BOUNDARY LAYER ANALYSIS

JOSEPH A. SCHETZ

Virginia Polytechnic Institute and State University

PRENTICE HALL, Upper Saddle River, New Jersey 07458

Original edition, entitled BOUNDARY LAYER ANALYSIS, 1st Edition, by SCHETZ, JOSEPH C., published by Pearson Education, Inc, publishing as Prentice Hall, Copyright © 1993.

KOREA edition published by PEARSON EDUCATION KOREA LTD, Copyright © 2008

This edition is manufactured in KOREA, and is authorized for sale only in KOREA.

ISBN 978-89-450-0624-0
ISBN 89-450-0624-9

This book is dedicated to all those who have helped me greatly as mentors over the years. These include, in particular, my high school physics and chemistry teacher at Dwight School, Charles Myron; my undergraduate thesis advisor at Webb Institute, Jens Holm; my Ph.D. advisor at Princeton University, Roger Eichhorn; and my boss in my first professional position at General Applied Science Laboratory, Antonio Ferri.

CONTENTS

PREFACE xiii

NOTATION xvii

1 INTRODUCTION TO VISCOUS FLOWS 1

1–1 The Importance of Viscous Phenomena 1

1–2 Conditions at a Fluid/Solid Boundary 2

1–3 Laminar Transport Processes 3

1–4 The Boundary Layer Concept 9

1–5 The Navier–Stokes Equations 11

1–6 Reduction of the Navier-Stokes to the Boundary Layer Equations 16

1–7 Separation and the Kutta Condition 19

1–8 Non-Newtonian Fluid Flow 22

1–9 Basic Notions of Turbulent Flow 23

Problems 28

2 INTEGRAL EQUATIONS AND SOLUTIONS FOR LAMINAR FLOW **30**

2–1 Introduction 30

2–2 The Integral Momentum Equation 31

2–3 Solution of the Integral Momentum Equation 35
2–3–1 The Pohlhausen Method, 35
2–3–2 The Thwaites-Walz Method, 38
2–3–3 Flows with Suction or Injection, 42

2–4 The Integral Energy Equation 43

2–5 Solution of the Integral Energy Equation 45
2–5–1 Unheated Starting-Length Problem, 45
2–5–2 Nonuniform Wall Temperature, 48

2–6 The Integral Species Conservation Equation 49

2–7 Solution of the Integral Species Conservation Equation 50

2–8 Relationship of Wall Friction, Heat Transfer, and Mass Transfer 51

2–9 Non-Newtonian Fluids 52
2–9–1 Power Law Fluid over a Flat Plate, 52
2–9–2 Bingham Plastic Fluid over a Flat Plate, 53

2–10 Discussion 54

Problems 54

3 DIFFERENTIAL EQUATIONS OF MOTION FOR LAMINAR FLOW **57**

3–1 Introduction 57

3–2 Conservation of Mass: The Continuity Equation 58

3–3 Conservation of Momentum: The Momentum Equation 60
3–3–1 Modeling of the Laminar Shear Stress, 62
3–3–2 Forms of the Momentum Equation for Laminar Flow, 63

3–4 Conservation of Energy: The Energy Equation 65
3–4–1 Modeling of the Laminar Heat Flux, 69

3–5 Conservation of Mass of Species: The Species Continuity Equation 70
3–5–1 Modeling of Laminar Diffusion, 70
3–5–2 Energy Transfer by Mass Transfer, 71
3–5–3 Physical Properties of Mixtures, 71

3–6 Transformation to Two-Dimensional Form 73

3–7 Mathematical Overview 73

Problems 75

4 EXACT AND NUMERICAL SOLUTIONS FOR LAMINAR, CONSTANT-PROPERTY, INCOMPRESSIBLE FLOWS **76**

4–1 Introduction 76

4–2 Parallel Flows 77
4–2–1 Steady Flow between Parallel Plates, 77
4–2–2 Unsteady Flow near an Infinite Plate, 78

4–3 Similar Solutions for the Velocity Field 79
4–3–1 Exact Solution for Flow over a Flat Plate, 81
4–3–2 Similar Solutions with Pressure Gradient, 85

4–4 Similar Solutions to the Low-Speed Energy Equation 87

4–5 Similar Solutions for Foreign Fluid Injection 89

4–6 Similar Solutions for Non-Newtonian Fluids 92
4–6–1 Power Law Fluid over a Flat Plate, 92
4–6–2 Bingham Plastic Fluid over a Flat Plate, 92

4–7 Numerically Exact Solutions 93
4–7–1 Numerical Analysis of the Linear Model Equation, 97
4–7–2 An Explicit Method for the Boundary Layer Equations, 103
4–7–3 Implicit Methods for the Boundary Layer Equations, 110
4–7–4 Transformations and Other Matters, 114
4–7–5 Finite Element Method, 118

Problems 124

5 COMPRESSIBLE LAMINAR BOUNDARY LAYERS **127**

5–1 Introduction 127

5–2 The Adiabatic Wall Temperature 128

5–3 The Reference Temperature Method 128

5–4 The Special Case of Prandtl Number Unity 129

5–5 The Recovery Factor for Nonunity Prandtl Number 132

5–6 Compressibility Transformations 134
5–6–1 Howarth-Dorodnitzin Transformation, 134
5–6–2 Crocco Transformation, 136
5–6–3 Levy-Lees Transformation, 136

5–7 Integral Method for Compressible Flow 137

5–8 Exact Solutions for Compressible Flow over a Flat Plate 138

5–9 Exact Stagnation Point Solutions 142

5–10 Flows with Mass Transfer 144
5–10–1 The Special Case of Pr = Le = 1, 144
5–10–2 Foreign Gas Injection through a Porous Wall, 146

5–11 Numerical Solutions of Compressible, Laminar Boundary Layer Flows 149

5–12 Real Gas Effects 157

5–13 Pressure Gradients and Separation in High-Speed Flows 164

5–14 Interactions in Hypersonic Flows 169

Problems 174

6 TRANSITION TO TURBULENT FLOW **176**

6–1 Introduction 176

6–2 A Simple Method Based on Re_θ 179

6–3 Hydrodynamic Stability Theory 179

6–4 The e^N Method 182

6–5 Other Prediction Methods 188

6–6 Selected Empirical Information 188
6–6–1 The Nature of Transition, 188
6–6–2 Effects of Free Stream Turbulence, 190
6–6–3 Effects of Pressure Gradients, 191
6–6–4 Effects of Roughness, 192
6–6–5 Bluff Bodies at Low Speeds, 194
6–6–6 Density-Stratified Flows, 196
6–6–7 Supersonic Flow, 197
6–6–8 Non-Newtonian Fluids, 198

Problems 200

7 WALL-BOUNDED, INCOMPRESSIBLE TURBULENT FLOWS **202**

7–1 Introduction 202

7–2 Engineering Requirements of Turbulent Analyses 202

7–3 Empirical Information on the Mean Flow as a Basis for Analysis 204
7–3–1 Flow over a Flat Plate, 204

7–3–2 Roughness Effects, 214
7–3–3 Injection and Suction through Porous Walls, 216
7–3–4 Flows with Axial Pressure Gradients, 220

7–4 Selected Empirical Turbulence Information 224

7–5 The Central Problem of the Analysis of Turbulent Flows 233

7–6 Mean Flow Turbulent Transport Formulations 237

7–7 Mean Flow Integral Methods 239

7–8 Mean Flow Models for the Eddy Viscosity and the Mixing Length 246
7–8–1 Models for the Inner Region, 247
7–8–2 Models for the Outer Region, 256
7–8–3 Composite Model for the Whole Boundary Layer, 262

7–9 Numerical Solution Methods for Mean Flow Formulations 264

7–10 Formulations Based on Turbulent Kinetic Energy 281

7–11 Formulations Based on Turbulent Kinetic Energy and a Length Scale 293

7–12 Formulations Based Directly on the Reynolds Stress 297

7–13 Direct Numerical Simulations 303

7–14 Boundary and Initial Conditions for Higher Order Models 304

7–15 Non-Newtonian Turbulent Flows 305

Problems 306

8 *INTERNAL FLOWS* **310**

8–1 Introduction 310

8–2 Laminar Internal Flows 312
8–2–1 Fully-Developed Flows, 312
8–2–2 Entrance Region Flows, 317
8–2–3 General Channel Flows, 322
8–2–4 Non-Newtonian Fluids, 324

8–3 Transition to Turbulent Flow 327

8–4 Turbulent Flows 328
8–4–1 Fully-Developed Flows, 328
8–4–2 Entrance Region Flows, 341
8–4–3 General Channel Flows, 350
8–4–4 Non-Newtonian Fluids, 353

8–5 High-Speed Flows 355

Problems 357

9 FREE SHEAR FLOWS **359**

9–1 Introduction 359

9–2 Laminar Free Shear Flows 361
9–2–1 The Simple Shear Layer, 361
9–2–2 Jets, 363
9–2–3 Wakes, 365

9–3 Mean Flow and Turbulence Data for a Round Jet in a Moving Stream 366
9–3–1 Introduction, 366
9–3–2 Constant-Density Flows, 366
9–3–3 Density Variations from Temperature Variations, 371
9–3–4 Density Variations from Composition Variations, 375

9–4 Mean Flow Turbulent Analyses 379
9–4–1 Similarity Analyses, 379
9–4–2 Analyses Based on Eddy Viscosity or Mixing-Length Models 386

9–5 Analyses Based on Turbulent Kinetic Energy 399

9–6 Analyses Based on Equations for Turbulent Kinetic Energy and a Length Scale 403

9–7 Analyses Based on the Reynolds Stress 408

9–8 Direct Numerical Simulations 411

9–9 Comments on the Use of Higher-Order Models for Free Mixing Flows 412

Problems 412

10 WALL-BOUNDED TURBULENT FLOWS WITH VARIABLE DENSITY AND HEAT AND MASS TRANSFER **414**

10–1 Introduction 414

10–2 Experimental Information 414
10–2–1 Mean Flow Data, 414
10–2–2 Turbulence Data, 426

10–3 Analysis 430
10–3–1 Boundary Layer Equations of Motion 430
10–3–2 Mean Flow Models for the Eddy Viscosity and Mixing Length, 433
10–3–3 Methods Based on Models for the Turbulent Kinetic Energy, 445

10–3–4 Methods Based on Models for Turbulent Kinetic Energy and a Length Scale, 445
10–3–5 Methods Based Directly on the Reynolds Stress, 449
10–3–6 Direct Numerical Simulations, 454

10–4 High-Speed Flows with Chemical Reactions 454
10–4–1 Approximate Solutions for Special Cases, 455

10–5 Viscous-Inviscid Interactions 457

Problems 460

11 THREE-DIMENSIONAL EXTERNAL BOUNDARY LAYER FLOWS 462

11–1 Introduction 462
11–1–1 Separation of Three-Dimensional Boundary Layers, 465

11–2 Laminar Three-Dimensional Boundary Layers 467
11–2–1 Boundary Layer Equations in Three Dimensions, 467
11–2–2 Special Cases, 470
11–2–3 Integral Methods, 475
11–2–4 Numerical Solution Techniques, 476

11–3 Transition to Turbulent Flow 480

11–4 Turbulent Three-Dimensional Boundary Layers 482
11–4–1 Introduction, 482
11–4–2 Mean Flow Observations, 487
11–4–3 Turbulence Data, 492
11–4–4 Integral Analyses, 495
11–4–5 Differential Methods, 499
11–4–6 High-Speed Flows, 502

11–5 Three-Dimensional Turbulent Jets 506
11–5–1 Introduction, 506
11–5–2 Experimental Data, 506
11–5–3 Analysis with a Mean Flow Turbulence Model, 509
11–5–4 Analysis with a Kϵ Turbulence Model, 508

Problems 510

APPENDIX A: LAMINAR THERMOPHYSICAL PROPERTIES OF SELECTED FLUIDS 512

APPENDIX B: COMPUTER CODES FOR STUDENTS 516

REFERENCES 558

INDEX 580

PREFACE

This book is intended as a text for courses in viscous fluid flow at high Reynolds numbers for advanced undergraduate and beginning graduate engineering students. Numerous homework problems ranging from the simple, but instructive, to the challenging are included. The emphasis is on understanding and analyzing flows of engineering interest. Thus, turbulent flows receive primary coverage, and a modern understanding of the physics of turbulent shear flows and turbulence models is discussed in detail. Further, the accurate analysis of practical flow problems, especially those involving turblence, requires the use of computerized methods, so numerical methods for the boundary layer equations are also treated in depth. Computer codes for a PC suitable for solving homework problems, including the use of turbulence models, are provided. Coverage of the older analytical methods is also included, to aid in developing understanding and because such methods are still widely used in preliminary design, especially for design optimization studies.

A second major theme of this volume is a concurrent treatment of the transfer of momentum, heat, and mass. This is primarily a book on viscous fluid dynamics, but the processes of convective heat and mass transfer are so closely connected to momentum transfer, that a unified presentation was deemed valuable. Such a presentation allows the scope of the book to span the entire range from low-speed to hypersonic flows. In addition, including mass transfer permits a discussion of viscous flows with chemical reactions.

An introductory coverage of simple non-Newtonian fluids of the *power law* and *Bingham plastic* types is also provided to orient the student to some of the important differences found with such fluids, compared to the more usual Newtonian fluids.

Chapter 1 is devoted to an introduction to the subject of viscous flow and why it is important to the engineer. The relevant physical phenomena and properties and dimensionless numbers are discussed. Then, the exact equations of motion for a constant-density fluid and the boundary layer assumptions are derived. The phenomena of flow separation and the *Kutta condition* are described next. The last section introduces the basic ideas associated with turbulent flows.

The subject matter of Chapter 2 is the approximate solution to laminar boundary layer problems, including heat and mass transfer, based on the integral forms of the equations. Worked examples and a computer code for the Thwaites-Walz method are presented.

Chapter 3 is concerned with the derivation of the boundary layer equations expressing conservation of total mass, momentum, energy, and mass of species, including variable density and properties.

The coverage in Chapter 4 begins with a few of the exact solutions for special cases of laminar flow. Then, *similar solutions* are discussed. The remainder of the chapter deals with numerical solutions of boundary layer problems using the finite difference method and the finite element method. Worked examples and computer codes are included.

Chapter 5 discusses high-speed laminar flows. Viscous heating, compressibility transformations, numerical solutions, and viscous-inviscid interactions are covered. Again, worked examples and a computer code are given.

Transition to turbulent flow is treated in Chapter 6. Hydrodynamic stability and the e^N method are described. Finally, selected empirical information is presented to illustrate the influence of roughness, pressure gradients, injection or suction, supersonic flow, and the like.

A detailed discussion of the nature of turbulent, wall-bounded flows and modern turbulence models is the focus of Chapter 7. The discussion is limited to constant-density flows for clarity. The problems of analyzing turbulent flow problems are emphasized. Integral and numerical methods are discussed, and extensive comparisons of predictions with experiment are provided. Worked examples and computer codes are presented.

The important subject of laminar and turbulent internal flows has been highlighted by devoting a separate chapter (Chapter 8) to that topic alone.

Chapter 9 covers free shear flows, such as wakes, jets, and shear layers. Both laminar and turbulent cases are considered over the range from low to high speeds. Worked examples and a computer code are provided.

Wall-bounded turbulent flows of variable density and properties are discussed in Chapter 10. Both low- and high-speed cases are included.

The last chapter (Chapter 11) contains a detailed presentation of current knowledge for truly three-dimensional boundary layer flows.

Throughout the book, separating, but not separated, flows are discussed, because the analysis of the latter cannot be handled with the usual boundary layer equations. Lastly, unsteady flows are not covered, and that is justified by the observation that those phenomena are usually omitted in courses at the level intended here.

The goal of this effort was to write a book appropriate for mechanical, aerospace, civil, ocean, and chemical engineering students. The treatment assumes that the student has taken at least one general undergraduate course in fluid mechanics and one in mathematics with partial differential equations. It also assumes that the student is computer literate. The simple codes included with the book are written in Fortran, and they are designed for use on PC's of the IBM PS/2 class. A course that would encompass the bulk of the material in this volume would likely be a two-semester graduate course. A good one-semester course at the undergraduate or graduate level can be formed with a judicious selection of all the material supplied. The selection of topics to be covered in a given course will clearly be strongly influenced by the subject area of the students. It is suggested, however, that all students receive some coverage of numerical solution techniques and turbulence models.

A number of people helped with the preparation of this volume and/or provided helpful suggestions. In particular, Ming Situ, Stanley Favin, Bernard Grossman, and Evangelos Hytopoulos helped greatly with the computer codes for the homework problems. Dr. Ephraim Sparrow provided many helpful suggestions. Finally, students in classes in which my earlier text and notes that formed the basis of this book were used have always been open with helpful comments. I thank them all. I also thank Frederick Billig and the Applied Physics Laboratory of Johns Hopkins University, as well as various government agencies that have supported my personal research in the boundary layer field.

Joseph A. Schetz
Blacksburg, VA

NOTATION

a	Speed of sound and amplification factor
A	Area or constant
$b_{1/2}$	Half-width
B_1, B_2, B'_1, B'_2	Constants
$B_{f,h}$	Injection/suction parameters
c	Average speed of molecules
c_i'	Fluctuating value of species concentration
c_i	Species concentration
C_i	Mean value of species concentration
C_1, C_2, C_μ, etc.	Constants
c_p	Specific heat at constant pressure
c_v	Specific heat at constant volume
c_r, c_i	Real and imaginary parts of the phase velocity
C_f	Skin friction coefficient
$\overline{C_f}$	Average skin friction coefficient
C_p	Pressure coefficient
C_D	Drag coefficient
D, d	Diameter and minor axis in three-dimensional jets
$\mathcal{D}$	Drag
D_h	Hydraulic diameter

D_{ij}	Binary diffusion coefficient
D_T	Turbulent diffusion coefficient
e	Internal energy
$E_1(k_1)$	Kinetic energy of axial fluctuations at wave number k_1
f	Friction factor
$f(\cdot)$	Function of $(\cdot)$
f_i	Body force vector
f_1, f_2, f_μ	Factors in $K\epsilon$ model
F_i	Element force vector
g	Acceleration of gravity, Clauser similarity variable, Johnson-King variable, and $G/(1 - G_w)$
G	Stagnation enthalpy ratio
h	Enthalpy
$\hslash$	Film coefficient
$\hslash_D$	Film coefficient for diffusion
h_1, h_3, h_c	Metric coefficients
H	Stagnation enthalpy
$H(\Lambda)$	Shape factor
i	$\equiv \sqrt{-1}$
j	Index
J	Integrated momentum flux
k	Thermal conductivity and average roughness size
k_m	Crocco-Lees mixing constant
k_T	Turbulent thermal conductivity
k_1	Wave number of fluctuations
K	Turbulent kinetic energy
$K(T)$	Equilibrium constant
K_1, K_2, etc.	Constants and geodesic curvatures
K_{ij}	Element stiffness matrix
ℓ	Turbulent length scale and arc length
ℓ_m	Mixing length
ℓ_1	Major axis in three-dimensional jets
L_m	Length scale in Johnson-King model
Le	Lewis number
Le_T	Turbulent Lewis number
m	Index along surface
$\dot{m}_i$	Diffusive mass flux of species i
M	Maximum value of index m and Mach number
M_{c1}	Convective Mach number
M_τ	Mach number in compressibility correction
n	Index across layer, frequency, and transverse streamline coordinate
n_y	y component of normal vector to surface
N	Maximum value of index n

Nu	Nusselt number
$\mathrm{Nu_{Diff}}$	Nusselt number for diffusion
N_{He}	Hedstrom number
N_{Pl}	Plasticity number
p	Pressure and exponent for power law fluids
p_i	Partial pressure of species i
p'	Fluctuating pressure
P	Mean pressure, scaled pressure, and perimeter
P_t	Total pressure
P_c, P_T, P_v	Power law decay exponents
Pr	Prandtl number
Pr_T	Turbulent Prandtl number
$\mathcal{P}$	Production of turbulent kinetic energy
q_i	Heat flux vector
q_T	Turbulent heat flux
q_w	Wall heat transfer rate
q_*	Friction velocity in three-dimensional flows
Q	Total velocity in three-dimensional flows
r	Radial coordinate, recovery factor, and element coordinate
$r_o(x)$	Body radius
$r_{1/2}$	Half-radius
R	Pipe radius, gas constant, and radius of curvature
Ri	Richardson number
Re	Reynolds number
R_o	Body nose radius of curvature
R_t	Turbulent Reynolds number
s, $\bar{s}$	Transformed streamwise and streamline coordinate and element coordinate
Sc	Schmidt number
Sc_T	Turbulent Schmidt number
St	Stanton number
$\mathrm{St_{Diff}}$	Stanton number for diffusion
$S(\Lambda)$	Shear parameter
t	Time
T	Static temperature
$\mathbf{T}_{x,y,z}$	Surface force vector
T_b	Bulk temperature
T^*	Reference temperature
T_t	Total (stagnation) temperature
T_0	Time period
$\bar{T}$	Mean temperature
T'	Fluctuating temperature
T_*	Heat transfer temperaure

T^+	$\equiv (T_w - \bar{T})/T_*$
u	Streamwise velocity
u_{ave}	Average velocity
u_i	Nodal velocity in FEM
$\bar{u}$	Transformed velocity and velocity from previous iteration in FEM
$\bar{\bar{u}}$	Mass-weighted mean streamwise velocity
u'	Fluctuating velocity
u_*	Friction velocity
u^+	$\equiv U/u_*$
u_c	Convective velocity
U	Mean streamwise and scaled velocity
U_p	Primary direction velocity in a channel
U_w	Velocity of moving wall
v	Transverse or radial velocity
v_w	Transverse velocity at the wall
v_i	Nodal velocity in FEM
$\bar{v}$	Transformed velocity and velocity from previous iteration in FEM
$\bar{\bar{v}}$	Mass-weighted mean transverse velocity
v_o^+	Dimensionless transverse velocity at the wall
v'	Fluctuating transverse velocity
V	Mean transverse, scaled and general velocity
V_o	Entrainment velocity
V_ϕ	Velocity from scalar potential in a channel
V_ψ	Velocity from vector potential in a channel
w	Spanwise velocity in three-dimensional flows
w'	Fluctuating spanwise velocity
W	Mean spanwise velocity in three-dimensional turbulent flows and channel half-height
W_i	Molecular weight
$W(y/\delta)$	Wake function
w_A	Reaction source term
x	Streamwise coordinate
$\bar{x}$	Transformed streamwise coordinate
x_1, x_3	Coordinates in three-dimensional flows
x_{PC}	Length of potential core
X_i	Mole fraction of species i
X	Scaled streamwise coordinate
y	Transverse coordinate
$\bar{y}$	Transformed transverse coordinate
$y^+ \equiv yu_*/\nu$	Transverse coordinate for the law of the wall
y_{max}	Length scale in Baldwin-Lomax model
Y	Transformed and scaled transverse coordinate

$Y_{1/2}$, $Z_{1/2}$ Half-widths in three-dimensional jets
z Spanwise coordinate in three-dimensional flows
z_A Dimensionless mass fraction
Z $\equiv k^m l^n$
$Z(p, T)$ Compressibility factor

Greek

α Wave number, amplification factor, and atom mass fraction
α_T $\equiv k_T/\rho c_p$
$\alpha(p)$ Friction law coefficient for power law fluids
β Pressure gradient parameter and wave number
β_w Wall streamline angle
β_ξ, β_ω Pressure gradient parameters in three dimensions
$\beta(p)$ Friction law function for power law fluids
$\overline{\chi}$ Hypersonic interaction parameter
ψ Planar stream function
Ψ Axisymmetric stream function
$\hat{\psi}$ Disturbance stream function
ϵ Dissipation of turbulent energy
$\epsilon_{n,m}$ Truncation error
ϵ_{xy} Strain
ρ Density
λ Pohlhausen pressure gradient parameter, pipe resistance coefficient, and second viscosity coefficient
λ^* Mean free path between molecules
Λ Thwaites-Walz pressure gradient parameter
τ, τ_{xy} Shear
τ_T Turbulent shear
τ_0 Yield stress for plastic fluid
Ω Intermittency and vorticity parameter
Ω_e Element area
Γ_e Element perimeter
μ Laminar viscosity
μ_a Apparent viscosity for non-Newtonian fluid
μ_{BP} Viscosity factor for Bingham plastic fluid
μ_{PL} Viscosity factor for power law fluid
μ_T Turbulent viscosity
μ_{Tx} Turbulent viscosity in the streamwise direction
μ_{Tz} Turbulent viscosity in the crossflow direction
κ Constant in the law of the wall and inverse of channel radius of curvature
κ_T Constant in the temperature law of the wall

ν	Laminar kinematic viscosity
ν_T	Turbulent kinematic viscosity
ϕ_i	Element interpolation function
ϕ	Amplitude function and dummy variable
$\phi_{1,2}$	Deformation angle
δ	Boundary layer thickness
δ_T	Thermal boundary layer thickness
δ_c	Concentration boundary layer thickness
δ^*, Δ^*	Displacement thickness
δ_k^*	Kinematic displacement thickness
δ_1^*, δ_2^*	Displacement thicknesses in three-dimensional flows
Δ_{PS}	Perry-Schofield length scale
Δ	Clauser integral boundary layer thickness
θ	Momentum thickness
θ_{11}, θ_{12}, θ_{21}, θ_{22}	Momentum thicknesses in three-dimensional flows
Θ, Θ_r	Excess temperature
ζ	$\equiv \delta_T/\delta$
Π	Wake parameter
ξ	Dummy variable and stretched time
η	Similarity variable
$\overline{\eta}$	Transformed transverse coordinate
ω	Dimensionless frequency, viscosity law exponent, and transformed lateral coordinate
ω_x, ω_y, ω_z	Components of vorticity
$\sigma_{K,\epsilon,\tau}$	Prandtl numbers for K, ϵ, τ
Γ	See Eq. (10–8)
γ	Ratio of specific heats and stability parameter in three-dimensional flows
ξ	Transformed streamwise coordinate

Subscripts

c	Values on the centerline
e	Values on the edge of the boundary layer
j	Initial values in a jet
t	Stagnation values
w	Wall values
∞	Conditions in the approach flow

CHAPTER 1

INTRODUCTION TO VISCOUS FLOWS

1-1 THE IMPORTANCE OF VISCOUS PHENOMENA

An applied fluid dynamicist is generally engaged in the analysis or design of a device that has a specific practical purpose (e.g., an airplane, a piping system, or a heat exchanger). In virtually all such situations, the cost of operation as well as performance is of primary importance, and this usually comes down to estimating the resistance of the device to fluid motion. For a vehicle, this resistance is called *drag*. For a piping system, one must determine the *pressure loss*. In other cases, we may speak simply of *frictional losses* or, in rotating electrical machines, *windage*. It should be clear, then, that the influence of fluid friction can seldom be disregarded in actual engineering practice, and the *inviscid fluid* assumption is too restrictive for most real situations. Thus, the main topic of this book is the dynamics of viscous fluids at conditions of interest to engineers.

A process that is closely allied to fluid friction is convection heat transfer. This subject is very important in cases ranging from the heat exchanger in an industrial boiler to a reentry vehicle. The close linkage that exists between fluid friction and convection phenomena—on physical grounds, on the basis of similar mathematical models, and on their practical importance—makes the concurrent study of these processes attractive, and that path will be followed here.

A third process related to viscous resistance and convection heat transfer is mass transfer with fluid motion. Examples include industrial drying operations and fuel droplet evaporation. Thus, we shall pursue the study of this process, along with viscous effects and convection heat transfer.

It has been stressed that the subject matter of interest is concerned with real phenomena. The reader will find it helpful, therefore, to employ his or her powers of observation and recollection while studying this material. Take every opportunity to note and learn from the examples of fluid flow that constantly surround you in the real world. Observe the cloud patterns, contrails from airplanes, and smoke plumes from industrial stacks in the sky. Watch the water flow in your bathtub or shower or behind your boat. You can easily find a myriad of other examples yourself. Try to relate them to the material under study. If you cannot, try to find a situation in your world that involves the subject of interest at the moment. This may all sound too abstract, but the student who tries to follow this advice will find it very helpful in gaining real understanding.

1–2 CONDITIONS AT A FLUID-SOLID BOUNDARY

The mathematical basis of modern fluid dynamics was developed mostly before the true molecular nature of matter was well understood. Thus, the fluid was considered a homogeneous, continuous medium, rather than consisting of rather widely spaced molecules as we now know is the case for gases at modest pressure, for example. With very few exceptions, such as the highly rarefied air at the outer edge of the atmosphere, it has been possible and convenient to retain that continuum mathematical formulation and apply it successfully to the real world by making certain assumptions. For our purposes here, the most important of these concern the conditions at the interface between the fluid and a solid surface. On the scale of an air molecule, even a smoothly machined surface looks like rough terrain. Each molecule striking the surface will have numerous collisions between the various peaks and valleys and other molecules and will exchange some of its momentum and thermal energy with the surface. It is clearly an awesome task to try to follow each molecule through its history and then sum up on a statistical basis to find the average momentum and thermal energy exchange at a given location on the surface. Experiments and analytical estimates have shown that an entirely adequate representation of physical reality is contained in the assumptions that, at a surface, a fluid loses all of its momentum relative to the surface in the interaction between the two and that the fluid comes to the same temperature as the surface. An essentially equivalent assumption is made for the case of mass transfer, but the matter can become somewhat more complicated. A detailed discussion is reserved until mass transfer is treated in depth in a later chapter.

The foregoing assumptions yield velocity and temperature profiles in the region near a solid surface resembling those shown in Fig. 1–1. Denoting the local velocity in the streamwise direction as u and the distance normal to the surface as y, the profile $u(y)$ at a given value of the streamwise coordinate x will look as shown. The velocity at the surface is taken to be zero (often called the *no-slip condition*). The no-slip condition leads to the clarifying notion that a surface is a *sink* for fluid

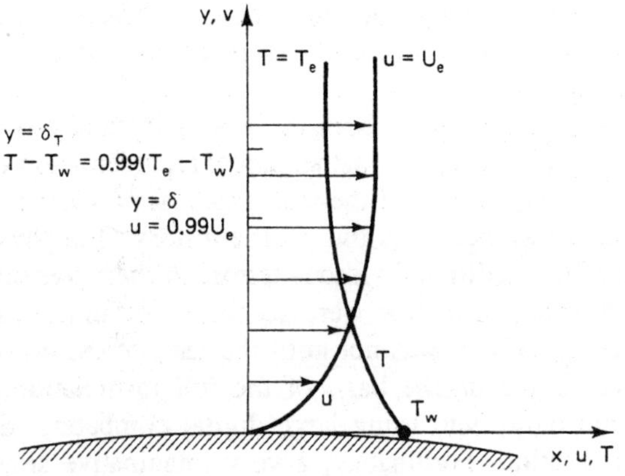

Figure 1–1 Typical velocity and temperature profiles in a boundary layer. Shown are the definition of the velocity and temperature boundary layer thicknesses.

momentum. The velocity at a distance far (on the scale of the viscous region) from the surface will have some value dependent on the free-stream velocity U_∞ and the shape of the body. Call this value U_e, to indicate conditions at the outer edge of the viscous region. Between the values of 0 and U_e, we expect the velocity profile to change smoothly due to the influence of fluid friction on the fluid in the layers just above and below any point of interest. The profile will blend into the edge value U_e only asymptotically, so it is common to identify the point where 99% of U_e is reached as the *thickness* δ of the viscous layer on the solid boundary. (Call this region from the surface to δ the *boundary layer*.)

In a similar way, the temperature profile will vary from $T = T_W$ at the wall to $T = T_e$ at the edge of the *thermal boundary layer* δ_T. We define δ_T as the point where $T - T_W = 0.99(T_e - T_W)$. Note that δ_T is not necessarily equal to δ; that is, the thermal boundary layer thickness does not have to be the same as the velocity boundary layer thickness.

For flows with mass transfer, a composition profile across the layer will resemble those for velocity and temperature just discussed, and there will be a *concentration boundary layer* of thickness δ_c.

1–3 LAMINAR TRANSPORT PROCESSES

The simplest class of flows in which viscous phenomena are important occurs when the streamlines form an orderly, roughly parallel pattern. The fluid in the viscous region may be thought of as proceeding along in a series of layers or laminates with smoothly varying velocity and temperature from laminate to laminate. Consider a deck of playing cards resting on a table, and push the first few cards slightly to one side. Friction from card to card will cause most of the rest of the cards to shift in the same direction, but by a lesser amount, down through the deck. The displaced pattern of the edge of the deck may be imagined to be a velocity profile, and the flow so represented would then have the laminated characteristic previously described. Viscous flows of this class are, therefore, called *laminar*. They are the simplest type of viscous flows, but unfortunately, only a fraction of flows of practical interest falls into this class. Nonetheless, some flows do, and this is a good place to begin the study of the details of viscous processes.

The reader is likely to be surprised at both the simplicity and crudity of the physical arguments that form the basis of the mathematical representation of laminar flows, especially in light of the complexity of the mathematical problem that results. Some perspective on this may be gained by noting a little history. The physical arguments and the formulation of the additional viscous terms, in their present form, to augment Euler's inviscid equations of motion were accomplished in the early 1800s by Navier (1823) and Stokes (1845). It was not until the late 1900s, however, that practical problems could be solved on the basis of the full formulation using the Navier-Stokes equations, and only then, using large digital computers. Such solutions are still by no means routine. Fortunately, a very imaginative simplification was developed just after the turn of the last century that can be used in many situations of engineering interest. We shall introduce that concept in the next section.

One can develop a suitable mathematical representation for the viscous terms by assembling some physical facts. First, a *fluid* will not sustain a stress as a result of a simple static displacement. The fluid will simply shift and then return to rest in a new stress-free state. That is the important basic mechanical difference between a fluid and a solid. If stresses in a fluid are to exist, they must result from relative motion, not position. Thus, in a fluid, stress can occur only in a region of variable velocity. We would also expect that a greater rate of variation would produce a greater stress. The simplest representation that meets these criteria is

$$\tau \sim \frac{\partial V}{\partial n} \qquad (1\text{–}1)$$

where τ is the viscous stress, V the velocity, and n the direction normal to the direction of the shear stress. Note that Eq. (1–1) is based on the rather bold assumption that only the first derivative is important. We might have included other terms proportional to, for example, $\partial^2 V/\partial n^2$, but we choose to try and keep the formulation as simple as possible.

The adequacy of Eq. (1–1) can be tested in the simple device shown in Fig. 1–2, called a *viscometer*. The no-slip condition at both solid surfaces and the small gap thickness produce the linear velocity profile shown. The frictional stress can be determined from the power required to turn the cylinder. It is found that many common fluids (e.g., air, water) obey the relation in Eq. (1–1) if we write

$$\tau = \mu \frac{\partial V}{\partial n} \qquad (1\text{–}2)$$

where μ is the *laminar coefficient of viscosity*, which depends strongly on the composition of the fluid. Equation (1–2) is referred to as *Stokes's law*. Fluids that obey this law are termed *Newtonian;* others are termed *non-Newtonian*. In non-Newtonian fluids, such as polymers, the shear stress depends nonlinearly on the rate of shear strain, as shown in Fig. 1–3. We will return to the subject of non-Newtonian fluids

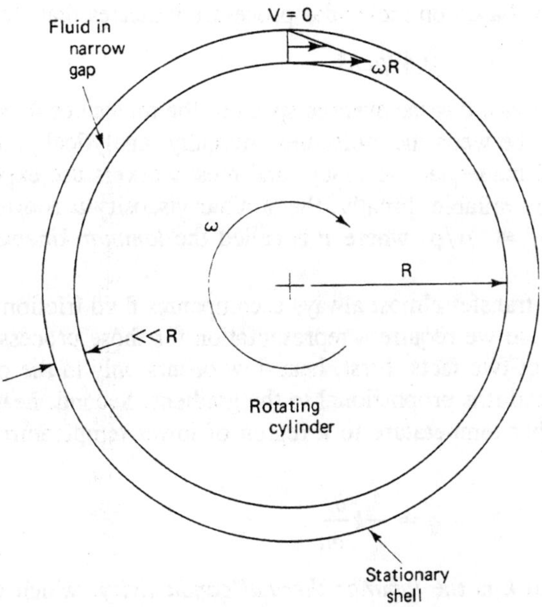

Figure 1–2 Schematic of a device for measuring fluid viscosity.

later, but, for now, consider the more common and simpler Newtonian case. The quantity μ is called a *physical property* of the fluid itself, since it does not depend on the state of fluid motion. Experiment shows that, essentially, $\mu = \mu(T)$ for a given fluid; the influence of pressure is negligible in most cases.

Looking at Eq. (1–2), we find that the dimensions of the relation suggest that

$$\mu \sim \text{density} \times \text{velocity} \times \text{length} \tag{1–3}$$

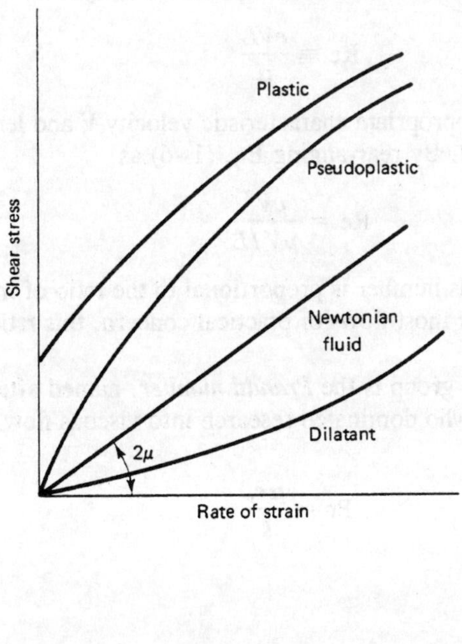

Figure 1–3 Variation of shear stress versus rate of strain for various types of fluids.

Indeed, an idealized analysis based on molecular processes indicates that, for gases,

$$\mu = 0.49\rho c\lambda^* \tag{1-4}$$

where ρ is the density of the gas, c is the average speed of the molecules ($c \sim \sqrt{T}$), and λ^* is the mean free path between the molecules. Actually, analytical predictions of μ have generally been of inadequate accuracy, and most workers use experimental data for μ when they are available. Finally, the laminar viscosity μ is often combined with the density as $\nu \equiv \mu/\rho$, where ν is called the *laminar kinematic viscosity*.

As stated earlier, heat transfer almost always accompanies fluid friction if there is a temperature difference, so we require a representation for those processes also. The physical basis consists of two facts. First, heat flow occurs only in the presence of a temperature gradient, and it is proportional to the gradient. Second, heat always flows from a region of higher temperature to a region of lower temperature. Thus, we are led to

$$q = -k\frac{\partial T}{\partial n} \tag{1-5}$$

where q is the heat flux and k is the *laminar thermal conductivity*, which depends strongly on fluid composition. Equation (1–5) is called *Fourier's law*. The minus sign ensures the correct direction of heat flow, with k always taken as positive. Experiment verifies the adequacy of Fourier's law and also shows that $k = k(T)$, again with a very weak dependence on pressure. The laminar thermal conductivity is thus also a physical property of the fluid and does not depend on fluid motion or the rate of heat transfer. Indeed, Eq. (1–5) also holds for solids.

There are two dimensionless groups involving the laminar viscosity and thermal conductivity that are useful. The first is the *Reynolds number*, named after a famous British fluid dynamicist of the late 1800s, Osborne Reynolds:

$$\text{Re} \equiv \frac{\rho VL}{\mu} \tag{1-6}$$

Here, one must select the appropriate characteristic velocity V and length scale L for a given flow being analyzed. By rearranging Eq. (1–6) as

$$\text{Re} = \frac{\rho V^2}{\mu V/L} \tag{1-6a}$$

we can see that the Reynolds number is proportional to the ratio of inertial forces to viscous forces in a flow. For most flows of practical concern, this ratio is very large, on the order of 10^3 to 10^8.

The second important group is the *Prandtl number*, named after a German researcher, Ludwig Prandtl, who dominated research into viscous flow in the first half of this century:

$$\text{Pr} = \frac{\mu c_p}{k} \tag{1-7}$$

Rearranging Eq. (1–7) to the form

$$\mathrm{Pr} = \frac{\mu/\rho}{k/(\rho c_p)} \tag{1–7a}$$

makes it plain that the Prandtl number is proportional to the ratio of the coefficient for diffusion of momentum to the coefficient for diffusion of heat. For most gases, Pr ~ 0.7. For liquids such as water, Pr ~ $O(10)$. For liquid metals, Pr << 1.

For the flow of a fluid consisting of a mixture of two (or more) different fluids with a nonuniform composition, there is a process of molecular diffusion akin to viscous shear and heat transfer. If the local concentration of a species is described by a *mass fraction* c_i defined as the mass of species i in a volume divided by the total mass of all fluids in that volume, the analog of Eqs. (1–2) and (1–5) for diffusion is

$$\dot{m}_i = -\rho D_{ij}\frac{\partial c_i}{\partial n} \tag{1–8}$$

where $\dot{m}_i$ is the local mass flux of species i relative to the local mass-averaged velocity of the fluid. Equation (1–8), called *Fick's law,* expresses the physical fact that a species diffuses from a region of higher concentration to a region of lower concentration. This simple expression neglects other possible processes of diffusion due to strong temperature or pressure gradients, which are often not important. The factor D_{ij} is called the *laminar binary diffusion coefficient* and is also a physical property of the fluids under consideration for laminar flows. In this book, we treat in detail only mixtures of two species, that is, *binary mixtures.*

The diffusion coefficient is combined with other properties to form two dimensionless groupings. The first is the *Schmidt number,*

$$\mathrm{Sc} = \frac{\mu}{\rho D_{ij}} \tag{1–9}$$

The second is the *Lewis number,*

$$\mathrm{Le} = \frac{\rho c_p D_{ij}}{k} \tag{1–10}$$

Obviously,

$$\mathrm{Pr} = \mathrm{Le}\cdot\mathrm{Sc} \tag{1–11}$$

Thus, the Schmidt number is the ratio of the coefficient for diffusion of momentum to the coefficient for diffusion of mass, and the Lewis number gives the corresponding ratio for mass transfer to heat transfer.

The determination of physical properties such as the laminar viscosity, thermal conductivity, and diffusion coefficient of a fluid comprises a separate field of endeavor, and the work is largely composed of painstaking experiments. The working fluid dynamicist must have readily available sources of this critical information. Some data for air and water are plotted in Figs. 1–4 and 1–5. Note that the Prandtl number is nearly constant for air. It is also nearly constant for most gases. Further

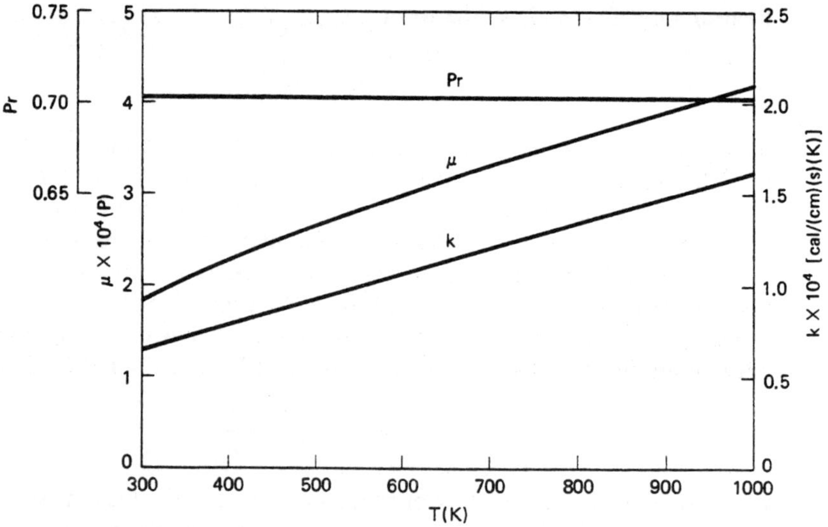

Figure 1–4 Laminar thermophysical properties for air at 1.0 atm.

tabulated data for a sampling of fluids are contained in Appendix A for the convenience of the reader. Appendix A also has tables of conversion factors that are helpful when data are to be obtained from a variety of sources and, hence, likely to be in a variety of systems of units.

It is also common to fit empirical information to a curve. A widely used formula for the viscosity of air is the Sutherland (1893) formula,

$$\mu = 0.1716\left(\frac{T}{273.1}\right)^{3/2}\frac{383.7}{T + 110.6} \tag{1–12}$$

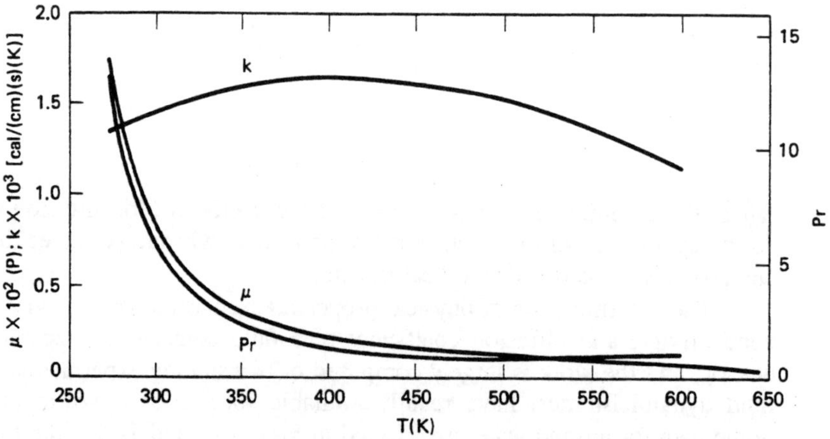

Figure 1–5 Laminar thermophysical properties for liquid water.

where T is in degrees Kelvin and μ is in millipoise (N · s/m² × 10⁴ gives mP). A simpler form, $\mu \sim T^{\omega}$($\omega \approx 0.76$ for air) is often employed.

1–4 THE BOUNDARY LAYER CONCEPT

Because the Navier-Stokes equations that result from the complete viscous additions to the inviscid equations are very difficult to solve for general flows, they sat largely unused for several decades and did not contribute significantly to the early development of, for example, airplanes or improved shiphulls. When faced with a difficult mathematical problem, it is natural to seek rigorous or at least reasonable simplifications. This is usually done by comparing the magnitudes of the various terms in the equation(s) to see whether any terms may be neglected, compared to those retained for conditions of interest. To make this procedure rational, it is necessary first to nondimensionalize and normalize all the independent and dependent variables, preferably to approximately order unity. For example, we may scale the local streamwise velocity u by the free-stream velocity U_∞ and the axial coordinate x by the length of the body, the diameter of a pipe, or some other length scale. When this is done for the Navier-Stokes equations, one is left with the factor 1/Re, multiplying all the viscous terms. Now, as we observed earlier, the Reynolds number is very large for most practical flows, so this factor would be roughly 10^{-3} to 10^{-8}.

Under such circumstances, it certainly seems safe to neglect all terms multiplied by 1/Re, since the other terms in the equations have coefficients of roughly unity, and we have sought to make all the terms themselves of order unity. This all seems plausible; the only problem is that one is, in this way, led back to the inviscid flow equations, which cannot predict frictional drag or heat transfer. The core of the problem arises because the terms that are neglected are the highest order derivatives. Depressing the order of the system from second order to first order means that only one boundary condition on u in the transverse or y-direction can be imposed. Thus, one cannot impose both the no-slip condition on the surface and merging into the free stream far from the surface. Given that choice, the only sensible decision is to forego the no-slip condition leading to zero wall shear and the usual inviscid flow analysis.

The matter rested in this seeming paradox until Prandtl's (1904) penetrating analysis. Prandtl reasoned that, for flows of high Reynolds number, viscous effects would be confined to a thin layer along a solid surface. Such a region must always exist, since the velocity must vary quickly from a relatively high value out in the flow down to zero on the body surface. This variation will produce large gradients and, hence, significant shear forces, even for low-viscosity (high Re) flows, by virtue of Eq. (1–2). Although the region in question may occupy only a small portion of a total flow field, it cannot be neglected, since all momentum, heat, and mass transfer to or from the surface must take place through it. Outside this boundary layer, the flow does behave as if it were inviscid. Thus was born the brilliant new idea of dividing the flow into two regions and using different simplified equations in each. For the outer region, we neglect viscous effects and use the inviscid Euler equations. For the inner, boundary layer region, we use a simplified version of the Navier-Stokes equations that still, however, retains some viscous terms. Of course, the solutions for the two regions must be suitably joined to produce a composite solution for the whole flow field. This idea is so powerful that it has been used in many other areas of science and mathematics besides fluid dynamics. The term *boundary layer analysis* is, therefore, widely used in a general way.

The single most important part of the boundary layer assumption is that the static pressure may be taken to be constant across the boundary layer. For many uses, the following physical derivation is adequate. We presume a thin viscous layer and a body or channel shape with modest surface curvature in the direction of flow. The reason for the latter restriction will become clearer in the next section. Since the layer is thin everywhere, the magnitude of the rate of growth of the layer, $d\delta/dx$, must be small. Thus, streamlines in the boundary layer must all be roughly parallel to the surface, and, by presumption,they must all have, at most, a modest curvature. This is the same as saying that the radius of curvature of the streamlines in the boundary layer R must be large. From elementary fluid mechanics, we have the relation

$$\frac{\partial p}{\partial n} = \frac{\rho V^2}{R} \tag{1–13}$$

which says that, for very large R, the variation in pressure normal to the streamlines (including the body surface itself) must be negligible.

The fact that the static pressure is constant across the thin boundary layer is important for two reasons. First, the Navier-Stokes equations have one equation expressing conservation of momentum for each coordinate direction. Thus, for a planar (x,y) formulation, we would have an x-momentum equation and a y-momentum equation. Under the boundary layer assumption, the y-momentum equation reduces to

$$\frac{\partial p}{\partial y} \approx 0 \tag{1–14}$$

a considerable simplification.

The second important result of taking the static pressure to be constant across the boundary layer concerns the joining of the outer, inviscid solution to the inner, boundary layer solution to produce a complete, composite solution. If the static pressure is constant across the layer, the static pressure predicted by the inviscid solution at the outer edge of the boundary layer determines the static pressure all across the boundary layer. One says that the inviscid solution *impresses* the static pressure on the boundary layer. This analysis can be carried further. From the point of view of the inviscid flow, the outer edge of the thin boundary layer is indistinguishable from the surface. Thus, an inviscid solution for the flow over the body or through the channel of interest is found first, and the predicted *inviscid pressure distribution on the surface*, $p(x)$, is taken as the *pressure in the boundary layer*. Also, Bernoulli's equation applied to the inviscid solution gives the *inviscid velocity on the surface*, which is *taken to be the outer edge velocity for the boundary layer*:

$$U_e \frac{dU_e}{dx} = -\frac{1}{\rho}\frac{dp}{dx} \tag{1–15}$$

From the point of view of the boundary layer, the outer edge of the boundary layer δ is far from the surface. Therefore, the boundary condition for the velocity in the boundary layer at the outer edge of the layer is written, not as $u(x,\delta) = U_e$, but rather, as

$$\lim_{y\to\infty} u(x, y) = U_e(x) \tag{1–16}$$

From the preceding discussion, it can be seen that the coupling between the outer, inviscid flow and the inner, boundary layer flow is all one way. The inviscid solution for flow over the body or through the channel can be found first, independently of the boundary layer. That solution determines $p(x)$ and $U_e(x)$ for the boundary layer solution. The boundary layer equations are then solved under the constraints given by $p(x)$ and $U_e(x)$.

The foregoing *heuristic* development is sufficient for understanding the essence of boundary layer theory, as described in the rest of this book. For completeness and for those interested in or curious about the mathematical details, the next two sections, giving a complete derivation for the steady, constant-density, constant-property case, is included. Those not interested can skip to Section. 1–7 without loss of continuity.

1–5 THE NAVIER-STOKES EQUATIONS

In some situations, the full viscous equations, usually called the Navier-Stokes equations, must be employed. Although the solution of those equations for general flows is beyond the scope of this book, it is worthwhile to include a derivation for laminar, constant-property cases.

Consider the surface forces acting on the fluid element shown in Fig. 1–6. The total surface forces acting on the faces perpendicular to the x-axis are denoted

$$T_x\,dy\,dz;\ \left(T_x + \frac{\partial T_x}{\partial x}dx\right) dy\,dz \tag{1–17}$$

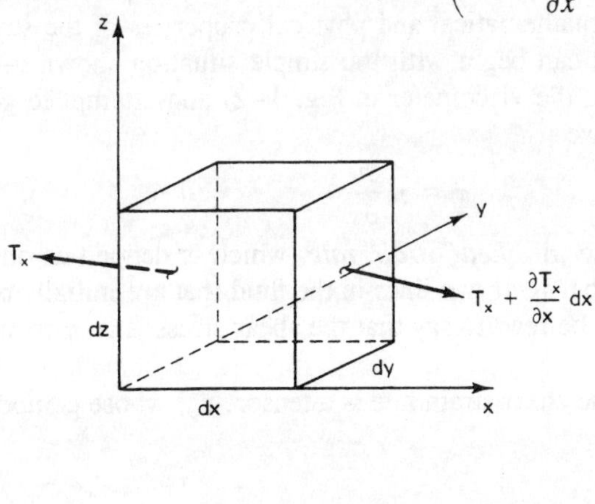

Figure 1–6 General surface forces acting on the sides of a fluid element perpendicular to the x-axis.

The net force from these is

$$\frac{\partial \boldsymbol{T}_x}{\partial x}\, dx\, dy\, dz \tag{1–18}$$

By analogy, the net forces on the other faces can be written

$$\frac{\partial \boldsymbol{T}_y}{\partial y}\, dy\, dx\, dz; \qquad \frac{\partial \boldsymbol{T}_z}{\partial z}\, dz\, dx\, dy \tag{1–19}$$

Now each of the force (per unit area, i.e., stress) vectors $\boldsymbol{T}_x$, $\boldsymbol{T}_y$, and $\boldsymbol{T}_z$ can be decomposed. We get

$$\begin{aligned} \boldsymbol{T}_x &= i\tau_{xx} + j\tau_{xy} + k\tau_{xz} \\ \boldsymbol{T}_y &= i\tau_{yx} + j\tau_{yy} + k\tau_{yz} \\ \boldsymbol{T}_z &= i\tau_{zx} + j\tau_{zy} + k\tau_{zz} \end{aligned} \tag{1–20}$$

The subscript notation—for example, on τ_{xy}—indicates quantities acting on the face perpendicular to the x-axis and in the y-direction. Taking moments of the forces about an arbitrary axis and setting the result equal to zero for a fluid particle in equilibrium leads directly to the result that

$$\begin{aligned} \tau_{xy} &= \tau_{yx} \\ \tau_{xz} &= \tau_{zx} \\ \tau_{yz} &= \tau_{zy} \end{aligned} \tag{1–21}$$

must hold in general.

A quantity with two subscripts (i.e., nine components) is called a *tensor*. Such a quantity follows in a logical sequence from a *scalar,* such as temperature, with only one component, to a *vector,* such as velocity, with three components, to a stress *tensor,* with nine components. With Eq. (1–21), we see that the stress consists of a *symmetrical tensor*. The diagonal components τ_{xx}, τ_{yy}, τ_{zz} are called *normal stresses,* and they clearly bear a close relation to the pressure. The off-diagonal components are viscous *shear stresses*.

The task now is to relate the viscous stresses to the dynamics of the flow. The major physical arguments necessary for this endeavor have been presented in Section 1–3. There are also some mathematical and physical properties of the stress tensor that will prove useful. One can begin with the simple situation shown in Fig. 1–7 (this is the same case as for the viscometer in Fig. 1–2) and attempt to generalize. For this simple case, we have

$$\tau = \mu \frac{\partial u}{\partial y} \tag{1–22}$$

The quantity $\partial u/\partial y$ is twice the *shear strain rate,* which is defined as one-half the rate of change of the angle between two lines in the fluid that are initially perpendicular. Thus, Eq. (1–22) can be read to say that the shear stress is proportional to the shear strain rate.

In the general case, the shear strain rate is a tensor, ϵ_{ij}, whose components can

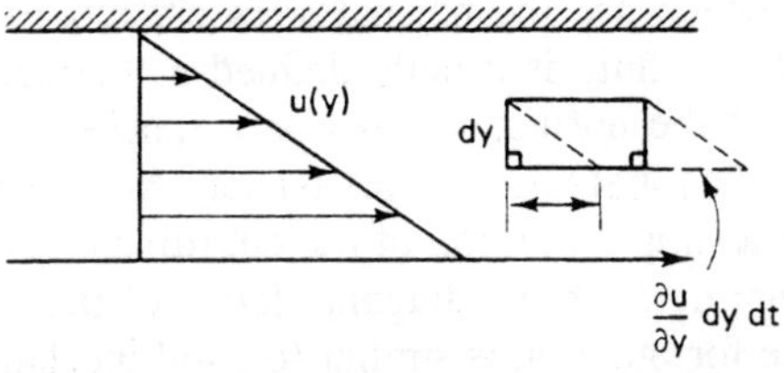

Figure 1–7 Velocity distribution of a viscous fluid in a narrow gap between a moving and a fixed surface.

be developed by considering Fig. 1–8 with ϵ_{xy} as an example. Clearly,

$$\epsilon_{xy} = \frac{1}{2}\left(\frac{d\phi_1}{dt} + \frac{d\phi_2}{dt}\right) = \frac{1}{2}\left(\frac{\partial v}{\partial x} + \frac{\partial u}{\partial y}\right) \tag{1–23}$$

and by analogy,

$$\epsilon_{yz} = \frac{1}{2}\left(\frac{\partial w}{\partial y} + \frac{\partial v}{\partial z}\right); \qquad \epsilon_{zx} = \frac{1}{2}\left(\frac{\partial u}{\partial z} + \frac{\partial w}{\partial x}\right) \tag{1–24}$$

As in solid mechanics, this tensor is symmetric. Thus, Eq. (1–22) is generalized to

$$\begin{aligned} \tau_{xy} &= \mu\left(\frac{\partial u}{\partial y} + \frac{\partial v}{\partial x}\right) \\ \tau_{yz} &= \mu\left(\frac{\partial v}{\partial z} + \frac{\partial w}{\partial y}\right) \\ \tau_{zx} &= \mu\left(\frac{\partial w}{\partial x} + \frac{\partial u}{\partial z}\right) \end{aligned} \tag{1–25}$$

for the shear stresses. The key assumption is that the viscous stress tensor is a linear function of the rate-of-strain tensor. This analysis leads directly to equations for a *Newtonian fluid*. It has also been assumed that the fluid is *isotropic*—that is, the process is independent of direction.

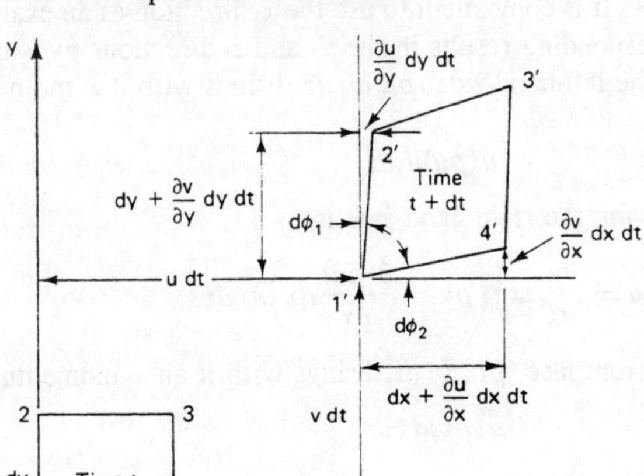

Figure 1–8 Distortion of a fluid element as it moves in a viscous shear flow.

The diagonal terms, which are the normal stresses, require special consideration. First, the pressure p in a fluid is usually defined as the negative (i.e., acting inward) of the mean value of the normal stresses on a sphere of unit radius, so the pressure must appear in the diagonal terms of the tensor. Also, we wish the final relation to be invariant under a simple rotation of the coordinate system. This requires the introduction of additional items in the diagonal terms of the tensor and a second viscosity coefficient, λ. The formulation is similar to solid mechanics, where we require two coefficients of elasticity for an isotropic solid. The final forms of the diagonal terms become

$$\tau_{xx} = 2\mu\frac{\partial u}{\partial x} + \lambda\left(\frac{\partial u}{\partial x} + \frac{\partial v}{\partial y} + \frac{\partial w}{\partial z}\right) - p$$

$$\tau_{yy} = 2\mu\frac{\partial v}{\partial y} + \lambda\left(\frac{\partial u}{\partial x} + \frac{\partial v}{\partial y} + \frac{\partial w}{\partial z}\right) - p \qquad (1\text{–}26)$$

$$\tau_{zz} = 2\mu\frac{\partial w}{\partial z} + \lambda\left(\frac{\partial u}{\partial x} + \frac{\partial v}{\partial y} + \frac{\partial w}{\partial z}\right) - p$$

But the average of the normal stresses must be equal to the negative of the pressure, whether there is motion (and therefore viscous normal stresses) or not. Thus,

$$-p = \frac{\tau_{xx} + \tau_{yy} + \tau_{zz}}{3} = -p + \frac{3\lambda + 2\mu}{3}\left(\frac{\partial u}{\partial x} + \frac{\partial v}{\partial y} + \frac{\partial w}{\partial z}\right) \qquad (1\text{–}27)$$

Obviously, the additional term on the right-hand side must be zero. For incompressible fluids, the quantity in parentheses is zero by mass conservation expressed as the continuity equation. For other fluids, it is common simply to assume that $\lambda = -2\mu/3$ (*Stokes's hypothesis*), although that relationship is hard to justify rigorously, except for monatomic gases. This completes the modeling of the stresses.

We turn to the fluid element in Fig. 1–9 to apply conservation of momentum. Clearly, it will be necessary here to balance forces with momentum changes in all three coordinate directions. It is convenient to use the x-direction as an example and then write down the corresponding results in the y- and z-directions by extension.

The mass entering the left-hand side, $\rho u\, dy\, dz$, brings with it x-momentum in the amount

$$u(\rho u)dydz \qquad (1\text{–}28)$$

and the x-momentum leaving the right-hand side is

$$\left(u + \frac{\partial u}{\partial x}dx\right)\left(\rho u + \frac{\partial(\rho u)}{\partial x}dx\right)dy\,dz \qquad (1\text{–}29)$$

The mass entering the bottom face, $\rho v\, dx\, dz$, brings with it an x-momentum of

$$u(\rho v)dx\,dz \qquad (1\text{–}30)$$

and the x-momentum leaving the top is

$$\left(u + \frac{\partial u}{\partial y}dy\right)\left(\rho v + \frac{\partial(\rho v)}{\partial y}dy\right)dx\,dz \qquad (1\text{–}31)$$

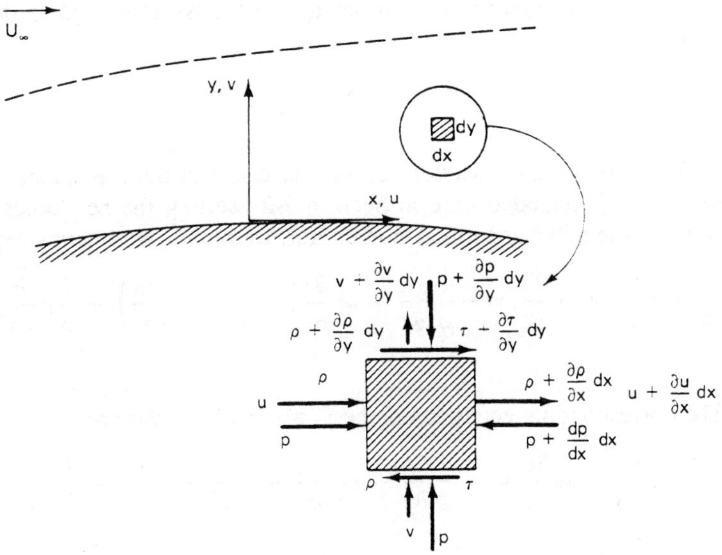

Figure 1–9 Differential volume for conservation of mass and momentum in the viscous layer.

The same procedure can be followed for the front and back faces. The net change in x-momentum in the volume is balanced by the resultant of the forces in the x-direction.

The primary fluid forces considered here are two surface forces: pressure and viscous forces. Volume (or body) forces, such as those due to gravity, will be neglected because they are small compared to inertia and the other forces for most conditions of practical interest. The summation of pressure forces in the x-direction is (see Fig. 1–9)

$$p\,dy\,dz - \left(p + \frac{\partial p}{\partial x}dx\right)dy\,dz \tag{1–32}$$

So the net pressure force is

$$\frac{\partial p}{\partial x}\,dx\,dy\,dz \tag{1–33}$$

Generalizing to include the net of all the surface forces acting in the x-direction, we obtain, after dividing by the volume $(dx)(dy)(dz)$,

$$\frac{\partial \tau_{xx}}{\partial x} + \frac{\partial \tau_{xy}}{\partial y} + \frac{\partial \tau_{xz}}{\partial z} \tag{1–34}$$

Using Eqs. (1–25) and (1–26) with $\lambda = -2\mu/3$, this becomes

$$-\frac{\partial p}{\partial x} + \mu\left(\frac{\partial^2 u}{\partial x^2} + \frac{\partial^2 u}{\partial y^2} + \frac{\partial^2 u}{\partial z^2}\right) + \frac{1}{3}\mu\frac{\partial}{\partial x}\left(\frac{\partial u}{\partial x} + \frac{\partial v}{\partial y} + \frac{\partial w}{\partial z}\right) \tag{1–35}$$

It is convenient to use conservation of mass expressed as the continuity equation

$$\frac{\partial u}{\partial x} + \frac{\partial v}{\partial y} + \frac{\partial w}{\partial z} = 0 \tag{1-36}$$

A derivation of this equation can be found in any fluid mechanics textbook. Also, a derivation is included here in Section 3-2. Setting the net forces in the x-direction equal to the net change in x-momentum, we obtain, after using Eq. (1–36),

$$u\frac{\partial u}{\partial x} + v\frac{\partial u}{\partial y} + w\frac{\partial u}{\partial z} = -\frac{1}{\rho}\frac{\partial p}{\partial x} + \nu\left(\frac{\partial^2 u}{\partial x^2} + \frac{\partial^2 u}{\partial y^2} + \frac{\partial^2 u}{\partial z^2}\right) + \frac{1}{3}\nu\frac{\partial}{\partial x}\left(\frac{\partial u}{\partial x} + \frac{\partial v}{\partial y} + \frac{\partial w}{\partial z}\right) \tag{1-37}$$

The corresponding equations in the y- and z-directions are

$$u\frac{\partial v}{\partial x} + v\frac{\partial v}{\partial y} + w\frac{\partial v}{\partial z} = -\frac{1}{\rho}\frac{\partial p}{\partial y} + \nu\left(\frac{\partial^2 v}{\partial x^2} + \frac{\partial^2 v}{\partial y^2} + \frac{\partial^2 v}{\partial z^2}\right) + \frac{1}{3}\nu\frac{\partial}{\partial y}\left(\frac{\partial u}{\partial x} + \frac{\partial v}{\partial y} + \frac{\partial w}{\partial z}\right) \tag{1-38}$$

and

$$u\frac{\partial w}{\partial x} + v\frac{\partial w}{\partial y} + w\frac{\partial w}{\partial z} = -\frac{1}{\rho}\frac{\partial p}{\partial z} + \nu\left(\frac{\partial^2 w}{\partial x^2} + \frac{\partial^2 w}{\partial y^2} + \frac{\partial^2 w}{\partial z^2}\right) + \frac{1}{3}\nu\frac{\partial}{\partial z}\left(\frac{\partial u}{\partial x} + \frac{\partial v}{\partial y} + \frac{\partial w}{\partial z}\right) \tag{1-39}$$

By virtue of Eq. (1–36), the quantity in parentheses in the last term of each of the foregoing momentum equations is zero for incompressible flow.

1–6 REDUCTION OF THE NAVIER-STOKES EQUATIONS TO THE BOUNDARY LAYER EQUATIONS

We next introduce the scaled variables $U = u/U_\infty$, $V = v/U_\infty$, $X = x/L$, $Y = y/L$, $P = p/\rho_{ref} U_\infty^2$, $\rho' = \rho/\rho_{ref}$, and $\mu' = \mu/\mu_{ref}$, into the two-dimensional, incompressible, steady Navier-Stokes equations to consider the simplest case (constant density, constant property flow, $\rho = \rho_{ref}$, $\mu = \mu_{ref}$) for illustration:

$$\frac{\partial U}{\partial X} + \frac{\partial V}{\partial Y} = 0 \tag{1-40}$$

$$U\frac{\partial U}{\partial X} + V\frac{\partial U}{\partial Y} = -\frac{\partial P}{\partial X} + \frac{1}{\text{Re}}\left(\frac{\partial^2 U}{\partial X^2} + \frac{\partial^2 U}{\partial Y^2}\right) \tag{1-41}$$

$$U\frac{\partial V}{\partial X} + V\frac{\partial V}{\partial Y} = -\frac{\partial P}{\partial Y} + \frac{1}{\text{Re}}\left(\frac{\partial^2 V}{\partial X^2} + \frac{\partial^2 V}{\partial Y^2}\right) \tag{1-42}$$

Where, $\text{Re} = \rho_{ref} U_\infty L/\mu_{ref}$. If one simply lets Re become very large, as is often the situation in real problems, the last terms on the right-hand side of the two momentum equations become negligible. This effect has serious consequences. The influence of the viscosity disappears entirely. The order of the differential equations is reduced by one, so some physically required boundary conditions must be dropped. The result is an inviscid analysis that cannot enforce the *no-slip condition* on the wall and that predicts zero frictional drag. This analysis is unacceptable in most circumstances.

The way out of the dilemma can be found by following the suggestions of Prandtl, as illustrated by a model problem. The behavior of a system of equations as the coefficient of the highest order derivative(s) tends to zero can be illustrated by a simple example developed by Prandtl himself to better understand the case of the Navier-Stokes equations as 1/Re goes towards zero as Re goes to very large values. The best kind of example is one where the full problem can be solved exactly, no matter what the value of the coefficients. Prandtl selected the simple spring-mass-damper dynamics problem described by

$$m\frac{d^2x}{dt^2} + a\frac{dx}{dt} + cx = 0 \tag{1–43}$$

The exact solution to this equation can be written

$$x(t) = C_1e^{\lambda_1 t} + C_2e^{\lambda_2 t} \tag{1–44}$$

where

$$\lambda_{1,2} = \frac{-a \pm \sqrt{a^2 - 4mc}}{2m} \tag{1–45}$$

Now let $m \to 0$, and the solution reduces to

$$\lim_{m\to 0} x(t) = C_2e^{-\frac{c}{a}t} \tag{1–46}$$

This is the same as the solution of the equation that results when one lets $m \to 0$ in Eq. (1–43) before solving it—namely,

$$a\frac{dx}{dt} + cx = 0 \tag{1–47}$$

Now consider a case with the initial conditions $x(0) = 0$ and $x'(0) = 1$. The reduced solution cannot satisfy both boundary conditions. It can, however, match the exact solution that uses both boundary conditions for large time. Call this solution $x_{\text{outer}}(t)$.

Up to this point, nothing creative has been done. But now we introduce a new independent variable $\xi \equiv t/m$, and transform the full equation to

$$\frac{d^2x}{d\xi^2} + a\frac{dx}{d\xi} + mcx = 0 \tag{1–48}$$

The fundamental idea behind this transformation is to stretch out the *thin* region for small time near $t = 0$, where rapid changes in $x(t)$ occur for small values of m.

Now, as $m \to 0$, the equation reduces to

$$\frac{d^2x}{d\xi^2} + a\frac{dx}{d\xi} = 0 \tag{1–49}$$

Note that the order of Eq. (1–49) has not been reduced, as was the case when we let $m \to 0$ in Eq. (1–43), so a solution to the simplified equation can still satisfy two boundary conditions. The exact solution to Eq. (1–49) is

$$x_{\text{inner}}(t) = C_3 e^{-a\xi} + C_4 \tag{1–50}$$

The first boundary condition requires $C_3 = -C_4$. The constant is determined further using the *method of matched asymptotic expansions*, which says that "the outer limit of the inner solution must equal the inner limit of the outer solution." [See Van Dyke (1964).] Thus, $C_4 = C_2$, as can be seen from Eqs. (1–46) and (1–50). Using $x'(0) = 1$ on the inner solution, $C_3 = -m/a = -C_4$. Finally,

$$x_{\text{inner}}(t) = \frac{m}{a}(1 - e^{-\frac{a}{m}t}) \tag{1–51}$$

The complete approximate solution to the original problem for small values of m is now a composite of the inner solution, valid near $t = 0$, and the outer solution,

$$x_{\text{outer}}(t) = \frac{m}{a}e^{-\frac{c}{a}t} \tag{1–52}$$

valid for large t. The composite solution compares favorably with the exact solution, valid for all t. (See Problem 1.7 at the end of the chapter.)

Now consider the reduction of the Navier-Stokes equations to the boundary layer equations using the same approach. Stretch the normal coordinate Y and the velocity in that direction, V, to expand the thin region for cases with high Re. Let the small parameter that will accomplish the stretching be represented by ϵ and remain unspecified for now. Take $u^* = U$, $P^* = P$, $v^* = V/\epsilon$, $x^* = X$, and $y^* = Y/\epsilon$, and the equations become

$$\frac{\partial u^*}{\partial x^*} + \frac{\partial v^*}{\partial y^*} = 0 \tag{1–53}$$

$$u^*\frac{\partial u^*}{\partial x^*} + v^*\frac{\partial u^*}{\partial y^*} = -\frac{\partial P^*}{\partial x^*} + \frac{1}{\text{Re}}\left(\frac{\partial^2 u^*}{\partial x^{*2}} + \frac{1}{\epsilon^2}\frac{\partial^2 u^*}{\partial y^{*2}}\right) \tag{1–54}$$

$$\epsilon u^*\frac{\partial v^*}{\partial x^*} + \epsilon v^*\frac{\partial v^*}{\partial y^*} = -\frac{1}{\epsilon}\frac{\partial P^*}{\partial y^*} + \frac{1}{\text{Re}}\left(\epsilon\frac{\partial^2 v^*}{\partial x^{*2}} + \frac{1}{\epsilon}\frac{\partial^2 v^*}{\partial y^{*2}}\right) \tag{1–55}$$

From the first momentum equation, Eq. (1–54), it can be seen that taking $\epsilon^2 = 1/\text{Re}$ simplifies the viscous terms by eliminating $\partial^2 u^*/\partial x^{*2}$ as Re becomes very large, but the order of the system, at least for y^*, is not reduced. Thus, two boundary conditions on y^*, at $y^*=0$ on the wall and $y^* \to \infty$, can be imposed, as is required physically. The continuity equation is unaffected. The result for the normal momentum equation, Eq. (1–55), is

$$\frac{1}{\sqrt{\mathrm{Re}}}\left(u^*\frac{\partial v^*}{\partial x^*}+v^*\frac{\partial v^*}{\partial y^*}\right)=-\sqrt{\mathrm{Re}}\frac{\partial P^*}{\partial y^*}+\frac{1}{\sqrt{\mathrm{Re}}}\left(\frac{1}{\mathrm{Re}}\frac{\partial^2 v^*}{\partial x^{*2}}+\frac{\partial^2 v^*}{\partial y^{*2}}\right) \tag{1–56}$$

Examining this equation, one can see that, as Re is allowed to become large, all the terms but the pressure gradient term become negligibly small. Thus, to order $O(\mathrm{Re}^{-1/2})$, the untransformed system of equations reduces to

$$\frac{\partial u}{\partial x}+\frac{\partial v}{\partial y}=0 \tag{1–57}$$

$$u\frac{\partial u}{\partial x}+v\frac{\partial u}{\partial y}=-\frac{1}{\rho}\frac{dp}{dx}+\nu\frac{\partial^2 u}{\partial y^2} \tag{1–58}$$

$$\frac{\partial p}{\partial y}\approx 0 \tag{1–59}$$

The reader should review the end of Section 1–4 to recall how these equations are to be used.

1–7 SEPARATION AND THE KUTTA CONDITION

The information developed so far can be used to explain the important phenomenon of *boundary layer separation,* and that will further lead to an explanation of the physical basis of the famous *Kutta condition,* widely used in inviscid aero- and hydrodynamics. When the shape of the surface of a body immersed in a flowing fluid or of a channel with fluid flowing through it is such that the static pressure increases rapidly in the streamwise direction, a dramatic and undesirable change in the flow pattern is often observed. Photographs and a schematic of the flow are given in Fig. 1–10. Streamlines in the boundary layer near the surface, which have been proceeding along generally following the shape of the surface, suddenly depart from the surface. We say that the flow *separates* from the surface. This occurrence is undesirable, because the body or channel shape was usually selected to produce a given static pressure variation for a specific purpose, such as to produce lift for a wing or to provide efficient deceleration of a flow in a diffuser. If the flow separates, the intended static pressure distribution will obviously be disrupted. Also, flow separation will disturb an intended heat or mass transfer distribution.

The process leading to separation can be made clear by a crude, but useful, argument. The development is crude, and not exact, because it uses Bernoulli's equation in the viscous boundary layer, even though that strictly applies only to inviscid flow. Our purpose at this point is, however, only to illuminate the phenomena involved, not to analyze the flow in detail, so some liberties may be taken. Bernoulli's equation states that

$$u\frac{du}{ds}=-\frac{1}{\rho}\frac{dp}{ds} \tag{1–60}$$

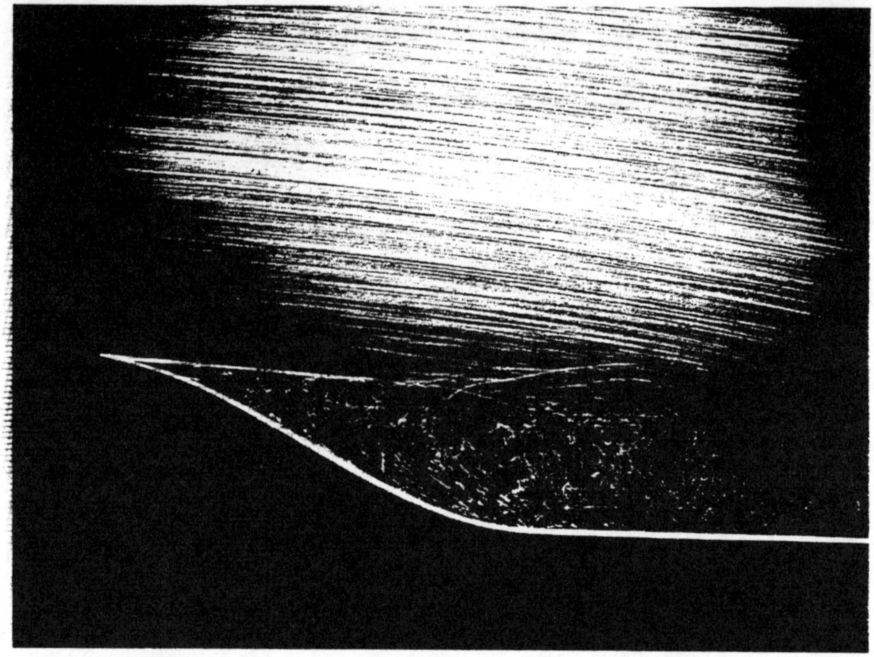

(A) Visualization of laminar separation from a curved wall by Werle (1982) at ONERA. Air bubbles in water

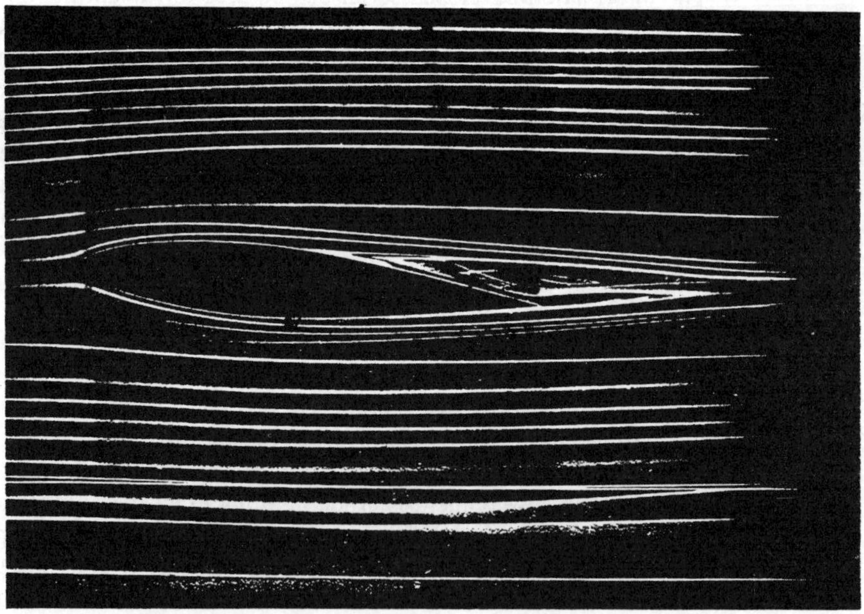

(B) Visualization of laminar separation from an airfoil at angle of attack by Werle (1982) at ONERA. Dye streaks in water.

Figure 1–10 Illustrations of the boundary layer separation process.

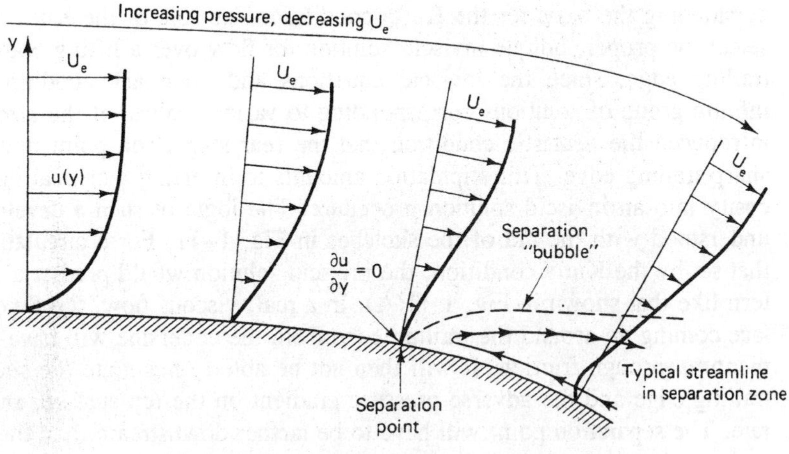

(C) Schematic illustration of separation.

Figure 1–10 *(cont.)*

along a streamline. In the inviscid flow outside the boundary layer, this equation will correctly relate the variation in the edge velocity, $U_e(x)$, to the pressure gradient. Note that for increasing pressure, the velocity decreases. For this exercise, assume also that Eq. (1–60) holds locally in the boundary layer. Now, the boundary layer assumption states that the static pressure is constant across the layer, so the pressure gradient (in the streamwise direction) must also be the same in the layer. Comparing two points—one at the edge of the layer and one well in the viscous layer—we may write

$$u\frac{du}{ds} = U_e\frac{dU_e}{ds} \tag{1–61}$$

from Eq. (1–60), since the pressure gradients are the same. This relation clearly shows that a given pressure gradient will produce a much larger change in velocity at a point in the lower velocity viscous layer than at a point in the inviscid flow at the edge. The ratio of the changes will be $U_e : u$. Since u approaches zero at the wall, a pressure gradient will produce a large Δu in a region where u is already small. The result is as shown by the successive profiles sketched in Fig. 1–10(C), leading finally to an actual reversal in the direction of flow. This reversed flow causes an effective obstacle to the upstream flow, so it *separates* from the original body surface to flow over the *separation bubble*. At the point of separation, $(\partial u/\partial y)_w = 0$, so the wall or skin friction is zero.

When a flow separates, the viscous region is no longer thin, even at high Reynolds numbers, and the boundary layer assumptions are no longer valid. Thus, the conditions for the applicability of the boundary layer assumptions must be a flow with high Reynolds number over streamlined, not bluff, shapes.

The phenomenon of separation is important in its own right, and further discussion of it will be presented later in this book. Separation is also important in understanding the basis for the Kutta condition. The role of the Kutta condition is to select the proper, unique inviscid solution for flow over a lifting body with a sharp trailing edge, since the inviscid equations and boundary conditions provide an infinite group of solutions corresponding to various values of the *circulation*. Kutta introduced the heuristic condition that the rear stagnation point must occur at the sharp trailing edge. This stipulation amounts to inserting a critical influence of viscosity into an inviscid solution procedure. The logic of such a development can be understood with the aid of the sketches in Fig. 1–11. For a circulation lower than that set by the Kutta condition, the inviscid solution would predict a streamline pattern like that shown in Fig. 1-11(A). In a real, viscous flow, the fluid near the surface coming up around the trailing edge from the underside will have lost some momentum through friction. It will then not be able to negotiate the sharp turn at the trailing edge and the adverse pressure gradient on the top surface, and it will separate. The separation point will have to be farther downstream than the inviscid stagnation point shown in the figure. In fact, the only stable location will be at the trailing edge, as shown in Fig. 1–11(B). This viscous process cannot be described by the inviscid analysis, so the Kutta condition must be asserted in order to make the inviscid analysis unique. Again, we see the great importance of even a small viscosity.

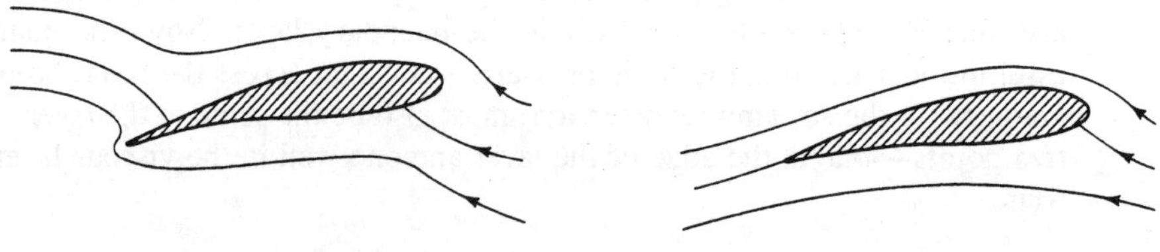

(A) Circulation less than required by Kutta Condition

(B) Circulation as required by Kutta Condition

Figure 1–11 Streamline patterns for flow past an airfoil predicted by inviscid fluid theory.

1–8 NON-NEWTONIAN FLUID FLOW

The general subject of non-Newtonian fluid flow is vast and complex. In this volume, the intent will be to provide an introduction to the subject and to describe some new phenomena that occur with such fluids, as opposed to the more conventional Newtonian fluids.

There are three broad classes of non-Newtonian fluids: time-independent fluids, where the shear depends only on the instantaneous rate of strain; time-dependent fluids, where the shear depends on the magnitude and duration of the strain; and viscoelastic fluids, which show a partial elastic recovery when the strain is removed.

Time-independent fluids are further subdivided into those with a yield stress and those without. Fluids with an initial yield stress are called *plastic fluids*, as shown in Fig. 1–3. The simplest model is a so-called *Bingham plastic*, which is described by an initial yield stress τ_o and a constant *plastic viscosity* μ_{BP}. Such a fluid will have a positive y-intercept and a constant positive slope on a plot such as Fig. 1–3. Thus,

$$\tau = \tau_o + \mu_{BP}\frac{\partial V}{\partial n} \tag{1–62}$$

The concept of an *apparent viscosity*, defined as

$$\mu_a = \mu_{BP} + \frac{\tau_o}{\partial V/\partial n} \tag{1–63}$$

is sometimes useful. Note that this concept implies a diminishing viscosity as the strain rate increases. Plastic fluids occur in some plastic melts, slurries, greases, some foodstuffs, and paper pulp, to give a few examples.

Plastic fluids without an initial yield stress are called *pseudoplastic fluids*, and most non-Newtonian fluids of practical interest fall in this category. Examples include many polymers, adhesives, paints, greases, and biological fluids. Figure 1–3 shows a typical curve of stress versus rate of strain. The simplest model for such fluids is the *power law model*, which states that

$$\tau = \mu_{PL}\left(\frac{\partial V}{\partial n}\right)^p \tag{1–64}$$

where p is usually less than unity. For this case, the apparent viscosity becomes

$$\mu_a = \mu_{PL}\,|\partial V/\partial n|^{p-1} \tag{1–65}$$

decreasing with increasing rate of strain. A *dilatant* fluid has p greater than unity, but such fluids are much less common.

Time-dependent non-Newtonian fluids usually exhibit *hysteresis;* that is, the curve of stress versus rate of strain will go up one path and down another with the same beginning and ending points as the loading is increased and then decreased. Viscoelastic fluids exhibit both viscous and elastic properties. These last two classes of fluids occur much less frequently in engineering practice, so they will not be discussed further here. (See Skelland (1967) for more information.)

1–9 BASIC NOTIONS OF TURBULENT FLOW

Most flows of practical concern are *turbulent*, and not *laminar*, so it is important to gain a clear understanding of the nature of turbulent flows early in the study of viscous phenomena. In this chapter, we deal only with the physical nature of turbulence. The analytical description of turbulent flows will be discussed in later chapters, after a solid foundation in the form of an analytical description of laminar flows has been laid.

(A) Re_D = 15,000. Dye streaks in water.

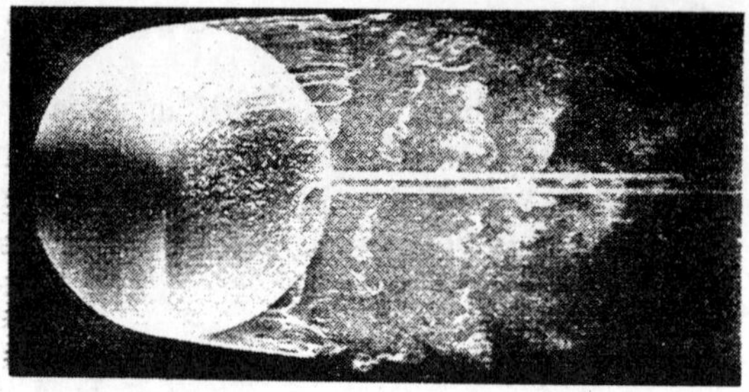

(B) Re_D = 15,000. Air bubbles in water.

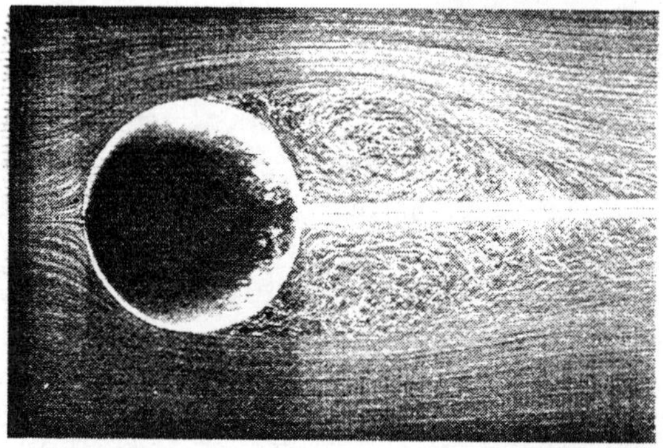

(C) Re_D = 30,000. Dye streaks in water.

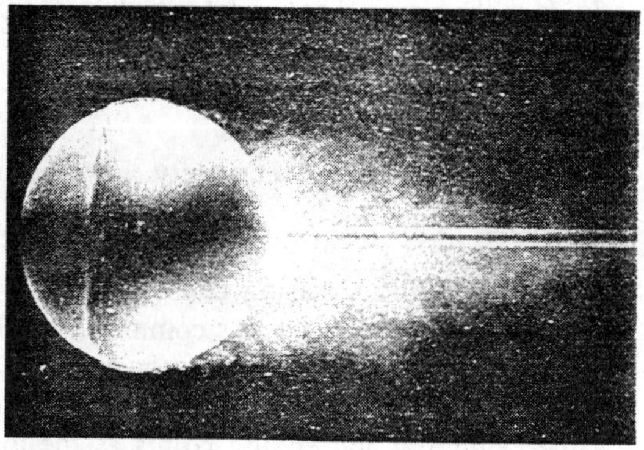

(D) RE_D = 30,000. Air bubbles in water.

Figure 1–12 Flow visualizations for the turbulent flow in the wake of a sphere at various Reynolds numbers by Werle (1982) at ONERA.

In Fig. 1–12, one sees beautiful visualizations of turbulent flow in the wake of a sphere by Werle (1982) at ONERA in France. It is worth the time to study these photographs carefully. Also, the reader should supplement these instantaneous examples with time-varying observations, the plume from a smokestack being a good example. The films, *Characteristics of Laminar and Turbulent Flow*[1] and *Turbulence*,[2] are both excellent and helpful. The flow visualizations shown in Fig. 1–12 reveal some of the most significant characteristics of turbulent flow. First, there is a general *swirling* nature of the flow, involving indistinct lumps of fluid called *eddies*. There is a very wide range in the size of the eddies occurring at the same time or at the same place in the turbulent region. Second, the instantaneous boundary between the turbulent region and the nonturbulent, outer, inviscid flow is sharp. The flow at a point near the average location of the edge of the boundary layer is intermittently either turbulent or not. Third, turbulence is basically an unsteady process. Any still picture, such as those shown in the figure, does not adequately convey the real situation. The unsteadiness may appear vaguely periodic, but a closer study will reveal that the flow is randomly unsteady. A harmonic analysis of the motion shows that fluctuations over a very wide range of effective frequencies (several orders of magnitude) are present. Fourth, turbulence is always three dimensional, even if the background flow is two dimensional. A final important point to be learned and remembered from visualizations of turbulent flows is that the irregular variations in the motion are not small with respect to either time or space. Some of the instantaneous velocity fluctuations in the wake behind a bluff body such as that shown in Fig. 1–12, for example, are of the order of magnitude of the free-stream velocity.

An instrument for measuring the fluctuating velocities in a turbulent flow is the *hot-wire anemometer*. This is a device that is simple in concept, and modern electronics has made it easy to use. The heart of the instrument is a very thin wire (say, 10^{-3} mm), stretched between two sharp prongs, that is electrically heated to a temperature slightly higher than the fluid. Fluid flowing past the wire will cool it by an amount depending directly on the instantaneous velocity. With very fine wires and suitable electronics, this device can follow even the very fast, small-scale components of turbulent motion. The output from a hot-wire anemometer displayed on an oscilloscope would look as shown in Fig. 1–13(A). If the probe were located near the outer edge of the turbulent zone, it would sense flow that is only intermittently turbulent, and the signal would look as shown in Fig. 1–13(B). The *intermittency* is defined as the fraction of the time that the flow is turbulent.

Looking at Fig. 1–13(A), one can identify a time-mean velocity,

$$U(x,y,z) \equiv \frac{1}{T_0}\int_0^{T_0} u(x,y,z,t)dt \qquad (1\text{–}66)$$

where T_0 is a time that is long compared to the longest periods of the fluctuations. To obtain some measure of the average magnitudes of the fluctuation, one can split the

[1] Produced by the University of Iowa.

[2] Produced by the Encyclopaedia Brittanica.

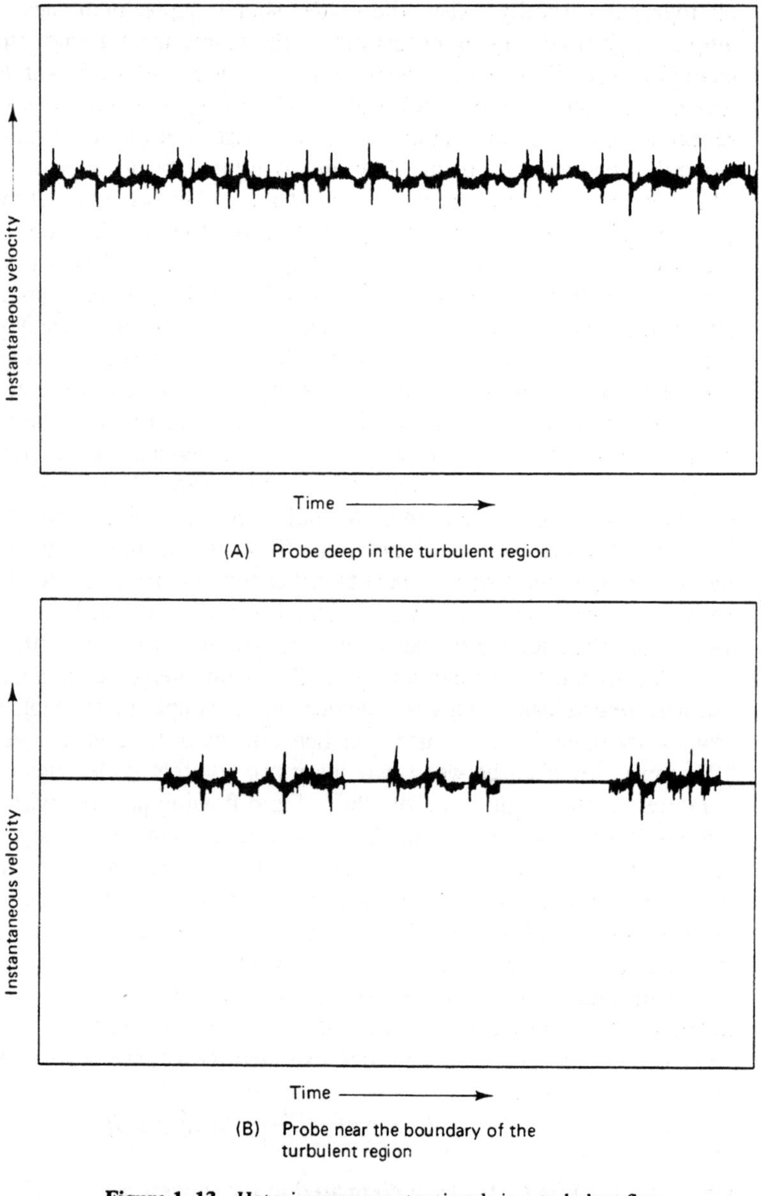

Figure 1–13 Hot-wire anemometer signals in a turbulent flow.

instantaneous velocity into a mean and a fluctuating part:

$$u(x,y,z,t) = U(x,y,z) + u'(x,y,z,t) \tag{1-67}$$

Obviously, the average value of u' is zero, so we take the root-mean-squared value, $\sqrt{(u')^2}$. The ratio of $\sqrt{(u')^2}$ to either U or U_∞ is called the *turbulence intensity*. Some profile results for the mean flow and the turbulence intensities in all three flow directions in a turbulent boundary layer are shown in Fig. 1–14.

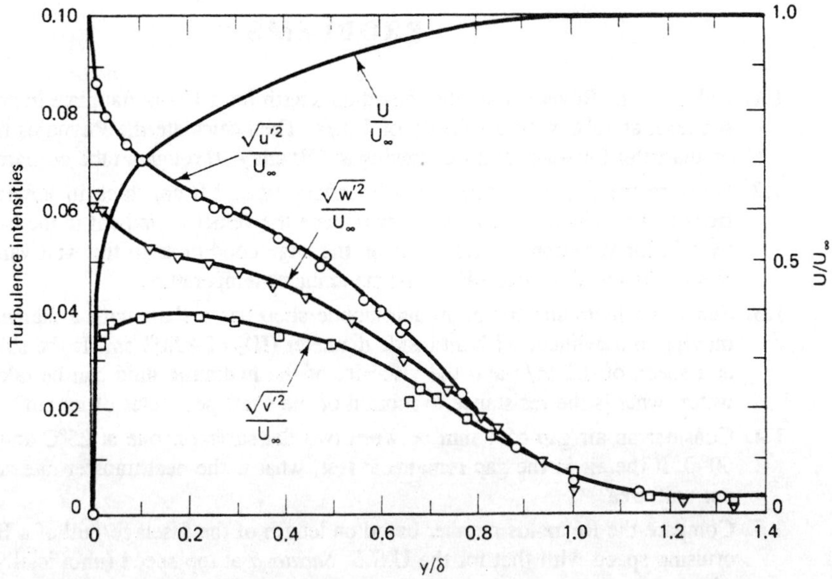

Figure 1–14 Measurements of the mean velocity and turbulent intensities in a boundary layer. (From Klebanoff, 1955).

One last conceptual matter is important: What is the difference between an unsteady laminar flow and a turbulent flow? And following that, can we speak of a *steady* turbulent flow? In principle, a velocity-time history such as that in Fig. 1–13(A) could be produced by an unsteady laminar flow. An example might be the flow in a laminar boundary layer near an oscillating surface. If the surface oscillations were controlled by a random number generator, and if they were at a high rate, the velocity near the surface could be sensed by a hot-wire anemometer, and the output might resemble that for a turbulent flow. The key to distinguishing between such a case and a genuinely turbulent flow is that the velocity fluctuations in the laminar case would be directly correlated with the oscillations of the surface (i.e., the boundary conditions). The fluctuations in a turbulent flow are truly random, and they are not correlated with the boundary conditions. This leads us to the apparent contradiction in the first terms contained in the phrase *steady turbulent flow*. The time-averaged, mean flow represented by U, for example, can indeed be steady. The mean flow behind the sphere in Fig. 1–12 will be constant with time (i.e., steady) if the cylinder is fixed and the approach flow is constant. To be precise, one says that such a flow is *steady in the mean*. A turbulent flow can also be *unsteady in the mean*. For the case of the wake of a sphere, the mean flow field would vary with time if the sphere were oscillated [i.e., $U = U(x,y,t)$]. However, any time variation of the mean component of a turbulent flow will be correlated with the time variations of the boundary conditions.

PROBLEMS

1.1. Calculate the Reynolds number based on length for a 1.0-m flat plate in an airstream at sea level at 25°C with a velocity of 3 m/s. Then calculate the Reynolds number based on diameter for water at 25°C flowing at 150 cm^3/s through a tube of diameter 1.0 cm.

1.2. Estimate the frictional stress in a boundary layer 1.0 cm thick in flowing water at a free-stream velocity of 5 m/s. Approximate the velocity gradient in the boundary layer by a linear variation in velocity from the edge conditions to the wall value. Compare this to the inertial force ρV^2. Assume ambient temperature.

1.3. Suppose a hydraulic lift in an automobile shop has a shaft with a diameter of 40 cm moving in a cylinder with an inside diameter (ID) of 40.02 cm. If the shaft is moving at a speed of 0.2 m/s and the viscosity of the hydraulic fluid can be taken as that of water, what is the resistance to motion of the shaft per meter of length?

1.4. Consider an air gap of 1 mm between two flat surfaces, one at 25°C and the other at 50°C. If the air in the gap remains at rest, what is the heat transfer rate across the gap per unit area?

1.5. Compare the Reynolds number based on length of the fuselage/hull of a Boeing 747 at cruising speed with that for the U.S.S. *Saratoga* at top speed (unofficially reported as 40 knots).

1.6. Suppose you are supplied with the test data shown in the following table (the data are for shear measurements made for three fluids as a function of varying rate of strain du/dy):

du/dy 1/sec	Shear, Fluid A lbf/ft^2	Shear, Fluid B lbf/ft^2	Shear, Fluid C lbf/ft^2
0.0	0.00	0.00	2.00
20.0	2.23	0.40	2.40
40.0	3.16	0.80	2.80
60.0	3.87	1.20	3.20
80.0	4.47	1.60	3.60
100.0	5.00	2.00	4.00

Plot the data and calculate the apparent viscosity μ_a for each. Can you characterize these three fluids by class?

1.7. Consider a damped spring-mass analog to the boundary layer flow situation, and solve for the exact solution and the composite *boundary layer* solution for a case with $m = 10^{-3}$, $a = 0.05$, and $c = 0.1$. Plot and discuss the results.

1.8. Consider Prandtl's damped spring-mass analog to the boundary layer flow situation. Determine the thickness of the *boundary layer* as a function of the small parameter m, keeping c and a constant.

1.9. Estimate the static pressure variation across a boundary layer 0.5 in thick over a wing section where the radius of curvature is 5.0 ft. The vehicle is flying at 10,000 ft, and the local inviscid velocity is 300 ft/s.

1.10. Suppose we wish to evaluate and then compare the magnitude of the terms in the x-momentum equation of the Navier-Stokes equations at a high Re in the middle of the boundary layer. For that purpose, assume that $u/U_e = 3/2(y/\delta) - 1/2(y/\delta)^3$, $\delta = 5.0x/(\mathrm{Re}_x)^{1/2}$, and $dp/dx = 0$. Take $x = 0.5$ ft, $U_e = 1.0$ ft/s, and $\mathrm{Re}_x = 5 \times 10^4$. Use the continuity equation to solve for $v(x,y)$, and then compare the terms in the x-momentum equation. Would the boundary layer assumption be justified?

1.11. Locate a clearly observable turbulent flow in your surroundings. Make a series of sketches showing the instantaneous and time-varying features of that flow.

CHAPTER 2

INTEGRAL EQUATIONS AND SOLUTIONS FOR LAMINAR FLOW

2–1 INTRODUCTION

Before undertaking the development of any analytical procedure, the analyst should ask what the desired result of the analysis is to be. A practicing engineer or applied scientist generally is not interested in the *elegance* of a solution, but rather needs the answer to a specific problem. If an elegant solution provides more information than is needed at an appreciable increase in the cost of obtaining the solution, compared with a cruder method, the cruder method is better on a cost-benefit basis.

What, then, is the output usually needed from a boundary layer analysis? Generally, the primary information required is skin friction drag, $C_f(x)$ $(\equiv \tau_w/(1/2\rho U_e^2))$. A closely related item would be the point, if any, where $\tau_w = C_f = 0$ occurs (i.e., the location of separation). If there is heating or cooling, the wall heat transfer rate, $q_w(x)$, is correspondingly sought. When there are concentration gradients, the mass transfer rate at the wall, $\dot{m}_{iw}$, is desired. The next type of information that might be useful is the transverse extent of the viscous region, $\delta(x)$ [and perhaps $\delta_T(x)$ and/or $\delta_c(x)$]. It may surprise the reader to learn that in the vast majority of practical cases, that much output is sufficient. Only rarely in a design calculation would one need to know as much detail as, for example, the streamwise velocity $u(x,y)$ at a given point in the boundary layer.

With the goal of predicting $C_f(x)$ and $\delta(x)$ [and perhaps $q_w(x)$ and $\dot{m}_{iw}(x)$], let us develop the simplest analytical method of achieving those ends. In this chapter, we consider only laminar flows. There are some cases of high Reynolds number flow where the Reynolds number in a substantial part of the flow is low enough for laminar flow to exist. Also, an understanding of laminar analyses is essential to attaining an understanding of the more widely applicable and more complicated turbulent analyses. To keep the derivation simple, it is helpful initially to adopt the additional restrictions of steady, incompressible (constant-density), and constant-property flow.

In formulating the appropriate analytical approach, it is important to note that all of the primary desired quantities [$C_f(x)$, $q_w(x)$, $\dot{m}_{iw}(x)$, $\delta(x)$, etc.] are functions only of the streamwise coordinate x. Apparently, the dependence on the transverse coordinate, y, is of lesser importance, and we may be able to take some liberties. Indeed, we will treat all the flow within the boundary layers at a given station x as one unit and apply conservation principles to it as a whole.

Before beginning, it is perhaps useful to digress a moment and say a few words about the philosophy of the foregoing approach. The engineering or physical science student generally takes this matter too lightly. By this point in his or her education, the student has seen conservation of mass, momentum, energy, and so on, applied to a variety of situations. It is taken for granted that there exist generally applicable conservation principles that can be rigorously used to produce an equation or equations that can faithfully describe real phenomena. The fact is, however, that this state of affairs exists only in the narrow branch of the spectrum of human endeavors known, roughly, as the natural sciences. It does not exist in law, medicine, history, sociology, or economics, to name just a few areas. Those subjects are purely empirical; there are no basic *first principles* that have any real generality. That is why the often-heard question, "If we can send men to the moon, why can't we control our economy?" is in truth nonsensical. That is also why Isaac Newton is commonly counted among the top 10 of the most important human beings who ever lived.

2–2 THE INTEGRAL MOMENTUM EQUATION

Consider the two-dimensional flow in Fig. 2–1. The main flow is from left to right and is bounded on the bottom by a rigid surface. From the discussion in Chapter 1, it is known that the velocity at the surface ($y = 0$) will be taken to be zero [i.e.,

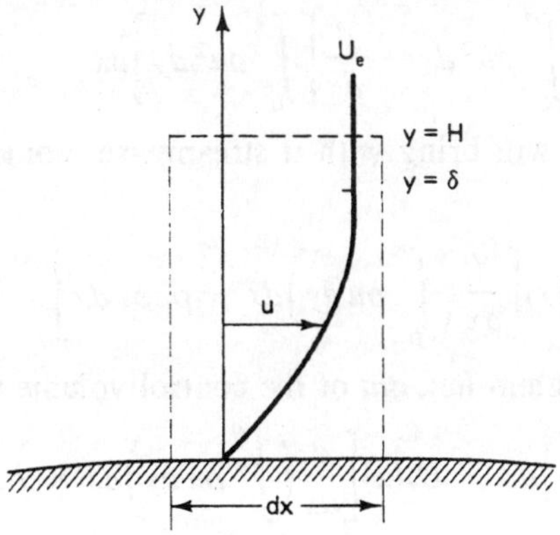

Figure 2–1 Control volume for integral momentum equation derivation.

$u(x, 0) = 0$]. We expect the velocity to vary smoothly from zero at the surface and to blend into the local boundary layer edge velocity $U_e(x)$ at $y = \delta$. We shall apply our conservation principles to a control volume that is finite in the transverse direction, $y = 0$ to $y = H$ [where $H > \delta(x)$], but differential in the streamwise direction dx. This approach is consistent with the notion of downplaying the importance of variations in the y-direction. The streamwise direction x is measured along the body surface in boundary layer analysis, and the transverse direction y is orthogonal to x.

The mass flow entering the left-hand side of the control volume is

$$\int_0^H \rho u \, dy \tag{2–1}$$

and the mass flow leaving the right-hand side can be expressed, using the first term of a *Taylor's series,* as

$$\int_0^H \rho u \, dy + \frac{d}{dx}\left[\int_0^H \rho u \, dy\right] dx \tag{2–2}$$

If the bottom is solid and the flow is steady, the mass in the control volume must remain constant, so the difference between the left- and right-hand side mass flows must come in or go out through the top. If the wall is porous and there is flow through it at v_w, then the quantity $\rho_w v_w$ must be subtracted from the preceding difference to obtain the flow through the top, namely,

$$\frac{d}{dx}\left[\int_0^H \rho u \, dy\right] dx - \rho_w v_w \, dx \tag{2–3}$$

The streamwise momentum flow in the left-hand side is

$$\int_0^H \rho u^2 \, dy \tag{2–4}$$

and that out of the right-hand side is

$$\int_0^H \rho u^2 \, dy + \frac{d}{dx}\left[\int_0^H \rho u^2 \, dy\right] dx \tag{2–5}$$

Any mass flow in the top will bring with it streamwise momentum in an amount proportional to $U_e(x)$

$$U_e(x)\left[\frac{d}{dx}\left(\int_0^H \rho u \, dy\right) dx - \rho_w v_w \, dx\right] \tag{2–6}$$

The net streamwise momentum flux *out* of the control volume then becomes

$$\frac{d}{dx}\left(\int_0^H \rho u^2 \, dy\right) dx - U_e(x)\left[\frac{d}{dx}\left(\int_0^H \rho u \, dy\right) dx - \rho_w v_w \, dx\right] \tag{2–7}$$

The change in x-momentum is balanced with the summation of forces in the x-direction on the fluid. Neglecting body forces such as gravity with respect to the inertia of the flow, we are left with pressure forces and shear forces. Since the boundary layer approximation is essentially $\partial p/\partial y \approx 0$ across the layer (see Chapter 1), p may be treated as $p(x)$ alone. The pressure force on the left-hand side is pH, and that on the right-hand side is

$$-\left(p + \frac{dp}{dx}dx\right)H \tag{2–8}$$

So the net streamwise pressure force on the volume is

$$-\left(\frac{dp}{dx}dx\right)H \tag{2–8a}$$

There is a laminar shear force on the bottom surface ($\tau_w = \mu(\partial u/\partial y)_{y=0}$) acting in the negative x-direction (i.e., retarding the motion of the fluid) and given by

$$-\tau_w\, dx = -\mu\frac{\partial u}{\partial y}\bigg|_{y=0} dx \tag{2–9}$$

But there is no such force on the top surface, since this surface is at, or above, the edge of the viscous layer (i.e., $(\partial u/\partial y)_{y=\delta} = 0$). Any small shear forces on the sides of the volume have no component in the x-direction.

All of the preceding information can be combined into one equation expressing conservation of momentum in the x-direction. In doing so, it is convenient to rewrite the second part of Eq. (2–7), using the rule for the derivative of a product, as

$$U_e(x)\frac{d}{dx}\left[\int_0^H \rho u\, dy\right]dx = \frac{d}{dx}\left[\int_0^H \rho u\, U_e(x)\, dy\right]dx - \frac{dU_e}{dx}\left[\int_0^H \rho u\, dy\right]dx \tag{2–10}$$

Setting Eq. (2–8a) plus Eq. (2-9) equal to Eq. (2–7), with Eq. (2–10), results in

$$-\tau_w - H\frac{dp}{dx} = -\rho\frac{d}{dx}\left[\int_0^H (U_e - u)u\, dy\right] + \frac{dU_e}{dx}\rho\left[\int_0^H u\, dy\right] + \rho_w v_w U_e \tag{2–11}$$

This equation can be further simplified by noting that, since the pressure is assumed constant across the boundary layer, the pressure can be related to the velocity in the inviscid flow just outside the boundary layer through Bernoulli's equation,

$$-\frac{1}{\rho}\frac{dp}{dx} = U_e\frac{dU_e}{dx} \tag{2–12}$$

Finally, we introduce new notation for the integrals in Eq. (2–11), rewritten slightly:

$$\theta \equiv \int_0^\delta \left(1 - \frac{u}{U_e}\right)\frac{u}{U_e}dy \tag{2–13}$$

$$\delta^* \equiv \int_0^\delta \left(1 - \frac{u}{U_e}\right) dy \tag{2–14}$$

Note that the last integral in Eq. (2–11) was combined with the dp/dx term using Eq. (2–12), to produce Eq. (2–14). Equations (2–13) and (2–14) represent more than notational convenience, and we shall attach physical meaning to θ and δ^* shortly. In the meantime, the equation becomes

$$\frac{d\theta}{dx} + \frac{1}{U_e}\frac{dU_e}{dx}(2\theta + \delta^*) - \frac{v_w}{U_e} = \frac{C_f}{2} \tag{2–15}$$

since $\rho = \rho_w$ with the assumptions in this chapter. Some people prefer to derive this same equation by integrating the boundary layer equations [Eqs. (1–57) through (1–59)] across the boundary layer.

This is a good point to step back and consider what has been produced. The equation looks tractable, since it is a relatively simple ordinary differential equation with x as the only independent variable. Our major desired output C_f appears directly as a dependent variable. Here, as with any boundary layer problem, $U_e(x)$ must be known beforehand from a prior inviscid solution (see Chapter 1), and v_w, if not zero, must be given as a boundary condition. But what about θ and δ^*? Inspection of Eqs. (2–13) and (2–14) reveals that they involve δ, one of our other desired outputs, but they also involve $u(y)$, which we do not know. Thus, we have one equation and three unknowns! But fluid dynamicists are an imaginative lot; there is a way out of this quandary.

First, however, look more closely at θ and δ^*. Equations (2–13) and (2–14) show that they are lengths related to δ, but how? The length δ^* is the simpler, so begin with it. The integrand is proportional to $(U_e - u)$, which is the difference between the actual viscous profile and that which would occur if the flow were inviscid; that is, $u = U_e$ for all y down to $y = 0$ (see Fig. 2–2). The difference $(U_e - u)$ is called the *velocity defect*. (Remember the term, as it is important in turbulent boundary layer analysis.) Clearly, there is less mass flow passing through the region $0 \le y \le \delta$ for the viscous profile than there would be if an inviscid profile existed. The difference is proportional to the integral of $(U_e - u)$. We can now ask whether it is possible to consider an adjusted inviscid profile that has the same mass flow near the body as the real viscous profile. Such a profile would have the shape shown by the dash/dot curve in Fig. 2–2, such that shaded area I is equal to shaded area II. That would satisfy the criteria, but how could such a profile occur? The answer is, only if the solid surface were *displaced* upward a distance δ^* such that

$$\rho U_e \delta^* = \rho \int_0^\delta (U_e - u) dy \tag{2–14a}$$

The length δ^* is, therefore, called the *displacement thickness*. A parallel argument can be made concerning the flow of momentum through the region $0 \le y \le \delta$, compared with that for an inviscid profile. It turns out that the surface must be

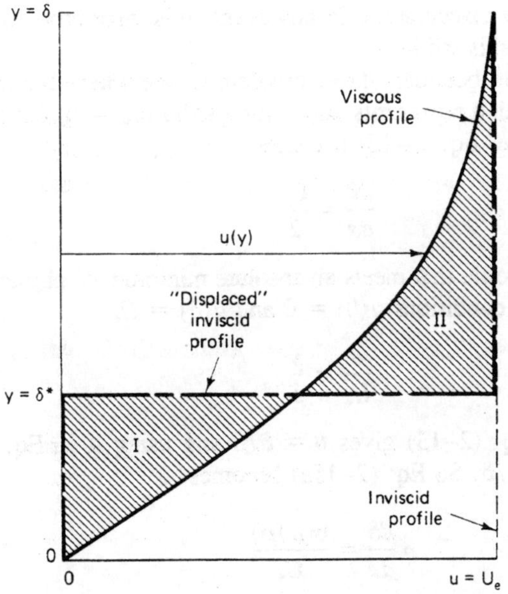

Figure 2–2 Schematic illustration of the definition of the displacement thickness δ^*.

shifted upward by an amount θ, called the *momentum thickness*. The two thicknesses are often also mentioned in reporting the results of experimental measurements, since it is difficult to accurately determine δ, the point where $u(x,y)$ blends smoothly into U_e because of scatter in the data. The integrals can be more accurately found, as the scatter tends to cancel out. In recent times, the same practice has developed in reporting numerical results, for analogous reasons.

2–3 SOLUTION OF THE INTEGRAL MOMENTUM EQUATION

2–3–1 The Pohlhausen Method

At this point, it is necessary to face directly the problem posed by one equation [Eq. (2–15)] and three unknowns: C_f, δ^*, and θ. Observe that if $u(y)$ were known, all three quantities would be defined by Eqs. (2-9), (2–13), and (2–14). Of course, that is foolish, because if we knew *u(y),* we would not be bothering with Eq. (2–15) at all. Looking more closely, however, we see that only

$$\frac{u}{U_e} = f\left(\frac{y}{\delta}\right) \tag{2–16}$$

that is, only the nondimensional *shape* of the profile is needed, not *u(y)* explicitly. If we are bold enough to guess the shape of the profile, the problem will reduce to one equation with one unknown, $\delta(x)$. One is emboldened by the fact that the profile will be integrated for δ^* and θ, so any inaccuracies will tend to cancel. That is correct, but it is also necessary to differentiate the assumed profile to get τ_w, and that operation tends to amplify inaccuracies. In any event, it is instructive to try a simple problem and see how all this works.

The simplest possible boundary layer problem is one where the inviscid solution is U_e = constant, which represents flow over a solid ($v_w = 0$) flat plate aligned with the flow. In that case, Eq. (2–15) becomes

$$\frac{d\theta}{dx} = \frac{C_f}{2} \tag{2–15a}$$

The crudest guess for a profile that meets an absolute minimum of physical criteria is a straight line joining the conditions $u(0) = 0$ and $u(\delta) = U_e$:

$$\frac{u}{U_e} = \frac{y}{\delta} \tag{2–17}$$

Using this equation in Eq. (2–13) gives $\theta = \delta/6$ and using it in Eq. (2-9) gives $\tau_w = \mu(\partial u/\partial y)_{y=0} = \mu U_e/\delta$. So Eq. (2–15a) becomes

$$\delta \frac{d\delta}{dx} = \frac{6(\mu/\rho)}{U_e} \tag{2–18}$$

which is one equation for the one unknown, δ, as promised. This is an easy equation to solve, since U_e = constant for this problem. Also, for a flat plate, the boundary layer thickness at the leading edge, $x = 0$, will be zero, so the single required initial condition is just $\delta(0) = 0$. The solution then is

$$\delta(x) = \sqrt{\frac{12(\mu/\rho)x}{U_e}} \tag{2–19}$$

which says that δ grows as $\sqrt{x}$. Rearranging terms, we can write Eq. (2–19) as

$$\frac{\delta}{x} = \sqrt{\frac{12\mu}{\rho U_e x}} = 3.46\text{Re}_x^{-1/2} \tag{2–19a}$$

where $\text{Re}_x = \rho U_e x/\mu$. The variation of $C_f(x)$ can be found by substituting Eq. (2–19) into the relation for $\tau_w = \mu(\partial u/\partial y)_{y=0}$ to yield

$$\tau_w = \frac{\mu U_e}{\delta} = \frac{\mu U_e}{\sqrt{12(\mu/\rho)x/U_e}} \tag{2–20}$$

or

$$C_f = \frac{\tau_w}{1/2\rho U_e^2} = 0.577\text{Re}_x^{-1/2} \tag{2–20a}$$

Now, how good is this crude solution? Experiment and elaborate exact analysis of this problem (presented in Chapter 4) indicate that

$$\frac{\delta}{x} = 5.0\text{Re}_x^{-1/2} \tag{2–21}$$

and

$$C_f = 0.664\text{Re}_x^{-1/2} \tag{2–22}$$

Thus, the simple analysis presented has achieved the correct dependence on Re_x and fairly good numerical values for the coefficients. Often, having the correct dependence on the flow parameters is sufficient for design.

Clearly, one could expect to do better quantitatively with a more realistic profile shape. Pohlhausen (1921b) used the function

$$\frac{u}{U_e} = a + b\left(\frac{y}{\delta}\right) + c\left(\frac{y}{\delta}\right)^2 + d\left(\frac{y}{\delta}\right)^3 + e\left(\frac{y}{\delta}\right)^4 \tag{2–23}$$

The coefficients a through e were determined by forcing satisfaction of a set of physically reasonable conditions on the profile. Pohlhausen used

$$\begin{array}{ll} y = 0\colon u = 0 & y = \delta\colon u = U_e \\ & \dfrac{\partial u}{\partial y} = 0 \\ \mu\dfrac{\partial^2 u}{\partial y^2} = \dfrac{dp}{dx} & \\ & \dfrac{\partial^2 u}{\partial y^2} = 0 \end{array} \tag{2–24}$$

The second condition at $y = 0$ can be developed by considering a small volume of fluid ($dx : dy : 1$) right on the wall. Since $u = v = 0$ on the wall, the momentum flux in and out of such a volume is negligible compared to the forces on the volume. Thus, equilibrium on the volume is achieved simply by a balance of the pressure force and the shear force. The net pressure force is

$$p(dy) - \left(p + \frac{dp}{dx}dx\right)dy \tag{2–25}$$

and the net shear force is

$$-\left(\mu\frac{\partial u}{\partial y}\right)dx + \mu\left(\frac{\partial u}{\partial y} + \frac{\partial}{\partial y}\left(\frac{\partial u}{\partial y}\right)dy\right)dx \tag{2–26}$$

So setting the sum of the forces to zero produces the stated condition. The result of applying these conditions to Eq. (2–23) is

$$\begin{gathered} a = 0, \qquad b = 2 + \frac{\lambda}{6}, \qquad c = -\frac{\lambda}{2}, \\ d = -2 + \frac{\lambda}{2}, \qquad e = 1 - \frac{\lambda}{6} \end{gathered} \tag{2–27}$$

where $\lambda = (\delta^2/\nu)dU_e/dx$ is the *Pohlhausen pressure gradient parameter*. Profiles for various values of λ are shown in Fig. 2–3. The profiles for *favorable* pressure gradients ($dp/dx < 0$ and $\lambda > 0$; i.e., no tendency toward separation) have a fuller shape than the flat-plate case. For *adverse* pressure gradients ($dp/dx > 0$ and

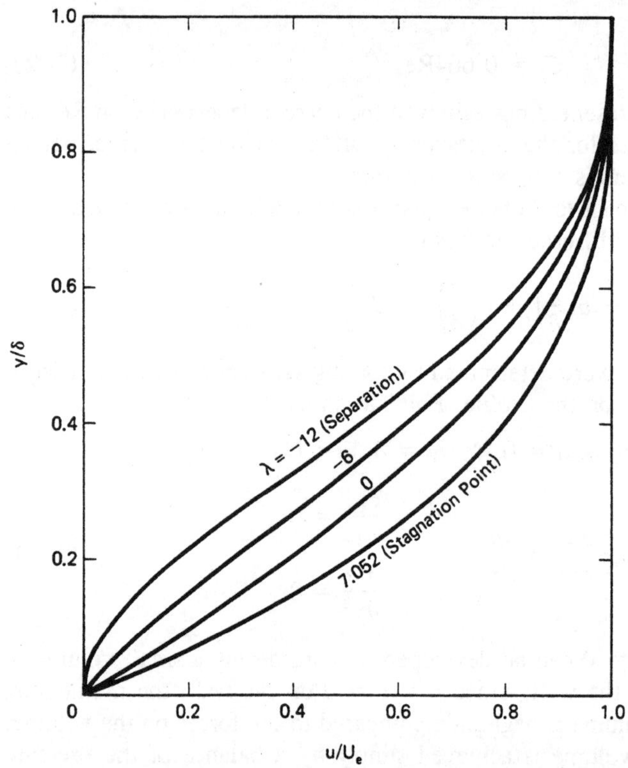

Figure 2–3 Laminar boundary layer velocity profiles by the Pohlhausen method for various values of the pressure gradient.

$\lambda < 0$), the fluid near the wall slows down rapidly, leading to an inflection point in the profile. Using Eq. (2–27), taking the first derivative of the profile, and evaluating at the wall, we find that $\tau_w = \mu(\partial u/\partial y)_{y=0} = 0$ (i.e., separation) when $\lambda = -12$. Many calculations have been made, and the results, when compared to experiment, indicate that this criterion for separation is rather crude. Attempts to improve the results by using other profiles have not been very successful. Nonetheless, the *momentum integral method,* as it is called, remains popular for obtaining simple approximate solutions. In modern times, it is especially useful in the beginning of design optimization iterations.

2–3–2 The Thwaites-Walz Method

Actually, the calculations with the Pohlhausen method are somewhat bothersome, but fortunately, the problem has been reformulated by Thwaites (1949) and Walz (1941) into a very convenient form. We multiply Eq. (2–15) by $U_e\theta/\nu$ and rearrange using $H \equiv \delta^*/\theta$ to get

$$\frac{\tau_w\theta}{\mu U_e} = \frac{U_e\theta}{\mu}\frac{d\theta}{dx} + \frac{\theta^2 dU_e/dx}{\nu}(H + 2) \tag{2–28}$$

The quantity H, called the *shape factor,* is a nondimensional function of the profile shape alone, as is the term on the left-hand side, which is commonly denoted S, the *shear correlation function*. With Eq. (2–28), it is more convenient to use $\Lambda = (\theta^2/\nu)dU_e/dx$ to parametize the profile shape, rather than λ from before. Now, saying that H and S are functions of the profile shape alone is the same as saying $H(\Lambda)$ and $S(\Lambda)$. With these, Eq. (2–28) can be rewritten as

$$U_e \frac{d}{dx}\left[\frac{\Lambda}{dU_e/dx}\right] = 2[S(\Lambda) - \Lambda(H(\Lambda) + 2)] = F(\Lambda) \tag{2–28a}$$

The Pohlhausen profile, or any other assumed profile shape, or the few available exact boundary layer solutions, or experimental data can be used to find the right-hand side. A good fit to the bulk of the available information is

$$F(\Lambda) = 0.45 - 6.0\Lambda \tag{2–29}$$

With this equation, Eq. (2–28a) can be integrated once and for all to yield

$$\theta^2(x) = \frac{0.45\nu}{U_e^6(x)}\int_0^x U_e^5(x')dx' + \theta^2(0)\left[\frac{U_e(0)}{U_e(x)}\right]^6 \tag{2–30}$$

This simple expression represents an approximate answer to all low-speed, steady, planar, laminar boundary layer problems! For a sharp-nosed body, $\theta(0) = 0$, and for a blunt-nosed body, $U_e(0) = 0$, so the last term drops out. For a blunt-nosed body at the forward stagnation point, we can write

$$\theta^2(0) = \frac{0.075\nu}{(dU_e/dx)_o} \tag{2–31}$$

The inviscid velocity gradient at the stagnation point can be taken to be $(dU_e/dx)_o = 2U_\infty/R_o$, where R_o is the radius of curvature at the stagnation point. Note that x is a running variable along the surface and that the variable of integration, x', runs from 0 up to the point of interest denoted by x.

A given problem is specified by $U_e(x)$ from the inviscid solution. The first step is to solve Eq. (2–30) for $\theta(x)$. Using $U_e(x)$, we then find $\Lambda(x) = (\theta^2/\nu)dU_e/dx$. We now also require correlations for $S(\Lambda)$ and perhaps $H(\Lambda)$. Thwaites (1949) gave the results listed in the first three columns of Table 2–1. At each point x along the surface, the appropriate S for $\Lambda(x)$ is found, and then the definition of S is unwound to yield

$$C_f = \frac{2\mu}{\rho U_e \theta}S(\Lambda) \tag{2–32}$$

The values for $S(\Lambda)$ and $H(\Lambda)$ in the table can be represented by the following curve fits:

$$0 \le \Lambda \le 0.1$$

$$S = 0.22 + 1.57\Lambda - 1.80\Lambda^2$$

$$H = 2.61 - 3.75\Lambda + 5.24\Lambda^2$$

TABLE 2–1 SHEAR AND SHAPE FUNCTIONS, THWAITES (1949)

Λ	$H(\Lambda)$	$S(\Lambda)$	δ/θ
0.10	2.28	0.359	8.5
0.080	2.34	0.333	8.2
0.064	2.39	0.313	8.2
0.048	2.44	0.291	8.1
0.032	2.49	0.268	8.1
0.016	2.55	0.244	8.0
0.0	2.61	0.220	8.0
-0.016	2.67	0.195	8.0
-0.032	2.75	0.168	8.0
-0.048	2.87	0.138	8.1
-0.064	3.04	0.104	8.1
-0.080	3.30	0.056	8.1
-0.084	3.39	0.038	8.2
-0.088	3.49	0.015	8.2
-0.090	3.55	0 (sep.)	8.2

$$-0.1 \le \Lambda \le 0$$

$$S = 0.22 + 1.402\Lambda + 0.018\Lambda/(\Lambda + 0.107)$$

$$H = 2.088 + 0.0731/(\Lambda + 0.14)$$

Approximate values for δ/θ are also included as the last column in Table 2–1.

Since the solutions from this procedure are based on a wider range of profile shapes than just that derived from Eq. (2–23), as well as on experiment, they are better than those obtained with the Pohlhausen procedure. The range of inaccuracy is generally less than 10% for favorable pressure gradients, but about 20 to 30% for strong, adverse pressure gradients.

The Thwaites-Walz method has been extended by Rott and Crabtree (1952) to the flow over an axisymmetric body described by the variation of the body radius $r_o(x)$. The result equivalent to Eq. (2–30) is

$$\theta^2(x) = \frac{0.45\nu}{r_o^2(x)U_e^6(x)} \int_0^x r_o^2(x')U_e^5(x')dx' \tag{2–30a}$$

Perhaps surprisingly, Rott and Crabtree show that the constants are the same as for the planar case. For a blunt-nosed body, the axisymmetric result becomes

$$\theta^2(0) = \frac{0.0563\nu}{(dU_e/dx)_0} \tag{2–31a}$$

The inviscid velocity gradient at the stagnation point can be taken to be $(dU_e/dx)_o = (3/2)U_\infty/R_o$.

A simple computer program for the Thwaites-Walz method for a personal computer (PC) is included in Appendix B.

Example 2–1. Application of the Momentum Integral Method

Consider the two-dimensional laminar flow of a fluid with a kinematic viscosity $\nu = 2.0 \times 10^{-4}$ m^2/s at $U_\infty = 10.0$ m/s over a surface that is a flat plate from the leading edge to $x = 1.0$ m. At that station, a ramp begins that produces an inviscid velocity distribution $U_e(x) = 10.5 - x/2$, m/s. This is an adverse pressure gradient, since U_e is decreasing, so that p increases. Calculate the boundary layer development over this surface up to $x = 2.0$ m. Does the flow separate?

Solution This problem can easily be treated with the Thwaites-Walz integral method. We use the code WALZ in Appendix B. Input data for the kinematic viscosity CNU = 0.0002, the free stream velocity is UINF = 10.0, and KASE = 0 for two-dimensional flow. Information for dx (DELX in the program) also must be given. We choose to make 41 steps (NMAX = 41) along the surface, from $x = 0$ to $x =$ 2.0 m ($dx = 0.05$ m). Finally, the inviscid velocity distribution is required. This kind of information for this particular case is included in the code in the loop DO 10 N = 1,NMAX. For other problems, this section of the code must be modified.

The primary output of the program is listed in the following table. The boundary layer thicknesses increase, the profile shape changes, and the skin friction coefficient decreases rapidly after the adverse pressure gradient begins, but the boundary layer does not separate by $x = 2.0$ m—i.e., $C_f > 0$.

A plot of some of these results is given in Figure 2–4. Note that the change in edge velocity is actually quite small but the effect on the output is significant. Observe the change in slope of the curves at $x = 1.0$ m.

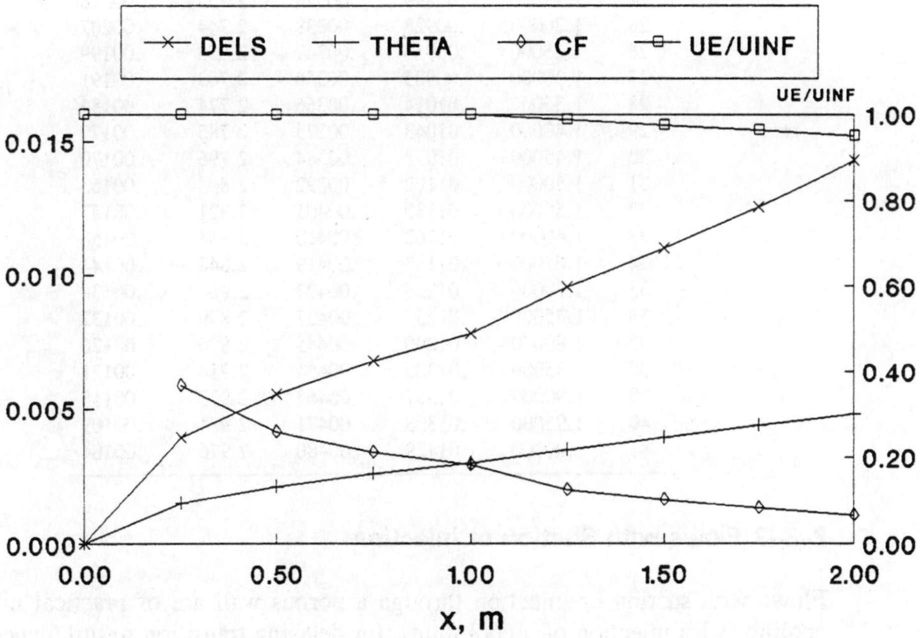

Figure 2–4 Streamwise variation of integral boundary layer properties for Example 2–1.

N	S	DELTS	THETA	H	CF
1	0.00000	.00000	.00000	2.610	—
2	0.05000	.00175	.00067	2.610	.01312
3	0.10000	.00248	.00095	2.610	.00928
4	0.15000	.00303	.00116	2.610	.00757
5	0.20000	.00350	.00134	2.610	.00656
6	0.25000	.00392	.00150	2.610	.00587
7	0.30000	.00429	.00164	2.610	.00536
8	0.35000	.00463	.00178	2.610	.00496
9	0.40000	.00495	.00190	2.610	.00464
10	0.45000	.00525	.00201	2.610	.00437
11	0.50000	.00554	.00212	2.610	.00415
12	0.55000	.00581	.00223	2.610	.00396
13	0.60000	.00607	.00232	2.610	.00379
14	0.65000	.00630	.00242	2.610	.00364
15	0.70000	.00655	.00251	2.610	.00351
16	0.75000	.00678	.00260	2.610	.00339
17	0.80000	.00700	.00268	2.610	.00328
18	0.85000	.00722	.00277	2.610	.00318
19	0.90000	.00743	.00285	2.610	.00309
20	0.95000	.00763	.00292	2.610	.00301
21	1.00000	.00783	.00300	2.610	.00293
22	1.05000	.00842	.00310	2.718	.00235
23	1.10000	.00870	.00319	2.726	.00225
24	1.15000	.00899	.00329	2.735	.00216
25	1.20000	.00928	.00338	2.744	.00207
26	1.25000	.00956	.00347	2.753	.00199
27	1.30000	.00985	.00356	2.763	.00191
28	1.35000	.01014	.00366	2.774	.00184
29	1.40000	.01043	.00375	2.785	.00177
30	1.45000	.01072	.00384	2.796	.00170
31	1.50000	.01102	.00392	2.808	.00163
32	1.55000	.01132	.00401	2.821	.00157
33	1.60000	.01162	.00410	2.834	.00150
34	1.65000	.01193	.00419	2.848	.00144
35	1.70000	.01225	.00427	2.864	.00138
36	1.75000	.01257	.00437	2.879	.00132
37	1.80000	.01290	.00445	2.896	.00126
38	1.85000	.01323	.00454	2.914	.00121
39	1.90000	.01357	.00463	2.933	.00115
40	1.95000	.01393	.00471	2.954	.00109
41	2.00000	.01429	.00480	2.976	.00104

2-3-3 Flows with Suction or Injection

Flows with suction or injection through a porous wall are of practical interest for cooling (with injection of a cool fluid), for delaying transition to turbulence by suction (see Chapter 6), and for preventing separation in an adverse pressure gradient by suction. Generally speaking, the momentum integral method is not accurate for these flows, especially those with injection, so a detailed discussion of such cases is reserved for later in this text, after more powerful methods have been introduced. However, Prandtl (1935) has presented a simple treatment of prevention of separation by suction that is illuminating. He considered the case of incipient separation, which, with the quartic profile given by Eq. (2–23), corresponds to $\lambda = -12$. For that profile, $\delta^*/\delta = 2/5$, $\theta/\delta = 4/35$, and $\tau_w = \mu(\partial u/\partial y)_{y=0} = 0$. Substituting into the integral momentum equation, Eq. (2–15), and making the assumption that $d\delta/dx = d\delta^*/dx = d\theta/dx = 0$ because the profile is assumed to be maintained constant by the suction, one gets the suction velocity needed to maintain incipient separation:

$$v_w = \frac{22}{35}\,\delta\,\frac{dU_e}{dx} \tag{2–33}$$

From Eqs. (2–24), and (2–12), we obtain

$$\left.\frac{\partial^2 u}{\partial y^2}\right|_w = \frac{1}{\mu}\frac{dp}{dx} = -\frac{U_e}{\nu}\frac{dU_e}{dx} \tag{2–34}$$

The assumed profile shape, Eqs. (2–23) and (2–27), gives

$$\left.\frac{\partial^2 u}{\partial y^2}\right|_w = 12\frac{U_e}{\delta^2} \tag{2–35}$$

at separation. Thus, combining Eqs. (2–34) and (2–35), we get

$$\delta = \sqrt{\frac{12\nu}{-dU_e/dx}} \tag{2–36}$$

and Eq. (2–33) becomes

$$v_w = -2.18\sqrt{-\nu\frac{dU_e}{dx}} \tag{2–37}$$

as the final expression for the suction velocity necessary to maintain incipient separation and prevent flow reversal.

2-4 THE INTEGRAL ENERGY EQUATION

Under the restrictions of planar, steady, constant-density, constant-property flow, there are still cases of important thermal energy, or heat, transfer. Consider air at 50°C flowing over a surface at 30 °C. There is significant heat transfer, but the range of temperature is small enough to neglect variations in density or physical properties, as can be seen by referring to the data in Appendix A. A *thermal boundary layer* will develop with a temperature profile varying from $T = T_w(x)$ at $y = 0$ to $T = T_e(x)$ at $y = \delta_T(x)$. The thickness δ_T of the thermal boundary layer may be thicker or thinner than that of the velocity boundary layer, depending on the physical properties and the particular problem of interest.

The physical properties combine to form the dimensionless *Prandtl number*,

$$\Pr \equiv \frac{\mu c_p}{k} \tag{2-38}$$

which represents the ratio of the coefficient for diffusion of momentum to the coefficient for diffusion of heat. Interestingly, most common gases have $\Pr \approx 0.7$, which means that heat will diffuse faster than momentum. All other things being equal, then, a thermal boundary layer in a gas flow is likely to be thicker than the corresponding velocity boundary layer in a ratio of roughly 1:0.7. The Prandtl number for common liquids (water, oil, and so on) is on the order of 10 to as high as 10^6. At the other extreme, the values for liquid metals are on the order of 10^{-2}. Obviously, the Prandtl number of the fluid is of great importance in boundary layer flows with heat transfer.

Approximate solutions at a level similar to those obtained for the velocity boundary layer in this chapter can be obtained by applying a simple statement of conservation of energy to a suitable control volume (see Fig. 2–5). Again, one could alternatively begin with the partial differential equation form of the energy equation to be derived in Chapter 3 and then integrate across the thermal boundary layer. For our purposes here, a suitable statement of conservation of energy can be written as

$$\text{thermal energy in} + \text{frictional heating} + \text{heat transfer at the surface} = \text{thermal energy out} \tag{2-39}$$

The thermal energy carried by the flow into the left-hand side of the control volume is

$$\rho c_p \int_0^H uT\,dy \tag{2-40}$$

and that carried out of the right-hand side is

$$\rho c_p \int_0^H uT\,dy + \frac{d}{dx}\left[\rho c_p \int_0^H uT\,dy\right]dx \tag{2-41}$$

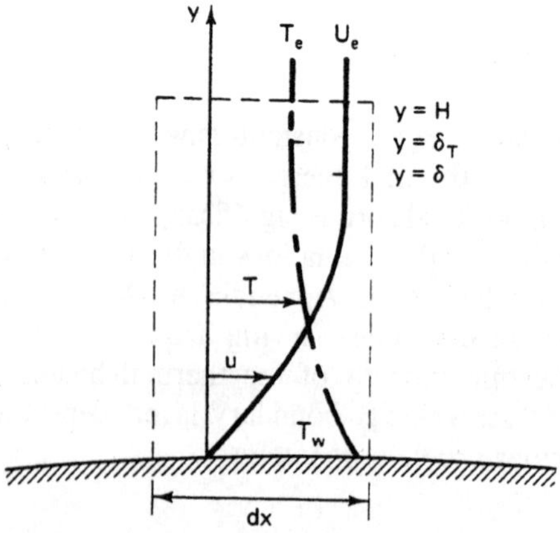

Figure 2–5 Control volume for derivation of the integral energy equation.

The mass flow through the top brings in thermal energy equal to

$$c_p T_e\left[\frac{d}{dx}\left(\int_0^H \rho u\, dy\right)dx - \rho_w v_w\, dx\right] \tag{2-42}$$

Since there is viscous shear throughout the boundary layer, $\partial u/\partial y \neq 0$, there is frictional heating. This can be expressed as

$$\int_0^H \mu\left[\frac{\partial u}{\partial y}\right]^2 dy\, dx \tag{2-43}$$

There is heat transfer $-k(\partial T/\partial y)_{y=0}$, at the surface and injected fluid carries in energy

$$\rho_w v_w c_p T_w\, dx \tag{2-44}$$

Substituting all of this into Eq. (2-39) results in the integral energy equation

$$\frac{d}{dx}\left[\int_0^H (T_e - T)u\, dy\right] + \frac{\nu}{c_p}\left[\int_0^H \left(\frac{\partial u}{\partial y}\right)^2 dy\right] - v_w(T_e - T_w) = \frac{k(\partial T/\partial y)_w}{\rho c_p} \tag{2-45}$$

Note again that under the assumptions in this chapter, $\rho_w = \rho$.

We are now in the same situation as we were regarding the integral momentum equation: Unless the shape of the profiles for u and T are specified, there are too many unknowns for this single equation.

2-5 SOLUTION OF THE INTEGRAL ENERGY EQUATION

It should not surprise the reader that most efforts to solve the integral energy equation closely follow the Pohlhausen method; that is, a polynomial assumption for the temperature profile is used, in conjunction with a polynomial for the velocity profile.

2-5-1 Unheated Starting-Length Problem

The method for solving the integral energy equation can be illustrated, and some useful results can be obtained at the same time, by considering a specific situation known as the unheated starting-length problem. The flow is as shown in Fig. 2-6

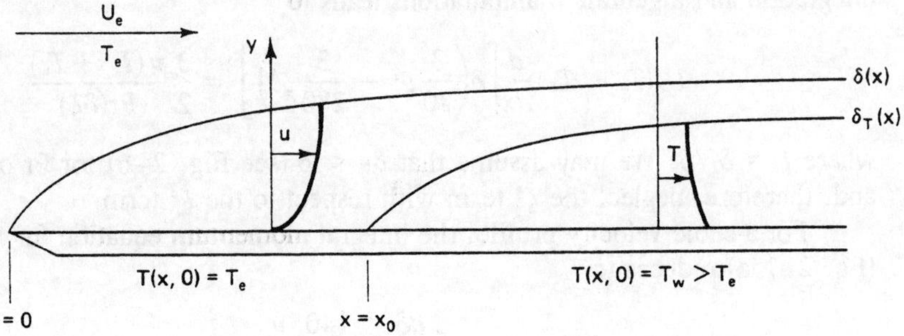

Figure 2-6 Schematic of the flow over a flat plate with an unheated starting length.

with a solid ($v_w = 0$) flat plate (U_e = constant, $dp/dx = 0$) whose initial portion $0 \le x \le x_o$ has the same wall temperature as the static temperature in the free stream (i.e., $q_w = 0$, since there is no temperature difference). For $x \ge x_o$, $T_w \ne T_e$, and there will be heat transfer at the surface. The velocity boundary layer will begin to grow at the leading edge of the plate ($x = 0$), but the thermal boundary layer will only begin at $x = x_o$. Also, it is assumed that the flow speed is low enough to neglect viscous heating with respect to the other energy transfer processes. Thus, the second integral in Eq. (2–45) is dropped.

The solution to this problem, as given by Eckert and Drake (1959), uses cubic polynomials for both the velocity and temperature profiles. For the velocity profile, the condition on $\partial^2 u/\partial y^2$ at $y = \delta$ is dropped from the group in Eq. (2–24). The resulting profile is

$$\frac{u}{U_e} = \frac{3}{2}\left(\frac{y}{\delta}\right) - \frac{1}{2}\left(\frac{y}{\delta}\right)^3 \tag{2–46}$$

The conditions imposed on the temperature profile are

$$\begin{aligned} &y = 0 : T = T_w \qquad && y = \delta_T : T = T_e \\ &\frac{\partial^2 T}{\partial y^2} = 0 && \frac{\partial T}{\partial y} = 0 \end{aligned} \tag{2–47}$$

In the same way as for the velocity profile, the second condition on $T(y)$ at $y = 0$ arises from consideration of a small volume of fluid on the wall. Since $u = v = 0$ on the wall, convection can be neglected. Thus, the net energy transfer through that volume,

$$\left(-k\frac{\partial T}{\partial y}\right)dx + k\left(\frac{\partial T}{\partial y} + \frac{\partial}{\partial y}\left(\frac{\partial T}{\partial y}\right)dy\right)dx \tag{2–48}$$

must be zero, leading to the stated condition. The resulting temperature profile is

$$\frac{T - T_w}{T_e - T_w} = \frac{3}{2}\left(\frac{y}{\delta_T}\right) - \frac{1}{2}\left(\frac{y}{\delta_T}\right)^3 \tag{2–49}$$

Substituting Eqs. (2–46) and (2–49) into Eq. (2–45) and performing the necessary integration and algebraic manipulations leads to

$$U_e(T_e - T_w)\frac{d}{dx}\left[\delta\left(\frac{3}{20}\zeta^2 - \frac{3}{280}\zeta^4\right)\right] = \frac{3}{2}\frac{\nu(T_e - T_w)}{\Pr(\delta\zeta)} \tag{2–50}$$

where $\zeta = \delta_T/\delta$. We may assume that $\delta_T < \delta$ (see Fig. 2–6) for Pr of order unity and, therefore, neglect the ζ^4 term with respect to the ζ^2 term.

For a cubic velocity profile, the integral momentum equation for flat-plate flow (Eq. 2–15a) reduces to

$$\delta\frac{d\delta}{dx} = \frac{140}{13}\frac{\nu}{U_e} \tag{2–51}$$

and the solution is

$$\delta^2 = \frac{280}{13}\frac{\nu x}{U_e} \tag{2–52}$$

Using these results in Eq. (2–50) yields

$$\frac{4}{3}x\frac{d\zeta^3}{dx} + \zeta^3 = \frac{13}{14}\left(\frac{1}{\mathrm{Pr}}\right) \tag{2–53}$$

The general solution to this first-order ordinary differential equation for ζ^3 is

$$\zeta^3 = cx^{-3/4} + \frac{13}{14}\left(\frac{1}{\mathrm{Pr}}\right) \tag{2–54}$$

The single required initial condition is $\zeta = 0$ (i.e., $\delta_T = 0$)at $x = x_o$, and the final solution may be written

$$\frac{\delta_T}{\delta} = \frac{1}{1.026\mathrm{Pr}^{1/3}}\left[1 - \left(\frac{x_0}{x}\right)^{3/4}\right]^{1/3} \tag{2–55}$$

For $x_o = 0$ (i.e., when the whole plate is heated),

$$\frac{\delta_T}{\delta} = \frac{1}{1.026\mathrm{Pr}^{1/3}} \tag{2–55a}$$

This expression turns out to be valid for Pr of order unity, despite our earlier assumption that $\delta_T < \delta$.

Of course, the heat transfer rate at the wall is generally of greater practical interest than $\delta_T(x)$. But now, having $\delta_T(x)$, we can evaluate $q_w = -k(\partial T/\partial y)_{y=0}$ using Eq. (2–49). It is common to express the wall heat transfer in terms of a *film coefficient* $\hbar$, defined by

$$q_w = -k\left(\frac{\partial T}{\partial y}\right)_{y=0} \equiv \hbar(T_w - T_e) \tag{2–56}$$

This expression defining $\hbar$ is known as *Newton's law of cooling* and is not a fundamental physical model. Unlike k, the quantity $\hbar$ is not a property of the fluid. Rather, *Fourier's law* [Eq. (1–5)] is the basic law used in heat transfer analysis. Only after the complete solution is obtained using Fourier's law is the answer usually rewritten in terms of the film coefficient $\hbar$. This is done partly by convention and partly because Newton's name is associated with $\hbar$. In any event, this dimensional quantity is usually incorporated into a dimensionless group known as the *Nusselt number*,

$$\mathrm{Nu} \equiv \frac{\hbar x}{k} \tag{2–57}$$

Thus, for the unheated starting-length problem,

$$\mathrm{Nu} = 0.332\mathrm{Pr}^{1/3}\left(\frac{\rho U_e x}{\mu}\right)^{1/2}\left[1 - \left(\frac{x_0}{x}\right)^{3/4}\right]^{-1/3} \tag{2–58}$$

In this instance, partly by good fortune, we get the correct coefficient, agreeing with the exact solution for the case in which $x_o = 0$.

Alternatively, the film coefficient can be put in the dimensionless form

$$\mathrm{St} \equiv \frac{h}{\rho U_e c_p} \tag{2–59}$$

known as the *Stanton number*. Then the solution becomes

$$\mathrm{St} = \frac{\mathrm{Nu}}{\mathrm{Re}\cdot\mathrm{Pr}} = 0.332\mathrm{Pr}^{-2/3}\mathrm{Re}^{-1/2}\left[1 - \left(\frac{x_0}{x}\right)^{3/4}\right]^{-1/3} \tag{2–60}$$

Comparing this result for the case with $x_o = 0$, i.e.,

$$\mathrm{St} = 0.332\mathrm{Pr}^{-2/3}\mathrm{Re}^{-1/2} \tag{2–60a}$$

with Eq. (2–22), we see that

$$\mathrm{St}\cdot\mathrm{Pr}^{2/3} = \frac{C_f}{2} \tag{2–61}$$

Equation (2–61) gives a useful, direct relation of skin friction to heat transfer called the *Reynolds analogy*. Although developed here under restrictive conditions, it is often assumed to hold for a wide range of flows.

2-5-2 Nonuniform Wall Temperature

Many practical problems involve cases where the wall temperature varies in a complicated fashion. The specific example of a step change in wall temperature just analyzed can be generalized to encompass essentially arbitrary variations of $T_w(x)$. The key to such generalization lies in the fact that Eq. (2–45) (and, for that matter, the corresponding exact boundary layer energy equation considered in Chapter 3 in the constant-density, constant-property case) is *linear* in the temperature. This means that the sum of any two (or more) solutions corresponding to two (or more) separate sets of boundary conditions represents an exact solution to the new problem specified by boundary conditions formed by the sum of the two (or more) sets of original boundary conditions. Thus, one says that the solutions may be *superimposed*. Consider the wall temperature variation shown in Fig. 2–7. The actual $T_w(x)$ shown by the solid curve can be approximated by the series of step changes drawn with dashes. In such a situation, the total heat flow at a location x is the sum of the contributions of the various step changes:

$$q_w = \sum_{1}^{N} h(x, \xi_i)(\Delta T_{wi}) \tag{2–62}$$

Here, $h(x,\xi_i)$ represents the film coefficient for heat transfer at x for a single step change ΔT_{wi}, located at $x = \xi_i$. Clearly, the quantity h is obtained from Eq. (2–58), writing ξ_i for x_o. The summation can be extended in the limit to an integral for a se-

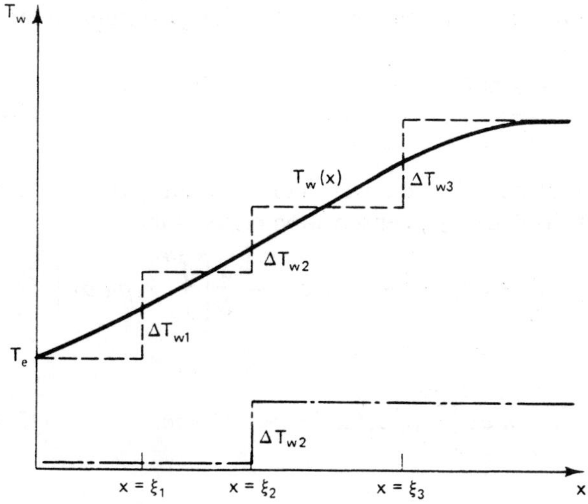

Figure 2–7 Variable wall temperature distribution approximated by a series of step changes.

ries of infinitesimal step changes:

$$q_w = \int_0^x h(x, \xi) \frac{dT_w}{d\xi} d\xi \tag{2–63}$$

A simple approximate procedure for carrying out the required calculations has been developed by Eckert et al (1957).

2–6 THE INTEGRAL SPECIES CONSERVATION EQUATION

Retaining the assumptions used throughout this chapter that the flow is planar and steady with constant density and constant physical properties, one can still envision some problems with significant mass transfer effects. The derivation of an integral equation for species concentration corresponding to the integral energy and momentum equations follows the same general lines as for those equations.

For simplicity, consider the fluid as a mixture of only two species (e.g., water vapor in air), denoted by subscripts 1 and 2. Referring to Fig. 2–1, we can write an integral mass conservation equation for fluid 1. For a steady flow, the difference between the mass of that fluid coming into the control volume through the left side and the mass of that fluid going out the right side must enter through the top and the bottom (if the wall is permeable to fluid 1). The mass of fluid 1 coming in the top can be found from the expression for the total mass coming in the top [Eq. (2–3)] by multiplying by the value of the mass fraction at the boundary layer edge, c_{1e}—that is,

$$c_{1e}\left[\frac{d}{dx}\left(\int_0^H \rho u \, dy\right)dx - \rho_w v_w \, dx\right] \tag{2–64}$$

Now recall the definition of the mass fraction, $c_i = \rho_i/\rho$. The difference between the mass of fluid 1 going out the right side and the mass of fluid 1 coming in the left side is

$$\frac{d}{dx}\left[\int_0^H c_1 \rho u \, dy\right] dx \tag{2-65}$$

If the mass flux by diffusion of fluid 1 at the wall into the volume is $\dot{m}_{1w}$ and that from injection through the wall is $\rho_w v_w c_{1w}$, conservation requires that

$$\dot{m}_{1w} + c_{1e}\frac{d}{dx}\left[\int_0^H \rho u \, dy\right] - c_{1e}\rho_w v_w + \rho_w v_w c_{1w} = \frac{d}{dx}\left[\int_0^H c_1 \rho u \, dy\right] \tag{2-66}$$

or

$$\frac{d}{dx}\left[\int_0^H (c_{1e} - c_1)\rho u \, dy\right] - \rho_w v_w (c_{1e} - c_{1w}) = -\dot{m}_{1w} \tag{2-67}$$

Note that $c_2 \equiv 1.0 - c_1$.

2–7 SOLUTION OF THE INTEGRAL SPECIES EQUATION

The simplest cases are those with a solid wall, so that $v_w = 0$. The conditions that can be imposed on the assumed species concentration profile follow directly from those used for the corresponding velocity and temperature profiles:

$$\begin{aligned} y = 0: \quad & c_1 = c_{1w} & \qquad y = \delta_c: \quad & c_1 = c_{1e} \\ & \frac{\partial^2 c_1}{\partial y^2} = 0 & & \frac{\partial c_1}{\partial y} = 0 \end{aligned} \tag{2-68}$$

Assuming a cubic profile, as was done for the temperature [see Eq. (2–49)], we get

$$\frac{c_1 - c_{1w}}{c_{1e} - c_{1w}} = \frac{3}{2}\left(\frac{y}{\delta_c}\right) - \frac{1}{2}\left(\frac{y}{\delta_c}\right)^3 \tag{2-69}$$

Using this equation with a cubic velocity profile for flow over a flat plate [see Eq. (2–46)], we can express the result as a *Nusselt number for diffusion:*

$$\text{Nu}_{\text{Diff}} \equiv \frac{\hbar_D x}{D_{12}} = 0.332\ \text{Re}_x^{1/2}\ \text{Sc}^{1/3} \tag{2-70}$$

Here, the *diffusion film coefficient* $\hbar_D$ is defined by

$$\dot{m}_{1w} = -\rho D_{12} \left.\frac{\partial c_1}{\partial y}\right|_{y=0} \equiv \rho \hbar_D (c_{1w} - c_{1e}) \tag{2-71}$$

and Sc ($\equiv \nu/D_{12}$) is the Schmidt number. Note the close relationship of Eq. (2–70) to Eq. (2–58) for $x_o = 0$ (i.e., the case of a constant-temperature plate). More will be made of this connection shortly.

Data for mass transfer from a flat plate in low-speed flow are scarce. In Fig. 2–8, the measurements of Christian and Kezios (1959) for naphthalene sublimation (at a low rate, so that $v_w \approx 0$) into air flowing axially over cylinders with $(\delta/R) << 1$ (so the effects of axisymmetry compared to a planar case are ≈2%) are compared with Eq. (2–70). The agreement is clear.

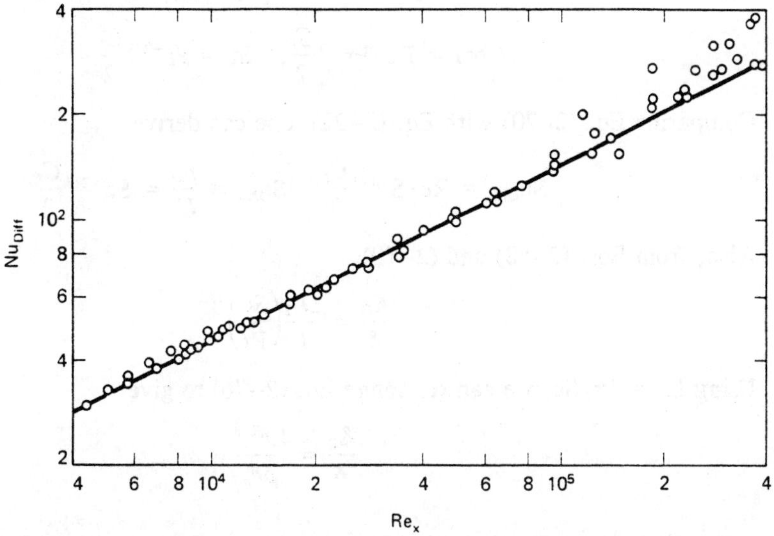

Figure 2–8 Mass transfer rate from a napthalene surface in an airstream compared to analysis. (From Christian and Kezios, 1959.)

2–8 RELATIONSHIP OF WALL FRICTION, HEAT TRANSFER, AND MASS TRANSFER

Earlier, we noted that the skin friction coefficient and the Stanton number, the Nusselt number, and the Nusselt number for diffusion are related. This observation can be formalized and generalized for cases where $v_w \approx 0$. First, the three processes of shear, heat transfer, and mass transfer are clearly related by virtue of the *laws* that are taken to govern each:

$$\tau = \mu \frac{\partial u}{\partial y}$$

$$q = -k \frac{\partial T}{\partial y} \tag{2–72}$$

$$\dot{m}_i = -\rho D_{ij} \frac{\partial c_i}{\partial y}$$

Also, the film coefficients for heat and mass transfer are defined in a similar manner:

$$q_w = h(T_w - T_e)$$
$$\dot{m}_{iw} = \rho h_D(c_{iw} - c_{ie}) \tag{2–73}$$

Comparing Eq. (2–58) with $x_o = 0$ with Eq. (2–22), we find that

$$\text{Nu} = \text{Re}\cdot\text{Pr}^{1/3}\frac{C_f}{2}; \quad \text{St} = \text{Pr}^{-2/3}\frac{C_f}{2} \tag{2–74}$$

Comparing Eq. (2–70) with Eq. (2–22), one can derive

$$\text{Nu}_{\text{Diff}} = \text{Re}\cdot\text{Sc}^{1/3}\frac{C_f}{2}; \quad \text{St}_{\text{Diff}} \equiv \frac{h_D}{U_e} = \text{Sc}^{-2/3}\frac{C_f}{2} \tag{2–75}$$

Also, from Eqs. (2–58) and (2–70),

$$\frac{h_D}{h} = \frac{D_{12}}{k}\left(\frac{\text{Sc}}{\text{Pr}}\right)^{1/3} \tag{2–76}$$

Using Le = Pr/Sc, we can rearrange Eq. (2–76) to give

$$\frac{h_D}{h} = \frac{\text{Le}^{2/3}}{\rho c_p} \tag{2–77}$$

2–9 NON-NEWTONIAN FLUIDS

The momentum integral method has been successfully applied to the laminar flow of some non-Newtonian fluids modeled as, for example, power law or Bingham plastic fluids. Indeed, the extension of the Pohlhausen method to these types of non-Newtonian fluids is surprisingly easy and quite informative.

2–9–1 Power Law Fluid over a Flat Plate

The biggest change that must be made to accommodate non-Newtonian fluids is in the relationship of the wall shear to the assumed velocity profile. Now, we must use [see Eq. (1-64)]

$$\tau_w = \mu_{\text{PL}}\left(\frac{\partial u}{\partial y}\right)^p_{y=0} \tag{2–78}$$

It is common again to employ the cubic velocity profile, Eq. (2–46) (see Acrivos et al., 1960). Substituting into Eq. (2–15a), the integral momentum equation for the case of a flat plate, gives

$$\frac{39}{280}\rho U_e^2\frac{d\delta}{dx} = \mu_{\text{PL}}\left(\frac{3}{2}\right)^p U_e^p\left(\frac{1}{\delta}\right)^p \tag{2–79}$$

This equation can be integrated as before, using $\delta(0) = 0$, to yield

$$\frac{\delta}{x} = \left[\frac{280(p+1)}{39}\left(\frac{3}{2}\right)^p\right]^{1/p+1}\left(\frac{x^p U_e^{2-p}\rho}{\mu_{\text{PL}}}\right)^{-1/p+1} \tag{2–80}$$

For a Newtonian fluid, $p = 1$ and $\mu_{PL} = \mu$, and the numerical factor for the same assumed velocity profile is $(280/13)^{1/2}$ (see Eq. (2–52). Also, where a Newtonian fluid result has $\text{Re}^{1/2}$, we now find the last term in Eq. (2–80). This is apparently a *Reynolds number for power law fluids*, which can be denoted Re_{PL}, i.e.,

$$\text{Re}_{\text{PL}} \equiv \frac{x^p U_e^{2-p}\rho}{\mu_{\text{PL}}} \tag{2–81}$$

Our primary interest is usually in the wall shear, which becomes

$$C_f(x) = \frac{\tau_w}{1/2\rho U_e^2} = 2\left[\left(\frac{3}{2}\right)\frac{39}{280(p+1)}\right]^{p/p+1}\text{Re}_{\text{PL}}^{-1/p+1} \tag{2–82}$$

For flow at the same edge velocity, a power law fluid will generally have a lower wall shear and a smaller boundary layer thickness than a Newtonian fluid with the same density and $\mu_{PL} = \mu$. This property has led to some interest in these fluids for drag reduction in water flows by adding a small amount of a power law fluid in solution.

2–9–2 Bingham Plastic Fluid over a Flat Plate

In the case of a Bingham plastic fluid, the integral momentum equation itself must also be modified. This is because $\tau_\delta \neq 0$ as $(\partial u/\partial y)_{y=\delta} = 0$ for such a fluid, which has a yield stress at zero velocity gradient (see Section 1-8). Thus, Eq. (2–15a) must be rewritten as [see Eq. (1-62)]

$$\frac{d\theta}{dx} = \frac{\tau_w - \tau_\delta}{\rho U_e^2} \tag{2–83}$$

Using the stress model for a Bingham plastic fluid, we can say that

$$\begin{aligned}\tau_w - \tau_\delta &= \left[\tau_o + \mu_{\text{BP}}\left(\frac{\partial u}{\partial y}\right)_{y=0}\right] - \left[\tau_o + \mu_{\text{BP}}\left(\frac{\partial u}{\partial y}\right)_{y=\delta}\right] \\ &= \mu_{\text{BP}}\left(\frac{\partial u}{\partial y}\right)_{y=0}\end{aligned} \tag{2–84}$$

so the effect appears to cancel out. But that is not quite true, as will be clear below in Eq. (2–87). The solution for the boundary layer thickness with a cubic profile assumption is, however, the same as for a Newtonian fluid, i.e. Eq. (2–52). It can be expressed here as

$$\frac{\delta}{x} = 4.64\ \text{Re}_{\text{BP}}^{-1/2} \tag{2–85}$$

where

$$\mathrm{Re}_{\mathrm{BP}} \equiv \frac{\rho U_e x}{\mu_{\mathrm{BP}}} \tag{2-86}$$

The yield stress does enter into the expression for the wall shear:

$$C_f(x) = \frac{\tau_w}{1/2\rho U_e^2} = \frac{\tau_o}{1/2\rho U_e^2} + \frac{0.646}{\mathrm{Re}_{\mathrm{BP}}^{1/2}} \tag{2-87}$$

2-10 DISCUSSION

In this chapter, we have presented some aspects of the more widely used approximate techniques for laminar boundary layer analysis. For the sake of relative simplicity, the coverage has been limited to steady, low-speed, constant-property cases in the planar geometry. Equivalent procedures have also been developed in the literature for cases relaxing some or all of these restrictions. Further material is given in subsequent chapters, and the interested reader can refer to older books, such as that by Schlichting (1968), for more complete historical coverage.

PROBLEMS

2.1. Assume that the velocity profile over a flat plate may be approximated as

$$\frac{u}{U_e} = \tanh\left(2.65\left(\frac{y}{\delta}\right)\right)$$

Calculate $\delta(x)$ and $C_f(x)$.

2.2. Air at 25°C flows over a flat plate at 30 m/s. How thick is the boundary layer at a distance of 3 cm from the leading edge? How big is the displacement thickness?

2.3. Consider a flat plate in a fluid medium that is at rest, except for a line *sink* located on the plate surface at $x = L$, as shown in the following sketch. Suppose the inviscid velocity field produced by the sink is $v_r = -C(L/r)$, where r is the radial distance from the sink to any point in the flow. Determine the momentum thickness distribution, $\theta(x)$.

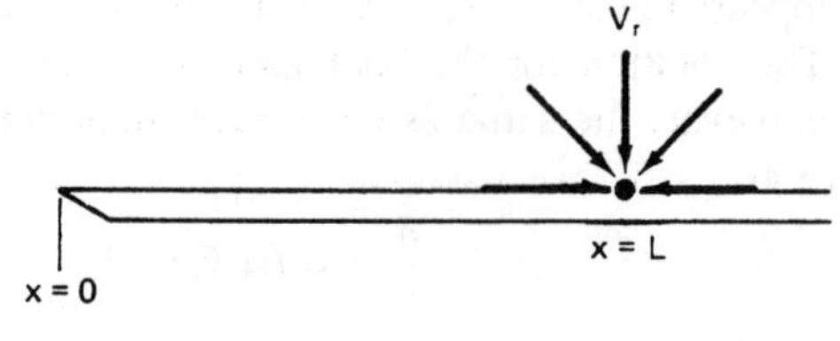

2.4. Suppose we have water at 20°C flowing over a body with a shape such that $U_e(x) = 2.0(1.0 - x)$ ft/s. Determine if and where separation will occur. Compare your results to the exact solution of Howarth (1938).

2.5. We wish to study the influence of the choice of velocity profile shape on the results of the Pohlhausen method. Use a cubic (third-order) velocity profile for the case of a flat plate, and compare the results for δ^*/δ and $\tau_w\delta/\mu U_e$ to the classical results. Also, how would the conditions for separation be changed with a cubic profile?

2.6. Water at 20°C is flowing over a surface such that $U_e(x) = 2.0 \sin^{1/5}(x/2)$ ft/s. What are the values of δ^* and C_f at $x = 1.0$ ft?

2.7. Air at 1.0 atm and room temperature flows over a plate at a rate of 3.0 m/s. Determine the mass flow (per unit width) of external stream fluid that enters the boundary layer between the leading edge and a station 50 cm from the leading edge.

2.8. Consider flow over a flat plate, and use the Thwaites-Walz method to predict δ, δ^*, θ, and C_f vs. x. Compare the results with the predictions of the Pohlhausen method and the exact solution in Eqs. (2–21) and (2–22).

2.9. What is $C_f(x)$ in the vicinity of the stagnation point for air at standard temperature and pressure (STP) flowing at 5.0 ft/s over a circular cylinder of diameter 1.0 in ? Use the Thwaites-Walz method.

2.10. Air at STP and 25.0 ft/s is flowing past a circular cylinder with a diameter of 0.5 in. Calculate δ, δ^*, and θ at the stagnation point.

2.11. Repeat Prob. 2.10, changing the body shape to a sphere.

2.12. Consider the general flow situation treated in Example 2–1. Everything is held fixed but the shape of the ramp. The free stream velocity variation over the ramp remains linear, but the rate of decrease is increased. What rate of free stream velocity decrease will produce separation by $x = 2.0$ m?

2.13. Consider again the general flow situation treated in Example 2–1. Everything is held fixed but the shape of the ramp. The free stream velocity variation over the ramp remains linear, but it increases at the same rate as the decrease in the original problem. Calculate the state of the flow in the boundary layer at $x = 2.0$ m.

2.14. We wish to study boundary layer development on an NACA 0010-35 airfoil at zero angle of attack. The geometry and inviscid velocity distribution are given in the following table. The radius of the leading edge is 0.272% of the chord, c. Compute the variation of $C_f(x)$ and $\theta(x)$ for air at 50.0 ft/s over a foil with a chord of 1.69 ft, assuming laminar flow. Use the Thwaites-Walz method.

x/c	y/c	U_e/U_∞
0.0	0.0	0.0
0.0125	0.0088	0.977
0.025	0.0127	1.016
0.05	0.0184	1.043
0.075	0.0229	1.059
0.1	0.0267	1.068
0.15	0.0329	1.083
0.2	0.0379	1.093
0.3	0.0448	1.102

x/c	y/c	U_e/U_∞
0.4	0.0488	1.109
0.5	0.050	1.111
0.6	0.0487	1.113
0.7	0.0439	1.107
0.8	0.0350	1.083
0.9	0.0210	1.021
0.925	0.0166	0.994
0.95	0.0118	0.957
0.975	0.0066	0.826
1.0	0.0010	0.691

2.15. Using a linear profile for the velocity [Eq. (2–17)] and a cubic profile for the temperature [Eq. (2–49)], develop an expression for Nu(x) equivalent to Eq. (2–58).

2.16. For a plate heated over its entire length, how is the average value of the film coefficient up to a station x related to the local value at that same station?

2.17. Water at 30°C flows over a flat plate at 0.5 m/s. The plate is heated to 40°C, starting at 1 cm from the leading edge. What is the total heat transfer from the plate up to a distance of 5 cm from the leading edge?

2.18. Hydrogen at 500°C and 15 atm flows over a plate at 2 m/s. The first 2 cm of the plate are maintained at that same temperature, the next 3 cm are at 600°C, and the rest of the plate is at 650°C. What is the heat transfer rate 6 cm from the leading edge?

2.19. Suppose we have a flat plate immersed in a liquid metal stream flowing at 0.5 ft/s with a fluid temperature of 60°F. The fluid has a specific gravity of 10.0, a viscosity the same as water at room temperature, and thermal conductivity and a Prandtl number the same as that of mercury. After a distance of 2.0 in, the plate is heated to 70°F for the rest of its length. What are the heat transfer rate and the thickness of the velocity and thermal boundary layers 3 in from the leading edge?

2.20. Compare the thermal boundary layer thickness to the velocity boundary layer thickness for (1) air at STP, (2) hydrogen at STP, (3) water at 70°F, (4) engine oil at 150°F, and (5) mercury at 75°F.

2.21. What is the value of the film coefficient for diffusion, $\hbar_D$, at a distance of 0.5 m for a mixture of CO_2 and air flowing over a flat plate at 3 m/s and 1 atm? Assume a small enough concentration of CO_2 such that all the properties of the mixture can be taken to be those for pure air. How does $\hbar_D$ compare with $\hbar$?

2.22. Consider flow at 4 ft/s of two different fluids over two flat plates 12 in long. One fluid is Newtonian with $\mu = 2 \times 10^{-3}$ lb_f s/ft^2. The other is a *power law fluid* with $\mu_{PL} = 2 \times 10^{-3}$ lb_f $s^{1/2}/ft^2$ and $p = 1/2$. The specific gravity of both fluids is 1.20. Calculate the boundary layer thickness and skin friction coefficient at the end of the plate for both cases.

2.23. Calculate the boundary layer flow in the first quadrant of a circular cylinder for a case of water at 68°F with $U_\infty = 10.0$ ft/s. and $R_o = 0.1$ ft. The inviscid flow solution for this flow is:

$$U_e(x) = 2.0 \sin(s/R_0)$$

where s is the surface distance from the stagnation point.

CHAPTER

3

DIFFERENTIAL EQUATIONS OF MOTION FOR LAMINAR FLOW

3–1 INTRODUCTION

In this chapter, attention is directed to the exact, laminar boundary layer form of the equations of motion. These are derived by applying the conservation principles to a differential volume of fluid, and coverage is restricted to planar or axisymmetric flows of perfect fluids (gases obeying the perfect gas law and incompressible liquids). The resulting system of equations, although mathematically complex, is capable of describing the fine details of the flow in the boundary layer. In light of the discussions in Chapter 2, the reader may well ask why it is worthwhile to go to this level of complexity. The answer is fourfold. First, there are some practical situations in which fine details or great precision regarding gross quantities such as C_f is required. The situation can be illustrated using a reentry vehicle as an example. Such a device is designed at the limits of performance and survivability, so the skin friction drag and wall heat transfer rate must be known quite accurately—indeed, to a precision of 1 to 2 percent. However, it is also important to know more detailed information about the profile, such as the maximum static temperature in the boundary layer, since that is crucial in predicting the communications *blackout* period. Second, it is important to have essentially exact, so-called *benchmark*, solutions to some problems, so that the adequacy of various approximate methods can be judged. Third, the widespread availability of large digital computers has put the solution of the exact equations within the reach of most practicing engineers. It is a matter of history in high technology that as soon as more accurate solutions become achievable, all workers want to have them. In the field of technology, there is little *nostalgia* for the old ways of doing things. Finally, the fourth and most important reason for studying the differential equations for the boundary layer is that approximate methods for turbulent flows have been less successful than for laminar flows. Thus, modern treatments of turbulent flows are based on differential formulations, and the study of the corresponding laminar cases provides a good background for undertaking the turbulent analyses.

The derivations will be presented in an apparently planar Cartesian coordinate system denoted by (x,y). However, the resulting boundary layer equations also apply to a so-called *body-fitted coordinate system,* in which x is measured along the body surface and y is locally normal to the surface. The equations apply as long as the boundary layer thickness is small compared to the local longitudinal radius of curvature, which is usually the case. If it is not, extra terms due to curvature must be added. These are neglected here.

3–2 CONSERVATION OF MASS: THE CONTINUITY EQUATION

The derivation of the continuity equation is relatively simple, and the student has probably seen it before. Note that this equation is not affected by an assumption of inviscid flow, compared to its form for a real, viscous fluid, since forces do not appear in the principle of conservation of mass.

Figure 3–1 shows the planar geometry of a differential volume of fluid in the boundary layer. The mass entering the left-hand side of the differential volume $(dx : dy : 1)$, assuming a unit depth in the z-direction out of the page, is

$$\rho u \, dy \tag{3–1}$$

and the mass leaving the right-hand side can be written

$$\left(\rho + \frac{\partial \rho}{\partial x} dx\right)\left(u + \frac{\partial u}{\partial x} dx\right) dy \tag{3–2}$$

Similarly, the mass flowing in through the bottom is

$$\rho v \, dx \tag{3–3}$$

and that out the top is

$$\left(\rho + \frac{\partial \rho}{\partial y} dy\right)\left(v + \frac{\partial v}{\partial y} dy\right) dx \tag{3–4}$$

Any net difference between the inflows and outflows must appear as an unsteady

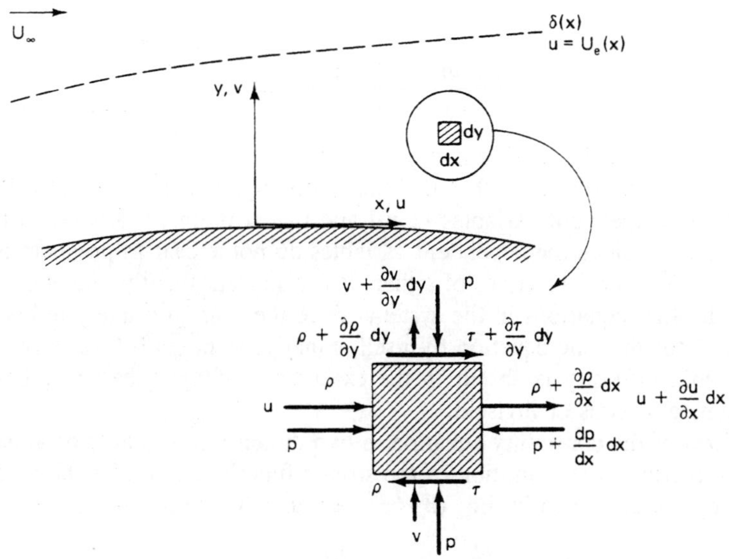

Figure 3–1 Differential volume for applying conservation of mass and momentum in the boundary layer.

change in the mass of fluid in the elemental volume which can be expressed as

$$\left(\frac{\partial \rho}{\partial t}\right) dxdy \tag{3–5}$$

Combining Eqs.(3–1) through (3–5) yields

$$\frac{\partial \rho}{\partial t} + \frac{\partial(\rho u)}{\partial x} + \frac{\partial(\rho v)}{\partial y} = 0 \tag{3–6}$$

For an axisymmetric flow, the distance from the axis r enters into the situation, and the corresponding equation becomes

$$\frac{\partial \rho}{\partial t} + \frac{1}{r}\frac{\partial(\rho u r)}{\partial x} + \frac{1}{r}\frac{\partial(\rho v r)}{\partial r} = 0 \tag{3–7}$$

For constant-density flow, Eqs. (3–6) and (3–7) simplify to

$$\frac{\partial u}{\partial x} + \frac{\partial v}{\partial y} = 0 \tag{3–6a}$$

and

$$\frac{1}{r}\frac{\partial(ur)}{\partial x} + \frac{1}{r}\frac{\partial(vr)}{\partial r} = 0 \tag{3–7a}$$

regardless of whether the flow is *steady or unsteady*.

It can easily be observed that the continuity equation is a partial differential equation [two independent variables (x,y)] and that it is linear, at least in the constant-density case, since the dependent variables do not appear as products of themselves or each other or derivatives of either. It is also clear that this equation must be coupled with other equations in the system, since there are too many unknowns (at least u and v) for this one equation by itself. Finally, as noted before, any term denoting the fluid viscosity is absent in the exact form of the equation, whether the fluid is assumed viscous or inviscid.

The form of the continuity equation in two dimensions—planar or axisymmetric—leads directly to the definition of the stream function. Consider the planar, incompressible case described by Eq. (3–6a). A scalar function ψ defined by

$$\frac{\partial\psi}{\partial y} = u; \quad -\frac{\partial\psi}{\partial x} = v \tag{3–8}$$

will always satisfy Eq. (3–6a), as can be seen by simple substitution. For the axisymmetric geometry, we can have another scalar function Ψ defined by

$$\frac{\partial\Psi}{\partial r} = ur; \quad -\frac{\partial\Psi}{\partial x} = vr \tag{3–9}$$

If the flow is compressible, one must add the restriction of steady flow, but then we can have

$$\frac{\partial\psi}{\partial y} = \rho u; \quad -\frac{\partial\psi}{\partial x} = \rho v \tag{3–10}$$

or

$$\frac{\partial\Psi}{\partial r} = \rho ur; \quad -\frac{\partial\Psi}{\partial x} = \rho vr \tag{3–11}$$

The concept of a stream function often proves useful in analysis, since it reduces the number of unknowns by one $[(u,v) \rightarrow \psi$ or $(u,v) \rightarrow \Psi]$, and the continuity equation is automatically solved, so the system of equations is correspondingly reduced by one.

3–3 CONSERVATION OF MOMENTUM: THE MOMENTUM EQUATION

Since the boundary layer approximation is equivalent to the statement that $\partial p/\partial y \approx 0$ across the layer, it is necessary to treat conservation of momentum only in the streamwise, x, direction in detail. Conservation of momentum in the y-direction is enforced by $\partial p/\partial y \approx 0$. Accordingly, consider again the elemental volume in Fig. 3–1, and note that the detailed derivation given here is only for the planar geometry. The mass entering the left-hand side, $\rho u\, dy$, brings with it x-momentum in the amount

$$u(\rho u)dy \tag{3–12}$$

and the x-momentum leaving the right-hand side is

$$\left(u + \frac{\partial u}{\partial x}\,dx\right)\left(\rho u + \frac{\partial(\rho u)}{\partial x}\,dx\right)dy \tag{3–13}$$

The mass entering the bottom face, $\rho v\, dx$ brings with it x-momentum of

$$u(\rho v)dx \tag{3–14}$$

and the x-momentum leaving the top is

$$\left(u + \frac{\partial u}{\partial y}\,dy\right)\left(\rho v + \frac{\partial(\rho v)}{\partial y}\,dy\right)dx \tag{3–15}$$

The x-momentum in the volume, $(\rho u)dx\, dy$, can also change in an unsteady fashion, and the change can be expressed simply as

$$\frac{\partial(\rho u)}{\partial t}\,dx\, dy \tag{3–16}$$

The net change in x-momentum in the volume is balanced by the resultant of the forces in the x-direction.

There are, in general, two kinds of fluid forces: *body forces* and *surface forces*. Body forces act on the bulk of the fluid in a volume as a whole. The most common example is gravity, but under special circumstances, magnetic or electric fields can also produce body forces. In most fluid flow problems, except those at very low speed, the influence of body forces is negligible compared to that of inertia and the other forces in the flow. We will carry along a generalized body force per unit volume, f_i, in the derivations in this section, but no problems involving body forces will be treated. The interested reader is referred to the chapter "Natural Convection" in the heat transfer test by Eckert and Drake (1972), for specific material on that subject.

The primary fluid forces in many practical problems are two surface forces: pressure and viscous forces. These forces influence the fluid in a volume by acting on the surface of the volume. The summation of pressure forces in the x-direction is (see Fig. 3–1)

$$p\, dy - \left(p + \frac{\partial p}{\partial x}\,dx\right)dy \tag{3–17}$$

We have a partial derivative for the pressure, $\partial p/\partial x$, not a total derivative, dp/dx, in order to include the unsteady case, i.e., $p(x, t)$. The boundary layer assumption,

$\partial p/\partial y \approx 0$, is retained. The only components of the viscous shear that produce forces in the x-direction are those acting on the top and bottom of the volume. Considering the typical shape of a velocity profile (see Fig. 2-1), the shear acting on the bottom surface tends to retard the flow, and it can be written

$$-\tau\, dx \tag{3–18}$$

The viscous shear on the top surface tends to accelerate the flow:

$$\left(\tau + \frac{\partial \tau}{\partial y}\, dy\right) dx \tag{3–19}$$

Setting the net change in x-momentum equal to the summation of fluid forces in the x-direction results in an equation that can be rearranged and simplified, after dividing through by $(dx)(dy)$, to give

$$u\left[\frac{\partial \rho}{\partial t} + \frac{\partial(\rho u)}{\partial x} + \frac{\partial(\rho v)}{\partial y}\right] + \left[\left(\frac{\partial u}{\partial x}\right)\left(\frac{\partial(\rho u)}{\partial x}\right)dx + \left(\frac{\partial u}{\partial y}\right)\left(\frac{\partial(\rho v)}{\partial y}\right)dy\right]$$
$$+ \rho\left[\frac{\partial u}{\partial t} + u\,\frac{\partial u}{\partial x} + v\,\frac{\partial u}{\partial y}\right] = -\frac{\partial p}{\partial x} + \frac{\partial \tau}{\partial y} + f_x \tag{3–20}$$

The group of terms in the first set of brackets should look familiar. It is the left-hand side of the continuity equation [see Eq. (3–6)]; thus, it is identically equal to zero. Looking at the terms in the second set of brackets, we see that they, and they alone, still contain the differential lengths dx and dy. Since Eq. (3–20) is a differential equation, it holds in the limit as dx and $dy \to 0$, so those terms disappear.

The boundary layer momentum equation then becomes

$$\frac{\partial u}{\partial t} + u\frac{\partial u}{\partial x} + v\frac{\partial u}{\partial y} = -\frac{1}{\rho}\frac{\partial p}{\partial x} + \frac{1}{\rho}\frac{\partial \tau}{\partial y} + \frac{f_x}{\rho} \tag{3–21}$$

This equation is valid for planar, unsteady, compressible boundary layer flow with body forces. Note that in this chapter, we have not yet had to say that the flow is laminar. Actually, Eq. (3–21) is equally valid for turbulent flow. The only distinction between the two cases will be in how the shear τ is *modeled*.

3–3–1 Modeling of the Laminar Shear Stress

The modeling of τ can be made clearer by first considering a restricted case—steady, constant-density flow, without body forces. The conclusions reached, however, are general and not influenced by those restrictions. The problem is specified by the appropriate form of the continuity equation

$$\frac{\partial u}{\partial x} + \frac{\partial v}{\partial y} = 0 \tag{3 6a}$$

and of the momentum equation

$$u\frac{\partial u}{\partial x} + v\frac{\partial u}{\partial y} = -\frac{1}{\rho}\frac{dp}{dx} + \frac{1}{\rho}\frac{\partial \tau}{\partial y} \tag{3–22}$$

and an equation of state, which, for this situation, is simply ρ = constant. Keeping in mind that the pressure $p(x)$ is *imposed* on the boundary layer by the inviscid solution, one can see that $p(x)$ is not really a dependent variable in this system. The actual dependent variables are u, v, τ, and ρ. But there are only three equations. Because one cannot hope to make the situation for ρ much simpler, and it is unreasonable to try to find some direct relation between u and v, attention is naturally directed to τ.

To close the system mathematically, the variable τ must be related to the other variables in a way that has some general validity, or the whole mathematical structure will be severely restricted. The process of finding the appropriate relation is called *modeling*. Fortunately, for laminar flow of a Newtonian fluid in a boundary layer, the simple expression

$$\tau = \mu \frac{\partial u}{\partial y} \tag{3-23}$$

is valid. In that case, $\mu = \mu(T)$ is a physical property of the fluid composition and temperature alone and is not dependent on the fluid motion. For non-Newtonian fluids, we have somewhat more complicated expressions, but the added complexity for *power law* or *Bingham plastic fluids*, to cite just two examples, is not great. The matter is vastly more complicated for turbulent flow, as we shall see in Chapter 7.

3-3-2 Forms of the Momentum Equation for Laminar Flow

Using Eq. (3-23) in Eq. (3-21) and dropping the body force terms yields

$$\rho\left[\frac{\partial u}{\partial t} + u\frac{\partial u}{\partial x} + v\frac{\partial u}{\partial y}\right] = -\frac{\partial p}{\partial x} + \frac{\partial}{\partial y}\left(\mu\frac{\partial u}{\partial y}\right) \tag{3-24}$$

Note that μ must be kept inside the derivative on the right-hand side if the temperature is varying, because $\mu = \mu(T)$.

If the flow is axisymmetric, the distance from the axis enters into the situation, and the equation becomes

$$\rho\left[\frac{\partial u}{\partial t} + u\frac{\partial u}{\partial x} + v\frac{\partial u}{\partial r}\right] = -\frac{\partial p}{\partial x} + \frac{1}{r}\frac{\partial}{\partial r}\left(\mu r\left(\frac{\partial u}{\partial r}\right)\right) \tag{3-25}$$

This equation is to be used with Eq. (3-7).

In most practical instances in the axisymmetric geometry, one is interested in the boundary layer on an axisymmetric body, as shown in Fig. 3-2. If an appropriate *body system of coordinates* is employed, the relevant forms of the momentum and continuity equations become

$$\rho\left[\frac{\partial u}{\partial t} + u\frac{\partial u}{\partial x} + v\frac{\partial u}{\partial y}\right] = -\frac{\partial p}{\partial x} + \frac{\partial}{\partial y}\left(\mu\frac{\partial u}{\partial y}\right) \tag{3-26}$$

and

$$r_0^j\frac{\partial \rho}{\partial t} + \frac{\partial(\rho u r_0^j)}{\partial x} + \frac{\partial(\rho v r_0^j)}{\partial y} = 0 \tag{3-27}$$

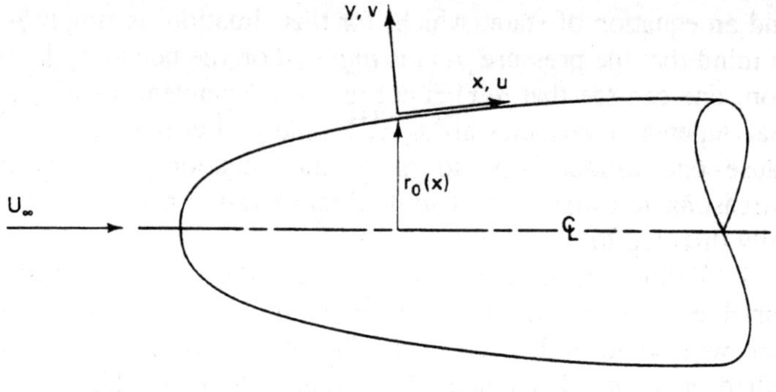

Figure 3–2 Schematic of the body-fitted coordinate system for the boundary layer on a body of revolution.

where $r_o(x)$ is the local body radius. These equations are valid for $\delta \ll r_o$ and no sharp corners; that is, d^2r_o/dx^2 must be well behaved. Note that Eq. (3–26) has the same form as the momentum equation for planar flow, Eq. (3–24). Thus, a convenient, unified presentation can be given. Equations (3–26) and (3–27) hold for flow over bodies in both the planar geometry ($j \equiv 0$) and the axisymmetric geometry ($j \equiv 1$).

If the flow has constant density, the momentum equation shows no explicit change in form; the coefficients involving ρ simply become constant. The continuity equation then becomes

$$\frac{\partial(ur_0^j)}{\partial x} + \frac{\partial(vr_0^j)}{\partial y} = 0 \tag{3–27a}$$

for both steady and unsteady flow. If the assumption of constant properties is also made, which is common with the constant-density assumption for gas flows, then the momentum equation becomes

$$\frac{\partial u}{\partial t} + u\frac{\partial u}{\partial x} + v\frac{\partial u}{\partial y} = -\frac{1}{\rho}\frac{\partial p}{\partial x} + \nu\frac{\partial^2 u}{\partial y^2} \tag{3–26a}$$

For steady flow, the first term on the left-hand side is dropped, and the pressure gradient term becomes $(-1/\rho)(dp/dx)$.

Another useful form of the momentum equation for planar or axisymmetric flows is developed by the introduction of a stream function as the dependent variable, in place of the two velocity components (u,v). For the case of a planar flow with constant density and constant properties, a stream function defined by Eq. (3–8) is substituted into Eq. (3–26a) to yield

$$\frac{\partial^2\psi}{\partial t\,\partial y} + \frac{\partial\psi}{\partial y}\frac{\partial^2\psi}{\partial x\,\partial y} - \frac{\partial\psi}{\partial x}\frac{\partial^2\psi}{\partial y^2} = -\frac{1}{\rho}\frac{\partial p}{\partial x} + \nu\frac{\partial^3\psi}{\partial y^3} \tag{3–28}$$

This is one equation for the one unknown, $\psi(x, y, t)$, with the equation of state, ρ = constant, implied.

For a *power law non-Newtonian fluid*, the stress model given by Eq. (1-64) must be used in Eq. (3–21). Dropping the body force terms then gives

$$\rho\left[\frac{\partial u}{\partial t} + u\frac{\partial u}{\partial x} + v\frac{\partial u}{\partial y}\right] = -\frac{\partial p}{\partial x} + \frac{\partial}{\partial y}\left(\mu_{\text{PL}}\left(\frac{\partial u}{\partial y}\right)^{p}\right) \tag{3–29}$$

For a *Bingham plastic non-Newtonian fluid,* the stress model given by Eq. (1–62) must be used in Eq. (3–21). But, the shear τ enters into Eq. (3–21) only as $\partial\tau/\partial y$, so the constant τ_o in Eq. (1-62) does not contribute. The result is that a Bingham plastic fluid is governed by the same form of the momentum equation as a Newtonian fluid, viz., Eq. (3–24).

3–4 CONSERVATION OF ENERGY: THE ENERGY EQUATION

The thermal energy in the flow can be expressed in terms of the internal energy, the enthalpy, or the temperature. Also, a static or a stagnation (or *total*) version of any of these three can be employed. For this reason, more different forms of the energy equation appear in the literature than do forms for any of the other equations. In any event, each of the foregoing dependent variables is a scalar, so only a single equation is required.

Conservation of energy is expressed by the first law of thermodynamics, which states that a change in energy of a system results from the difference between heat transfer and work. In Fig. 3–3, consider each process separately. As before, the

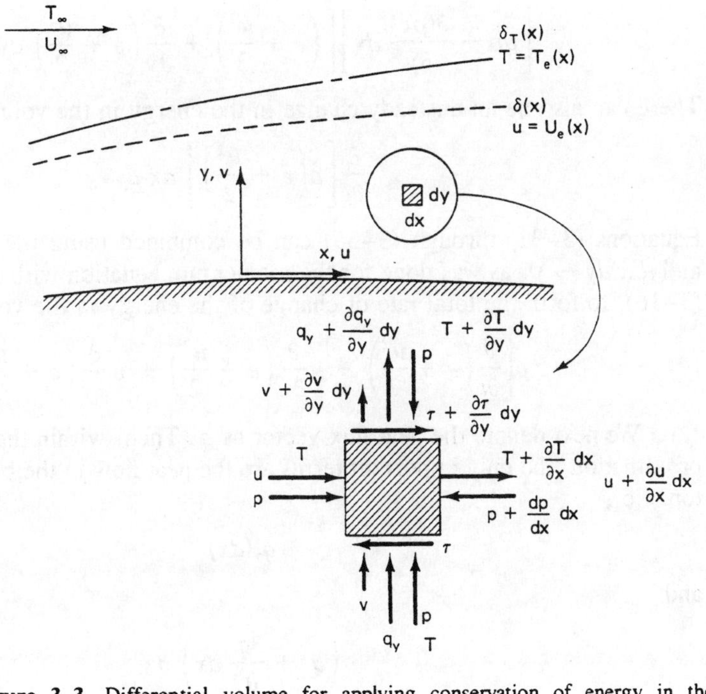

Figure 3–3 Differential volume for applying conservation of energy in the boundary layer.

boundary layer assumptions on the flow are $\partial p/\partial y \approx 0$, $\partial u/\partial y >> \partial u/\partial x$, and $u >> v$, and the equivalent assumption for the temperature field is $\partial T/\partial y >> \partial T/\partial x$. Changes in the energy of the flow are expressed in terms of changes in the complete internal energy (thermal $e(T)$ plus kinetic energy) as

$$de + d\left(\frac{u^2}{2}\right) = c_v dT + d\left(\frac{u^2}{2}\right) \tag{3–30}$$

for a perfect gas. In this section, we consider a fluid of only one chemical species.

The mass flow into the left-hand side brings in energy at a rate

$$\rho u\left(e + \frac{u^2}{2}\right)dy \tag{3–31}$$

The mass flow out of the right-hand side carries energy out at a rate

$$\left[\rho u + \frac{\partial(\rho u)}{\partial x}\,dx\right]\left[\left(e + \frac{u^2}{2}\right) + \frac{\partial}{\partial x}\left(e + \frac{u^2}{2}\right)dx\right]dy \tag{3–32}$$

For the bottom and top surfaces, one obtains

$$\rho v\left(e + \frac{u^2}{2}\right)dx \tag{3–33}$$

and

$$\left[\rho v + \frac{\partial(\rho v)}{\partial y}\,dy\right]\left[\left(e + \frac{u^2}{2}\right) + \frac{\partial}{\partial y}\left(e + \frac{u^2}{2}\right)dy\right]dx \tag{3–34}$$

There can also be an unsteady change in the energy in the volume, expressed as

$$\frac{\partial}{\partial t}\left[\rho\left(e + \frac{u^2}{2}\right)\right]dx\,dy \tag{3–35}$$

Equations (3–31) through (3–35) can be combined using the continuity equation and $dx, dy \rightarrow 0$, as was done for the momentum equation with Eqs. (3–12) through (3–16), to form the total rate of change of the energy in the volume:

$$\rho\left[\frac{\partial}{\partial t}\left(e + \frac{u^2}{2}\right) + u\frac{\partial}{\partial x}\left(e + \frac{u^2}{2}\right) + v\frac{\partial}{\partial y}\left(e + \frac{u^2}{2}\right)\right] \tag{3–36}$$

We next denote the heat flux vector as q_i. Then, within the boundary layer approximation, the only important terms are the heat flow in the bottom and out of the top, i.e.,

$$q_y(dx)$$

and (3–37)

$$\left(q_y + \frac{\partial q_y}{\partial y}\,dy\right)dx$$

So the net heat flow rate to the volume is

$$-\frac{\partial q_y}{\partial y}\,dy\,dx \tag{3–37a}$$

Work can be done on the system by all the forces that are acting on it. Neglecting body forces, one need only be concerned with pressure and viscous forces. Work is a force acting through a distance, so for a flowing system, the work per unit time becomes a force times a velocity. The pressure forces do net work on the side surfaces in the amount

$$-\frac{\partial(pu)}{\partial x}\,dx\,dy \tag{3–38}$$

and on the top and bottom surfaces in the amount

$$-\frac{\partial(pv)}{\partial y}\,dy\,dx \tag{3–39}$$

In both of these equations, one must carefully reckon the signs, considering the orientation of the pressure force and the velocity components to the surfaces in question.

With the boundary layer approximation, we neglect shear on the sides compared with shear on the top and bottom surfaces, so we have, for the work done by frictional forces,

$$-\tau u\,dx + \left(\tau u + \frac{\partial(\tau u)}{\partial y}\,dy\right)dx \tag{3–40}$$

Combining Eqs. (3–38) through (3–40), we obtain the net work per unit time:

$$\left(\frac{\partial(\tau u)}{\partial y} - \frac{\partial(pu)}{\partial x} - \frac{\partial(pv)}{\partial y}\right)dx\,dy \tag{3–41}$$

Now we combine Eq. (3–36), which is the rate of increase of total energy, with Eq. (3–37a), which is the rate at which heat is transferred into the element, and Eq. (3–41), which is the rate at which the stresses on the boundaries do work on the volume, through the first law, to get

$$\rho\left[\frac{\partial}{\partial t}\left(e + \frac{u^2}{2}\right) + u\frac{\partial}{\partial x}\left(e + \frac{u^2}{2}\right) + v\frac{\partial}{\partial y}\left(e + \frac{u^2}{2}\right)\right]$$
$$= -\frac{\partial q_y}{\partial y} + \frac{\partial(\tau u)}{\partial y} - \frac{\partial(pu)}{\partial x} - \frac{\partial(pv)}{\partial y} \tag{3–42}$$

This is a correct, but inconvenient, form of the energy equation, so we will make some substitutions.

First, we replace the internal energy with the enthalpy, defined as

$$h \equiv e + \frac{p}{\rho}; \qquad dh = c_p dT \tag{3–43}$$

The operator on the left-hand side of Eq. (3–42) can be written in a shorthand form as

$$\rho\left[\frac{\partial(\cdot)}{\partial t} + u\frac{\partial(\cdot)}{\partial x} + v\frac{\partial(\cdot)}{\partial y}\right] = \rho\frac{D(\cdot)}{Dt} \tag{3–44}$$

Using this with Eq. (3–43), one can write

$$\rho\frac{De}{Dt} = \rho\frac{D}{Dt}\left(h - \frac{p}{\rho}\right) = \rho\frac{Dh}{Dt} - \frac{Dp}{Dt} + \frac{p}{\rho}\frac{D\rho}{Dt} \tag{3–45}$$

Now the continuity equation, Eq. (3–6), can be expanded and rewritten with this notation as

$$\frac{D\rho}{Dt} = -\rho\left(\frac{\partial u}{\partial x} + \frac{\partial v}{\partial y}\right) \tag{3–6b}$$

Using this form and rearranging Eq. (3–45), we obtain

$$\rho\frac{De}{Dt} = \rho\frac{Dh}{Dt} - \frac{\partial(pu)}{\partial x} - \frac{\partial(pv)}{\partial y} - \frac{\partial p}{\partial t} \tag{3–46}$$

Thus, Eq. (3–42) can be recast to read

$$\rho\left[\frac{\partial}{\partial t}\left(h + \frac{u^2}{2}\right) + u\frac{\partial}{\partial x}\left(h + \frac{u^2}{2}\right) + v\frac{\partial}{\partial y}\left(h + \frac{u^2}{2}\right)\right] = -\frac{\partial q_y}{\partial y} + \frac{\partial(\tau u)}{\partial y} + \frac{\partial p}{\partial t} \tag{3–47}$$

The second step necessary to achieve the form of the energy equation desired for many uses is to remove the $u^2/2$ terms. For this purpose, consider the momentum equation, Eq. (3–21), and drop the body force term and multiply through by the streamwise velocity component u, to get

$$\rho\left[\frac{\partial(u^2/2)}{\partial t} + u\frac{\partial(u^2/2)}{\partial x} + v\frac{\partial(u^2/2)}{\partial y}\right] = -u\frac{\partial p}{\partial x} + u\frac{\partial \tau}{\partial y} \tag{3–48}$$

where we have used $u\,\partial u/\partial(\cdot) = \partial(u^2/2)/\partial(\cdot)$. This equation is often called the *mechanical energy* equation. Now, we subtract Eq. (3–48) from Eq. (3–47) and obtain

$$\rho\left[\frac{\partial h}{\partial t} + u\frac{\partial h}{\partial x} + v\frac{\partial h}{\partial y}\right] = u\frac{\partial p}{\partial x} - \frac{\partial q_y}{\partial y} + \frac{\partial p}{\partial t} + \tau\frac{\partial u}{\partial y} \tag{3–49}$$

or, using Eq. (3–43),

$$\rho c_p\left[\frac{\partial T}{\partial t} + u\frac{\partial T}{\partial x} + v\frac{\partial T}{\partial y}\right] = u\frac{\partial p}{\partial x} + \frac{\partial p}{\partial t} - \frac{\partial q_y}{\partial y} + \tau\frac{\partial u}{\partial y} \tag{3–49a}$$

This is the desired result for many cases—the exact, boundary layer energy equation written in terms of the static temperature. It is valid for both laminar and turbulent flow, since the shear τ and now also the heat transfer rate q_y have not been *modeled*.

3–4–1 Modeling of the Laminar Heat Flux

For laminar flow of many common fluids, a suitable approximation is *Fourier's law*,

$$q_y = -k\frac{\partial T}{\partial y} \tag{1–5a}$$

where $k = k(T)$ is a physical property of the fluid composition and temperature. Note that q_y does not depend on the motion of the fluid. Perhaps surprisingly, the *same expression* has been found to describe the behavior of *most non-Newtonian fluids*. Using this expression and the Newtonian expression for shear in laminar flow, we obtain, for the energy equation,

$$\rho c_p\left[\frac{\partial T}{\partial t} + u\frac{\partial T}{\partial x} + v\frac{\partial T}{\partial y}\right] = \frac{\partial}{\partial y}\left(k\frac{\partial T}{\partial y}\right) + \mu\left(\frac{\partial u}{\partial y}\right)^2 + u\frac{\partial p}{\partial x} + \frac{\partial p}{\partial t} \tag{3–50}$$

For constant-property flow, the energy equation can be written

$$\rho c_p\left[\frac{\partial T}{\partial t} + u\frac{\partial T}{\partial x} + v\frac{\partial T}{\partial y}\right] = k\frac{\partial^2 T}{\partial y^2} + u\frac{\partial p}{\partial x} + \frac{\partial p}{\partial t} + \mu\left(\frac{\partial u}{\partial y}\right)^2 \tag{3–51}$$

If the fluid is also incompressible, we can write

$$\rho c_v\left[\frac{\partial T}{\partial t} + u\frac{\partial T}{\partial x} + v\frac{\partial T}{\partial y}\right] = k\frac{\partial^2 T}{\partial y^2} + \mu\left(\frac{\partial u}{\partial y}\right)^2 \tag{3–52}$$

The apparent simplification achieved in Eq. (3–52) over (3–51) results from starting with Eq. (3–42), subtracting the mechanical energy equation, Eq. (3–48), and then using the continuity equation for incompressible flow, Eq. (3–6a).

Finally, it is important to note that Eq. (3–50) [or (3–51) or (3–52)] holds for flow over axisymmetric bodies if (x,y) is chosen as in Fig. 3–2, in the same way as for the momentum equation.

For general axisymmetric flows, the exact forms are

$$\rho c_p\left[\frac{\partial T}{\partial t} + u\frac{\partial T}{\partial x} + v\frac{\partial T}{\partial r}\right] = \frac{1}{r}\frac{\partial}{\partial r}\left(kr\frac{\partial T}{\partial r}\right) + \mu\left(\frac{\partial u}{\partial r}\right)^2 + u\frac{\partial p}{\partial x} + \frac{\partial p}{\partial t} \tag{3–53}$$

and

$$\rho c_v\left[\frac{\partial T}{\partial t} + u\frac{\partial T}{\partial x} + v\frac{\partial T}{\partial r}\right] = \frac{k}{r}\frac{\partial}{\partial r}\left(r\frac{\partial T}{\partial r}\right) + \mu\left(\frac{\partial u}{\partial r}\right)^2 \tag{3–54}$$

For most cases in the low-speed, constant-property regime, the viscous heating terms (the last term in Eqs. (3–51), (3–52), and (3–54)) are negligible. On the other hand, they are very important in high-speed flows, as will be discussed in Chapter 5.

For non-Newtonian fluids, it was noted earlier that the heat transfer model is the same as for Newtonian fluids. That gives the energy equation the same form, except for the viscous heating term, since it involves τ (see Eq. (3–49a). If viscous heating is important in a particular case involving a non-Newtonian fluid, a relevant shear model, e.g., Eq. (1–62) or Eq. (1–64), must be used.

3–5 CONSERVATION OF MASS OF SPECIES: THE SPECIES CONTINUITY EQUATION

Application of the principle of conservation of mass to a species i requires only a slight generalization of the development for the conservation of the total mass in Section 3–2. The net mass flow of species i through the sides (the difference between the flow into the left side and the flow out of the right side) is

$$\frac{\partial}{\partial x}(\rho u c_i)\, dx\, dy \tag{3–55}$$

For the flows into the bottom and out of the top, we must now allow for a mass flux of species i by diffusion as well as by convection. Denote the diffusive flux as $\dot{m}_i$. The net mass flow of species i in the y-direction is then

$$\frac{\partial}{\partial y}(\rho v c_i + \dot{m}_i)\, dy\, dx \tag{3–56}$$

The diffusive flux in the x-direction is neglected with the boundary layer assumption, in the same way as for shear and heat transfer.

Any accumulation of mass of species i in the volume must appear as an unsteady term

$$\frac{\partial}{\partial t}(\rho c_i)\, dx\, dy \tag{3–57}$$

Applying conservation of mass of species i and assuming no production or consumption of species by chemical reactions, one obtains, after using the continuity equation (for the total mass, Eq. (3–6)),

$$\rho\left[\frac{\partial c_i}{\partial t} + u\frac{\partial c_i}{\partial x} + v\frac{\partial c_i}{\partial y}\right] = -\frac{\partial \dot{m}_i}{\partial y} \tag{3–58}$$

This equation holds for laminar or turbulent flow; the diffusive flux must still be modeled.

3–5–1 Modeling of Laminar Diffusion

Assuming a binary mixture and neglecting thermal and pressure diffusion, Fick's law gives

$$\dot{m}_i = -\rho D_{ij}\frac{\partial c_i}{\partial y} \tag{3–59}$$

Thus, Eq. (3–58) becomes

$$\rho\left[\frac{\partial c_i}{\partial t} + u\frac{\partial c_i}{\partial x} + v\frac{\partial c_i}{\partial y}\right] = \frac{\partial}{\partial y}\left(\rho D_{ij}\frac{\partial c_i}{\partial y}\right) \tag{3–60}$$

If the fluid has constant density and constant properties, the equation simplifies to

$$\frac{\partial c_i}{\partial t} + u\frac{\partial c_i}{\partial x} + v\frac{\partial c_i}{\partial y} = D_{ij}\frac{\partial^2 c_i}{\partial y^2} \tag{3–61}$$

For axisymmetric flows, we have

$$\rho\left[\frac{\partial c_i}{\partial t} + u\frac{\partial c_i}{\partial x} + v\frac{\partial c_i}{\partial r}\right] = \frac{1}{r}\frac{\partial}{\partial r}\left(\rho D_{ij} r\frac{\partial c_i}{\partial r}\right) \tag{3–62}$$

There is much less work in the literature and generally less need for material on mass transfer with non-Newtonian fluids.

3–5–2 Energy Transfer by Mass Transfer

Mass transfer can contribute to energy transfer, even if the fluid is at a constant temperature. This phenomenon is a consequence of the fact that different species generally have different specific heats (e.g., c_p). For a laminar flow, the result is an additional term in Eq. (3–47), the energy equation expressed as a function of stagnation enthalpy:

$$-\frac{\partial}{\partial y}\left[\left(\frac{1}{\mathrm{Le}} - 1\right)\rho D_{ij} \sum_i h_i\frac{\partial c_i}{\partial y}\right] \tag{3–63}$$

Here,

$$h_i = \int c_{pi}\, dT + h_i^o \tag{3–64}$$

where h_i^o is the heat of formation of species i at the reference temperature for enthalpy. One can see how this term appears by noting the occurrence of the diffusional mass flux, modeled as in Eq. (3–59), times the relevant enthalpy h_i summed over all species. The energy equation for problems with mass transfer is usually written in terms of the stagnation enthalpy.

3–5–3 Physical Properties of Mixtures

The general-purpose computer code of Svehla and McBride (1973), used for most of the data on physical properties in Appendix A, will also calculate properties of mixtures of Newtonian fluids. One should not suppose that a property (e.g., viscosity) of a mixture is a linear function of the concentration between the property values of each single constituent. Some results for mixture viscosity and thermal conductivity for H_2 in air are shown in Fig. 3–4, from Eckert et al. (1958).

For mixtures of only two constituents, the following approximate rules may be used. For the viscosity of gases, the equation of Buddenberg and Wilke (1949) applies:

$$\mu = \frac{X_1^2}{(X_1^2/\mu_1) + 1.385(X_1X_2RT/pW_1D_{12})} + \frac{X_2^2}{(X_2^2/\mu_2) + 1.385(X_2X_1RT/pW_2D_{21})} \tag{3–65}$$

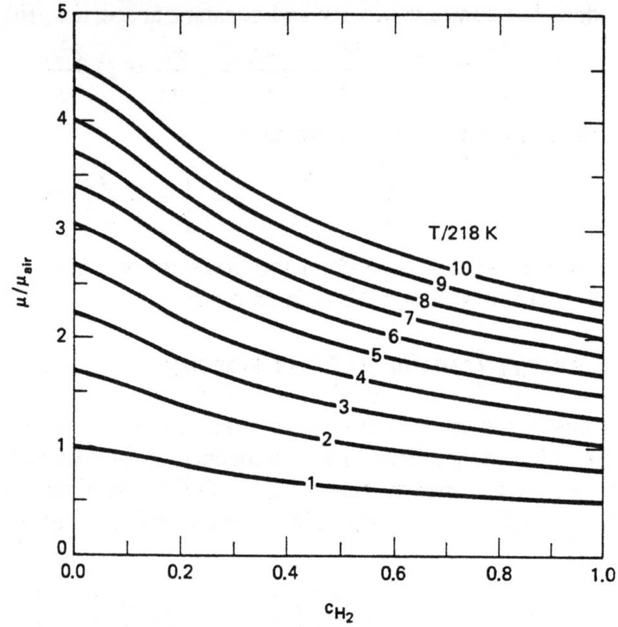

(A) Mixture viscosity

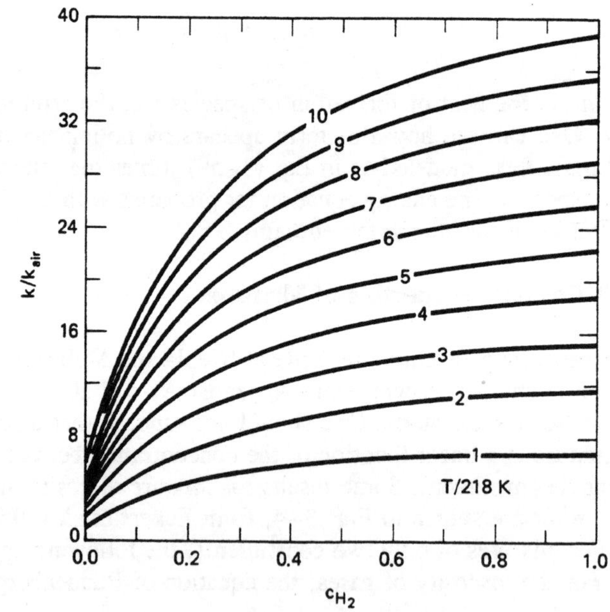

(B) Mixture thermal conductivity

Figure 3–4 Variation of mixture viscosity and thermal conductivity for H_2 in air. (From Eckert et al., 1958.)

If the fluid has constant density and constant properties, the equation simplifies to

$$\frac{\partial c_i}{\partial t} + u\frac{\partial c_i}{\partial x} + v\frac{\partial c_i}{\partial y} = D_{ij}\frac{\partial^2 c_i}{\partial y^2} \qquad (3\text{–}61)$$

For axisymmetric flows, we have

$$\rho\left[\frac{\partial c_i}{\partial t} + u\frac{\partial c_i}{\partial x} + v\frac{\partial c_i}{\partial r}\right] = \frac{1}{r}\frac{\partial}{\partial r}\left(\rho D_{ij} r\frac{\partial c_i}{\partial r}\right) \qquad (3\text{–}62)$$

There is much less work in the literature and generally less need for material on mass transfer with non-Newtonian fluids.

3–5–2 Energy Transfer by Mass Transfer

Mass transfer can contribute to energy transfer, even if the fluid is at a constant temperature. This phenomenon is a consequence of the fact that different species generally have different specific heats (e.g., c_p). For a laminar flow, the result is an additional term in Eq. (3–47), the energy equation expressed as a function of stagnation enthalpy:

$$-\frac{\partial}{\partial y}\left[\left(\frac{1}{\text{Le}} - 1\right)\rho D_{ij}\sum_i h_i\frac{\partial c_i}{\partial y}\right] \qquad (3\text{–}63)$$

Here,

$$h_i = \int c_{pi}\,dT + h_i^o \qquad (3\text{–}64)$$

where h_i^o is the heat of formation of species i at the reference temperature for enthalpy. One can see how this term appears by noting the occurrence of the diffusional mass flux, modeled as in Eq. (3–59), times the relevant enthalpy h_i summed over all species. The energy equation for problems with mass transfer is usually written in terms of the stagnation enthalpy.

3–5–3 Physical Properties of Mixtures

The general-purpose computer code of Svehla and McBride (1973), used for most of the data on physical properties in Appendix A, will also calculate properties of mixtures of Newtonian fluids. One should not suppose that a property (e.g., viscosity) of a mixture is a linear function of the concentration between the property values of each single constituent. Some results for mixture viscosity and thermal conductivity for H_2 in air are shown in Fig. 3–4, from Eckert et al. (1958).

For mixtures of only two constituents, the following approximate rules may be used. For the viscosity of gases, the equation of Buddenberg and Wilke (1949) applies:

$$\mu = \frac{X_1^2}{(X_1^2/\mu_1) + 1.385(X_1X_2RT/pW_1D_{12})} + \frac{X_2^2}{(X_2^2/\mu_2) + 1.385(X_2X_1RT/pW_2D_{21})} \qquad (3\text{–}65)$$

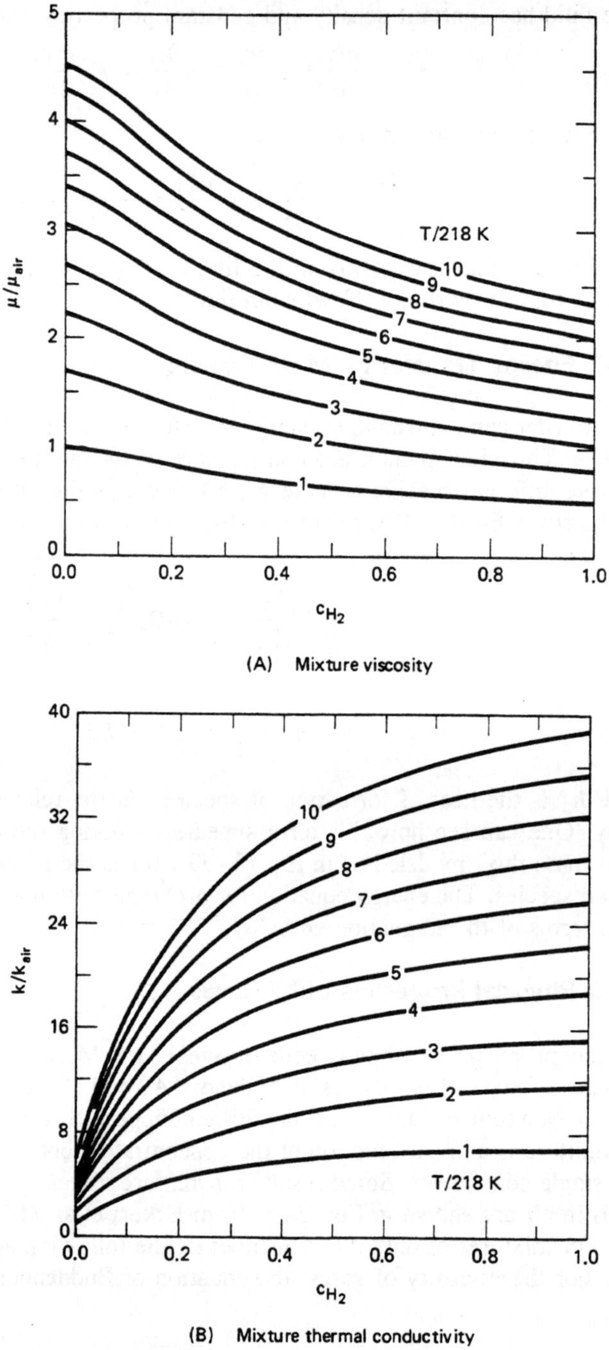

Figure 3–4 Variation of mixture viscosity and thermal conductivity for H_2 in air. (From Eckert et al., 1958.)

Here, X_i is the mole fraction of species i. For the thermal conductivity of gases, one can use

$$k = \frac{1}{2}\left[(k_1X_1 + k_2X_2) + \frac{k_1k_2}{(X_1\sqrt{k_2} + X_2\sqrt{k_1})^2}\right] \tag{3–66}$$

There are no corresponding simple expressions for liquid mixtures, so one must refer to published data.

3–6 TRANSFORMATION TO TWO-DIMENSIONAL FORM

Under some circumstances, the calculation of the boundary layer flow over an axisymmetric body can be reduced to the calculation of the boundary layer flow over an equivalent two-dimensional body. The transformation that accomplishes this reduction for cases where the effects of transverse curvature are negligible ($\delta/r_o << 1$) was introduced by Mangler (1945). It will be illustrated here for constant-density, constant-property flows. The compressible version is also available. The axisymmetric equations are Eqs. (3–26a) and (3–27a) with $j = 1$. If we introduce new, transformed variables $(\bar{x}, \bar{y})$, defined as

$$\bar{x} = \frac{1}{L^2}\int_0^x r_o^2\, dx', \qquad \bar{y} = \frac{r_o y}{L}$$
$$\bar{u} = u, \qquad \bar{v} = \frac{L}{r_o}\left(v + \frac{uy}{r_o}\frac{dr_o}{dx}\right), \qquad \bar{U}_e(\bar{x}) = U_e(x) \tag{3–67}$$

the equations become

$$\frac{\partial \bar{u}}{\partial \bar{x}} + \frac{\partial \bar{v}}{\partial \bar{y}} = 0$$
$$\bar{u}\frac{\partial \bar{u}}{\partial \bar{x}} + \bar{v}\frac{\partial \bar{u}}{\partial \bar{y}} = -\frac{1}{\rho}\frac{dp}{d\bar{x}} + \nu\frac{\partial^2 \bar{u}}{\partial \bar{y}^2} \tag{3–68}$$

This transformation allows the direct development of some simple results to be given later. It is also useful in transforming the shape of computational regions in the numerical solution of boundary layer problems.

3–7 MATHEMATICAL OVERVIEW

Before leaving this chapter, which presents the derivation of the exact, boundary layer forms of the equations of motion that will be used throughout the remainder of the book, it is appropriate—indeed, essential—to consider the nature of the mathematical problem that has been created. For purposes of discussion, the set of equations consisting of Eqs. (3–26), (3–27), (3–50), and (3–60), together with an appropriate equation of state, usually

$$p = \frac{\rho RT}{W}; \qquad W = \left(\sum_i c_i/W_i\right)^{-1} \tag{3-69}$$

can be said to represent the system. To keep these considerations simple, we restrict ourselves to fluids of only one species here.

There are four dependent variables—u, v, ρ, and T [$p(x)$ must be given from the inviscid solution]—and four equations. There are three independent variables—x, y, and t—so the system consists primarily [except for Eq. (3–69)] of partial differential equations. The equations also have variable coefficients, and more than that, Eqs. (3–26) and (3–50) are nonlinear on the left-hand sides. The nonlinearity is evident from the existence of terms such as $u\,\partial u/\partial x = \partial(u^2/2)/\partial x$, which clearly have a nonlinear form for the dependent variables. It has the same general consequences as the occurrence of terms such as yy' and y^2 in an ordinary differential equation for $y(x)$: Methods that rely on superposition for solving the equation cannot be used. Accordingly, the powerful techniques of classical analysis, such as *Fourier series* and *Laplace transforms,* are excluded from use in boundary layer theory. The exception to this situation is the case of the energy equation for incompressible flow, Eq. (3–52). There, the momentum and continuity equations can be solved for (u,v) uncoupled from the energy equation, since ρ = constant is known *a priori*. The energy equation is rendered linear with known, variable coefficients, and it can be solved taking advantage of that fact.

Let us now turn our attention to the matter of boundary conditions and initial conditions. Looking at the system of equations, one can see that it is second order in y (second derivatives with respect to y are present) and first order in x and t. The mathematical requirement is matched well with physical intuition. One would expect to impose conditions on the body surface, $y = 0$, and at the edge of the boundary layer, $y = \delta$ or δ_T. Actually, from the point of view of the boundary layer, $y = \delta$ is the same thing as the asymptotic limit $y \to \infty$, since the boundary layer flow will blend asymptotically into the outer inviscid flow. One also would expect to specify the flow at the leading edge or the front stagnation point of a body (i.e., $x = 0$). Finally, for an unsteady flow, the state of the flow at the beginning of the period of interest, $t = t_o$, must be specified. In an unsteady case, the boundary conditions on x and y can also be arbitrary, but given, functions of time.

The form of the equations and the required boundary conditions produce a natural *front-to-back* (in the streamwise direction) orientation to the mathematical problem. Such a system is mathematically characterized as *parabolic* and is solved by *downstream marching* procedures, whether analytical or numerical techniques are used.

At the risk of confusing the reader, we observe here that this physically satisfying state of affairs occurs only within the boundary layer approximation for viscous flows. For steady flows at low Reynolds number or over bluff bodies, the Navier-Stokes equations must be used, and they are not *parabolic,* but *elliptic*. Thus, they require two boundary conditions on x, since they are second order in x (see Eqs. (1-

41) and (1-42)). This means that one must specify the condition of the flow not only in front of the body, which is generally a straightforward task, but also somewhere behind the body—for example, in the wake. But how is one to know the velocity profile in the wake *before* solving the problem? Obviously, no exact information will be available, and some approximate formulation must be employed. This difficulty with the *downstream boundary condition* is one of the major problems now under study in viscous flow research. Those matters, however, are beyond the scope of this book.

PROBLEMS

3.1. For incompressible, constant-property, planar, unsteady flow, write the energy equation using a stream function.

3.2. Consider a flat plate of infinite extent in the x-direction. One can then say that $u \neq u(x)$. How does such a condition affect Eqs. (3–6a) and (3–26a)?

3.3. For the same situation as in Problem 3.2, what is the effect of a porous surface at $y = 0$ with suction [i.e., $v(0, x) = v_w < 0$]? With injection ($v_w > 0$)?

3.4. Calculate and plot the variation in the viscosity of a CO_2-air mixture from 0 to 100% at 400°K.

3.5. Calculate and plot the variation in the thermal conductivity of an H_2-air mixture from 0 to 100% at 300°K. Find some exact data in the literature for a point or two, and compare with your prediction.

3.6. Consider a release of H_2 from a slit across a uniform stream of N_2 at 1 m/s at 400°K and 0.1 atm. The H_2 exits in the same direction and at the same velocity as the N_2. Estimate the rate of spread of the H_2 plume. (*Hint:* Note the effect of a uniform velocity on the steady form of Eq. (3–60).)

3.7. Consider the steady, planar flow of a constant-density *power law* fluid, and write the governing equations in terms of a stream function.

3.8. For a constant-density flow, the continuity equation can be taken as Eq. (3–6a). Then, $v(x, y)$ can be calculated if $u(x, y)$ is known. For that purpose, assume the velocity profile in Eq. (2–23), with Eq. (2–27) and Eq. (2–21) for flow over a flat plate. How does $v(x, y)$ behave at the boundary layer edge?

3.9. A number of years ago, Richard von Mises introduced an unusual transformation into boundary layer analysis. He proposed transforming the independent variables from (x, y) to (x, ψ) where ψ is the stream function defined in Eq. (3–8). What is the form of the momentum equation after this transformation is made?

3.10. After applying the von Mises transformation described in Problem 3.9, make the further change of replacing the usual dependent variable $u(x, y)$ by the stagnation pressure, defined as $P_t = p + \rho u^2/2$.

CHAPTER 4

Exact and Numerical Solutions for Laminar, Constant-Property, Incompressible Flows

4–1 INTRODUCTION

The mathematical formulation of boundary layer problems, as developed in Chapter 3, is quite complex, so there is only a small group of cases that admit to an exact analytical solution, even if the discussion is restricted to steady, incompressible, constant-property flow. Only a fraction of this small class of problems has any practical application. We will treat a few representative cases in the next section and the most widely applicable cases in Chapter 8, which deals with internal flows.

The next group of exact solutions that exists is for those restrictive classes of flows for which helpful simplifications of the equations are possible and rigorous, but whose solutions must usually be obtained numerically. Fortunately, relatively simple methods can be used for this group. Sections 4–3 through 4–6 deal with some of these flows. These solutions have some practical and scientific value, but they do not apply, for example, to the flow over general bodies of engineering interest.

Until quite recently, books dealing with the subject would have a chapter consisting of detailed coverage of the material just described, together with many other exact solutions of even less practical value. However, the current widespread availability of the large digital computer has put *numerically exact* solutions of rather general boundary layer problems within the reach of all engineers and applied scientists. Thus, the main body of this chapter is devoted to numerical methods. Much of that material will also carry over directly into the treatment of turbulent flows.

4–2 PARALLEL FLOWS

Parallel flows are characterized by conditions that render $u(y)$ alone. The usual assumption is flat surfaces that are effectively infinite in the $\pm x$ direction. We shall consider two such cases in detail in this section. For the moment, we explore the consequences of taking $u(y)$ alone in the equations of motion. This condition is equivalent to $\partial u/\partial x = 0$, which leads to $\partial v/\partial y = 0$ from the continuity equation, Eq. (3–6a). Integrating, we get v = constant. If there is a solid surface, $v_w = 0$, and that leads to $v = 0$ everywhere. With $\partial u/\partial x = 0$ and $v = 0$, the momentum equation, Eq. (3–26a), becomes

$$\frac{\partial u}{\partial t} = -\frac{1}{\rho}\frac{dp}{dx} + \nu\frac{\partial^2 u}{\partial y^2} \tag{4–1}$$

This is a single equation for the single unknown u. More than that, a great mathematical simplification has been achieved, since the equation is now linear.

4–2–1 Steady Flow between Parallel Plates

The physical problem of interest here is the flow between two very large, flat parallel plates separated by a small distance $2W$ under steady conditions. The flow far from the edges of the plate will not be affected by the faraway edges, so the distance from the edge, x, will not enter. Equation (4–1) accordingly reduces to

$$0 = -\frac{1}{\rho}\frac{dp}{dx} + \nu\frac{d^2 u}{dy^2} \tag{4–2}$$

Note that the equation has been reduced all the way to a linear ordinary differential equation from the original nonlinear partial differential momentum equation. The linear equation can easily be integrated twice to obtain $u(y)$. There are two common cases. The first is where the lower plate at $y = -W$ is at rest and the upper plate at $y = +W$ is moving in the x-direction at a velocity U_w, and there is no pressure gradient. This situation is called *Couette flow*, and the solution is simply

$$\frac{u}{U_w} = \frac{1}{2}\left(1 + \frac{y}{W}\right) \tag{4–3}$$

That is, the velocity varies linearly from $u = 0$ at $y = -W$ to $u = U_w$ at $y = +W$. This is the flow in the viscometer in Fig. 1–2.

The second case has both plates at rest, but there is a throughflow driven by a pressure gradient dp/dx that is constant with x. The solution is

$$u = -\frac{W^2}{2\mu}\frac{dp}{dx}\left(1 - \left(\frac{y}{W}\right)^2\right) \tag{4–4}$$

which is a parabolic velocity profile symmetric about $y = 0$.

Since the system is linear, the two solutions can be superimposed to give the following general solution for a moving upper plate with pressure gradient:

$$\frac{u}{U_w} = \frac{1}{2}\left(1 + \frac{y}{W}\right) - \frac{W^2}{2\mu U_w}\frac{dp}{dx}\left(1 - \left(\frac{y}{W}\right)^2\right) \tag{4-5}$$

If the pressure is decreasing in the $+x$-direction, the flow is all in the $+x$-direction. If the pressure is increasing in the $+x$-direction at a sufficient rate, reversed flow can occur near the lower, stationary wall. The situation is illustrated in Fig. 4-1.

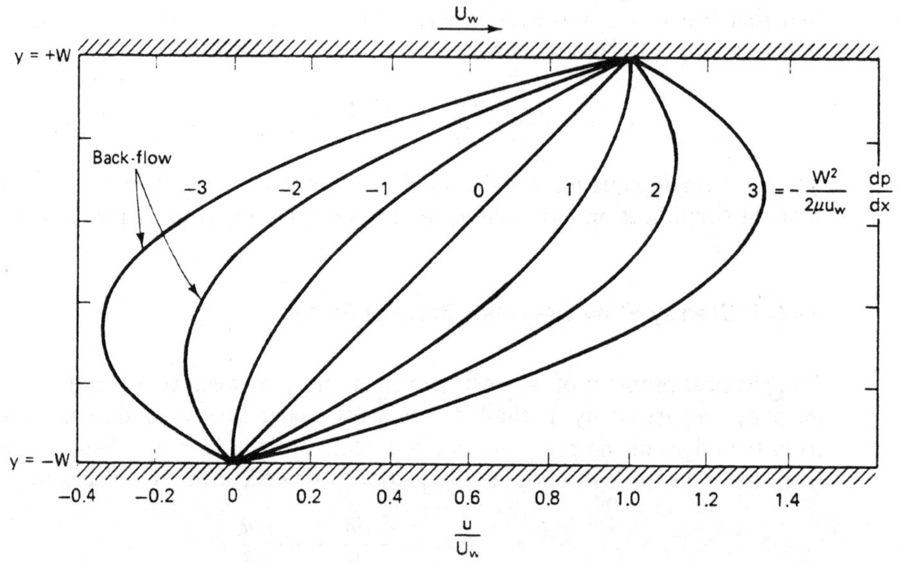

Figure 4-1 Couette flow between two parallel flat walls.

4-2-2 Unsteady Flow near an Infinite Plate

Here, we are concerned with the unsteady flow above a single infinite plate in the $\pm x$-direction, with the fluid above the plate unbounded in the $+y$-direction and initially at rest. Such flows are called *Stokes flows*. Equation (4-1) is reduced in those cases to

$$\frac{\partial u}{\partial t} = \nu \frac{\partial^2 u}{\partial y^2} \tag{4-6}$$

Again, there are two common examples. The first has the plate suddenly moving in the $+x$-direction with a velocity U_w at time $t = 0$. Since Eq. (4-6) is a linear partial differential equation, and since we have simple geometry and simple boundary and initial conditions, the solution can be obtained by classical methods. Transform techniques yield

$$\frac{u}{U_w} = \operatorname{erfc}\left(\frac{y}{\sqrt{4\nu t}}\right) \tag{4-7}$$

where erfc(x) is defined as the *complementary error function* of argument x that varies smoothly from a value of unity when the argument is zero to zero as the argument goes to infinity. Erfc is a tabulated function that can be found in standard mathematical tables. Defining the boundary layer thickness as the point where the velocity is within 1 percent of the value far from the surface (in this case, u goes to 0 far above the plate), we get

$$\delta \approx 4\sqrt{\nu t} \tag{4–8}$$

This solution has an important property relevant to our later analysis of more general boundary layer flows. Note that the two independent variables y and t appear only in the combination $y/\sqrt{4\nu t}$. If that were known beforehand, Eq. (4–6) could have been rewritten as an ordinary differential equation in terms of a single independent variable $\eta = y/\sqrt{4\nu t}$ and a single dependent variable $F(\eta) = u/U_w$, namely,

$$F'' + 2\eta F' = 0 \tag{4–9}$$

The boundary conditions can also be written in terms of η alone as $f(0) = 1$ and $\lim_{\eta\to\infty} F(\eta) = 0$. Thus, this problem could have been treated as being governed by an ordinary differential equation, rather than the generally more complicated situation posed by a partial differential equation. In such a case, one says that the profiles are all *similar*, since the profile $u(y)$ at any given time t can be scaled to look exactly like that at any other time using the factor $\sqrt{4\nu t}$ on y.

The second physical problem of interest under this category is the case where the plate oscillates in the $\pm x$-direction in a harmonic fashion, e.g., $u(0, t) = U_w \cos(nt)$. Standard mathematical methods yield the solution

$$\frac{u}{U_w} = e^{-\sqrt{(n/2\nu)}y} \cos\left(nt - \sqrt{\frac{n}{2\nu}}y\right) \tag{4–10}$$

This is clearly a damped oscillation that decays in amplitude with increasing distance above the surface. Also, there is a phase lag of $y\sqrt{n/2\nu}$ between those fluid particles at a height y and those on the surface. Some representative profiles are shown in Fig. 4–2.

This solution is of some utility in its own right, but it has gained far more fame as a key element in the development of a widely used turbulence model. That application will be discussed in Chapter 7.

4–3 SIMILAR SOLUTIONS FOR THE VELOCITY FIELD

There is a second class of flow problems that admits to relatively simple exact solutions. Again, the conditions are such that simplifications can rigorously be made in the equations of motion. This class of problems is somewhat more general than that typified by parallel flows, but the simplifications of the equations are not as great, so that the solutions are harder to obtain.

If one considers steady, planar, constant-density, constant-property flows, the

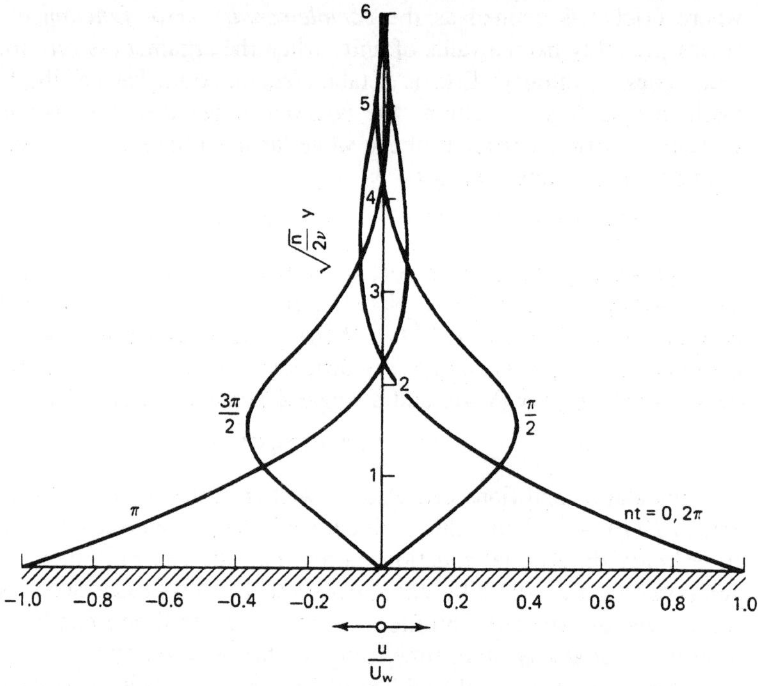

Figure 4–2 Velocity distribution in the neighborhood of an oscillating wall.

mathematical problem is posed by Eq. (3–6a) and the steady, constant-property form of Eq. (3–26a):

$$\frac{\partial u}{\partial x} + \frac{\partial v}{\partial y} = 0 \tag{3–6a}$$

$$u\frac{\partial u}{\partial x} + v\frac{\partial u}{\partial y} = -\frac{1}{\rho}\frac{dp}{dx} + \nu\frac{\partial^2 u}{\partial y^2} \tag{3–26a}$$

The difficulty is that these are nonlinear partial differential equations. For the class of problems in the preceding section, the terms on the left-hand side of the momentum equation disappear, and the system becomes linear. Those problems are specifically chosen such that this happens, but that cannot be expected to occur in general.

On the other hand, is it possible to find a class of problems such that the partial differential equations reduce rigorously to ordinary differential equations, even if nonlinear? The reader has likely seen a comparable maneuver before with the method of *separation of variables*. There, one hopes to find cases where

$$u(x, y) = X(x) \cdot Y(y) \tag{4–11}$$

giving two ordinary differential equations rather than one partial differential equation. Unfortunately, that procedure will not work for boundary layer problems. So what do we do then? Certainly, problems where the x- or y-dependence would disappear altogether cannot be very general. The last possibility is the introduction of one new independent variable—say, η—that is a function of (x, y) such that

$$u(x, y) = u[\eta(x, y)] \tag{4–12}$$

Of course, not only the equations, but the boundary conditions, must be compatible with this representation. Surely, that will not be possible for all flow problems, but perhaps there are some of this type.

The development of these solutions proceeded logically from two roots. First, it had been known for some time from experiment that flat-plate boundary layer profiles remained unchanged along the plate if plotted as u/U_e versus y/δ. We say that each normalized profile is *similar* to every other. Further, experiment showed that $\delta \sim \sqrt{x}$, so profiles plotted as u/U_e versus $y/\sqrt{x}$ would also be *similar*.

The second root was the solution of the case of a suddenly started plate, discussed in Section 4–2–2. A crude heuristic approximation to Eq. (3–26a) for steady, constant-property flow over a flat plate ($dp/dx \equiv 0$) would be

$$U_e \frac{\partial u}{\partial x} \approx \nu \frac{\partial^2 u}{\partial y^2} \tag{4–13}$$

The streamwise variable can be combined with U_e to produce a *pseudotime*, x/U_e. Solutions to problems governed by this equation would then often involve $y/\sqrt{x}$, since solutions to Eq. (4–6) often involve $y/\sqrt{t}$. This line of reasoning is not rigorous, but it also strongly suggests that a promising trial form for $\eta(x, y)$ would be $\eta \sim y/\sqrt{x}$.

4–3–1 Exact Solution for Flow over a Flat Plate

The exact solution for flow over a flat plate was originally produced by Blasius (1908), who was a doctoral student under Prandtl. It is convenient to use the stream function formulation, since then only one equation for one unknown $\psi(x, y)$ is involved. That equation is, from Eq. (3–28) for steady flow with $dp/dx = 0$,

$$\frac{\partial \psi}{\partial y}\frac{\partial^2 \psi}{\partial y\, \partial x} - \frac{\partial \psi}{\partial x}\frac{\partial^2 \psi}{\partial y^2} = \nu \frac{\partial^3 \psi}{\partial y^3} \tag{4–14}$$

We choose to write the assumption in Eq. (4–12) as

$$\frac{u}{U_e} = Af'(\eta) \tag{4–15}$$

with $\eta = By/\sqrt{x}$, where A and B are constants to be determined later to simplify the final equation. Now, it cannot be expected that $\psi = \psi(\eta)$ alone, since the value of ψ at any point y above the plate determines the total mass flow between $y = 0$ and the given y. We know that the boundary layer grows proportionally to $\sqrt{x}$, so it can be conjectured that

$$\psi \sim \sqrt{x}\, f(\eta) = C\sqrt{x}\, f(\eta) \tag{4–16}$$

where C is another *convenience* constant. The form of $f(\eta)$ in Eqs. (4–15) and (4–16) can be seen to be correct by reexamining the definition of the stream function, Eq. (3–8), and noting that $\eta \sim (y/\sqrt{x})$. Substituting into Eq. (4–14) and collecting terms, we find it convenient to select A, B, and C so that

$$\begin{aligned} \eta &= y\sqrt{\frac{U_e}{2\nu x}} \\ \frac{u}{U_e} &= f'(\eta) \\ \psi &= \sqrt{2\nu U_e x}\, f(\eta) \\ v &= \sqrt{\frac{\nu U_e}{2x}}\,(\eta f' - f) \end{aligned} \tag{4–17}$$

giving the final equation

$$f''' + ff'' = 0 \tag{4–18}$$

This equation surely has a simple appearance, although it is important to note the remaining nonlinearity in the term ff''.

Thus far, this looks like good progress, but the boundary conditions must still be examined. Indeed, it is the boundary conditions that determine which problems admit to the foregoing solution and which do not. That is also the case with the method of separation of variables in other contexts. We begin with the usual boundary conditions written in terms of (u, v) and (x, y):

$$\begin{aligned} &y = 0,\ x \geq 0: && u(x, 0) = v(x, 0) = \psi(x, 0) = 0 \\ &y \to \infty,\ \text{all } x: && u(x, y) \to U_e \\ &x = 0,\ y > 0: && u(0, y) = U_e \end{aligned} \tag{4–19}$$

The last condition imposes a uniform approach stream with velocity U_e. All the conditions must be cast in terms of $\eta \sim y/\sqrt{x}$, even though we will then be unable to tell the difference between $y \to \infty$ and $x \to 0$ with $y > 0$. This behavior is accommodated in Eq. (4–19), but it is worth observing that a *uniform approach flow* is *required*. Not all flat-plate boundary layer problems have similar solutions. Finally, the boundary conditions of interest here can be written in terms of η as

$$\begin{aligned} &\eta = 0: && f(0) = f'(0) = 0 \\ &\eta \to \infty: && f'(\eta) \to 1.0 \end{aligned} \tag{4–20}$$

Blasius solved Eq. (4–18) with boundary conditions given by Eq. (4–20) using inner and outer series expansions that are joined within the boundary layer. In modern times, this type of problem is generally treated by a so-called *shooting method*. The difficulty to be handled is that we have a *two-point boundary value problem*. The third-order system in Eq. (4–18) has the required three boundary conditions in Eq. (4–20), but they are not all given at the same value of the independent variable, η. There are two values of the dependent variable specified at $\eta = 0$

and one at $\eta \to \infty$. That is perfectly acceptable from a mathematical standpoint, and it is required from a physical standpoint, but it puts the problem in a more complicated form than the more common arrangement. The usual way to solve a third-order problem numerically is to recast the single third-order equation into three first-order equations and solve each, starting from an initial value. Following that procedure, we rewrite Eq. (4–18), using $g_1 = f$, $g_2 = f'$, and $g_3 = f''$, as

$$g_1' = g_2, \qquad g_2' = g_3, \qquad g_3' = -g_1 g_3 \tag{4–21}$$

The boundary conditions then become

$$g_1(0) = 0, \qquad g_2(0) = 0, \qquad \lim_{\eta\to\infty} g_2(\eta) = 1.0 \tag{4–22}$$

Usually, one would expect to have $g_3(0) = f''(0)$, which would be given. Thus, it is convenient to recast the original problem and guess trial values of $g_3(0) = f''(0)$ and then produce trial solutions and iterate until $\lim_{\eta\to\text{Large}} g_2 \approx 1.0$, within some tolerance. In this way, one is *shooting* at $g_2 = 1.0$ for large η. The simple system of first-order equations can be solved by a number of well-known procedures, such as the *Runge-Kutta method*, and convenient packaged routines are available in any computer center. The overall iterative procedure can easily be accomplished with *Newton's method*.

The solution is shown in Fig. 4–3 and Table 4–1. It is common to take the

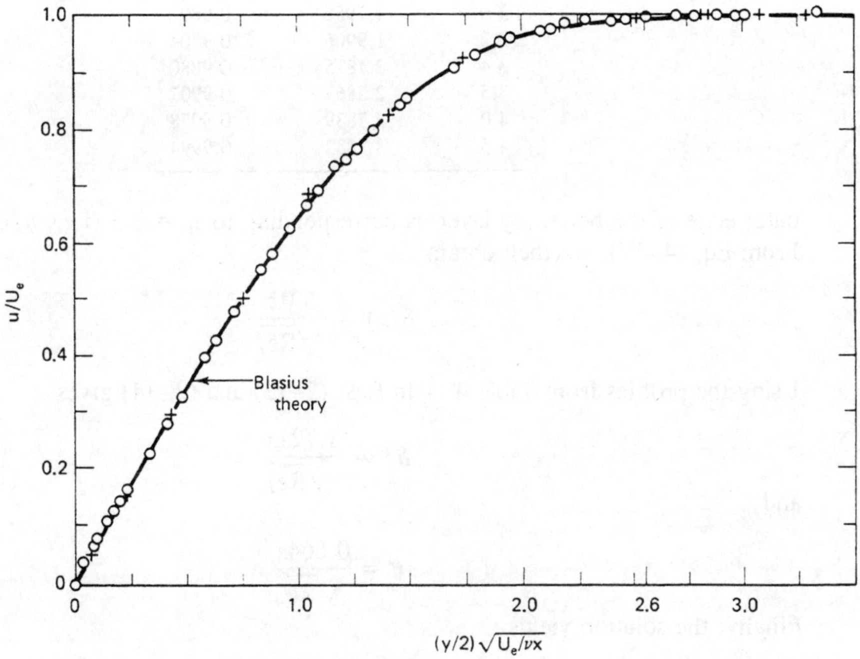

Figure 4–3 Comparison between Blasius solution and experiments of Nikuradse for laminar flow over a flat plate. (From Nikuradse, 1942.)

TABLE 4–1 TABULATED BLASIUS SOLUTION FOR LAMINAR FLOW ON A FLAT PLATE

η	$f(\eta)$	$f'(\eta)$
0.0	0.0000	0.0000
0.1	0.0024	0.0470
0.2	0.0094	0.0939
0.3	0.0211	0.1408
0.4	0.0376	0.1876
0.5	0.0586	0.2342
0.6	0.0844	0.2806
0.7	0.1147	0.3265
0.8	0.1497	0.3720
0.9	0.1891	0.4167
1.0	0.2330	0.4606
1.2	0.3337	0.5453
1.4	0.4507	0.6244
1.6	0.5830	0.6967
1.8	0.7289	0.7611
2.0	0.8868	0.8167
2.2	1.0550	0.8633
2.4	1.2315	0.9011
2.6	1.4148	0.9306
2.8	1.6033	0.9529
3.0	1.7956	0.9691
3.2	1.9906	0.9804
3.4	2.1875	0.9880
3.5	2.2863	0.9907
4.0	2.7839	0.9978
4.5	3.2832	0.9994

outer edge of the boundary layer as corresponding to $\eta = 3.5$ (i.e., $u/U_e = 0.99$). From Eq. (4–17), we then obtain

$$\delta(x) = \frac{5.0x}{\sqrt{\text{Re}_x}} \tag{4–23}$$

Using the profiles from Table 4–1 in Eqs. (2–13) and (2–14) gives

$$\delta^* = \frac{1.721x}{\sqrt{\text{Re}_x}} \tag{4–24}$$

and

$$\theta = \frac{0.664x}{\sqrt{\text{Re}_x}} \tag{4–25}$$

Finally, the solution yields

$$C_f = \frac{\sqrt{2}\,f''(0)}{\sqrt{\text{Re}_x}} = \frac{0.664}{\sqrt{\text{Re}_x}} \tag{4–26}$$

Flow over a flat plate with suction or injection through the wall (i.e., $v_w \neq 0$) can also be treated as a similar solution for a restricted variation $v_w(x)$. That solution is discussed in Section 4–5–1, in conjunction with mass transfer.

4-3-2 Similar Solutions with Pressure Gradient

Earlier, it was stated that the similarity technique would not work for general body shapes [i.e., essentially arbitrary $U_e(x)$], and one should not expect it to, on purely physical grounds. This statement can be made clear by considering boundary layer development on an airfoil as an example. From the stagnation point to about the maximum thickness, there will be a favorable pressure gradient ($dp/dx < 0$) with full profiles (see Fig. 2–3 for $\lambda > 0$). Around the maximum thickness ($dp/dx \approx 0$), the profiles will resemble those for $\lambda \approx 0$. After that, there is an adverse pressure gradient ($dp/dx > 0$, $\lambda < 0$), and the profiles tend to develop an inflection point. Clearly, all these profiles will not be *similar* in shape. However, there are some special cases with pressure gradients (i.e., $U_e(x) \neq$ constant) that do admit to solutions of this type. By substituting a fairly general assumed form into the momentum equation, collecting terms, and then insisting that all explicit dependence on x alone disappear, it emerges that cases in which $U_e(x) = U_1 x^m$ will serve. These are special cases corresponding to inviscid flow over a wedge with an opening angle $\beta\pi = 2m\pi/(m + 1)$. The momentum equation takes the form

$$f''' + ff'' - \frac{2m}{m+1}[(f')^2 - 1] = 0 \tag{4-27}$$

with

$$\eta = y\sqrt{\frac{m+1}{2}\frac{U_e}{\nu x}}$$
$$\psi = \sqrt{\frac{2}{m+1}}\sqrt{\nu U_e x}\, f(\eta) \tag{4-28}$$

This case was derived by Falkner and Skan (1930), and the solutions shown in Fig. 4–4 were obtained by Hartree (1937). The case $\beta = m = 0$ is the flat-plate problem. Values of $\beta < 0$ represent *adverse* pressure gradients, and $\beta = -0.1988$ corresponds to separation (see Fig. 4–4). Cases with $\beta > 0$ represent *favorable* gradients, and $\beta = m = 1$ is the case of planar flow at a stagnation point. Some of the gross features of the solutions are tabulated in Table 4–2. It is clear that a favorable pressure gradient leads to higher skin friction, thinner layers, and fuller profiles, compared to flow over a flat plate. Conversely, an adverse pressure gradient leads to lower skin friction, thicker layers, and less full profiles. Only a rather modest adverse pressure gradient ($\beta = -0.1988$) is needed to induce separation.

The planar stagnation point solution ($\beta = m = 1$) is useful in a general way, since it provides the initial conditions for nonsimilar boundary layer calculations over blunt-nosed bodies of general shapes. In that case, $U_e(x) = U_1 x$, and using this information in Eq. (4–28) together with the results in Table 4–2 gives

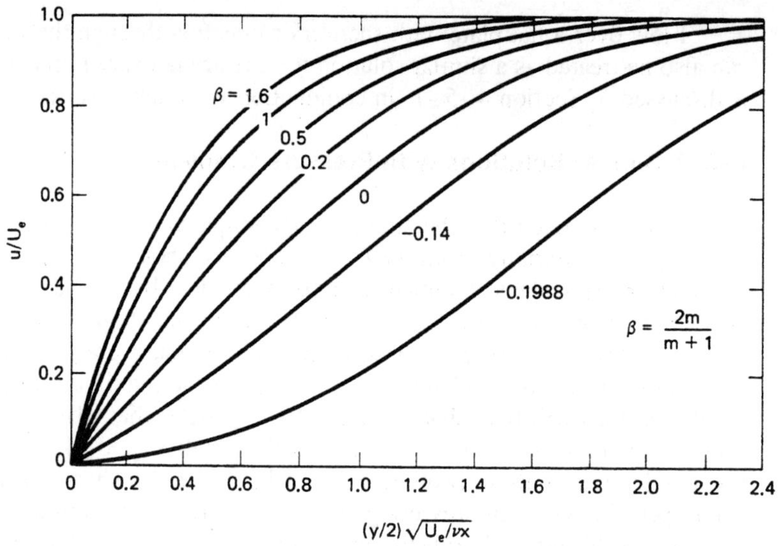

Figure 4–4 Velocity profiles for laminar flow over wedges of various angles. (From Hartree, 1937.)

TABLE 4–2 TABULATION OF SOME IMPORTANT RESULTS FOR FALKNER-SKAN WEDGE FLOWS

m	β	$C_f(\text{Re}_x)^{1/2}$	$\delta^*/x(\text{Re}_x)^{1/2}$	$\theta/x(\text{Re}_x)^{1/2}$	$\delta/x(\text{Re}_x)^{1/2}$
1	1	2.465	0.648	0.292	2.4
1/3	1/2	1.515	0.985	0.429	3.4
0.111	0.20	1.025	1.320	0.548	4.2
0	0	0.664	1.721	0.664	5.0
−0.024	−0.05	0.559	1.879	0.701	5.4
−0.065	−0.14	0.328	2.334	0.788	5.8
−0.0904	−0.1988	0	3.427	0.868	7.1

$$\delta^2(0) = \frac{0.17\nu}{(dU_e/dx)_0} \tag{4–29}$$

Equation (4–29) can be compared with the approximate result for $\theta(0)$ in Eq. (2–31a). Here also, one can take $(dU_e/dx)_0 = 2U_\infty/R_0$, where R_0 is the nose radius. The corresponding case in axisymmetric flow—i.e., the stagnation point on a blunt-nosed body of revolution—is $\beta = 1/2$ and $m = 1/3$, with $(dU_e/dx)_0 = (3/2)U_\infty/R_0$.

It is important to remember that in similar solutions, *all profiles* for any given case (a given m) have the *same* nondimensional *shape*. Thus, the separation profile for $\beta = -0.1988$ does not develop at some axial station along the body starting from a more usual profile upstream. Rather, the *profiles* at *every axial station* along the body are *separated*. These comments should help the reader grasp just how *special*, and not general, these cases really are.

4–4 SIMILAR SOLUTIONS TO THE LOW-SPEED ENERGY EQUATION

For steady, low-speed, constant-density, constant-property flows, the exact energy equation becomes

$$u\frac{\partial T}{\partial x} + v\frac{\partial T}{\partial y} = \frac{\nu}{\Pr}\frac{\partial^2 T}{\partial y^2} \tag{4–30}$$

Comparing this equation with the momentum equation for steady flow over a flat plate,

$$u\frac{\partial u}{\partial x} + v\frac{\partial u}{\partial y} = \nu\frac{\partial^2 u}{\partial y^2} \tag{4–31}$$

we can anticipate that there will be some situations where *similar* solutions also exist for the energy equation.

For a flat plate with a constant wall temperature and a uniform temperature approach flow, the boundary conditions are

$$\begin{aligned} y = 0,\ x \geq 0: &\quad T = T_w = \text{constant} \\ y \to \infty,\ \text{all } x: &\quad T = T_e = \text{constant} \\ x = 0,\ y > 0: &\quad T = T_e = \text{constant} \end{aligned} \tag{4–32}$$

The temperature is more conveniently written in terms of a so-called *excess temperature*

$$\Theta(x, y) \equiv \frac{T - T_w}{T_e - T_w} \tag{4–33}$$

which ranges from zero to unity. Using this equation and the definitions in Eq. (4–17), we transform Eq. (4–30) into

$$\Theta'' + \Pr \cdot f\Theta' = 0 \tag{4–34}$$

This equation is *linear* because $f(\eta)$ is known from the previous solution of the momentum equation, and it is simply a variable coefficient here.

The solution to Eq. (4–34) can be written in integral form as

$$\Theta = \frac{\displaystyle\int_0^\eta e^{-\int_0^\eta \Pr\cdot f\,d\eta}\,d\eta}{\displaystyle\int_0^\infty e^{-\int_0^\eta \Pr\cdot f\,d\eta}\,d\eta} \tag{4–35}$$

The actual solution is obtained by numerical evaluation of the integrals, since $f(\eta)$ is available only as an array of numerical values. Some temperature profiles plotted as

$$1 - \Theta = \frac{T - T_e}{T_w - T_e} \tag{4–36}$$

are given in Fig. 4–5. When compared with the velocity profile in Fig. 4–3 corresponding to a flat plate ($\beta = 0$), they show an important effect of the Prandtl number. For Pr = 1.0, the nondimensional profiles are the same. For Pr > 1, the temperature profiles are *fuller* and $\delta_T < \delta$. For Pr < 1, the reverse situation holds. (Remember that Pr = 0.7 for air and most gases.) The result for the heat transfer coefficient is

$$\mathrm{Nu}_x \equiv \frac{hx}{k} = 0.332\ \mathrm{Pr}^{1/3}\sqrt{\mathrm{Re}_x} \tag{4–37}$$

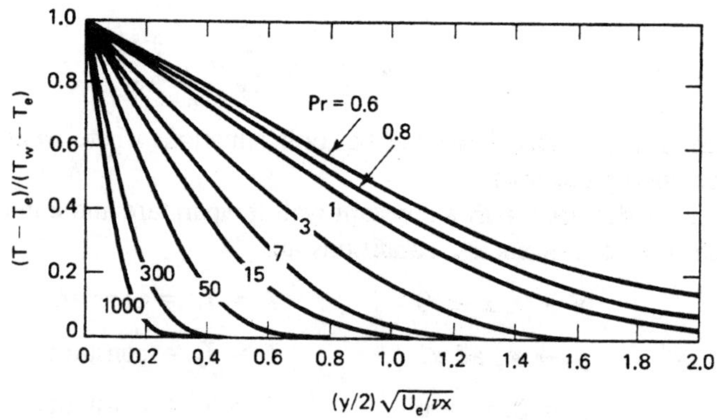

Figure 4–5 Temperature profiles for laminar flow over a constant-temperature flat plate. (From Eckert and Drewitz, 1940.)

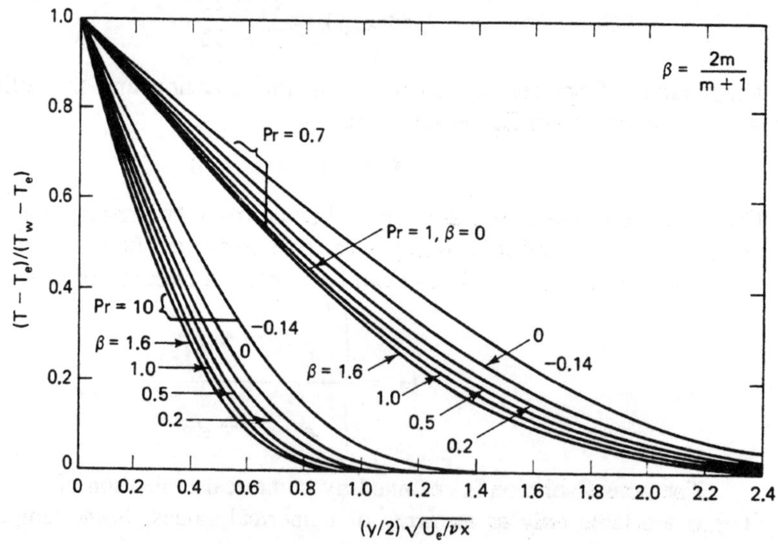

Figure 4–6 Temperature profiles for laminar flow over constant-temperature wedges. (From Eckert, 1942.)

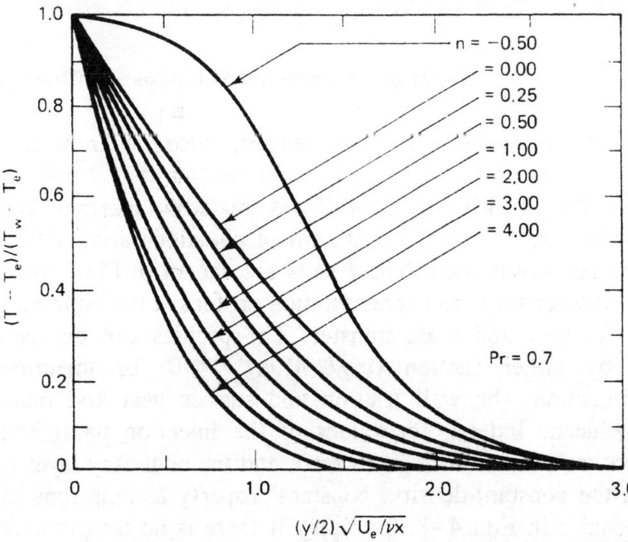

Figure 4–7 Temperature profiles for laminar flow over variable-temperature flat plates. (From Levy, 1952.)

There are also similar solutions to Eq. (4–30) corresponding to the *wedge* flows $U_e \sim x^m$ for the momentum equation. These were found by Fage and Falkner (1931) for $U_e \sim x^m$ and/or $(T_w - T_e) \sim x^n$. Some of these are presented in Figs. 4–6 and 4–7. The profiles in Fig. 4–6 show the effect of pressure gradient ($m \neq 0$) and Prandtl number for the case of a constant wall temperature ($n = 0$), and those in Fig. 4–7 show the effect of variable wall temperature ($n \neq 0$) for a flat plate ($m = 0$) at Pr = 0.7. Comparing the two, one can note that the effect of a variable wall temperature is larger than that of a pressure gradient. The special case with $n = -1/2$ serves to emphasize that point. Figure 4–7 shows that such a case has $(\partial T/\partial y)_w = 0$, and thus, $q_w = 0$, even though $T_w \neq T_e$.

4–5 SIMILAR SOLUTIONS FOR FOREIGN FLUID INJECTION

Under the assumptions of constant density and constant properties taken for this chapter, only rather restricted cases of injection of a foreign fluid can be treated. More general cases are discussed in Chapter 5. Here, we must assume not only that the properties are independent of temperature and pressure, but that the properties of the two fluids are essentially equal. Taking this all together, we find that the complete system of equations to be used in such a case for a steady flow over a flat plate is

$$\begin{aligned}
&\frac{\partial u}{\partial x} + \frac{\partial v}{\partial y} = 0 \\
&u\frac{\partial u}{\partial x} + v\frac{\partial u}{\partial y} = \nu\frac{\partial^2 u}{\partial y^2} \\
&u\frac{\partial T}{\partial x} + v\frac{\partial T}{\partial y} = \frac{\nu}{\text{Pr}}\frac{\partial^2 T}{\partial y^2} \\
&u\frac{\partial c_i}{\partial x} + v\frac{\partial c_i}{\partial y} = \frac{\nu}{\text{Sc}}\frac{\partial^2 c_i}{\partial y^2}
\end{aligned} \tag{4–38}$$

Actually, only one species equation—say, for c_1—is needed, since we know that $c_1 + c_2 \equiv 1.0$.

A problem governed by Eq. (4–38) with $v_w \neq 0$ was treated by Hartnett and Eckert (1957). In a case where $v_w \sim x^{-1/2}$, this system of equations admits to a similar solution, and that property was used. The results are shown in Figs. 4–8, 4–9, and 4–10 for velocity, temperature, and concentration profiles in the boundary layer and film coefficients for heat and mass transfer. The profiles can be seen to be greatly influenced by either suction $[v_w/U_e)(\mathrm{Re}_x)^{1/2} < 0]$ or injection $[v_w/U_e)(\mathrm{Re}_x)^{1/2} > 0]$. For injection, the wall friction and surface heat and mass transfer are dramatically reduced. Indeed, for values of the injection parameter greater than 0.62, the surface transfer rates all go to zero, and the boundary layer is said to be *blown off*. With the constant-density, constant-property assumptions of this chapter, the velocity profiles in Fig. 4–8 also apply if there is no temperature difference—and thus, no heat transfer—and only a single species—and thus, no mass transfer.

The transverse mass flux of the air (species 2) at any point is composed of a part due to a *diffusive velocity* and a part due to the *convective velocity* v. Thus,

$$\rho_2 v_2 = -\rho D_{12}\frac{\partial c_2}{\partial y} + \rho c_2 v \tag{4–39}$$

or

$$v_2 = -\frac{D_{12}}{c_2}\frac{\partial c_2}{\partial y} + v \tag{4–40}$$

Since no air penetrates into the wall, $v_2(x, 0) \equiv 0$, and

$$v_w = \frac{D_{12}}{c_{2w}}\frac{\partial c_2}{\partial y}\bigg|_w \tag{4–41}$$

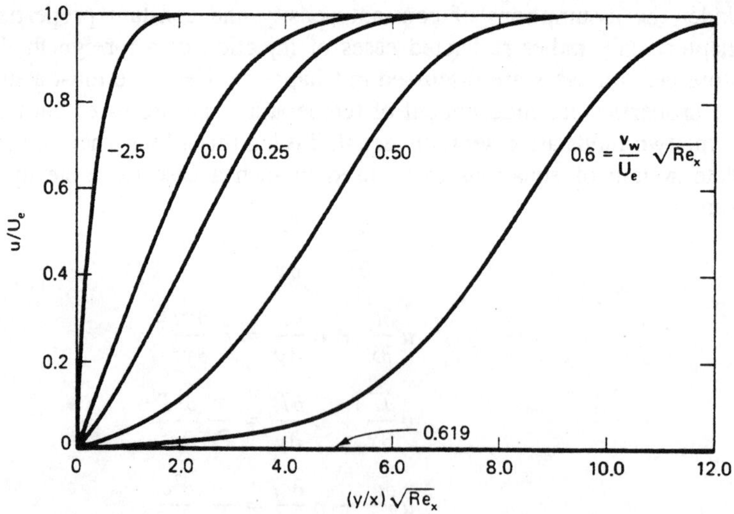

Figure 4–8 Velocity profiles for fluid injection from a flat plate. (From Hartnett and Eckert, 1957.)

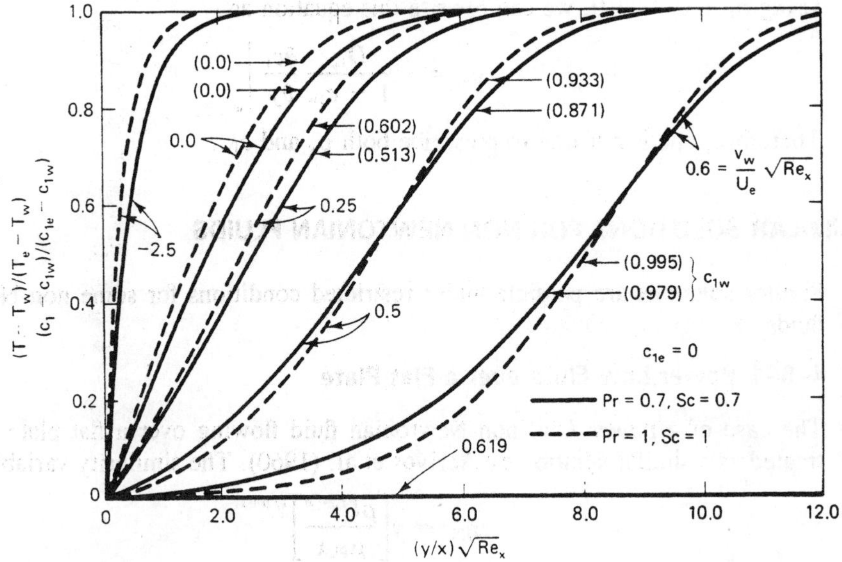

Figure 4–9 Temperature and mass fraction profiles for fluid injection from a flat plate. (From Hartnett and Eckert, 1957.)

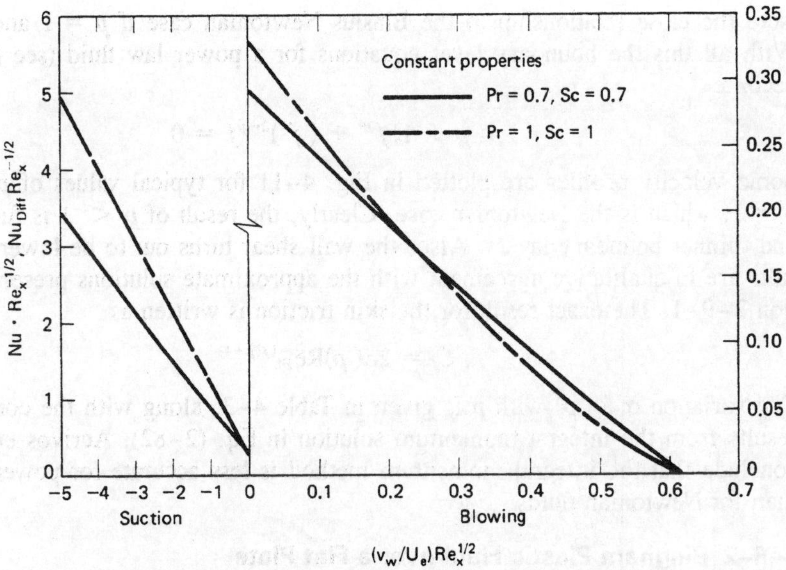

Figure 4–10 Nondimensional heat and mass transfer rates for fluid injection from a flat plate. (From Hartnett and Eckert, 1957.)

Using $c_1 + c_2 \equiv 1.0$, we can rewrite this equation as

$$v_w = -\frac{D_{12}}{1 - c_{1w}} \frac{\partial c_1}{\partial y}\bigg|_w \tag{4–42}$$

Therefore, one is not free to prescribe both v_w and c_{1w}.

4–6 SIMILAR SOLUTIONS FOR NON-NEWTONIAN FLUIDS

Similar solutions are possible under restricted conditions for some non-Newtonian fluids.

4–6–1 Power Law Fluid over a Flat Plate

The case of a *power law*, non-Newtonian fluid flowing over a flat plate has been treated as a similar solution by Acrivos et al. (1960). The similarity variable used is

$$\eta_{\mathrm{PL}} \equiv y\left[\frac{\rho U_e^{2-p}}{\mu_{\mathrm{PL}} x}\right]^{1/p+1} \tag{4–43}$$

and the velocity is expressed as

$$\frac{u}{U_e} = f'(\eta_{\mathrm{PL}}) \tag{4–44}$$

Note the close relationship to the Blasius Newtonian case if $p = 1$ and $\mu_{\mathrm{PL}} = \mu$. With all this the boundary layer equations for a power law fluid (see Eq. (3–29) become

$$p(p + 1)f''' + (f'')^{2-p}f = 0 \tag{4–45}$$

Some velocity profiles are plotted in Fig. 4–11 for typical values of p, including $p = 1$, which is the Newtonian case. Clearly, the result of $p < 1$ is fuller profiles and thinner boundary layers. Also, the wall shear turns out to be lower. These results are in qualitative agreement with the approximate solutions presented in Section 2–9–1. The exact result for the skin friction is written as

$$C_f = 2c(p)\mathrm{Re}_{\mathrm{PL}}^{-1/(p+1)} \tag{4–46}$$

The variation of $c(p)$ with p is given in Table 4–3, along with the corresponding results from the integral momentum solution in Eq. (2–82). Acrivos et al. (1960) conclude that the integral momentum method is less accurate for power law fluids than for Newtonian fluids.

4–6–2 Bingham Plastic Fluid over a Flat Plate

Since shear stress appears in the momentum equation only as $\partial\tau/\partial y$, substituting the shear model for a *Bingham plastic* (Eq. (1–62)) results in the same equation as for a Newtonian fluid with μ_{BP} written in place of the usual μ. Thus, the Blasius solution given in Section 4–3–1 can be used for a *Bingham plastic* case. The important dif-

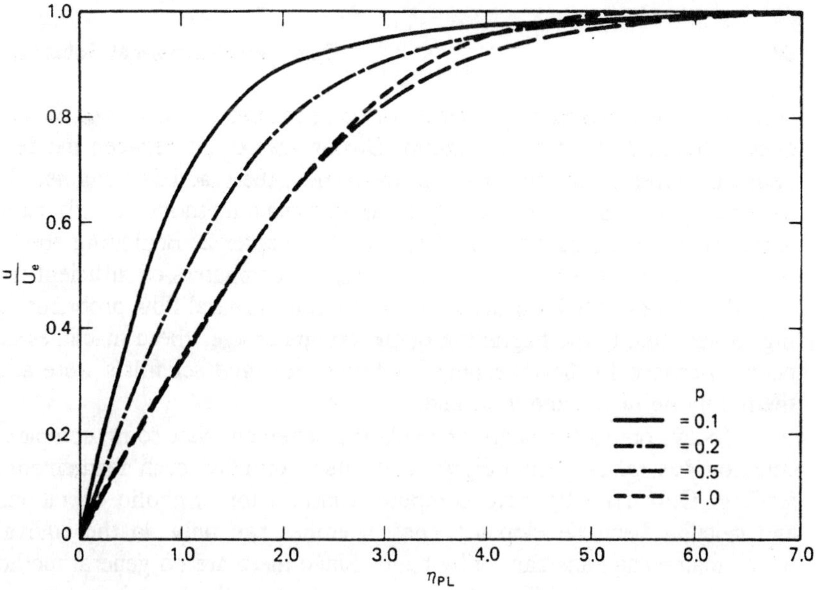

Figure 4–11 Velocity profiles for power law fluid over a flat plate. (From Acrivos et al., 1960.)

TABLE 4–3 EXACT SOLUTION FOR COEFFICIENT IN SKIN FRICTION RELATION (EQ. 4–46) FOR POWER LAW FLUID OVER A FLAT PLATE

p	$c(p)$, Exact	$c(p)$, Eq. (2–82)
0.05	1.017	0.926
0.1	0.969	0.860
0.2	0.873	0.747
0.3	0.733	0.655
0.5	0.576	0.518
1.0	0.332	0.323

ference between the case of a *Bingham plastic* fluid and that of a Newtonian fluid emerges in the wall shear stress or skin friction results, where an extra term due to the yield stress τ_0 arises:

$$C_f = \frac{\tau_0}{1/2\rho U_e^2} + \frac{0.664}{\sqrt{\text{Re}_x}} \tag{4–26a}$$

These results are all quite similar to those obtained with the momentum integral method in Chapter 2.

4–7 NUMERICALLY EXACT SOLUTIONS

Up to this point, the reader has seen representative examples of all the classes of exact solutions of the boundary layer equations for low-speed flows. Clearly, the scope of problems covered by these solutions is extremely limited, and they do not cover cases of such fundamental interest to the engineer as flow over an airplane wing, over a ship hull, or through a pump. During the period between the development of boundary layer theory in 1904 and the 1960s, the practicing engineer had little use for exact solutions and had to rely on approximate methods, usually based on the integral momentum equation, as described in Chapter 2. Beginning about 1960, however, engineers began to gain access to digital computers of sufficient size and speed to undertake essentially exact solutions to more general flow problems. It is interesting to note that at the beginning of the computer age, these machines had generally been purchased for bookkeeping, and engineers and scientists were allowed to use them only on nights and weekends.

Today, one often hears or reads the statement that some complex differential equation was *solved* with a digital computer. Actually, such a statement can be misleading. Only recently have computer routines for symbolic operations in algebra and calculus been developed.[1] These routines can only do the equivalent of that which mathematicians can do by hand. Since there are no general methods for solving nonlinear partial differential equations such as the boundary layer equations exactly by hand, the computer cannot do so either. The real power of the digital computer is that it can do *arithmetic* very rapidly on a very large scale, without errors. But boundary layer problems are not arithmetic problems; they are differential equation problems. So how can the computer be used to solve them? The answer is simple: One must reduce the solution of the differential equations to an essentially arithmetic problem. But the only way to do that involves undoing one of the great intellectual developments in all of human thought—the *limit process*. The reader can recall his or her introduction to the calculus, involving the conceptually difficult idea

$$\left.\frac{dF}{dx}\right|_{x=x_0} \equiv \lim_{\Delta x\to 0}\frac{F(x_0+\Delta x)-F(x_0)}{\Delta x} \tag{4–47}$$

The extension of this idea leads to differential equations. If the process is reversed, a differential equation becomes a series of algebraic relations between the values of the dependent variables at various specific values of the independent variables separated by small *finite differences* (e.g., Δx and Δy).

Consider the situation shown in Fig. 4–12. The region of interest in the flow is overlaid with a grid with small, but finite spacing (Δx, Δy). We will be concerned with the values of the dependent variables at the intersections of the grid lines, called *node* points. Clearly, there will be many such points, and a simple, unambiguous notation is required to identify them. It is convenient to use an integer pair (n, m) for this purpose. The index n identifies a location in the x-direction. The initial station, $x = x_i$, is denoted by $n = 1$, and n runs up to a value of $n = N$ at the last station, $x = x_f$. For the transverse coordinate, pick $m = 1$ at $y = 0$ and $m = M$ at $y = y_{max}$, where y_{max} must clearly be greater than $\delta(x)$. Thus,

$$x = x_i + (n-1)\Delta x \tag{4–48}$$

$$y = (m-1)\Delta y \tag{4–49}$$

[1] The article, "Computer Algebra," by R. Pavelle, M. Rothstein, and J. Fitch, in *Scientific American*, December 1981, provides interesting background information.

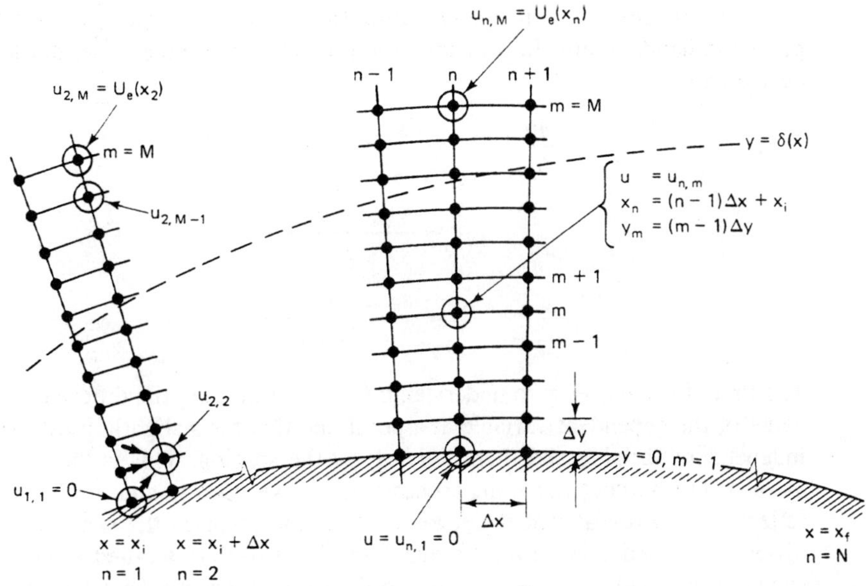

Figure 4–12 Schematic of the grid system and notation for the finite difference solution of boundary layer problems.

A pair of specific integer values (n, m) then locates a specific node point in the region of interest, and a dependent variable is written, for example, as $u(x, y) \rightarrow u_{n,m}$.

Before proceeding, we should note that the analyst must pick the values of (N, M) for a given problem. This is the same as picking $(\Delta x, \Delta y)$. The choice must be based on a good estimate of the rapidity of variation expected in the dependent variables with (x, y). Looking at Fig. 4–3, for example, one might say that such a simple velocity profile could be adequately described by 15 to 20 points across the layer. In a thin boundary layer, $\partial u/\partial y \gg \partial u/\partial x$, so we would certainly expect to be able to use a $\Delta x > \Delta y$. The choice of $(\Delta x, \Delta y)$ is also strongly influenced by the boundary conditions for a given problem. If a problem involves a rapidly varying $U_e(x)$, a smaller Δx will be required. All of this discussion may seem rather nonspecific, but it represents the true state of affairs. The analyst must make these choices; the computer, in general, cannot. Only experience can make the task easier, but most general-purpose computer codes have helpful suggestions for the novice in the user's manual. Finally, it is always prudent to run a sample calculation with a smaller grid size. One is presuming that the finite differences chosen are small enough that the numerical solution obtained approaches the solution to the differential equation, which is based on the limit as $(\Delta x, \Delta y) \rightarrow 0$. This whole business is of more than abstract interest. Indeed, it could hardly be of more practical concern, since it involves money—sometimes a lot of it. Obviously, the smaller the choices for $(\Delta x, \Delta y)$ the bigger and more expensive will be the calculations. From the practical side, then, any $(\Delta x, \Delta y)$ smaller than is really required to approximate the exact solution to the differential equations as closely as needed is too small.

Let us return now to manipulating the differential equations so that the computer can handle them. In principle, the procedure is simple. Consider a first partial derivative

$$\left.\frac{\partial u}{\partial y}\right|_{x_n, y_m} \approx \frac{\Delta u}{\Delta y} = \frac{u_{n,m+1} - u_{n,m}}{\Delta y} \tag{4–50a}$$

$$\approx \frac{u_{n,m} - u_{n,m-1}}{\Delta y} \tag{4–50b}$$

$$\approx \frac{u_{n,m+1} - u_{n,m-1}}{2(\Delta y)} \tag{4–50c}$$

The first of these is easy to understand: One is just taking the difference between the value of the dependent variable at a point and that for a slightly greater value of the independent variable and then dividing by the spacing. This is the same as in Eq. (4–47), but without the limit process; hence we use $\approx$, and not $=$. A moment's reflection will reveal that the second expression involves the same ideas. Equation (4–50a) is called a *forward* difference, and Eq. (4–50b) is called a *backward* difference. In the limit as $\Delta y \to 0$, they are the same thing, but here the limit process is not involved, and they are not the same. They represent two separate approximations to $\partial u/\partial y$. Each of these approximations has a *one-sided* nature, so the *central* difference scheme of Eq. (4–50c) is preferred in many instances.

An approximation for the second partial derivative follows directly along the same line of thought:

$$\left.\frac{\partial^2 u}{\partial y^2}\right|_{x_n, y_m} = \left.\frac{\partial}{\partial y}\left(\frac{\partial u}{\partial y}\right)\right|_{x_n, y_m} \approx \frac{\Delta(\Delta u)}{(\Delta y)^2}$$

$$\approx \frac{(u_{n,m+1} - u_{n,m}) - (u_{n,m} - u_{n,m-1})}{(\Delta y)^2} \tag{4–51}$$

$$\approx \frac{u_{n,m+1} - 2u_{n,m} + u_{n,m-1}}{(\Delta y)^2}$$

Even-order derivatives come out directly as *central* differences.

It may look now as if we are ready to proceed and solve all possible boundary layer problems. All that appears necessary is to insert approximations such as Eqs. (4–50) and (4–51) into the boundary layer equations and develop a computer code to perform the required calculations. But two important matters remain. The first is an analysis of the behavior of the solution of the finite difference equations compared to the behavior of the solution of the original differential equations and also of the behavior of the finite difference solutions themselves. Unfortunately, this analysis cannnot be rigorously performed for a nonlinear system. Instead, one is forced to analyze an idealized related linear *model* equation and carry over the results by *analogy*. This is obviously perilous, and a conservative attitude is necessary. The situation will be discussed in the next section. The second matter not yet discussed is boundary and initial conditions. Perhaps surprisingly, that is a simple item that emerges naturally as a part of the discussion.

4–7–1 Numerical Analysis of the Linear Model Equation

The linear *model* parabolic equation to be treated is Eq. (4–13) [or its close relative, Eq. (4–6)]. Equation (4–13) is known as the *heat equation,* and its numerical solution has been thoroughly analyzed. The discussion here follows that in the text by Carslaw and Jaegar (1959), with adaptations. We restate the equation here for convenience.

$$U_e \frac{\partial u}{\partial x} = \nu \frac{\partial^2 u}{\partial y^2} \tag{4–13}$$

For the term $\partial u/\partial x$, it appears that three choices, following from Eq. (4–50a), Eq. (4–50b), or Eq. (4–50c), are available. Actually, that is not so. For this parabolic system, the method of solution is to start from a given initial profile, $u(x_i, y)$, and "*march*" downstream. At $x = x_i$, the approximation for $\partial u/\partial x$ corresponding to Eq. (4–50b) would involve information at $x = x_i - \Delta x$, which is unavailable. The same thing holds for Eq. (4–50c). Clearly, then, a *forward* difference must be used:

$$\left.\frac{\partial u}{\partial x}\right|_{x_n, y_m} \approx \frac{u_{n+1,m} - u_{n,m}}{\Delta x} \tag{4–52}$$

One could arrive at the same conclusion by arguing that it is necessary to step forward from the initial station in order to begin to march downstream. Only an expression such as Eq. (4–52) will involve information at a downstream location and none from upstream.

Substituting Eqs. (4–51) and (4–52) into Eq. (4–13) gives

$$U_e \frac{(u_{n+1,m} - u_{n,m})}{\Delta x} = \nu \frac{(u_{n,m+1} - 2u_{n,m} + u_{n,m-1})}{(\Delta y)^2} \tag{4–53}$$

This can be rearranged into an *explicit* relation for $u_{n+1,m}$, the velocity in the boundary layer at a downstream station, in terms of information at the initial station, viz.,

$$u_{n+1,m} = Q(u_{n,m+1} + u_{n,m-1}) - (2Q - 1)u_{n,m} \tag{4–53a}$$

where $Q \equiv \nu(\Delta x)/(U_e(\Delta y)^2)$ involves only the parameters of the problem and the step sizes. Equation (4–53a) is called an *explicit finite difference formulation.*

The method of solution can be described by referring to Fig. 4–12. We begin at $x = x_i$ with a known initial profile, written here as $u_{1,m}$ for all $1 \le m \le M$. The boundary conditions are $u = 0$ on the wall (or $u_{n,1} = 0$) and $u = U_e$ for large y (or $u_{n,M} = U_e$). The velocity at $x = x_i + \Delta x$ and $y = \Delta y$, the first interior point on the new profile, can be found using Eq. (4–53a). The result is

$$u_{2,2} = Q(u_{1,3} + u_{1,1}) - (2Q - 1)u_{1,2} \tag{4–53b}$$

where $u_{1,1} = 0$, since it is on the wall, i.e., at $m = 1$. The values $u_{1,3}$ and $u_{1,2}$ are known, since they are on the given initial profile. Notice how information at three points on the known profile feeds the value at one point on the new profile [see Fig. 4–13(A)].

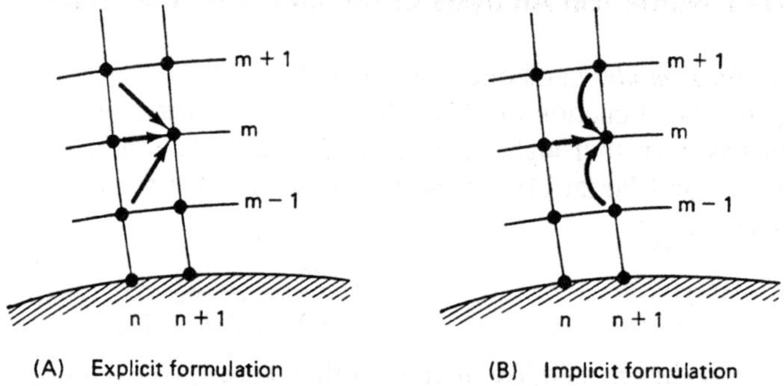

Figure 4–13 Schematic illustration of the flow of information in marching downstream for the explicit and implicit finite difference formulations.

The velocity at the next point up on the profile, $x = x_i + \Delta x$ and $y = 2(\Delta y)$, can be found from

$$u_{2,3} = Q(u_{1,4} + u_{1,2}) - (2Q - 1)u_{1,3} \tag{4–53c}$$

Here, $u_{1,4}$, $u_{1,2}$, and $u_{1,3}$ are all known, since they are on the initial profile. The process is continued up to the next-to-last y point, $x = x_i + \Delta x$ and $y = (M - 2)(\Delta y)$, where we have

$$u_{2,M-1} = Q(u_{1,M} + u_{1,M-2}) - (2Q - 1)u_{1,M-1} \tag{4–53d}$$

Note that $u_{1,M} = U_e(x_i)$ and $u_{2,M} = U_e(x_i + \Delta x)$, by applying the top boundary condition. In this way, a complete profile is calculated at $x = x_i + \Delta x$. Observe how easily the boundary and initial conditions entered into the calculations. The wall boundary condition is used directly in the calculation of all points that are one grid spacing up from the wall (i.e., $m = 2$). The outer boundary condition is used directly in the calculation of all points that are one grid spacing down from the top boundary (i.e., $m = M - 1$). The initial conditions are used going from $n = 1$ to $n = 2$.

With the foregoing result, one step in the downstream marching process is completed, since the solution at $x = x_i + \Delta x$ (i.e., $u_{2,m}$) is now known. The rest of the process is conceptually simple. Consider the newly calculated profile at $x = (x_i + \Delta x)$ to be known in the same sense as the given initial profile at $x = x_i$, and step downstream to $x = (x_i + \Delta x) + \Delta x$ in exactly the same fashion as from x_i to $(x_i + \Delta x)$. But then, if you can make two steps, you can certainly make another, and another, and so on, until $x = x_f = x_i + (N - 1)\Delta x$ is reached. Plainly, many, many simple, repetitive numerical operations are involved. It is possible to apply this procedure by hand, but it surely would involve more time than any rational person would want to spend. The method is, however, perfectly suited to a digital computer.

Having seen how to achieve a numerical approximation to a model parabolic problem, it is natural now to inquire how good the resulting solution might be. It happens that if one blindly applied this procedure to a variety of situations, seemingly good solutions in some cases and obvious nonsense in others would be obtained. The difficulty can be traced to two important items: *round-off error* and *stability*.

Round-off errors come from the finite number of digits that must be handled in any computer. Clearly, each number must be rounded off at the last digit that can be retained. This statement may sound like a truism, and indeed it is, but its implications are wide ranging. The problem arises because many, many separate calculations are to be made for any given problem. If the errors accumulate a little at each step, it is possible to reach the point where the errors dominate the solution. Such a situation is termed an *unstable* procedure. If the errors tend to cancel each other, the total error by the end of the problem remains small, and the procedure is called *stable*. *Truncation errors* arise in a different way. Equation (4–50a) can be rearranged and viewed as the first term in a series expansion for $u(y + \Delta y)[u_{n,m+1}]$ from $u(y)[u_{n,m}]$, involving $\partial u/\partial y$. A complete series expansion would also have terms with $(\Delta y)^2(\partial^2 u/\partial y^2)$ and so on. Thus, Eq. (4–50a) represents a *truncated* series. For small Δy, it can be an accurate approximation, but the neglected terms will contribute small errors that can accumulate or tend to cancel in the same way as the *round-off* errors.

The stability question can be analyzed mathematically following the procedure developed by von Neumann (1952) for the *heat equation,* applied here to its equivalent, the *model* equation for the boundary layer equations. The procedure relies on the fact that this equation is linear and uses *Fourier analysis*. The solution of some physical problems described by the heat equation, and thus also of the model equation, can be written in the form

$$u(x, y) = \xi(x)e^{i\beta y} \tag{4–54}$$

We therefore adopt that convenient form for our analysis. In it, β is a wave number of a periodic function. Since the equation is *linear*, the behavior of an initial *error*, represented by Eq. (4–54), superimposed on a background solution can be analyzed either with or without the background solution present. Clearly, it is more convenient to do so without the background solution. Also, we seek a general result, not the behavior of an error with respect to the solution of a particular problem. Further, since the equation is linear, we can proceed with any single value of β and be assured that the corresponding result for any other β or combination of β's can be found by superposition.

In finite difference form, Eq. (4–54) becomes

$$\begin{aligned} u_{n,m} &= \xi_n e^{i\beta(m-1)\Delta y} \\ u_{n,m+1} &= \xi_n e^{i\beta m(\Delta y)} \\ u_{n,m-1} &= \xi_n e^{i\beta(m-2)\Delta y} \\ u_{n+1,m} &= \xi_{n+1} e^{i\beta(m-1)\Delta y} \end{aligned} \tag{4–55}$$

where $\xi_n = \xi((n-1)\Delta x + x_i)$. Substituting into Eq. (4–53a), we get

$$\xi_{n+1}e^{i\beta(m-1)\Delta y} = Q(\xi_n e^{i\beta m(\Delta y)} + \xi_n e^{i\beta(m-2)\Delta y}) - (2Q-1)\xi_n e^{i\beta(m-1)\Delta y} \quad (4\text{–}56)$$

Dividing through by $e^{i\beta(m-1)\Delta y}$ and then by ξ_n, we obtain

$$\begin{aligned}\frac{\xi_{n+1}}{\xi_n} &= Q(e^{i\beta(\Delta y)} + e^{-i\beta(\Delta y)}) - (2Q-1)\\ &= 2Q\cos[\beta(\Delta y)] - (2Q-1)\end{aligned} \quad (4\text{–}57)$$

the last step involving trigonometric identities. In any event, for stability, we want $|\xi_{n+1}| \leq |\xi_n|$ (i.e., the errors are not to grow). That condition is satisfied for $Q \leq 1/2$, no matter what the value of β, as a brief examination of Eq. (4–57) will show. This is not an abstract result, as the working of Problem 4.14 demonstrates. The practical consequences of such a restriction can be seen by rewriting it, using the definition of Q, as

$$\Delta x \leq \frac{1}{2}\frac{U_e(\Delta y)^2}{\nu} \quad (4\text{–}58)$$

It often happens that Eq. (4–58) forces the use of a much smaller Δx than would otherwise seem required to describe adequately the variations of the flow in the x-direction. The result is more calculation steps and, therefore, greater computer cost with no attendant improvement in the solution. Note that one should be cautious about choosing a smaller value of Δy than is necessary, since, from Eq. (4–58), that immediately implies an even smaller value of Δx.

It is possible for a stable procedure that *converges* to a solution as the step sizes are decreased to converge to the solution of a different differential equation than the original one. That is, the truncation error may not decrease to zero as Δx and Δy do. By using a Taylor's series, an expression for the truncation error can be developed. For the present explicit method, the truncation error $\epsilon_{n,m}$ is found to be

$$\epsilon_{n,m} = U_e\frac{\Delta x}{2}\left(\frac{\partial^2 u}{\partial x^2}\right) - \nu\frac{(\Delta y)^2}{12}\left(\frac{\partial^4 u}{\partial y^4}\right) + 0((\Delta x)^2) + 0((\Delta y)^4) \quad (4\text{–}59)$$

Thus, $\epsilon_{n,m} \to 0$ as $\Delta x, \Delta y \to 0$, as desired.

A way out of the restriction set forth in Eq. (4–58) can be found with some increase in conceptual complexity. We begin by considering again the finite difference approximation to the second derivative with respect to y [Eq. (4–51)]. The approximation for the second derivative should apply to the whole region between x and $x + \Delta x$. Equation (4–51) is written at the beginning of this strip-shaped region. One can just as well use

$$\left.\frac{\partial^2 u}{\partial y^2}\right|_{n+1,m} \approx \frac{u_{n+1,m+1} - 2u_{n+1,m} + u_{n+1,m-1}}{(\Delta y)^2} \quad (4\text{–}60)$$

which is written at the downstream end of the region. In this case, the finite difference represented is formulated using as-yet unknown information at the downstream station that we are stepping toward, rather than known information at the station where we are. In the limit as $\Delta x \to 0$, the distinction would be lost, but we are not letting $\Delta x \to 0$ here, and the distinction is real. Substituting Eqs. (4–52) and (4–60) into Eq. (4–13) results in

$$-Qu_{n+1,m-1} + (1 + 2Q)u_{n+1,m} - Qu_{n+1,m+1} = u_{n,m} \tag{4–61}$$

This is an expression for the new velocity at one step Δx downstream along the mth grid line in terms of the known value before the step on that same grid line and the new values obtained after a step Δx on the $(m + 1)$th and $(m - 1)$th grid lines. The information flows as shown in Fig. 4–13(B). Thus, Eq. (4–61) is an *implicit* relation for one unknown in terms of some of the other unknowns. It is, therefore, called an *implicit finite difference* formulation. Fortunately, we have one equation like Eq. (4–61) for each unknown, so the system is closed. The solution of a system of linear algebraic equations such as this for all the unknowns at each step Δx downstream generally involves inverting or decomposing a matrix of the coefficients of the unknowns. Since there may be a large number (approximately 10 to 100) of grid points across the layer, the matrix can be quite large, and that is what deterred the use of this type of formulation, even in the early days of the availability of digital computers. However, the matrix is tridiagonal—that is, all the entries off the main diagonal and one diagonal line to each side are zero. This is because the equation for each unknown involves only two other unknowns—the ones on each side of it—and none of the others. The necessary manipulations are, therefore, not difficult or expensive with modern machines and methods.

With the added complexity of the *implicit* formulation, the reader may well question the utility of the approach. The key lies in the fact that a stability analysis of this formulation proceeding in exactly the same way as for the explicit formulation reveals that the implicit method is unconditionally stable (i.e., it is stable for any value of Q). The only restrictions on the choices of Δx and Δy come from considerations of profile shapes, boundary conditions, and so on, as discussed earlier. Again, things come down to a matter of cost, and the implicit method is simply cheaper to run for a given problem than the explicit method.

The truncation error for the implicit method is found to be

$$\epsilon_{n,m} = -U_e \frac{(\Delta x)}{2}\left(\frac{\partial^2 u}{\partial x^2}\right) - \nu \frac{(\Delta y)^2}{12}\left(\frac{\partial^4 u}{\partial y^4}\right) + 0((\Delta x)^2) + 0((\Delta y)^4) \tag{4–62}$$

which again shows that $\epsilon_{n,m} \to 0$ as Δx, $\Delta y \to 0$.

Since the finite difference representation for $\partial^2 u/\partial y^2$ must hold from n to $n + 1$, a logical extension of the simple explicit method ($\partial^2 u/\partial y^2$ evaluated at n) and the simple implicit method ($\partial^2 u/\partial y^2$ evaluated at $n + 1$) is to take the average of the two, giving $\partial^2 u/\partial y^2$ evaluated at $n + \frac{1}{2}$, i.e., halfway across the strip over which one is stepping. Intuitively, such a technique makes good sense. Further motivation for it is found upon comparing Eqs. (4–59) and (4–62) for the truncation errors. Clearly, taking the average of the two will cause the cancellation of the term in $\epsilon_{n,m}$ involving Δx. This technique is called the *Crank-Nicolson method*. It is also an implicit method, since it involves unknowns at $n + 1$, and a stability analysis shows it to be unconditionally stable. Because it is stable and has improved accuracy (smaller truncation errors) for a given Δx, it is quite popular. It does, however, involve more computations per step than the simple implicit method.

Earlier, we stated that an implicit method involves solving a system of equations whose matrix of coefficients is tridiagonal. Here, we shall show how the system can be solved directly in a simple manner. Referring back to Eq. (4–61) and Fig. 4–13(B), one can see that the algebraic problem at each step from n to $n + 1$ in the x-direction will involve solving a system of equations for $u_{n+1,m}$ from $m = 2$ to $m = M - 1$ of the form

$$\begin{aligned}
A_{22}u_2 + A_{23}u_3 + A_{24}u_4 + A_{25}u_5 + \cdots + A_{2,M-1}u_{M-1} &= B_2 \\
A_{32}u_2 + A_{33}u_3 + A_{34}u_4 + A_{35}u_5 + \cdots + A_{3,M-1}u_{M-1} &= B_3 \\
&\vdots \\
A_{M-1,2}u_2 + A_{M-1,3}u_3 + \cdots + A_{M-1,M-1}u_{M-1} &= B_{M-1}
\end{aligned} \tag{4–63}$$

First, one can ask what happened to u_1 and u_M, and what about the boundary conditions? On a surface, $u_1 = 0$, and u_M is known as $u_M = U_e(n + 1)$. The boundary conditions are implemented by first taking $A_{21}u_1$ from its apparent place at the extreme left of the first equation in the matrix and putting it into B_2, since that term involves all known quantities. Next, the term $A_{M-1,M}u_M$ is taken from its apparent place at the right end of the left-hand side of the last equation in the matrix and is put into B_{M-1} on the right-hand side, since it is known.

Now, the system of interest here is *tridiagonal,* since Eq. (4–61) has only three of the unknowns for any given m. Thus, in the first equation in the preceding matrix, A_{24} to $A_{2,M-1}$ are all identically zero. In the second equation, A_{35} to $A_{3,M-1}$ are zero. And the last equation will have $A_{M-1,2}$ to $A_{M-1,M-3}$ zero. For the model equation, all the coefficients on the main diagonal are simply $(1 + 2Q)$, and all the coefficients on the diagonals above and below the main diagonal are simply $-Q$ (see Eq. (4–61). The matrix of coefficients will thus have a lot of zeroes, and one says that such a matrix is *sparse*. The system might still be large, but fortunately, there is a convenient method of direct solution for the tridiagonal case known as the *Thomas algorithm*. According to this method, we rewrite the system for a generic interior point m as

$$A_m u_{m-1} + B_m u_m + C_m u_{m+1} = D_m \tag{4–64}$$

and introduce X_m and Y_m by

$$u_m = X_m u_{m+1} + Y_m \tag{4–65}$$

where

$$X_m \equiv \frac{-C_m}{B_m + A_m X_{m-1}}, \qquad Y_m \equiv \frac{D_m - A_m Y_{m-1}}{B_m + A_m X_{m-1}} \tag{4–66}$$

The reader can substitute these two expressions into the original equation and confirm that it is satisfied. That is,

$$u_m = \frac{-C_m u_{m+1}}{B_m + A_m X_{m-1}} + \frac{D_m - A_m Y_{m-1}}{B_m + A_m X_{m-1}} \tag{4-67}$$

will satisfy the original equation. Now consider the situation for $m = 2$. We have

$$A_2 u_1 + B_2 u_2 + C_2 u_3 = D_2 \longrightarrow B_2 u_2 = -C_2 u_3 + (D_2 - A_2 u_1) \tag{4-68}$$

and

$$u_2 = X_2 u_3 + Y_2 \tag{4-69}$$

Comparing these two expressions, we see that the following relationship must hold:

$$\frac{B_2}{1} = \frac{-C_2}{X_2} = \frac{D_2 - A_2 u_1}{Y_2} \tag{4-70}$$

We also can write

$$X_2 = \frac{-C_2}{B_2 + A_2 X_1}, \qquad Y_2 = \frac{D_2 - A_2 Y_1}{B_2 + A_2 X_1} \tag{4-71}$$

Comparing Eqs. (4–70) and (4–71), we see that we must have

$$X_1 = 0, \qquad Y_1 = u_1 = 0 \tag{4-72}$$

This leads to the final, direct procedure: We solve for X_m and Y_m from $m = 2$ to $m = M - 1$, using Eq. (4–66), starting from $X_1 = Y_1 = 0$. We then solve for u_m from $m = M - 1$ down to $m = 2$ with Eq. (4–65), using the top boundary condition for u_M, i.e., $u_M = U_e(n + 1)$.

4-7-2 An Explicit Method for Solving the Boundary Layer Equations

From the discussion in Section 4–7–1, it should not be surprising to find that the earliest finite difference codes for the boundary layer equations were based on an explicit formulation. We will examine the explicit method of Wu (1961) for illustrative purposes. The method is presented here in a planar, incompressible form for simplicity and clarity.

The differential equations to be treated are the continuity equation, Eq. (3–6a), and the momentum equation, Eq. (3–26a), without the unsteady term in the latter:

$$\begin{gathered} \frac{\partial u}{\partial x} + \frac{\partial v}{\partial y} = 0 \\ u\frac{\partial u}{\partial x} + v\frac{\partial u}{\partial y} = -\frac{1}{\rho}\frac{dp}{dx} + \nu\frac{\partial^2 u}{\partial y^2} \end{gathered} \tag{4-73}$$

The parabolic momentum equation is solved with an explicit, "downstream-marching" procedure, and the continuity equation is viewed as an auxiliary equation for determining $v(x, y)$.

The explicit, finite difference form of the momentum equation is not difficult to derive by analogy with the explicit treatment of the model equation, Eq. (4–13). Comparing the two differential equations, we see that there is great similarity between the first and last terms in the momentum equation and the two terms in the model equation. Thus, those two terms in the momentum equation can be treated as they were for the model equation. The pressure gradient term is *nonhomogeneous*, since the pressure gradient is specified from the inviscid solution. The extra term with v in the momentum equation is not difficult to treat, since $\partial u/\partial y$ is not in the downstream direction. Thus, we can use a central difference to try for better accuracy. With all of this in mind, the explicit, finite difference form of the momentum equation can be written

$$u_{n,m}\left(\frac{u_{n+1,m} - u_{n,m}}{\Delta x}\right) + v_{n,m}\left(\frac{u_{n,m+1} - u_{n,m-1}}{2(\Delta y)}\right) = -\frac{1}{\rho}\left(\frac{p_{n+1} - p_n}{\Delta x}\right) + \nu\left(\frac{u_{n,m+1} - 2u_{n,m} + u_{n,m-1}}{(\Delta y)^2}\right) \tag{4–74}$$

This equation provides an explicit relation for calculating the streamwise velocity after a downstream step $u_{n+1,m}$, in terms of known information at the station before the step. We have

$$u_{n+1,m} = \frac{\nu(\Delta x)}{u_{n,m}(\Delta y)^2}(u_{n,m+1} + u_{n,m-1}) - \left(2\left(\frac{\nu(\Delta x)}{u_{n,m}(\Delta y)^2}\right) - 1\right)u_{n,m} - \frac{1}{\rho u_{n,m}}(p_{n+1} - p_n) - \frac{v_{n,m}}{u_{n,m}}\left(\frac{\Delta x}{\Delta y}\right)\left(\frac{u_{n,m+1} - u_{n,m-1}}{2}\right) \tag{4–74a}$$

or

$$u_{n+1,m} = Q_{n,m}(u_{n,m+1} + u_{n,m-1}) - (2Q_{n,m} - 1)u_{n,m} - \frac{1}{\rho u_{n,m}}(p_{n+1} - p_n) - \frac{v_{n,m}}{u_{n,m}}\left(\frac{\Delta x}{\Delta y}\right)\left(\frac{u_{n,m+1} - u_{n,m-1}}{2}\right) \tag{4–74b}$$

where

$$Q_{n,m} \equiv \frac{\nu(\Delta x)}{u_{n,m}(\Delta y)^2} \tag{4–75}$$

Observe the great similarity of Eq. (4–74b) to Eq. (4–53a). In Eq. (4–74b), there are two additional terms: one from the pressure gradient and the other in the convective derivative involving $v_{n,m}$. Also, note that the factor involving the step sizes, $Q_{n,m}$, [Q in Eq. (4–53a)] is no longer a constant across the layer at any x-station, since it now contains $u_{n,m}$ rather than U_e.

An important matter was handled rather easily here. In contrast to the linear model equation, the differential momentum equation is nonlinear. The finite difference form of the momentum equation [Eq. (4–74)] is, however, linear! That was accomplished by the simple device of *lagging* the u in the coefficient of $\partial u/\partial x$, i.e., evaluating it at n, where it is known, and not at $n + 1$, where it is not known. The same device is used for the $v\,\partial u/\partial y$ term. Again, the frequent profound difference between numerics and mathematics can be seen.

The finite difference form of the simpler, differential continuity equation proves a little more difficult to develop, as a result of the added complication of coupling between two equations, which we have not considered before. A straightforward formulation would appear to be

$$\frac{u_{n+1,m} - u_{n,m}}{\Delta x} + \frac{v_{n,m+1} - v_{n,m-1}}{2(\Delta y)} = 0 \tag{4–76}$$

This equation is to be used to calculate the transverse velocity after a downstream step $v_{n+1,m}$. It is easy to see, however, that none of those quantities appears in the equation, so this formulation will not serve. If the $\partial v/\partial y$ term is approximated in terms of quantities after the step, i.e.,

$$\frac{\partial v}{\partial y} \approx \frac{v_{n+1,m+1} - v_{n+1,m-1}}{2(\Delta y)} \tag{4–77}$$

an equation for v at every other grid point on the new profile results. It would then be necessary to interpolate for the values at the missing points. To obtain good accuracy, the use of a very small step size Δy would be required, since $\partial v/\partial y$ is large near the wall. The final scheme developed and actually employed involved a backward difference for $\partial v/\partial y$ at the downstream station $n + 1$:

$$\left.\frac{\partial v}{\partial y}\right|_{n+1,m-1/2} \approx \frac{v_{n+1,m} - v_{n+1,m-1}}{\Delta y} \tag{4–77a}$$

This equation can be interpreted as a central difference about a location halfway between two grid points $(n + 1, m)$ and $(n + 1, m - 1)$, that is, at the point $(n + 1, m - 1/2)$, shown as a cross in Fig. 4–14. The expression used for $\partial u/\partial x$ in Eq. (4–76) can be viewed as a central difference around $(n + 1/2, m)$, shown as a triangle in the figure. That point is rather far from $(n + 1, m - 1/2)$, so for consistency and symmetry, it is now necessary to write a corresponding form for $\partial u/\partial x$:

$$\begin{aligned}\left.\frac{\partial u}{\partial x}\right|_{n+1/2,m-1/2} &= \frac{1}{2}\left[\left.\frac{\partial u}{\partial x}\right|_{n+1/2,m} + \left.\frac{\partial u}{\partial x}\right|_{n+1/2,m-1}\right] \\ &\approx \frac{1}{2}\left[\frac{u_{n+1,m} - u_{n,m}}{\Delta x} + \frac{u_{n+1,m-1} - u_{n,m-1}}{\Delta x}\right]\end{aligned} \tag{4–78}$$

With this form, the continuity equation is written as

$$\frac{1}{2}\left[\frac{u_{n+1,m} - u_{n,m}}{\Delta x} + \frac{u_{n+1,m-1} - u_{n,m-1}}{\Delta x}\right] + \frac{v_{n+1,m} - v_{n+1,m-1}}{\Delta y} = 0 \tag{4–79}$$

which permits the accurate determination of $v_{n+1,m}$, starting at $m = 2$, using the boundary condition on v at the wall $(m = 1)$, and proceeding up to $m = M$.

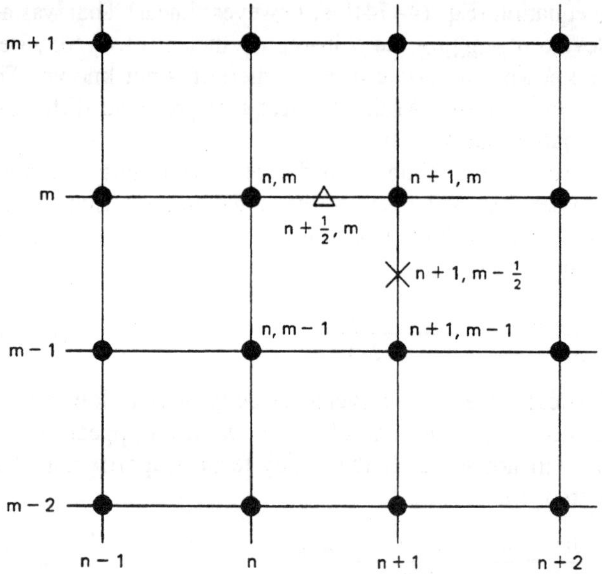

Figure 4–14 Schematic of the locations of effective centers for the differencing scheme of Wu.

The next matter of importance is stability, since the technique being discussed is an explicit method. The original system of equations is nonlinear, so the method of von Neumann cannot be rigorously applied. The *circuit analogy* method of Karplus (1958) can be applied to this nonlinear system on an *ad hoc* basis. The analogy comes from the fact that the voltage or current distribution of a network of electrical resistors arranged in a regular pattern can be described by an equation of the form

$$a(\phi_{n,m+1} - \phi_{n,m}) + b(\phi_{n,m-1} - \phi_{n,m}) + c(\phi_{n+1,m} - \phi_{n,m}) + d(\phi_{n-1,m} - \phi_{n,m}) = 0 \tag{4–80}$$

where a is positive and ϕ is any dependent variable. It can be proven that if (1) all the coefficients are positive or (2) some are negative, and the sum is negative, then the system is stable. Since all the finite difference approximations discussed so far can be put in the form of Eq. (4–80), the analogy can be used to infer stability of the finite difference formulation.

For the stated formulation, the equations can be rearranged to give

$$\begin{aligned} a &= -\frac{\Delta x}{2(\Delta y)} v_{n,m} + \frac{\nu(\Delta x)}{(\Delta y)^2} \\ b &= \frac{\Delta x}{2(\Delta y)} v_{n,m} + \frac{\nu(\Delta x)}{(\Delta y)^2} \\ c &= -u_{n,m} \\ d &= 0 \end{aligned} \tag{4–81}$$

where

$$a > 0 \qquad \textit{if } v_{n,m} < 0 \tag{4–82}$$

or

$$\Delta y < \frac{2v}{v_{n,m}} \qquad \textit{if } v_{n,m} > 0 \tag{4–83}$$

Since $u_{n,m} > 0$, $c < 0$, and the sum of the coefficients must be negative for stability; that is,

$$\frac{2\nu(\Delta x)}{(\Delta y)^2} - u_{n,m} < 0 \tag{4–84}$$

or

$$\frac{\nu(\Delta x)}{u_{n,m}(\Delta y)^2} < \frac{1}{2} \tag{4–84a}$$

leading to

$$\Delta x < \frac{1}{2}\frac{u_{n,m}(\Delta y)^2}{\nu} \tag{4–85}$$

This result is nearly identical to that rigorously derived for the *model* equation in Sec. 4–6–1, $\nu(\Delta x)/(U_e(\Delta y)^2) < 1/2$. The important difference is that in Eq. (4–84a), the $u_{n,m}$ in the denominator is a variable that is always less than U_e, so this requirement is more restrictive than that for *the model equation*. Indeed, the value of $u_{n,m}$ at the first grid point out from the wall will generally be a small fraction of U_e, and that sets the restriction of Δx for the whole layer. Earlier, we found the condition in Eq. (4–83) that must be met to keep $a > 0$. That condition must still be met.

Once an accurate solution for the velocity components is obtained, various other quantities are of interest to the analyst. Certainly, the most important is

$$\tau_w = \mu\left(\frac{\partial u}{\partial y}\right)_{y=0} \approx \frac{\mu}{2(\Delta y)}(4u_{n,2} - u_{n,3}) \tag{4–86}$$

This is a three-point formula using $u_{n,1} = 0$. The boundary layer thickness, δ, can be found by searching the array $u_{n,m}$ for the value of m, and thus y, corresponding to $u_{n,m} = 0.99U_e$. It may be necessary to interpolate between two grid points to obtain an accurate value. The integral thicknesses, δ^* and θ, can be found by integrating the velocity profiles in the definitions in Eqs. (2–13) and (2–14), using the well-known *trapezoidal rule* or *Simpson's rule*.

The foregoing method is simple and quite successful. Even with the step-size restrictions described, very accurate calculations can be made in a very short time for planar, incompressible, constant-property flows. A comparison of numerical solutions obtained on a PC using this method and the exact solution of the so-called *Howarth problem* ($U_e = U_0 - Ax$) is shown in Fig. 4–15. The Howarth problem gives a flow that is a nonsimilar, adverse pressure gradient flow leading to separation, so it is a severe test.

A simple computer code for a PC is included in Appendix B for use in working the problems at the end of the chapter.

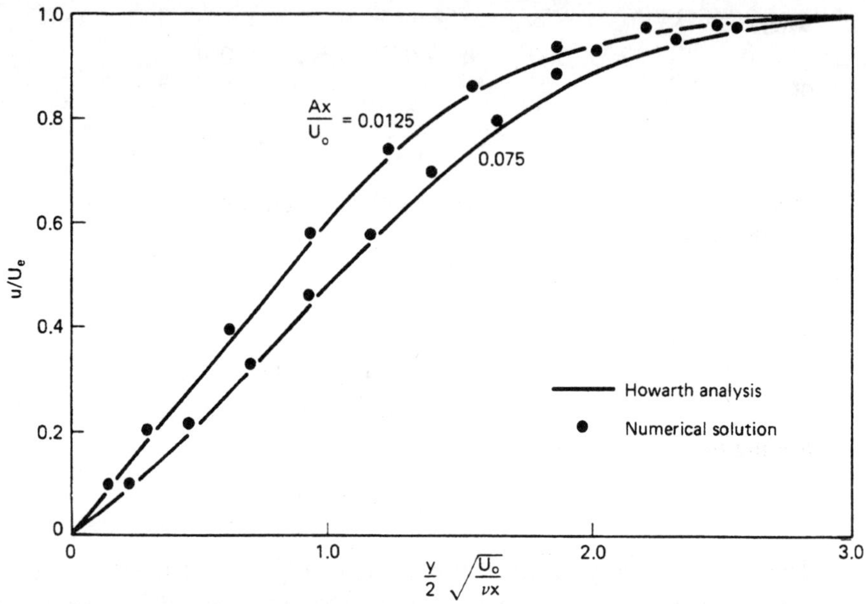

Figure 4–15 Numerical solution of the Howarth problem by the explicit method. (From Grossman and Schetz, 1985.)

Example 4–1. Application of Explicit Numerical Method

Consider the two-dimensional laminar flow of a fluid with a kinematic viscosity $\nu = 2.0 \times 10^{-4}$ m²/s at $U_\infty = 10.0$ m/s over a surface that is a flat plate from the leading edge to $x = 1.0$ m. At that station, a ramp begins that produces an inviscid velocity distribution $U_e(x) = 10.5 - x/2$ m/s. This distribution represents an adverse pressure gradient, since U_e is decreasing, so that p increases. Calculate the boundary layer development over this surface up to $x = 2.0$ m. Does the flow separate?

Solution In Chapter 2, we treated this flow problem with the Thwaites-Walz integral method, using the code WALZ in Appendix B. Now we have more powerful tools of analysis available. The finite difference methods permit solution for the details of the flow, as well as for the integral thicknesses and skin friction. We apply the code ILBLE in Appendix B to the problem.

Since the first part of the problem is flow over a flat plate, the Blasius solution can be used to obtain "initial" conditions at $x = 1.0$. Thus, the numerical calculation will begin at XI = 1.0 and go to XF = 2.0. We must again provide input data for the kinematic viscosity as CNU = 0.0002 and the free stream velocity as UINF = 10.0. The Blasius solution at $x = 1.0$ gives the boundary layer thickness, DELT = 0.0224, the displacement thickness, DELTS = 0.00771, the momentum thickness, THETA = 0.00297, and the skin friction coefficient, CF = 0.00297. We choose 21 points across the initial boundary layer (DY = 0.00112) and add about 80 points above that to give

MMAX = 100. One is not free to pick DX or NMAX for this explicit method, since the step size must be controlled by the stability criterion.

Initial profiles for U and V are required. For simplicity, we adopt the cubic velocity profile [Eq. (2-46)] for U and V = 0 as adequate approximations to the exact Blasius solution, which would require tabular input for the profiles.

Finally, the inviscid velocity distribution is required. This kind of information for this particular case is included directly in the code following the determination of DX and then the new X at each step downstream. Note that in the code, all velocities, including UE, are normalized with UINF.

For other problems, the relevant sections of the code must be modified to accommodate the new input information.

Excerpts from the station output are listed in the following table. With the stability criterion, small step sizes DX are needed, and the output for every single step fills nearly 20 pages. Note, for future reference, that about 600 steps in the x-direction were required with this method. The boundary layer thicknesses increase and the skin friction coefficient decreases rapidly in the adverse pressure gradient, but the boundary layer does not separate by x = 2.0 m—i.e., $C_f > 0$.

X	DELT	DELTS	THETA	CF	UE	DUEDX
1.0000	0.0224	0.00771	0.00297	0.00297	1.0000	−0.0500
1.0505	0.0224	0.00889	0.00320	0.00221	0.9975	−0.0500
1.1007	0.0235	0.00926	0.00330	0.00205	0.9950	−0.0500
1.1491	0.0235	0.00958	0.00339	0.00194	0.9926	−0.0500
1.2000	0.0246	0.00990	0.00348	0.00184	0.9900	−0.0500
1.2496	0.0258	0.01021	0.00356	0.00175	0.9875	−0.0500
1.3003	0.0269	0.01052	0.00365	0.00167	0.9850	−0.0500
1.3497	0.0269	0.01082	0.00374	0.00160	0.9825	−0.0500
1.4001	0.0280	0.01112	0.00383	0.00153	0.9800	−0.0500
1.4502	0.0291	0.01142	0.00393	0.00146	0.9775	−0.0500
1.5000	0.0291	0.01173	0.00400	0.00140	0.9750	−0.0500
1.5502	0.0302	0.01204	0.00408	0.00134	0.9725	−0.0500
1.5999	0.0302	0.01235	0.00417	0.00128	0.9700	−0.0500
1.6498	0.0313	0.01267	0.00425	0.00122	0.9675	−0.0500
1.6997	0.0325	0.01300	0.00434	0.00116	0.9650	−0.0500
1.7501	0.0325	0.01333	0.00443	0.00110	0.9625	−0.0500
1.8001	0.0336	0.01368	0.00451	0.00105	0.9600	−0.0500
1.8501	0.0347	0.01403	0.00460	0.00099	0.9575	−0.0500
1.8999	0.0347	0.01439	0.00468	0.00093	0.9550	−0.0500
1.9499	0.0358	0.01476	0.00477	0.00088	0.9525	−0.0500
2.0000	0.0370	0.01515	0.00485	0.00082	0.9500	−0.0500

The velocity profiles at the end of the computation region, x = 2.0, are given in the next table. Only every other point, up to just beyond the boundary layer, is listed to conserve space. Note that by this station, there are about 45 points across the boundary layer. That number is probably more than is needed for accuracy, and it makes DY small and then DX especially small through the stability criterion [see Eq. (4-85)]. This matter will be discussed further in Section 4-7-4.

M	Y	U	V
1	0.0	0.0	0.0
3	0.0022	0.0475	0.0001
5	0.0045	0.1065	0.0003
7	0.0067	0.1761	0.0006
9	0.0090	0.2546	0.0011
11	0.0112	0.3397	0.0017
13	0.0134	0.4284	0.0024
15	0.0157	0.5175	0.0032
17	0.0179	0.6033	0.0040
19	0.0202	0.6825	0.0048
21	0.0224	0.7523	0.0056
23	0.0246	0.8107	0.0063
25	0.0269	0.8569	0.0069
27	0.0291	0.8913	0.0074
29	0.0314	0.9153	0.0079
31	0.0336	0.9308	0.0082
33	0.0358	0.9401	0.0084
35	0.0381	0.9453	0.0086
37	0.0403	0.9479	0.0088
39	0.0426	0.9491	0.0089
41	0.0448	0.9497	0.0090
43	0.0470	0.9499	0.0091
45	0.0493	0.9500	0.0093
47	0.0515	0.9500	0.0093

Plots of some of these results are shown in Figure 4–16. Comparing these with the results for the same problem by the Thwaites-Walz method used in Example 2–1, one can see that the agreement for θ is quite good, that for C_f is good, but that for δ^* is not so good. This is perhaps not surprising, since the Thwaites-Walz method is really based on determining θ, with the other quantities obtained after the fact. The predicted velocity profile shows the effects of the adverse pressure gradient, with an inflection point developing near the wall. Of course, the integral method does not yield a prediction for the profile that could be compared with the numerical prediction.

4–7–3 Implicit Methods for Solving the Boundary Layer Equations

An early implicit procedure for solving the boundary layer equations was developed by Parr (1963). Parr used

$$\frac{\partial^2 u}{\partial y^2} \approx \frac{1}{2}\left[\frac{u_{n+1,m+1} - 2u_{n+1,m} + u_{n+1,m-1}}{(\Delta y)^2} + \frac{u_{n,m+1} - 2u_{n,m} + u_{n,m-1}}{(\Delta y)^2}\right] \tag{4–87}$$

which is the Crank-Nicolson method. For the other derivatives, he took

$$\frac{\partial u}{\partial x} \approx \frac{(u_{n+1,m} - u_{n,m})}{\Delta x} \tag{4–88}$$

and

$$\frac{\partial u}{\partial y} \approx \frac{1}{2}\left(\frac{u_{n+1,m+1} - u_{n+1,m-1}}{2(\Delta y)} + \frac{u_{n,m+1} - u_{n,m-1}}{2(\Delta y)}\right) \tag{4–89}$$

Parr found that an unconditionally stable procedure also resulted for the boundary

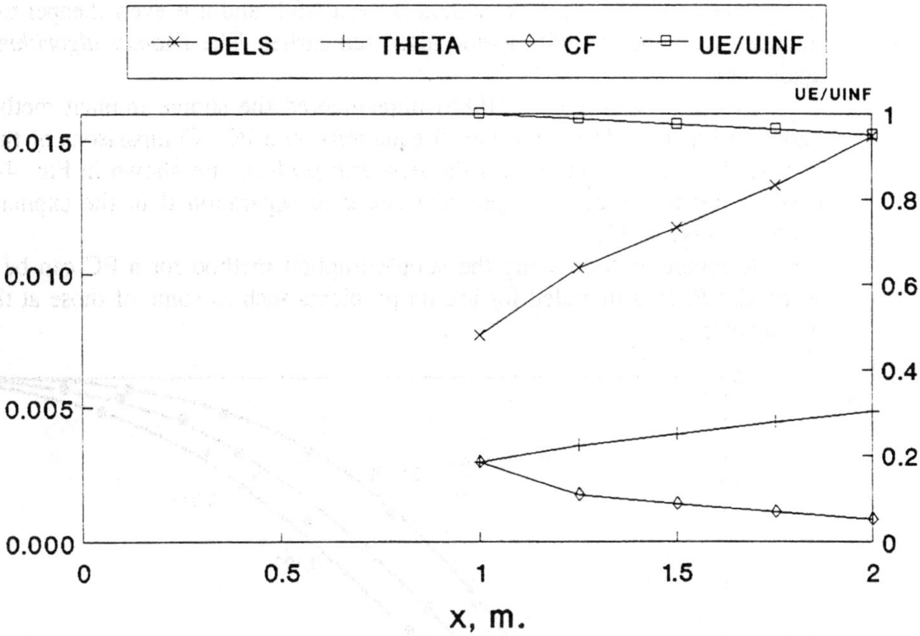

(A) Streamwise variation of integral boundary layer properties.

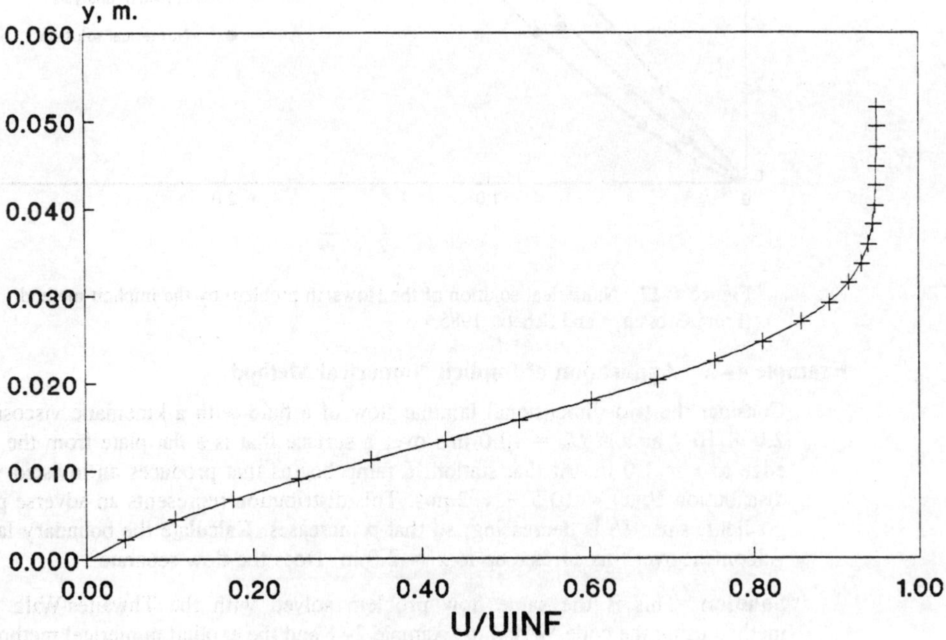

(B) Velocity profile at x = 2.0 m.

Figure 4–16 Predicted results for Example 4–1.

layer equations. This *implicit* method is successful, and it is even cheaper to run than the corresponding explicit method discussed earlier. The *Thomas algorithm* is useful again here.

Grossman and Schetz (1985) implemented the simple implicit method (analogous to Eq. (4–61) for the model equation) on a PC. Comparisons of the predictions and the exact solution for the *Howarth problem* are shown in Fig. 4–17. The implicit method is able to approach closer to separation than the explicit method shown in Fig. 4–15.

A computer code using the simple implicit method for a PC can be found in Appendix *B*. It is intended for use on problems such as some of those at the end of the chapter.

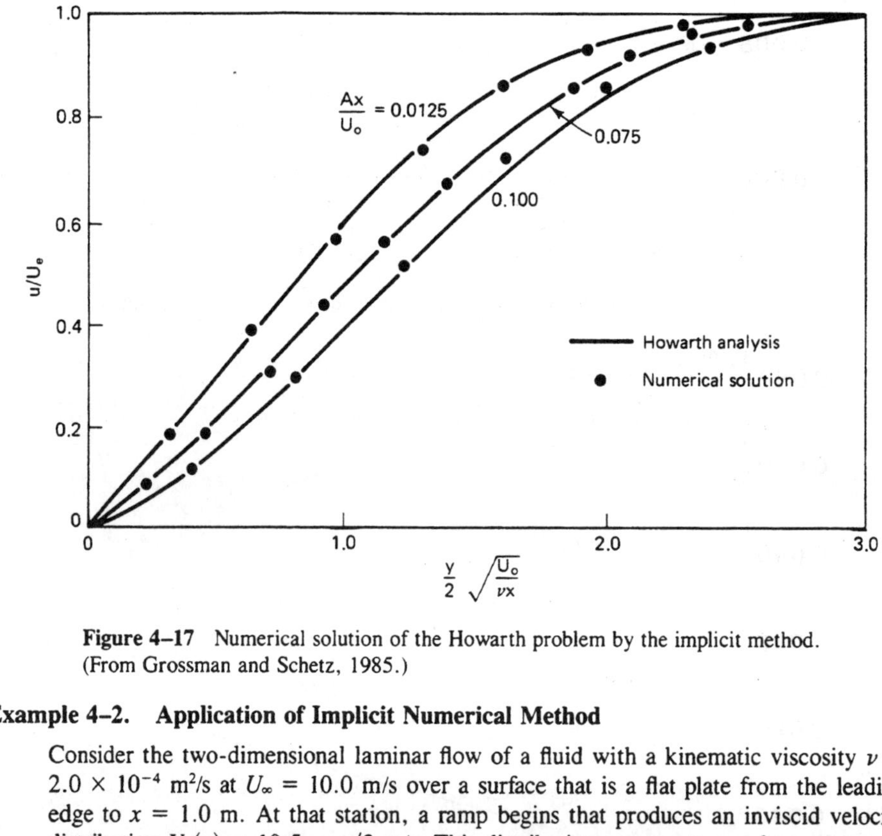

Figure 4–17 Numerical solution of the Howarth problem by the implicit method. (From Grossman and Schetz, 1985.)

Example 4–2. Application of Implicit Numerical Method

Consider the two-dimensional laminar flow of a fluid with a kinematic viscosity $\nu = 2.0 \times 10^{-4}$ m²/s at $U_\infty = 10.0$ m/s over a surface that is a flat plate from the leading edge to $x = 1.0$ m. At that station, a ramp begins that produces an inviscid velocity distribution $U_e(x) = 10.5 - x/2$ m/s. This distribution represents an adverse pressure gradient, since U_e is decreasing, so that p increases. Calculate the boundary layer development over this surface up to $x = 2.0$ m. Does the flow separate?

Solution This is the same flow problem solved with the Thwaites-Walz integral method using the code WALZ in Example 2–1 and the explicit numerical method using code ILBLE in Example 4–1. Here, we apply the implicit numerical method in code ILBLI in Appendix B to the problem, for comparison in terms of accuracy of the predictions and the computational effort required.

Many, but not all, parts of the input are the same as for the explicit calculation. Since the first part of the problem is flow over a flat plate, the Blasius solution can be used to obtain "initial" conditions at $x = 1.0$. Thus, the numerical calculation will begin at XI = 1.0 and go to XF = 2.0. We must again provide input data for the kinematic viscosity as CNU = 0.0002 and the free stream velocity as UINF = 10.0. The Blasius solution at $x = 1.0$ gives the boundary layer thickness, DELT = 0.0224, the displacement thickness, DELTS = 0.00771, the momentum thickness, THETA = 0.00297, and the skin friction coefficient, CF = 0.00297. We choose 21 points across the initial boundary layer (DY = 0.00112) and add about 80 points above that to give MMAX = 100. Since the implicit method is unconditionally stable, no stability criterion need be followed, and we can select the streamwise step size DX and the number of points NMAX based only on considerations of accuracy. We select NMAX = 41 to give DX = 0.025, which is about the size of the initial boundary layer thickness.

Initial profiles for U and V are required. For simplicity, we adopt the cubic velocity profile [Eq. (2–46)] for U and V = 0 as adequate approximations to the exact Blasius solution, which would require tabular input for the profiles.

Finally, the inviscid velocity distribution is required. This kind of information for this particular case is included directly in the code following the determination of the new X at each step downstream in the loop DO 5 N = 1, NMAX. Again, UE is normalized with UINF in the code.

For other problems, the relevant sections of the code must be modified to accommodate the new input information.

The station output at every other station is listed in the following table. Again, the boundary layer thicknesses increase and the skin friction coefficient decreases rapidly in the adverse pressure gradient, but the boundary layer does not separate by $x = 2.0$ m—i.e., $C_f > 0$.

N	X	DELT	DELTS	THETA	CF
1	1.0000	0.0224	0.00771	0.00297	0.00297
3	1.0500	0.0213	0.00863	0.00315	0.00234
5	1.1000	0.0224	0.00900	0.00323	0.00215
7	1.1500	0.0235	0.00937	0.00333	0.00202
9	1.2000	0.0246	0.00970	0.00343	0.00190
11	1.2500	0.0258	0.01002	0.00352	0.00181
13	1.3000	0.0258	0.01034	0.00361	0.00173
15	1.3500	0.0269	0.01064	0.00370	0.00165
17	1.4000	0.0280	0.01094	0.00379	0.00158
19	1.4500	0.0280	0.01125	0.00388	0.00151
21	1.5000	0.0291	0.01155	0.00397	0.00145
23	1.5500	0.0302	0.01186	0.00405	0.00138
25	1.6000	0.0302	0.01217	0.00414	0.00132
27	1.6500	0.0314	0.01248	0.00423	0.00126
29	1.7000	0.0325	0.01280	0.00431	0.00121
31	1.7500	0.0325	0.01312	0.00440	0.00115
33	1.8000	0.0336	0.01345	0.00448	0.00109
35	1.8500	0.0347	0.01379	0.00457	0.00104
37	1.9000	0.0347	0.01414	0.00465	0.00098
39	1.9500	0.0358	0.01450	0.00474	0.00093
41	2.0000	0.0370	0.01487	0.00482	0.00088

The predicted velocity profiles at the end of the computation region, $x = 2.0$, are given in the next table. Only every other point, up to just beyond the boundary layer, is listed to conserve space.

M	Y	U	V
1	0.0	0.0	0.0
3	0.0022	0.0502	0.0001
5	0.0045	0.1119	0.0003
7	0.0067	0.1841	0.0006
9	0.0090	0.2649	0.0011
11	0.0112	0.3518	0.0016
13	0.0134	0.4418	0.0023
15	0.0157	0.5313	0.0031
17	0.0179	0.6167	0.0038
19	0.0202	0.6947	0.0046
21	0.0224	0.7626	0.0053
23	0.0246	0.8187	0.0060
25	0.0269	0.8625	0.0066
27	0.0291	0.8947	0.0070
29	0.0314	0.9170	0.0074
31	0.0336	0.9314	0.0077
33	0.0358	0.9401	0.0080
35	0.0381	0.9451	0.0081
37	0.0403	0.9477	0.0083
39	0.0426	0.9490	0.0084
41	0.0448	0.9496	0.0085
43	0.0470	0.9499	0.0087
45	0.0493	0.9500	0.0088
47	0.0515	0.9500	0.0089

Some simple plots of these predictions are given in Figure 4–18. Looking at these and comparing them with the tables and plots for Example 4–1 with the explicit method, one can see that the results are in good agreement. With the explicit method, the velocity profile is a little less full near the wall, leading to a slightly lower skin friction coefficient. Also, the explicit method predicts a higher velocity v at the boundary layer edge. It is important to note that the implicit method involves much less computational effort, needing only 41 streamwise steps, compared with about 600 streamwise steps for the explicit method.

4–7–4 Transformations and Other Matters

With the material in Sections 4–7–2 and 4–7–3, the reader has already seen the primary numerical methods used for boundary layer problems. Further extensions of these basic methods involve the treatment of compressible fluids (which are discussed in Chapter 5), turbulent cases (which are discussed in later chapters), more efficient procedures for implementing the basic methods, and prior transformation of

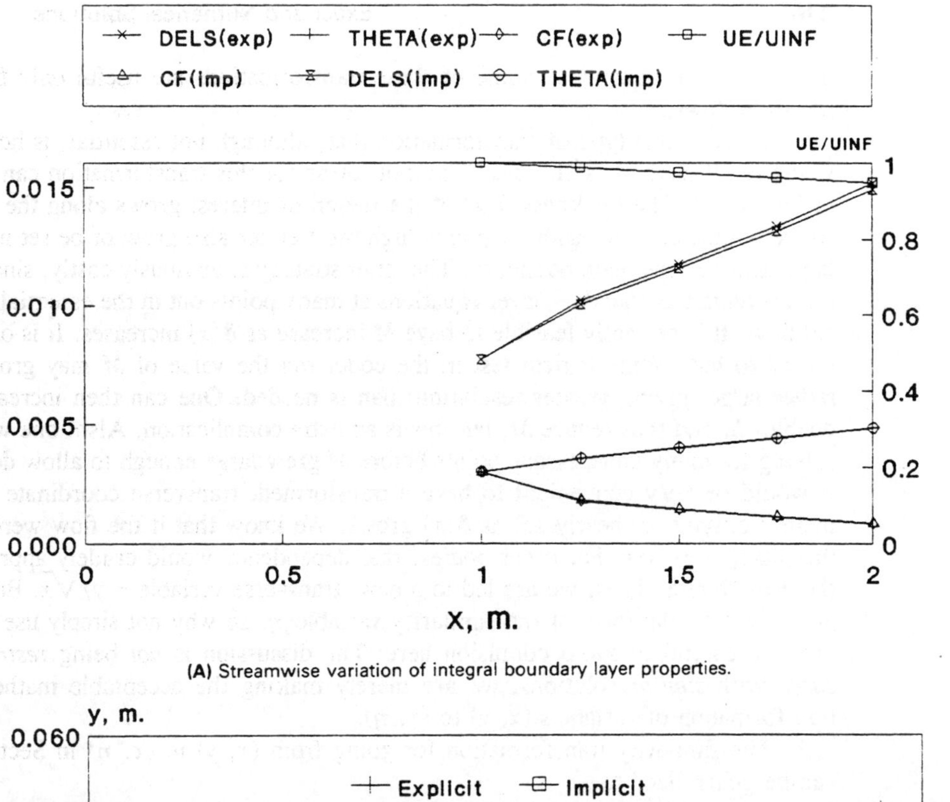

(A) Streamwise variation of integral boundary layer properties.

(B) Velocity profiles at $x = 2.0$ m.

Figure 4–18 Predicted results for Example 4–2.

the differential equations. Some of these transformations are useful only for compressible flows.

There is one type of transformation that, although not essential, is helpful for virtually all methods and cases. The motivation for this transformation can be seen in Fig. 4–12. The thickness $\delta(x)$ of the region of interest grows along the surface, so the computational region M points high must either also grow or be set much too large near the upstream boundary. The latter strategy is obviously costly, since it implies solving the boundary layer equations at many points out in the essentially inviscid flow. It is perfectly feasible to have M increase as $\delta(x)$ increases. It is only necessary to have some logical test in the code, but the value of M may grow to be rather large, giving greater resolution than is needed. One can then increase (say, double) Δy and thus reduce M, but this is an extra complication. Also, one would be solving for many unnecessary points before M grew large enough to allow doubling. It would be very convenient to have a transformed, transverse coordinate that remains constant, or nearly so, as $\delta(x)$ grows. We know that if the flow were over a flat plate, $\delta \sim \sqrt{x}$. For other bodies, that dependence would crudely approximate the growth rate. Thus, we are led to a new, transverse variable $\sim y/\sqrt{x}$. But this is no more than the form of the similarity variable η, so why not simply use it? One must be careful to avoid confusion here: The discussion is *not* being *restricted* to *cases* with *similar* solutions; we are merely making the acceptable mathematical transformation of variables (x, y) to (x, η).

The similarity transformation for going from (x, y) to (x, η) in Section 4–3 can be generalized to

$$s = \int_0^x U_e\,dx', \qquad \eta = \frac{yU_e}{\sqrt{2\nu s}} \tag{4–90}$$

In terms of these variables, the equation for *nonsimilar* boundary layer flows becomes

$$f''' + ff'' - \frac{2s}{U_e}\frac{dU_e}{ds}((f')^2 - 1) = 2s\left(f'\left(\frac{\partial f'}{\partial s}\right) - f''\left(\frac{\partial f}{\partial s}\right)\right) \tag{4–91}$$

The pressure gradient parameter becomes

$$\beta \equiv \frac{2s}{U_e}\frac{dU_e}{ds} \tag{4–92}$$

For nonsimilar cases, we are not restricting the form of the inviscid velocity distribution, $U_e(s)$, as was required to obtain similar solutions. Also, the right-hand side of the equation is not forced to be equal to zero by insisting on $f(\eta)$ alone. Here, the general form $f(s, \eta)$ is retained with

$$\frac{u}{U_e} = \left(\frac{\partial f}{\partial \eta}\right)_s, \qquad \psi = \sqrt{2\nu s}\,f(s, \eta) \tag{4–93}$$

Also,

$$v = -U_e\sqrt{2\nu s}\left(\left(\frac{\partial\eta}{\partial s}\right)\left(\frac{\partial f}{\partial\eta}\right) + \left(\frac{\partial f}{\partial s}\right) + \left(\frac{f}{2s}\right)\right) \tag{4-94}$$

The boundary conditions are

$$f(s, 0) = 0, \qquad f'(s, 0) = 0, \qquad \lim_{\eta\to\infty} f'(s, \eta) = 1 \tag{4-95}$$

for a solid wall. If there is injection or suction, the boundary condition on the stream function becomes

$$f(s, 0) = f_w = -\frac{1}{\sqrt{2\nu s}}\int_0^s v_w\, ds \tag{4-96}$$

For numerical solution by the methods described earlier for the untransformed equations, it is more convenient to rewrite these transformed equations with $F(s, \eta) = u/U_e$ rather than with a stream function. Doing so puts Eq. (4-91) in the form

$$F'' + \beta(1 - F^2) - VF' - 2sFF_s = 0 \tag{4-97}$$

where V is a transformed normal velocity that must be found from the transformed continuity equation

$$2sF_s + V' + F = 0 \tag{4-98}$$

which can be written

$$V = V_w - \int_0^\eta (2sF_s + F)\, d\eta \tag{4-99}$$

This integration can be accomplished numerically by the *trapezoidal rule*. The numerical solution of Eq. (4-97) will be illustrated for the simple implicit method discussed earlier. The streamwise derivative $\partial F/\partial s$ is approximated as a forward difference. The transverse derivatives, $F' = \partial F/\partial\eta$ and F'', are evaluated at the downstream station we are stepping towards. In that case, Eq. (4-97) can be approximated as

$$A_m F_{n+1,m-1} + B_m F_{n+1,m} + C_m F_{n+1,m+1} = D_m \tag{4-100}$$

where

$$A_m = -\frac{1}{(\Delta\eta)^2} - \frac{V_{n,m}}{2(\Delta\eta)}$$

$$B_m = \frac{2}{(\Delta\eta)^2} + \beta F_{n,m} + 2s\frac{F_{n,m}}{\Delta s}$$

$$C_m = -\frac{1}{(\Delta\eta)^2} + \frac{V_{n,m}}{2(\Delta\eta)}$$

$$D_m = \beta - 2s\frac{F_{n,m}^2}{\Delta s}$$

Clearly, this system can be solved by the *Thomas algorithm*.

A thorough discussion of all these matters can be found in Blottner (1975). A typical computer code for solving boundary layer problems by the methods described here might have a *flowchart* like the one shown in Fig. 4–19.

4–7–5 Finite Element Method

The numerical treatment of physical problems by what has become known as the *finite element method* (FEM) is based upon a rather different idea than the *finite difference methods* (FDM) that have been discussed so far. In FDM, one introduces approximations to the derivatives in the differential equation(s) [see Eq. (4–52), for example], and that reduces the differential equation(s) to a system of algebraic equations for the unknowns at each grid point. In FEM, one introduces approximations to the dependent variables themselves, not the derivatives. The region of interest is again subdivided into small regions in what may look like a grid familiar in FDM, but the small regions are called *finite elements*, and they need not be regular in shape. Here, we shall restrict the discussion to rectangular elements for simplicity. We consider approximating the variation of the dependent variable(s) over an element by some preselected functional form called an *interpolation function* in terms of values of the dependent variable(s) at the *nodes* of the element. If this assumed function is substituted into the exact differential equation(s), it will give an error or *residual*, since it is not, in general, a solution to the equation(s). The idea is to minimize that error in some weighted sense over the whole region, in order to produce a close approximation to the solution of the exact differential equation(s). This is called the *method of weighted residuals*, an approach the reader has likely seen before—for example, in the study of vibrations, where it is quite popular. Again, a system of algebraic equations for the unknowns at the nodes will result.

The earliest attempts to treat boundary layer equations by what might broadly be called FEM date back a number of years [see Schetz (1963), Schetz and Jannone (1965), and Schetz (1966)]. With the limited computer power available at that time, it was necessary to use very large elements—on the order of one-half the boundary layer thickness. To achieve acceptable accuracy with such large elements, very good interpolation functions that mimicked the behavior of the real solutions were required. The functions used were developed from solutions to the linearized form of the momentum equation [see Eq. (4–13)] for the problem of interest. These functions were then distorted by adding factors to the argument. The simplest type of error minimization condition, called *collocation*, was used where the weighting function was just a *Dirac delta function*. Very good qualitative behavior of the solutions

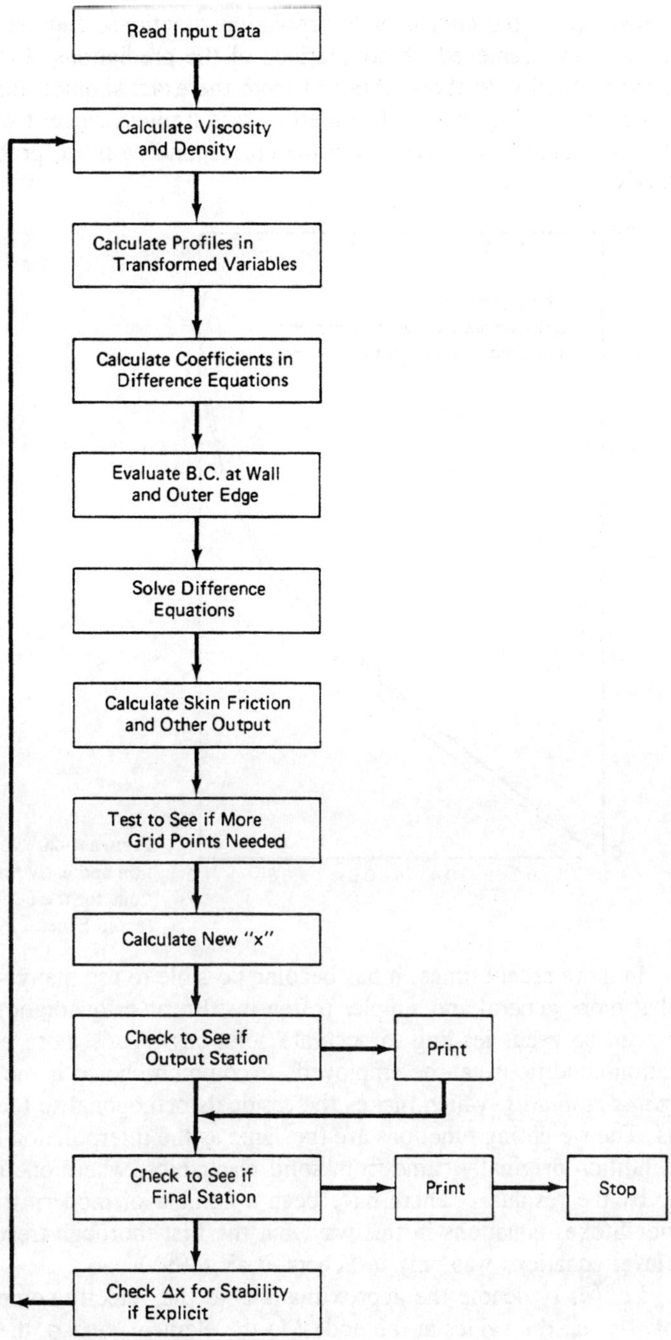

Figure 4–19 Flowchart for a typical boundary layer computer code.

was ensured by the choice of interpolation functions, and reasonable quantitative agreement was achieved. A comparison of the predictions obtained with the finite element method with those obtained from the exact solution for the flat plate problem is shown in Fig. 4–20. The method was limited, since it was a combination of analytical methods (the solution of the corresponding linear problem) and numerical methods.

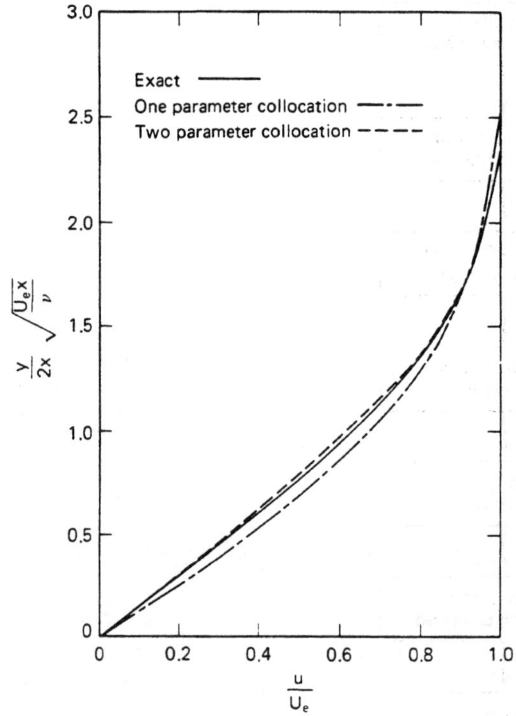

Figure 4–20 Comparison of exact solution and early finite element approximations for the laminar flat plate problem. (From Schetz, 1966.)

In more recent times, it has become possible to use many very small elements, so that more general and simpler (often just linear or quadratic) interpolation functions can be used, leading to accurate solutions. Also, more powerful error minimization conditions can be employed. A common choice is the *Galerkin method of weighted residuals,* which makes the residuals orthogonal to the interpolation functions. The weighting functions are the same as the interpolation functions. This kind of condition originally came from solid mechanics, where one minimized the work done by the residuals. There have been a number of modern treatments of the full Navier-Stokes equations in this way, but the first thorough treatment of the boundary layer equations was only in Schetz et al. (1991).

Let (u, v) denote the approximations to the velocity components over an element, (u_i, v_i) the values at the node i in the element, and ϕ_i the interpolation functions over an element. Then the finite element approximation to the velocities over an element can be written

$$u(x, y) = \sum_{i=1}^{N} u_i \phi_i(x, y)$$
$$v(x, y) = \sum_{i=1}^{n} v_i \phi_i(x, y) \qquad (4\text{–}101)$$

If these relations are substituted into the momentum equation, an error or residual, R, will be produced. When the Galerkin error minimization condition is applied, a matrix problem for each element results of the form

$$[K]_e[u]_e = [F]_e \qquad (4\text{–}102)$$

where

$$K_{ij} = \int_{\Omega_e} \left(\bar{u}\frac{\partial \phi_j}{\partial x}\phi_i + \bar{v}\frac{\partial \phi_j}{\partial y}\phi_i + \nu\frac{\partial \phi_j}{\partial y}\frac{\partial \phi_i}{\partial y} \right) dx\, dy \qquad (4\text{–}103)$$

$$F_i = -\int_{\Omega_e} \frac{1}{\rho}\frac{dp}{dx}\phi_i\, dx\, dy + \int_{\Gamma_e} \nu\frac{\partial u}{\partial y}\phi_i n_y\, ds \qquad (4\text{–}104)$$

The last term in these two equations is the result of applying the *divergence theorem* to the viscous term in the momentum equation. The matrix M_{ij} is called the *stiffness matrix* and the vector F_i is called the *force vector*, holdover terms from solid mechanics, where Galerkin-type methods were originally developed. The quantity n_y is the y-component of the surface normal. Also, $\bar{u}$ and $\bar{v}$ are the approximations to the velocity components from the previous iteration. Since the algebraic system to be solved is nonlinear, an iterative procedure is appropriate. We will return to the continuity equation later.

The simplest element that can be used is the four-node, quadrilateral element with a linear interpolation function for the velocity in each direction. Such a function is called a *bilinear interpolation function*. Now, consider an element, as shown in Fig. 4–21. The origin of the local coordinate system is at the center of the element. With respect to this system, face 1–4 corresponds to $r = -1$, face 2–3 corresponds to $r = 1$, face 1–2 corresponds to $s = -1$, and face 3–4 corresponds to $s = 1$. The solution over the element can then be expressed as

$$u(r, s) = a_1 + a_2 r + a_3 s + a_4 rs \qquad (4\text{–}105)$$

We require that the solution be equal to the nodal values at the nodes, so that the constants a_i can be determined in terms of the nodal values as

$$u_1 = u(-1, -1) = a_1 - a_2 - a_3 + a_4$$
$$u_2 = u(1, -1) = a_1 + a_2 - a_3 - a_4$$
$$u_3 = u(1, 1) = a_1 + a_2 + a_3 + a_4 \qquad (4\text{–}106)$$
$$u_4 = u(-1, 1) = a_1 - a_2 + a_3 - a_4$$

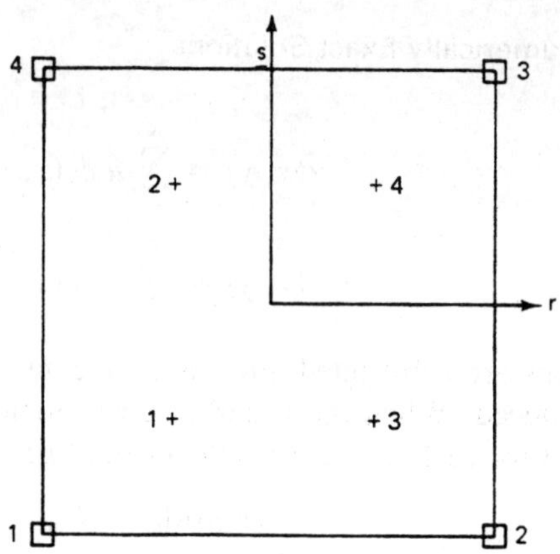

(A) Geometry and numbering convention for nodal (□) and quadrature (+) points

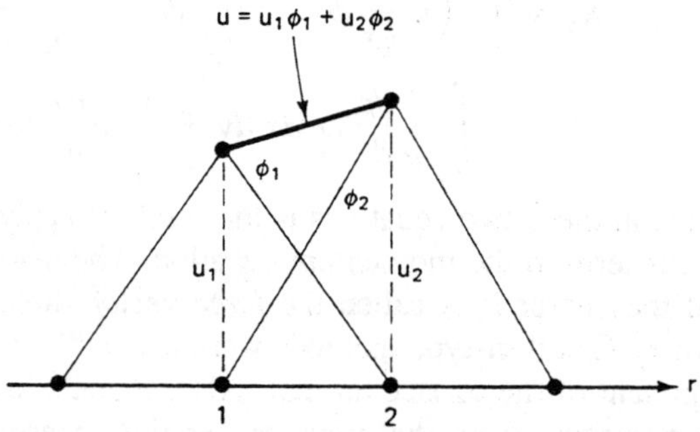

(B) Illustration of the construction of the approximation to the velocity

Figure 4–21 Schematics of the bilinear element for the finite element method.

This system can be inverted, and the results for the constants as a function of the nodal values can be inserted into Eq. (4–105) to give

$$u(r, s) = [\phi_1 \quad \phi_2 \quad \phi_3 \quad \phi_4]\begin{bmatrix} u_1 \\ u_2 \\ u_3 \\ u_4 \end{bmatrix} \tag{4–107}$$

where

$$\phi = \begin{bmatrix} \frac{1}{4}(1 - r)(1 - s) \\ \frac{1}{4}(1 + r)(1 - s) \\ \frac{1}{4}(1 + r)(1 + s) \\ \frac{1}{4}(1 - r)(1 + s) \end{bmatrix} \tag{4–108}$$

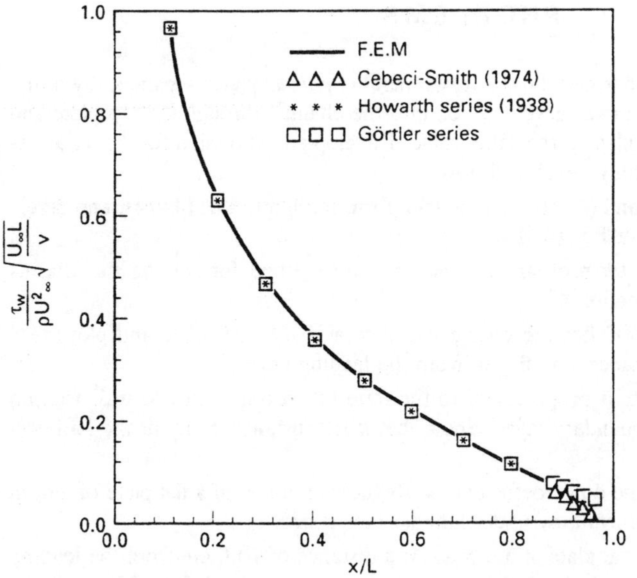

Figure 4–22 Comparison of FEM prediction using a bilinear element with prior solutions for the Howarth problem. (From Schetz, Hytopoulos, and Gunzburger, 1991.)

With this definition of an element, the numerical integration of the integrals in Eqs. (4–103) and (4–104) can be accomplished exactly with a 2×2 quadrature. The K-matrix is 4×4, and the F-vector has four components for each element. Since the problem under discussion is parabolic, and since one is "marching" downstream from an initial condition, the nodal values at 1 and 4 are known from the upstream solution. Therefore, they can be applied as boundary conditions at the element level. Doing so reduces the matrix of unknowns to 2×2. Finding the solution for the column of elements at a given streamwise station requires the assembly of the element stiffness matrices into a global matrix and the assembly of the element force vectors into a global force vector. This procedure leads to a tridiagonal system of equations for the nodal unknowns on the downstream side of the column of elements. That system can be solved using the *Thomas algorithm* described earlier.

The boundary conditions are simply the no-slip condition on the wall and the edge velocity, $U_e(x)$, on the top boundary of the computational region.

After the solution for u is obtained on a given column, the solution for v is obtained from the continuity equation, rewritten as

$$v(x, y) = -\int_0^y \frac{\partial u}{\partial x}\, dy \tag{4–109}$$

The integral is evaluated numerically using the *trapezoidal rule* and backward differences for $\partial u/\partial x$, formed with the values of u on the upstream and downstream faces of the element. It is interesting that the result of these operations is identical to Eq. (4–79), which was derived in a much different way.

Some results for the Howarth problem are compared with the analytical solution using nine terms in the series, as presented by Howarth in Fig. 4–22. The good agreement can be observed.

PROBLEMS

4.1. Consider the axial laminar flow between two parallel *porous* plates separated by a distance $2W$. There is a transverse velocity v_w into the channel through the top plate and out through the bottom plate at the same value. Develop an expression for the axial velocity profile $u(y)$ for fully developed flow.

4.2. Substitute Eqs. (4–15) and (4–16) and the trial form for η into Eq. (4–14), and determine A, B, and C to give Eq. (4–18).

4.3. Write and test a computer program for the *shooting* method for solving the Blasius problem described in Section 4–3–1.

4.4. For air at 1 atm and 25°C flowing over a flat plate at 5 m/s, calculate and plot the v velocity profile at a distance $x = 0.3$ m from the leading edge.

4.5. Show that $(\delta^*/\tau_w)dp/dx$ is proportional to the ratio of pressure force to wall friction force on the fluid in a boundary layer. Show that it is constant for any of the Falkner-Skan *wedge* flows.

4.6. Find an expression for the drag coefficient vs. Reynolds number of a flat plate of length L and width W in a uniform flow under laminar conditions.

4.7. Water is flowing over a flat plate at 3.0 m/s. At a distance of 10.0 cm from the leading edge, what is the distance above the plate to the point where $u = 1.5$ m/s? What is the value of v at that point? What is the thickness of the boundary layer and the displacement thickness?

4.8. Hydrogen at 400°K and 0.1 atm is flowing over a flat plate at 2.0 m/s. How large are the boundary layer and momentum thicknesses at 150.0 cm from the leading edge? What is the value of the skin friction coefficient?

4.9. Consider flow problems described by the approximate model equation

$$U_e \frac{\partial u}{\partial x} + V_w \frac{\partial u}{\partial y} = \nu \frac{\partial^2 u}{\partial y^2}$$

with boundary conditions $u(x, 0) = 0$, $u(0, y) = U_e$, $\lim_{y \to \infty} u = U_e$, $U_e =$ constant, and $V_w = f(x)$. Determine what variations $V_w(x)$ permit similar solutions.

4.10. Calculate $\beta(x)$ for the flow described in Problem 2.14. What does this calculation imply about the utility of similarity methods for this type of flow problem? Can you estimate the likelihood of separation?

4.11. Consider airflow over a flat plate for two cases: one with $T_w =$ constant and the other with $(T_w - T_e) \sim x^2$. Compare the thickness of the thermal boundary layers for the two cases at the same station x.

4.12. Consider ambient air flowing over a flat plate at 3.0 m/sec. At a station 1.0 m from the leading edge, what are the values of u and v at the edge of the velocity boundary layer? If the plate is heated to 325°K, what is the temperature at the same location?

4.13. Write and test a computer program for one axial step Δx, using the explicit method and the *model* equation, as described in Section 4–7–1. Take the initial profile, at $x = x_i$, to be the cubic polynomial in Eq. (2–46), with $U_e =$ constant $= 3$ m/s, $\delta(x_i) = 0.5$ cm, and the fluid to be air at 1 atm and 25°C. Use $\Delta y = \delta/20$ and $\Delta x = \delta/2$.

4.14. To test the importance of the stability criteria developed in Section 4–7–1, examine the behavior of an initial error represented by $u_{1,1} = 1.0$ and $u_{1,m} = 0$ for all $m \neq 1$.

Using Eq. (4–53a), make calculations for $n = 2, 3$ and $m = 1, 2, 3, 4, 5$, with $Q = 0.25$, 0.5, and 0.6. Take the flow to be symmetrical about the line $m = 1$.

4.15. Redo Problem 4.14, using the implicit formulation in Eq. (4–61).

4.16. Take a physical problem that can be described by the *model* equation, and use an implicit finite difference method to obtain the solution after one step in the x-direction. The initial profile is a linear variation from 0.0 at $y = 0$ ($M = 1$) up to a constant value of 1.0 beginning at the edge, which is located at the third grid point ($M = 3$). Use $Q = 1.0$ and solve *by hand.*

4.17. Perform a stability analysis of the implicit finite difference formulation for the *model* equation, Eq. (4–61), using the method used for the explicit formulation in Section 4–7–1.

4.18. Perform a stability analysis of the simple, explicit finite difference formulation for the *model* equation, Eq. (4–53a), using the Karplus method.

4.19. Derive a finite difference formulation of the *model* equation, Eq. (4–13), based on the Crank-Nicolson method.

4.20. Develop a finite difference formulation for the *model* equation, Eq. (4–13), using a central difference for $\partial u/\partial x$. This method could work after the first step away from the initial profile. Perform a stability analysis by a method of your choice.

4.21. Suppose we have water at 20°C flowing over a body with a shape such that $U_e(x) = 2.0(1.0 - x)$ ft/s. Determine the solution up to separation, using the explicit finite difference method. Compare your results to the exact solution of Howarth (1938).

4.22. Do Problem 4.21, using the implicit finite difference method.

4.23. Consider the flow problem treated in Example 4–2. Perform a *grid dependence study* by making calculations with a larger and smaller dx and dy and comparing the results with each other and those given in the example.

4.24. Consider flow problems described by the approximate model equation

$$U_e\frac{\partial u}{\partial x} + V_w\frac{\partial u}{\partial y} = \nu\frac{\partial^2 u}{\partial y^2}$$

with boundary conditions $u(x, 0) = 0$, $u(0, y) = U_e$, $\lim_{y\to\infty} u = U_e$, $U_e =$ constant, and $V_w =$ constant. We wish to analyze a case for a liquid ($\mu = 100\ \mu_{H2O}$) flow at 20°C at 0.5 m/sec over the surface with $V_w = -.005$ m/sec. The initial station has $\delta =$ 0.5 cm with a linear velocity profile. Use a Crank-Nicolson procedure to determine the velocity profile 2.0 cm further downstream. You may use a very crude Δy to save time.

4.25. Assume that the following equation is a suitable approximate equation for boundary layer problems with suction or injection [i.e., $V_w(x) \neq 0$]:

$$U_e\frac{\partial u}{\partial x} + V_w\frac{\partial u}{\partial y} = \nu\frac{\partial^2 u}{\partial y^2}$$

Develop an explicit finite difference approximation for the solution of this equation. Perform a stability analysis and state the results.

4.26. Test the accuracy of the explicit and implicit numerical procedures for solving boundary layer equations by computing a simple case for which an exact solution is available. Select a fluid and realistic values for the free stream velocity and the length of a flat plate (to give an appropriate value for Re_L), and make numerical calculations using both approaches. Compare the predictions for the u and v velocity profiles at the end of the plate with the Blasius solution.

4.27. Suppose we wish to analyze a flow problem as shown in the sketch for Problem 2.3 for air at STP and $C = 0.5$ m/s. Use an explicit numerical procedure to find $\theta(x)$ and $C_f(x)$ for $0 \leq x \leq 0.8L$. Compare your results with the predictions of the Thwaites-Walz method.

4.28. Suppose we wish to analyze a flow problem as shown in the sketch for Problem 2.3 for air at STP and $C = 0.5$ m/s. Use an implicit numerical procedure to find $\theta(x)$ and $C_f(x)$ for $0 \leq x \leq 0.8L$. Compare your results with the predictions of the Thwaites-Walz method.

4.29. For the flow described in Problem 2.14, calculate the variation of $C_f(x)$, $\delta(x)$, $\delta^*(x)$, and $\theta(x)$, assuming laminar flow. Use whatever numerical method you prefer.

4.30. Consider a flow similar to that in Example 4.2, except that the edge velocity between $x = 1.0$ and $x = 2.0$ decreases quadratically with an increasing slope, rather than linearly with a constant slope. If the edge velocity at $x = 2.0$ decreases to the same value as in the example and the slope at $x = 1.0$ is taken to be zero, how does the boundary layer at that station differ? Use the implicit numerical method.

CHAPTER

5

COMPRESSIBLE LAMINAR BOUNDARY LAYERS

5-1 INTRODUCTION

For a steady, compressible laminar flow over a planar or axisymmetric body, the equations to be treated are the steady forms of Eqs. (3–26), (3–27), and (3–50) or an energy equation in another form, such as in terms of h [and Eq. (3–60) if there is more than one species]. Looking at this system and limiting the discussion to a single species for the moment, one sees that the influence of compressibility is first contained directly in the density terms in the continuity equation, Eq. (3–27), and more passively as a variable coefficient in the momentum, Eq. (3–26), and energy, Eq. (3–50) equations. The second influence of compressibility is to produce temperature variations that are too large to permit the assumption of constant properties μ and k. Actually, the main concern here is with the viscosity, since we may use the definition of the Prandtl number to write

$$k = \frac{\mu c_p}{\mathrm{Pr}} \tag{5–1}$$

The advantage of this equation will become clearer if the reader recalls that the Prandtl number is nearly constant for most gases over a wide range of temperature (see Fig. 1–4 and the tables in Appendix A). Further, it is common to use the energy equation written in terms of the enthalpy, Eq. (3–49), in compressible flow problems. Then one writes

$$-\frac{\partial q_y}{\partial y} = \frac{\partial}{\partial y}\left(k\frac{\partial T}{\partial y}\right) = \frac{\partial}{\partial y}\left(\frac{k}{c_p}\frac{\partial h}{\partial y}\right) = \frac{\partial}{\partial y}\left(\frac{\mu}{\mathrm{Pr}}\frac{\partial h}{\partial y}\right) \tag{5-2}$$

In the simplest cases, therefore, one may say that the added complexity with compressible, laminar boundary layer problems is centered on variable ρ and μ and various, but essentially constant, values of Pr. We know that $\mu = \mu(T)$, and an equation of state will give, in general, $\rho = \rho(T, p)$. However, the pressure is constant across the layer, so we really have to contend only with density variations produced by temperature variations in the boundary layer for single-species gas flows.

5-2 THE ADIABATIC WALL TEMPERATURE

By Eq. (2-56), the wall temperature for adiabatic conditions (no heat transfer) would appear to be just $T_w = T_e$. That equation (2-56) is, however, suitable only to low-speeds cases. At high speeds, where compressibility becomes important, a new phenomenon occurs. By definition, a high-speed flow has appreciable kinetic energy, and that kinetic energy can be dissipated into heat by friction within the boundary layer. This process is represented by the term $\mu(\partial u/\partial y)^2$ in the energy equation [e.g., Eq. (3-50)]. The kinetic energy in the flow is simply the difference between the total (stagnation) and static temperatures:

$$T_{\mathrm{kinetic}} = T_t - T = \frac{V^2}{2c_p} \tag{5-3}$$

If the wall is adiabatic, the temperature that the wall attains at steady equilibrium will depend on how much of the kinetic energy is *recovered* on the wall. This quantity is expressed as a *recovery factor* r, defined as

$$T_{\mathrm{aw}} = T_e + r\frac{U_e^2}{2c_p} \tag{5-4}$$

The value of r is generally less than, but near, unity for gases. More will be said about r later. In any event, a film coefficient for high-speed flow must be based on

$$q_w = \hbar(T_w - T_{\mathrm{aw}}) \tag{5-5}$$

and not on Eq. (2-56).

5-3 THE REFERENCE TEMPERATURE METHOD

In view of the discussion in Section 5-1, it is reasonable to ask whether there might be some value of the temperature—say, between T_e and T_w—at which the density and physical properties of the fluid could be evaluated and used in the available constant-density, constant-property solutions to provide an adequate approximation to the actual, variable-density, variable-property flow. The answer, at least for gases, is yes, and that value of the temperature is called the *reference temperature* T^*. Here we quote the relation due to Eckert (1956):

$$T^* = T_e + 0.5(T_w - T_e) + 0.22(T_{aw} - T_e) \tag{5–6}$$

No equivalent relation with the same high degree of accuracy of results has been developed for liquids. Also, for lower speed flows that may still involve large temperature differences, the last term in Eq. (5–6) disappears.

With this concept, all the solutions in Chapters 2 and 4 can be carried over to compressible flow problems. For example, we may rewrite Eq. (2–60a) as

$$\text{St}^* = 0.332(\text{Pr}^*)^{-2/3}(\text{Re}^*)^{-1/2} \tag{5–7}$$

where the superscript * denotes that all physical properties and the density are to be evaluated at $T = T^*$. This formula can then be used to give the film coefficient $\hbar$ that can be used with Eq. (5–5) for high-speed flow over a flat plate.

5–4 THE SPECIAL CASE OF PRANDTL NUMBER UNITY

We saw in Section 4–4–1 that the case Pr = 1 was a special case giving identical nondimensional profiles for velocity and temperature in low-speed flow. The matter can be pursued further for high-speed flows. Although this is a special situation not generally found in the real world, many gases have Pr ≈ 0.7 (see Appendix A), which is not far from unity, and the results of studying the case Pr = 1 are quite informative.

For our purposes here, we take the energy equation in terms of the stagnation enthalpy $(h + u^2/2)$ as in Eq. (3–47), with the heat transfer term written as in Eq. (5–2) and the shear as $\mu(\partial u/\partial y)$. Also, we restrict the discussion to planar, steady flow. After some rearranging, we arrive at

$$\left(\rho u \frac{\partial}{\partial x} + \rho v \frac{\partial}{\partial y}\right)\left(h + \frac{u^2}{2}\right) = \frac{\partial}{\partial y}\left[\mu\left(\frac{\partial}{\partial y}\left(\frac{h}{\text{Pr}} + \frac{u^2}{2}\right)\right)\right] \tag{5–8}$$

Clearly, if Pr = 1, a solution to this equation is

$$h + \frac{u^2}{2} = \text{constant} \tag{5–9}$$

The constant can be evaluated from a boundary condition, say, at the wall, where $u = 0$. A suitable, simple case will be an insulated wall with no heat flow, and the result is

$$h + \frac{u^2}{2} = h_{aw} = \text{constant} = h_e + \frac{U_e^2}{2} \tag{5–10}$$

or

$$T + \frac{u^2}{2c_p} = T_{aw} = \text{constant} = T_e + \frac{U_e^2}{2c_p} \tag{5–10a}$$

for cases where c_p can be assumed constant. This result is known as the *Busemann* (1931) *energy integral.*

The foregoing analysis was generalized by Crocco (1932). We start here with the steady form of the energy equation in terms of the static enthalpy h as in Eq. (3–49), with the heat transfer term as in Eq. (5–2) and the laminar shear as usual, to give

$$\rho\left[u\frac{\partial h}{\partial x} + v\frac{\partial h}{\partial y}\right] = u\frac{dp}{dx} + \frac{\partial}{\partial y}\left(\frac{\mu}{\text{Pr}}\frac{\partial h}{\partial y}\right) + \mu\left(\frac{\partial u}{\partial y}\right)^2 \tag{5–11}$$

Next, we substitute the general assumption $h = h(u)$ into this equation to yield

$$\rho\frac{dh}{du}\left(u\frac{\partial u}{\partial x} + v\frac{\partial u}{\partial y}\right) = \frac{\partial}{\partial y}\left(\frac{\mu}{\text{Pr}}\frac{dh}{du}\frac{\partial u}{\partial y}\right) + u\frac{dp}{dx} + \mu\left(\frac{\partial u}{\partial y}\right)^2 \tag{5–11a}$$

where we have used

$$\begin{aligned}\frac{\partial h}{\partial x} &= \frac{dh}{du}\frac{\partial u}{\partial x}\\ \frac{\partial h}{\partial y} &= \frac{dh}{du}\frac{\partial u}{\partial y}\end{aligned} \tag{5–12}$$

The first term on the right-hand side of Eq. (5–11a) can be expanded with the product rule to yield

$$\frac{dh}{du}\frac{\partial}{\partial y}\left(\frac{\mu}{\text{Pr}}\frac{\partial u}{\partial y}\right) + \frac{\mu}{\text{Pr}}\frac{d^2h}{du^2}\left(\frac{\partial u}{\partial y}\right)^2 \tag{5–13}$$

The last term in Eq. (5–13) comes from continuing the logic shown in Eq. (5–12). Now using Eq. (5–13), we take Eq. (5–11a) and combine it with the steady form of the momentum equation, Eq. (3–24), multiplied by dh/du, which results in

$$-\frac{dp}{dx}\left[\frac{dh}{du} + u\right] + \frac{dh}{du}\left[\frac{\partial}{\partial y}\left(\mu\frac{\partial u}{\partial y}\right) - \frac{\partial}{\partial y}\left(\frac{\mu}{\text{Pr}}\frac{\partial u}{\partial y}\right)\right] = \frac{\mu}{\text{Pr}}\left(\frac{\partial u}{\partial y}\right)^2\left[\frac{d^2h}{du^2} + \text{Pr}\right] \tag{5–14}$$

There are two cases that admit to simple solutions by inspection. Both must have $\text{Pr} = 1$.

The first case arises for

$$\frac{dh}{du} + u = 0 \tag{5–15}$$

With this expression, Eq. (5–14) reduces to

$$\frac{d^2h}{du^2} + 1 = 0 \tag{5–16}$$

which, when integrated, results in

$$h = -\frac{u^2}{2} + \text{constant} \tag{5-9a}$$

equivalent to the earlier result, Eq. (5-9). By differentiating and evaluating at the wall, where $u = 0$, we obtain

$$\left.\frac{\partial h}{\partial y}\right|_w = -u_w \left.\frac{\partial u}{\partial y}\right|_w = 0 \tag{5-17}$$

This condition implies no heat transfer, because

$$\left.\frac{\partial h}{\partial y}\right|_w = c_p \left.\frac{\partial T}{\partial y}\right|_w = -\frac{c_p}{k} q_w \tag{5-18}$$

It is important to observe, however, that this solution holds for arbitrary dp/dx.

The second case emerges under the restriction of $dp/dx \equiv 0$, which again leads to Eq. (5-16) from Eq. (5-14), but without Eq. (5-15). Simple integration in this case yields

$$h = -\frac{u^2}{2} + C_1 u + C_2 \tag{5-19}$$

The constants, C_1 and C_2, can be found from the boundary conditions at the boundary layer edge and on the wall, i.e.,

$$\begin{aligned} h(U_e) &= h_e \\ h(0) &= h_w \end{aligned} \tag{5-20}$$

The final result is

$$h(u) = -\frac{u^2}{2} + \left(h_e + \frac{U_e^2}{2} - h_w\right)\frac{u}{U_e} + h_w \tag{5-21}$$

or, if c_p can be assumed constant,

$$T = -\frac{u^2}{2c_p} + \left(T_e + \frac{U_e^2}{2c_p} - T_w\right)\frac{u}{U_e} + T_w \tag{5-21a}$$

Differentiating Eq. (5-21) and evaluating at the wall gives

$$\begin{aligned} \left.\frac{\partial h}{\partial y}\right|_w &= -u_w \left.\frac{\partial u}{\partial y}\right|_w + \left(\frac{h_e + U_e^2/2 - h_w}{U_e}\right)\left.\frac{\partial u}{\partial y}\right|_w \\ &= \left(\frac{h_e + U_e^2/2 - h_w}{U_e}\right)\left.\frac{\partial u}{\partial y}\right|_w \end{aligned} \tag{5-22}$$

which is not zero. This case thus allows heat transfer into and out of the wall under the restriction of a zero pressure gradient.

The relation in Eq. (5–21a) can be used to show an important point about very high-speed boundary layer flows. Consider the case of flow over a constant-pressure region of a body moving at $M_e = 8$ in the atmosphere at sea level, with $T_e = 300°\text{K}$. Using isentropic relations, we find that $T_e/T_{t,e} = 0.0725$. Thus, $T_{t,e} = 4{,}140°\text{K}$, if it is assumed for the purposes of this illustrative example only that air behaves as a perfect gas under these conditions. The maximum tolerable surface temperature for most practical applications is about 1,000°K. We now seek the maximum value of the static temperature in the boundary layer. The working of Problem 5.3 at the end of the chapter will show that the maximum static temperature occurs where

$$\left.\frac{u}{U_e}\right|_{T=T_{\max}} = \frac{T_{t,e} - T_w}{U_e^2/c_p} \tag{5–23}$$

For the hypothetical case under discussion, we find that $T_{\max} = 1{,}640°\text{K}$, which is much higher than either T_e or T_w. That is, there is a static *temperature peak* in the boundary layer for flows with a high Mach number. This information is very important for reentry vehicles, since it leads to dissociation and ionization of the air in the boundary layer, producing the communications *blackout* familiar to all avid observers of the space program.

Finally, we note from Eq. (5–10a) that the adiabatic wall temperature T_{aw} for $\text{Pr} = 1$ equals $T_{t,e}$. Consequently, for $\text{Pr} = 1$, the recovery factor r is also unity, as can be seen by comparing Eq. (5–10a) with Eq. (5–4).

5–5 THE RECOVERY FACTOR FOR NONUNITY PRANDTL NUMBER

A useful, rather general result can be found by analyzing a simple, seemingly restrictive problem—the *plate thermometer*. This is the case of an adiabatic, flat plate suspended in a high-speed fluid stream. The temperature measured by a thermometer (e.g., a thermocouple) embedded in the plate will be, by definition, the adiabatic wall temperature T_{aw}. Knowing this, Pr, and $T_{t,e}$, we can find $r = r(\text{Pr})$.

Pohlhausen (1921a) treated this problem under the constant-property assumption, but including, of course, the frictional heating term, $\mu(\partial u/\partial y)^2$, in the energy equation, or there would be no important recovery process. We use the steady form of Eq. (3–51) with $dp/dx = 0$, written as

$$u\frac{\partial T}{\partial x} + v\frac{\partial T}{\partial y} = \frac{\nu}{\text{Pr}}\frac{\partial^2 T}{\partial y^2} + \frac{\nu}{c_p}\left(\frac{\partial u}{\partial y}\right)^2 \tag{5–24}$$

The boundary conditions are

$$\begin{aligned} y = 0,\ x \geq 0&: \quad \frac{\partial T}{\partial y} = 0 \\ y \to \infty,\ \text{all } x&: \quad T(x, y) \to T_e \end{aligned} \tag{5–25}$$

It is helpful to introduce a dimensionless *excess temperature*

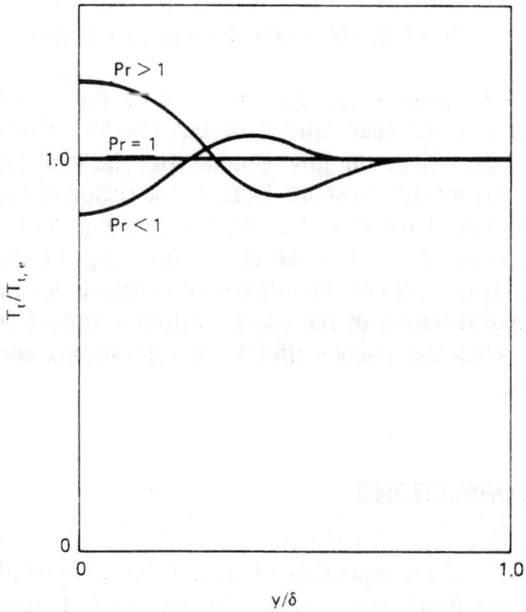

Figure 5–1 Sketches of total temperature profiles in a high-speed boundary layer for Prandtl number greater than, equal to, and less than unity.

$$\Theta_r \equiv \frac{T - T_e}{U_e^2/(2c_p)} \tag{5–26}$$

The similarity variable η and the representation of the velocity field in terms of $f(\eta)$ from Sections 4–3–1 and 4–4–1 can also be used here because of the constant-property assumption. With both of these, and using Eq. (5–26), we see that Eq. (5–24) becomes

$$\Theta_r'' + \text{Pr} \cdot f\Theta_r' + \frac{\text{Pr}}{2}(f'')^2 = 0 \tag{5–27}$$

The boundary conditions are

$$\begin{aligned} \Theta_r'(0) &= 0 \\ \lim_{\eta\to\infty} \Theta_r(\eta) &= 0 \end{aligned} \tag{5–28}$$

The solution of Eq. (5–27) subject to Eq. (5–28) can be found directly by the method of *variation of parameters*. Our primary interest here is in the value of the temperature at the wall, from which the recovery factor may be found. The result is

$$\begin{aligned} r &= \frac{\text{Pr}}{2}\int_0^\infty \exp\left(-\text{Pr}\int_0^\eta f\,d\eta\right)\left[\int_0^\eta (f'')^2 \exp\left(\text{Pr}\int_0^\eta f\,d\eta\right) d\eta\right] d\eta \\ &\approx \sqrt{\text{Pr}} \quad \textit{for } 0.5 \le \text{Pr} \le 5.0 \end{aligned} \tag{5-29}$$

The key result, $r = \sqrt{\text{Pr}}$, for laminar flow has been found to hold under much more general conditions than those assumed for this simple flow problem. Indeed, it is reputed to have been a classified secret on both sides during World War II.

Sketches of typical total temperature profiles across a boundary layer on an adiabatic wall for $\text{Pr} > 1$, $\text{Pr} = 1$, and $\text{Pr} < 1$ are shown in Fig. 5–1. In Section

5–4, it was seen that T_t = constant = $T_{t,e}$ for Pr = 1 [see Eq. (5–10a)]. For Pr < 1, the recovery factor r is less than unity [see Eq. (5–29)]. Consequently, $T_{aw} = T(x, 0) = T_t(x, 0) < T_{t,e}$. Since the flow is adiabatic, the total thermal energy in the boundary layer must remain constant. Thus, if one region of the flow has $T_t < T_{t,e}$, some other region must have $T_t > T_{t,e}$, as shown in Fig. 5–1. This picture is reversed for Pr > 1, where $T_{aw} > T_{t,e}$. All of the foregoing is a result of the fact that the Prandtl number is the ratio of the diffusion coefficients for momentum and thermal energy. This redistribution of the energy within a viscous region can produce remarkable results when the viscous effects are very strong, such as in a vortex or a separation bubble.

5–6 COMPRESSIBILITY TRANSFORMATIONS

With the increased complexity of the equations of motion for compressible (variable-density), variable-property flows, it was natural to seek ways of rigorously extending the material at hand for constant-density, constant-property flows to those cases. Ways were sought to transform a compressible boundary layer problem into an equivalent incompressible problem. The existing solutions could then be transformed back to give a solution for the original compressible problem. This quest ended in success, with some restrictions, and we will discuss two early examples and then the latest of the various transformations.

5–6–1 Howarth-Dorodnitzin Transformation

The first *compressibility transformation* seems to have been developed independently by Howarth (1948) and Dorodnitsyn (1942). The basic idea is to introduce a distortion of the transverse coordinate y using the density as a *weighting factor:*

$$Y \equiv \int_0^y \frac{\rho}{\rho_e}\, dy' \tag{5–30}$$

The streamwise coordinate x is left undistorted. The velocity components are now written in terms of a compressible stream function [see Eq. (3–10)] as

$$u = \frac{\partial \psi}{\partial Y}; \qquad v = -\frac{\rho_e}{\rho}\left[\frac{\partial \psi}{\partial x} + u\left(\frac{\partial Y}{\partial x}\right)_y\right] \tag{5–31}$$

Substitution into the steady form of the momentum equation [Eq. (3–24)] for flow over a flat plate, $dp/dx = 0$, results in

$$\frac{\partial \psi}{\partial Y}\frac{\partial^2 \psi}{\partial x\, \partial Y} - \frac{\partial \psi}{\partial x}\frac{\partial^2 \psi}{\partial Y^2} = \frac{\partial}{\partial Y}\left(\frac{\rho\mu}{\rho_e^2}\frac{\partial^2 \psi}{\partial Y^2}\right) \tag{5–32}$$

This equation in the (x, Y)-plane is very similar to that for constant-density, constant-property flow in the (x, y)-plane [Eq. (4–14)]. The equivalence becomes complete if one takes $\rho\mu$ = constant = $\rho_e\mu_e$. Since, for a given pressure, $\rho \propto 1/T$, it follows that $\mu \propto T$, which is a stronger variation with temperature than is actually observed for air and other common gases (see, for example, Eq. (1–12) and Fig. 1–4). Nonetheless, this approach does provide a method for treating compressible boundary layer flows. One can, for example, take the Blasius solution (Section 4–3–1) to be applicable in the (x, Y)-plane. The inversion of the transformation requires evaluating

$$y = \int_0^Y \frac{\rho_e}{\rho}\, dY' \tag{5–33}$$

which, in turn, requires knowledge of $\rho(x, Y)$. In general, this information must come from a solution of the energy equation in the (x, Y)-plane to obtain $T(x, Y)$. It is not necessary, but one can also make the additional assumptions that permit the use of a Crocco integral, Eq. (5–10a), for an adiabatic wall, or Eq. (5–21a), both for Pr = 1. This procedure gives $T(x, Y)$ directly in terms of the known $u(x, Y)$. The results for adiabatic wall cases are shown in Fig. 5–2. Note that the attendant high temperature and thus low density near the wall (small y) stretch out the inner, nearly linear portion of the Blasius profile at high Mach numbers. Since the velocity at the wall must be zero, there is clearly some region near the wall where the flow is subsonic, even though the edge Mach number outside the boundary layer is supersonic. For an adiabatic case, this region can be large, since the temperature and thus the speed of sound near the wall are high. For example, for the case of $M_e = 4$ in Fig. 5–2, the flow is subsonic up the point where $u/U_e \approx 0.4$, which is at $y/\delta \approx 0.4$. This matter is important for shock–boundary layer interactions, discussed later in the chapter.

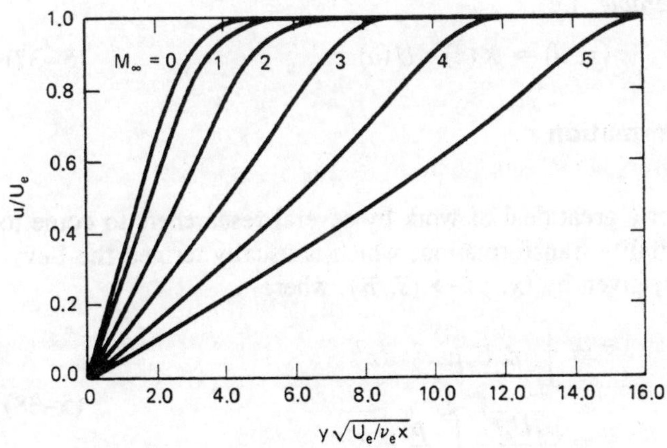

Figure 5–2 Velocity profiles on an insulated flat plate for laminar flow with Prandtl number unity and viscosity proportional to temperature. (From Crocco, 1946.)

5-6-2 Crocco Transformation

Another important compressibility transformation was developed by Crocco (1946). Here, the viscous stress $\tau = \mu \partial u/\partial y$ and the static enthalpy h are taken to be the dependent variables, and x and u are taken to be the independent variables. A primary motivation for this apparently strange change of variables comes from the fact that the usual momentum equation is rendered nonlinear by the term $u\partial u/\partial x$. Thus, making u an independent variable removes the source of the nonlinearity.

The steady form of the continuity equation, Eq. (3-6), the steady form of the momentum equation, Eq. (3-24), and the steady form of the energy equation written in terms of the static enthalpy, Eq. (5-11), transform to

$$u\frac{\partial}{\partial x}\left(\frac{\rho\mu}{\tau}\right) + \frac{\partial^2 \tau}{\partial u^2} - \frac{\partial p}{\partial x}\frac{\partial}{\partial u}\left(\frac{\mu}{\tau}\right) = 0 \tag{5-34}$$

and

$$(1 - \Pr)\frac{\partial h}{\partial u}\tau\frac{\partial \tau}{\partial u} + \tau^2\left(\frac{\partial^2 h}{\partial u^2} + \Pr\right) - \Pr(\rho\mu)u\frac{\partial h}{\partial x} + \mu \Pr\frac{\partial p}{\partial x}\left(\frac{\partial h}{\partial u} + u\right) = 0 \tag{5-35}$$

After a solution is obtained, one can invert the transformation to find

$$y(x, u) = \int_0^u \frac{\mu(h)\, du'}{\tau(x, u')} \tag{5-36}$$

Again, the appeal of the somewhat unrealistic assumption $\rho\mu = C$ is apparent in Eqs. (5-34) and (5-35): Without that assumption, this transformation played a very important role in early numerical solutions, as described in a later section.

For the case $\partial p/\partial x = 0$, the momentum equation is very much simplified. Indeed, the variables are *separable*, i.e.,

$$\tau(x, u) = X(x) \cdot U(u) \tag{5-37}$$

5-6-3 Levy-Less Transformation

At this point, we jump over a great deal of work by several researchers to come to the most general compressibility transformation, which is usually termed the Levy-Lees (1956) transformation, given by $(x, y) \to (\overline{s}, \overline{\eta})$, where

$$\begin{aligned} \overline{s} &\equiv \int_0^x \rho_e U_e \mu_e r_0^{2j}\, dx' \\ \overline{\eta} &\equiv \frac{\rho_e U_e r_0^j}{\sqrt{2\overline{s}}} \int_0^y \frac{\rho}{\rho_e}\, dy' \end{aligned} \tag{5-38}$$

Here, r_0 is the local body radius (see Fig. 3–2). This transformation clearly contains elements of the Howarth-Dorodnitsyn transformation, the Blasius similarity variable, and the Mangler transformation. Thus, it will be useful for compressible, similar flows. It is also useful for compressible, nonsimilar flows that are treated by numerical methods, since it keeps the growth of the computational region in the transverse direction under control (see Section 4–7–4). Indeed, in recent times, the latter utility is by far the greatest advantage of the Levy-Lees transformation. In numerical solutions one does not make the assumption $\rho\mu$ = constant.

The results of the Levy-Lees transformation on the two main operators in the boundary layer equations are

$$\rho u \frac{\partial(\cdot)}{\partial x} + \rho v \frac{\partial(\cdot)}{\partial y} = \rho U_e^2 \rho_e \mu_e r_0^{2j}\left(f' \frac{\partial(\cdot)}{\partial \bar{s}} - f' \frac{\partial \bar{\eta}}{\partial \bar{s}} \frac{\partial(\cdot)}{\partial \bar{\eta}} - \frac{f}{2\bar{s}} \frac{\partial(\cdot)}{\partial \bar{\eta}}\right)$$

and (5–39)

$$\frac{\partial}{\partial y}\left([\cdot]\frac{\partial(\cdot)}{\partial y}\right) = \frac{\rho U_e^2 r_0^{2j}}{2\bar{s}} \frac{\partial}{\partial \bar{\eta}}\left(\rho[\cdot]\frac{\partial(\cdot)}{\partial \bar{\eta}}\right)$$

where $u(\bar{s}, \bar{\eta}) = U_e(\bar{s})f'(\bar{s}, \bar{\eta})$.

5-7 INTEGRAL METHOD FOR COMPRESSIBLE FLOW

The Thwaites-Walz integral method (see Section 2–3–2) has been extended to compressible flows by Cohen and Reshotko (1956a) under restrictions that permit the use of the compressibility transformations to obtain a direct correspondence with incompressible cases. Cohen and Reshotko assumed $\rho\mu$ = constant, Pr = 1, and treated both an adiabatic wall and cases with heat transfer. These are restrictive assumptions, but the resulting method provides such a simple, powerful tool for treating high-speed laminar cases that it should be a part of every analyst's arsenal.

The inviscid solution over the body of interest must provide $U_e(x)$ and thus $p_e(x)$ which can also be used to obtain $M_e(x)$ and $T_e(x)$. The result corresponding to Eq. (2–30) is

$$\theta_i^2(X) = \frac{A\nu_0}{U_{e,i}^B(X)} \int_0^X U_{e,i}^{B-1}(X')\, dX' \tag{5–40}$$

with $A \approx 0.45$ and B depending on the heat transfer rate through $S_w = T_w/T_{t,e} - 1$. For no heat transfer ($T_w = T_{aw} = T_{t,e}$ with Pr = 1.0), $S_w = 0$ and $B \approx 6$. For other values of S_w, they provide a curve for B. In this equation, the quantities with a subscript correspond to the transformed, incompressible variables. Those and the transformed x coordinate are defined as

$$X = \int_0^x \frac{a_e}{a_t}\frac{p_e}{p_t}\, dx'; \qquad U_{e,i} = \frac{a_t}{a_e} U_e; \qquad \theta = \frac{p_t}{p_e}\frac{a_e}{a_t}\theta_i \tag{5–41}$$

where, the variable-density form of the momentum (and displacement) thickness is implied, i.e.,

$$\theta \equiv \int_0^\infty \left(1 - \frac{u}{u_e}\right)\frac{\rho u}{\rho_e u_e}\,dy; \qquad \delta^* \equiv \int_0^\infty \left(1 - \frac{\rho u}{\rho_e u_e}\right)dy \tag{5-42}$$

Also, ν_0 denotes the kinematic viscosity evaluated at $T_{t,e}$ and $p_{t,e}$. After finding θ_i, one must find Λ_i in order to proceed to solve for δ^* and τ_w as before. Now

$$\Lambda_i(X) = \frac{\theta_i^2(X)}{\nu_o}\frac{dU_{e,i}}{dX} \tag{5-43}$$

The same tabular values for $H(\Lambda_i)$ and $S(\Lambda_i)$ as for the low-speed case (Table 2-1) can be used here also for the adiabatic wall case. For heat transfer, they give tabular values as a function of S_w. The result is

$$\frac{\delta^*}{\theta} = H(\Lambda_i) + \frac{\gamma - 1}{2}M_e^2(H(\Gamma_i) + 1) \tag{5-44}$$

and

$$\tau_w = \frac{S(\Lambda_i)\mu_w U_e}{\theta}\left(\frac{T_e}{T_w}\right) \tag{5-45}$$

This method is easy to apply and yields reasonable accuracy, so it is useful for rapid estimating purposes.

5–8 EXACT SOLUTIONS FOR COMPRESSIBLE FLOW OVER A FLAT PLATE

The first exact solutions for high-speed airflow over a flat plate with a realistic viscosity law and Pr $\neq$ 1 were obtained by Crocco (1946). He used the transformation of his own devising (see Section 5–6–2) and reduced the problem to the simultaneous solution of two ordinary differential equations. He made other simplifying assumptions and then solved the problem by the method of successive approximations without the aid of a modern computer. With all of this, however, the solutions were essentially exact, and their attainment was a landmark event at the time.

Very extensive calculations for flow over a flat plate using the Crocco method were presented by Van Driest (1952). He used the Sutherland viscosity law and Pr = 0.75, and both adiabatic wall and heat transfer (with various T_w/T_e) cases were studied over a wide range of Mach numbers, $0 \leq M_e \leq 20$. Actually, the results at the higher Mach numbers cannot be viewed as exact, since they were obtained assuming a perfect gas. This assumption is untenable for practical cases of high M_e,—certainly for $M_e > 6$—since dissociation becomes important. Still, Van Driest's results, as a whole, are worthy of careful study, since they cover such wide ranges of the important parameters. In Fig. 5–3, one can see the variation of the average skin friction coefficient with Mach number. Note the strong effect of T_w/T_e at low Mach

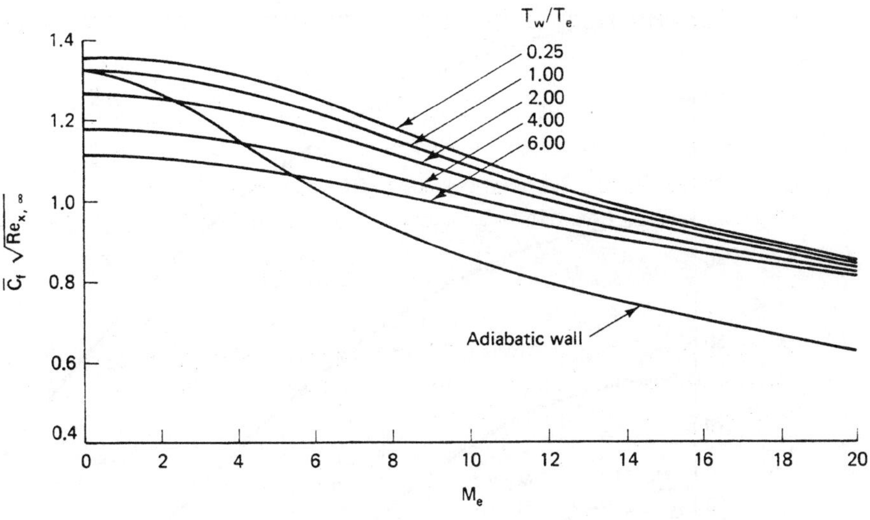

Figure 5–3 Mean skin friction coefficient for laminar flow over a flat plate with Pr = 0.75 and the Sutherland viscosity law. (From Van Driest, 1952.)

number and the important difference between adiabatic and heat transfer cases. Velocity profiles for adiabatic cases are shown in Fig. 5–4. They are generally similar to those given in Fig. 5–2, where Pr = 1. Again, there is an extended linear portion of the profile near the wall, and the dimensionless boundary layer thickness grows rapidly with edge Mach number. Temperature profiles for the same adiabatic cases are plotted in Fig. 5–5.

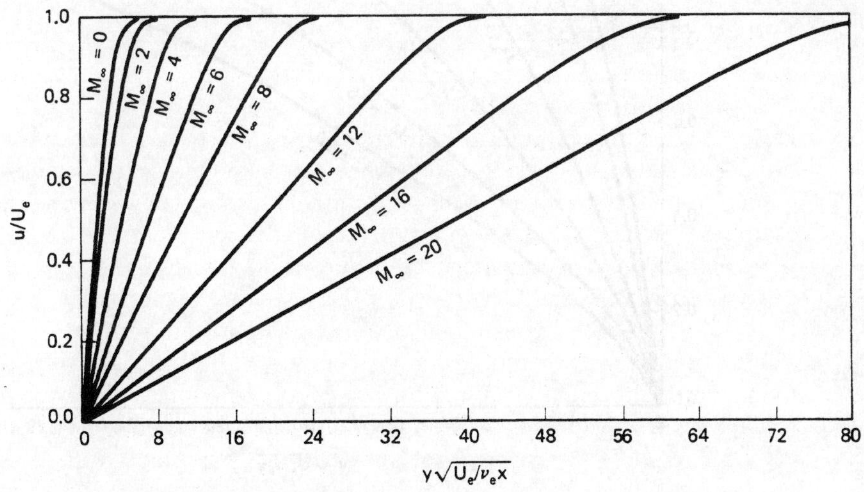

Figure 5–4 Velocity profiles on an insulated flat plate for laminar flow with Pr = 0.75 and the Sutherland viscosity law. (From Van Driest, 1952.)

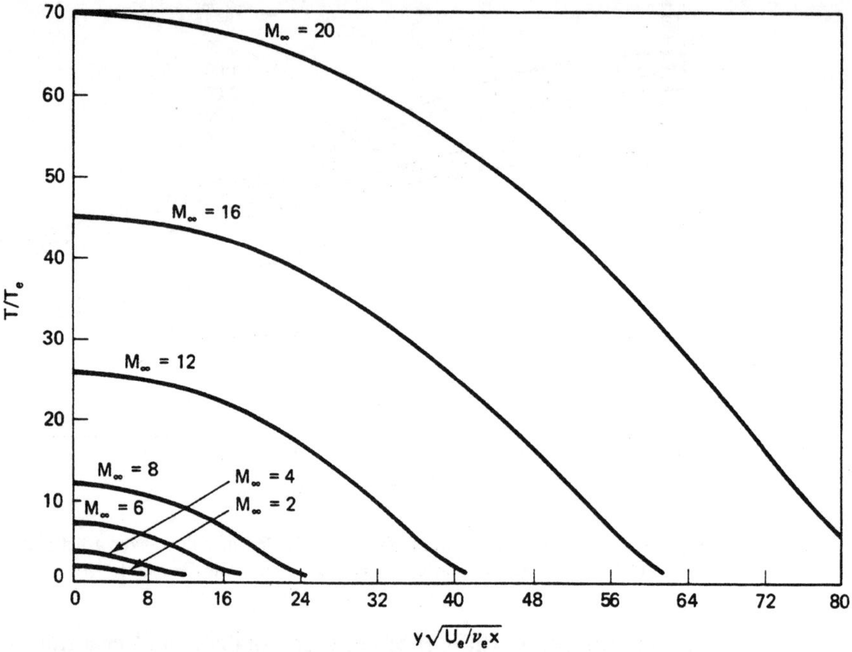

Figure 5–5 Temperature profiles on an insulated flat plate for laminar flow with Pr = 0.75 and the Sutherland viscosity law. (From Van Driest, 1952.)

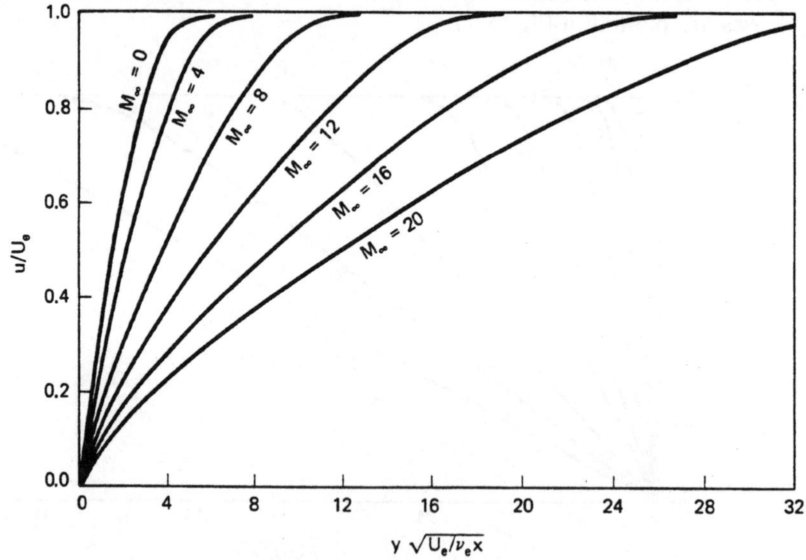

Figure 5–6 Velocity profiles on a flat plate for laminar flow with $T_w/T_e = 1.0$, Pr = 0.75, and the Sutherland viscosity law. (From Van Driest, 1952.)

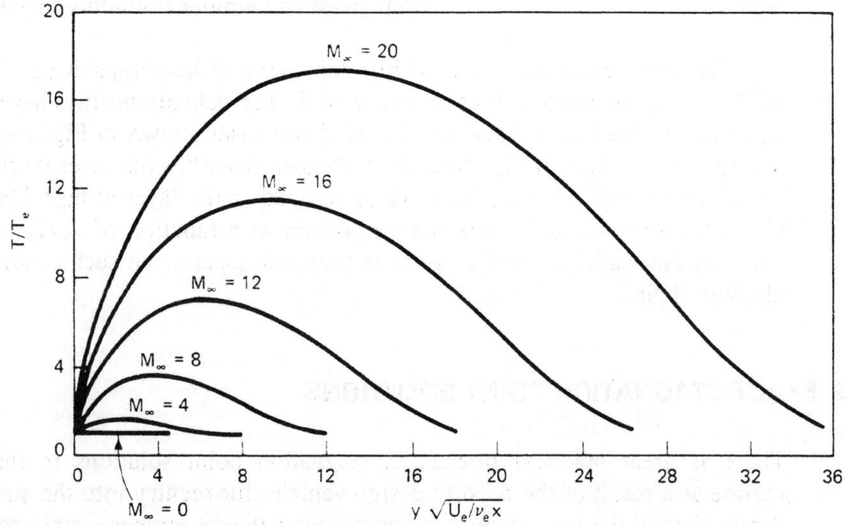

Figure 5–7 Temperature profiles on a flat plate for laminar flow with T_w/T_e = 1.0, Pr = 0.75, and the Sutherland viscosity law. (From Van Driest, 1952.)

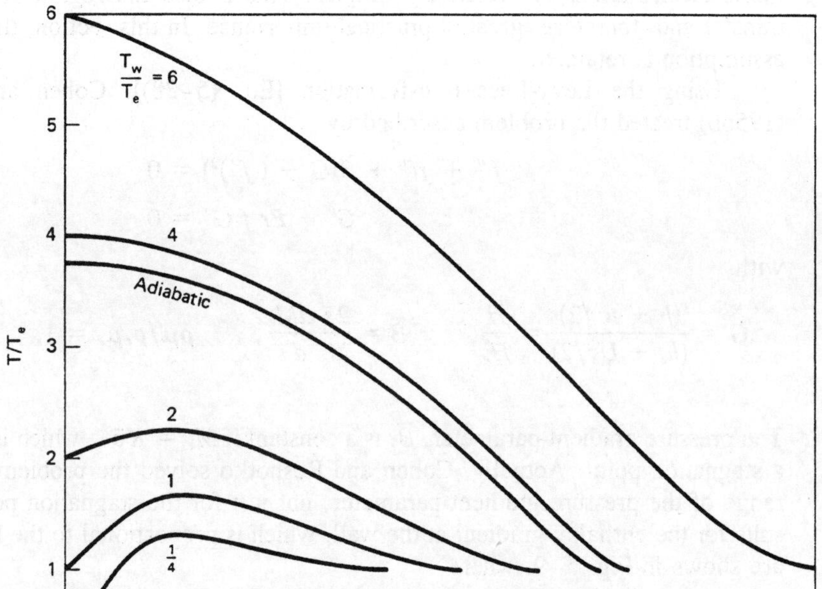

Figure 5–8 Temperature profiles on a flat plate for laminar flow at M_e = 4 with various T_w/T_e, Pr = 0.75, and the Sutherland viscosity law. (From Van Driest, 1952.)

Velocity and temperature profiles for a case of heat transfer (to the wall) with $T_w/T_e = 1.0$ are given in Figs. 5–6 and 5–7. The velocity profiles have more curvature near the wall than those for the adiabatic cases shown in Fig. 5–4. The static temperature profiles in Fig. 5–7 show the peaks within the layer discussed earlier. Clearly, very high temperatures can be reached in the layer at high Mach numbers. Figure 5–8 shows static temperature profiles as a function of T_w/T_e for $M_e = 4.0$. For very hot walls, $T_w \geq T_{aw}$, and the peaks disappear, but such conditions are unlikely in flight.

5–9 EXACT STAGNATION POINT SOLUTIONS

There is great practical interest in stagnation point solutions in the high-speed regime as a result of the need to design vehicles for reentry into the atmosphere that can withstand the very high heat transfer rates that accompany rapid deceleration of blunt bodies by drag. It is fortunate that the flow at the stagnation point and in its immediate vicinity can be treated as a *similar* solution, so the mathematics and numerical calculations are relatively simpler. This is also the region of highest heat transfer and, therefore, greatest practical importance. In this section, the perfect gas assumption is retained.

Using the Levy-Lees transformation [Eq. (5–38)], Cohen and Reshotko (1956b) treated the problem described by

$$f''' + ff'' + \bar{\beta}(G - (f')^2) = 0 \tag{5–46}$$

$$G'' + Pr\, f\, G' = 0 \tag{5–47}$$

with

$$G = \frac{(h + u^2/2)}{(h_e + U_e^2/2)} = \frac{H}{H_e}, \qquad \bar{\beta} = \frac{2\bar{s}}{M_e}\frac{dM_e}{d\bar{s}}, \qquad \rho\mu/\rho_e\mu_e = 1, \qquad \mathrm{Pr} = 1 \tag{5–48}$$

The pressure gradient parameter, $\bar{\beta}$, is a constant if $M_e = K\bar{s}^m$, which is the case for a stagnation point. Actually, Cohen and Reshotko solved the problem for a whole range of the pressure gradient parameter, not just for the stagnation point. The results for the enthalpy gradient at the wall, which is proportional to the heat transfer, are shown in Fig. 5–9, where

$$q_w = \frac{1}{\mathrm{Pr}_w}\frac{U_e\rho_w\mu_w r_0^j}{\sqrt{2\bar{s}}}(h_w - (h_e + U_e^2/2))(g')_w \tag{5–49}$$

and $g = G/(1 - G_w)$. The dependence of the heat transfer on the wall on the free stream stagnation enthalpy ratio is much stronger than on the pressure gradient parameter, especially for strongly cooled walls.

Eckert and Tewfik (1960) took a mean value for $(g')_w$, introduced a dependence on Pr_w to agree with low-speed solutions, replaced the free stream stagnation enthalpy with the recovery enthalpy, and suggested evaluating the properties at the

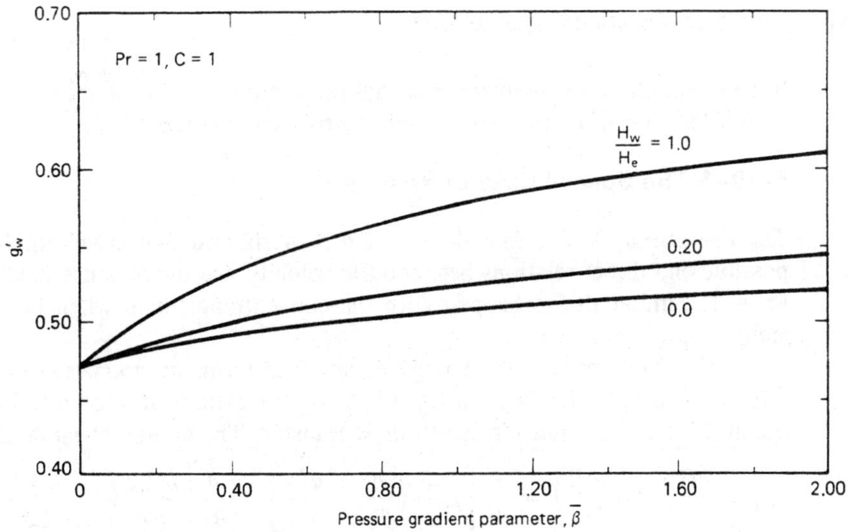

Figure 5–9 Heat transfer parameter as a function of the pressure gradient parameter for high-speed laminar flow. (From Cohen and Reshotko, 1956b.)

reference temperature T^*, resulting in

$$q_w = \frac{0.35}{\mathrm{Pr}^{*2/3}} \frac{U_e \rho^* \mu^* r_0^j}{\sqrt{\bar{s}^*}} (h_w - h_{aw}) \tag{5–50}$$

Introducing a film coefficient based on enthalpy, $\hbar_i$, leads to a Stanton number relation

$$\mathrm{St}^* = \frac{0.35 \mu^* r_0^j}{\sqrt{\bar{s}^*}} \tag{5–51}$$

For a two-dimensional stagnation point, this equation becomes

$$\mathrm{St}^* = \frac{0.5}{\mathrm{Pr}^{*2/3} \sqrt{\rho^*(C_o x) x / \mu^*}} = \frac{0.5}{\mathrm{Pr}^{*2/3}\, \mathrm{Re}_x^{*1/2}} \tag{5–52}$$

For an axisymmetric stagnation point, the expression has the same form, but the constant becomes 0.7. To apply Eq. (5–52) to an actual body, one must relate the C_o in $U_e = C_o x$ to the body nose radius. This can be done in a simple fashion using *Newtonian impact theory,* which gives the following very good approximation for blunt, axisymmetric bodies:

$$C_o \approx \frac{U_\infty}{R_0} \sqrt{2 \frac{\rho_\infty}{\rho_e}} \tag{5–53}$$

The resulting predictions for heat transfer at the nose are in good agreement with experiment. Note that the heat transfer is a strong function of the nose radius of the body, $q_w \propto 1/\sqrt{R_0}$. A large nose radius, such as on most reentry bodies, thus reduces the heat transfer.

5–10 FLOWS WITH MASS TRANSFER

In this section, we extend the material in Section 4–5 to the more general case of two fluids with different and variable thermophysical properties.

5–10–1 The Special Case of Pr = Le = 1

The discussions in Sections 4–4–1 and 5–4 showed that great simplifications are possible and direct relations between the velocity and temperature fields result when Pr = 1. Similar things happen with the concentration field when Le = Sc = 1 as well.

With two species, the energy equation in terms of stagnation enthalpy, $H = h + u^2/2$, is extended beyond Eq. (5–8) by the extra term shown in Eq. (3–63) to account for energy transfer due to mass transfer. The energy equation becomes

$$\left(\rho u \frac{\partial H}{\partial x} + \rho v \frac{\partial H}{\partial y}\right) = \frac{\partial}{\partial y}\left[\mu \frac{\partial}{\partial y}\left(\frac{h}{\text{Pr}} + \frac{u^2}{2}\right)\right] + \frac{\partial}{\partial y}\left[\frac{\mu\,\text{Le}}{\text{Pr}}\left(1 - \frac{1}{\text{Le}}\right)\sum_i h_i \frac{\partial c_i}{\partial y}\right] \tag{5–54}$$

The species equation is

$$\rho u \frac{\partial c_i}{\partial x} + \rho v \frac{\partial c_i}{\partial y} = \frac{\partial}{\partial y}\left(\frac{\mu\,\text{Le}}{\text{Pr}} \frac{\partial c_i}{\partial y}\right) \tag{5–55}$$

and the momentum equation remains as the steady form of Eq. (3–24).

Following the logic in Section 5–4 for deriving $h(u)$, we postulate here $c_i = c_i(u)$. Then the diffusive term on the right-hand side of Eq. (5–55) becomes

$$\begin{aligned}\frac{\partial}{\partial y}\left(\frac{\mu\,\text{Le}}{\text{Pr}} \frac{\partial c_i}{\partial y}\right) &= \frac{\partial}{\partial u}\left(\frac{\mu\,\text{Le}}{\text{Pr}} \frac{d c_i}{du} \frac{\partial u}{\partial y}\right)\frac{\partial u}{\partial y} \\ &= \frac{\mu\,\text{Le}}{\text{Pr}}\left(\frac{\partial u}{\partial y}\right)^2 \frac{d^2 c_i}{du^2} + \frac{\partial}{\partial y}\left(\frac{\mu\,\text{Le}}{\text{Pr}} \frac{\partial u}{\partial y}\right)\frac{d c_i}{du}\end{aligned} \tag{5–56}$$

The left-hand side of Eq. (5–55) becomes

$$\frac{d c_i}{du}\left(\rho u \frac{\partial u}{\partial x} + \rho v \frac{\partial u}{\partial y}\right) \tag{5–57}$$

Substituting for the term in parentheses in Eq. (5–57) from the momentum equation and using Eq. (5–56), we obtain, for Eq. (5–55) as a whole,

$$\frac{d c_i}{du}\left(\frac{\partial}{\partial y}\left(\mu \frac{\partial u}{\partial y}\right) - \frac{dp}{dx}\right) = \frac{\mu\,\text{Le}}{\text{Pr}}\left(\frac{\partial u}{\partial y}\right)^2 \frac{d^2 c_i}{du^2} + \frac{\partial}{\partial y}\left(\frac{\mu\,\text{Le}}{\text{Pr}} \frac{\partial u}{\partial y}\right)\frac{d c_i}{du} \tag{5–58}$$

For Pr = Le = 1 and $dp/dx \equiv 0$, this equation reduces to $d^2 c_i/du^2 = 0$, which admits to the solution

$$c_i = D_1 u + D_2 \tag{5–59}$$

where the constants D_1 and D_2 are to be determined from the boundary conditions at the wall and the boundary layer edge. Thus,

$$c_i = (c_{ie} - c_{iw})\frac{u}{U_e} + c_{iw} \tag{5–60}$$

Taking the derivative at the wall gives

$$\left.\frac{\partial c_i}{\partial y}\right|_w = \frac{(c_{ie} - c_{iw})}{U_e}\left.\frac{\partial u}{\partial y}\right|_w \tag{5–61}$$

which shows that one must have mass transfer at the wall for nonzero skin friction ($(\partial u/\partial y)_w \neq 0$), as in Fig. 5–10(C); that is, an impermeable wall will not admit this solution.

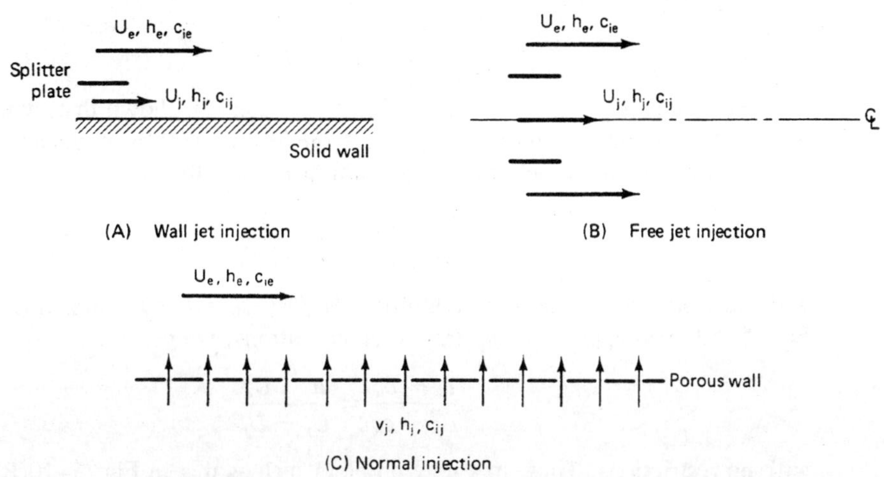

Figure 5–10 Schematics of some flows with injection of a foreign fluid.

It is also possible to seek solutions of the form $c_i = c_i(H)$ using the species equation and the energy equation. The result for Pr = Le = 1 is

$$c_i = D_3 H + D_4 \tag{5–62}$$

Evaluating the derivative at the wall, we obtain

$$\left.\frac{\partial c_i}{\partial y}\right|_w = D_3\left.\frac{\partial H}{\partial y}\right|_w = D_3\left.\frac{\partial h}{\partial y}\right|_w \tag{5–63}$$

so an impermeable wall must also be insulated—i.e., no mass transfer implies no heat transfer.

Looking back at Eq. (5–54), one can see that for Le = 1, the energy equation with mass transfer takes the same form as that without mass transfer, that is, Eq. (5–8). Thus, the relations for $h(u)$ developed in Section 5–4 hold in such cases also.

The utility of the various generalized Crocco integrals can be illustrated by considering two sample mixing flows, shown in Figs. 5–10(A) and (B). For the case in Fig. 5–10(A), with an impermeable wall, we cannot use $c_i(u)$, but we can use

$c_i(H)$ for an impermeable wall, as long as the wall is also insulated. Applying the initial conditions in the wall jet and the external flow to Eq. (5–62) gives

$$\frac{c_i - c_{ie}}{c_{ij} - c_{ie}} = \frac{H - H_e}{H_j - H_e} \tag{5–64}$$

On the other hand, we could choose to use the relation $h(u)$ in Eq. (5–19) if $dp/dx = 0$ and the wall is not insulated. Then, applying the conditions in the wall jet and the external flow results in

$$H = (H_e - h_w)\frac{u}{U_e} + h_w \tag{5–65}$$

with the restriction that

$$\frac{H_e - h_w}{H_j - h_w} = \frac{U_e}{U_j} \tag{5–66}$$

For the case given in Fig. 5–10(B), it can easily be shown that, with $dp/dx = 0$, Eq. (5–19) applies, since $(\partial u/\partial y)_{y=0} = (\partial h/\partial y)_{y=0} = 0$ by symmetry. Using the initial conditions in the jet and the external flow leads to

$$\frac{H - H_e}{H_j - H_e} = \frac{u - U_e}{U_j - U_e} \tag{5–67}$$

with no restrictions. Since, in addition, $(\partial c_i/\partial y)_{y=0} = 0$ by symmetry, $c_i(u)$ as in Eq. (5–59) also applies. Using the initial conditions, we get

$$\frac{c_i - c_{ie}}{c_{ij} - c_{ie}} = \frac{u - U_e}{U_j - U_e} \tag{5–68}$$

with no restrictions. Thus, in a flow problem such as that in Fig. 5–10(B), it would be necessary only to solve the momentum equation for $u(x, y)$, since the enthalpy and species fields can be found directly from Eqs. (5–67) and (5–68) if one is willing to accept the restriction Pr = Le = 1. (Such an approximation is quite reasonable for turbulent jet mixing flows, no matter what species are involved.)

5–10–2 Foreign Gas Injection through a Porous Wall

Exact solutions for injection of a light gas (H_2) through a porous wall into a high-speed airstream without restrictive assumptions on the physical properties and with variable density have been published by Eckert et al. (1958). Such a case is of practical interest for cooling a surface in a high-speed flow. The solutions obtained, however, are restricted to a *similar* distribution of injection (i.e., $v_w \sim x^{-1/2}$), and that is unrealistic from a practical viewpoint. Nonetheless, some useful conclusions can be drawn from the results.

Results for a flat plate treated earlier by Van Driest without injection are shown in Fig. 5–11 for various values of c_{1w} (H_2 concentration at the wall) corresponding to various dimensionless injection rates f_w through a generalized form of Eq. (4–42),

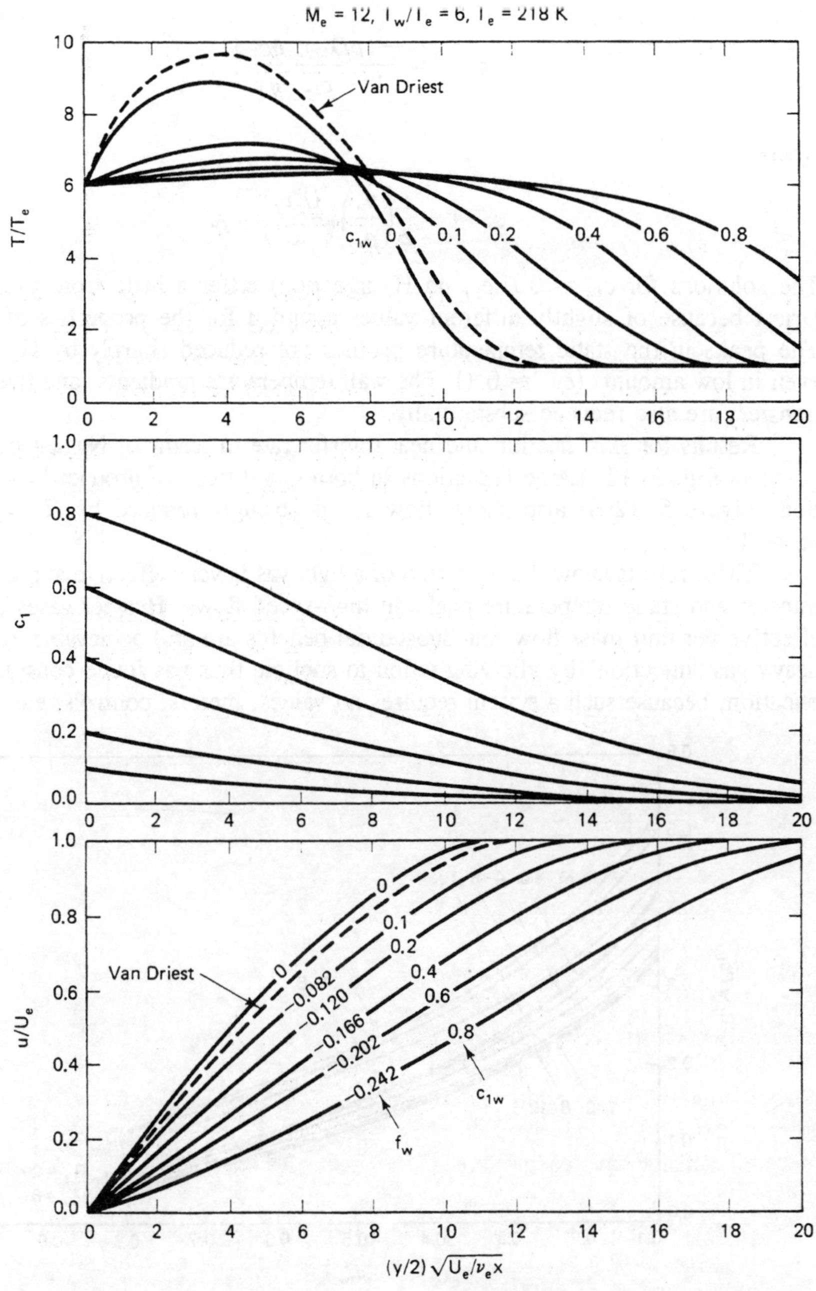

Figure 5–11 Velocity, temperature, and mass fraction profiles for H_2 injection into high-speed airflow over a flat plate. (From Eckert et al., 1958.)

i.e.,

$$\rho_w v_w = \frac{-(\rho D_{12})_w}{1 - c_{1w}} \frac{\partial c_1}{\partial y}\bigg|_w \tag{5-69}$$

where

$$v_w = -\frac{1}{2}\left(\frac{\rho_w}{\rho_e}\right)\left(\frac{U_e \nu_e}{x}\right)^{1/2} f_w \tag{5-70}$$

The solutions for $c_{1w} = 0$ (i.e., no H_2 injection) differ a little from those of Van Driest because of slightly different values assumed for the properties of pure air. The peaks in the static temperature profiles are reduced sharply by H_2 injection, even in low amounts ($c_{1w} \approx 0.1$). The wall temperature gradients, and thus the heat transfer, are also reduced substantially.

Results for skin friction and heat transfer rate in terms of Nusselt number are given in Fig. 5–12. Large reductions in both quantities are produced by H_2 injection. Figure 5–12(B) also shows how T_{aw} is strongly reduced by H_2 injection at $M_e = 12$.

These results show that injection of a light gas is very effective at reducing heat transfer and static temperature peaks in high-speed flows. Heavier gases are not as effective per unit mass flow, but substantial benefits are still achievable. The use of heavy gas "injection" by allowing teflon to sublime thus has found considerable application, because such a system requires no values, meters, controls, etc.

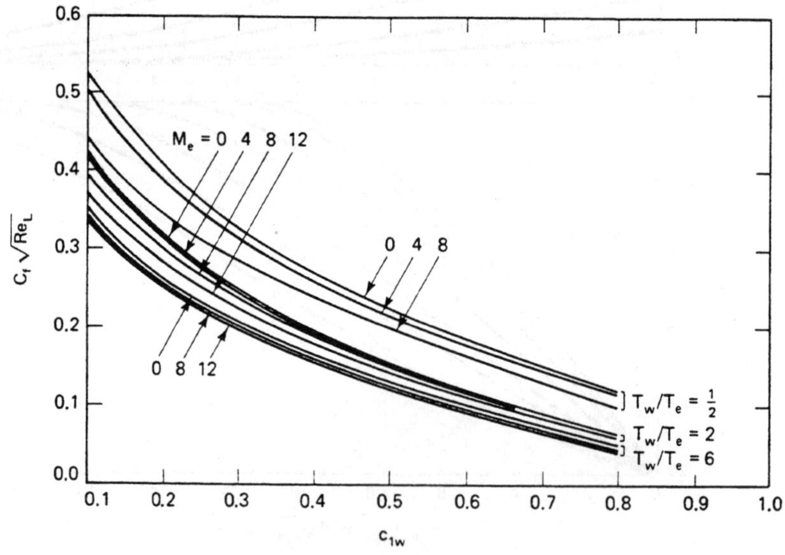

(A) Skin friction coefficient

Figure 5–12 Skin friction and heat transfer for H_2 injection into an airstream over a flat plate. (From Eckert et al., 1958.)

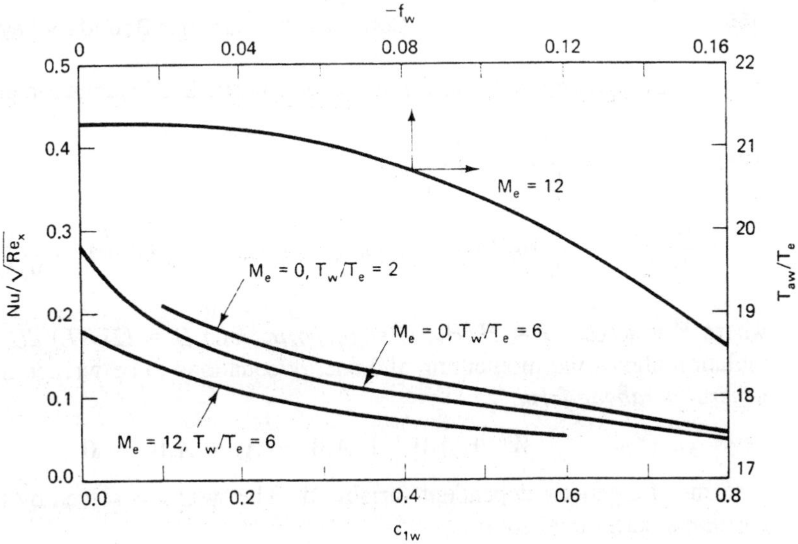

(B) Dimensionless heat transfer rate

Figure 5–12 *(continued)*

5–11 NUMERICAL SOLUTIONS OF COMPRESSIBLE, LAMINAR BOUNDARY LAYER FLOWS

In principle, the numerical solution of compressible boundary layer problems follow directly from that of incompressible problems. The main differences are the direct coupling with a now nonlinear energy equation and the necessity for calculating the variable physical properties. The evaluation and use of variable physical properties presents no real difficulties within a numerical solution procedure. The properties are evaluated in subroutines, using either a table-lookup procedure or curve fits. Since the equations are solved by considering a number of node points across the layer, it is no problem to employ the appropriate variable properties point to point. At high Mach number ($M_e \approx 6$ or greater), one generally also has to deal with non-perfect gas behavior and chemical reactions. Those matters are discussed in a detailed Review by Blottner (1970), but they are beyond the scope of this section. Some material on so-called *real gas behavior* is given in a later section of this chapter.

Many workers use the equations of motion for nonsimilar flows in Levy-Lees coordinates, rewritten in a slightly different form suggested by Blottner (1970). The continuity equation becomes

$$2\bar{s}F_{\bar{s}} + V' + F = 0 \tag{5–71}$$

The momentum equation is

$$2\bar{s}FF_{\bar{s}} + VF' = \beta\left(\frac{\rho_e}{\rho} - F^2\right) + (CF')' \tag{5–72}$$

The energy equation is written in terms of a normalized stagnation enthalpy G as

$$2\bar{s}FG_{\bar{s}} + VG' = \frac{C}{\mathrm{Pr}}G'' + \left(\frac{C}{\mathrm{Pr}}\right)'G' + \frac{U_e^2}{H_e}\left(\left(1 - \frac{1}{\mathrm{Pr}}\right)(C'FF' + C(F')^2 + CFF'') + CFF'\left(1 - \frac{1}{\mathrm{Pr}}\right)'\right) \tag{5–73}$$

where $F = u/U_e$, $G = H/H_e$, $C = \rho\mu/\rho_e\mu_e$, and $\beta = (2\bar{s}/U_e)\, dU_e/d\bar{s}$. This formulation allows the momentum and energy equations to be written in the so-called *general parabolic form*,

$$W'' + A_1W' + A_2W + A_3 + A_4W_{\bar{s}} = 0 \tag{5–74}$$

in terms of a generic dependent variable W. The system is solved by the *Thomas algorithm* at each axial step.

The value of V is then determined by numerical integration of

$$V = V_w - \int_0^{\eta} (2\bar{s}F_{\bar{s}} + F)\, d\eta \tag{5–75}$$

or by a finite difference treatment of the continuity equation.

For multicomponent flows, the diffusion equation(s) can also be written in the general parabolic form and then solved by the same procedures.

Virtually all modern methods are based on an implicit finite difference formulation, and the well-developed method of Flugge-Lotz and Blottner (1962) has been selected here as a representative example. This is a *fully implicit* method; that is, the second derivatives with respect to y are evaluated at the downstream station toward which the solution is marching. First, to demonstrate the adequacy of the method, one can compare the results obtained with the finite difference method to the exact results from Low (1955) for the relatively simple case of a *similar*, compressible boundary layer flow with $\mu \sim T$. A *similar* profile is used as the initial condition for the finite difference calculations, and the results after 10 downstream steps are shown in Fig. 5–13. Note that the ordinate is *not* a similarity variable. Excellent agreement with the appropriate *similar* profile at this downstream station can be seen.

The effects of employing various grid spacings Δx and Δy are displayed in Figs. 5–14 and 5–15. Clearly, good accuracy can be obtained for a case such as this with about 20 points across the boundary layer. Nearly doubling the number of points (decreasing $(\Delta y/L)\sqrt{\mathrm{Re}}$ from 1.0 to 0.6) produces no significant changes. Observe the expanded scales on the ordinates. Also, once a suitable Δx has been chosen, reducing it further by a factor of 10 achieves nothing useful.

Anderson and Lewis (1971) have used the Blottner method in Levy-Lees coordinates to treat both laminar and turbulent problems with the assumption of perfect gas behavior or local chemical equilibrium. A comparison of predicted results and data for a high-Mach-number, high-temperature laminar flow over a flat plate is

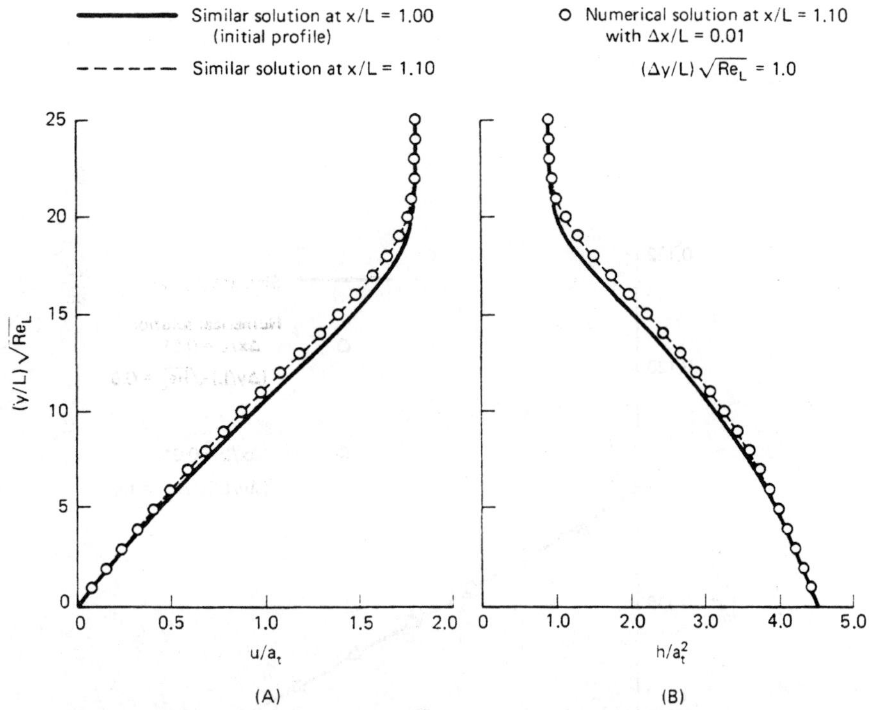

Figure 5–13 Comparison of similar solution and nonsimilar numerical solution from Flügge-Lotz and Blottner (1962) for laminar flow over a flat plate at Mach 3.0 and $T_w/T_{aw} = 2.0$.

shown in Fig. 5–16. It is interesting to note that the difference between the predictions assuming a perfect gas and those assuming a real gas was only 3 percent under the given conditions. Schetz et al. (1982) extended this method to cases with mass transfer.

Finally, it is important to assess the applicability of these elaborate methods to more complex cases than flow over a flat plate, no matter how high the Mach number. A case with a complicated geometry and a strongly nonsimilar pressure distribution is hypersonic flow over a flat-nosed cylinder with rounded corners. That case was treated by Smith and Clutter (1965), and the results are compared to shock tube experiments by Kemp et al. (1959) in Fig. 5–17 in terms of the wall heat transfer. Very good agreement can be seen. These results are representative of the level of precision that the designer can expect for high-speed laminar boundary layer predictions. Compressible laminar boundary layer flows treated by the *finite element method* are discussed in Hytopoulos, Schetz, and Gunzburger (1992).

A simple computer code for a PC based on the implicit finite difference method for compressible laminar boundary layer flows is presented in Appendix *B*.

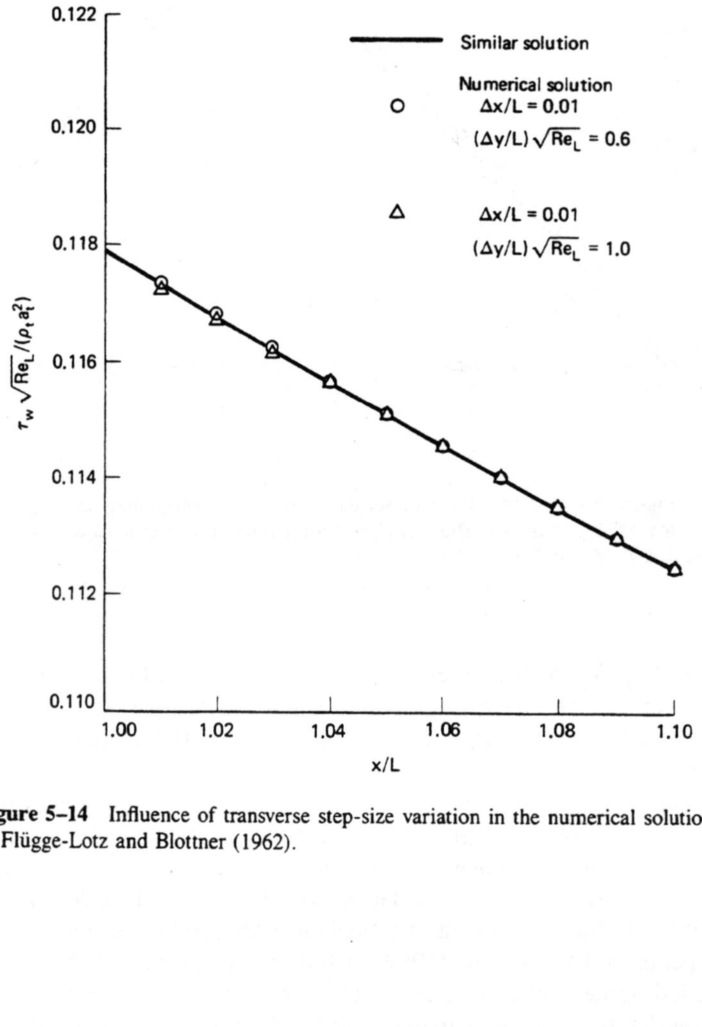

Figure 5–14 Influence of transverse step-size variation in the numerical solution of Flügge-Lotz and Blottner (1962).

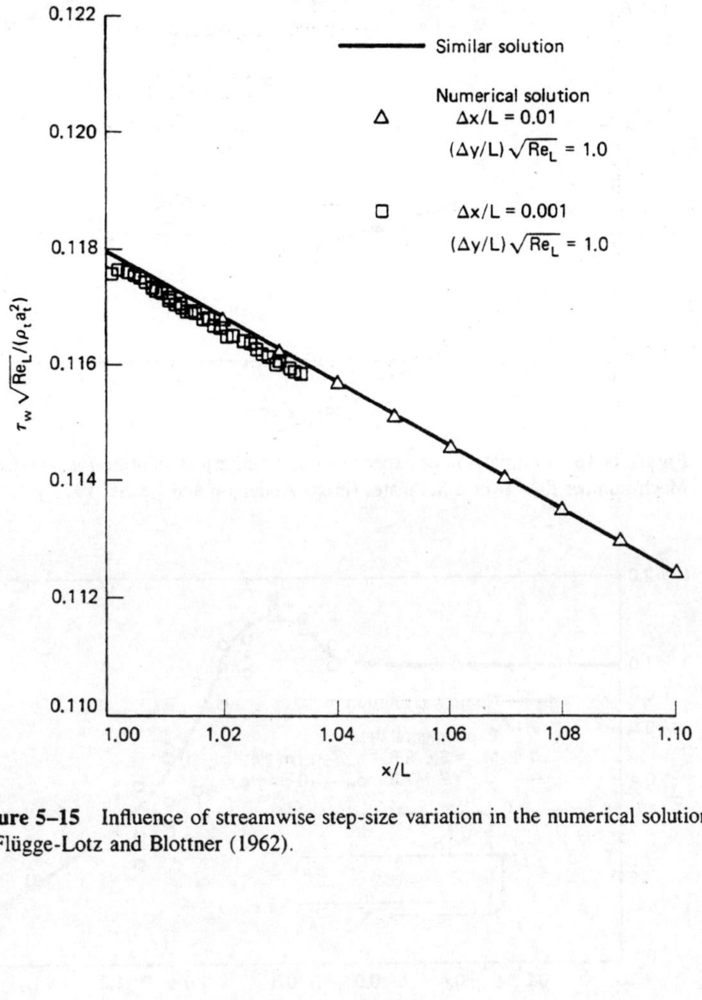

Figure 5–15 Influence of streamwise step-size variation in the numerical solution of Flügge-Lotz and Blottner (1962).

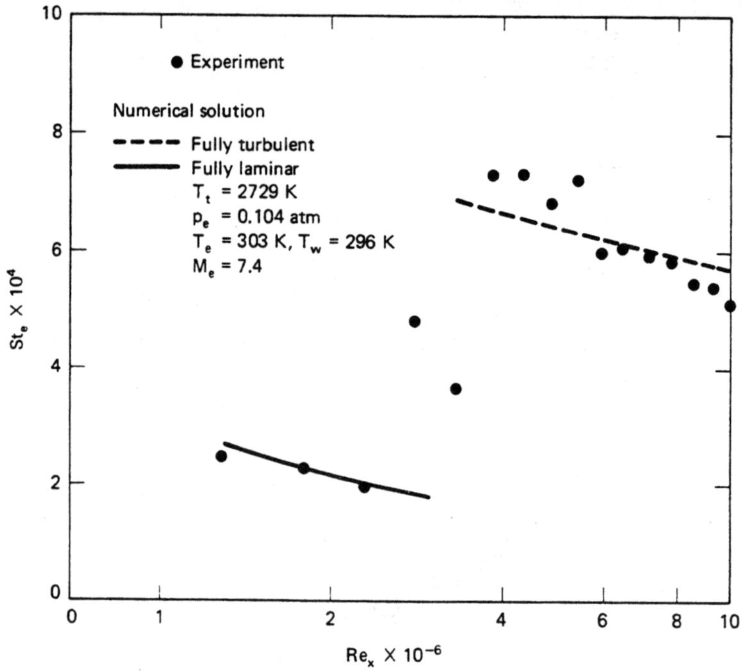

Figure 5–16 Comparison of experiment and numerical solution for laminar high-Mach-number flow over a flat plate. (From Anderson and Lewis, 1971.)

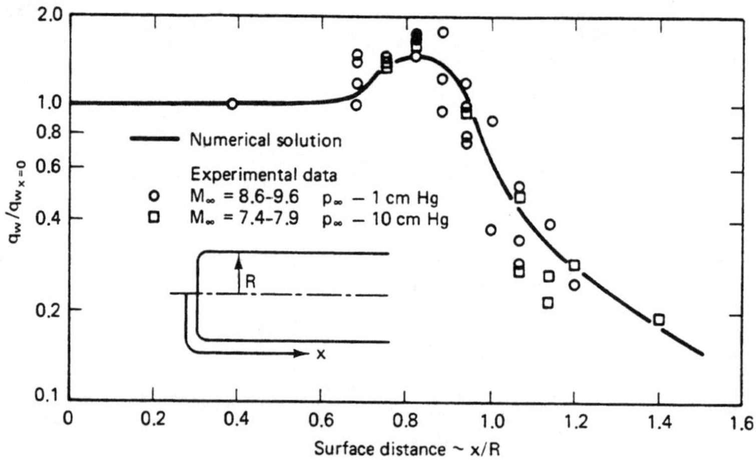

Figure 5–17 Comparison of calculated and measured heat transfer on a flat-nosed cylinder in high-Mach-number laminar flow. (From Smith and Clutter, 1965.)

Example 5–1. High-Speed Laminar Boundary Layer Problem

Use a numerical method to solve for the development of a laminar boundary layer on a flat plate 1.0 m long in a Mach 4.0 air flow. The free stream temperature is 300°K, and the free stream pressure is 0.01 atm. The wall temperature is also 300°K.

Solution This flow problem can be treated with the code CLBL in Appendix B. We start at the leading edge with XI = 0.0 and go to XF = 1.0. Then, input TINF = 300.0 and XMIF = 4.0. The corresponding free stream velocity can be calculated from the definition of the speed of sound as UINF = 1,390.0 m/s. At these free stream conditions, the kinematic viscosity is CNUINF = 0.00157 m^2/s. We select NMAX = 101, NMAX = 200, and DY = 0.0002 for a fine grid to resolve the thin boundary layer near the leading edge. Finally, we set TW = TE = 1.0.

A listing of the output at the various stations is as follows:

N	X	CF	ST
6	0.050	0.00341	0.00167
11	0.100	0.00198	0.00113
16	0.150	0.00160	0.00094
21	0.200	0.00139	0.00084
26	0.250	0.00125	0.00076
31	0.300	0.00115	0.00070
36	0.350	0.00107	0.00066
41	0.400	0.00101	0.00062
46	0.450	0.00095	0.00059
51	0.500	0.00091	0.00056
56	0.550	0.00087	0.00054
61	0.600	0.00083	0.00052
66	0.650	0.00080	0.00050
71	0.700	0.00077	0.00048
76	0.750	0.00075	0.00047
81	0.800	0.00073	0.00046
86	0.850	0.00071	0.00044
91	0.900	0.00069	0.00043
96	0.950	0.00067	0.00042
101	1.000	0.00065	0.00041

Some of the output is plotted in Figure 5–18. The results for CF and ST (C_f and St respectively) indicate the applicability of *Reynolds analogy* to this kind of flow. Note the temperature peak in the boundary layer and the shape of the velocity profile near the wall. These features were discussed earlier. Since this is a case with a relatively cool wall (T_w/T_e = 1.0), the so-called linear region of the velocity profile near the wall is not so prominent. The conditions of this problem result in Re = 885,350 at the end of the plate. The assumption of laminar flow for that Re at the given Mach number is justified, as will be discussed in Chapter 6.

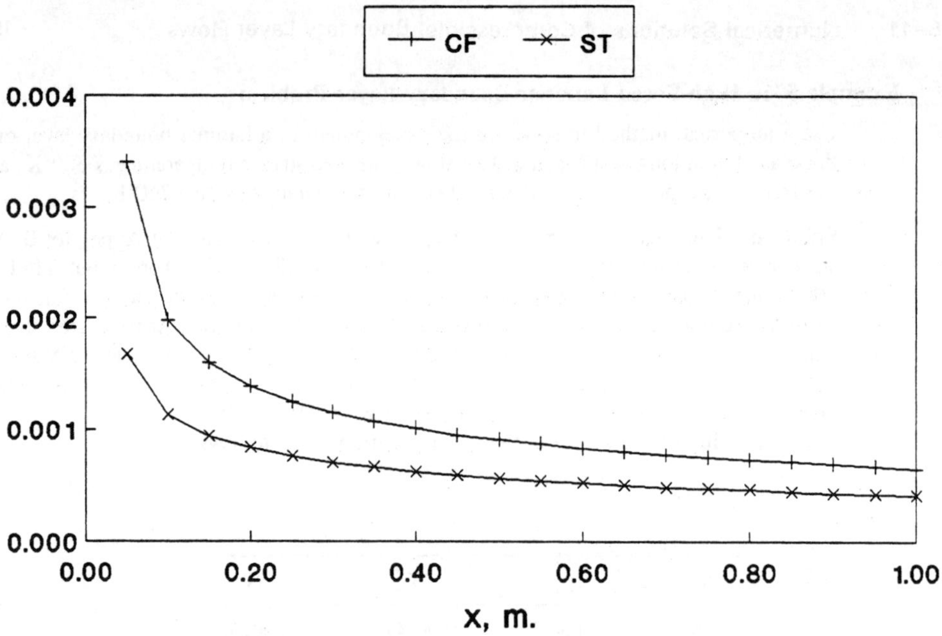

(A) Skin friction and Stanton number

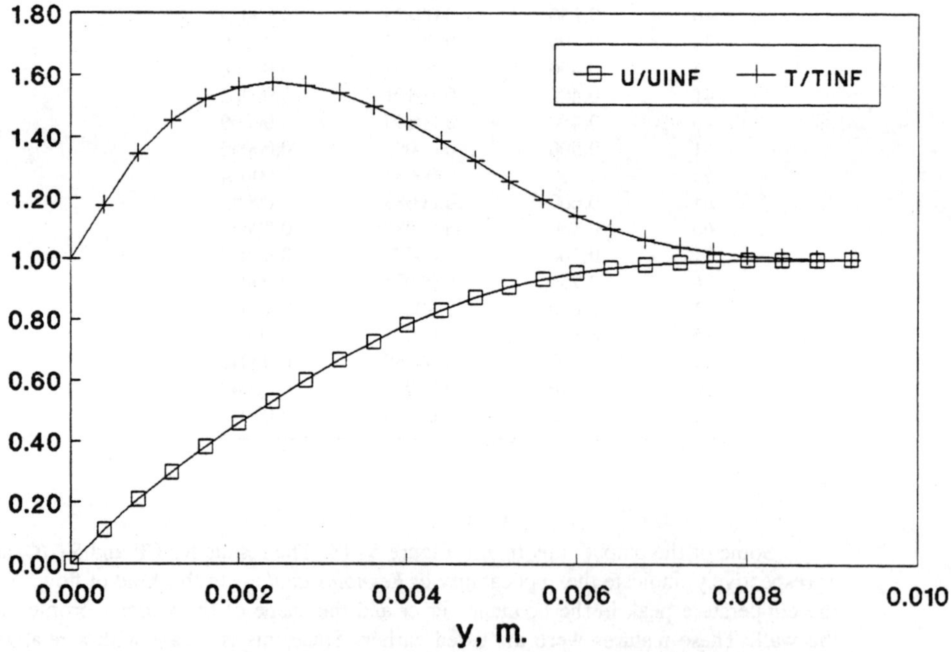

(B) Velocity and temperature profiles at x = 1.0 m.

Figure 5–18 Predicted results for Example 5–1.

5–12 REAL GAS EFFECTS

Up to this point, we have assumed that the gaseous working fluid could be adequately represented as a perfect gas, often with constant specific heats. At some conditions, those assumptions begin to break down, and it is necessary to consider so-called *real gas effects*. Since mass transfer has been treated in this text, these important effects can be discussed. The general subject is very broad and complex, and we shall concentrate here on the first-order effects for simple gases, such as N_2 and O_2 and their products.

The equation of state for a perfect gas is generalized as

$$\frac{p}{\rho(R/W)T} = Z(p, T) \tag{5–76}$$

where $Z(p, T)$ is the *compressibility factor*. Clearly, $Z = 1$ corresponds to a perfect gas, and deviations from unity indicate nonperfect or real gas effects. These effects occur in two regimes. At high pressures and low temperatures, intermolecular forces become important, and other equations of state have been developed. Of much greater practical importance to the aerodynamicist is the regime of high temperatures and low or modest pressures, where dissociation and even ionization of molecules become significant.

For simplicity, consider a *generic* diatomic gas A_2 which can approximately model the behavior of air or any other atmosphere that consists of a mixture of simple gases. This idea was first proposed by Lighthill (1957). The dissociation reaction for such a gas may be written

$$A_2 \rightleftarrows A + A \tag{5–77}$$

and the mass fractions of atoms and molecules are denoted, respectively, by

$$\frac{\rho_A}{\rho} = c_A = \alpha$$

and

$$\frac{\rho_{A_2}}{\rho} = c_{A_2} = 1 - \alpha \tag{5–78}$$

The quantity α is often called the *degree of dissociation*. Each species can be assumed to behave as a perfect gas, so

$$p_A = \rho_A(R/W_A)T, \qquad p_{A_2} = \rho_{A_2}(R/W_{A_2})T \tag{5–79}$$

Now,

$$p = p_A + p_{A_2} = \rho RT\frac{(1 + \alpha)}{2W_A} \tag{5–80}$$

where $2W_A$ is the molecular weight of the molecules. The enthalpy of each species is found using Eq. (3–64), and the mixture enthalpy is just

$$h = \alpha h_A + (1 - \alpha)h_{A_2} \tag{5–81}$$

The foregoing chemical reaction can be treated in three ways. The first and simplest model asserts that the reaction is proceeding so slowly at the local temperature and pressure, that there is no change in composition due to chemistry in the time scale of interest for the flow problem. The composition is said to remain *frozen*. The next most complicated model takes the reaction as proceeding very rapidly, so that the local composition comes instantly to the value it would have if a long time were available. The composition is then in *equilibrium* with the local temperature and pressure. Obviously, the static temperature and pressure are the relevant quantities for considerations of chemistry. The most complicated situation is the general case where neither extreme is adequate, and one must treat the actual time rate of change of composition described by *chemical kinetics*. A detailed treatment at that level is beyond the scope of this text; the reader can refer to Vincenti and Kruger (1965) or other available volumes for a discussion. The composition for the equilibrium limit is found, using the *law of mass action*, to be

$$\frac{4\alpha^2}{(1-\alpha^2)} = \frac{K(T)}{p} \tag{5-82}$$

where $K(T)$ is the *equilibrium constant* for the relevant reaction. For the dissociation of O_2, a plot is shown in Fig. 5–19. Such results can be generalized with only slight approximation to show the degree of dissociation for a generic diatomic gas, as in Fig. 5–20, where θ_D is the ratio of the heat of dissociation to $R/2W_A$. Note the strong effect of pressure at a given temperature.

In the presence of chemical reactions, the diffusion equations must be expanded to include a term for the chemical production of species. Thus, for steady laminar flow, Eq. (3–60) becomes, in the notation of this section,

$$\rho u \frac{\partial \alpha}{\partial x} + \rho v \frac{\partial \alpha}{\partial y} = \frac{\partial}{\partial y}\left(\rho D_{12} \frac{\partial \alpha}{\partial y}\right) + w_A \tag{5-83}$$

The extra term w_A is the chemical production term for atoms. The continuity and momentum equations remain unchanged; the necessary changes to the energy equation were already made to accommodate mass transfer in Section 3–5–2. In the notation of this section, the energy equation, in terms of the stagnation enthalpy H, is

$$\rho u \frac{\partial H}{\partial x} + \rho v \frac{\partial H}{\partial y} = \frac{\partial}{\partial y}\left(\frac{\mu}{\Pr}\frac{\partial H}{\partial y} + \mu\left(1 - \frac{1}{\Pr}\right)\frac{\partial u^2/2}{\partial y}\right) - \frac{\partial}{\partial y}\left(\left(\frac{1}{\text{Le}} - 1\right)\rho D_{12}(h_A - h_{A_2})\frac{\partial \alpha}{\partial y}\right) \tag{5-84}$$

Also,

$$h_A = \int_0^T c_{P_A}\, dT + h_A^0, \qquad h_{A_2} = \int_0^T c_{P_{A_2}}\, dT \tag{5-85}$$

The heat of dissociation is included in the energy equation in the term $(h_A - h_{A_2})$.

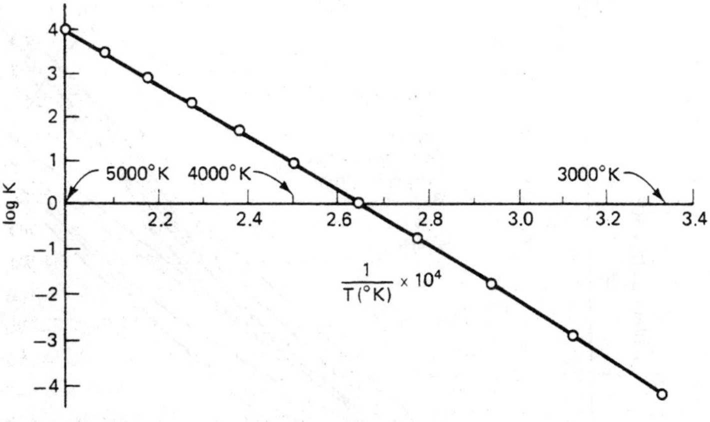

Figure 5-19 Equilibrium constant for O_2 dissociation vs. temperature T. (From Woolley, 1955.)

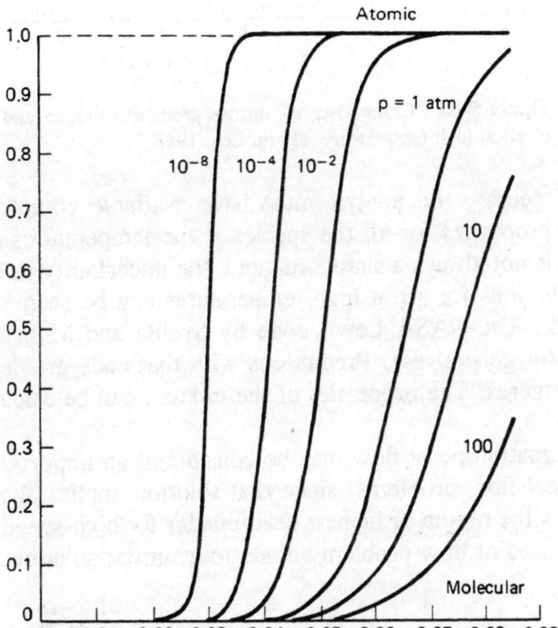

Figure 5-20 Degree of dissociation for a diatomic gas vs. temperature T. (From Woolley, 1955.)

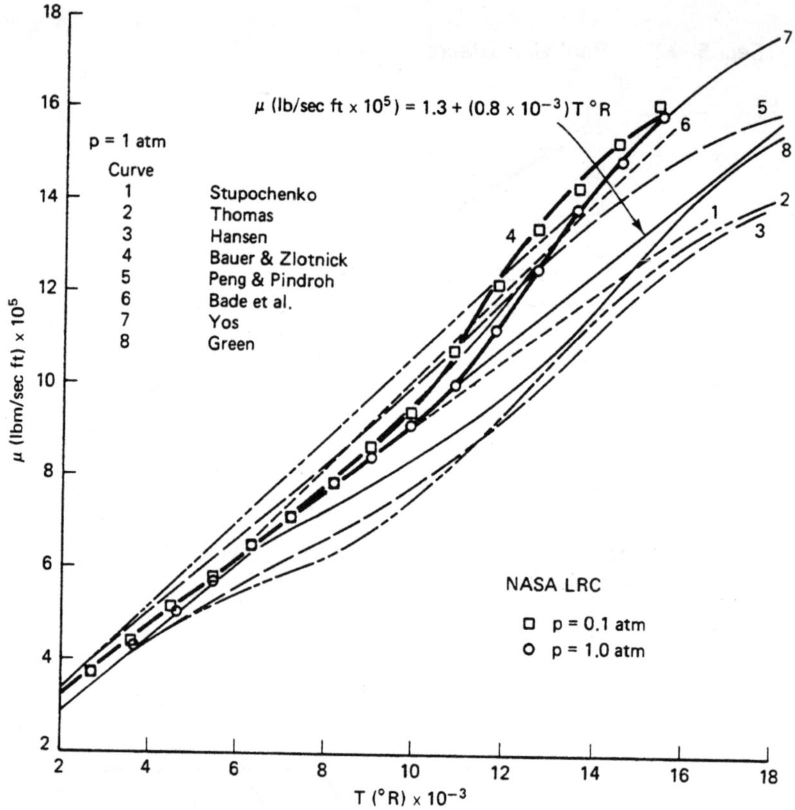

Figure 5–21 Comparison of various predictive models and data for the viscosity of air at high temperature. (From Lee, 1988.)

Of course, the analyst must have available complete information regarding physical properties for all the species at the temperatures and pressures of interest, and that is not always a simple matter. The uncertainty in both the data and theoretical predictions for air at high temperatures can be seen in the plots in Figs. 5–21 and 5–22. The NASA Lewis code by Svehla and McBride (1973) can be recommended for general use. Predictions with that code are denoted by (+) and (×) in the two figures. The properties of the mixture can be calculated as indicated in Section 3–5–3.

Stagnation point flow may be considered an important representative example of practical flow problems, since that solution applies at the nose of blunt bodies, and that is the region of highest heat transfer for high-speed vehicles. Recall that this stagnation point flow problem admits to a similar solution. If we introduce

$$z_A = \frac{\alpha}{\alpha_e}, \qquad z_{A_2} = \frac{1-\alpha}{1-\alpha_e} \tag{5–86}$$

the diffusion equation for this case becomes, in Levy-Lees coordinates,

$$\left(\frac{C}{Sc} z_A'\right)' + f z_A' = -\frac{w_A/\alpha_e}{(j+1)(dU_e/dx)_0} \tag{5–87}$$

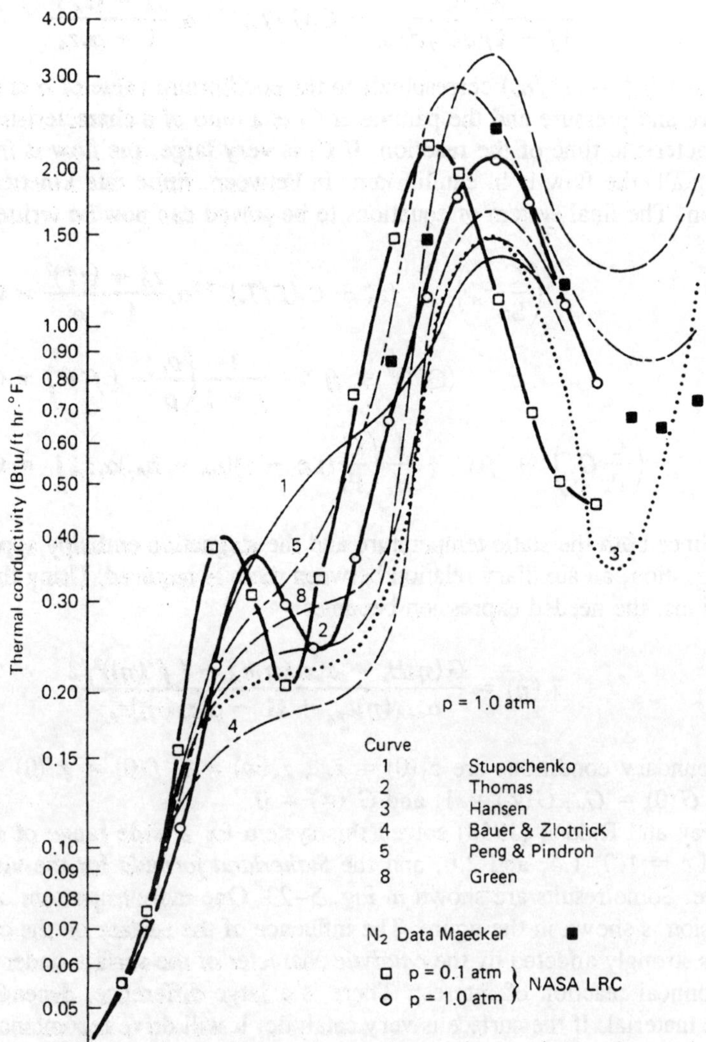

Figure 5–22 Comparison of various predictive models and data for the thermal conductivity of air at high temperature. (From Lee, 1988.)

Fay and Riddell (1958) developed an approximate kinetics scheme for the dissociation reaction and arrived at

$$\frac{w_A/\alpha_e}{(j+1)(dU_e/dx)_0} = C_1(T/T_e)^{-3.5}\,\alpha_e\frac{z_A^2-(z_A^*)^2}{1+\alpha_e z_A} \tag{5-88}$$

where α^* ($z_A^* = \alpha^*/\alpha_e$) corresponds to the equilibrium value of α at the local temperature and pressure and the parameter C_1 is a ratio of a characteristic flow time to a characteristic time of the reaction. If C_1 is very large, the flow is frozen. If C_1 is very small, the flow is in equilibrium. In between, finite rate kinetics controls the reaction. The final system of equations to be solved can now be written

$$\left(\frac{C}{\mathrm{Sc}}z_A'\right)' + fz_A' + C_1(T/T_e)^{-3.5}\alpha_e\frac{z_A^2-(z_A^*)^2}{1+\alpha} = 0 \tag{5-89}$$

$$(Cf'')' + ff'' + \frac{1}{j+1}\left(\frac{\rho_e}{\rho} - (f')^2\right) = 0 \tag{5-90}$$

$$\left(\frac{C}{\mathrm{Pr}}G'\right)' + fG' + \frac{1}{H_e}\left(\frac{C}{\mathrm{Pr}}(\mathrm{Le}-1)(h_A - h_{A_2})\alpha_e z_A'\right)' = 0 \tag{5-91}$$

Since both the static temperature and the stagnation enthalpy appear in the energy equation, an auxiliary relation between them is required. Using the appropriate definitions, the needed expression becomes

$$T(\eta) = \frac{G(\eta)H_e - \alpha_e z_A(\eta)h_A^0 - (f'(\eta))^2/2}{\alpha_e z_A(\eta)c_{p_A} + (1-\alpha_e z_A(\eta))c_{p_{A_2}}} \tag{5-92}$$

The boundary conditions are $z_A(0) = z_{Aw}$, $z_A(\infty) = 1$; $f(0) = f'(0) = 0$, $f'(\infty) = 1$; and $G(0) = G_w$, $G(\infty) = 1$, and $G'(\infty) = 0$.

Fay and Riddell (1958) solved this system for a wide range of C_1 with Pr = 0.71, Le = 1.0, 1.4, and 2.0, and the *Sutherland formula* for the viscosity of the mixture. Some results are shown in Fig. 5–23. One more important concept in this discussion is shown in the figure: The influence of the surface on the chemical reactions is strongly affected by the *catalytic* character of the surface under discussion on the chemical reaction of interest. There is a large difference, depending upon the surface material. If the surface is very catalytic, it will drive recombination of atoms to completion on the surface, even if conditions in the flow indicate a frozen flow. The effect on the heat transfer rate can be large, as can be seen on the left-hand side of the figure. If conditions in the flow indicate local chemical equilibrium, the catalytic effect is small for the usual case of walls at a modest temperature that favors recombination. The effects of a catalytic wall were found to be important in predicting heat transfer to the space shuttle. Favorable comparison between predictions and experiment, such as that shown in Fig. 5–24, demonstrate the adequacy of the whole analysis.

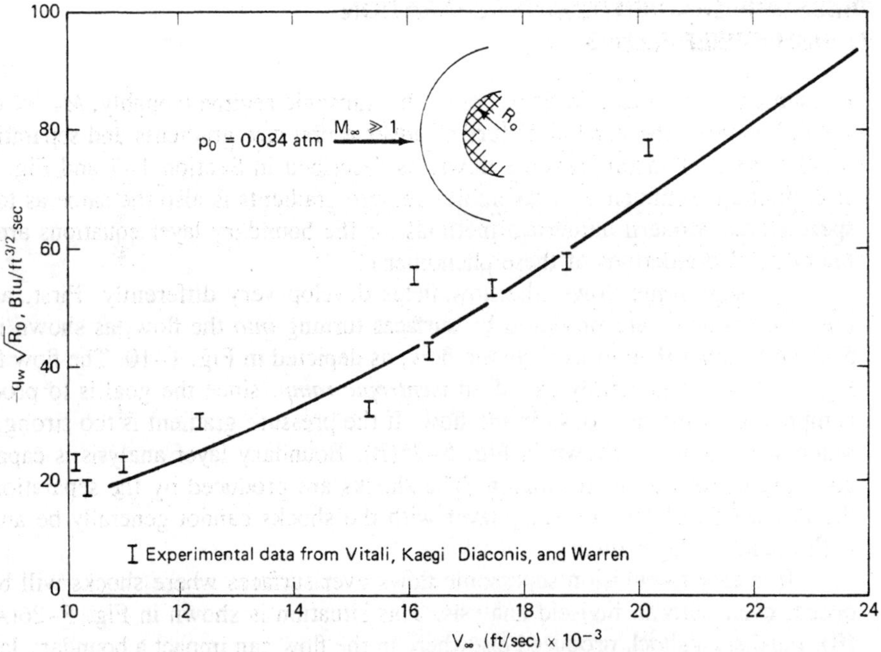

Figure 5–23 Variation of stagnation point heat transfer parameter with a dissociation reaction-rate parameter in laminar high-speed flow. (From Fay and Riddell, 1958.)

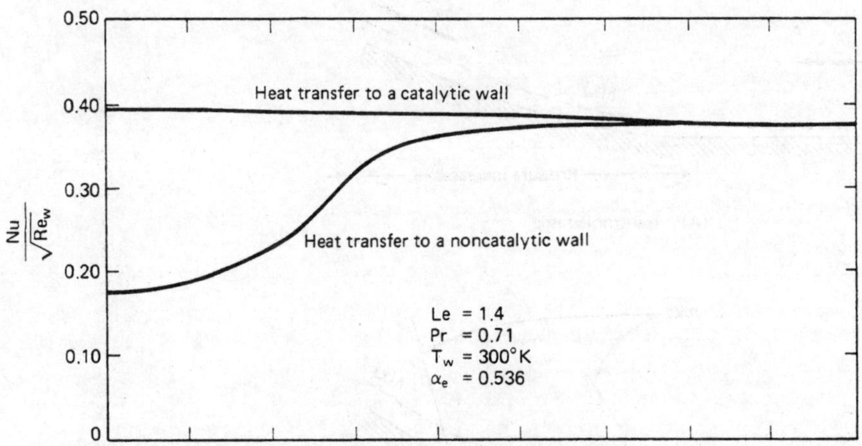

Figure 5–24 Comparison of predictions with experiment for stagnation point heat transfer to a catalytic surface in a high-Mach-number laminar flow. (From Fay and Riddell, 1958.)

5–13 PRESSURE GRADIENTS AND SEPARATION IN HIGH-SPEED FLOWS

For subsonic compressible flows below the transonic regime (roughly, $M_\infty < 0.8$ for slender bodies,) the general picture of adverse pressure gradients and separation resembles that for incompressible flows, as described in Section 1–7 and Fig. 1–10. The qualitative influence of favorable pressure gradients is also the same as for low-speed cases. Modern numerical methods for the boundary layer equations are capable of good predictions of these phenomena.

For supersonic flows, the flow fields develop very differently. First, adverse pressure gradients are produced by surfaces turning *into* the flow, as shown in Fig. 5–25(A), rather than *away from* the flow, as depicted in Fig. 1–10. The flow field in Fig. 5–25(A) is generally called an *isentropic ramp,* since the goal is to produce a compression without shocks in the flow. If the pressure gradient is too strong, separation will occur, as shown in Fig. 5–25(B). Boundary layer analysis is capable of good predictions up to separation. The shocks are produced by the separation, and the interaction of the boundary layer with the shocks cannot generally be analyzed with boundary layer theory.

It is easy to envision supersonic flows over surfaces where shocks will be produced, even with an inviscid analysis. This situation is shown in Fig. 5–26(A) and (B). Further, a shock produced elsewhere in the flow can impact a boundary layer on a surface, as illustrated in Fig. 5–26(C). Also, for subsonic approach flows near Mach 1.0, the presence of a body usually accelerates the flow to locally supersonic

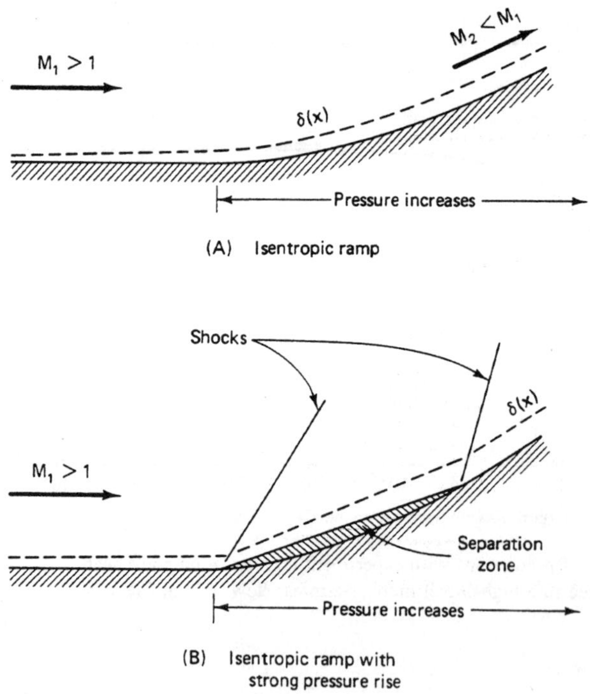

Figure 5–25 Sketches of typical flow fields for laminar boundary layers on supersonic compression surfaces where the inviscid flow is shock free.

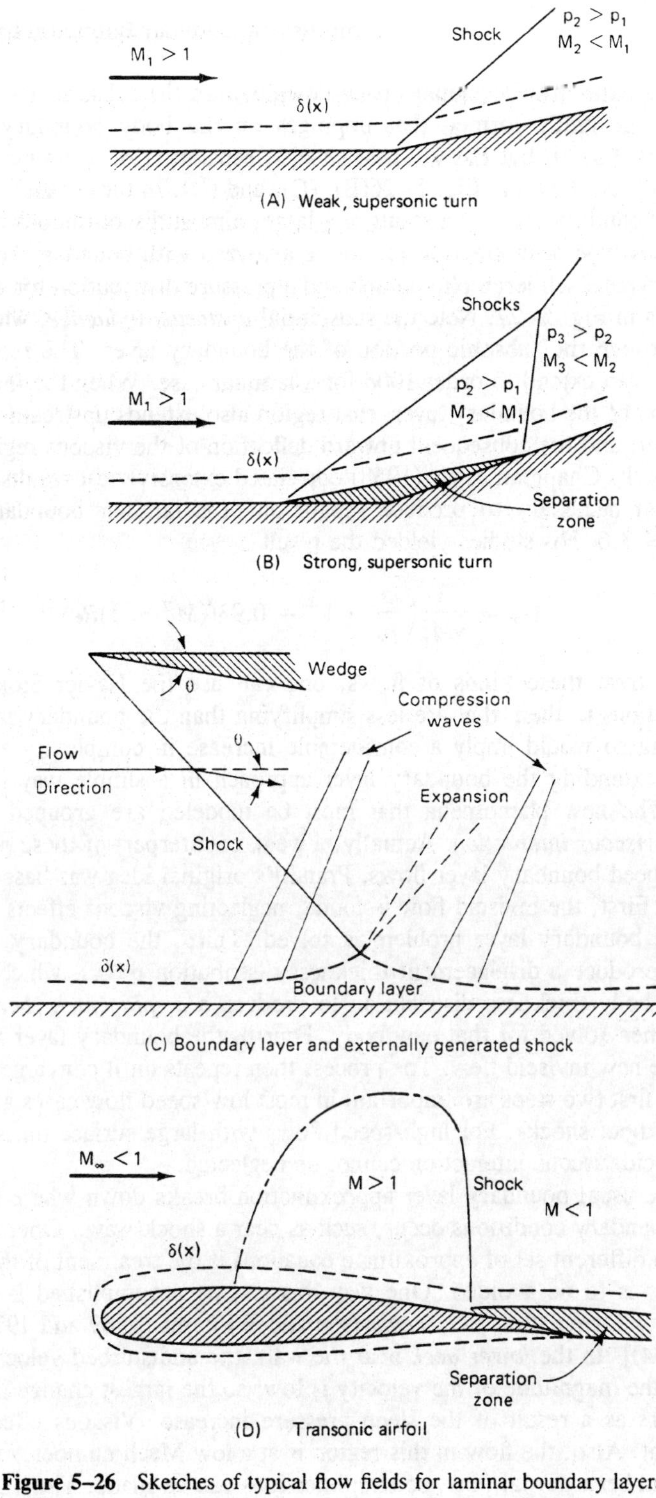

Figure 5–26 Sketches of typical flow fields for laminar boundary layers on high-speed compression surfaces where the inviscid flow contains shocks.

speeds. As the flow decelerates back down toward the subsonic free stream value, a shock is generally formed that impinges on the body boundary layer (see Fig. 5–26(D)). For all but the weakest shocks, separation of a laminar boundary layer will result, as shown in Fig. 5–26(B), (C), and (D). In these cases, the shocks separate the boundary layer as a result of a large, almost discontinuous pressure rise, but again, this type of interaction cannot be analyzed with boundary layer theory.

A typical schlieren photograph and a pressure distribution for a supersonic case are given in Fig. 5–27. Note the substantial *upstream influence,* which is communicated through the subsonic portion of the boundary layer. The region of upstream influence can extend to order 100δ for a laminar case. When the main shock causes separation of the boundary layer, that region also extends upstream of the shock impact point, and the consequent upward deflection of the viscous region produces another shock. Chapman et al. (1958) correlated extensive test results for the pressure coefficient necessary to produce separation of a laminar boundary layer for $1.1 \leq M_e \leq 3.6$. His studies yielded the result

$$C_p \equiv \frac{2}{\gamma M_e^2}\left(\frac{p}{p_e} - 1\right) = 0.93((M_e^2 - 1)Re)^{-1/4} \tag{5-93}$$

To treat these kinds of flows, one can use the Navier-Stokes equations or modifications to them that are less simplifying than the boundary layer assumption. But doing so would imply a considerable increase in complexity and cost. Or one can try extending the boundary layer approach in a simple way to meet the new needs. The new phenomena that must be modeled are grouped under the term *inviscid-viscous interaction*. Actually, a weak counterpart of these phenomena exists in low-speed boundary layer flows. Prandtl's original idea was based on an iterative scheme. First, the inviscid flow is found, neglecting viscous effects altogether. Second, the boundary layer problem is solved. Third, the boundary layer solution is used to produce a displacement thickness distribution $\delta^*(x)$, which is added to the original body thickness distribution to produce a *displacement body*. The inviscid flow is then solved for that new body. Fourth, the boundary layer flow is resolved, using the new inviscid flow. The process then repeats until convergence. In practice, only the first two steps are important in most low-speed flow cases and in high-speed flows without shocks. For high-speed flows with large surface turns and/or shocks, the inviscid-viscous interaction cannot be neglected.

The usual boundary layer approximation breaks down where severe gradients in the boundary conditions occur, such as near a shock wave. One, therefore, needs to seek a different set of approximate equations if the treatment of the Navier-Stokes equations is to be avoided. One way this can be accomplished is in terms of the *triple deck* formulation [see Stewartson and Williams (1969 and 1973) and Stewartson (1974)]. In the *lower deck* near the wall, the undisturbed velocity profile is linear and the magnitude of the velocity is low, so the largest change in relative velocity occurs as a result of the steep pressure increase. Viscous effects remain very important. Also, the flow in this region is at a low Mach number, even with a high-Mach-number external flow, so simplifications can be made. The *main deck* consists

(A) Schlieren photograph

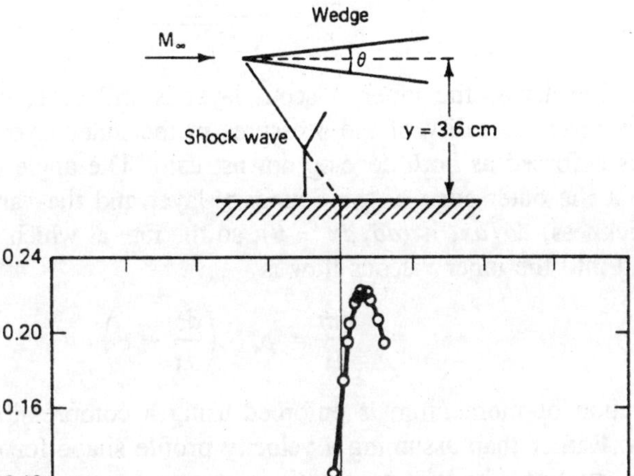

(B) Pressure distribution on the wall

Figure 5–27 Experimental observations for shock impingement on a supersonic laminar boundary layer. (From Liepmann, Roshko, and Dhawan, 1952.)

of the rest of the boundary layer. The pressure gradient can be taken as larger than the viscous stress gradients, so an inviscid assumption can be made (at least for laminar flows), although the flow is rotational. The *upper deck* is the inviscid, irrotational flow outside the boundary layer, and the pressure change in it is directly proportional to the local streamline angle. A more detailed consideration of the *triple deck* method is beyond the scope of this book. The interested reader can refer to the articles by Stewartson and Williams cited earlier and the review by Adamson and Messiter (1980) for more complete coverage.

Another, simpler approximate formulation that was introduced before is worthy of discussion. A first key paper was published by Crocco and Lees (1952), and that has been the basis of much subsequent work. The main idea of the approach is illustrated in Fig. 5–28. There is an inner, viscous region of height δ with a nonuniform velocity profile. This region interacts with an outer, supersonic, inviscid (actually taken to be isentropic) region that makes a small angle θ with the surface. A central ingredient in the approach is direct coupling of the inner, viscous flow to the outer, inviscid supersonic flow, using the pressure-flow angle relation from linearized supersonic theory, i.e.,

$$\frac{dp}{p} = \frac{\gamma M_e^2}{\sqrt{M_e^2 - 1}}\, d\theta \tag{5–94}$$

The pressure across the inner, viscous layer is still taken to be constant, as in a boundary layer. Equality of the pressures in the inner layer and the outer inviscid stream is enforced as both develop downstream. The angle between the streamline angle θ at the outer edge of the boundary layer and the variation of the boundary layer thickness, $d\delta/dx$, is $(d\delta/dx - \theta)$, so the rate at which external stream fluid is entrained into the inner viscous flow is

$$\frac{d\dot{m}}{dx} = \rho_e U_e\left(\frac{d\delta}{dx} - \theta\right) \tag{5–95}$$

Conservation of momentum is enforced using a compressible integral momentum equation. Rather than assuming a velocity profile shape for use in the integral momentum equation, the whole analysis is recast in terms of a *mixing coefficient* k_m, defined by

$$\frac{d\dot{m}}{dx} = k_m \rho_e U_e \tag{5–96}$$

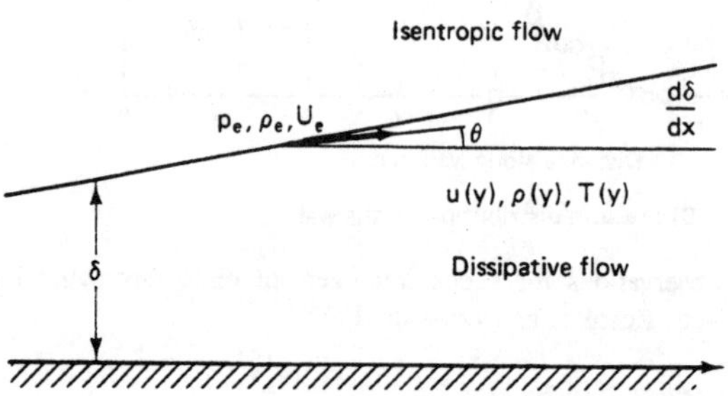

Figure 5–28 Schematic of the flow field modeled in the Crocco-Lees mixing theory.

A great simplification results with the assumption of a constant mixing coefficient, and some justifications for assuming that are provided. The analysis was restricted to an adiabatic wall and Pr = 1 so that the stagnation temperature could be taken to be constant across the inner layer, satisfying conservation of energy. The approach has been successfully extended to turbulent cases.

An important feature of Crocco and Lees's formulation is the possible occurrence of a *viscous throat* in the inner viscous layer when the average Mach number across that layer becomes unity. Requiring a smooth passage of the inner flow through the *saddle-point singularity* at this viscous throat serves to make the solutions unique. The whole phenomenon can be likened to the passage through Mach 1.0 in the throat of a converging-diverging nozzle, assuming inviscid flow. The matter can be illustrated using one-dimensional equations for the inner region, written in terms of averaged variables over the region. Allowing entrainment into the inner region and neglecting friction on the wall for simplicity, we can combine the energy and momentum equations to read

$$\frac{dM^2}{dx} = \frac{-2M^2\left(1 + \frac{(\gamma - 1)}{2}M^2\right)}{(1 - M^2)}\left[\frac{1}{A}\frac{dA}{dx} - \left((1 + \gamma M^2) - \frac{\gamma M^2 U_e}{V}\right)\frac{1}{\dot{m}}\frac{d\dot{m}}{dx}\right] \tag{5–97}$$

Clearly, at $M = 1$, the numerator must go to zero. The viscous throat does not occur at the minimum area ($dA/dx = 0$), as in the inviscid case due to the extra term involving viscous entrainment. This kind of condition can occur in the reattachment region of a shock-boundary layer problem with separation. Laminar cases encounter the phenomenon only for a strong negative pressure gradient upstream of shock impingement. It also occurs for subsonic slot injection (see Fig. 5–10(A)) into a supersonic flow. One is then not free to specify both the injection rate and the static pressure in the slot. More details on both of these situations will be provided in the discussion of the turbulent case in a later chapter.

For the laminar shock–boundary layer interaction case, the formulation was refined by Lees and Reeves (1964). These researchers avoided the use of the entrainment rate model by adding the integral form of the mechanical energy equation to the formulation. It was then necessary to assume a family of velocity profile shapes. Some comparisons between predictions and experiment are shown in Figs. 5–29 and 5–30. The case in Fig. 5–29 is for a wedge mounted on a wall, such as in Fig. 5–26(B), and the case in Fig. 5–30 is for shock impingement on a boundary layer on a flat wall, such as in Fig. 5–26(C). The good agreement can be noted.

5–14 INTERACTIONS IN HYPERSONIC FLOWS

The hypersonic flow regime can be crudely defined as encompassing flow problems with a free stream Mach number greater than about 5.0. Some matters that are important in that regime have already been discussed here, e.g., real gases. In this section, we extend the material in the previous section on inviscid-viscous interactions

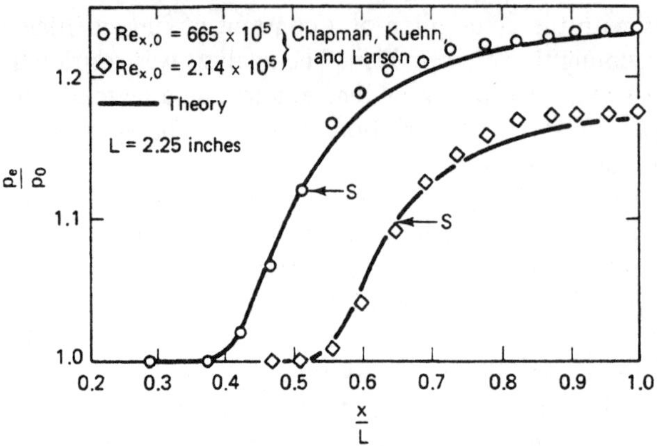

Figure 5–29 Comparison of theory with experiment for wall pressure distribution upstream of a disturbance to a Mach 2.0 laminar boundary layer. (From Lees and Reeves, 1964.)

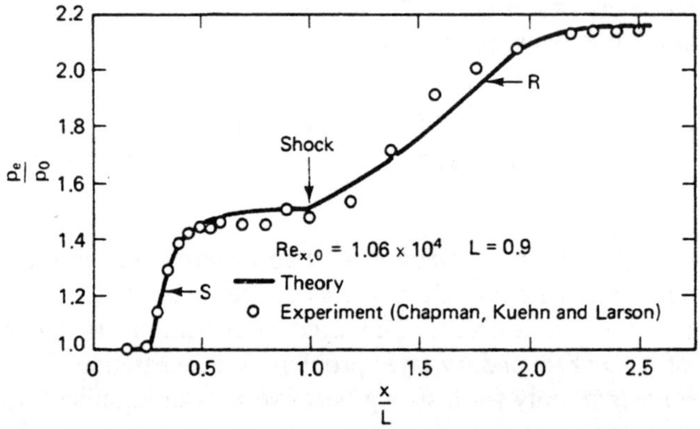

Figure 5–30 Comparison of theory with experiment for wall pressure distribution with shock impingement on a Mach 2.45 laminar boundary layer. (From Lees and Reeves, 1964.)

to the hypersonic regime. There are three kinds of such interactions in general hypersonic flow problems: pressure interactions, entropy interactions, and vorticity interactions. Pressure interactions are most important for slender bodies, and the flat plate case will be used to illustrate that situation. Vorticity interactions predominate for blunt bodies, and the stagnation point flow will be used to illustrate that situation.

The pressure interaction is important in hypersonic flows for three reasons. First, as the Mach number increases, the boundary layer thickness becomes relatively greater (see Fig. 5–2). Second, practical hypersonic flows often occur during conditions characterized by a low Reynolds number (high altitude, for example), and a low Reynolds number leads to thicker boundary layers. Third, small deflection angles lead to large pressure changes in the hypersonic regime. Consider the flat plate case to be representative of slender bodies. Near the leading edge, the rate of growth of the displacement thickness is large (recall that $\delta^* \sim x^{1/2}$). The growth produces a *displaced body* that looks rather blunt in the inviscid flow, leading to a shock and a large pressure rise. In turn, the development of the boundary layer is affected. This area is called the *strong interaction* regime. Further downstream, the rate of growth of the displacement thickness is less, and we have the *weak interaction* regime. The case of flat plate flow has been analyzed by Hayes and Probstein (1959), and the results for the pressure rise can be expressed in terms of the *hypersonic interaction parameter*

$$\bar{\chi} \equiv \frac{M_\infty^3\sqrt{C}}{\sqrt{Re}}; \qquad C \equiv \frac{\rho_w \mu_w}{\rho_e \mu_e} \tag{5-98}$$

The results are compared with data in Fig. 5–31. Note that large pressure rises are possible in the strong regime that is usually taken to be $\bar{\chi} > 3$. Generally, skin friction and heat transfer are the primary quantities of interest to the boundary layer analyst, and both are increased by the interaction. The increase in skin friction is shown in Fig. 5–32, which has measurements and the predictions of an analysis by Li and Nagamatsu (1953).

We turn now to the stagnation point region of a blunt body in hypersonic flow. The shock wave in front of the body will be very strong, since the Mach number is high, and the density behind the shock is nearly constant at the strong shock limit.

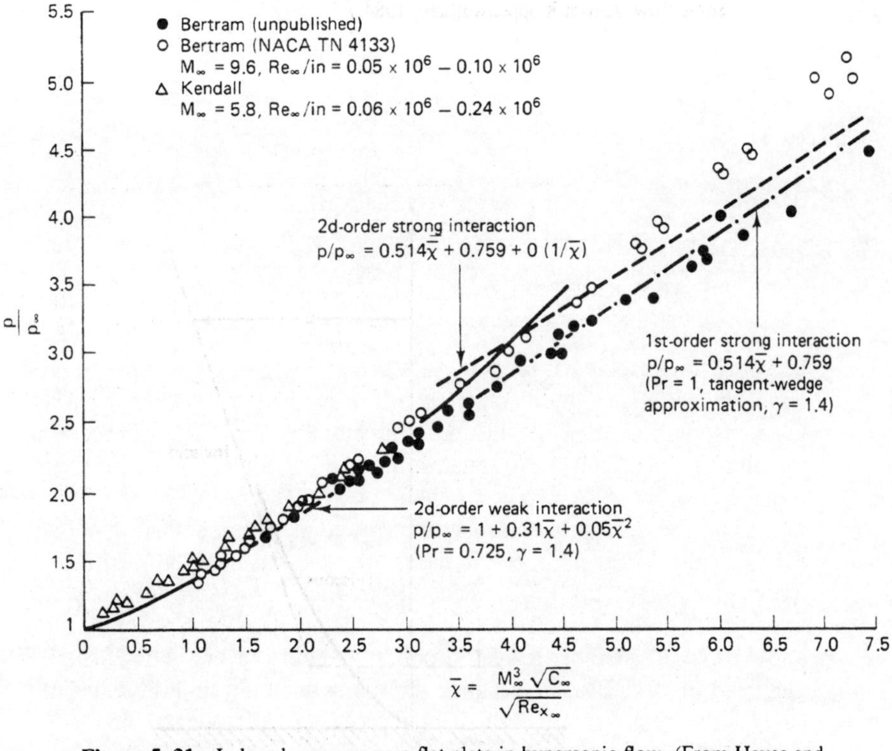

Figure 5–31 Induced pressure on a flat plate in hypersonic flow. (From Hayes and Probstein, 1959.)

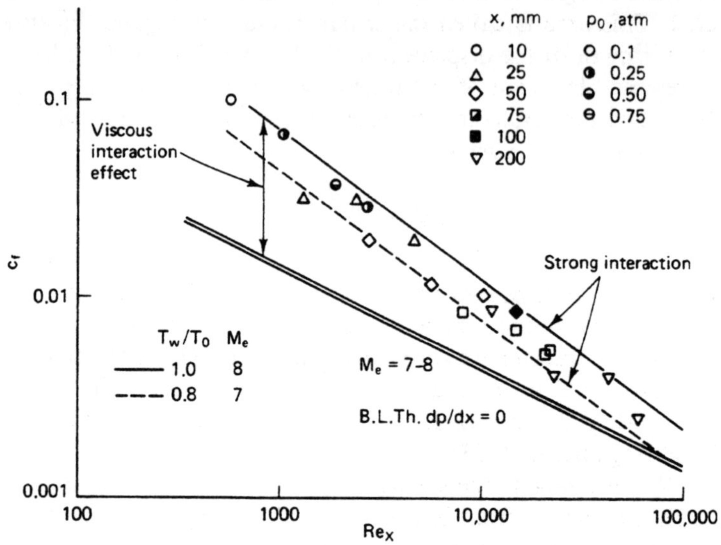

Figure 5–32 Effect of viscous interaction on skin friction for a flat plate in hypersonic flow. (From Koppenwallner, 1984.)

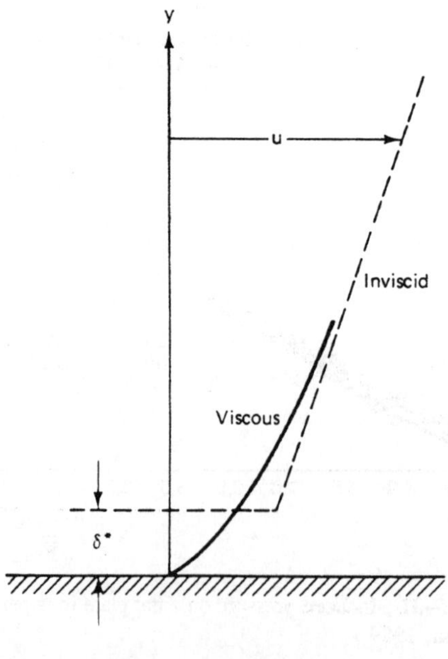

Figure 5–33 Velocity profile for a boundary layer with free stream vorticity.

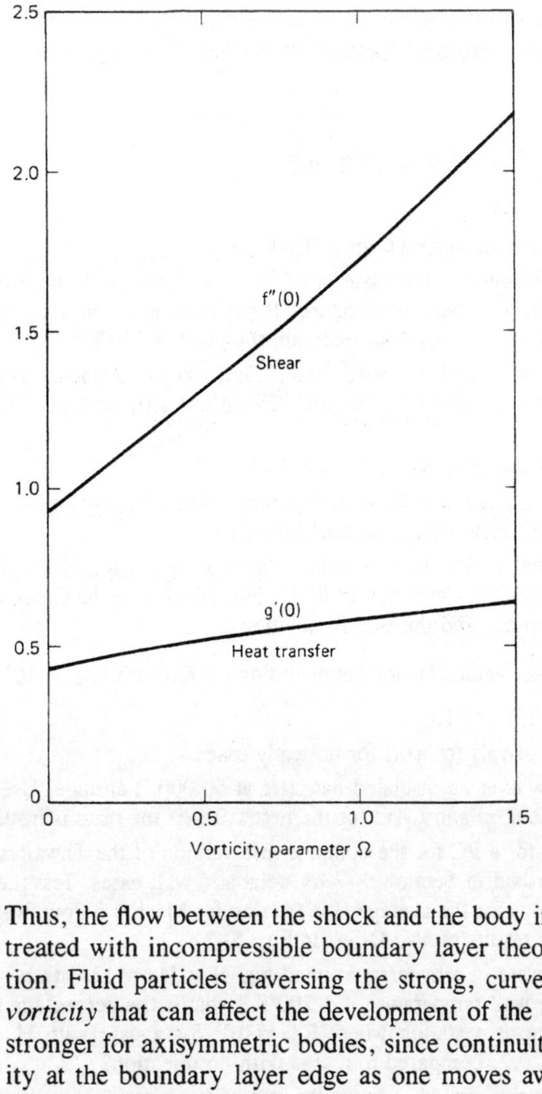

Figure 5–34 Effect of vorticity parameter on skin friction and heat transfer at an axisymmetric stagnation point. (From Kemp, 1959.)

Thus, the flow between the shock and the body in the stagnation point region can be treated with incompressible boundary layer theory to a good degree of approximation. Fluid particles traversing the strong, curved bow shock acquire a *free stream vorticity* that can affect the development of the boundary layer. The effect is much stronger for axisymmetric bodies, since continuity leads to an increase in the vorticity at the boundary layer edge as one moves away from the stagnation point. The analysis of the stagnation point region uses Eq. (5–46) with $G = 1.0$ and $\bar{\beta} = 1/2$ for incompressible flow at an axisymmetric stagnation point. The boundary condition on f' as η goes to infinity must be replaced to account for the vorticity at the boundary layer edge. The velocity profile can be expected to look as shown in Fig. 5–33, where the boundary layer profile should have a constant, non-zero slope at the boundary layer edge. Thus, the appropriate outer boundary condition can be written as $f''(\infty) = \Omega$, where Ω is a *vorticity parameter* defined as

$$\Omega \equiv \frac{1}{\left(\frac{8}{3}\right)^{3/4}\left(\frac{\gamma - 1}{\gamma + 1}\right)^{7/4}} \sqrt{\frac{\mu_e}{2\rho_e U_\infty R_o}} \tag{5–99}$$

Kemp (1959) treated this problem, with the results shown in Fig. 5–34. Note that large increases in skin friction and heat transfer are found as the vorticity parameter increases.

PROBLEMS

5.1. Plot T_{kinetic}/T_t versus V for air and CO_2 up to 3,000 m/s.

5.2. Consider the flow of H_2 over a flat plate at $M_\infty = 2.0$, $p_\infty = 10^{-1}$ atm, and $T_{t,\infty} = 600°K$. What is the adiabatic wall temperature? If the plate is 20 cm long and 10 cm wide, how much cooling is necessary to maintain the plate at 300°K?

5.3. Prove that for Pr = 1, the maximum static temperature in the boundary layer over a flat plate occurs where u is given by Eq. (5–23). (*Hint:* Differentiate with respect to u.)

5.4. Plot $T(u)$ for the numerical example in Section 5–4.

5.5. Find the adiabatic wall temperature for a surface with a local Mach number of 0.85 at an altitude of 10,000 m, assuming a *standard* atmosphere.

5.6. Find u/U_e versus y for a flat-plate boundary layer in air with $M_\infty = 3.0$, $p_\infty = 10^{-1}$ atm, $T_{t,\infty} = 300°K$, $T_w/T_\infty = 1.0$, and $\delta = 1$ cm. (*Hint:* Use the Crocco integral, the Howarth transformation, and the Blasius solution.)

5.7. Plot $\frac{\delta}{x}\sqrt{\text{Re}_x}$ and $\frac{\delta_T}{x}\sqrt{\text{Re}_x}$ versus M_e for flat-plate flow of air with $\text{Re}_L = 10^5$ and with an adiabatic wall and $T_w/T_\infty = 1.0$.

5.8. Rederive the Crocco integrals for $h(u)$ for unsteady cases.

5.9. Consider Mach 2.0 flow over an insulated flat plate at 60,000 ft altitude. Use the integral method to calculate $\theta(x)$ and $C_f(x)$ for the first 1.0 ft if the plate is insulated.

5.10. Write a computer code for a PC for the compressible version of the Thwaites-Walz integral method, as described in Section 5–7 for adiabatic wall cases. Test the code by comparing the predictions for the average skin friction for Mach 4.0 flow over an insulated flat plate with the results of Van Driest in Fig. 5–3.

5.11. Consider a porous flat plate in a hypersonic wind tunnel at $M_e = 12$ with $p_t = 30$ atm and $T_e = 200°K$. If the wall temperature is 1,200°K, what is the percentage reduction in the heat transfer rate at a station where $\text{Re}_x = 10^5$ for a case with H_2 injection sufficient to give $c_{1w} = 0.2$, compared to a case with no injection?

5.12. For the conditions of Problem 5.11, what is the rate of H_2 injection required?

5.13. The atmosphere on a remote planet is mostly hydrogen. Consider a probe with surfaces like a flat plate flying at Mach 3.0 in the planet's atmosphere. The static temperature is 500°K, and the static pressure is 0.1 atm. The flat plate surfaces are 30 cm long and 20 cm wide, and the plate temperature must be kept at 300°K. What is the drag and heat transfer on each plate?

5.14. What strength of impinging shock wave is required to separate a laminar boundary layer at $M_e = 2.0$ at $\text{Re} = 10^5$?

5.15. Estimate the pressure rise on an adiabatic flat plate with $M_e = 12$, where $\text{Re}_x = 10^5$. Is this a *strong* or *weak* interaction?

5.16. Use the compressible numerical method to treat the compressible version of the *Howarth problem,* defined as a case with an external Mach number variation of $M_e(x) = M_{eo}(1 - x/L)$ with an insulated wall. Consider a case with $M_{eo} = 2.0$, and compare your prediction for the separation point with the value of $(x/L)_{\text{sep.}} \approx 0.09$ obtained by an early numerical solution. What is the effect of Mach number in the range $0 \leq M_{eo} \leq 4$? What is the effect of wall cooling in the range $0.25 \leq T_w/T_e \leq 2.0$ at $M_{eo} = 3.0$?

CHAPTER 6

TRANSITION TO TURBULENT FLOW

6-1 INTRODUCTION

The material in Chapters 1 to 5 completed the study of two-dimensional external laminar boundary layer flows at the level intended for this book. Internal flows will be covered in Chapter 8, and three-dimensional external laminar flows are discussed in Chapter 11. Chapters 7, 10, and, 11 are concerned with external turbulent boundary layer flows. In this chapter, one of the most important and probably the most difficult subject in boundary layer work is addressed: the prediction of transition from laminar to turbulent flow. In Chapter 1, some features of turbulent flows were introduced, and it has been stressed that most flows of engineering interest are turbulent. But how is the fluid dynamicist to be certain that a flow will be laminar or turbulent under a given set of conditions? This question is of more than casual interest, because the levels of skin friction and heat transfer, for example, are quite different in the two cases (see Chapters 4 and 7, for example). This difference is largely because of the greatly increased mixing due to the random motion of the turbulent eddies. If the skin friction is strongly influenced, one can guess that separation is strongly influenced, and that is, indeed, the case. Thus, the items of principal interest to the analyst or designer depend critically on whether a given flow is laminar or turbulent. Common experience tells us that flows with high Reynolds numbers are more likely to be turbulent, but such a crude statement is hardly satisfactory for careful design. The designer needs accurate methods for predicting transition, but unfortunately, these are hard to achieve. In Sections 6–2 to 6–4, attempts at the analytical prediction of transition and their limitations are discussed. The remainder of the chapter is concerned with selected empirical information on the subject. This coverage is an accurate reflection of the current state of the art in the area, where much more trust is placed in empiricism than in analysis.

Before examining some of the analytical prediction techniques, it is informative to study briefly the earliest organized experimental investigations of transition—Reynolds's (1883) classical pipe flow experiments. These can be supplemented by the reader's own observations of transition phenomena, in a water faucet efflux as the valve is opened (increasing the Reynolds number) or in the smoke plume from a cigarette resting in an ashtray, to cite two common examples. Reynolds's experiments were conducted in an apparatus shown schematically in Fig. 6–1. As the flow

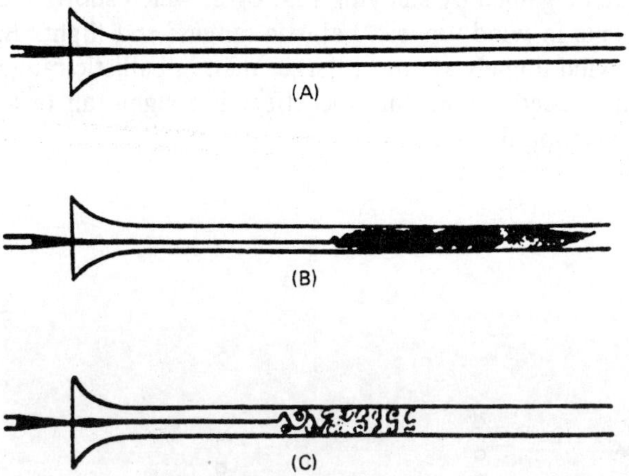

(A)–(C) Reynolds's sketches of flow observation: (A) low speed (low Re_D); (B) higher speed (higher Re_D); (C) spark illumination

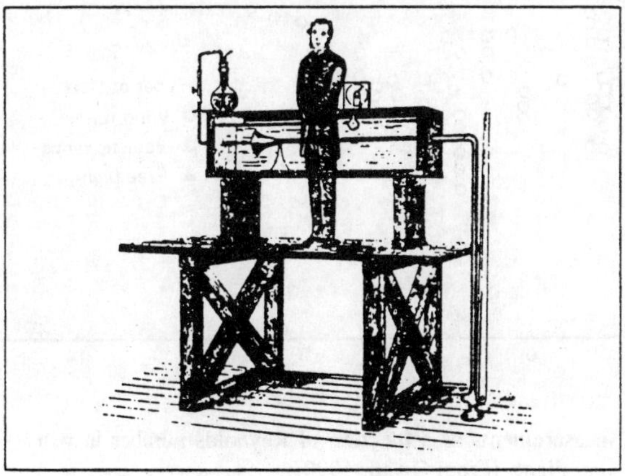

(D) Reynolds's experimental apparatus

Figure 6–1 The classic pipe flow experiments of Reynolds (1883).

rate through the glass pipe was increased, the behavior of the dye stream changed from that shown in Fig. 6–1(A) to that in Fig. 6–1(B), indicating a transition from laminar to turbulent flow. By increasing the velocity in a pipe of fixed size D with the same fluid (constant ρ and μ), Reynolds was increasing the dimensionless grouping $\rho VD/\mu \equiv Re_D$, named after him. He found that the value for transition, called the *critical* Reynolds number, was approximately 2,300. Later studies have shown that this critical value is very sensitive to the inlet conditions. By careful tailoring of the inlet flow, it has been possible to maintain laminar flow up to $Re_D \approx 20{,}000$. A value of about 2,300 is commonly used in engineering practice, where the inlet conditions are seldom "clean" in the laboratory sense, but this extreme sensitivity to the conditions of the experiments persists throughout all transition work. It contributes to the apparent *scatter* in the empirical information, but actually one is seeing the effects of small but important differences in seemingly identical experiments. Some appreciation of this matter can be gained by studying Fig. 6–2, which shows the results of transition measurements in wind tunnels, ballistic ranges, and flight. Background disturbances in most wind tunnels are much larger than in ballistic ranges or flight, especially at supersonic speeds. Note that the effect is roughly an order of magnitude variation in the transition Re.

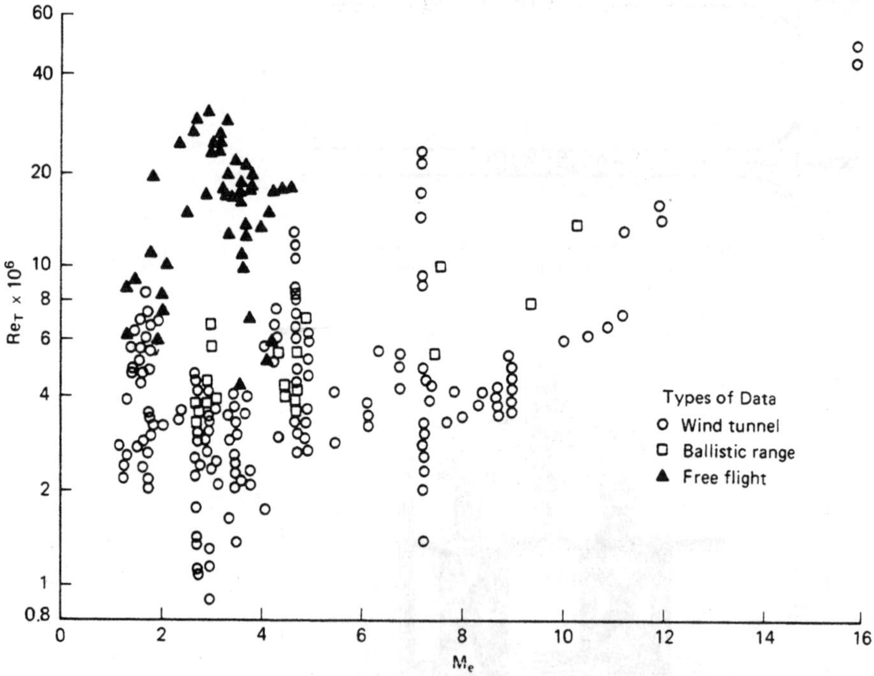

Figure 6–2 Transition measurements as a function of Reynolds number in wind tunnels, ballistic ranges, and flight. (From Owen, 1990.)

6–2 A SIMPLE METHOD BASED ON Re_θ

A crude but very simple method for roughly predicting the location of transition was presented by Michel (1952). He succeeded in correlating the transition location for low-speed flows on the basis of only the Reynolds number based on momentum thickness, $Re_\theta \equiv \rho U_e \theta/\mu$. Michel found that

$$\mathrm{Re}_{\theta,\,\mathrm{trans}} \approx 2.9\,\mathrm{Re}_{x,\,\mathrm{trans}}^{0.4} \tag{6–1}$$

where $\mathrm{Re}_x \equiv \rho U_e x/\mu$. The required distribution of $\theta(x)$ can be calculated by the Thwaites-Walz method (see Section 2–3–2). A calculation of $\theta(x)$ can be expressed as Re_θ vs. Re_x and inserted into Eq. (6–1). When the relation in this equation is satisfied, transition at that Re_x is indicated. The results are reliable to about ± 30–50 percent.

To see how the procedure works, consider the simple case of a flat plate. From Eq. (4–25), we can write

$$\mathrm{Re}_\theta = 0.664\,\mathrm{Re}_x^{0.5}$$

Inserting this equation into Eq. (6–1) yields

$$0.664\,\mathrm{Re}_{x,\,\mathrm{trans}}^{0.5} \approx 2.9\,\mathrm{Re}_{x,\,\mathrm{trans}}^{0.4} \tag{6–1a}$$

which can be solved to give $\mathrm{Re}_{x,\,\mathrm{trans}} \approx 2.5 \times 10^6$.

6–3 HYDRODYNAMIC STABILITY THEORY

Visual observations of turbulence clearly indicate that the transition process is the end result of the growth of initially small, probably random disturbances in the flow, unless there are large background disturbances that bypass this process. For an example view the film *Characteristics of Laminar and Turbulent Flow* produced by the University of Iowa. Small disturbances due, for example, to noise or slight vibrations of solid surfaces are always present in the background of any flow. Apparently, under some conditions in the flow, these disturbances are damped out, whereas under other conditions, they are amplified. From Reynolds's experiments, one can expect that the value of a suitable Reynolds number will describe the boundary between the two cases. Recall from Chapter 1 that the Reynolds number expresses the ratio of inertial forces to viscous forces in a flow. Thus, increasing the Reynolds number means decreasing the relative importance of viscous forces and viscous damping. This situation bears a close resemblance to the vibration of mechanical systems, which had been analyzed rather thoroughly by the early 1900s. Thus, it should not be surprising that attempts were made to apply the same general methods to the question of the stability of a flow to small disturbances. These attempts were quite successful and yet unsuccessful at the same time. (This will become clear shortly.) Also, the mathematics of the analyses proved very complex and interesting (to mathematicians). For these reasons, we give only an overview of the theory here.

For the interested reader, there are whole books devoted to the subject, and Malik's (1990) review is informative.

In general, one wishes to analyze the unsteady three-dimensional behavior of a small disturbance in a background laminar boundary layer flow. This is obviously a difficult undertaking in the general case, and some simplifying assumptions are required. First, there is a helpful assist from a theorem by Squire (1933), which proved that, in low-speed flow, two-dimensional disturbances are always less stable than three-dimensional disturbances. Thus, one need only consider the two-dimensional case to find the minimum instability conditions. The principal simplifying assumptions are that (1) the disturbance quantities remain small compared to the baseline flow (if $u = u_0 + \hat{u}$, then $\hat{u} \ll u_0$, etc.), (2) the baseline flow is a function only of the transverse coordinate [i.e., $u_0 = u_0(y)$ alone, etc.], and (3) the disturbance can be written in terms of a stream function in the form

$$\psi(x, y, t) = \phi(y) \exp[i\alpha(x - ct)] \tag{6–2}$$

Here, $i \equiv \sqrt{-1}$, $\phi(y)$ is a complex amplitude function, α is the wave number of the disturbance ($= 2\pi/\text{wavelength}$), and c is the complex phase velocity ($= c_r + ic_i$). Looking at Eq. (6–2), one can see that stability with respect to time depends on the value of αc_i. For $\alpha c_i < 0$, there is damping (i.e., the disturbance decreases with time); $\alpha c_i = 0$, a neutral condition is achieved, and for $\alpha c_i > 0$, amplification of the disturbance and thus instability results. Perhaps it is helpful to note that the form assumed in Eq. (6–2) implies that

$$\begin{aligned} \hat{u} &= \frac{\partial \psi}{\partial y} = \frac{\partial \phi}{\partial y} \exp[i\alpha(x - ct)] \\ \hat{v} &= -\frac{\partial \psi}{\partial x} = -i\alpha\phi(y) \exp[i\alpha(x - ct)] \end{aligned} \tag{6–3}$$

With this information, the unsteady Navier-Stokes equations of motion for a laminar, incompressible, constant-property flow can be simplified to

$$(u_0 - c)(\phi'' - \alpha^2\phi) - u_0''\phi = \frac{-i}{\alpha \operatorname{Re}}(\phi^{(iv)} - 2\alpha^2\phi'' + \alpha^4\phi) \tag{6–4}$$

This is the famous Orr-Sommerfeld equation, and it proved difficult to solve for general profile shapes $u_0(y)$ before the advent of the large digital computer. For that reason, much of the earliest work concentrated on cases with a linear profile ($u_0 \sim y$, i.e., $u_0'' \equiv 0$) and/or inviscid flow ($\operatorname{Re} \to \infty$).

Various calculations for the general case have been presented. Lin (1945) and others used asymptotic theory, and in more recent times, direct numerical solutions have been obtained, such as those of Wazzan et al. (1968). Some of these results are shown in Figs. 6–3 and 6–4. The results of the theory have been confirmed by experiment, as far as the onset of instability is concerned. Schubauer and Skramstad (1947) placed a thin magnetic ribbon in a laminar, flat-plate boundary layer and then excited the ribbon externally to vibrate, producing disturbances at various known wavelengths. Theory and experiment agreed rather well, as shown in the figures.

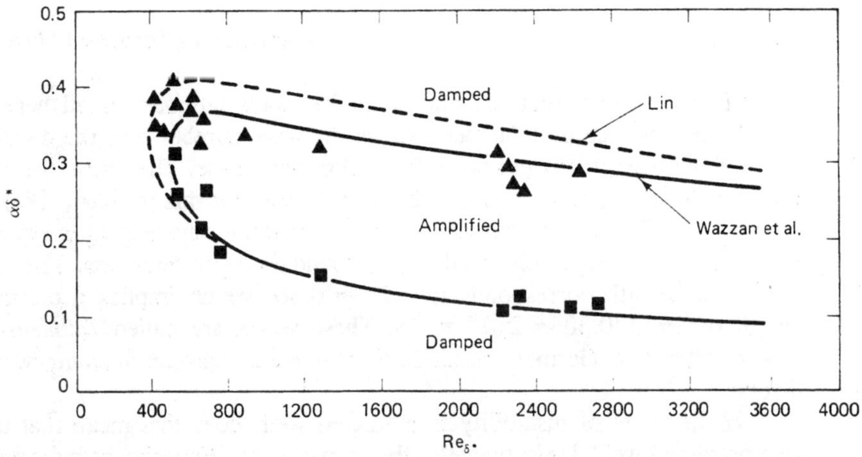

Figure 6–3 Comparison of the predictions of the hydrodynamic stability theory by Lin (1945) and Wazzan et al. (1968) for the wavelength of neutral disturbances for flow over a flat plate with the measurements of Schubauer and Skramstad (1947).

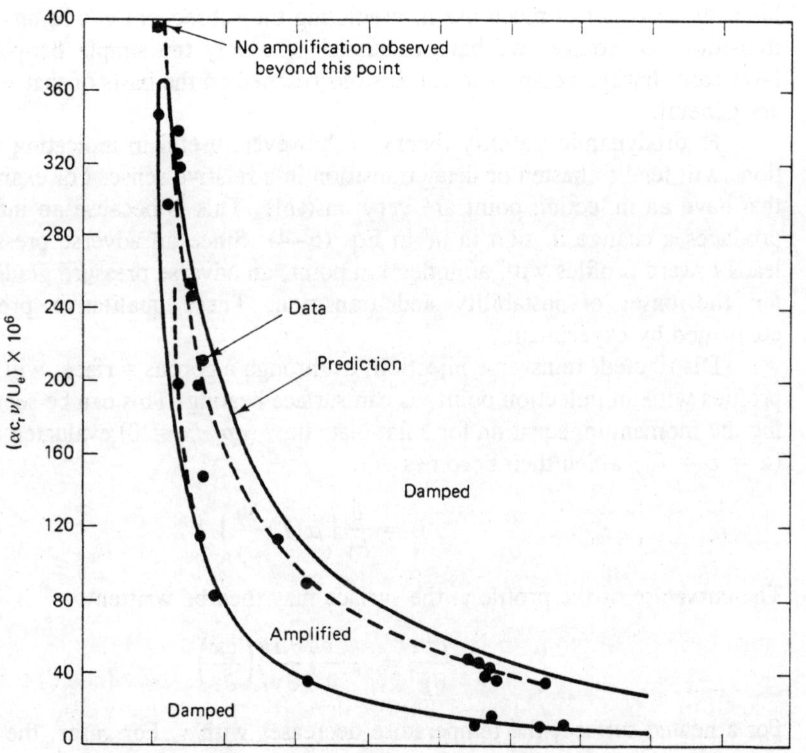

Figure 6–4 Comparison of the prediction of the hydrodynamic stability theory by Lin (1945) for the frequency of neutral disturbances for flow over a flat plate with the measurements of Schubauer and Skramstad (1947).

It can be seen that the maximum Reynolds number (based here on δ^*) at which a disturbance is stable depends on the wave number α of the disturbance, but that number is generally unknown for background noise. Therefore, the only rational choice is to focus on the worst case, which gives, for this problem, $(\mathrm{Re}_{\delta^*})_{\mathrm{crit}} = 520$ for $\alpha\delta^* \approx 0.30$. It is informative to observe that the wavelengths of typical unstable disturbances are large compared to the boundary layer thickness. The smallest unstable wavelength corresponds to $\alpha\delta^* \approx 0.30$, which implies a minimum wavelength of $2\pi\delta^*/0.30 \approx 21\delta^* \approx 7\delta$. These waves are called *Tollmein-Schlichting waves,* after two German researchers who made substantial contributions to the theory.

If the onset of instability is predicted well, does that mean that transition is also predicted well? Unfortunately, the answer is no. Experiment indicates that transition on a flat plate in a relatively clean external flow (a low free stream turbulence level of about 0.1%) occurs at about $(\mathrm{Re}_x)_{\mathrm{trans}} \approx 2.5 \times 10^6$. Using the Blasius solution (Eq. (4-24)), we can see that this translates to $(\mathrm{Re}_{\delta^*})_{\mathrm{trans}} \approx 2{,}700$, which is much higher than the instability condition, at 520. The difficulty is simply that *instability* and *transition* are *not the same thing*. Instability is merely the very early precursor of ultimate transition to turbulent flow, and knowing the conditions for instability alone is not of much use in predicting the subsequent conditions ($\mathrm{Re}_{x,\,\mathrm{trans}}$) for transition. Of course, we have discussed here only the simple flat-plate boundary layer case, but the pessimistic conclusions reached on the basis of that single example are general.

Hydrodynamic stability theory is, however, useful in indicating which conditions will tend to hasten or delay transition in a relative sense. For example, profiles that have an inflection point are very unstable. This is because an inflection point produces a change in sign in u_0'' in Eq. (6–4). Since an adverse pressure gradient leads toward profiles with an inflection point, an adverse pressure gradient will hasten the onset of instability and transition. These qualitative predictions are confirmed by experiment.

Distributed, transverse injection, as through a porous surface, will also produce profiles with an inflection point, as can surface heating. This can be seen by examining the momentum equation for a flat-plate flow ($dp/dx = 0$) evaluated at the surface ($u = v = 0$), which then becomes

$$0 = \frac{\partial}{\partial y}\left(\mu(T)\frac{\partial u}{\partial y}\right) \tag{6–5}$$

The curvature of the profile at the surface may then be written

$$\frac{\partial^2 u}{\partial y^2} = -\frac{1}{\mu}\left(\frac{\partial \mu}{\partial y}\right)\left(\frac{\partial u}{\partial y}\right) \tag{6–6}$$

For a heated surface, the temperature decreases with y. For gases, the viscosity increases with temperature, and $\partial\mu/\partial y < 0$, leading to $(\partial^2 u/\partial y^2)_w > 0$. Near the outer edge of the layer, the curvature is always negative, so surface heating in a gas

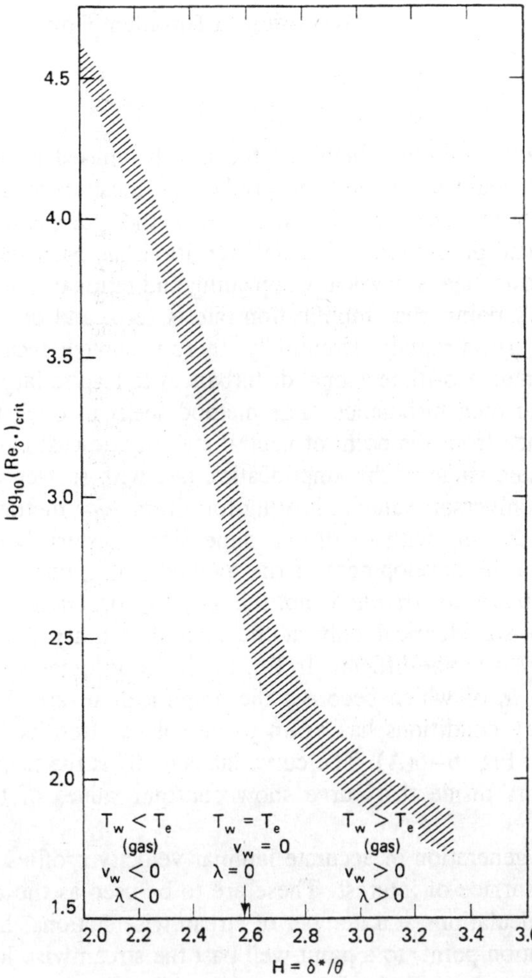

Figure 6–5 Minimum Reynolds number for instability in low-speed flow of gases (heating/cooling effects), as influenced by pressure gradient, injection/suction, and heating/cooling. (From Stuart, 1963.)

flow implies an inflection point in the layer and decreased stability. Since, for liquids, the viscosity generally decreases with temperature, this whole picture is reversed.

All of these influences on profile shape can be expressed in terms of the shape parameter $H \equiv \delta^*/\theta$. Adverse pressure gradients increase H (see Table 2–1), as does injection or surface heating for gases. Favorable pressure gradients, suction, and surface cooling in gas flows reduce H. The effect of variations in H on stability is shown for gases in Fig. 6–5 from Stuart (1963). The effects of pressure gradient and suction or injection are the same for gases and liquids, but the effects of heating and cooling are reversed. These results have led to the design of airfoils with the maximum thickness as far back as possible (e.g., in the famous P–51 *Mustang*), to delay the beginning of the adverse pressure gradient and thus transition. Also, suction remains a topic of study in the quest to delay transition and thus reduce total drag on aircraft with *laminar flow control*.

6–4 THE e^N METHOD

For many years, hydrodynamic stability theory rested largely unused by working fluid dynamicists, because it could not be used to predict the location of transition for a given flow problem. More recently, a clever heuristic method for extending the theory to produce approximate predictions of actual transition has been developed by Jaffe et al. (1970). The basic idea is physically appealing and rational, but not rigorous. At the neutral stability point, the amplification rate is zero, and beyond that point, the amplification rate grows rapidly. Eventually, there is enough accumulated amplification of the initial small two-dimensional disturbances to lead to large three-dimensional disturbances and then turbulence. The method seeks to track the amplification rate of a disturbance from the point of neutral stability downstream along the surface, until the integrated value of the amplification rate with surface distance reaches a certain, hopefully universal, value indicating transition. The method is not rigorous, since it deals directly only with small two-dimensional disturbances.

There are three steps in the development of the method. First, one must deal with the *spatial* behavior of *small disturbances,* not the *temporal* or *timewise* behavior treated before. The two are identical only at the neutral stability point. The derivation of the spatial analysis is not difficult. In Eq. (6–2), α becomes a complex number, the imaginary part α_i of which becomes the amplification rate. Equation (6–4) with suitable boundary conditions has again to be solved. Results for flow over a flat plate are shown in Fig. 6–6(A). The curve labeled "0" is the neutral stability curve, and the contours inside that curve show constant values of the amplification.

The second step is the generation of accurate laminar velocity profiles at various stations along the body surface of interest. These are to be used as the baseline profiles $u_0(y)$ for stability calculations at a number of streamwise stations, from the leading edge (or front stagnation point) to a point well past the streamwise location, where transition may be expected on the basis of some cruder estimation technique or data correlations. Obviously, accurate profiles are required, or else the stability calculations based on them will be in error. Also, since calculations at a number of stations must be made, efficiency is important. A modern, implicit finite difference procedure is recommended for this purpose.

The third step is to implement the method and see whether comparisons with experiment yield any universal value for the integrated amplification rate that can be used as a predictor of transition in later calculations. If A_0 is the amplitude of a fixed frequency at the neutral stability point, and if A is the amplitude at some greater x, then

$$\frac{A}{A_0} = e^N \tag{6–7}$$

where

$$N = -\int_{x_n}^{x} \alpha_i(x')\,dx' \tag{6–8}$$

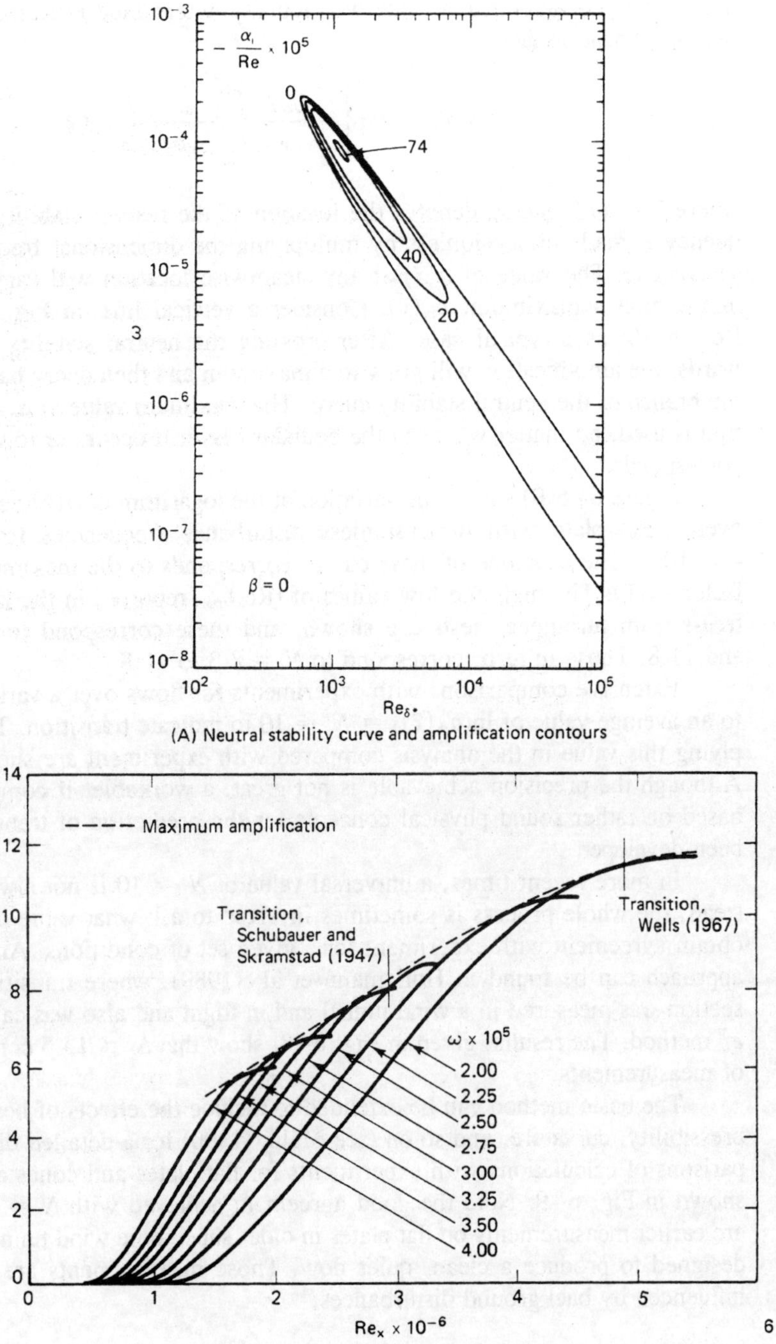

Figure 6–6 Calculations of the stability of flat-plate boundary layer flow. (From Jaffe et al., 1970.)

The integrated amplification rate was actually determined in terms of a dimensionless amplification factor

$$a(\bar{x}) = \exp\left[-\frac{U_\infty L}{\nu}\int_{\bar{x}_n}^{\bar{x}} \frac{\alpha_i}{U_e\delta/\nu}\frac{U_e}{U_\infty}\,d\bar{x}\right] \tag{6-9}$$

where $\bar{x} = x/L$ and $\bar{x}_n$ denotes the location of the neutral stability point. The frequency is made dimensionless by multiplying the dimensional frequency by ν/U_e^2, yielding ω. The value of $a(\bar{x})$ at any steamwise location will vary with frequency and exhibit a maximum, $a_m(\bar{x})$. Consider a vertical line on Fig. 6–6(A) at, say, $\mathrm{Re}_{\delta^*} = 10^3$ as a typical case. After crossing the neutral stability curve going upwards, the amplification will grow to a maximum and then decay back to zero on the top branch of the neutral stability curve. The maximum value of a_m at each axial station is used, no matter where in the boundary layer it occurs or to what frequency it corresponds.

Figure 6–6(B) shows the variation of the logarithm of $a(\bar{x})$ versus Re_x for flow over a flat plate with dimensionless disturbance frequencies from 2×10^{-5} to 4×10^{-5}. The envelope of these curves corresponds to the maximum amplification factor $a_m(\bar{x})$. The high and low values of $(\mathrm{Re}_x)_{\mathrm{trans}}$ reported in the literature for low-free-stream turbulence tests are shown, and these correspond to $\ln(a_m(\bar{x})) = 8.3$ and 11.8. Those in turn, correspond to $N = 8.3$ to 11.8.

Extensive comparisons with experiments for flows over a variety of bodies led to an average value of $\ln(a_m(\bar{x})) = N \approx 10$ to indicate transition. The results of applying this value in the analysis compared with experiment are shown in Fig. 6–7. Although the precision achievable is not great, a workable, if complicated, method based on rather sound physical concepts for the prediction of transition has clearly been developed.

In more recent times, a universal value of $N \approx 10$ is not always retained. Instead, the whole process is sometimes inverted to ask what value of N is needed to obtain agreement with experiment for a given set of conditions. An example of this approach can be found in Horstmann et al. (1989), where transition on an airfoil section was measured in a wind tunnel and in flight and also was calculated with the e^N method. The results, given in Fig. 6–8, show that $N = 13.5$ correlated both sets of measurements.

The basic method can be extended to include the effects of heat transfer, compressibility, curvature, and so on (see Malik (1990) for a detailed discussion). Comparisons of calculations with experiments for flat plates and cones at March 3.5 are shown in Fig. 6–9. Note the good agreement achieved with $N = 10$. Also shown are earlier measurements on flat plates in older supersonic wind tunnels that were not designed to produce a clean, quiet flow. Those measurements are clearly strongly influenced by background disturbances.

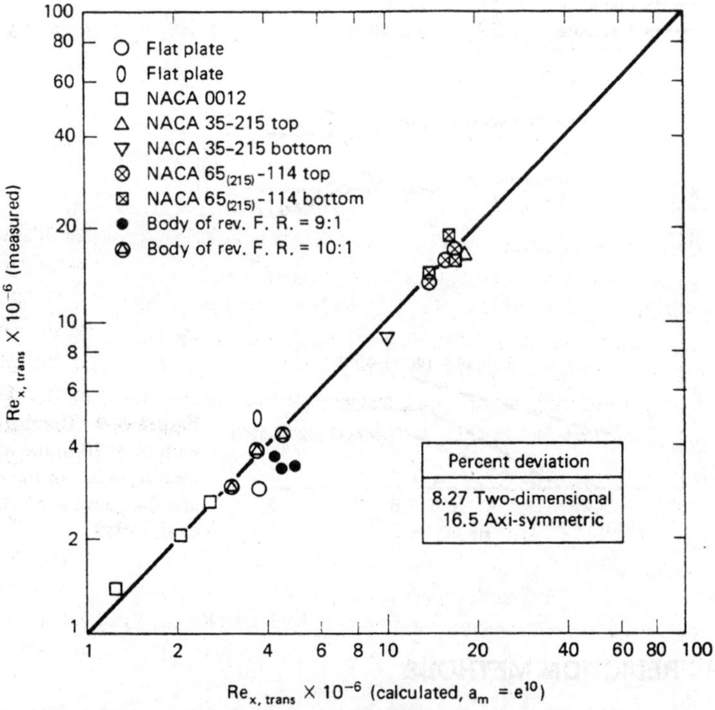

Figure 6–7 Comparison of prediction of transition by the e^{10} method with experiment. (From Jaffe et al., 1970.)

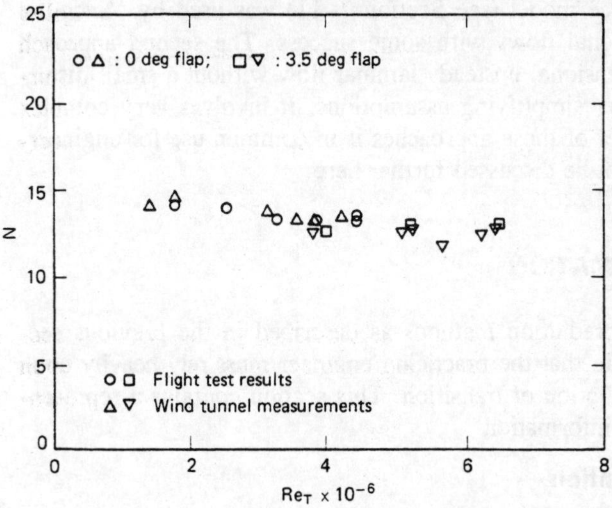

Figure 6–8 Calculated N factor at transition on a wing section for flight and wind tunnel experiments (From Horstmann et al., 1989.)

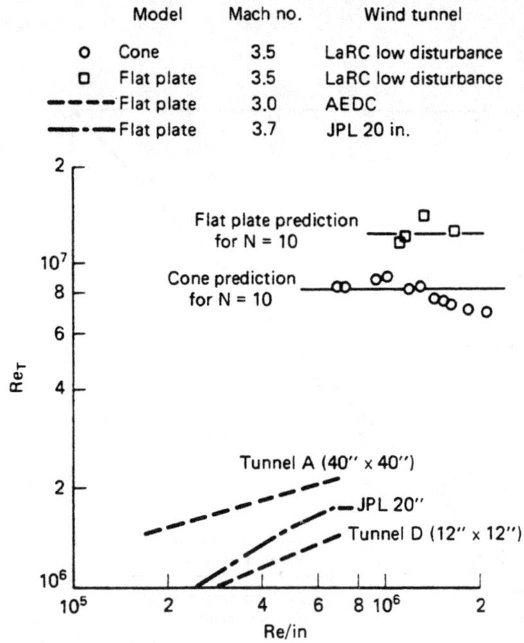

Figure 6–9 Comparison of predictions with $N = 10$ in the e^N method with measurements of transition on cones and flat plates at Mach 3.5 (from Chen et al., 1989.)

6–5 OTHER PREDICTION METHODS

There are two other, more elaborate, approaches to the prediction of transition. The first uses turbulence models of the type to be described in detail later in this book. For example, the $K\epsilon$ turbulence model (see Section 7–11) was used by Vancoillie (1984) for computing transitional flows with some success. The second approach deals directly with three-dimensional unsteady laminar flow without a small-disturbance assumption or any other simplifying assumptions. It involves very complex and costly calculations. Neither of these approaches is in common use for engineering purposes, and they will not be discussed further here.

6–6 SELECTED EMPIRICAL INFORMATION

With the state of analytical prediction methods as described in the previous sections, the reader can appreciate that the practicing engineer must rely heavily upon empirical information for prediction of transition. This section contains a representative sample of the available information.

6–6–1 The Nature of Transition

The first thing that is important to understand about transition is that it is not a single, abrupt event in the flow, but rather a process involving several steps that occur over a region of the flow. Some researchers identify as many as seven steps, going from instability through large three-dimensional disturbances and finally to turbulence. The process can be seen in the excellent detailed photographs of flow over a flat plate from Werle (1982), shown in Fig. 6–10, where the flow is from left to

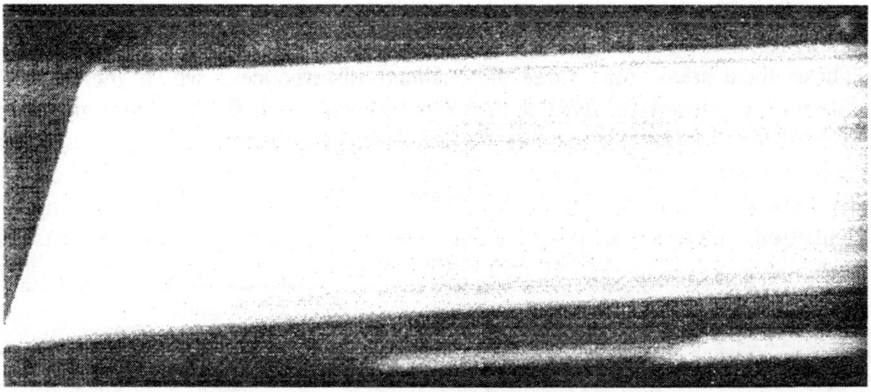

(A) $Re_L = 20{,}000$; Stable.

(B) $Re_L = 100{,}000$; Disturbances grow.

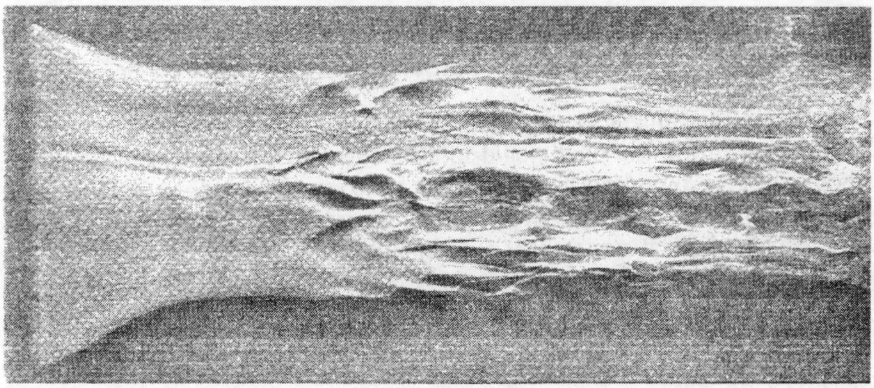

(C) $Re_L = 100{,}000$ with 1 deg. angle of attack to produce larger disturbances; Transition.

Figure 6–10 Photographs of dye streaks in a boundary layer on a flat plate in a water flow showing the growth of disturbances leading to transition. (From Werle, 1982, at ONERA.)

right. After the neutral stability point, two-dimensional disturbances begin to grow. These then break into three-dimensional disturbances, which themselves break down. Soon, localized *spots* of turbulence appear, and, finally, these merge together to produce fully turbulent flow. The length of the transition region, Δx_{tr}, as a function of Mach number and the Reynolds number at transition, has been correlated by Potter and Whitfield (1962), as shown in Fig. 6–11. Note that the length of the transition region is a sizable fraction of the total length up to the end of transition.

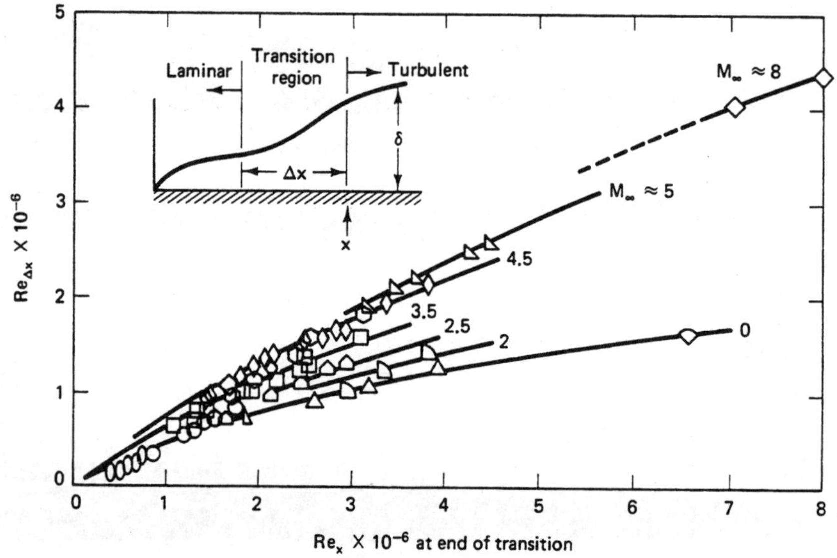

Figure 6–11 Correlation of the length of the transition region with the Reynolds number for transition and the Mach number. (From Potter and Whitfield, 1962.)

6–6–2 Effects of Free Stream Turbulence

Intuitively, one might expect that the level and nature of the background disturbances present in the flow would influence transition, and that is indeed the case. Some data for low-speed flow over a flat plate are given in Fig. 6–12, together with the results of a theory by Van Driest and Blumer (1963). This theory is founded on the simple physical notion that transition will correspond to a critical value of a Reynolds number based on the vorticity in the boundary layer given by

$$\mathrm{Re}_{\mathrm{vort}} \equiv \frac{(\partial u/\partial y)y^2}{\nu} \tag{6–10}$$

This special Reynolds number represents the ratio of a so-called inertial stress $\rho y^2(\partial u/\partial y)^2$ to the local shear stress $\tau = \mu(\partial u/\partial y)$. For flow over a flat plate, Van Driest and Blumer give

$$\frac{1{,}690}{\mathrm{Re}_{x,\,\mathrm{trans}}^{1/2}} = 1.0 + 19.6\,\mathrm{Re}_{x,\,\mathrm{trans}}^{1/2}\left(\frac{\overline{(u')^2}}{U_e^2}\right) \tag{6–11}$$

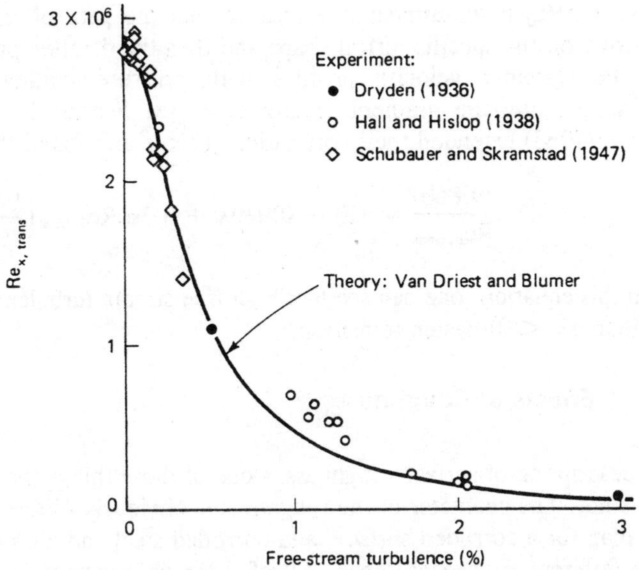

Figure 6–12 Comparison of prediction with experiment for the effect of free stream turbulence on transition for flow over a flat plate. (From Van Driest and Blumer, 1963.)

6–6–3 Effects of Pressure Gradients

Figure 6–13 shows a collection of wind tunnel data for the chordwise location of transition on various airfoils near zero angle of attack as a function of the chord Reynolds number, $Re_c = \rho U_\infty c/\mu$. The transition location always moves forward as

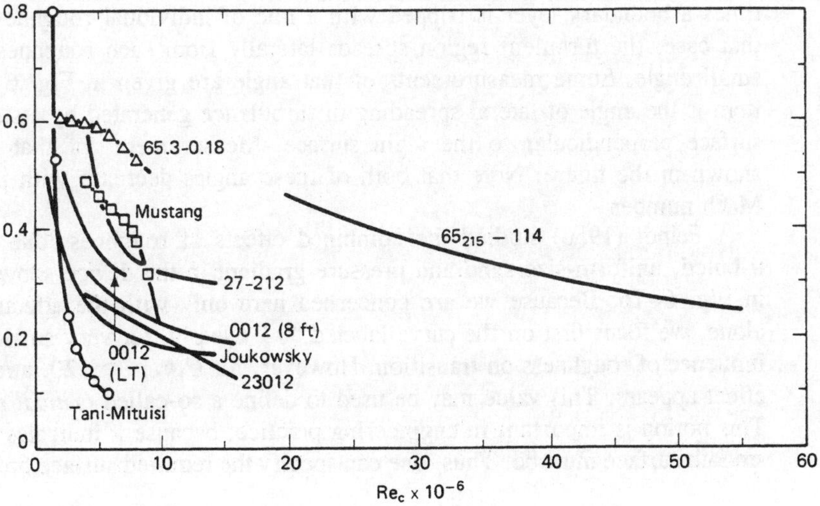

Figure 6–13 Collection of transition location data on airfoils. (From Lin, 1959.)

the chord Reynolds number is increased, but the path of movement depends significantly on the specific airfoil shape and thus the detailed pressure variation.

For laminar velocity profiles with pressure gradients described by the Pohlhausen pressure gradient parameter λ [see Section 2–3–1], Van Driest and Blumer (1963) extended their correlation scheme and found that

$$\frac{9{,}860}{\mathrm{Re}_{\delta,\,\mathrm{trans}}} = 1.0 - 0.049\lambda + 3.36\,\mathrm{Re}_{\delta,\,\mathrm{trans}}\left(\frac{\overline{(u')^2}}{U_e^2}\right) \tag{6–12}$$

From this equation, one can see that high free stream turbulence or adverse pressure gradients ($\lambda < 0$) hasten transition.

6–6–4 Effects of Roughness

The description of surface roughness is one of those things that sounds simple but really is not. The character of the roughness is obviously different for a machined surface than for a corroded surface, and corroded steel and aluminum surfaces are also quite different from each other. All of these observations have led to considerable difficulty in describing the influence of roughness in terms of any single parameter, such as the average roughness height k. Most careful studies have been done either for a single, isolated, two-dimensional roughness element—usually a wire on the surface—or for the simple distributed roughness produced by uniform-size sand glued to the surface.

The studies of single roughness elements were conducted because there has always been a great interest in *tripping* a normally laminar boundary layer in a lower Re wind tunnel or towing tank experiment, in order to simulate flight at a much higher Re. Dryden (1953) collected data for the influence of single two-dimensional roughness, as plotted in Fig. 6–14 in terms of $(\mathrm{Re}_{x,\,\mathrm{trans}})_{\mathrm{rough}}/(\mathrm{Re}_{x,\,\mathrm{trans}})_{\mathrm{smooth}}$. Sometimes a boundary layer is tripped with a line of individual roughness elements. In that case, the turbulent region spreads laterally from each roughness element at a small angle. Some measurements of that angle are given in Fig. 6–15. A related item is the angle of lateral spreading of turbulence generated by a sidewall or other surface perpendicular to the main surface. Measurements of that angle are also shown in the figure. Note that both of these angles decrease with increasing edge Mach number.

Feindt (1956) studied the combined effects of roughness due to evenly distributed, uniform-size sand and pressure gradient in the device shown in the insert in Fig. 6–16. Because we are concerned here only with the effects of roughness alone, we focus first on the curve labeled "0." Clearly, for very small k, there is no influence of roughness on transition. However, for $U_1 k/\nu \geq 120$, a rapidly growing effect appears. This value may be used to define a so-called *critical roughness size*. This notion is important in engineering practice, because it indicates how smooth a smooth surface must be. Thus, one can specify the required surface finish on a ration-

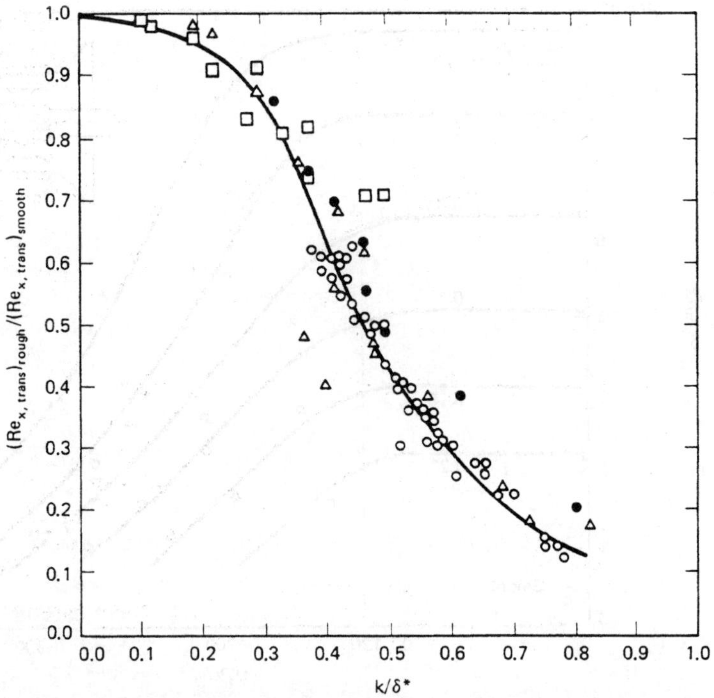

Figure 6–14 Ratio of the transition Reynolds number for flow over a flat plate with a single roughness element to that for a clean plate. (Collected from various sources by Dryden, 1953.)

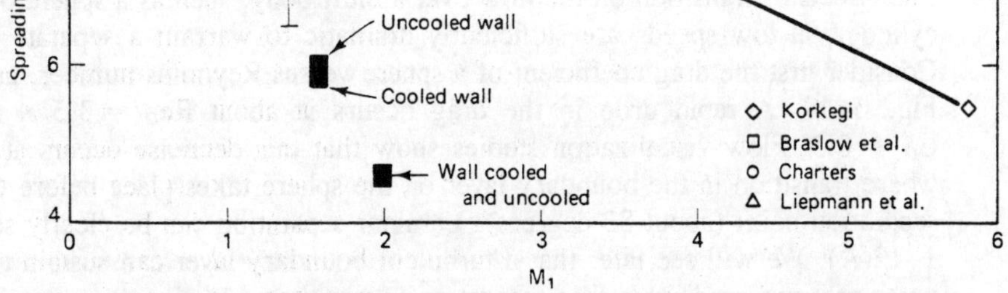

Figure 6–15 Angle of lateral spreading of turbulence from roughness elements and sidewalls. (From Wilson, 1966.)

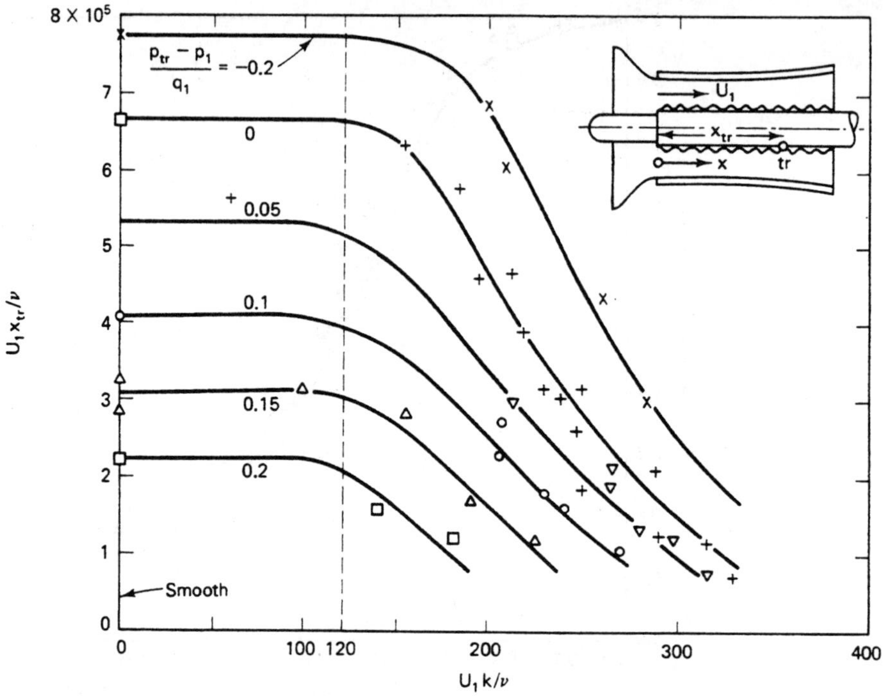

Figure 6–16 Effect of roughness due to evenly distributed, uniform-size sand grains and pressure gradient on transition for low-speed flow. (From Feindt, 1956.)

al basis, from the point of view of transition. The figure also shows the effect of both adverse and favorable pressure gradients.

6–6–5 Bluff Bodies at Low Speeds

The effects of transition on the flow over a bluff body, such as a sphere or a circular cylinder, at low speeds are sufficiently dramatic to warrant a separate discussion. Consider first the drag coefficient of a sphere versus Reynolds number, as plotted in Fig. 6–17. A rapid drop in the drag occurs at about $Re_D \approx 3.5 \times 10^5$, where $C_D \approx 0.3$. Flow visualization studies show that this decrease occurs at conditions where transition in the boundary layer on the sphere takes place before the laminar separation point (about 82 degrees). Laminar separation can be clearly seen in Fig. 1–12(A). We will see later that a turbulent boundary layer can sustain a larger adverse pressure gradient before separation occurs than a laminar one can. This means that the flow remains attached further along the surface of the sphere, keeping the pressure distribution close to the inviscid case (drag = 0) longer. The situation is displayed in Fig. 6–18. Thus, even though the local skin friction drag is higher in

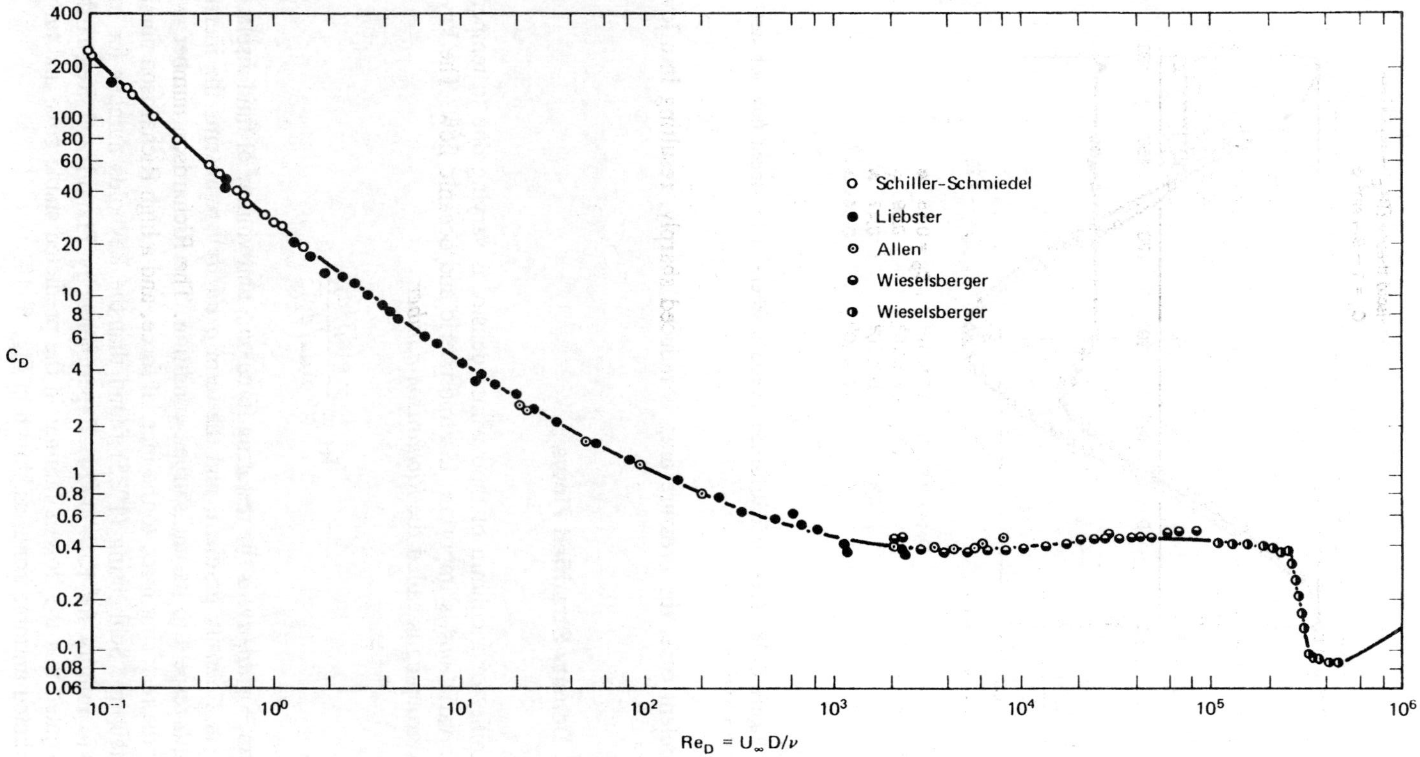

Figure 6–17 Experimental data from various researchers for the drag coefficient of a sphere in low-speed flow as a function of Reynolds number.

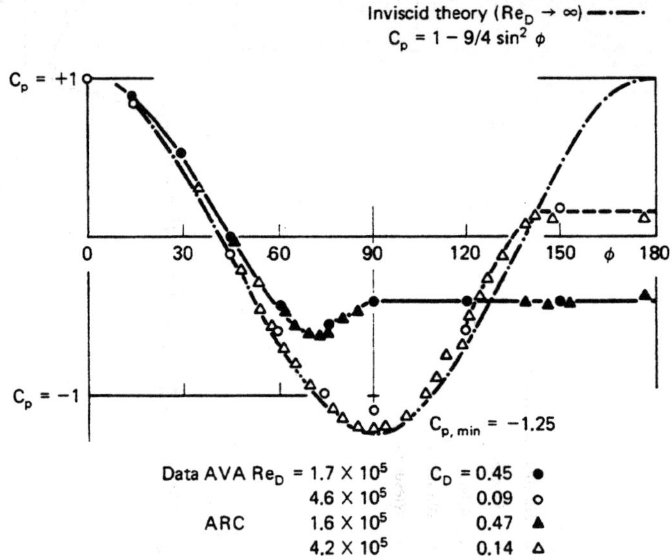

Figure 6–18 Pressure distribution around a sphere in low-speed flow for various Re_D.

the turbulent case, the pressure drag is reduced sharply, resulting in a lower total drag.

6–6–6 Density-Stratified Flows

The stability of a column of fluid whose density is varying due to temperature or pressure variations is important in atmospheric and oceanic flow. The key dimensionless grouping is called the *Richardson number*

$$\text{Ri} \equiv \frac{-g(d\rho/dy)}{\rho(du/dy)_w^2} \tag{6–13}$$

The term $-g(d\rho/dy)$ is the restoring force on a unit volume of fluid displaced a distance y in a density gradient, and the term $\rho(du/dy)^2$ represents the inertia of the fluid with respect to its undisturbed condition. The Richardson number is thus the ratio of the restoring force to the inertial force, and a high Richardson number indicates stability. Schlichting (1935) found that the Reynolds number for instability tended to infinity for Ri = 1/24. Experiments by Prandtl and Reichardt (1934) demonstrated that the flow is laminar in the predicted stable zone and turbulent in the predicted unstable zone, as shown in Fig. 6–19.

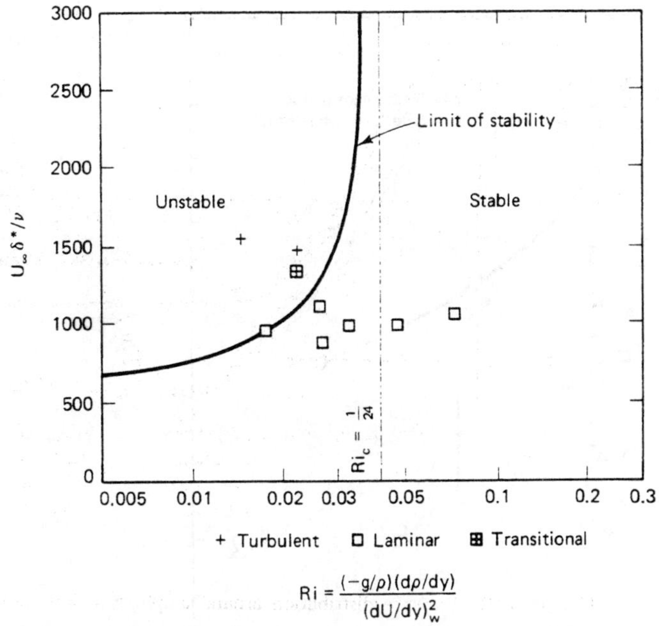

Figure 6–19 Density-stratified flow over a flat plate as a function of Richardson number. (From Schlichting, 1935.)

6–6–7 Supersonic Flow

The effect of increasing the Mach number is to decrease the transition Reynolds number, at least up to $M_\infty \approx 4$, as shown in Fig. 6–20 for flow over insulated cones. The data, shown as squares and triangles, may have been influenced by flow irregularities and high free stream turbulence. The general effect of T_w near $T_w/T_{aw} \approx 1.0$ is shown in Fig. 6–21 for Mach 3.0. Data for that value and higher Mach numbers at lower values of T_w/T_{aw} on slightly blunted cones are given in Fig. 6–22. Compressible hydrodynamic stability theory indicates that wall cooling is stabilizing up to about Mach 4.0 and destabilizing at higher Mach numbers, where the second mode of the disturbance becomes important.

The effect of isolated, two-dimensional roughness is shown in Fig. 6–23 for two Mach numbers. For high Mach numbers, it is virtually impossible to induce transition by *tripping* with a wire. The predicted effect of evenly distributed, uniform roughness is indicated in Fig. 6–24 as function of T_w/T_e at one moderate Mach number. These results show that cooling a rough surface reduces the Reynolds number (based on free stream conditions) for transition. The effect is explained by studying more closely the flow near a roughness element on the surface. A cooler fluid implies a lower kinematic viscosity and, hence, a higher unit Reynolds number based on conditions in that region of the flow.

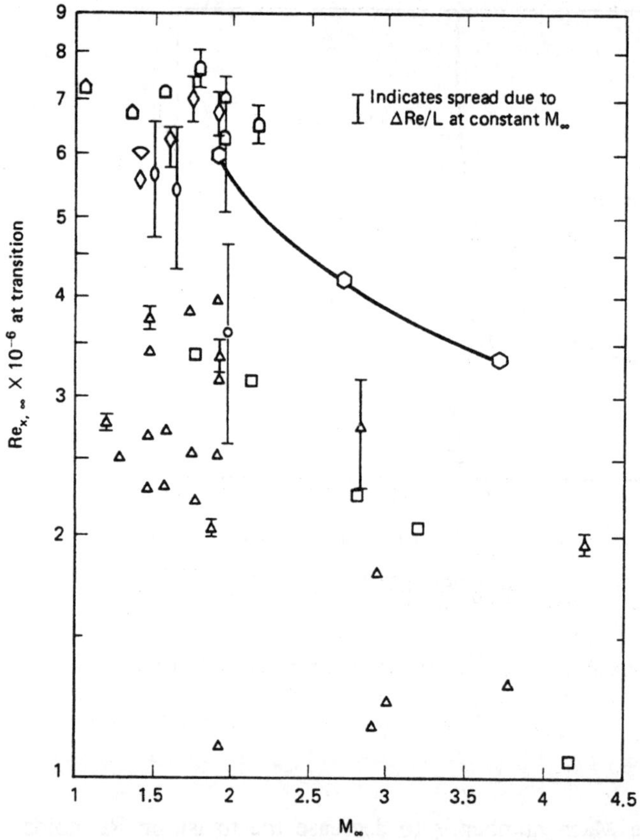

Figure 6–20 Experimental data for the effect of Mach number on transition on sharp, insulated cones. (Data from many sources collected by Wilson, 1966.)

It is interesting to mention here the phenomenon of *relaminarization*. In high-speed flows, it is possible to produce severe pressure gradients. For example, a rapid-expansion nozzle will produce a very strong, favorable pressure gradient that can actually cause a turbulent boundary layer to revert to a laminar state. Back et al. (1969) found that this occurred for flows with

$$\frac{\nu_e}{U_e^2}\frac{dU_e}{dx} > 2 \times 10^{-6} \tag{6–14}$$

6–6–8 Non-Newtonian Fluids

There is little organized information in the literature on transition in flows of non-Newtonian fluids. This is partly because there are so many different kinds of non-Newtonian fluids and partly because the apparent viscosity of most non-Newtonian fluids of practical interest is so high (low apparent Re $= \rho VL/\mu_a$), that most flows remain laminar.

Some theoretical studies of the stability of non-Newtonian fluids are summarized in a review article by Pearson (1976). Zahorski (1982) indicates a destabilizing tendency for viscoelastic non-Newtonian fluids compared to Newtonian fluids, all

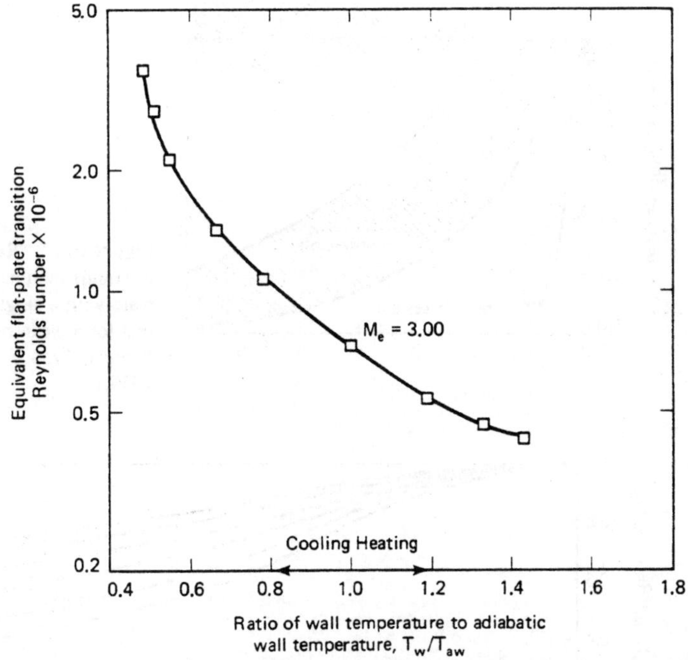

Figure 6–21 Effect of wall temperature on transition at Mach 3.0. Data from cones converted to data for equivalent flat plate. (From Jack and Diaconis, 1955.)

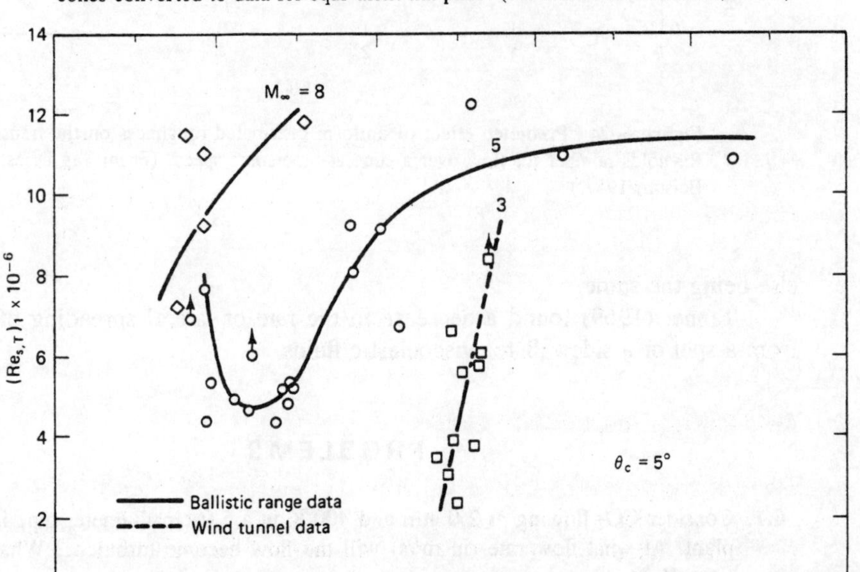

Figure 6–22 Effect of Mach number and wall temperature ratio on transition for cones. (From Sheetz, 1969.)

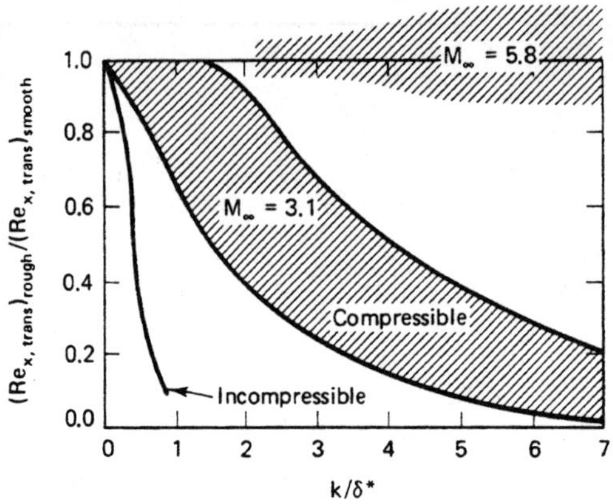

Figure 6–23 Ratio of the transition Reynolds number for flow over a flat plate with a single roughness element to that for a clean plate. (M = 3.1 from Brinich, 1954; M = 5.8 from Korkegi, 1956.)

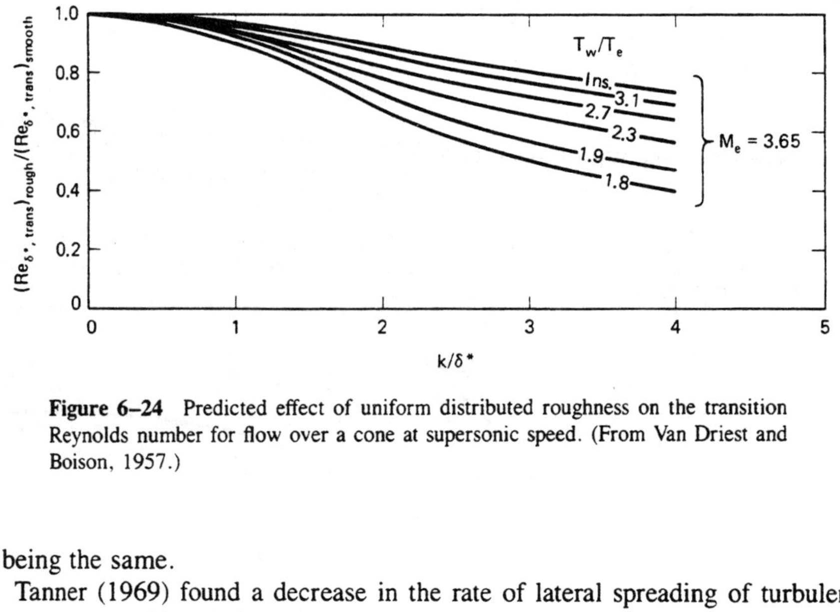

Figure 6–24 Predicted effect of uniform distributed roughness on the transition Reynolds number for flow over a cone at supersonic speed. (From Van Driest and Boison, 1957.)

else being the same.

Tanner (1969) found a decrease in the rate of lateral spreading of turbulence from a spot or a sidewall for viscoelastic fluids.

PROBLEMS

6.1. Consider CO_2 flowing at 2.0 atm and 400°K in a 5.0-cm-diameter pipe in a chemical plant. At what flow rate (in m^3/s) will the flow become turbulent? What if the fluid were H_2?

6.2. For flow over a flat plate, what is the Reynolds number based on x for instability if the disturbance has a wavelength $= 10\delta$?

6.3. What is the transition Reynolds number based on x for flow over an NACA 0012 airfoil at $M_\infty \approx 0$? What is the length of transition Δx, expressed in terms of a Reynolds number?

6.4. Compare the prediction of the method of Michel with that of Van Driest and Blumer for the transition Reynolds number for flow over a flat plate. Assume negligible free stream turbulence.

6.5. For water flow over a flat plate at 300°K and 3 m/s, compare the length to transition for free stream turbulence levels of 2.0% and 0.1%. What if the surface had a wire with $k/\delta^* = 0.2$?

6.6. Consider flow over two surfaces, each in an external stream with 1% free stream turbulence. The first surface is a flat plate ($\lambda = 0$). The other has a pressure gradient such that $\lambda = -4$. Compare the transition Reynolds number Re_δ for the two.

6.7. Water at 3.0 m/s is flowing over a surface with a negligible pressure gradient. What surface roughness can be allowed to have the surface behave as if it is smooth, as far as transition is concerned? How would your answer change if the pressure gradient were a strong adverse one?

6.8. Write and test a computer code for the Michel method of predicting transition, coupled with the Thwaites-Walz integral method for treating laminar boundary layers.

6.9. Air at standard temperature and pressure is flowing past a 25-cm-diameter ball on the top of a flagpole. What is the velocity for the decrease in drag due to transition? How much will the pressure at the rear of the sphere change?

6.10. For the ARDC model atmosphere, what value of the velocity gradient corresponds to the stability limit at 1,000 m?

6.11. Compare the transition Reynolds number for flow over an insulated cone at Mach 1.5, 3.0, 4.5, and 6.0. What would be the effect of roughness on the surface with $k/\delta^* = 2.0$ at Mach 3.0 and 6.0?

6.12. A test engineer wishes to trip the boundary layer on an insulated flat plate that is to be tested in a wind tunnel at Mach 3.0 with $T_t = 600$°K and $p_t = 2.0$ atm. Use the compressible integral method to calculate $\delta^*(x)$ up to $x = 30$ cm. How large a trip is needed at that station to reduce the transition Reynolds number by a factor of 4?

CHAPTER 7

WALL-BOUNDED, INCOMPRESSIBLE EXTERNAL TURBULENT FLOWS

7–1 INTRODUCTION

In this chapter, we begin the study of the most important part of boundary layer theory from a practical point of view—turbulent flows. The general subject is vast and complex, so it is prudent to start with a portion that one can grasp and then build on. First, we restrict the chapter to constant-density, constant-property (laminar thermophysical properties) flows. Second, only *wall-bounded, external turbulent flows* are considered. These are flows with a rigid (not necessarily impermeable) surface boundary on one side of the turbulent, viscous region. Those flows with rigid boundaries all around the flow are discussed in a later chapter, on internal flows. Flows without any such boundaries are called *free turbulent flows,* and they are also the subject of a separate chapter. The reason for the latter important distinction will become clear shortly.

7–2 ENGINEERING REQUIREMENTS OF TURBULENT ANALYSES

In Chapter 1, we indicated that turbulent flow is necessarily unsteady and three dimensional. Moreover, the frequencies of the unsteadiness and the size of the scales of the motion have been found to span several orders of magnitude. Reflection on these facts will yield a perspective on the magnitude of the problem of attempting to analyze all the details of a turbulent flow, even a very simple turbulent flow such as that over a flat plate. One example of a specific aspect of this general situation may be helpful. If the scales of the turbulent motion span orders of magnitude in size, how are the grid spacings (Δx, Δy) in a numerical method to be chosen intelligently? Obviously, results will be very crude if the grid spacings are larger than the scales of a significant part of the motion. We thus incur a requirement for about 10^3 or 10^4 grid points across a boundary layer and 10^6 or 10^7 total grid points. (The basis for these estimates will be made clear subsequently.) It is safe to conclude, then, that the task of producing a general analysis for all the details of a practical turbulent flow is hopeless at the present time. The computational cost of a complete simulation varies as Re^3, so only flows with very low Re can be treated at all at this time, even if great amounts of computer time are expended. It is unlikely that the picture will change anytime soon. Some knowledgeable workers estimate that it will be late in the last half of the next century before direct calculations of turbulent flows become common in engineering practice, even with very optimistic projections of improvements in computers and numerical methods.

Given this bleak assessment, is the designer to be left completely without analytical tools? Of course, the answer is *no,* and the opening that can be exploited comes from the limited design information required from a turbulent analysis. Referring back to the logic first presented in Chapter 2, we can state that the first requirement of a boundary layer analysis is a prediction of $C_f(x)$, including any points where $C_f = 0$ (i.e., where separation occurs). Second, one may be interested in $\delta(x)$ [and perhaps $\delta^*(x)$ and $\theta(x)$]. Other quantities, such as detailed velocity and temperature profiles, are of rapidly diminishing utility for design in most cases.

Before proceeding further, we must pause and ask what $C_f(x)$, for example, means in a turbulent flow. Consider a typical engineering application. For a frictional resistance prediction for a ship moving at constant speed, do we need, or even want, the instantaneous fluctuating value of the wall shear at a point on the surface? Obviously, we do not; we want to know the steady power plant size and propeller performance required to propel the ship. Even if the ship's speed were not constant, one would not need the fluctuating value of $C_f(x)$. The designer would need any unsteady variation in $C_f(x)$ that is correlated with the unsteady changes in the initial or boundary conditions. (Reread the end of Section 1–8, and rethink the distinction between a *steady turbulent flow* and an *unsteady turbulent flow*.) The needs of the analyst may be seen as most commonly corresponding to the *time-averaged* or *mean* quantities [e.g., $U(x, y)$] introduced in Section 1–8. Such quantities can be steady or unsteady and two dimensional or three dimensional. (Only the fluctuating part of the turbulent motion is always three dimensional.) Also, only the *size* of the scales of the fluctuating part of the motion spans orders of magnitude; the *range* of scales of variation in the mean flow is much smaller.

With all of this in mind, it can be seen that a turbulent analysis aimed only at predicting the mean flow [$C_f(x)$, $\delta(x)$, etc., and perhaps $U(x, y)$, etc.] might well serve the main purposes of the designer and might also be within the range of modern analytical or numerical capability. This notion has been taken as defining the scope of the text. Lest the reader relax at this point, it should be noted that only limited success at achieving reliable, efficient predictions of turbulent flows has been attained to date, and then only with the expenditure of great amounts of ingenuity and intellectual energy.

It is important to note in passing, however, that there are some practical situations where a knowledge of the detailed fluctuating character of the motion is essential. An example is the ignition and combustion of mixing streams of a fuel and oxidant (e.g., H_2 and air). The rate of conversion of chemical to thermal energy in the flame depends on the instantaneous local temperature in a very nonlinear fashion. Thus, in order to predict the steady heat release at a point in the flame, one cannot base the rates on $\overline{T}(x, y)$. Instead, it would be necessary to calculate the rates with $T'(x, y, z, t)$ and then take the time average of the results. Neither is it sufficient to use a quantity such as $\sqrt{\overline{(T')^2}}$, because of the fact that *the average of a quantity raised to a power is not the same as the power of the average of the quantity*. This type of situation is at the forefront of current research in turbulence, and a more detailed discussion is outside the scope of the book.

Before beginning the development of mean flow analyses, it is helpful to consider some of the available empirical information. That will serve as a solid basis for the analytical development. In order to allow treatment of the mean flow without considering all the details of the instantaneous fluctuating motion, it will be necessary to draw heavily on the experimental data base. The intent will be to develop a *semiempirical analysis*. Neither *first principles* alone nor *empirical correlations* alone would do the job. Instead, the analysis will fall somewhere in between these extremes: Empirical information will be used within a framework of the equations of motion. In the sections that follow, we will collect information from the empirical data base and put it in as compact a form as possible.

7-3 EMPIRICAL INFORMATION ON THE MEAN FLOW AS A BASIS FOR ANALYSIS

7-3-1 Flow over a Flat Plate

In a complex field, it is wise to begin with the simplest physical situation that is representative of the more general cases. For boundary layers, that case is flow over a flat plate ($dP/dx = dU_e/dx = 0$). If one compares measurements of the velocity profiles over a flat plate in laminar (see Fig. 4–3) and turbulent cases (see U/U_∞ in Fig. 1–4), a striking difference, illustrated in Fig. 7–1, becomes apparent: When plotted against y/δ, the mean turbulent velocity profile *appears* to intersect the wall ($y = 0$) at a greater value than $U = 0$. Actually, it proceeds down toward $y/\delta = 0$, decaying slowly, and then drops quickly to $U = 0$ at $y/\delta = 0$ over a very small range of y/δ. This behavior can be likened to that which would result for a laminar boundary layer composed of two layers of two different fluids: a high viscosity fluid in the outer region of about $0.05 \le y/\delta \le 1.00$ and a much lower viscosity fluid near the wall $0 \le y/\delta \le 0.05$. Since a simple force balance will show that the

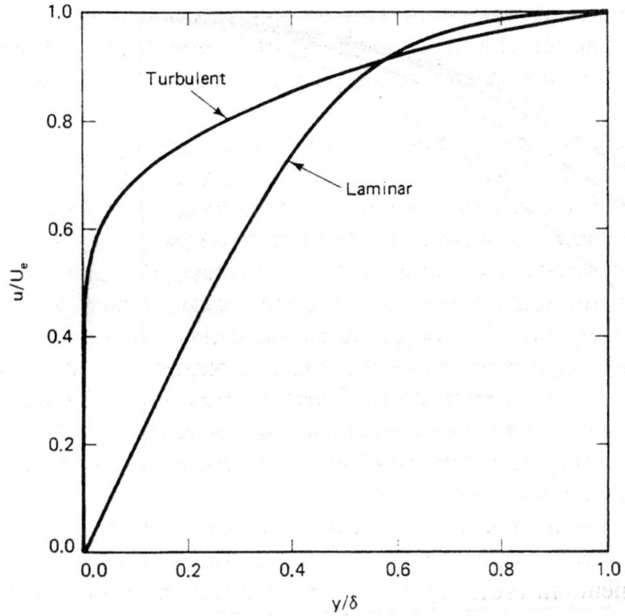

Figure 7–1 Comparison of the shapes of typical laminar and turbulent boundary layer velocity profiles.

shear at the interface must be continuous, we can write

$$\mu_1 \left(\frac{\partial u_1}{\partial y} \right)_{\text{interface}} = \mu_2 \left(\frac{\partial u_2}{\partial y} \right)_{\text{interface}} \tag{7–1}$$

So the gradient of the velocity profile must change sharply at the interface in the ratio μ_2/μ_1. This little development is more than of just casual interest. We shall see later that it contains the germ of one of the most important ideas in wall-bounded turbulent boundary layers.

There is another important difference between laminar and turbulent boundary layer profiles on a flat plate. For laminar flow, all flat-plate profiles are the same on a plot of U/U_e versus y/δ, no matter what the fluid, the Reynolds number, the roughness size, and so on. By contrast, for turbulent flows, one finds the situation shown in Fig. 7–2. Changes in the Reynolds number and/or the roughness change C_f, and that influences the shape of the profile in terms of U/U_e versus y/δ.

One consequence of the behavior shown in Fig. 7–2 is directly important in analysis: For laminar flow, the ratio $\delta^*/\theta \equiv H$, called the *shape parameter*, may be taken to describe the shapes of the profiles that occur under different flows. For example, a laminar flat-plate profile always has $H = 2.6$. This behavior was used in Section 6–3 and also in the integral analyses of Chapter 2. For turbulent flows, the concept is simply not valid, although it has often been used. From the figure, it can be seen that, for a turbulent flat-plate boundary layer, H depends on C_f. Thus, H cannot be used to indicate unequivocally the shape of the profiles to be found for turbulent boundary layer flows for a given body (implying a specific dP/dx).

It is very important to develop the most powerful and comprehensive correlating variables possible, so that the available empirical information can be presented

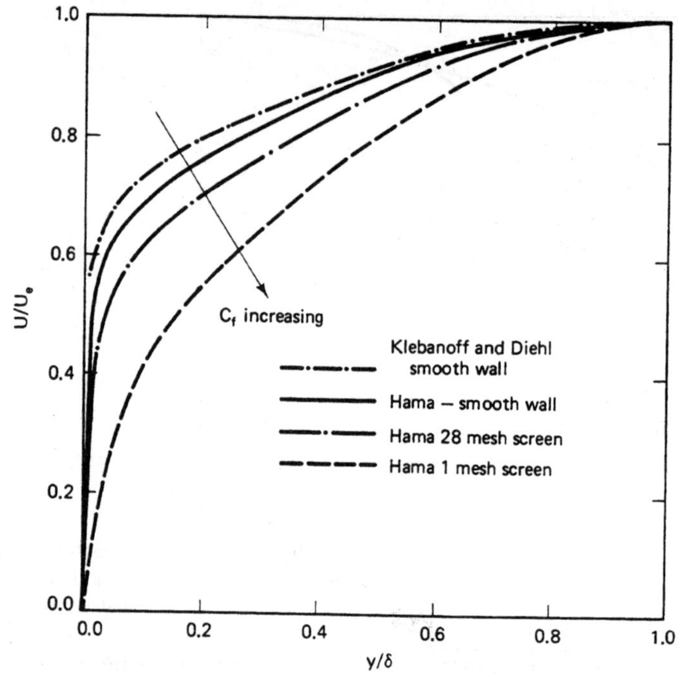

Figure 7–2 Turbulent boundary layer velocity profiles on smooth and rough flat plates. (From Clauser, 1956.)

and studied in as compact a form as possible. The problem is complex, and there is a lot of data, so it is obviously helpful to consider the data in the simplest form, involving the fewest groupings of variables and parameters, in order to understand as much as possible about the problem. Clearly, U/U_e and y/δ are not sufficient to collapse all the data, even for simple turbulent flat-plate flow, and we will have to look deeper. Many workers contributed to the final results that we will develop, but the presentation by Clauser (1956) is the clearest and most comprehensive, and the general outline of his approach is followed here.

The curves in Fig. 7–2 cannot be collapsed into a single curve by multiplying by any single scaling factor. However, if the *velocity defect*, defined as $(1 - U/U_e)$, is considered, a factor proportional to $1/\sqrt{C_f}$ will be seen to correlate all the curves. The velocity defect is the local decrease in velocity below the boundary layer edge velocity. Looking at the figure, one can see that a greater C_f leads to a greater velocity defect. This makes good physical sense, because a greater C_f means there is a greater relative retarding force acting on the fluid in the boundary layer, implying a greater loss in velocity (and, therefore, momentum). Since the profiles are expressed in terms of velocities, it is convenient to introduce a derived velocity called the *friction velocity* u_* as a normalizing quantity to express the influence of C_f. The wall shear divided by the density has the units of velocity squared, so one can write

$$u_* \equiv \sqrt{\frac{\tau_w}{\rho}} \tag{7–2}$$

which leads to

$$\frac{u_*}{U_e} = \sqrt{\frac{\tau_w}{\rho U_e^2}} = \sqrt{\frac{C_f}{2}} \tag{7–2a}$$

Thus, the proper choice for the ordinate of turbulent, flat-plate boundary layer profiles can be taken to be $(U/U_e - 1)/\sqrt{C_f/2} = (U - U_e)/u_*$. It suffices to retain y/δ as the other axis, because y is the relevant independent variable and δ is a suitable reference scale. The success of this choice of coordinates is shown in Fig. 7–3, a plot commonly called a *defect law plot;* the statement $(U - U_e)/u_* = f(y/\delta)$ is called the *defect law*.

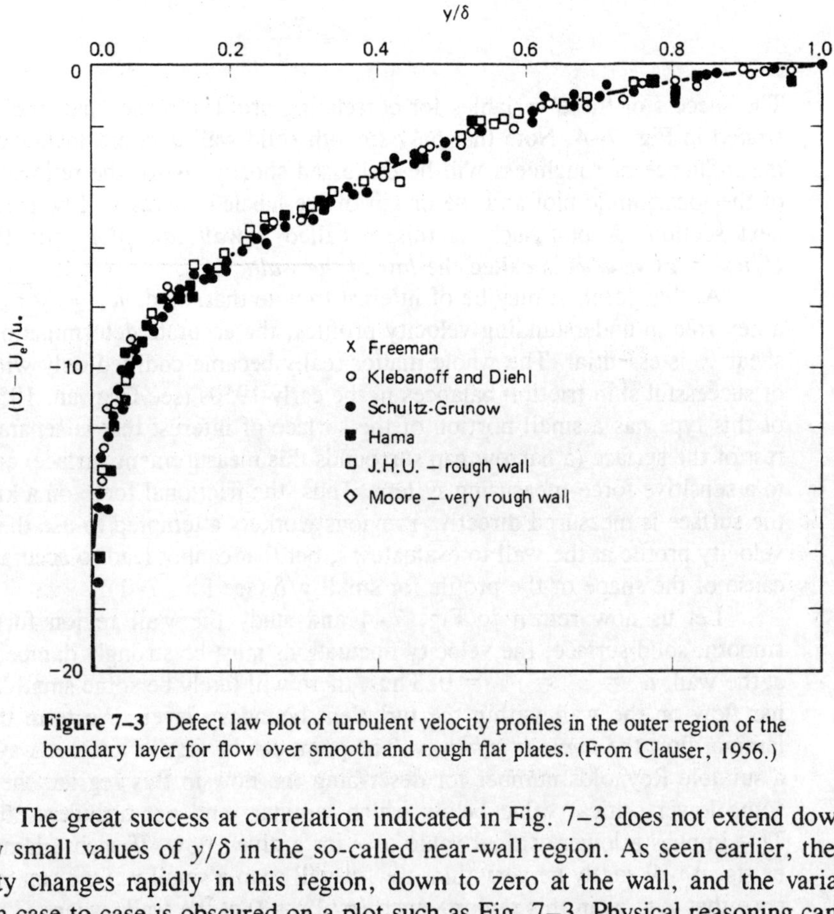

Figure 7–3 Defect law plot of turbulent velocity profiles in the outer region of the boundary layer for flow over smooth and rough flat plates. (From Clauser, 1956.)

The great success at correlation indicated in Fig. 7–3 does not extend down to very small values of y/δ in the so-called near-wall region. As seen earlier, the velocity changes rapidly in this region, down to zero at the wall, and the variation from case to case is obscured on a plot such as Fig. 7–3. Physical reasoning can tell us that the previous correlating variables should not work near the wall. We saw that in the outer part of the boundary layer, it was the local value of the mean velocity relative to the edge velocity $(U - U_e)$ that was important. The value U of the local mean velocity relative to the wall value of zero did not enter into the situation directly. In the innermost part of the layer near the wall, however, that measure will surely become important. It seems plausible to retain the friction velocity as a scaling quantity for velocity, especially in the wall region, so the trial grouping $(U/u_* \equiv u^+)$ can be selected. For the outer part of the boundary layer, location in the layer y was reckoned with respect to δ as y/δ. In the inner, wall-dominated region, it can be reasoned that the value of δ will be less important. The inner region occupies only a small portion ($\approx$10%) of the total height δ, and changes at far-removed points near $y = \delta$ will only be weakly felt in the inner region. It remains then to render the length y dimensionless in some other way that is more appropriate to the near-wall region. Earlier, we saw how to make a length dimensionless by introducing a Reynolds number. A little thought applied to the present situation leads to a Reynolds number of the form

$$y^+ \equiv \frac{yu_*}{\nu} \tag{7–3}$$

The success of these variables for correlating profiles in the inner region is demonstrated in Fig. 7–4. Note that only smooth solid wall data are included in this plot; the influence of roughness will be discussed shortly. Also, the rationale for the use of the logarithmic plot and the origin of the labeled curves will be presented in the next section. A plot such as this is called a *wall law plot,* and the statement $U/u_* = g(yu_*/\nu)$ is called the *law of the wall.*

At this point, it may be of interest to note that, since $u_* \equiv \sqrt{\tau_w/\rho}$ plays such a key role in understanding velocity profiles, the accurate determination of the wall shear τ_w is essential. The whole matter really became codified only with the advent of successful skin friction balances in the early 1950s (see Dhawan, 1953). A device of this type has a small portion of the surface of interest that is separated from the rest of the surface (a narrow gap surrounds this measurement surface) and connected to a sensitive force-measuring system. Thus, the frictional force on a known area of the surface is measured directly. Previous workers attempted to use the slope of the velocity profile at the wall to evaluate τ_w, but that cannot lead to accurate results because of the shape of the profile for small y/δ (see Fig. 7–1).

Let us now return to Fig. 7–4 and study the wall region further. Near a smooth, solid surface, the velocity fluctuations must be strongly damped, since right at the wall, $u' = v' = w' = 0$. Thus, there will likely be some small layer of laminar flow on the wall within the turbulent boundary layer. We term this region of laminar flow the *laminar sublayer*. The point can be argued another way. If yu_*/ν is a suitable Reynolds number for describing the flow in this region, there should be some lower critical value below which laminar, and not turbulent, flow prevails. This implies a lower critical value of y for a given flow. If such a laminar sublayer exists, it will surely be very thin, i.e., it will span a small y. One can therefore suppose that $\tau = \tau_w$ in this region, since $\partial\tau/\partial y = 0$ at the wall, where $U = V = 0$ for a flat plate, as indicated by Eq. (3–22) [see Fig. 7–5 for an illustration of $\tau(y)$]. This assumption leads to

$$\tau = \mu\frac{\partial U}{\partial y} = \tau_w \neq F(y) \tag{7–4}$$

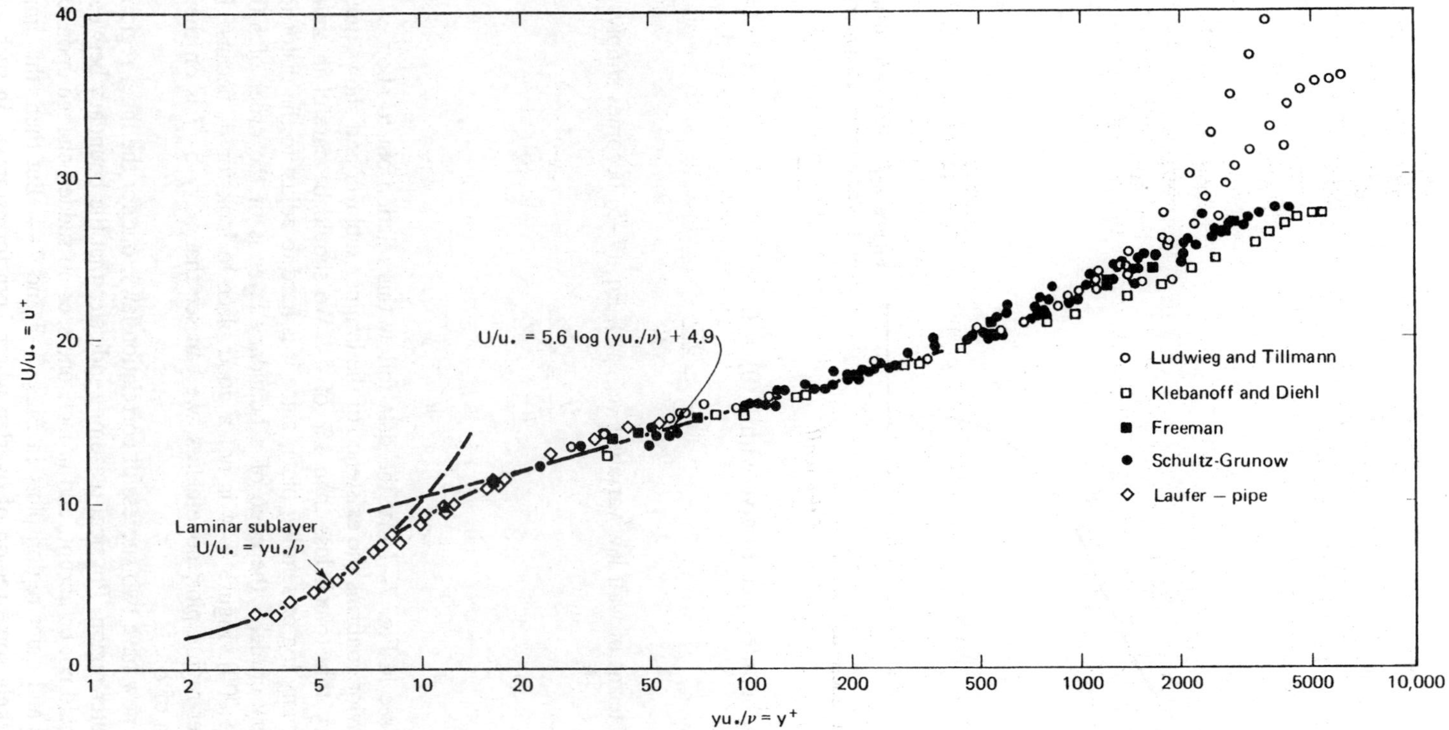

Figure 7–4 Universal wall law plot for turbulent boundary layers on smooth, solid surfaces. (From Clauser, 1956.)

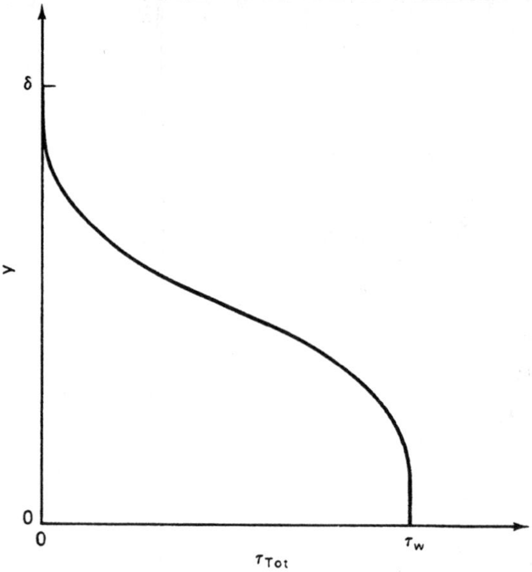

Figure 7–5 Schematic of the variation of the shear across a boundary layer.

which can be integrated to give with $U(0) = 0$

$$U = \frac{y\tau_w}{\mu} \tag{7–5}$$

Recast in terms of wall law variables (u^+, y^+), Eq. (7–5) becomes simply

$$\frac{U}{u_*} = \frac{yu_*}{\nu} \tag{7–5a}$$

or

$$u^+ = y^+ \tag{7–5b}$$

One can see in Fig. 7–4 that the data follow this relation out to about $y^+ \approx 5–7$. This behavior confirms the existence of the laminar sublayer and shows that it is, indeed, very thin (much less than 1% of δ). We should be careful in studying the figure (or any log or semilog plot), since it is hard to adjust to the distortion of the scale. Very crudely, the edge of a boundary layer is of the order of 5,000, i.e., $\delta^+ = O(5{,}000)$. Figure 7–4 is not a good place to look for δ, because Fig. 7–4 is an inner region plot; nonetheless, we can see that $y^+ \approx 5–7$ is on the order of 1/1,000th of δ^+.

We now have two successful correlation laws, one for the inner region and one for the outer region. Physical reasoning indicates that the boundary between the regions should not be abrupt, and in fact, one sees no sudden change upon examining the inner and outer region plots in Figs. 7–3 and 7–4. But then, this implies that there is likely some region of overlap where both forms apply. In such an *overlap region,* we may say that both

$$\frac{U - U_e}{u_*} = f\left(\frac{y}{\delta}\right) \tag{7–6}$$

and

$$\frac{U}{u_*} = g\left(\frac{yu_*}{\nu}\right) \tag{7–7}$$

hold. These equations can be rewritten, respectively, as

$$\frac{U}{u_*} = f\left(\frac{y}{\delta}\right) + \frac{U_e}{u_*} \tag{7–6a}$$

and

$$\frac{U}{u_*} = g\left[\left(\frac{y}{\delta}\right)\left(\frac{\delta u_*}{\nu}\right)\right] \tag{7–7a}$$

In the overlap region, the two must be equal for all y. Since y is the independent variable, it may vary arbitrarily within the overlap region, and the dependent variable, U/u_*, must track exactly with both Eqs. (7–6a) and (7–7a). For a given boundary layer profile, U_e/u_* and $\delta u_*/\nu$ are constants, (i.e., they do not vary with y). Thus, a constant multiplicative factor on the argument inside the function g must have the same effect as a different constant additive factor outside the function f. The logarithm is the only function that displays such a property, because the logarithm of a product can be written as the sum of two logarithms. Thus, we may propose, for the *overlap region only,* that

$$\frac{U - U_e}{u_*} = A \log\left(\frac{y}{\delta}\right) + B \tag{7–6b}$$

and

$$\frac{U}{u_*} = A \log\left(\frac{yu_*}{\nu}\right) + C \tag{7–7b}$$

The *wall law plot* was already plotted on a log scale in Fig. 7–4, and the *defect law plot* from Fig. 7–3 can be replotted as in Fig. 7–6. Clearly, there is an overlap region, since the log laws follow the data, and that region is not small (it spans roughly $30 \approx y^+ \approx 300$).

Examining the data so plotted, Clauser (1956) proposed that $A = 5.6$, $B = -2.5$, and $C = 4.9$. Other workers have contemplated different lines drawn through the data, leading to slightly different values of the constants, but we will use Clauser's values throughout. Also, some write the equations with a natural logarithm and use a constant κ, which is related to A as $\kappa = \ln(10)/A$, so that

$$\frac{U}{u_*} = \frac{1}{\kappa} \ln\left(\frac{yu_*}{\nu}\right) + C \tag{7–7c}$$

For $A = 5.6$, $\kappa = 0.41$.

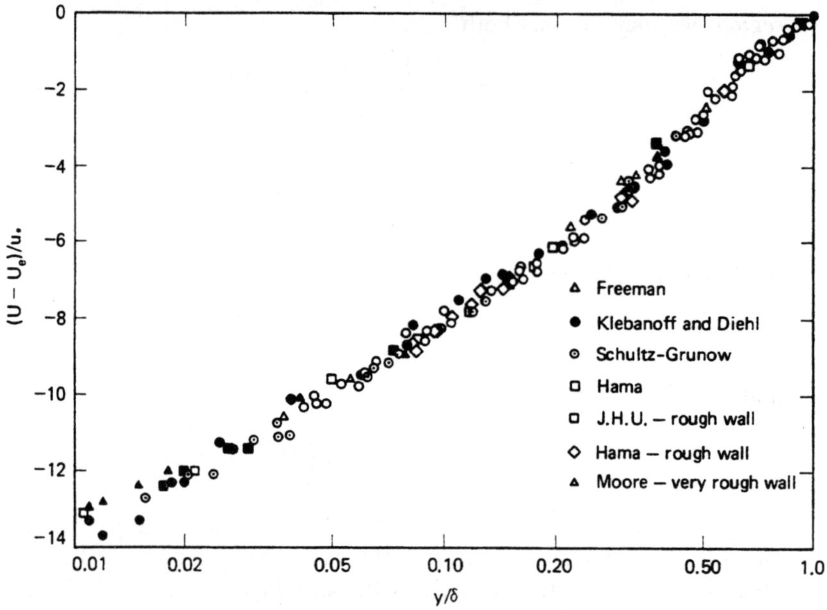

Figure 7–6 Logarithmic plot of the defect law for flat-plate turbulent boundary layers. (From Clauser, 1956.)

Coles (1956) introduced the notion of a *law of the wake* to describe the velocity profiles in the outer region. He used the fact that the deviation of the velocity above the logarithmic law, when normalized with the maximum value of that deviation at the outer edge of the layer, $y = \delta$, is a function of y/δ alone. He correlated that function as a *wake function* $W(y/\delta)$, which is selected to have $W(0) = 0$ and $W(1) = 2$. Coles's wake function is given by

$$\frac{U/u_* - \left(\dfrac{1}{\kappa}\ln(yu_*/\nu) + C\right)}{U_e/u_* - \left(\dfrac{1}{\kappa}\ln(\delta u_*/\nu) + C\right)} \equiv \frac{1}{2}W\left(\frac{y}{\delta}\right) \tag{7–8}$$

shown schematically in Fig. 7–7. Coles (1956) then proposed that

$$W\left(\frac{y}{\delta}\right) = 2\sin^2\left(\frac{\pi}{2}\frac{y}{\delta}\right) \tag{7–8a}$$

Finally, the *law of the wake* can be written as

$$\frac{U}{u_*} = \frac{1}{\kappa}\ln\left(\frac{yu_*}{\nu}\right) + C + \frac{\Pi}{\kappa}W\left(\frac{y}{\delta}\right) \tag{7–9}$$

giving a composite profile that is valid for both the overlap layer and the outer region. Here, $\Pi = -\kappa B/2$ is a *wake parameter*. For $B = -2.5$, $\Pi = 0.51$ [Coles (1956) recommended 0.55].

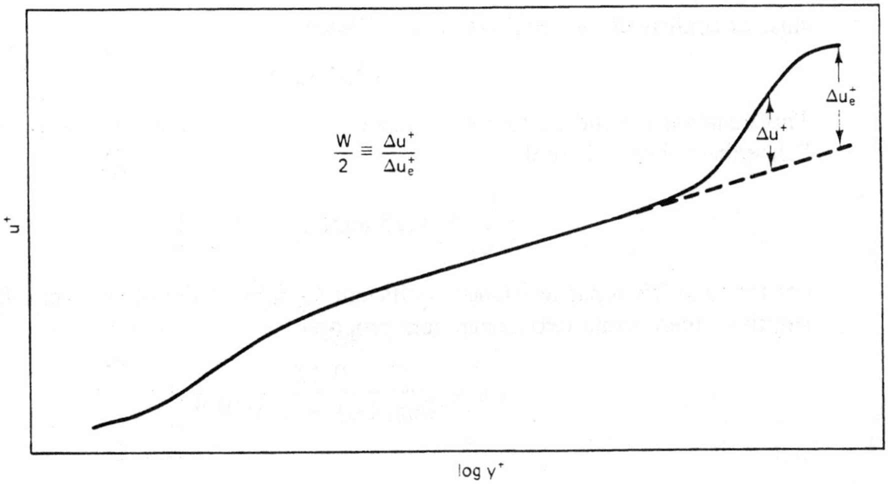

Figure 7–7 Schematic of Coles's wake function.

The whole turbulent boundary layer can now be viewed as consisting of four main layers, each with its own length and velocity scales after Kovaszny (1967). The innermost layer is the laminar sublayer, which has a length scale ν/u_* (see Eq. (7–3)) and a velocity scale u_*. The inner region has a length scale y and a velocity scale u_*. The outer region's length scale is δ, and its velocity scale is U_e. Corrsin and Kistler (1955) introduced a fourth layer, called the *superlayer*, which is a thin, corrugated interfacial region between turbulent and nonturbulent fluid at the outer edge. The velocity scale in this region can be taken to be the *entrainment velocity*, which is proportional to the rate at which free stream fluid is drawn into the viscous region. This velocity is also the relative speed of advance of the turbulent interface with respect to the external, nonturbulent fluid. Kovaszny (1967) proposed that the entrainment velocity be defined as

$$V_0 \equiv U_e \frac{d}{dx}(\delta - \delta^*) \tag{7–10}$$

He then used this equation to form a characteristic length scale for the superlayer as $10\nu/V_0$.

A skin friction law can be found by eliminating U and y from Eqs. (7–6b) and (7–7b) by subtraction to give

$$\frac{U_e}{u_*} = A \log\left(\frac{\delta u_*}{\nu}\right) + C - B \tag{7–11}$$

which can be rewritten as

$$\sqrt{\frac{2}{C_f}} = A \log\left(\text{Re}_\delta \sqrt{\frac{C_f}{2}}\right) + C - B \tag{7–12}$$

Other simpler, explicit formulas have also been derived, either by approximating Eq. (7–12) or by fitting empirical data directly. One of the simplest is due to Blasius, as confirmed by Schultz-Grunow (1940):

$$C_f = 0.0456(\mathrm{Re}_\delta)^{-1/4} \tag{7–13}$$

This equation is valid up to approximately $\mathrm{Re}_x = 10^7$. Many people use the famous Schoenherr (1932) formula:

$$\frac{1}{\sqrt{C_f}} = 4.15 \log(\mathrm{Re}_x C_f) + 1.7 \tag{7–14}$$

For the total frictional resistance coefficient C_D for turbulent flow over a flat plate of length L, these same two researchers proposed

$$C_D = \frac{0.427}{(\log(\mathrm{Re}_L) - 0.407)^{2.64}} \tag{7–13a}$$

and

$$\frac{1}{\sqrt{C_D}} = 4.13 \log(\mathrm{Re}_L C_D) \tag{7–14a}$$

7–3–2 Roughness Effects

Surface roughness has an important direct effect on the flow in the inner, wall-dominated region. It has only an indirect effect on the outer flow, increasing C_f, and that effect is included in the outer region scaling with u_*. The precise description of surface roughness patterns and their effects on turbulent flows cannot be based on any single parameter, such as an average roughness size k. Research is still being performed, and a detailed discussion of the matter is not within the scope of this book. For our purposes, we will proceed as if k alone were sufficient, except for a few comments. In the wall region, one would suppose that k enters into the situation dimensionlessly, as $ku_*/\nu \equiv k^+$. For a roughness size k^+ well within the laminar sublayer, the influence of k^+ can be expected to be small, so one might say that surfaces with $k^+ < 5$ (approximately) are *smooth,* except for unusual roughness patterns. For surfaces with $k^+ > 10–12$ (approximately), the laminar sublayer begins to disappear. Experiment shows that the influence of roughness on the *wall law plot* is to shift the logarithmic portion of the smooth wall curve down and to the right, corresponding to an increase in C_f. An increase in C_f increases u_*, which increases y^+ and decreases u^+. The slope of the logarithmic region, however, does not change. The factor A (and hence, κ) in the logarithmic law is not influenced by processes in the sublayer, so it will also not be influenced by modest roughness. This result can be represented as a downward shift $\Delta U/u_*$ at a fixed yu_*/ν. This shift is a function of k^+, as shown in Fig. 7–8. The *law of the wall* is then written as

$$\frac{U}{u_*} = A \log\left(\frac{yu_*}{\nu}\right) + C - \frac{\Delta U}{u_*} \tag{7–15}$$

For large $k^+ (k^+ \geq 70)$, where the laminar sublayer disappears, the flow is said

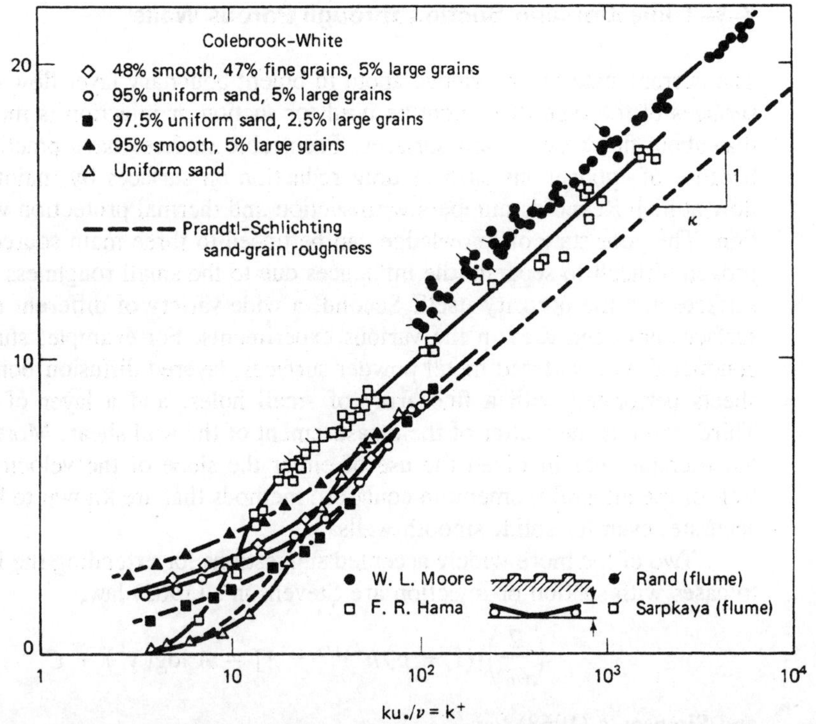

Figure 7–8 Downward shift of the logarithmic portion of the wall law as a result of roughness. (From Hama, 1954.)

to be *fully rough,* and the inner layer must be independent of viscosity. For the latter to happen, one can see from Eq. (7–15) that $\Delta U/u_*$ must have the form

$$\frac{\Delta U}{u_*} = A \log\left(\frac{ku_*}{\nu}\right) + D \tag{7–16}$$

so that the viscosity cancels out. In that case, the curves in Fig. 7–8 must have the slope $A = 5.6$ for large k^+, which they do. The quantity D in Eq. (7–1b) is the *y-intercept* in the figure.

The new *law of the wall* for fully rough surfaces, Eqs. (7–15) and (7–16), can be combined with the *defect law*, Eq. (7–6b) (which holds for smooth and rough surfaces) as before to determine a skin friction law. The result is

$$\sqrt{\frac{2}{C_f}} = A \log\left(\frac{\delta}{k}\right) - B + C - D \tag{7–17}$$

The quantity D depends somewhat on the character of the roughness, even for large k^+, as can be seen in Fig. 7–8. For small k^+, the situation is more complex, with the type of roughness strongly influencing the flow, in addition to k^+.

7-3-3 Injection and Suction through Porous Walls

The current state of knowledge about turbulent boundary layer flow over permeable surfaces of the type that might be used for suction or injection is much poorer than that about flows over solid surfaces. Permeable surfaces have practical importance because of applications such as drag reduction on surfaces by maintaining laminar flow at high Reynolds numbers with suction and thermal protection with fluid injection. The poor state of knowledge can be traced to three main sources. First, it has proven difficult to separate the influences due to the small roughness of most porous surfaces and the porosity itself. Second, a wide variety of different types of porous surfaces has been used in the various experiments. For example, studies have been conducted with sintered metal powder surfaces, layered diffusion-bonded screening, sheets perforated with a fine array of small holes, and a layer of small spheres. Third, there is the matter of the measurement of the wall shear. Most of the work in the literature has involved the use of either the slope of the velocity profile at the wall or the integral momentum equation, methods that are known to be generally inaccurate, even for solid, smooth walls.

Two of the more widely accepted suggestions for extending the logarithmic law to cases with suction or injection are Stevenson's (1963) law,

$$\left(\frac{2}{v_0^+}\right)[(1 + v_0^+ u^+)^{1/2} - 1] = A \log(y^+) + C \tag{7-18}$$

and Simpson's (1968) law,

$$\left(\frac{2}{v_0^+}\right)[1 + v_0^+ u^+)^{1/2} - (1 + 11.0v_0^+)^{1/2}] = A \log\left(\frac{y^+}{11.0}\right) \tag{7-19}$$

where $v_0^+ \equiv v_w/u_*$. One of these seems to agree best with some data and the other with other data. Both equations have been used to attempt to correlate independent sets of data. In Fig. 7–9, both laws are applied to data obtained from a porous surface of layered screening examined by Scott et al. (1964). In Fig. 7–10, the two laws were applied to a porous surface of sintered metal examined by Schetz and Nerney (1977). For Fig. 7–9, Stevenson's law appears best, but for Fig. 7–10, Simpson's law looks best, at least as far as collapsing all the data into a narrow band is concerned. Actually, the data of Schetz and Nerney (1977) showed a behavior more like that observed over rough solid surfaces, as may be seen in Fig. 7–11. For all cases, a shift of $\Delta U/u_* \approx 3.5$ was found in the range $0 \leq v_0^+ \leq 0.1$. The wall shear in tests by Schetz and Nerney was measured directly with a floating element balance.

A series of experiments was reported in Kong and Schetz (1981, 1982) in which six surfaces were studied under the same conditions in the same apparatus: (1) smooth, solid; (2) rough, solid (sandpaper); (3) sintered metal; (4) bonded screening, porous; (5) bonded screening with a solid sheet bonded to the back; and (6) a thin sheet with 0.15-mm holes on a 0.625-mm center-to-center square pattern. The sandpaper roughness was in the same nominal range ($k^+ \approx 5–7$) as the sintered and screening surfaces. A typical wall law plot is shown in Fig. 7–12. The effect of uni-

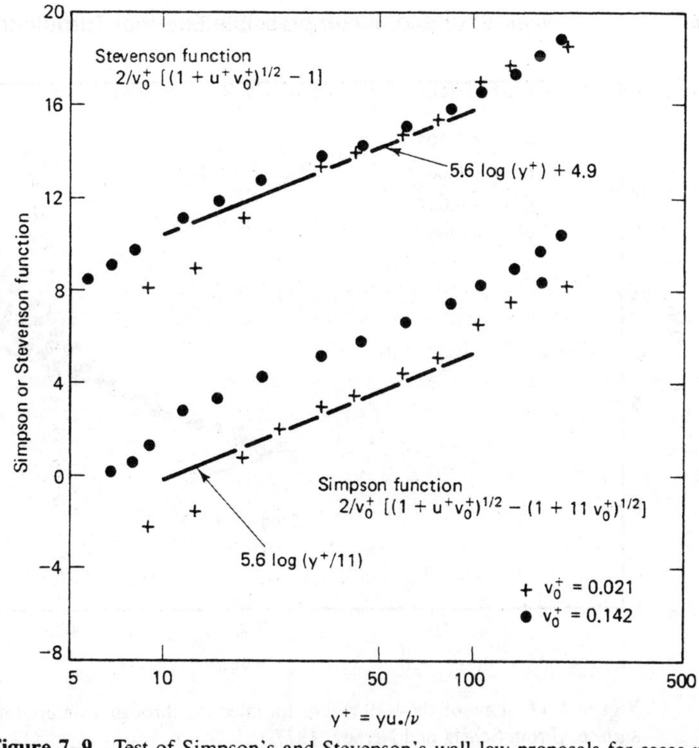

Figure 7–9 Test of Simpson's and Stevenson's wall law proposals for cases with injection or suction against the data of Scott et al. (1964). (From Schetz and Favin, 1971.)

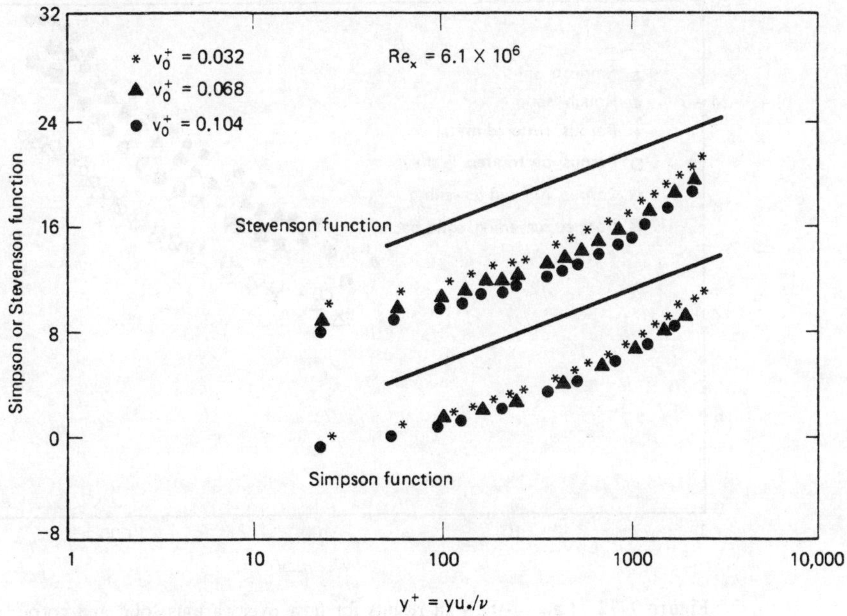

Figure 7–10 Test of Simpson's and Stevenson's wall law proposals against the data of Schetz and Nerney (1977).

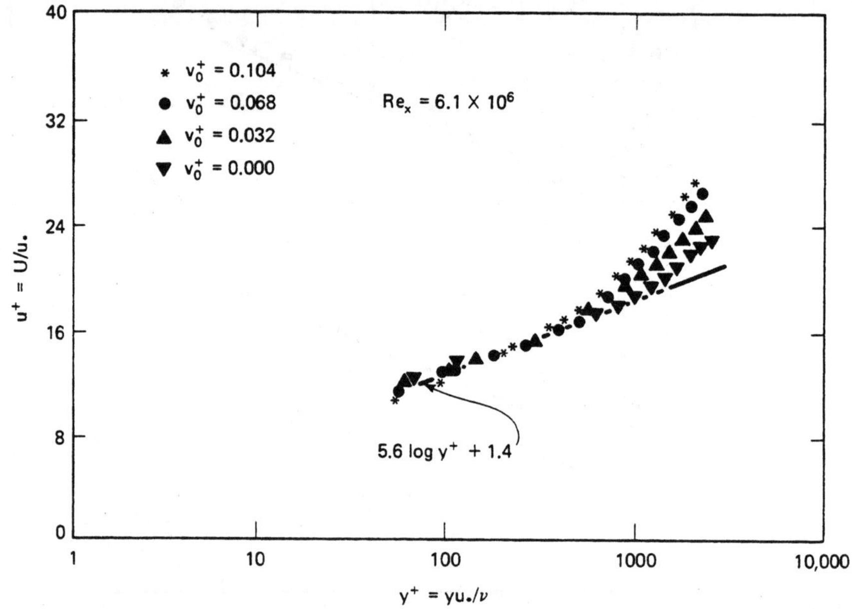

Figure 7–11 Law-of-the-wall results for injection through a sintered metal porous surface. (From Schetz and Nerney, 1977.)

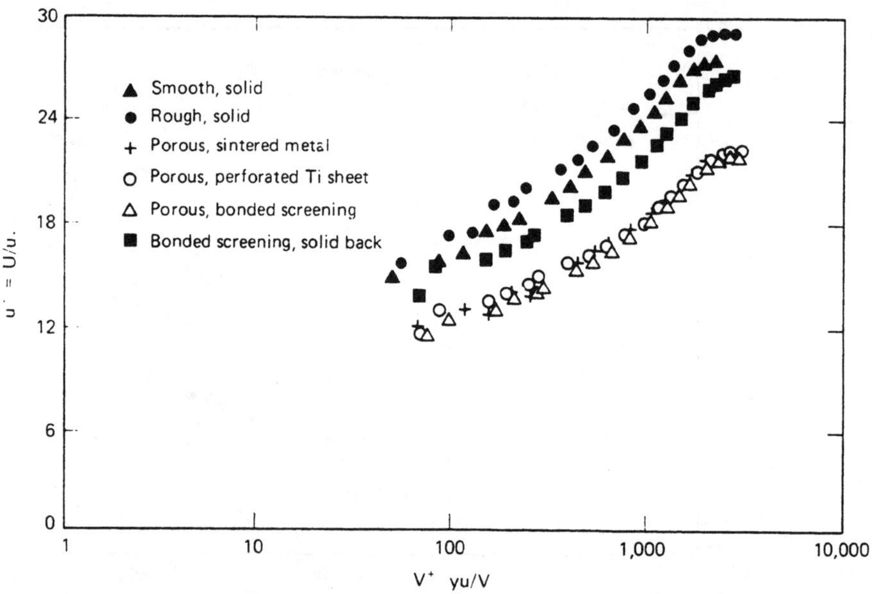

Figure 7–12 Law-of-the-wall results for flow over various solid and porous surfaces. (From Kong and Schetz, 1982.)

form roughness on a solid surface can be seen in the sandpaper-roughened and solid-backed screening cases, where a small shift $\Delta U/u_*$ was found in the logarithmic region, as would be expected for k^+ values of this magnitude (see Fig. 7–8). To see the effect of porosity without any interference from different surface roughness patterns, it is clearest to compare the results of the "smooth" perforated titanium wall with those of the smooth, solid wall, or those of the porous screen wall with those of the solid screen wall. The comparisons reveal that the effect of porosity is to shift the logarithmic region of the wall law downward by an amount $\Delta u^+ \approx 3$–4 from the results with the solid wall. The combined effects of a small amount of roughness and porosity can be seen by comparing the results of the sintered metal, porous wall with those of the smooth, solid wall, or those of the porous, rough screening wall with those of the smooth, solid wall. The downward shift of the logarithmic region of the wall law by the combined effects of a small amount of roughness and porosity is approximately the sum of the individual effects.

In general, skin friction is increased by suction and decreased by injection, as shown in Fig. 7–13. However, it is important to observe that the skin friction over a real, porous surface is increased, compared with flow over a smooth solid surface at the same conditions, even for $v_w \equiv 0$. The initial increase in C_f at $v_w = 0$ was found to be approximately the sum of about a 15–20% increase each from roughness and porosity. The smoother perforated sheet has a lower initial increase at $v_w = 0$, but the decrease as v_w increases is less than for the sintered surface. This is presumably because the injection is more uniform on the small scale for the sintered material than for the individual small holes of the perforated sheet.

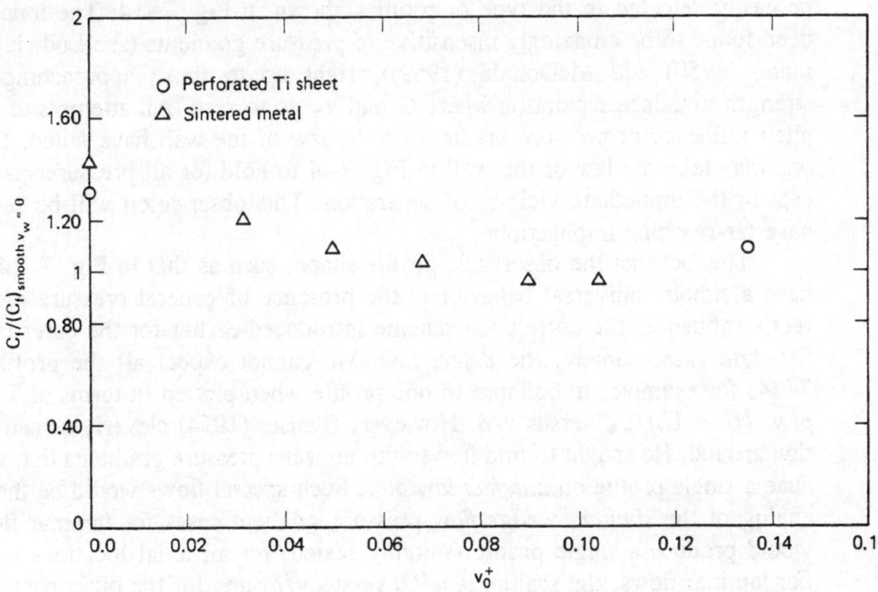

Figure 7–13 Reduction of skin friction with injection through two porous surfaces. (From Schetz and Nerney (1977) and Collier and Schetz, 1983.)

7–3–4 Flows with Axial Pressure Gradients

We have suceeded in constructing a more or less tidy picture of the mean flow for the simple case of a flat plate, but it is now necessary to turn to more general cases. For example, consider the flow over an airfoil. Each case will have a different pressure gradient, dP/dx, depending on the shape of the section. More than that, every airfoil will have a region of favorable pressure gradient near the front, a region of small pressure gradient near the maximum thickness, and then a region of adverse pressure gradient towards the rear. Can we really expect to collapse all the profiles along the surface into a single shape, as with the *defect law* for flow over a flat plate? After all, such a collapse is not possible even for laminar flows with pressure gradients, where only the highly restrictive *similar, wedge flow* solutions exhibit the said behavior, and wedge flows can have only a completely adverse or completely favorable pressure gradient along the whole surface. Measurements of boundary layer profiles along the top and bottom surfaces of an airfoil are shown in Fig. 7–14, and they confirm the reservations stated. The profile shapes change markedly along the surface, and no simple scaling will bring them into coincidence. The same behavior is observed for flows over axisymmetric bodies and in planar or axisymmetric channels of varying cross section. The simple fact is that general body or channel shapes produce general pressure gradients which lead to turbulent velocity profiles whose shapes do not fall into any one simple *family*.

The foregoing assessment holds in general, and the issue will have to be confronted directly. There is, however, one simplifying experimental result that cannot be easily detected in the type of profiles shown in Fig. 7–14: The inner flow has been found to be amazingly insensitive to pressure gradients (see Ludwieg and Tillmann (1950) and McDonald (1969)), right up to those approaching sufficient strength to induce separation where C_f and u_* go to zero. All attempts to find an explicit influence of pressure gradient on the law of the wall have failed. In essence, one may take the law of the wall in Fig. 7–4 to hold for all pressure gradients, except in the immediate vicinity of separation. This observation will be seen later to have far-reaching implications.

The fact that the observable profile shape, such as that in Fig. 7–14, does not have a simple, universal behavior in the presence of general pressure gradients directly influences the correlation scheme introduced earlier for the outer layer of the flat-plate case, namely, the *defect law*. We cannot expect all the profiles in Fig. 7–14, for example, to collapse to one profile when plotted in terms of a *defect law plot*, $(U - U_e)/u_*$ versus y/δ. However, Clauser (1954) cleverly turned this question around. He sought to find flows with nonzero pressure gradients that would produce a single profile on a *defect law plot*. Such special flows would be the turbulent analog of the *similar, wedge flow* pressure gradient cases for laminar flows. Each would produce a single profile (suitably scaled) for all axial locations in the flow. For laminar flows, the scaling is u/U_e versus y/δ, and for the outer part of a turbulent boundary layer, the scaling is $(U - U_e/u_*$ versus y/δ. If the profile shape on a suitable plot is viewed as an *output* and some variable or parameter involving the

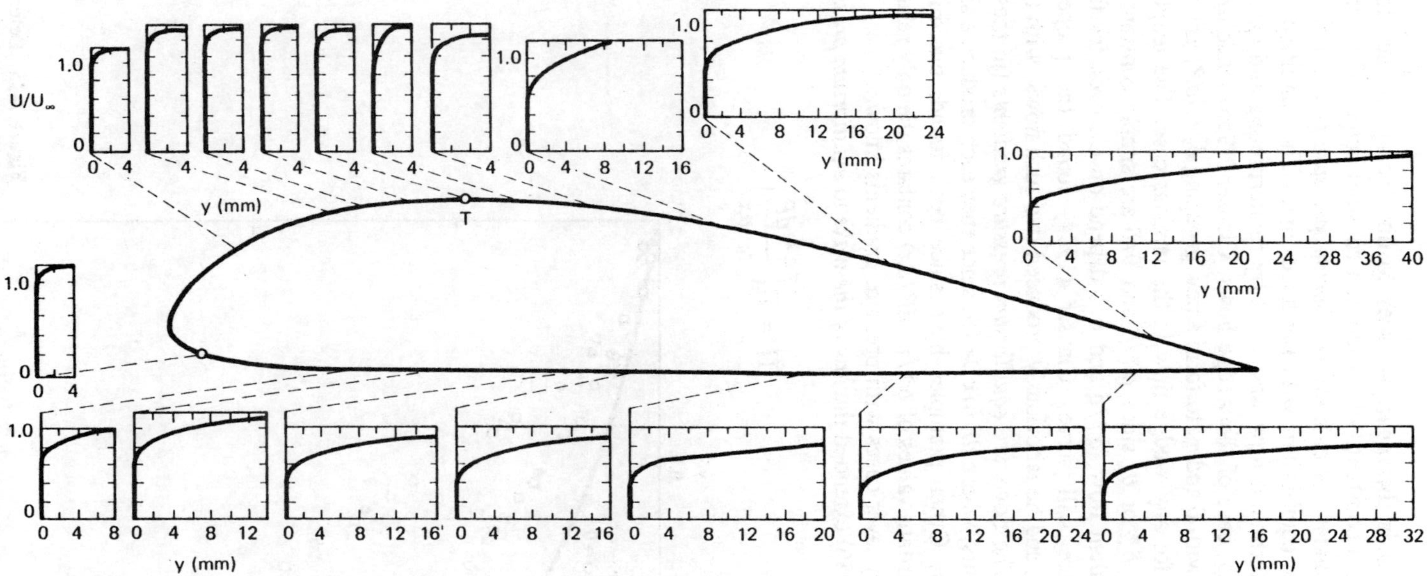

Figure 7–14 Boundary layer profiles along an airfoil. $C_L = 0.4$, $\mathrm{Re}_L = 4.3 \times 10^6$. Pressure side turbulent, suction side turbulent beyond T. (From Stüper, 1934.)

pressure gradient is the *input,* we seek to find the appropriate form of the input which, when held constant, will result in a constant output. Clauser called this class of turbulent boundary layer flows *equilibrium pressure gradient flows,* since the output would be in equilibrium with the fixed input. The condition for such pressure gradients in a turbulent flow was found by reinterpreting and extending the corresponding laminar case of the wedge flows. Clauser (1956) made dP/dx dimensionless by scaling with δ and τ_w to form a new grouping $\delta/\tau_w(dP/dx)$, which was found to be constant for any wedge flow in the laminar case. The length scale could as well have been δ^* or θ, since δ^*/δ and θ/δ are simply constants for any wedge flow. For turbulent flow, δ^*/δ and θ/δ depend on C_f, even for flat-plate flow, so only one choice will serve. Clauser (1954) found the proper choice to be $\delta^*/\tau_W(dP/dx)$, and he laboriously produced two such flows, with the results shown in Fig. 7–15. The flows are *equilibrium pressure gradients* (in these cases, adverse pressure gradients, where $dP/dx > 0$), since they each produce a single profile on a *defect law plot*. Other examples have since been found, but this class of flows specified by constant values of $\delta^*/\tau_w(dP/dx)$ contains the only relatively simple turbulent boundary layer cases with pressure gradients known.

Coles (1956) extended the *law of the wake* to *equilibrium pressure gradients* by letting

$$\Pi = \Pi\left(\frac{\delta^*}{\tau_w}\frac{dP}{dx}\right) \tag{7–20}$$

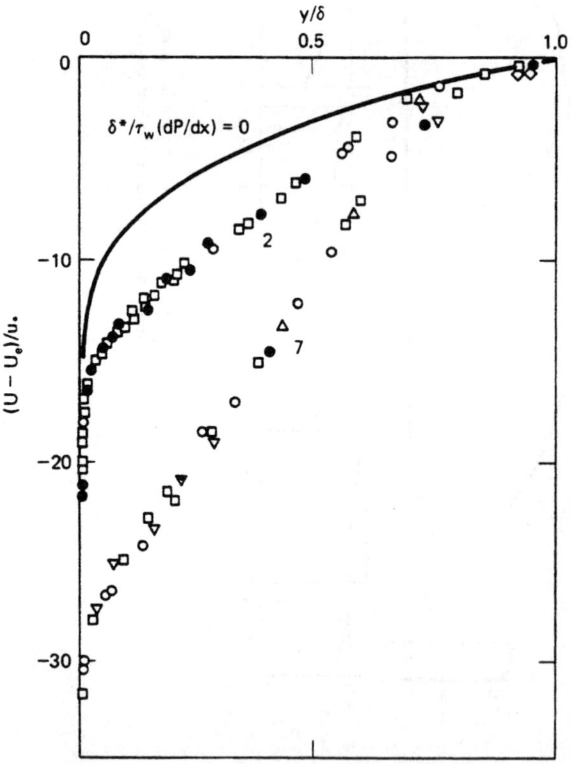

Figure 7–15 Defect law plot of velocity profiles for equilibrium pressure gradient flows. (From Clauser, 1956.)

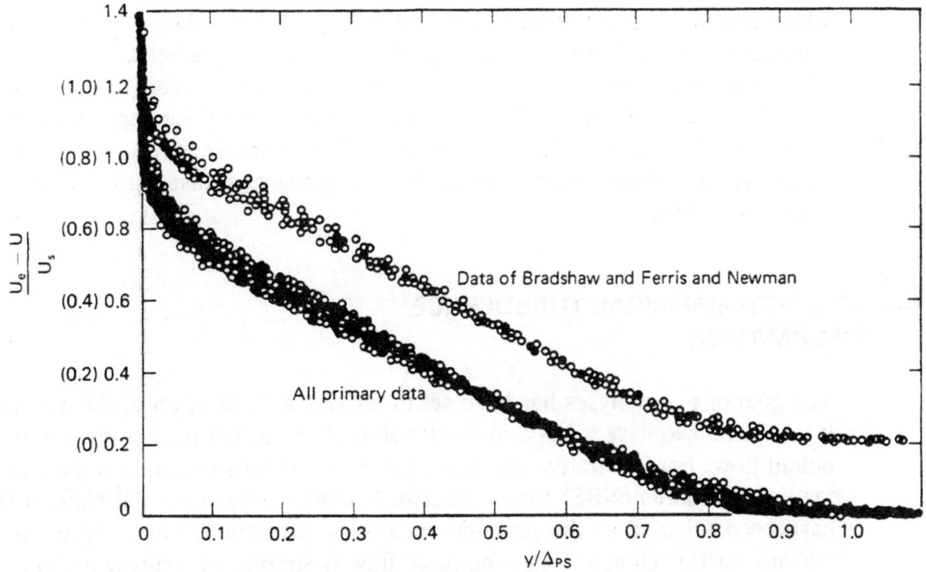

Figure 7–16 Universal velocity defect law plot for adverse pressure gradient flows. (From Perry and Schofield, 1973.)

White (1974) has proposed the specific expression

$$\Pi = 0.8\left[\frac{\delta^*}{\tau_w}\frac{dP}{dx} + 0.5\right]^{3/4} \tag{7–21}$$

This enables the development of a skin friction law for these cases by evaluating the *law of the wake,* Eq. (7–10), at $y = \delta$, where $U = U_e$. We obtain

$$\sqrt{\frac{2}{C_f}} = A \log\left(\mathrm{Re}_\delta \sqrt{\frac{C_f}{2}}\right) + C + \frac{2\Pi}{\kappa} \tag{7–22}$$

where an equation such as Eq. (7–21) is implied.

For very strong adverse pressure gradients, the shear profile shape changes, and the maximum shear does not remain on the wall. This phenomenon can be understood by recalling that the shear on the wall goes to zero at separation, while the shear above the wall remains finite (see Fig. 1–10(C)). When $\tau_{max}/\tau_w > 3/2$, a modified outer region law is required, and Perry and Schofield (1973) introduced

$$\frac{U_e - U}{U_s} = F\left(\frac{y}{\Delta_{PS}}\right) \tag{7–23}$$

The new scaling velocity U_s is defined by

$$\left(\frac{U_s}{U_e}\right)^{3/2} = 13.53\frac{u_{m\tau}}{U_e}\sqrt{\frac{\delta^*}{y_m}} \tag{7–24}$$

where $u_{m\tau} = (\tau_{max}/\rho)^{1/2}$ and y_m is y at τ_{max}. The new scale length is taken to be $\Delta_{PS} = 2.86\delta^* U_e/U_s$. The success of this correlation is demonstrated in Fig. 7–16, which contains a great number of different experiments. Also, it is not necessary to introduce a restriction such as an equilibrium pressure gradient. The important role of the maximum shear in the boundary layer will be seen again later in this chapter.

Adverse pressure gradients strong enough to produce separation are beyond the scope of this book, since the boundary layer approximations do not apply in such cases. The interested reader can find a comprehensive discussion of the subject in Simpson (1989).

7-4 SELECTED EMPIRICAL TURBULENCE INFORMATION

The goal of our analyses has been set to provide a description of the mean flow, but it is essential to have a basic understanding of the actual fluctuating nature of a turbulent flow. For illustrative purposes, the excellent data obtained at the National Bureau of Standards (NBS) for a turbulent boundary layer on a flat plate in the 1950s have been selected as the principal data to be presented. There were two primary reasons for this choice: First, the basic flow is simple, yet representative; second, a complete range of measurements is available for the same flow.

Some of the NBS data from Klebanoff (1955) has already been given in Fig. 1–4. Figure 7–17 shows more detailed profiles in the wall region for the three turbulence intensities. Notice that the axial intensity ($\sqrt{\overline{u'^2}}/U_e$) is the largest intensity, but the out-of-plane intensity ($\sqrt{\overline{w'^2}}/U_e$) is larger than the transverse intensity ($\sqrt{\overline{v'^2}}/U_e$). This is a result of the fact that the flow is freer to fluctuate side to side than up and down, where the wall exerts a direct restraint. It is also of interest to have an idea of the variations about the rms averages. Blackwelder and Kaplan (1976) have determined that the probability of finding $u' \approx \sqrt{\overline{u'^2}}$ is about 0.3 and that for finding $u' \simeq 3\sqrt{\overline{u'^2}}$ is 0.01.

Turbulence intensity profiles for flow over a rough flat plate from Corrsin and Kistler (1954) are plotted in Fig. 7–18. The roughness was in the form of corrugated paper, so it was two dimensional and approximately sinusoidal. The height of the roughness was 0.2 cm, which was sufficient to put the flow into the *fully rough* regime. The turbulence intensity levels found are much higher than those for the smooth wall, but the relative ordering of the three components is the same.

The outer edge of a turbulent boundary layer is characterized by an unsteady, ragged *turbulent front*. A hot-wire anemometer placed at $y = \delta$ (recall that δ is a *mean* flow quantity) will sense alternately turbulent and then inviscid flow. This is reflected in a quantity Ω called the *intermittency*, or the fraction of the time that the flow is turbulent. Some measurements are shown in Fig. 7–19. Note that the intermittency remains significantly less than unity rather far down into the boundary layer, indicating that the transverse motion of the turbulent front is on a scale which is an appreciable fraction of δ.

The transport of x-momentum across the boundary layer by the turbulence is through the correlation $-\overline{u'v'}$. When formed as $-\rho\overline{u'v'}$, this quantity has the di-

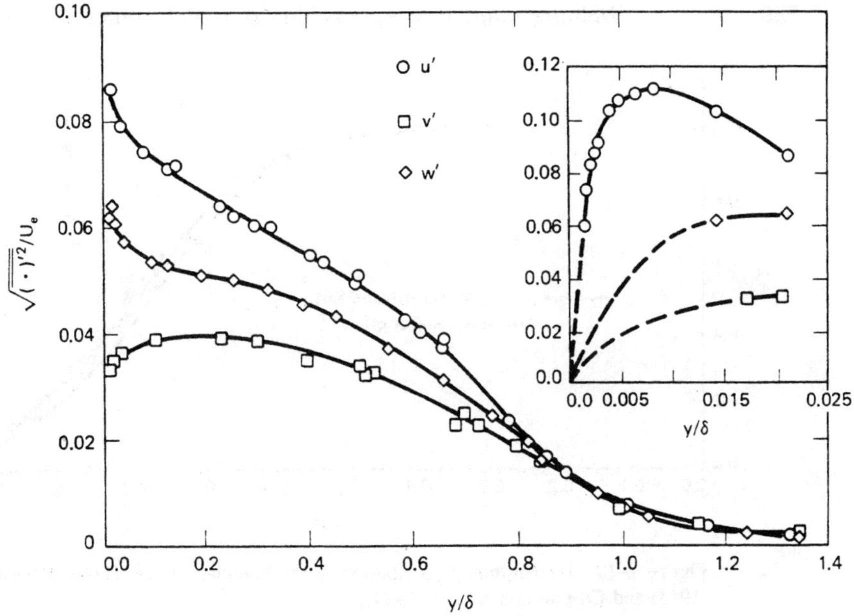

Figure 7–17 Turbulence intensity profiles for flow over a flat plate. (From Klebanoff, 1955.)

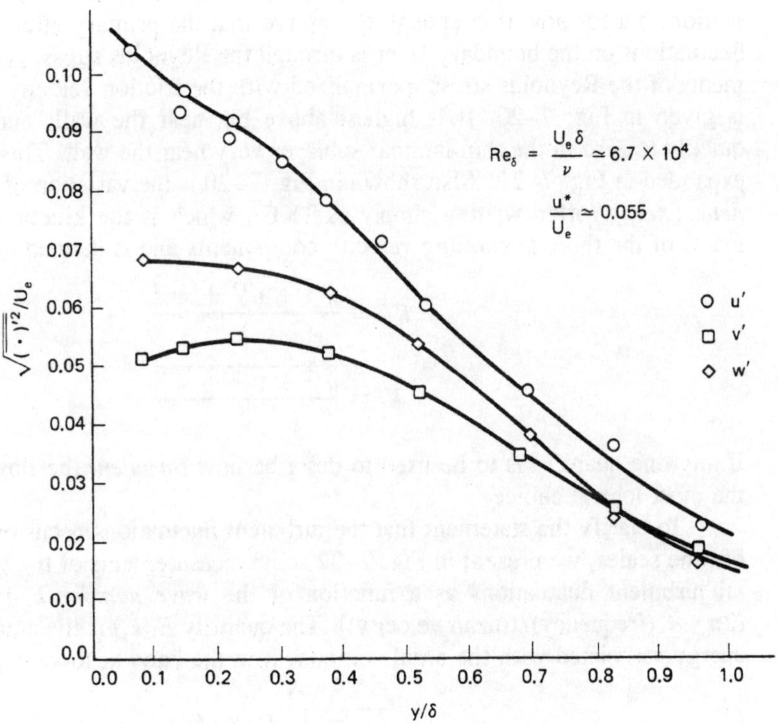

Figure 7–18 Turbulent intensity profiles for flow over a *fully rough* flat plate. (From Corrsin and Kistler, 1954.)

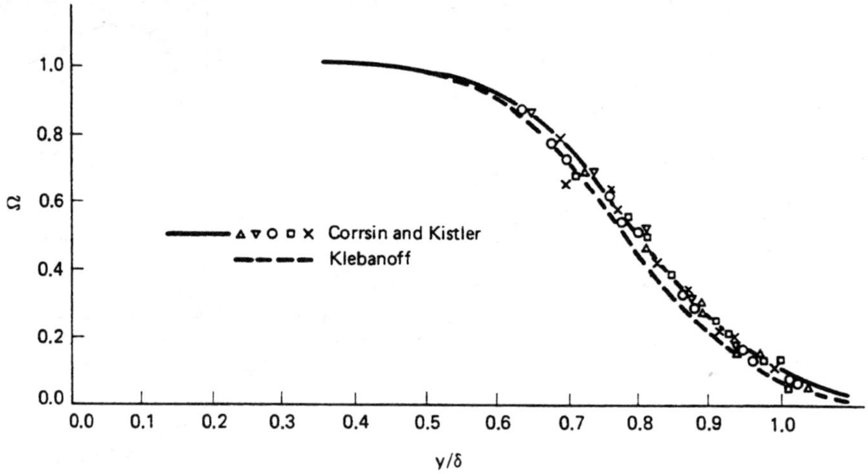

Figure 7–19 Intermittency distribution in the boundary layer. (From Klebanoff (1955) and Corrsin and Kistler, 1954.)

mensions of a stress, and it is called the *turbulent shear stress* or the *Reynolds stress,* after Osborne Reynolds. This concept will be derived more directly in a later section, but for now it is enough to observe that the primary effect of the turbulent fluctuations on the boundary layer is through the Reynolds stress. A plot of measurements of the Reynolds stress, normalized with the friction velocity across the layer, is given in Fig. 7–20. It is highest above but near the wall, and then drops off quickly to zero in the thin laminar sublayer very near the wall. This region is shown expanded in Fig. 7–21. Also shown in Fig. 7–20 is the variation of the *turbulent kinetic energy* (often written simply as TKE), which is the kinetic energy (per unit mass) of the three *fluctuating* velocity components and is defined as

$$K' \equiv \frac{u'^2 + v'^2 + w'^2}{2}$$
$$K \equiv \frac{\overline{u'^2 + v'^2 + w'^2}}{2} \tag{7–25}$$

If any one quantity is to be used to describe how *turbulent* the flow is, the TKE is the most logical choice.

To clarify the statement that the turbulent fluctuations occur over a wide range of time scales, we present in Fig. 7–22 some measurements of the *spectra* of the axial turbulent fluctuations as a function of the *wave number* k_1 [wave number $\equiv$ $(2\pi) \times$ (frequency)/(mean velocity)]. The quantity $E_1(k_1)$ is the amount of turbulent energy associated with the axial component in the band k_1 to $k_1 + dk_1$. Thus,

$$\overline{u'^2} = \int_0^\infty E_1(k_1)dk_1 \tag{7–26}$$

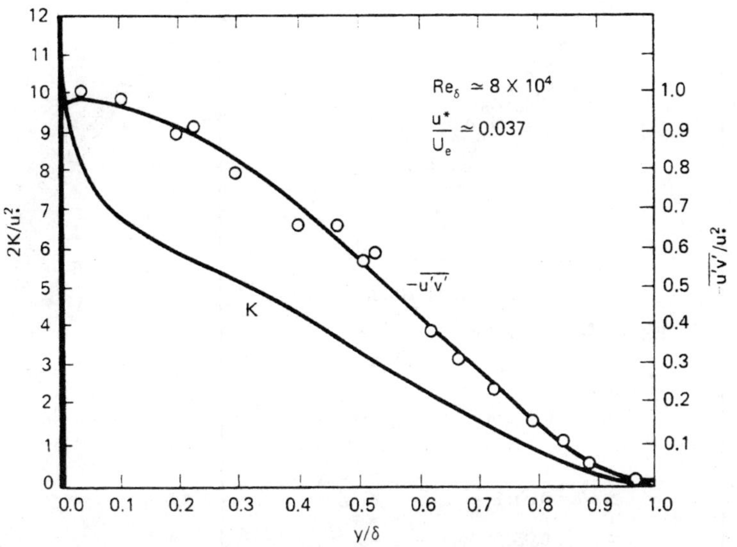

Figure 7–20 Profiles of Reynolds stress and turbulent kinetic energy in a turbulent boundary layer on a flat plate. (From Klebanoff, 1955.)

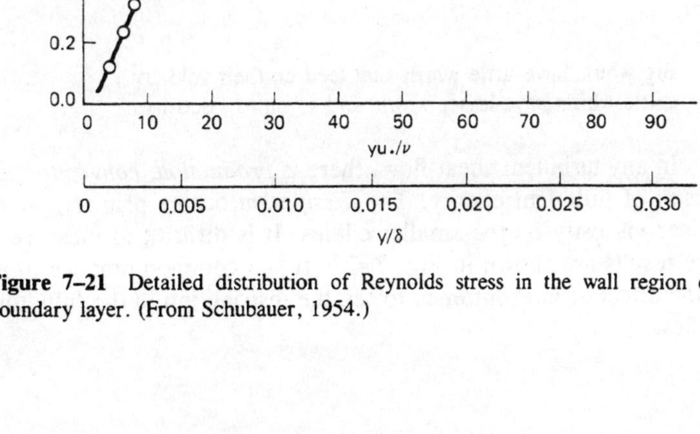

Figure 7–21 Detailed distribution of Reynolds stress in the wall region of a boundary layer. (From Schubauer, 1954.)

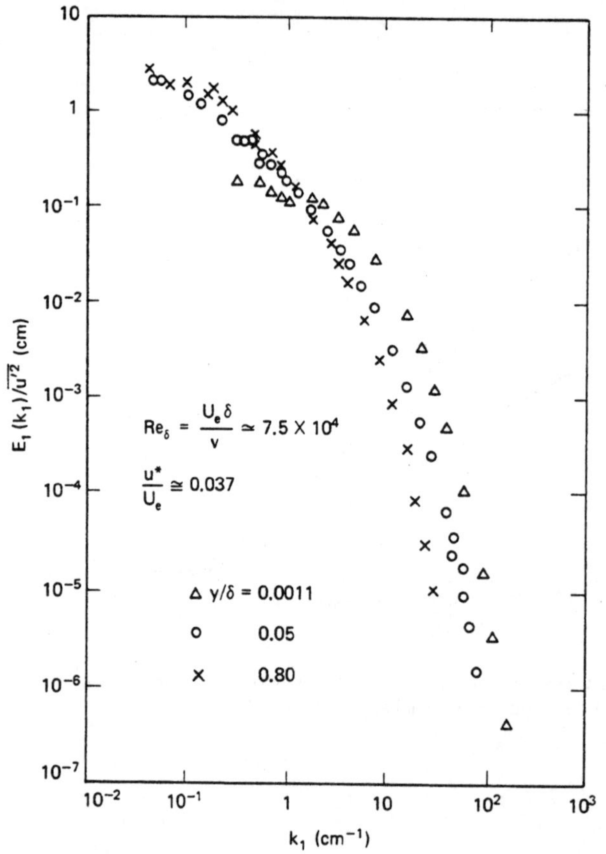

Figure 7–22 Spectra of the axial turbulence intensity in the boundary layer on a flat plate. (From Klebanoff, 1955.)

Observe first that there is appreciable motion occurring over about five decades in wave number. A second point is that the energy in the larger eddies (lower wave numbers) decreases as the wall is approached (look at the data for $y/\delta = 0.0011$), since the rigid wall inhibits larger scale motions. One says that there is an *energy cascade* flowing from the larger scales to the smaller scales and ultimately to dissipation by the viscosity into heat. The energy for the large-scale motion is extracted from the external stream. There is a parody on a children's rhyme that illustrates the point:

> Big whirls have little whirls that feed on their velocity.
> Little whirls have lesser whirls and so on to viscosity.

In any turbulent shear flow, there is *production, convection, diffusion,* and *dissipation* of turbulent energy. The *dissipation* occurs primarily due to the action of laminar viscosity on the smaller eddies. It is difficult to measure this quantity, but some results are shown in Fig. 7–23. It is a common practice to adjust such results for the effect of intermittency to see the distribution in the fully turbulent regions of the flow.

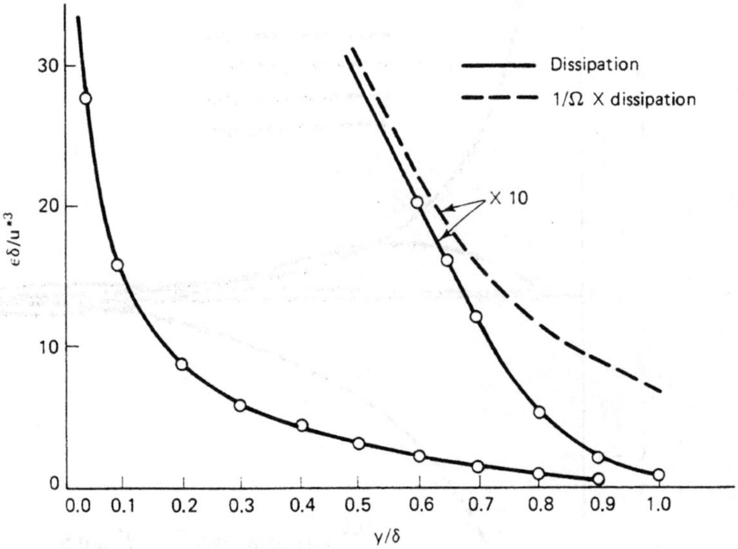

Figure 7–23 Distribution of the dissipation of turbulent kinetic energy across a flat-plate boundary layer. (From Klebanoff, 1955.)

The distributions of all the aforementioned processes across the layer can be plotted in the form of a turbulent kinetic energy balance, as in Fig. 7–24. Except near the outer edge, the main contributions to the overall balance are from *production* and *dissipation,* which nearly counterbalance each other.

For flow over permeable surfaces, the state of knowledge of turbulent quantities is again weaker than for solid surfaces, as was the case for the mean flow. Some data for axial and normal turbulence intensities and Reynolds stress are shown in Figs. 7–25, 7–26, and 7–27 for the six surfaces (three solid and three permeable) discussed earlier for the mean flow. The rough solid surfaces show clear increases above the smooth solid wall values for all three quantities. The effects of porosity per se can be seen by comparing the solid to the porous screening or the smooth solid to the smooth perforated sheet. There appears to be an identifiable effect of the porosity, and it is largest for the normal turbulence intensity and the Reynolds stress. That makes sense on physical grounds, since the normal turbulence component will not be as strongly damped by a porous surface as by a solid surface.

In addition to the results of measurements of the various quantities that reflect the fluctuating nature of the turbulent flow, a physical picture of the processes that lead to these data is helpful. Even such quantities as the turbulent intensities are averages over time, and one needs some grasp of the behavior of separate, unsteady turbulent events to understand the true physics of turbulent flow. The brief outline here is based mainly on the recent review of Kline and Robinson (1990). The idea of distinct eddies is not really clear or particularly helpful, because there is a large range of eddy sizes present essentially at the same time at the same place in a turbulent flow. It is more helpful to think of fluid *clumps* and *vortex lines*. A clump of

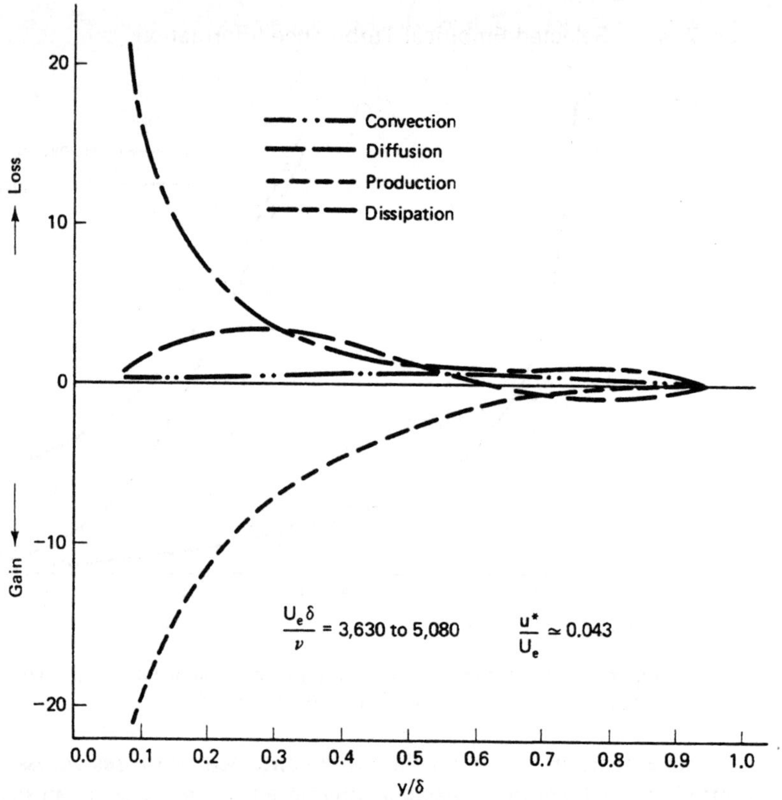

Figure 7–24 Turbulent energy balance in the boundary layer on a smooth flat plate. (From Klebanoff, 1955.)

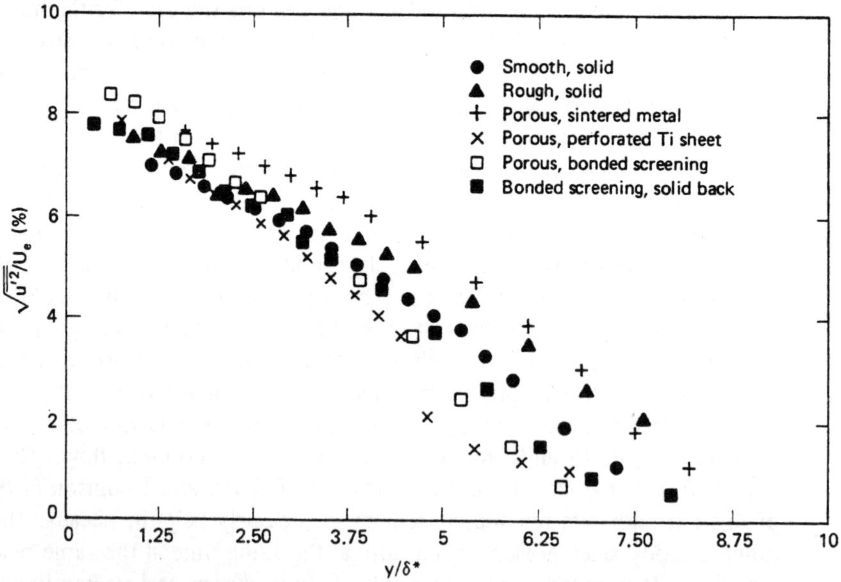

Figure 7–25 Axial turbulence intensity profiles over several solid and porous surfaces. (From Kong and Schetz, 1982.)

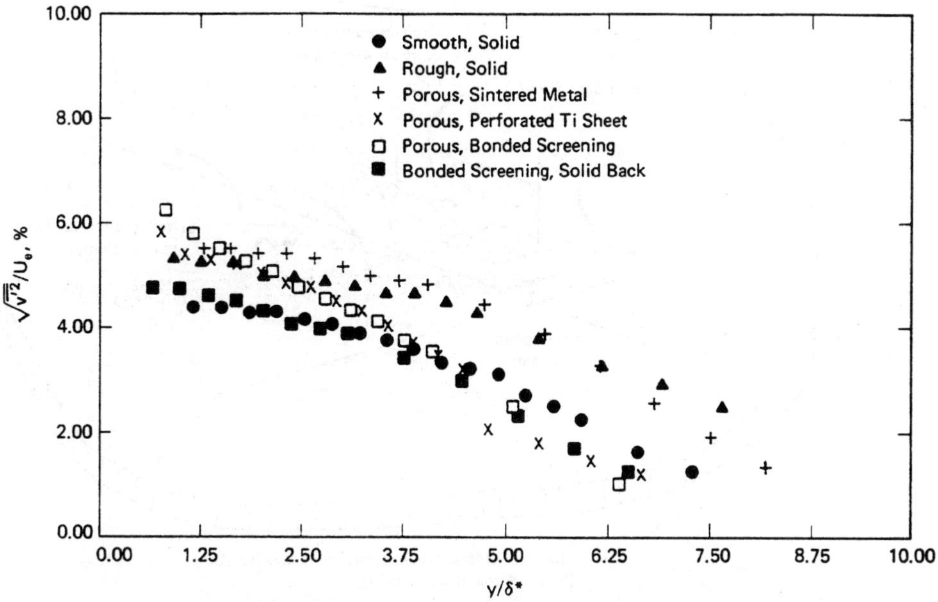

Figure 7–26 Normal turbulence intensity profiles over several solid and porous surfaces. (From Kong and Schetz, 1982.)

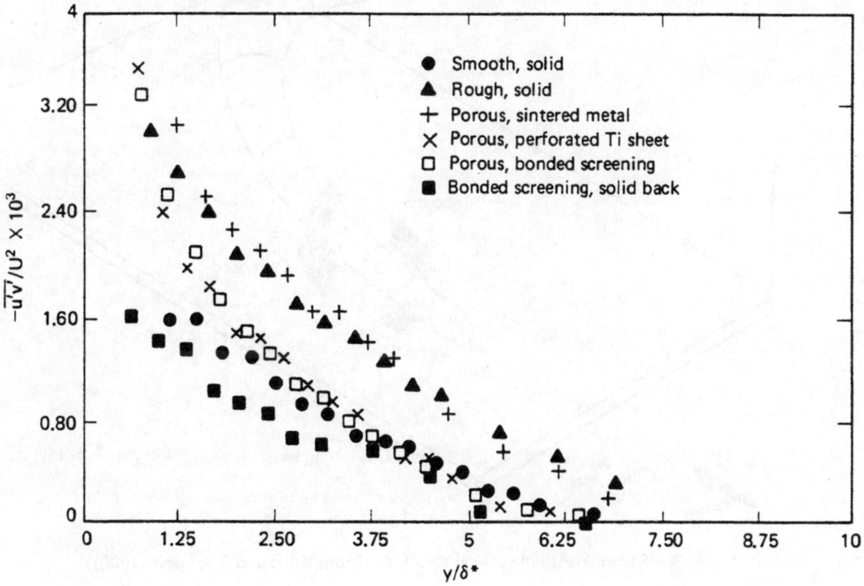

Figure 7–27 Reynolds stress profiles over several solid and porous surfaces. (From Kong and Schetz, 1982.)

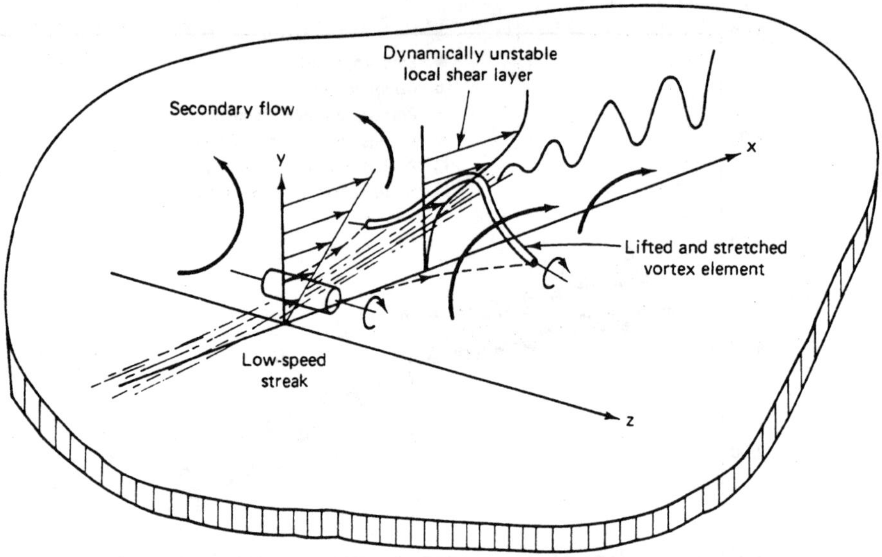

(A) Model of streak breakup that precedes bursting
(From Kline et al. (1967))

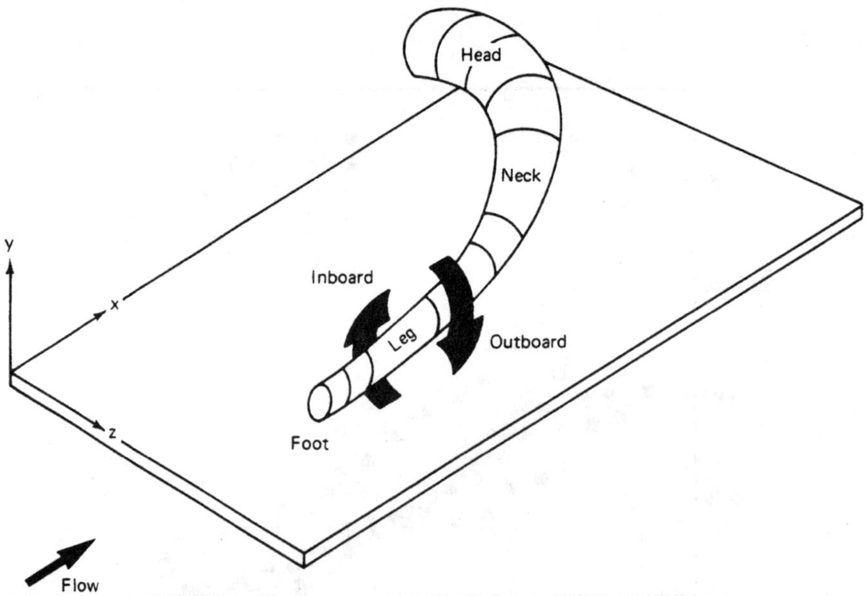

(B) Hook-shaped vortical structure (From Kline and Robinson (1990))

Figure 7–28 Vortical structures in wall region of turbulent boundary layers.

fluid may be said to move from one location to another in a turbulent flow. We will use that notion shortly to describe the Reynolds stress. A turbulent flow can be crudely envisioned as a superposition of vortex lines, so consider a vortex element of diameter d, length L, and angular velocity ω that results in a mass proportional to $\rho L d^2$, a kinetic energy proportional to $\rho L d^4 \omega^2$, and an angular momentum proportional to $\rho L d^4 \omega$. Let this vortex element be stretched but maintain constant mass and angular momentum (a good assumption for low diffusion). As L increases, the diameter will decrease and the angular velocity will increase, leading to an increase in the kinetic energy. Thus, work done on the ends of the element to stretch it will increase the kinetic energy in the element. This leads to a picture of the energy cascade mentioned earlier. Energy flows from the large-scale motions to the small-scale motions within them as a result of stretching of the vortex lines in the small-scale motions by the large-scale motions. In a two-dimensional (*in the mean*) turbulent flow, the vortex lines are mostly across the boundary layer. But the stretching must be in that direction also, so the turbulence must be three dimensional. Most of the mean vorticity is concentrated near the wall, due to the shape of the mean velocity profile (see Fig. 7–2). This leads to another consequence of vortex stretching: the randomly occurring *lift-up* of vortex elements near the wall, followed by sudden oscillation and *bursting* and *ejection* into the outer region. These processes are sketched in Fig. 7–28(A) from Kline et al (1967). More recent studies point to the more frequent occurrence of the hook-shaped structures shown in Fig. 7–28(B), where right and left hooks appear randomly. Rao et al. (1971) determined the mean burst time period to be about $5\delta/U_e$. Kline et al. (1967) found that the ejected fluid from a burst had a streamwise velocity of about $0.8U$. Thus, the ejection process is responsible for some portion of the Reynolds stress, since a clump of ejected fluid has $u' < 0$ and $v' > 0$, leading to a correlation of $u'v'$ of less than zero. Kim et al. (1971) suggest that virtually all turbulence production is a result of the bursting or ejection process, especially near the wall.

The wall region shows a periodic structure in both time and space. ***Streaks*** of low-momentum fluid with a length on the order of δ and a width on the order of $30\nu/u_*$ occur in a periodic pattern in the streamwise and spanwise directions. A horizontal shear layer is formed on top of these streaks at about $y^+ \approx 40$, where the local velocity profile shows an inflection point. The streaks are ended with the ejection of low-momentum fluid into the outer region. The ejected fluid is replaced by an inward *sweep* of high-momentum fluid from the edge of *buffer region,* near $y^+ \approx 40$. Again, this sweep-ejection process produces the major part of the Reynolds stresses.

7–5 THE CENTRAL PROBLEM OF THE ANALYSIS OF TURBULENT FLOWS

If it is assumed that the unsteady, laminar equations contain the essence of turbulent motions, then unsteady versions of Eqs. (1–36) through (1–39) will serve for a constant-density, constant-property (laminar properties) case. This assumption is neither as controversial nor as helpful as it may seem. It is not controversial because the derivation of these equations did not in any way preclude the existence of a randomly fluctuating motion. It is not particularly helpful because it really brings us no simplification. One is still confronted with a three-dimensional, randomly unsteady motion with very large ranges in physical scales and frequencies, resulting in a truly massive computational problem.

Earlier, we said that the analyst is usually interested only in the time-averaged, mean motion. How does that general goal fit into a formulation based on an unsteady system of equations? The answer is simple, in principle: Take the average over time. Using the division of the motion into mean and fluctuating parts, as, for example,

$$u(x, y, z, t) = U(x, y) + u'(x, y, z, t) \tag{1-67}$$

for a planar, steady (*in the mean*) case, we can substitute into the equations and then take the average over time, term by term. We will work here with the two-dimensional boundary layer form of the equations for clarity of the development, since we anticipate an analysis for two-dimensional mean flow in a boundary layer. A few simple rules must also be followed. If f and g are any two fluctuating variables and F and G are their mean values, and if a bar over a quantity $\overline{(\cdot)}$ denotes the operator of taking the time mean, then

$$\begin{aligned} &\overline{f'} \equiv 0, && \overline{F} \equiv F \\ &\overline{f + g} = F + G, && \overline{F \cdot g} = F \cdot G \\ &\overline{\frac{\partial f}{\partial s}} = \frac{\partial F}{\partial s}, && \overline{\int f\, ds} = \int F\, ds \end{aligned} \tag{7-27}$$

Substituting into the continuity equation Eq. (3–6a), we obtain

$$\begin{aligned} 0 = \frac{\partial u}{\partial x} + \frac{\partial v}{\partial y} &= \frac{\partial(U + u')}{\partial x} + \frac{\partial(V + v')}{\partial y} \\ &= \left(\frac{\partial U}{\partial x} + \frac{\partial V}{\partial y}\right) + \left(\frac{\partial u'}{\partial x} + \frac{\partial v'}{\partial y}\right) \end{aligned} \tag{7-28}$$

Taking the time mean of the whole equation yields

$$\frac{\partial U}{\partial x} + \frac{\partial V}{\partial y} = 0 \tag{7-29}$$

which says that the continuity equation for the mean flow has the same form as for a laminar flow. Note, however, that Eq. (7–29) substituted into Eq. (7–28) also implies that

$$\frac{\partial u'}{\partial x} + \frac{\partial v'}{\partial y} = 0 \tag{7-30}$$

Thus, the instantaneous fluctuations satisfy a continuity equation of the same form.

The nonlinear terms in the momentum equation, Eq. (3–26a), must be treated somewhat more carefully. It is helpful to note that the convective derivative (the nonlinear terms) can be rewritten as

$$
\begin{aligned}
u\frac{\partial u}{\partial x} + v\frac{\partial u}{\partial y} &= \frac{\partial(u^2)}{\partial x} + \frac{\partial(uv)}{\partial y} \\
&= u\frac{\partial u}{\partial x} + v\frac{\partial u}{\partial y} + u\left(\frac{\partial u}{\partial x} + \frac{\partial v}{\partial y}\right)
\end{aligned}
\tag{7–31}
$$

The last term is zero by virtue of the continuity equation [Eq. (7–28)]. We now substitute in terms of mean and fluctuating quantities:

$$
\begin{aligned}
\frac{\partial(u^2)}{\partial x} + \frac{\partial(uv)}{\partial y} &= \frac{\partial(U + u')^2}{\partial x} + \frac{\partial((V + v')(U + u'))}{\partial y} \\
&= \frac{\partial(U^2 + 2Uu' + u'^2)}{\partial x} + \frac{\partial(UV + Vu' + Uv' + u'v')}{\partial y}
\end{aligned}
\tag{7–32}
$$

Taking the time average results in

$$
\frac{\partial(U^2)}{\partial x} + \frac{\partial(UV)}{\partial y} + \frac{\partial(\overline{u'^2})}{\partial x} + \frac{\partial(\overline{u'v'})}{\partial y}
\tag{7–32a}
$$

Using the time-averaged continuity equation in the same manner as in Eq. (7–31), we can rewrite Eq. (7–32a) in more familar form as

$$
U\frac{\partial U}{\partial x} + V\frac{\partial U}{\partial y} + \frac{\partial(\overline{u'^2})}{\partial x} + \frac{\partial(\overline{u'v'})}{\partial y}
\tag{7–32b}
$$

The other terms in the momentum equation can be treated easily:

$$
\begin{aligned}
\overline{\frac{\partial u}{\partial t}} &= \overline{\frac{\partial(U + u')}{\partial t}} = \frac{\partial U}{\partial t} \\
\overline{\frac{\partial p}{\partial x}} &= \overline{\frac{\partial(P + p')}{\partial x}} = \frac{\partial P}{\partial x} \\
\overline{\mu\frac{\partial^2 u}{\partial y^2}} &= \mu\overline{\frac{\partial^2(U + u')}{\partial y^2}} = \mu\frac{\partial^2 U}{\partial y^2}
\end{aligned}
\tag{7–33}
$$

Note that nothing interesting happens to the viscous term.

Putting these all together, we can write the time-averaged momentum equation for turbulent flow in a boundary layer as

$$
\frac{\partial U}{\partial t} + U\frac{\partial U}{\partial x} + V\frac{\partial U}{\partial y} + \frac{\partial(\overline{u'^2})}{\partial x} = -\frac{1}{\rho}\frac{\partial P}{\partial x} + \nu\frac{\partial^2 U}{\partial y^2} + \frac{1}{\rho}\frac{\partial(-\rho\overline{u'v'})}{\partial y}
\tag{7–34}
$$

If the flow is steady (*in the mean*), the first term drops out. It is a common assumption to take $\partial(\overline{u'^2})/\partial x \ll \partial(\overline{u'v'})/\partial y$, and this is confirmed by experiment in many flows. Finally, the simplest system of equations that can possibly represent the mean motion of a turbulent flow in a boundary layer may be written

$$\frac{\partial U}{\partial x} + \frac{\partial V}{\partial y} = 0 \tag{7–29}$$

$$U\frac{\partial U}{\partial x} + V\frac{\partial U}{\partial y} = -\frac{1}{\rho}\frac{dP}{dx} + \nu\frac{\partial^2 U}{\partial y^2} + \frac{1}{\rho}\frac{\partial(-\rho\overline{u'v'})}{\partial y} \tag{7–34a}$$

Far from any rigid surface, the laminar shear term may be assumed small compared with momentum transport by turbulence and dropped. Certainly, the term involving $-\rho\overline{u'v'}$ cannot be dropped, or the system would collapse completely back to that for laminar flow. That is the only term remaining that shows any influence of the true fluctuating nature of the turbulent flow.

As with any boundary layer problem, in the problem specified by Eqs. (7–29) and (7–34a), the pressure field $P(x)$ is taken to be *imposed* on the shear flow by an external, inviscid flow, so that we would expect to solve these two equations for the two unknowns (U, V). However, we notice an additional unknown term, the last on the right-hand side of Eq. (7–34a). This is a momentum transfer term that plays the same role as simple Newtonian shear in a laminar flow, so that the grouping $-\rho\overline{u'v'} = \tau_T$ is termed the *turbulent shear stress* or, as we said earlier, *Reynolds stress*. Note that this turbulent *shear* term came, not from the laminar viscous terms, but from the inviscid convective derivative terms. Working with the framework of Eqs. (7–29) and (7–34a), we must relate this term to the other independent or dependent variables in order to complete the problem from a mathematical standpoint. Since these other variables are all mean quantities, such a relationship or model is termed a *mean flow model*. The development of models of that type will be the subject of Sections 7–6 and 7–8.

The reader must understand that there is no easy way out of the quandary just stated. The matter will be pursued to more complex formulations at the end of the chapter, but there will always be more unknowns than equations. The extra unknown(s) will always be turbulent quantities, such as $-\rho\overline{u'v'}$, and it will always be necessary to insert extra, generally semiempirical relations in order to *close* the system mathematically. This is called the *closure problem* for turbulent flows. It might be easier to grasp how we arrived at this perplexing state if one reflects on the simple fact that really nothing substantive has been done, either by asserting definitions of mean flow quantities, as by equations such as Eq. (1–67), or by manipulating the equations of motion, or by taking the average over time, as has been done here. To this point in developing the mathematical formulation, nothing has been said about the real *nature* of turbulent flows. The equations are still waiting for us to do so, before the formulation can be complete.

Prior to proceeding with the task of developing models (i.e., *modeling* turbulent shear), it is useful to develop a physical description of the process of momentum transfer by the fluctuations. Consider the sketch in Fig. 7–29. If a fluid clump from a height $y_0 + \Delta y$ moves down to y_0, it will tend to retain its mean velocity $U(y_0 + \Delta y)$, which will appear as a positive fluctuation on the mean velocity at y_0 [i.e., $U(y_0)$]. The instantaneous mass flow through the plane $y = y_0$ will be $-\rho v'$, and its axial velocity will be $U(y_0) + u'$, so it will transport an increment of axial momentum across the plane $y = y_0$ in the amount $(-\rho u'v')$. If the clump came from below, extending this logic leads to the same result. Averaging over time, we find that the *net* transport of axial momentum in the transverse direction by the velocity fluctuations is $-\rho\overline{u'v'}$.

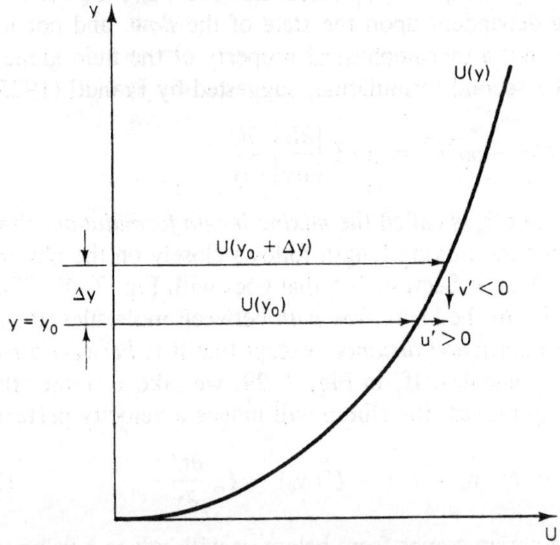

Figure 7–29 Schematic illustration of the evolution of the Reynolds stress.

7–6 MEAN FLOW TURBULENT TRANSPORT FORMULATIONS

In general, to develop a mean flow turbulent analysis, it is first necessary to model the turbulent shear at each point across the layer. Recalling the discussion of Section 7–5, this means that a relation

$$\tau_T \equiv -\rho\overline{u'v'} = f(x, y; U, V, P \text{ or } U_e, \mu, \rho) \tag{7–35}$$

must be sought so that the system posed by Eqs. (7–29) and (7–34a) can be closed. Clearly, this relation must be based on empirical information, so we are led back to the material in Section 7–3 for mean flow data. Note that no turbulent quantities appear on the right-hand side of Eq. (7–35). Perhaps it will be helpful at this point to state clearly the goal of this part of the *modeling* efforts, using the inner region as an example. We wish to find a relationship, schematically like that of Eq. (7–35), that will, when used in the equations of motion (Eqs. (7–29) and (7–34a)), faithfully reproduce the solid curve through the data of Fig. 7–4. An analysis of this type is often called *semiempirical*. An *empirical* approach would employ curves simply fitted to the data, without recourse to any equations based on first principles that govern the problem.

The stated goal has been accomplished within two related formulations. The first is the *eddy viscosity formulation* introduced by Boussinesq (1877), which takes the form

$$\tau_T = -\rho \overline{u'v'} = \mu_T \frac{\partial U}{\partial y} \tag{7–36}$$

by analogy with laminar flow [see Eq. (1–2)]. Here, however, μ_T, the *eddy viscosity*, must be expected to be dependent upon the state of the *flow*, and not just the state of the *fluid*. Thus μ_T is not a thermophysical property of the fluid alone, as is μ, the laminar viscosity. The second formulation, suggested by Prandtl (1925), is

$$\tau_T = -\rho \overline{u'v'} = \rho \ell_m^2 \left|\frac{\partial U}{\partial y}\right| \frac{\partial U}{\partial y} \tag{7–37}$$

where ℓ_m is a *mixing length*, so this is called the *mixing length formulation*. Prandtl's development of the concept of the mixing length follows closely on the physical interpretation of $-\rho\overline{u'v'}$ at the end of Section 7–5 that goes with Fig. 7–29. The *mixing length* ℓ_m is crudely similar to the *mean free path* between molecules λ^*, in that it is taken as some *effective interaction distance*, except that it is *between turbulent clumps* rather than between molecules. If, in Fig. 7–29, we take ℓ_m rather than Δy as the distance the fluid clump moves, the clump will induce a velocity perturbation

$$u'(y_0) = U(y_0 + \ell_m) - U(y_0) \approx \ell_m \frac{\partial U}{\partial y} \tag{7–38}$$

with $v'(y_0) < 0$. If the fluid clump comes from below, it will induce a velocity perturbation

$$u'(y_0) = U(y_0 - \ell_m) - U(y_0) \approx -\ell_m \frac{\partial U}{\partial y} \tag{7–39}$$

with $v' > 0$. By continuity, $v' \sim -u'$, and this relation, together with Eqs. (7–38) and (7–39), leads directly to Eq. (7–37). We write

$$\left|\frac{\partial U}{\partial y}\right| \frac{\partial U}{\partial y}$$

rather than $(\partial U/\partial y)^2$ to ensure the proper sign for τ_T in relation to that of $\partial U/\partial y$. An additional relationship for ℓ_m as a function of the independent variables or parameters or of the mean dependent variables is still required. This relationship will also be a function of the state of the *flow*, and not the *fluid*, just as was the eddy viscosity μ_T.

At this stage, it is appropriate to make some observations about the formulations in Eqs. (7–36) and (7–37). First, both the *eddy viscosity* and *mixing length formulations* are what are called *gradient transport formulations*, since they employ a gradient (here $\partial U/\partial y$) of some relevant dependent variable to represent a transport process. Thus, our representations of the laminar transport of momentum, $\tau = \mu(\partial u/\partial y)$, and the thermal energy, $q = -k(\partial T/\partial y)$, are also gradient transport formulations. There is however, a special weakness of such formulations for turbulent shear flows that is most easily seen for the case of a profile such as that sketched in Fig. 7–30. Profiles of this type occur, for example, with high-velocity tangential injection through an upstream slot in the wall into a turbulent boundary layer. Since the mean velocity profile $U(y)$ has a maximum, it must have a point where $\partial U/\partial y = 0$. Both formulations imply that $\tau_T = -\rho\overline{u'v} = 0$ at such a point, and that is not borne out by experiment for unsymmetrical *maxima* or *minima*. There is no simple way to repair this basic weakness of these formulations, but one can take some small comfort in the fact that most boundary layer flows do not involve profiles with a maximum or minimum. For a symmetrical profile, $\partial U/\partial y = 0$ on the plane of symmetry, and $-\rho\overline{u'v'} = 0$ there also. The second general observation to be made about the two formulations is that they are essentially equivalent. Looking at Eqs. (7–36) and (7–37), one can obviously write

$$\mu_T = \rho\,\ell_m^2 \left|\frac{\partial U}{\partial y}\right| \tag{7–40}$$

Thus, if a relationship for μ_T, for example, is known, a corresponding relationship for ℓ_m can be found from Eq. (7–40), or vice versa.

Specific models for the eddy viscosity and the mixing length will be developed in Section 7–8. Before that, however, we shall look at integral methods for analyzing turbulent boundary layers.

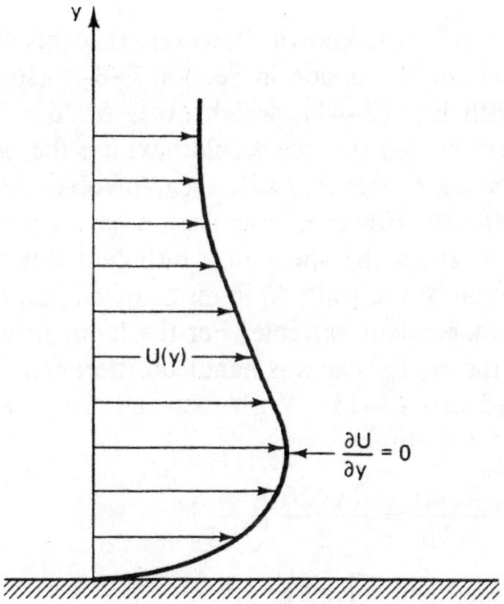

Figure 7–30 Illustration of a turbulent boundary layer velocity profile with a velocity maximum (i.e., $\partial U/\partial y = 0$).

7–7 MEAN FLOW INTEGRAL METHODS

There are methods for turbulent flow analysis that follow along the lines of the integral methods for laminar flows discussed in Chapter 2. They bear roughly the same relation to turbulent analyses based on differential formulations as is the case for laminar flows. Such methods are generally simple to apply, and they yield results that are sometimes adequate for engineering work. For purposes of illustration, we describe two of these methods here—an older one that is suitable only for very simple problems and a newer one due to Moses (1969) that can be used for more general flows. In each case, the reader should be alert to how the necessary modeling of the turbulent shear enters into the analysis.

The earliest application of an integral method to a case of turbulent flow was by Prandtl (1927) for the flat-plate problem. The momentum integral equation, Eq. (2–15), originally derived in Chapter 2 for steady, laminar flows, holds here for the mean flow of a steady (*in the mean*) turbulent flow if the dependent variables are all interpreted as mean values. Thus, for a solid flat plate, we have

$$\frac{d\theta}{dx} = \frac{C_f}{2} \tag{2–15a}$$

if, in the definition of θ, Eq. (2–13), we understand that the profile is described by $U(y)$, the mean velocity.

As with the laminar problem, an assumed profile shape is needed. The mean velocity profile for a turbulent flow over a flat plate can be crudely represented by

$$\frac{U}{U_e} = \left(\frac{y}{\delta}\right)^{1/n} \tag{7–41}$$

with $n \approx 7$ for $\mathrm{Re}_x \approx 10^6$ to 10^7. It is known, however, that this cannot be correct in general because of the data and discussion in Section 7–3–1 (see also Fig. 7–2). Nonetheless, we proceed with Eq. (7–41), which gives $\delta^*/\delta = 1/8$ and $\theta/\delta = 7/72$ for $n = 7$. It might be expected that we would next use the profile to find the wall shear, as in the laminar case, but the wall shear involves $(\partial U/\partial y)_w$, which, from Eq. (7–41), goes to infinity. However, one should have anticipated a need to insert some extra information about the shear in a turbulent flow anyway. In this method, we need only the shear at the wall, C_f in terms of δ [i.e., $C_f(\mathrm{Re}_\delta)$], since δ is to emerge as the primary dependent variable. For the laminar analysis, δ played the same role, even though the wall shear was handled differently. For the required $C_f(\mathrm{Re}_\delta)$, Prandtl (1927) used Eq. (7–13). With that and the just-determined θ/δ substituted into Eq. (2–15a), we obtain

$$\frac{7}{72}\frac{d\delta}{dx} = \frac{0.0456}{2}\left(\frac{\nu}{U_e\delta}\right)^{1/4} \tag{7–42}$$

This equation can be rearranged and solved to give

$$\frac{\delta(x)}{x} = \frac{0.375}{\mathrm{Re}_x^{1/5}} \tag{7–43}$$

where $\delta(0) = 0$ was used. This result indicates that a turbulent boundary layer grows much faster ($\delta \sim x^{4/5}$) than a laminar one, which has $\delta/x \sim \mathrm{Re}_x^{-1/2}$ ($\delta \sim x^{1/2}$). The result is rather crude, first, because it uses a profile shape that is not based on the best variable groupings for a turbulent boundary layer, and second, because it completely neglects transition and the region of laminar flow at the beginning of the plate.

Unfortunately, there is no more that can be done with this method. For example, if one wishes to solve problems with pressure gradients, it would be necessary to have $C_f(\mathrm{Re}_\delta, dP/dx)$ for general forms of dP/dx. If that much information were actually available, one would not need this type of analysis.

An integral method for turbulent boundary layers that has much greater capability than the one just examined and that is still simple to apply has been developed by Moses (1969). This method uses a generalized form of the momentum integral equation, Eq. (2–15), written

$$\frac{U(y_1/\delta)}{U_e}\frac{d}{dx}\left[\mathrm{Re}_\delta\int_0^{y_1/\delta}\frac{U}{U_e}d\left(\frac{y}{\delta}\right)\right]-\frac{1}{U_e}\frac{d}{dx}\left[U_e\mathrm{Re}_\delta\int_0^{y_1/\delta}\left(\frac{U}{U_e}\right)^2d\left(\frac{y}{\delta}\right)\right]$$
$$=\frac{\tau_w-\tau(y_1/\delta)}{\mu U_e}-\frac{(y_1/\delta)\,\mathrm{Re}_\delta}{U_e}\frac{dU_e}{dx} \qquad (7\text{–}44)$$

where U is now understood for u from before and where y_1/δ can have any value from 0 to 1.0. For $y_1/\delta = 1.0$, Eq. (7–44) becomes the usual *momentum integral* equation. For other values of y_1/δ, separate, additional equations are produced. One is then applying conservation of mass and momentum to a control volume $dx : y_1 : 1$, with $y_1 < \delta$, where before, a control volume $dx : H : 1$, with $H \geq \delta$, was used. The biggest difference in the result is that the shear at $y = y_1$ appears now, while the shear at $y = H$ is zero and does not appear. Moses obtained two equations, with $y_1/\delta = 1.0$ and 0.3. The latter value corresponds roughly to dividing the loss of momentum across the layer in half.

For the required assumption of a velocity profile shape, Moses used an approximate form of the *law of the wake*. He rewrote Eq. (7–9) using the definition of $W(y/\delta)$, Eq. (7–8), and $u^*/U_e = \sqrt{C_f/2}$, and then he approximated the *wake function* as

$$W\left(\frac{y}{\delta}\right) = 2\left[3\left(\frac{y}{\delta}\right)^2 - 2\left(\frac{y}{\delta}\right)^3\right] \qquad (7\text{–}45)$$

The final result is

$$\frac{U}{U_e} = 1 + \frac{\sqrt{C_f/2}}{\kappa}\ln\left(\frac{y}{\delta}\right) - \frac{2\Pi}{\kappa}\sqrt{\frac{C_f}{2}}\left[1 - \frac{1}{2}W\left(\frac{y}{\delta}\right)\right] \qquad (7\text{–}46)$$

For the laminar sublayer, Moses used another polynomial, and he noted that the choice of profile for the thin sublayer is not important for integral methods for flows with a high Reynolds number. That choice does not affect the values of the integrals greatly. This composite profile shape is much better and sounder than taking $U/U_e = f(y/\delta)$ as in Eq. (7–41).

A model for the turbulent shear is introduced with an eddy viscosity model [see Eq. (7–36)]. For this integral method, it is only necessary to model the turbulent shear at a single point $y/\delta = y_1/\delta$, since it is only necessary to represent $\tau(y_1/\delta)$ in Eq. (7–44). This is a strength of the formulation, since turbulence modeling is introduced at a much more fundamental level than simply inputting the skin friction, which really is the desired output. Moses used the completely empirical eddy viscosity model

$$\left.\frac{\mu_T + \mu}{\rho U_e \theta}\right|_{y_1/\delta} = 0.0225 + \frac{125}{\mathrm{Re}_\delta} \tag{7-47}$$

based on a direct curve fit of some data by Coles (1962). With all of this, the system consists of two ordinary differential equations—Eq. (7–44) evaluated at $y_1/\delta = 1.0$ and 0.3—together with the velocity profile, Eq. (7–46) [and Eq. (7–45) and the sublayer profile], for two unknowns, $\delta(x)$ (written as Re_δ) and C_f. The equations can be solved numerically using one of the programmed routines for solving ordinary differential equations available in any computer center.

Moses (1969) has applied his method to a number of experimental cases, and he has achieved the same good level of agreement as any of the current integral methods. Generally, the agreement is almost as good [for gross quantities such as C_f and $\delta(x)$] as that achieved by the best differential methods to be presented later. In order to display the performance of the various calculation methods, three adverse pressure gradient flows of increasing complexity have been selected for discussion. All of the methods perform well for flat plate flows (with zero pressure gradient), and the cases with favorable pressure gradients are much less challenging. The three cases we will consider are (1) the equilibrium case $[(\delta^*/\tau_w)(dP/dx) = 2]$ of Clauser (1956) discussed earlier, (2) the classic experiment of Schubauer and Klebanoff (1950), and (3) the airfoil flow of Newman (1951). All three flows are sufficiently well documented to permit unambiguous calculations and comparisons with the data. The flows are three of the more complex flows that were chosen as test cases for calculation methods at the 1968 AFOSR-IFP-Stanford Conference (see Kline et al., 1969). The predictions of Moses (1969) are compared with the data for these three cases in terms of $\theta(x)$ and $C_f(x)$ in Figs. 7–31 to 7–33. It can be seen that the agreement for the Clauser equilibrium case is poor, while it is good for the Schubauer and Klebanoff and Newman cases until separation is approached. It is a sad fact that the agreement with the Clauser experiments should be poor. Those famous experiments that played such a critical role in developing our current understanding of turbulent boundary layer flows have since been found to be tainted by unintentional three-dimensional effects.

To illustrate the performance of the various models and methods, we will also use a few of the cases from another conference at Stanford in 1981 that featured an organized effort at comparisons of predictions with experiment for complex turbulent flows (see Kline et al., 1982). The low-speed case chosen for inclusion is the adverse pressure gradient experiment of Samuel and Joubert (1974). A comparison of their data with a prediction from the Moses method is shown in Fig. 7–34. The generally good agreement can be noted.

A simple computer code for a PC implementing the Moses method may be found in Appendix B.

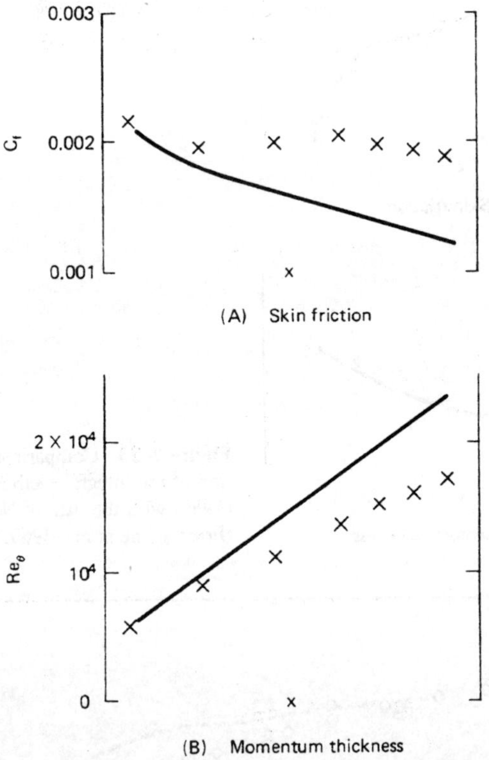

Figure 7–31 Comparison of the prediction of the integral method of Moses (1969) with the data of Clauser (1956). (From Kline et al., 1969.)

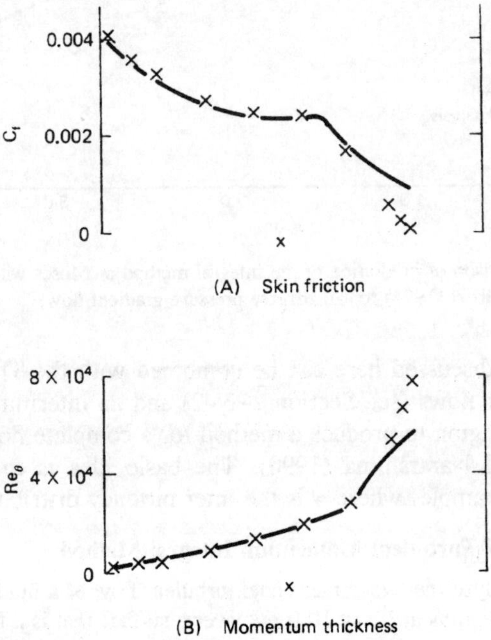

Figure 7–32 Comparison of the prediction of the integral method of Moses (1969) with the data of Schubauer and Klebanoff (1950). (From Kline et al., 1969.)

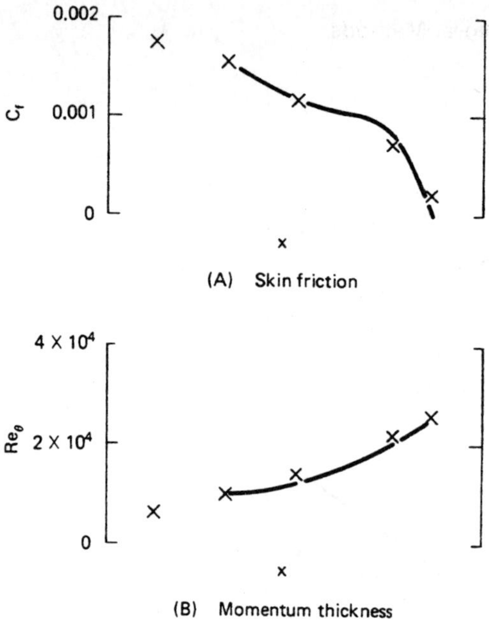

Figure 7–33 Comparison of the prediction of the integral method of Moses (1969) with the data of Newman (1951). (From Kline et al., 1969.)

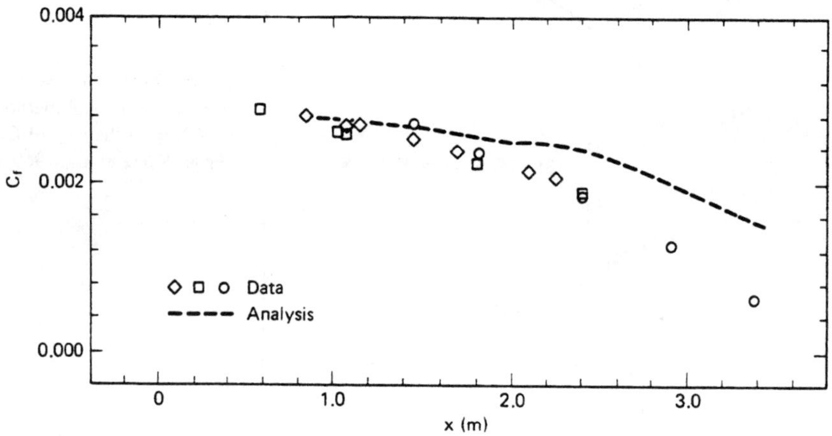

Figure 7–34 Comparison of prediction of the integral method of Moses with the data of Samuel and Joubert (1974) for an adverse pressure gradient flow.

The integral method discussed here can be combined with the Thwaites-Walz integral method for laminar flows (see Section 2–3–2) and an intermittency distribution for the transitional region to produce a method for a complete flow following the suggestion of Dey and Nararshima (1990). The basic idea is to take $\delta^* = (1 - \gamma)\delta^*_{\text{lam}} + \gamma\delta^*_{\text{turb}}$, for example, where γ is the intermittency distribution.

Example 7–1. Application of Turbulent Momentum Integral Method

Suppose we wish to analyze the two-dimensional turbulent flow of a fluid with a kinematic viscosity $\nu = 10^{-5}$ m²/s at $U_\infty = 10.0$ m/s over a surface that is a flat plate from the leading edge to $x = 5.0$ m. At that station, a ramp begins that produces an inviscid velocity distribution $U_e(x) = 15.0 - x$ m/s. This is a rather strong adverse pressure gradient, since U_e is decreasing rapidly, so that p increases rapidly. Calculate the boundary layer development over this surface up to $x = 7.0$ m. Does the flow separate?

Solution This problem can be solved with the Moses integral method, so we can use the code MOSES in Appendix B. Input data CNU = 0.00001 for the kinematic viscosity must be provided. The first part of the flow is a flat plate, so the simple integral solution can be used to calculate δ, θ, and C_f at $x = 5.0$ m [XX(1) = 5.0], to be used for initial conditions for the calculation of the rest of the flow. Thus, we start with CF = 0.002665 and RTHETA = 8336.3. Information for dx also must be given and is specified by choosing to make 21 steps (NMAX = 21) along the surface from $x = 5.0$ to $x = 7.0$ m, a not unreasonable figure ($dx = 0.10$ m).

Finally, the inviscid velocity distribution is required. This kind of information for this particular case is included in the code as UE(N) and DUE(N) and in the loop DO 10 N = 2,NMAX. For other problems, this section of the code and other relevant sections must be modified.

The primary output is listed in the following table. The boundary layer thicknesses increase, the profile shape changes, and the skin friction coefficient decreases rapidly in this relatively strong adverse pressure gradient. The boundary layer does not separate by $x = 7.0$ m—i.e., $C_f > 0$—but the boundary layer thickness nearly doubles, and C_f is reduced by about 40 percent. Compare the behavior of this turbulent boundary layer to the laminar case in Example 2–1 in a much weaker pressure gradient.

A plot of the output is given in Figure 7–35.

X	UE	CF	DELTA	THETA	RTHETA	H
5.00	10.000	0.00266	0.0772	0.00834	8336.4	1.337
5.10	9.900	0.00260	0.0792	0.00876	8667.5	1.347
5.20	9.800	0.00253	0.0814	0.00919	9004.7	1.356
5.30	9.700	0.00247	0.0837	0.00964	9348.9	1.365
5.40	9.600	0.00241	0.0862	0.01010	9700.5	1.374
5.50	9.500	0.00235	0.0888	0.01059	10060.4	1.383
5.60	9.400	0.00230	0.0916	0.01109	10429.0	1.392
5.70	9.300	0.00224	0.0945	0.01162	10807.2	1.401
5.80	9.200	0.00219	0.0976	0.01217	11195.7	1.410
5.90	9.100	0.00213	0.1007	0.01274	11595.2	1.419
6.00	9.000	0.00208	0.1041	0.01334	12006.6	1.428
6.10	8.900	0.00203	0.1075	0.01397	12430.7	1.438
6.20	8.800	0.00198	0.1111	0.01462	12868.5	1.448
6.30	8.700	0.00192	0.1149	0.01531	13320.9	1.458
6.40	8.600	0.00187	0.1188	0.01603	13789.1	1.469
6.50	8.500	0.00182	0.1229	0.01679	14274.1	1.481
6.60	8.400	0.00176	0.1272	0.01759	14777.3	1.492
6.70	8.300	0.00171	0.1317	0.01843	15299.9	1.506
6.80	8.200	0.00165	0.1364	0.01932	15843.6	1.519
6.90	8.100	0.00160	0.1413	0.02026	16409.8	1.534
7.00	8.000	0.00154	0.1465	0.02125	17000.4	1.550

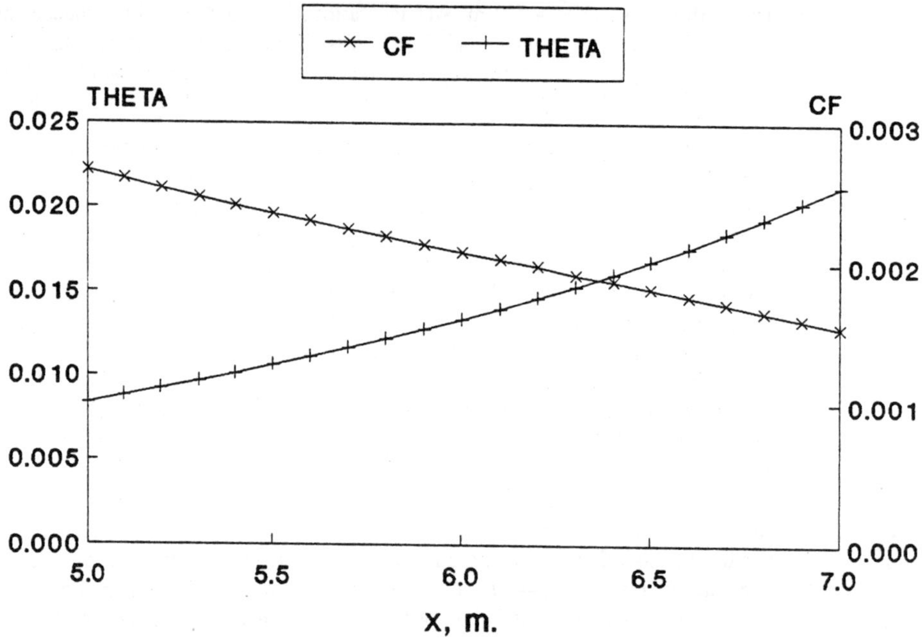

Figure 7–35 Streamwise variation of integral boundary layer properties for Example 7–1.

7–8 MEAN FLOW MODELS FOR THE EDDY VISCOSITY AND THE MIXING LENGTH

In order to exploit the availability of computerized methods for solving the differential equations of motion for the mean flow in the boundary layer, it is necessary to have a detailed representation of either the eddy viscosity or the mixing length across the entire layer. Using either Eq. (7–36) or Eq. (7–37) (with the required detailed auxiliary representations mentioned before) in Eq. (7–34a), we can solve the system of Eqs. (7–29) and (7–34a) by rather direct extensions of the methods developed in Chapter 4. The aims of employing a differential formulation rather than an integral method are to provide better accuracy for C_f, to afford more detailed information [e.g., $U(x, y)$ in the layer] and, it is hoped, to work with a more fundamental formulation that requires less of the prior knowledge that is important to the success of integral methods.

It can be expected that the modeling effort will take place in two stages: one for the inner, wall-dominated part of the layer and one for the outer part of the layer. We can then anticipate piecewise models for either the eddy viscosity or the mixing length to describe the variation over the whole layer. It is easiest to begin with the inner region for smooth, solid surfaces, since we have, in the *wall law plot* of Fig. 7–4, a compact presentation of empirical information that holds for both zero and nonzero pressure gradients.

7–8–1 Models for the Inner Region

For each region, we begin the modeling effort with the best available experimental information. For the inner region, the *law of the wall* states that

$$u^+ = g(y^+) \tag{7–48}$$

where the form of $g(y^+)$ is described by the data in Fig. 7–4. Analytical expressions have been derived for two parts of this region. For the sublayer (roughly, $0 \leq y^+ \leq 5–7$), there is

$$u^+ = y^+ \tag{7–49}$$

and for the overlap region (*roughly*, $30 \leq y^+ \leq 300$), there is

$$u^+ = A \log(y^+) + C = \frac{1}{\kappa} \ln(y^+) + C \tag{7–50}$$

Recall now the philosophy to be used to develop semiempirical models. We seek expressions for the eddy viscosity and/or mixing length that can be substituted into the equations of motion [Eqs. (7–29) and (7–34a)] to produce solutions for U, V that mimic the behavior of mean flow data when plotted on the wall law and defect law plots. That task sounds reasonable until one questions the mechanics of the process in detail. We cannot just try to guess a long series of expressions and then test them to see whether one works; the process is much too indirect. Instead, the logic must somehow be inverted to permit a more direct process for determining suitable models. Toward that end, we write the mean flow momentum equation as in Eq. (7–34a) and use $-\rho\overline{u'v'} = \tau_T$ as a more compact notation. Then this equation can be integrated with respect to y^+ using Eq. (7–48), following Coles (1955), to give

$$\frac{\tau + \tau_T}{\tau_w} = 1 + \frac{y}{\tau_w}\frac{dP}{dx} + \frac{\nu}{u_*^2}\frac{du_*}{dx}\int_0^{y^+} g^2(y^+)\, dy^+ \tag{7–51}$$

Careful study shows that the last two terms can safely be neglected, compared to unity, up through the entire inner region in virtually all flows. Thus, the equations of motion for the inner region become simply

$$\tau + \tau_T = \mu\frac{\partial U}{\partial y} - \rho\overline{u'v'} = \tau_w \tag{7–52}$$

This is a powerful result. The nonlinear partial differential equations describing turbulent boundary layer development have been integrated over the inner region without specifying a turbulence model. Equation (7–52) can now be used to find a turbulence model by a direct approach.

The simplest part of the inner region to treat is the logarithmic region, where $\tau_T \gg \tau$. (We write τ for the laminar shear and τ_T for the turbulent shear.) The laminar shear can be neglected in the logarithmic region, since it is also the inner part of the outer region where turbulent processes predominate (see Fig. 7–21). Thus, Eq.

(7–52) becomes

$$\tau_T = \tau_w \tag{7–53}$$

Using first the *eddy viscosity formulation*, Eq. (7–36), in Eq. (7–53) results in

$$\mu_T \frac{\partial U}{\partial y} = \tau_w \tag{7–54}$$

From Eq. (7–50), the velocity gradient $\partial U/\partial y$ in this region is

$$\frac{\partial U}{\partial y} = u_* \frac{1}{\kappa y} \tag{7–55}$$

Noting that $u_* \equiv \sqrt{\tau_w/\rho}$, substituting Eq. (7–55) into Eq. (7–54), and solving for μ_T produces

$$\mu_T = \kappa \rho u_* y \tag{7–56}$$

which is the desired result—an *eddy viscosity model* for the logarithmic part of the inner region. The process of finding a turbulence model has thus been inverted to a direct procedure. In retrospect, one might have been able to guess the form of this result by noting that viscosity ~ *density* × *velocity* × *length* and picking u^* as the appropriate characteristic velocity and y as the appropriate characteristic length scale for the inner region.

This development can be repeated with the *mixing length formulation*. We use Eq. (7–37) in Eq. (7–53) to obtain

$$\rho \ell_m^2 \left| \frac{\partial U}{\partial y} \right| \frac{\partial U}{\partial y} = \tau_w \tag{7–57}$$

We have an expression for $\partial U/\partial y$ in Eq. (7–55). Upon substituting this expression into Eq. (7–57), we get a *mixing length model* for the logarithmic part of the inner region:

$$\ell_m = \kappa y \tag{7–58}$$

Note that Eq. (7–58) could have been found from Eq. (7–56) using Eq. (7–40), or vice versa. Most people find it easier to think about a mixing length than an eddy viscosity. The result in Eq. (7–58) looks reasonable from a physical viewpoint, since the turbulent eddies probably are larger on the average as one moves away from the wall, and the interaction distance between clumps, i.e., the mixing length, should grow in the same proportions.

The models just developed for the logarithmic region cannot apply all the way to the wall. They both go to zero at the wall, as is required, but the eddy viscosity and the mixing length must both be zero in the laminar sublayer a short distance above the wall. Also, a look at Fig. 7–4 reveals that something important happens around $y^+ \approx 30$–50, where the profile departs from the log law. These models must be made to describe that behavior. The modification of the models for the inner part of the inner region ($y^+ < 50$) has been accomplished by two workers using two philosophically different approaches. Van Driest (1956a) used *deductive* logic, and Reichardt (1951b) used *inductive* logic. Van Driest tried to represent the physics of the damping of the turbulent eddies by the rigid wall by likening that flow to the laminar flow near an oscillating wall in a fluid otherwise at rest. That problem had been solved by Stokes (1851) (see Section 4–2–2), who showed that the motion diminishes with distance from the wall according as $\exp(-y/A)$, where A is a factor depending on the frequency of oscillation and the kinematic viscosity of the fluid [see Eq. (4–10)]. Van Driest reasoned that, if the plate is fixed and the fluid oscillates (crudely speaking, as in the eddies), then the factor $[1 - \exp(-y/A)]$ should be applied to the fluid oscillation to model the damping effect of the wall. He used the *mixing length formulation* and took ℓ_m to be decreased near the wall by the aforementioned factor. Thus, using the model for the logarithmic portion of the inner region, Eq. (7–58), we can write a model for the whole inner region as

$$\tau + \tau_T = \tau_w = \mu \frac{\partial U}{\partial y} + \rho\kappa^2 y^2 \left[1 - \exp\left(\frac{-y}{A}\right)\right]^2 \left|\frac{\partial U}{\partial y}\right| \frac{\partial U}{\partial y} \tag{7–59}$$

or, in dimensionless terms,

$$\frac{\tau + \tau_T}{\tau_w} = 1 = \frac{du^+}{dy^+} + \kappa^2 (y^+)^2 \left[1 - \exp\left(\frac{-y^+}{A^+}\right)\right]^2 \left(\frac{du^+}{dy^+}\right)^2 \tag{7–59a}$$

Both τ and τ_T must now be included, since laminar processes grow in importance as the wall is approached (see Fig. 7–21). It remains to find the factor A^+, which is a dimensionless *effective frequency* of the turbulent fluctuations. Van Driest (1956a) accomplished that task by algebraically solving Eq. (7–59a) to produce an ordinary differential equation for $u^+(y^+)$, viz.,

$$\frac{du^+}{dy^+} = \frac{2}{1 + \sqrt{1 + 4\kappa^2 (y^+)^2 [1 - \exp(-y^+/A^+)]^2}} \tag{7–60}$$

which can be numerically solved with $u^+(0) = 0$ for $u^+(y^+)$. But, of course, we know $u^+(y^+)$; it is defined by the data in Fig. 7–4. Van Driest (1956a) simply tried various values and found that $A^+ = 26$ with $\kappa = 0.40$ produces an excellent representation of the data, as shown in Fig. 7–36. For any particular set of wall law constants A (and hence, κ) and C, A^+ changes slightly. With $C = 4.9$ and $\kappa = 0.41$, as adopted throughout this volume, a value of $A^+ = 24$ is more appropriate. Given all of this, the Van Driest *mixing length model* for the complete inner region is

$$\ell_m = \kappa \left[1 - \exp\left(\frac{-yu_*}{A^+\nu}\right)\right] y \tag{7–61}$$

An equivalent eddy viscosity model can be found using Eq. (7–40):

$$\mu_T = \rho\kappa^2 y^2 \left[1 - \exp\left(\frac{-yu_*}{A^+\nu}\right)\right]^2 \left|\frac{\partial U}{\partial y}\right| \tag{7–62}$$

It will be helpful to keep this whole procedure in mind for later reference. An eddy viscosity or mixing length model for the inner region is thus determined, including

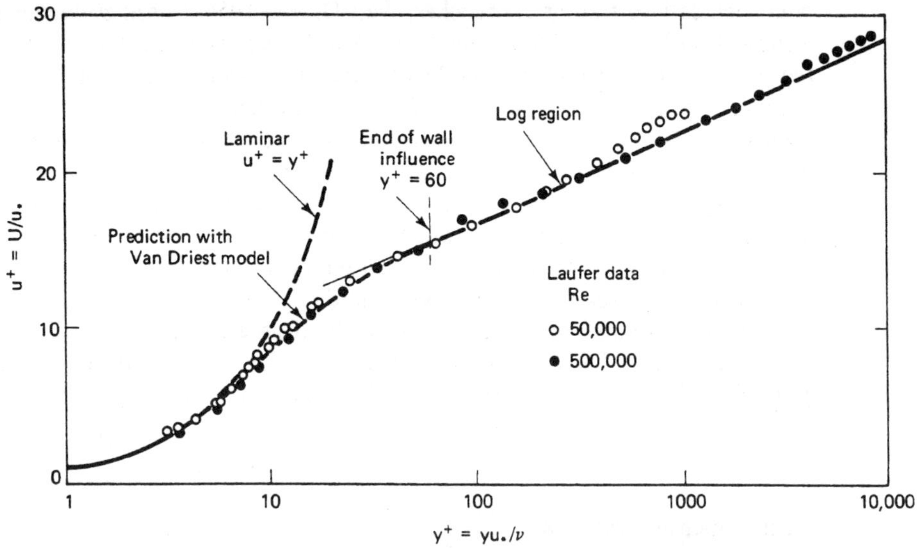

Figure 7–36 Comparison of prediction based on the Van Driest model with experiment for the wall law region. (From Van Driest, 1956a.)

the constants in it (κ, A^+, etc.), such that it produces a function $u^+(y^+)$ that goes through the data on a wall law plot.

Reichardt (1951b) approached the task of modifying a model for the logarithmic region to represent the wall effect quite differently. He simply sought a curve that would mimic the experimentally observed variation in $u^+(y^+)$ between the logarithmic region and the wall, where, he reasoned, the eddy viscosity should decay according as $\mu_T \sim y^3$, based on the continuity equation. The form he selected was

$$\mu_T = \kappa\rho\nu\left[\left(\frac{yu_*}{\nu}\right) - y_a^+ \tanh\left(\frac{yu_*}{\nu y_a^+}\right)\right] \tag{7-63}$$

Here, the constant y_a^+ is a dimensionless length scale of the order of the laminar sublayer thickness chosen to obtain the best fit with the experimental $u^+(y^+)$ in Fig. 7–4, especially where merging into the log law occurs. Actually, y_a^+ corresponds more closely to the intersection of Eqs. (7–5b) and (7–7b) on a wall law plot, as in Fig. 7–4. For the Clauser (1956) wall law constants, $y_a^+ = 9.7$ with $\kappa = 0.41$. Again, Eq. (7–40) can be used to yield the corresponding *mixing length model*.

Despite the fact that the development of the foregoing two models is very different, the models are, to all intents and purposes, equal as far as their ability to reproduce the experimentally observed $u^+(y^+)$. That, after all, was the only goal of their development. At the mean flow level of analysis, we regard Fig. 7–4 as representing all we know about turbulent flow. Some workers prefer the use of the Reichardt model in numerical solution procedures because it does not require evaluating $|\partial U/\partial y|$ as a factor times $\partial U/\partial y$ [cf. Eqs. (7–62), (7–63), and (7–36)].

Earlier, we observed that the major effect of roughness is to shift the logarithmic region of the *law of the wall* downward by $\Delta u^+(k^+)$; roughness does not change the slope of $u^+(y^+)$ (see Section 7–3–2). One can, therefore, attempt to modify the Van Driest and Reichardt models for roughness. Indeed, Van Driest (1956a) did so by adding an extra term to the wall damping factor in Eq. (7–61):

$$\ell_m = \kappa\left[1 - \exp\left(\frac{-y^+}{26}\right) + \exp\left(\frac{-60y^+}{26k^+}\right)\right]y \tag{7–64}$$

He did not, however, check the resulting $u^+(y^+, k^+)$ against the Hama data for $\Delta u^+(k^+)$. Further study has shown that Eq. (7–64) cannot be made to predict those data even if the factor 60 is adjusted (see Schetz and Nerney, 1977). In fact, the Van Driest model has proven very difficult to extend to any more general flows than the case of the smooth, solid wall that Van Driest himself studied.

The Reichardt model was extended to roughness cases by Schetz and Nerney (1977). One can note that the only *free* quantity in the Reichardt model is y_a^+, which is a length scale of the order of the thickness of the laminar sublayer. Now, it is reasonable to expect the thickness of the laminar sublayer and, thus, also y_a^+ to decrease with increasing roughness. Schetz and Nerney (1977) have determined the variation $y_a^+(k^+)$ such that the correct shift Δu^+ is produced for the Prandtl-Schlichting uniform sand roughness data, as an example. The required function y_a^+ is shown in Fig 7–37. Separate functions $y_a^+(k^+)$ have to be developed for other specific types of roughness. That, however, is not difficult, as long as $\Delta u^+(k^+)$ is available for the type of roughness of interest.

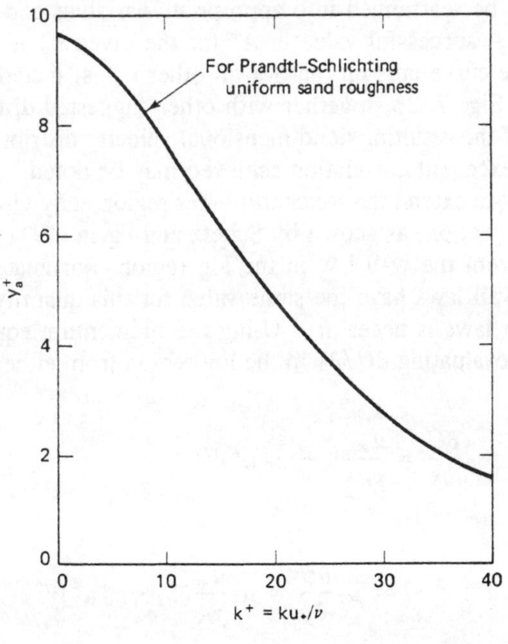

Figure 7–37 Variation of the Reichardt sublayer parameter to account for uniform roughness. (From Schetz and Nerney, 1977.)

A number of researchers have attempted to extend the Van Driest eddy viscosity model for the inner region to cases with suction or injection by seeking an appropriate form for $A^+(v_0^+)$. Most have done so *heuristically*, without recourse to any law of the wall, and did not check to see whether the proposed model produced $u^+(y^+, v_0^+)$ that went through an appropriate law of the wall. It is actually a simple matter to carry out the necessary calculations correctly, once a specific form of the law of the wall for flows with injection or suction has been chosen. Schetz and Favin (1971) performed this task using the Stevenson law, Eq. (7–18).

Near the wall, the momentum equation can be written

$$\tau + \tau_T = \tau_w + \rho v_w U \tag{7–65}$$

or

$$\frac{\tau + \tau_T}{\tau_w} = 1 + v_0^+ u^+ \tag{7–65a}$$

This relationship leads to an equation corresponding to Eq. (7–60), the one used by Van Driest to find $A^+(0) = 26$, which is again an ordinary differential equation that can be solved to give $u^+(y^+)$. The equation is

$$\frac{du^+}{dy^+} = \frac{2(1 + v_0^+ u^+)}{1 + \sqrt{1 + 4\kappa^2(1 + v_0^+ u^+)(y^+)^2[1 - \exp(-y^+/A^+)]^2}} \tag{7–66}$$

with $u^+(0) = 0$. The universal constant κ is generally taken to be 0.40 to 0.43, and Van Driest used 0.40 to derive the value $A^+(0) = 26$. For a given v_0^+, a trial value of A^+ can be input into Eq. (7–66), and $u^+(y^+)$ can easily be produced numerically. This numerical result can be rearranged into appropriate variables and plotted on a Stevenson wall law plot. A successful value of A^+ for the given v_0^+ is a value such that the log portion of the curve falls on that for all other v_0^+'s, including $v_0^+ \equiv 0$. The results are shown in Fig. 7–38, together with other suggested distributions. A Stevenson wall law plot of the resulting nondimensional velocity distribution is given in Fig. 7–39, where the excellent correlation achieved may be noted.

It is also not difficult to extend the Reichardt inner region eddy viscosity model to flows with suction or injection, as shown by Schetz and Favin (1971). In particular, one needs $(\partial U/\partial y)$ from the wall law in the log region. Fortunately, both the Stevenson and Simpson wall laws have the same value for this quantity, so that no choice between these two laws is necessary. Using the momentum equation in the form of Eq. (7–65a) and evaluating $\partial U/\partial y$ in the log region from either Eq. (7–18) or Eq. (7–19) as

$$\frac{\partial U}{\partial y} = \frac{u_*}{y\kappa}(1 + v_0^+ u^+)^{1/2} \tag{7–67}$$

we can write

$$\tau_w(1 + v_0^+ u^+) = \tau_T = \mu_T \frac{\partial U}{\partial y} = \mu_T \frac{u_*}{y\kappa}(1 + v_0^+ u^+)^{1/2} \tag{7–68}$$

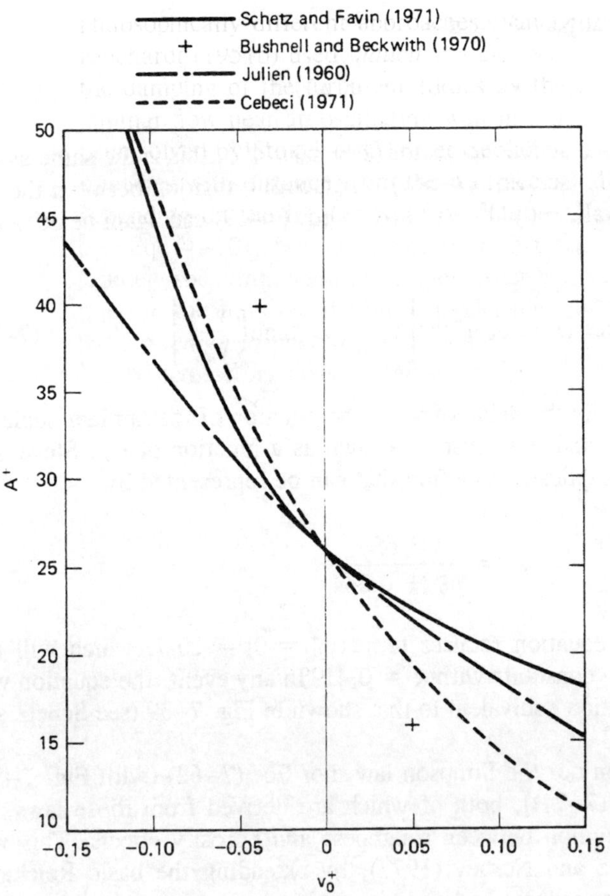

Figure 7–38 Various proposals for the Van Driest damping factor for injection or suction.

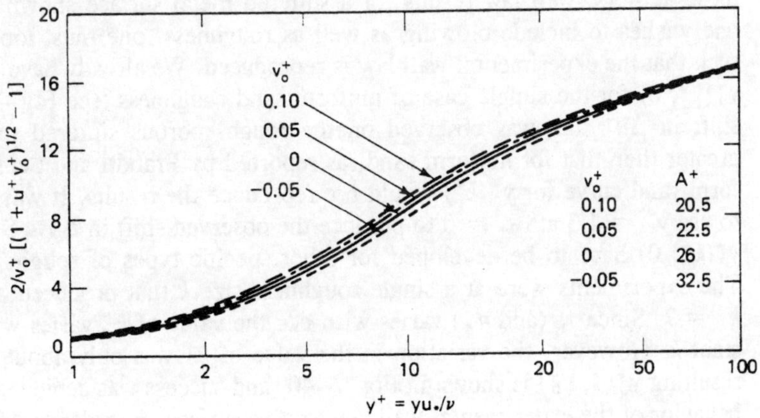

Figure 7–39 Stevenson wall law plot of the predictions of the Van Driest model, as extended by Schetz and Favin (1971).

This equation can be rearranged to yield

$$\mu_T = \kappa(\rho u_* y)(1 + v_0^+ u^+)^{1/2} \tag{7–69}$$

in the log region. But for the additional factor $(1 + v_0^+ u^+)^{1/2}$, this is the same as the earlier result for solid walls [see Eq. (7–56)]. A smooth transition between the required behavior near the wall, $\sim(y^+)^3$, and that in Eq. (7–69) can again be achieved by a hyperbolic tangent:

$$\mu_T = \kappa\rho\nu(1 + v_0^+ u^+)^{1/2}\left[y^+ - y_a^+ \tanh\left(\frac{y^+}{y_a^+}\right)\right] \tag{7–70}$$

The value of y_a^+ is essentially the value of y^+ at the junction of the laminar sublayer extended and the log law, and this must be known as a function of v_0^+. Stevenson (1963) has suggested a complicated function that can be represented by

$$y_a^+ = \frac{3.65}{v_0^+ + 0.344} \tag{7–71}$$

Note, however, that this equation reduces to $y_a^+(v_0^+ = 0) = 10.7$, which will not match the Clauser wall law constants with $\kappa = 0.41$. In any event, the equation will result in a wall law correlation equivalent to that shown in Fig. 7–39 (see Schetz and Favin, 1971).

Neither the Stevenson nor the Simpson law, nor Eq. (7–62) (with Fig. 7–38) or Eq. (7–70) [with Eq. (7–71)], both of which are derived from those laws, attempts to make any distinction between roughness and porosity effects. This was done, however, by Schetz and Nerney (1977), by extending the basic Reichardt model to cases with both roughness and porosity and injection or suction, based on these authors' wall law results for a sintered metal surface shown in Fig. 7–10. If one wishes to include blowing as well as roughness, one must look for $y_a^+(k^+, v_0^+)$ such that the experimental wall law is reproduced. We already have given a complete $y_a^+(k^+, 0)$ for the simple case of uniform sand roughness (see Fig. 7–37). Since the shift in Δu^+ that was observed on the rough, porous sintered metal surface was greater than that for uniform sand, as reported by Prandtl and Schlichting, the uniform sand curve for $y_a^+(k^+)$ would not reproduce the results. It was found necessary to use $y_a^+ = 4.3$ at $v_{0+} = 0$ to produce the observed shift in Δu^+. Separate functions $y_a^+(k^+, 0)$ have to be developed for other specific types of rough, porous surfaces. The experiments were at a single roughness size k that produced a nearly constant $k^+ = 7$. Since τ_w (and u_*) varies with v_W, the value of k^+ varies with v_0^+ for a constant k. However, the variation in the value of k^+ was only about 20 percent. The resulting $y_a^+(7, v_0^+)$ is shown in Fig. 7–40, and success was achieved in matching the behavior of the experimental wall law for this surface. In order to determine the general function $y_a^+(k^+, v_0^+)$, further injection experiments with other types of porous surfaces will have to be conducted.

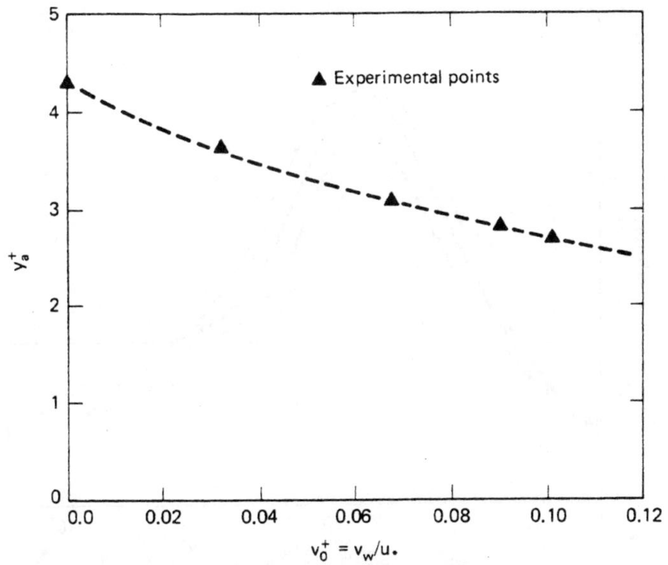

Figure 7–40 Variation in the Reichardt sublayer parameter to reproduce the wall law observed for injection through a sintered metal surface. (From Schetz and Nerney, 1977.)

We come now to the important matter of flows with pressure gradients over smooth, solid surfaces. It is well to begin the discussion by recalling that the law of the wall has been found to be insensitive to pressure gradient effects. This fact is important, because the law of the wall is the only sound basis for developing mean flow turbulence models for the inner region. Despite that, several workers (e.g., Cebeci and Smith, 1969) have introduced modifications to inner region eddy viscosity and mixing length models to account for pressure gradient effects that change the resulting $u^+(y^+)$ as a function of pressure gradient when they should not. Granville (1989) has discussed this matter in some depth. Following Reichardt, he invokes a condition that the eddy viscosity should vary as y^3. The Van Driest formula [Eq. (7–62)] gives a dependence on y^4. Granville (1989) proposes that

$$\frac{\mu_T}{\mu} = \kappa y^+(\tau/\tau_w)[1 - \exp(-y^+\sqrt{1 + ap^+}/24)^2] \tag{7–72}$$

where $p^+ = \alpha(\nu/\rho u_*^3)\, dP/dx$, with α slightly less than unity, and $a = 14.5$ for $p^+ > 0$ and $a = 18.0$ for $p^+ < 0$, and $\kappa = 0.40$. Note that the exponent in the Van Driest damping function has been changed so that its argument is squared. Thus, this model is no longer *deductive* but rather, *inductive,* since the direct physical basis of the *Stokes problem* solution used in the original Van Driest model has been lost. But, the model does meet the aforementioned criterion according to which the law of the wall is to remain essentially unaffected by pressure gradients (at least in the log law portion), as shown in Fig. 7-41.

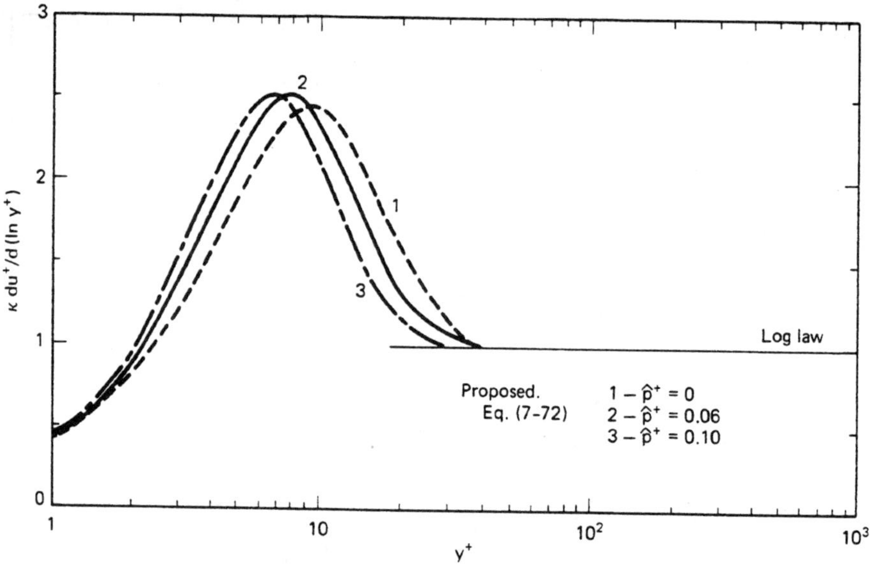

Figure 7–41 Variation in slope $\kappa\, du^+/d(\ln y^+)$ with normal distance y for proposed formula. (From Granville, 1989.)

7–8–2 Models for the Outer Region

For the outer region, there are again different approaches to model development, but in this case there is one approach leading to a *mixing length model* and a very different approach leading to an *eddy viscosity model.*

A number of researchers, including Prandtl himself, have used a simple physical argument leading to a *mixing length model.* They reason that the mixing length must be proportional to some length scale, on dimensional grounds. Further, in the outer region, the distance from the wall y is less significant than the size of the viscous region, δ. Also, the mixing length should be a constant in the outer part of the layer, since the explicit effects of the wall are negligible, and the eddies are insensitive to transverse locations in such a turbulent zone. In this way, one arrives at $\ell_m \sim \delta$. The proportionality constant is on the order of one-tenth, and a value of 0.09 is generally accepted. Thus,

$$\ell_m = 0.09\delta \tag{7–73}$$

This same model, with the same constant, is presumed to apply for all flows, either with or without pressure gradients. The model has been used successfully by Spalding and his co-workers at Imperial College (see Patankar and Spalding, 1967), as well as by others. Some comparisons of predictions with experiment will be given in a later section.

In light of the discussions in Sections 7–3–1 and 7–3–4, it should not be a surprise that Clauser (1956) developed an *eddy viscosity model* for the outer region by using his view of the generalized *defect law plot*. It is simplest to discuss the flat-plate case initially. First, Clauser (1956) retained the assumption that the turbulent transport coefficient should be constant across the outer region, but he used that assumption with an *eddy viscosity* rather than a *mixing length*. Thus, he took

$$\mu_T = \mu_T(x) \neq f(y) \tag{7–74}$$

for the outer region. Note that Eq. (7–40) shows that a constant mixing length implies a varying eddy viscosity and vice versa, since $\partial U/\partial y$ is not constant. Second, Clauser used the fact that a turbulent velocity profile *appears* to intersect the wall ($y = 0$) at a nonzero value of the velocity [$U(0) \approx U_e/2$]. Putting these ideas together, he proposed to treat the outer part of the turbulent boundary layer as if it were *pseudolaminar*, with some augmented viscosity $\mu_T(x)$ and with a *slip* velocity at $y = 0$. Using

$$\frac{U}{U_e} = g'\left(\frac{y}{\delta}\right) \tag{7–75}$$

in the mean flow momentum equation, we obtain

$$gg'' + \left(\frac{\mu_T}{\rho U_e \delta (d\delta/dx)}\right) g''' = 0 \tag{7–76}$$

For actual laminar flows, Blasius used $y/\sqrt{x}$, but the results gave $\delta \sim \sqrt{x}$, so y/δ amounts to the same thing. Rescaling both variables to an $f(\eta)$, one can obtain

$$f''' + ff'' = 0 \tag{7–76a}$$

which is the same as the equation used by Blasius for the real laminar flat-plate problem (see Section 4–3–1). Clauser (1956) solved this equation—this time for various values of the *slip* velocity $U(0)/U_e$—with the results shown in Fig. 7–42. The problem now becomes finding a method to rescale the coordinates so that the profiles will all collapse into one curve (or nearly so) resembling a *defect law plot*. In that way, these *pseudolaminar* profiles would mimic the behavior of the real turbulent profiles. Clauser (1956) tried two choices: making the curves coincide at $y/\delta = 0$ and making them have the same slope at $y/\delta = 0$. In both cases, the area above the curve(s) is kept equal. For the second choice, the coordinates are

$$\frac{U - U_e}{U_e\sqrt{(\eta)_{\delta^*}\, U'(0)/U_e}} \quad \text{and} \quad \eta\sqrt{\frac{U'(0)/U_e}{(\eta)_{\delta^*}}} \tag{7–77}$$

where $(\eta)_{\delta^*}$ is (η) at $y = \delta^*$. The result is shown in Fig. 7–43, which indeed resembles a *defect law plot*. The process of achieving complete equivalence between the two plots leads to the desired *eddy viscosity* model. Clauser (1956) reasoned on dimensional grounds that $\mu_T \sim$ *density* × *velocity* × *length*. For the characteristic velocity, he made the obvious choice of u_*. For the characteristic length, he chose to use a new integral thickness Δ that he developed from the turbulent *defect law plot*, defined as

$$\frac{\Delta}{\delta} \equiv -\int_0^1 \frac{(U - U_e)}{u_*}\, d\left(\frac{y}{\delta}\right) \tag{7–78}$$

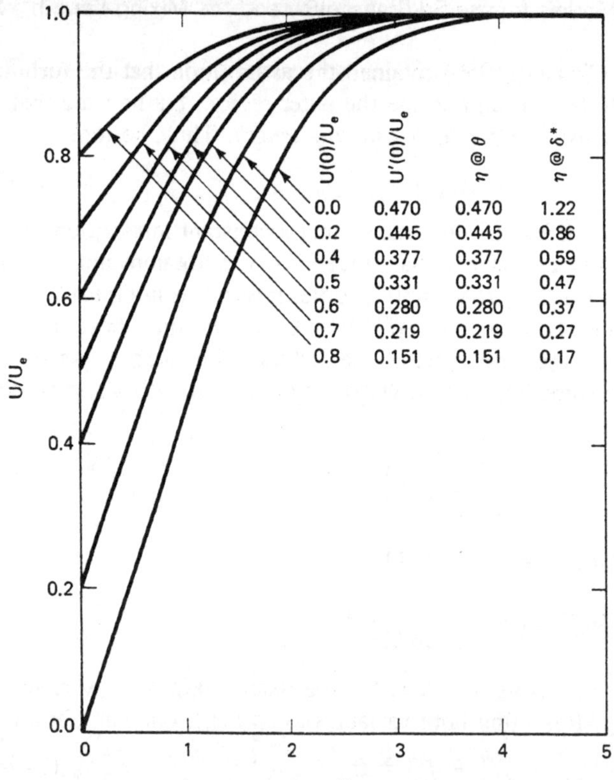

Figure 7–42 Solutions of the Blasius equation for various values of *slip* velocity at the wall. (From Clauser, 1956.)

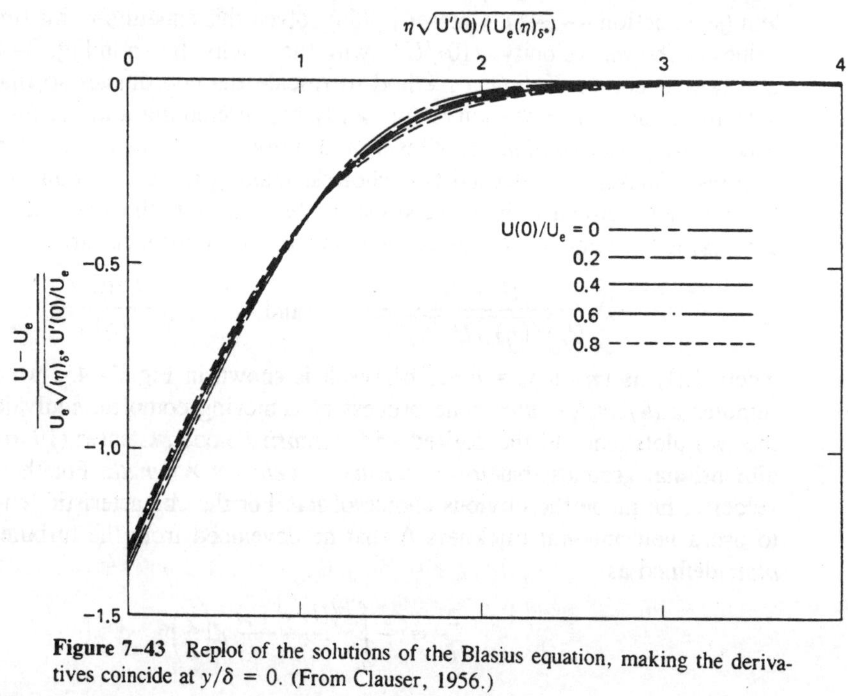

Figure 7–43 Replot of the solutions of the Blasius equation, making the derivatives coincide at $y/\delta = 0$. (From Clauser, 1956.)

For a flat plate, $\Delta/\delta = 3.6$. Thus, Clauser's proposed eddy viscosity model has the form

$$\mu_T = C\rho u_* \Delta \tag{7-79}$$

But, by definition,

$$\Delta = \frac{U_e}{u_*}\delta^* \tag{7-80}$$

[cf. Eqs. (7–78) and (2–14)], so the model becomes

$$\mu_T = C\rho U_e \delta^* \tag{7-79a}$$

For the *pseudolaminar* profile in Fig. 7–43, the wall shear can be evaluated using Eq. (7–79a) for the viscosity. The viscosity of the outer region can be used at the wall, because the whole layer is modeled by these *pseudolaminar* profiles with slip at the wall. We have

$$\begin{aligned}\frac{\tau_w}{\rho} = \frac{\mu_T}{\rho}\frac{\partial U}{\partial y}\bigg|_{y=0} &= \frac{\mu_T}{\rho}\frac{U'(0)}{U_e}(\eta)_{\delta^*}\frac{U_e}{\delta^*} \\ &= CU_e^2\frac{U'(0)}{U_e}(\eta)_{\delta^*}\end{aligned} \tag{7-81}$$

Thus,

$$u_* \equiv \sqrt{\frac{\tau_w}{\rho}} = U_e\sqrt{C(\eta)_{\delta^*}U'(0)/U_e} \tag{7-82}$$

and, using Eq. (7–80), we obtain

$$\Delta = \frac{\delta^*}{\sqrt{C(\eta)_{\delta^*}U'(0)/U_e}} \tag{7-83}$$

The correct coordinates for a *defect law plot* are $(U - U_e)/u_*$ and y/δ. The *pseudolaminar* coordinates of Eq. (7–77) can be related to the defect law coordinates, using Eqs. (7–82) and (7–83), by

$$\begin{aligned}\frac{U - U_e}{U_e\sqrt{(\eta)_{\delta^*}U'(0)/U_e}} &= \sqrt{C}\left(\frac{U - U_e}{u_*}\right) \\ \eta\sqrt{\frac{U'(0)/U_e}{(\eta)_{\delta^*}}} &= \frac{1}{\sqrt{C}}\left(\frac{y}{\Delta}\right) = \frac{1}{3.6\sqrt{C}}\left(\frac{y}{\delta}\right)\end{aligned} \tag{7-84}$$

Finally, the constant C can be chosen to make the *pseudolaminar* profiles in Fig. 7–43 *fit* the data on the real turbulent *defect law plot*, Fig. 7–3. Clauser (1956) chose $C = 0.018$, with the results shown in Fig. 7–44. Thus, the well-known *Clauser eddy viscosity model* becomes

$$\mu_T = 0.018\rho U_e \delta^* \tag{7-85}$$

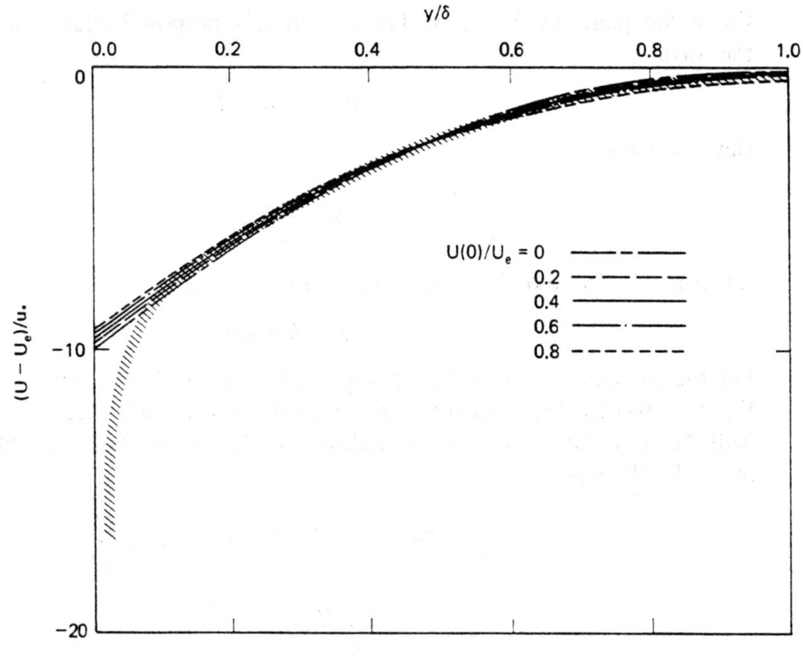

Figure 7–44 Comparison of turbulent data with calculated *pseudolaminar* profiles on a defect law plot. (From Clauser, 1956.)

Next, Clauser (1956) went back and established an equivalence between *laminar, wedge-type* pressure gradient flows and turbulent, *equilibrium pressure gradient* flows. The general procedure is as previously described for the flate-plate case. Happily, the same form for the eddy viscosity, Eq. (7–79a), suffices. This may be taken as a demonstration of the soundness of the approach leading to the model. The equivalence achieved is shown in Fig. 7–45. Clauser also found that the constant in Eq. (7–79a) depended weakly, if at all, on $(\delta^*/\tau_w)dP/dx$ (see Fig. 7–46). He chose $C = 0.018$ for all cases. Some other researchers looking at Fig. 7–46 have since concluded that $C = 0.0168$ is a better choice.

A word of caution is in order here. The Clauser model has been developed only for *equilibrium pressure gradient* flows. If it is applied to flows that are not in equilibrium [i.e., $(\delta^*/\tau_w)dP/dx \neq$ constant], one can expect poorer predictions. The model is, however, used very generally without much regard to this basic restriction, except to view results based on it with caution for flows that are strongly out of equilibrium. This is mainly because there is no other known eddy viscosity model that is more generally applicable.

In numerical solutions, it is sometimes difficult to determine the boundary layer thickness with high accuracy. This is because there is some "scatter" in the numerical "data," and the exact location at which $U = 0.99U_e$ is hard to find precisely. The matter is not important in itself, but, if, as in the mixing length model, such information enters directly into subsequent calculations, there is cause for concern.

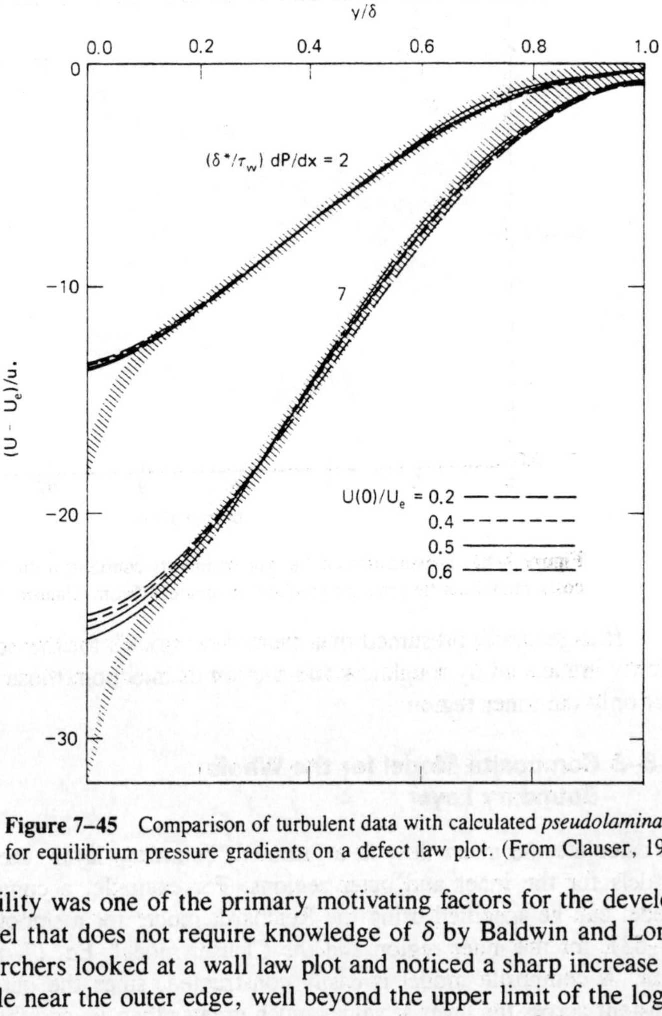

Figure 7–45 Comparison of turbulent data with calculated *pseudolaminar* profiles for equilibrium pressure gradients on a defect law plot. (From Clauser, 1956.)

That possibility was one of the primary motivating factors for the development of a simple model that does not require knowledge of δ by Baldwin and Lomax (1978). These researchers looked at a wall law plot and noticed a sharp increase in the slope of the profile near the outer edge, well beyond the upper limit of the log region (see Fig. 7–4). The value of $y = y_{MAX}$, where the maximum of the slope occurs, was taken as a measure of the width of the boundary layer. The quantity to be maximized is defined for two-dimensional boundary layer flow as

$$F(y) = y|\partial U/\partial y|(1 - e^{-y^+/A^+}) \tag{7–86}$$

This relation was developed by noting that

$$\frac{\partial u^+}{\partial(\ln y^+)} \propto y\frac{\partial U}{\partial y} \tag{7–87}$$

It was found that the ratio of $y_{MAX}F_{MAX}$ to $U_e\delta^*$ is reasonably constant over a wide range of conditions. Thus, the final expression to be used is

$$\mu_T = 0.018(1.60)\rho y_{MAX}F_{MAX} \tag{7–88}$$

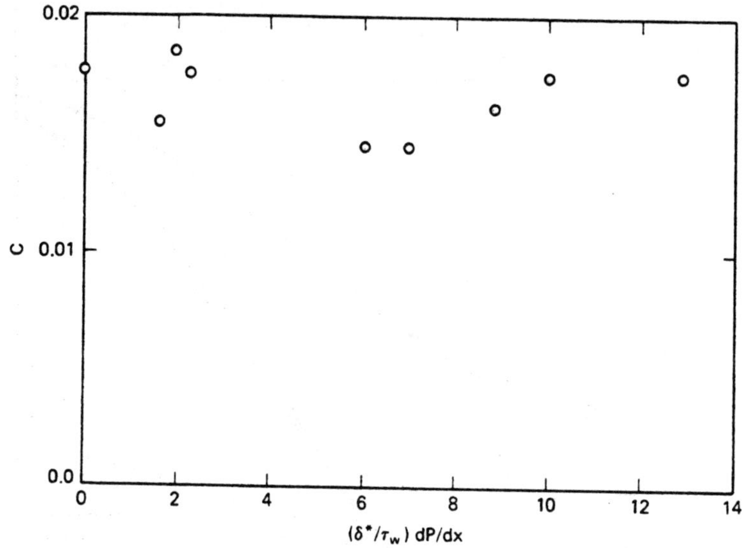

Figure 7–46 Dependence of the proportionality constant in the Clauser eddy viscosity model on the pressure gradient parameter. (From Clauser, 1956.)

It is generally presumed that mean flow models for the outer region are not explicitly influenced by roughness and suction or injection; those processes directly affect only the inner region.

7–8–3 Composite Model for the Whole Boundary Layer

To calculate the mean flow in a turbulent boundary layer, the analyst must choose models for the inner and outer regions. For example, a composite eddy viscosity model can be specified using the Reichardt model for a smooth, solid surface, Eq. (7–63), for the inner region and the Clauser model, Eq. (7–85), for the outer region. A composite model is easily constructed, since the outer region models are constant across the layer at values much greater than μ, and the inner region models all begin with μ_T or ℓ_m at zero (at $y = 0$), which then grows with increasing y [see, for example, Eq. (7–56)]. One simply first calculates the constant μ_T (or ℓ_m), based on the outer region model, and then uses the inner model for all values of y (starting from $y = 0$ and proceeding upwards) for which μ_T (or ℓ_m) is less than $(\mu_T)_{\text{outer}}$ [or $(\ell_m)_{\text{outer}}$]. The result for the outer region model is used for all other points. Plots of the variation of typical composite models for the mixing length and eddy viscosity across the whole boundary layer are given in Figs. 7–47 and 7–48.

All of the preceding may sound complicated, and indeed it is, if done by hand. However, these detailed models are used only within computerized procedures for numerical solutions. In such cases, one has a subroutine in the overall code that performs the operations required to construct the composite model. The computer time and, thus, cost of such calculations is minimal, since only simple arithmetic is involved.

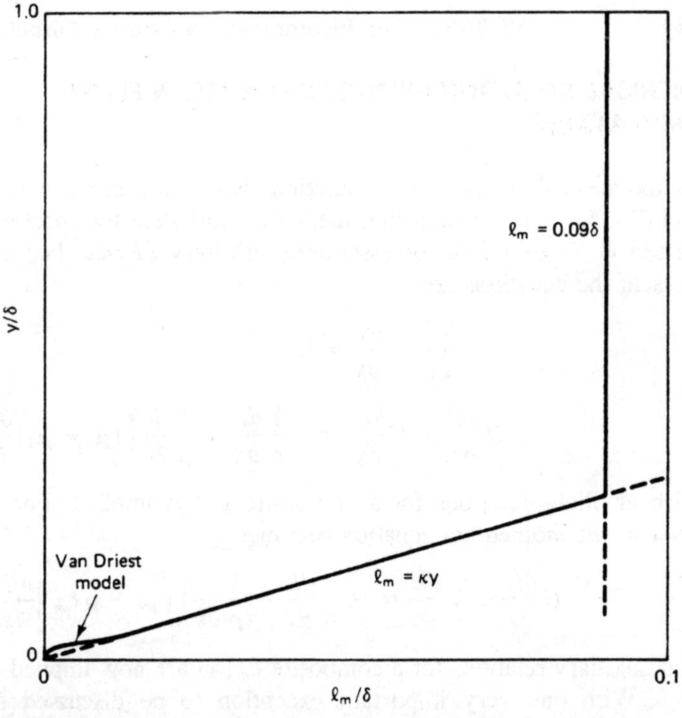

Figure 7–47 Schematic of the composite mixing length model for the whole turbulent boundary layer.

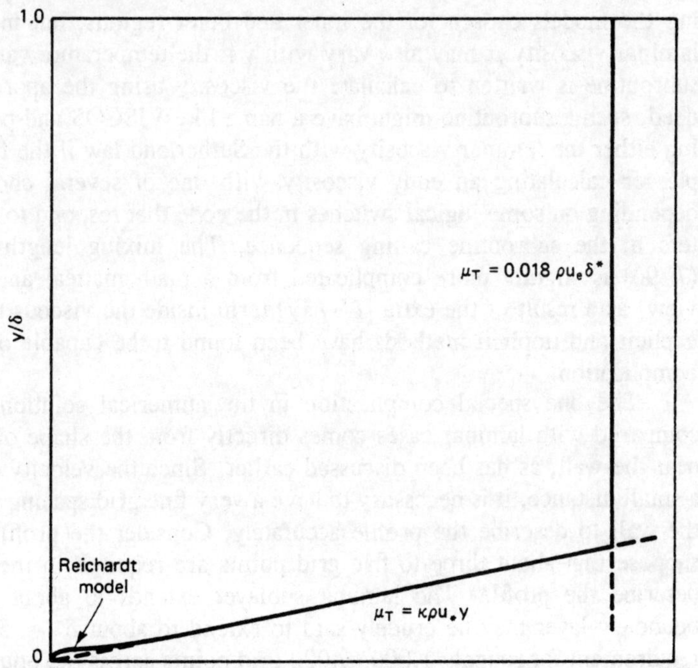

Figure 7–48 Schematic of the composite eddy viscosity model for the whole turbulent boundary layer.

7-9 NUMERICAL SOLUTION METHODS FOR MEAN FLOW FORMULATIONS

At the mean flow level of formulation, one is concerned with solving Eqs. (7–29) and (7–34a) with a composite, mean flow turbulent transport model of the type discussed in Section 7–8 for essentially arbitrary dP/dx. For an eddy viscosity approach, the equations are

$$\frac{\partial U}{\partial x} + \frac{\partial V}{\partial y} = 0$$
$$U\frac{\partial U}{\partial x} + V\frac{\partial U}{\partial y} = -\frac{1}{\rho}\frac{dP}{dx} + \frac{1}{\rho}\frac{\partial}{\partial y}\left((\mu + \mu_T)\frac{\partial U}{\partial y}\right) \tag{7–89}$$

with auxiliary relations for a composite $\mu_T(y)$ implied. For a mixing length approach, the momentum equation becomes

$$U\frac{\partial U}{\partial x} + V\frac{\partial U}{\partial y} = -\frac{1}{\rho}\frac{dP}{dx} + \frac{1}{\rho}\frac{\partial}{\partial y}\left(\left(\mu + \rho\ell_m^2\left|\frac{\partial U}{\partial y}\right|\right)\frac{\partial U}{dy}\right) \tag{7–90}$$

and auxiliary relations for a composite $\ell_m(y)$ are now implied.

With one very important exception to be discussed shortly, solving Eq. (7–89) or Eq. (7–90) numerically is really no more complicated than numerically solving the laminar boundary layer problems discussed in Chapters 4 and 5. For the turbulent case, the eddy viscosity, for example, will necessarily vary with y, following the models chosen for the inner and outer regions, but in a laminar case, the laminar viscosity μ may also vary with y if the temperature varies. In either case, a subroutine is written to calculate the viscosity using the appropriate relations. Indeed, such a subroutine might have a name like VISCOS and be capable of calculating either the laminar viscosity with the Sutherland law if the fluid is air, for example, or calculating an eddy viscosity with one of several choices for models, all depending on some logical switches in the code that respond to the value of parameters in the subroutine calling sequence. The mixing length formulation in Eq. (7–90) is slightly more complicated from a mathematical and numerical point of view, as a result of the extra $|\partial U/\partial y|$ term inside the viscous term. However, both explicit and implicit methods have been found to be capable of handling this extra complication.

The one special complication in the numerical solution of turbulent layers compared with laminar cases comes directly from the shape of the velocity profile near the wall, as has been discussed earlier. Since the velocity changes rapidly over a small distance, it is necessary to have a very fine grid spacing Δy in the region near the wall to describe the profile accurately. Consider the profiles in Fig. 7–4, and suppose that about three to five grid points are required in the laminar sublayer to describe the profile. The laminar sublayer extends to about $y^+ \approx 5–7$, and the boundary layer may be crudely said to extend to about $\delta^+ \approx 5{,}000$. In this way, a requirement for roughly 1,000–5,000 grid points across the boundary layer is apparently derived. This case is to be contrasted with laminar cases, where about 25 grid points are commonly used. Roughly speaking, computer cost varies with the number of grid points, so a cost factor on the order of 10^2 to 10^3 could be expected.

The resolution of this difficulty is a good example of the value of thinking about the physics of the situation before simply attacking a problem on the computer by brute force. A study of velocity profiles such as those presented earlier in this chapter [e.g., in Fig. 7–2] reveals that while the velocity changes very rapidly near the wall, it changes rather slowly with y over the rest of the profile. Thus, a very small Δy (compared to δ) is required near the wall, but a much larger Δy would be adequate in the outer part of the layer. This difference leads to the idea that a variable, increasing Δy (with increasing y) would be very useful. The result would be extra complexity in the numerical analysis of the differential boundary layer equations of motion beyond that discussed in Section 4–7, but that is a one-time cost in analysis and programming, and the reduction in computer execution cost per run is so dramatic that the use of a variable grid spacing is compelling. The most common scheme is based on

$$(\Delta y)_{m+1} = k(\Delta y)_m \tag{7–91}$$

with $1.05 \leq k \leq 1.10$. This is obviously a geometrically increasing sequence. With such a scheme, and with $k = 1.09$, about 80–100 grid points across the layer have been found to be satisfactory for producing accurate results for skin friction. Even with very strong distortion of the grid, the total number of grid points across a turbulent boundary layer cannot be reduced much below these values. Blottner (1975) has studied the matter, with the result shown in Fig. 7–49. To achieve a 1-percent level of uncertainty in C_f requires about 30 points with strong grid distortion and/or advanced numerical methods such as the Keller box scheme.

Turbulent boundary layer calculations require the use of a large number of grid points no matter what choice of Δy is used, so efficiency of calculation is always very important. The implicit method of Blottner is strongly recommended here. A very convenient code based on the Blottner method and Eq. (7–91) with many options for the user's easy choice of eddy viscosity models (and even of the constants involved) was developed by Miner et al. (1975). This code can also treat laminar and compressible flows.

To begin to demonstrate the utility of numerical solutions of a mean flow formulation in predicting turbulent boundary layers, some comparisons from the work of Smith et al. (1965) have been chosen. This work employed the Van Driest eddy viscosity model (with $\kappa = 0.40$) for the inner region and the Clauser model for the outer region (with $C = 0.0168$). A so-called *intermittency factor* also multiplied the outer eddy viscosity in an attempt to account for the effects of the intermittency near the outer edge of the layer. The intermittency factor is an empirical function that decays towards zero as y approaches δ and is taken from the measurements of Klebanoff (1955), as shown in Fig. 7–19. The use of such a factor has only a small effect on the results. Figures 7–50 and 7–51 show predictions versus experiment for a flat-plate flow on a *wall law plot* and a *defect law plot*, respectively. Clearly, excellent agreement was achieved.

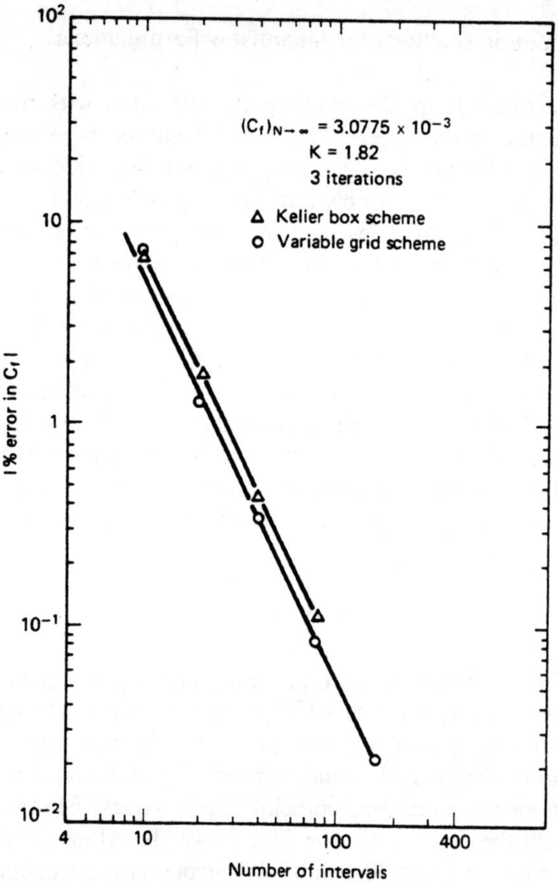

Figure 7–49 Accuracy of skin friction for turbulent boundary layer at $Re_x = 1.88 \times 10^6$ with various numbers of intervals in η-direction. (From Blottner, 1975.)

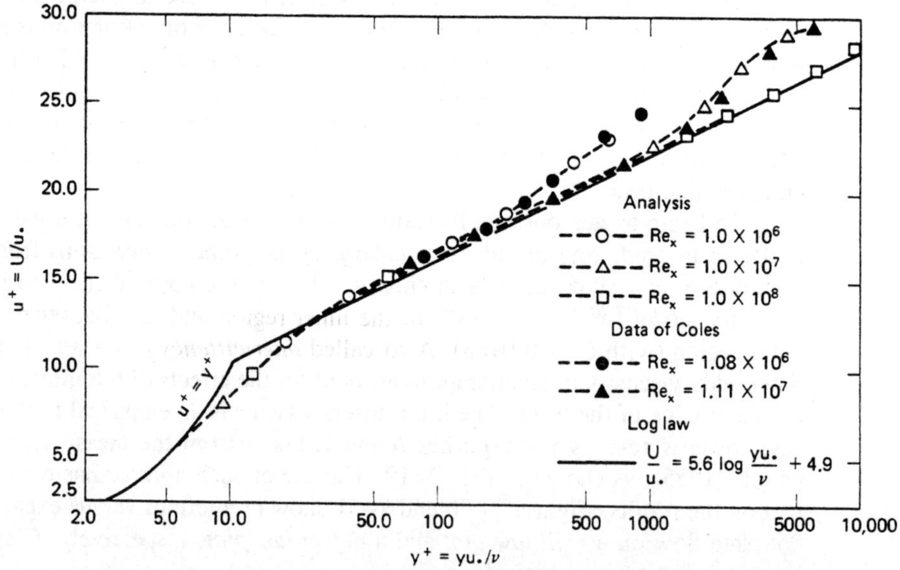

Figure 7–50 Comparison of predictions of Smith et al. (1965) with Van Driest and Clauser eddy viscosity models and data for the wall region.

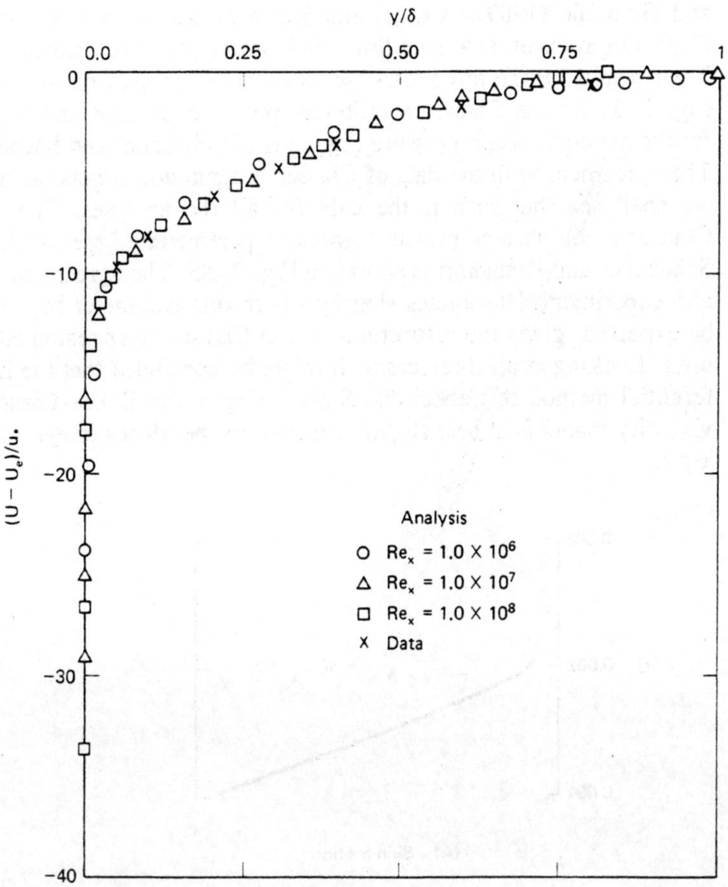

Figure 7–51 Comparison of predictions of Smith et al. (1965) with Van Driest and Clauser eddy viscosity models and data for the outer region of flat-plate boundary layers.

For the 1968 Stanford competition mentioned earlier, Cebeci and Smith (1969) extended their inner region model in an effort to account for strong pressure gradients. They proposed modifying the Van Driest damping factor to read

$$A = 26\nu\left(\frac{\tau_w}{\rho} + \frac{y}{\rho}\frac{dP}{dx}\right)^{-1/2} \tag{7–92}$$

In later work, they also suggested that

$$A = 26(\nu/u_*)\left[1 - 11.8(\nu U_e/u_*^3)\frac{dU_e}{dx}\right]^{1/2} \tag{7–93}$$

One can question this approach, since the *law of the wall* is known to be virtually unaffected by pressure gradients almost to separation, and the Van Driest model with $A^+ = 26$ faithfully reproduces that law. (See the discussion in Section 7–8–1 and Granville (1989).) Cebeci and Smith do not show how their model with Eq. (7–92) reproduces data from flows with strong pressure gradients on a *wall law plot*. Their results for the three test cases chosen for comparison in this book are given in Fig. 7–52 for the Clauser equilibrium pressure gradient and Figs. 7–53 and 7–54 for the nonequilibrium pressure gradients of Schubauer and Klebanoff and Newman. The agreement with the data of Clauser is again not as good as one might hope, but we shall see that such is the case for all the analyses. The strong variation in Clauser's equilibrium pressure gradient parameter $(\delta^*/\tau_w)dP/dx$ for the case of Schubauer and Klebanoff is shown in Fig. 7–55. The agreement between prediction and experiment deteriorates sharply where that parameter begins to vary, as might be expected, given the restrictions on the Clauser outer region eddy viscosity model used. Looking at all three cases, it might be concluded that the more elaborate, differential method of Cebeci and Smith using a Van Driest–Clauser composite eddy viscosity model is at best slightly superior to the Moses integral method for predicting C_f.

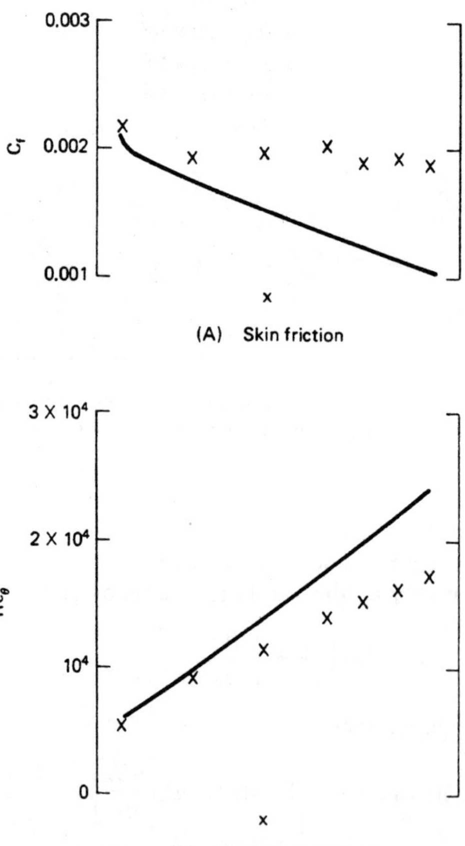

Figure 7–52 Comparison of the prediction of Cebeci and Smith (1969) using a Van Driest–Clauser eddy viscosity model with the data of Clauser (1956). (From Kline et al., 1969.)

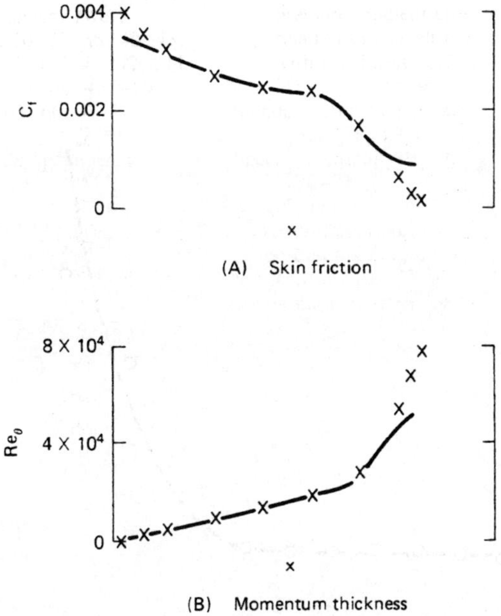

Figure 7–53 Comparison of the prediction of Cebeci and Smith (1969) using a Van Driest–Clauser eddy viscosity model with the data of Schubauer and Klebanoff (1950). (From Kline et al., 1969.)

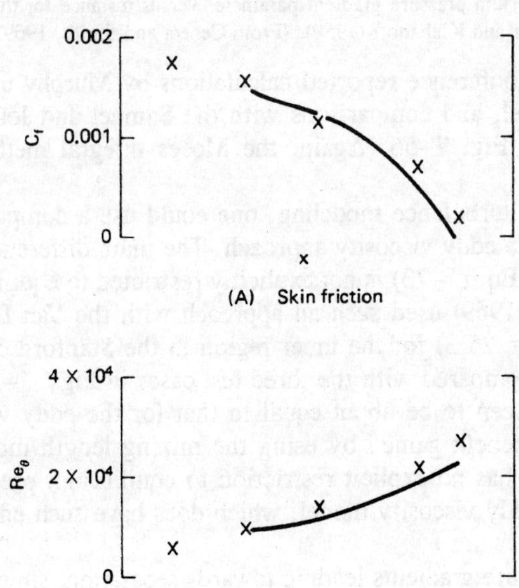

Figure 7–54 Comparison of the prediction of Cebeci and Smith (1969) using a Van Driest–Clauser eddy viscosity model with the data of Newman (1951). (From Kline et al., 1969.)

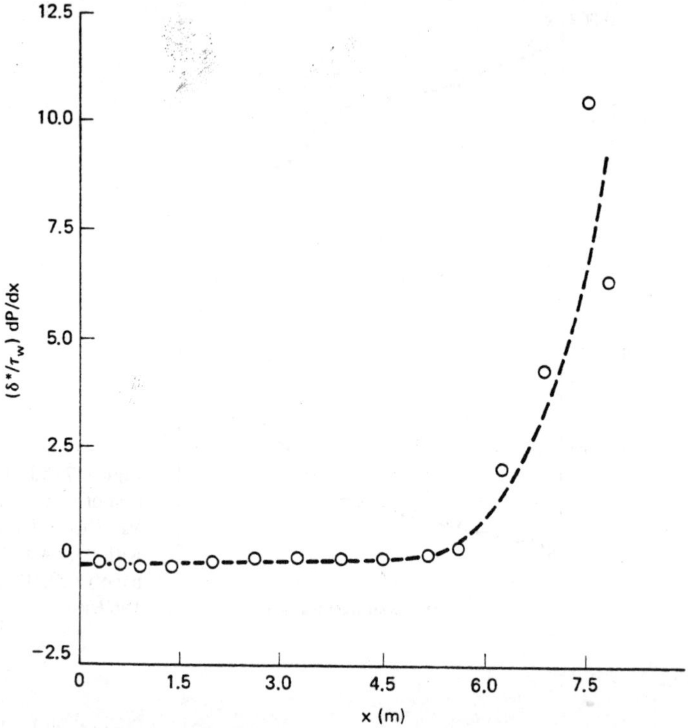

Figure 7–55 Equilibrium pressure gradient parameter versus distance for the experiment of Schubauer and Klebanoff (1950). (From Cebeci and Smith, 1969.)

The 1981 Stanford conference reported calculations by Murphy using a composite eddy viscosity model, and comparisons with the Samuel and Joubert (1974) experiment are plotted in Fig. 7–56. Again, the Moses integral method does as well.

At the same level of turbulence modeling, one could use a composite mixing length model rather than an eddy viscosity approach. The main difference would be in the outer region, where Eq. (7–73) is not explicitly restricted to equilibrium pressure gradients. Ng et al. (1969) used such an approach with the Van Driest model (with $\kappa = 0.435$ and $A^+ = 25.3$) for the inner region in the Stanford competition, and their predictions are compared with the three test cases in Figs. 7–57 to 7–59. The performance can be seen to be about equal to that for the eddy viscosity approach. There is no real benefit gained by using the mixing length model, which, from its crude derivation, has no explicit restriction to equilibrium pressure gradients, compared with an eddy viscosity model, which does have such an explicit restriction.

For very strong pressure gradients leading towards separation, simple eddy viscosity and mixing length models overpredict the measured values. This is illustrated in Figs. 7–60 and 7–61, from Simpson et al. (1981). There are no known mean flow models that can predict this behavior. Clearly, that is the reason the predictions deteriorate as separation is approached.

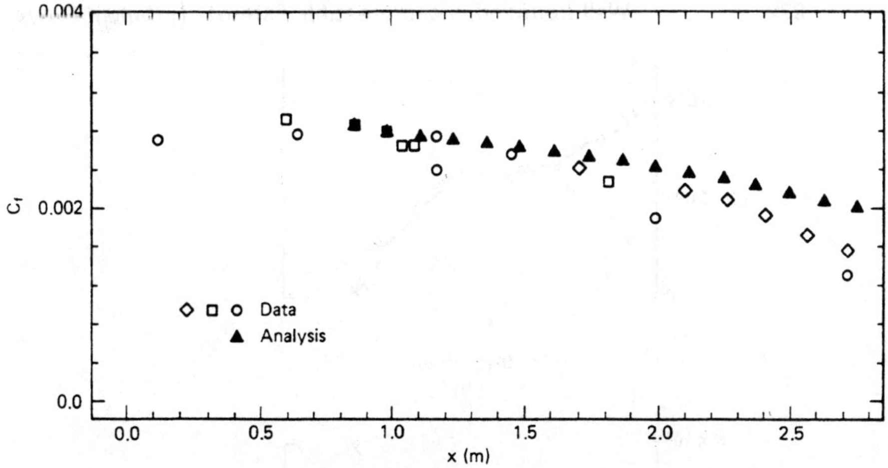

Figure 7–56 Comparison of the prediction of Murphy using a composite eddy viscosity model with the data of Samuel and Joubert (1974) for an adverse pressure gradient flow. (From Kline et al., 1982.)

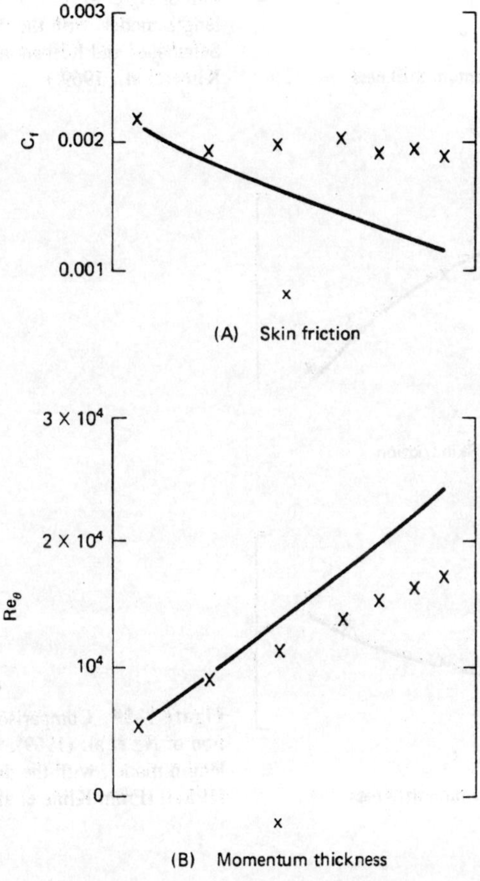

Figure 7–57 Comparison of the prediction of Ng et al. (1969), using a mixing length model, with the data of Clauser (1956). (From Kline et al., 1969.)

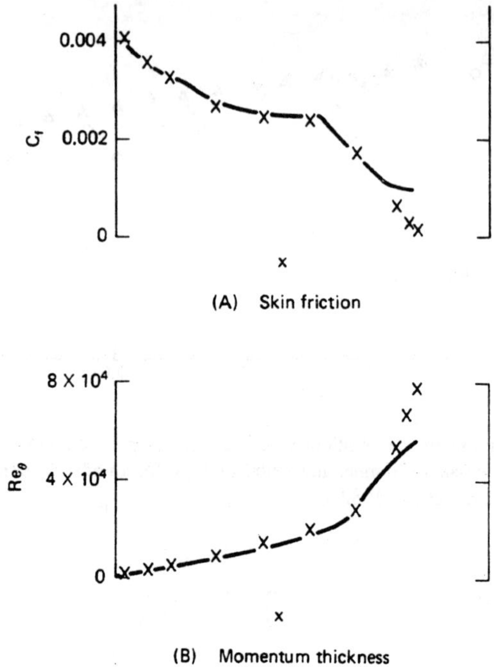

Figure 7–58 Comparison of the prediction of Ng et al. (1969), using a mixing length model, with the data of Schubauer and Klebanoff (1950). (From Kline et al., 1969.)

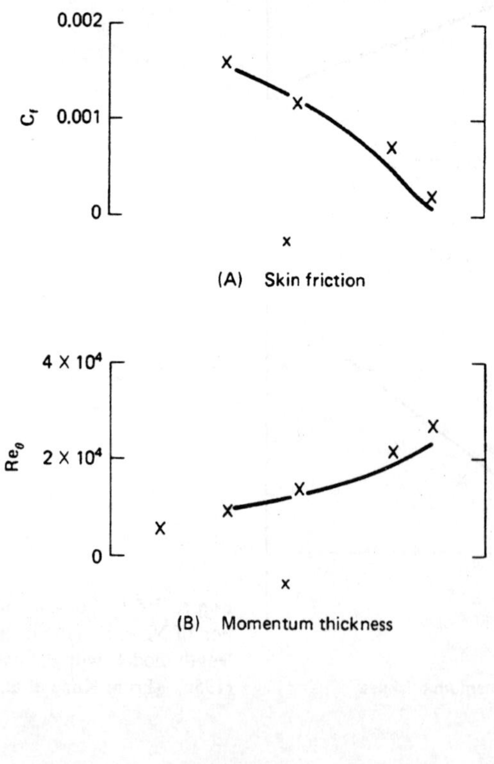

Figure 7–59 Comparison of the prediction of Ng et al. (1969), using a mixing length model, with the data of Newman (1951). (From Kline et al., 1969.)

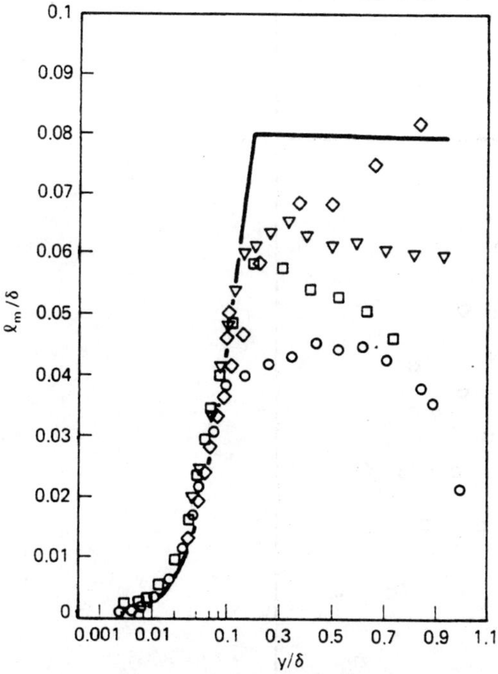

(A) Upstream of separation

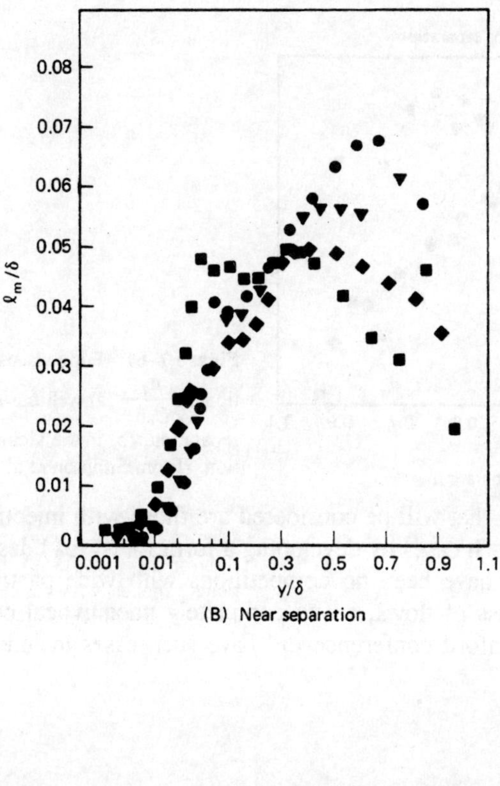

(B) Near separation

Figure 7–60 Mixing length distributions, ℓ_m/δ vs. y/δ, (a) well upstream from separation, (b) in the vicinity of separation. (From Simpson et al., 1981.)

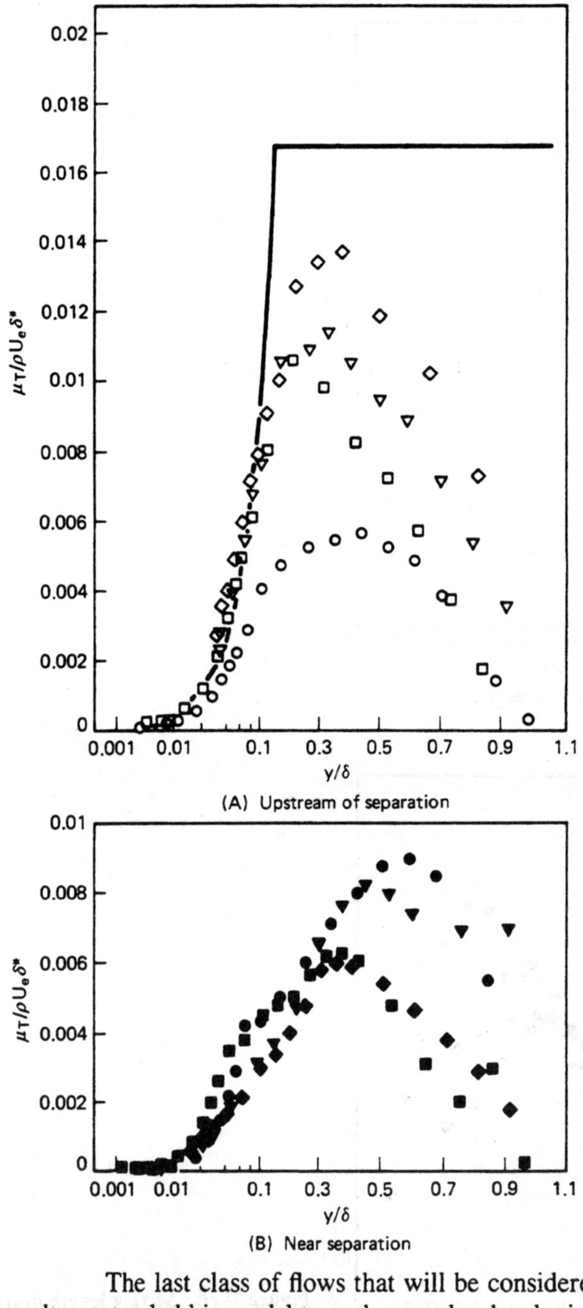

Figure 7–61 Eddy viscosity distributions, $\frac{\mu_T}{\rho U_e \delta^*}$ (a) well upstream from separation, (b) in the vicinity of separation. (From Simpson et al., 1981.)

The last class of flows that will be considered are those with injection. Cebeci has extended his model to such cases by developing a form for $A^+(v_0^+)$ described earlier (see Fig 7–38). There have been no competitions with wide participation attempting to predict this class of flows, so no completely unequivocal comparisons are possible. The 1981 Stanford conference did have such cases included, but few researchers treated them, and, of those that did, none used a simple eddy viscosity or mixing length model. Some comparisons of predictions with data are, however, available from original works. A prediction from Cebeci and Mosinskis (1971) is compared with one of the experimental cases of Simpson (1968) in Fig. 7–62. Predictions from Schetz and Favin (1971) using the extended Reichardt model of Eq. (7–70) together with Eq. (7–71) are compared with two cases from Simpson (1968)—one with injection and one with suction—in Fig. 7–63. The performance of the two formulations is judged to be roughly equal.

The preceding comparisons demonstrate a reasonably good capability for predicting the development of turbulent boundary layers with numerical solutions of mean flow formulations using eddy viscosity or mixing length models. A simple computer code for a PC is in Appendix B. A second code utilizing a stretched grid is also in Appendix B.

There is one more matter that needs to be addressed and that concerns the interconnected issues of the boundary conditions at the wall and the grid spacing near

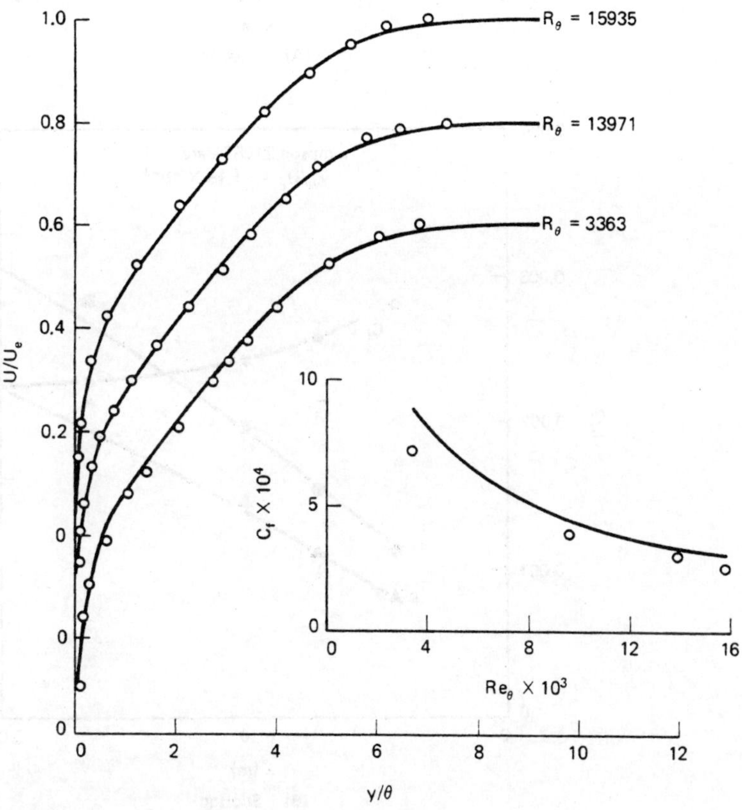

Figure 7–62 Comparison of predictions of Cebeci and Mosinskis (1971) based on an eddy viscosity formulation for flow with injection with the data of Simpson (1968). $v_w/U_e = 0.00784$.

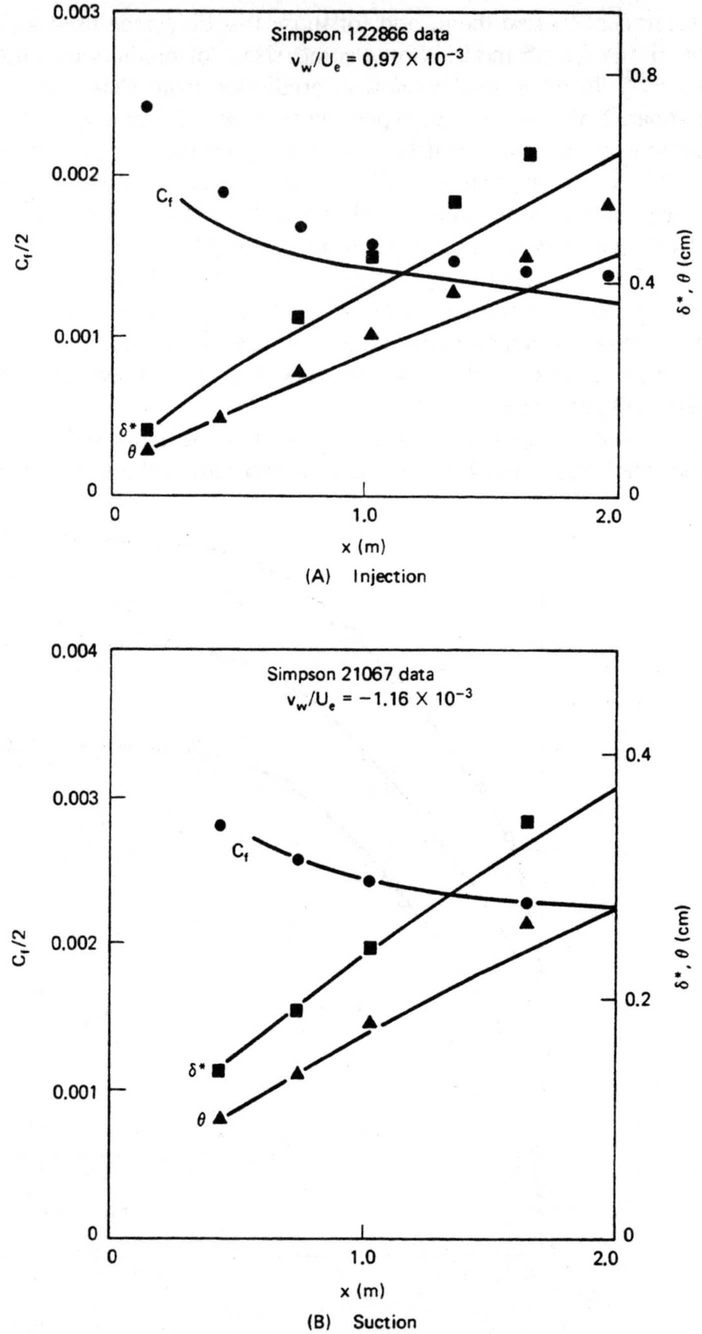

Figure 7–63 Comparison of predictions of Schetz and Favin (1971) using the extended Reichardt eddy viscosity model for cases with injection and suction with the data of Simpson (1968).

the wall. There would appear to be no issue with regard to the boundary conditions on the wall. One should simply enforce the *no-slip condition* as $U = V = 0$ on $y = 0$, and that is what was done in the foregoing calculations. However, attempts to relieve the necessity for very fine grid spacing near the wall and thus reduce computational cost have involved modifying the wall boundary conditions. The most widely used approach is the *wall function* method developed by the Spalding group [see Patankar and Spalding (1966)]. In that approach, one presumes the accuracy of the log law, Eq. (7–7c), and uses the law to deduce an artificial boundary condition on $U \neq 0$, to be enforced at some point above the actual wall. A point in the range $30 < y^+ < 100$ is suggested [see Rodi, 1980]. Once a value of y^+ is selected to be the boundary location in the calculation, Eq. (7–7c) is used to determine the boundary value of U to impose at that location. Sometimes, continuity of shear at the match point is also required, and that serves to determine u_*. This method results in a saving in the cost of computation, compared to integrating down to the wall, but the actual level of the cost of computation is so low for two-dimensional calculations that it is hard to justify the extra approximations in the approach. Also Wahls et al. (1989) found that the basic method performed poorly in strong adverse pressure gradients. As a more appealing alternative, Caille and Schetz (1991) have used an integral conservation-of-momentum condition on an inner strip, with a numerical solution on a coarse grid for the outer region.

EXAMPLE 7–2. Numerical Solution of a Turbulent Boundary Layer with Mean-flow Models

Let us compare the predicted development of a turbulent boundary layer on a flat plate to $x = 6.0$ ft, starting from initial conditions at $x = 5.0$ ft, with an eddy viscosity model and a mixing length model. The fluid has a kinematic viscosity $\nu = 0.00001$, and the edge velocity is 10.0 ft/s. We then compare the predicted results with those from the empirical Blasius skin friction law and the simple integral method.

Solution The code ITBL in Appendix B can be used to solve this problem. The first input data needed are XI = 5.0 and XF = 6.0. Then, we require CNU = 0.00001 and UINF = 10.0. Since this code uses an implicit method, the number of steps in the streamwise direction (IMAX) and, thus, DX can be selected free of any stability criterion. Choose NMAX = 101 for DX = 0.01 as a rather fine grid. The issue of the grid size in the y-direction is complicated for a uniform grid spacing. (Reread the beginning of Section 7–9.) On most PCs, a maximum of about 550 points can be put across the layer, due to limitations on *RAM*. That value was used for MMAX here. A total of 401 points was put across the initial boundary layer thickness of DEL = 0.0856 ft, obtained from the integral method (DY = 0.0002145). The Blasius skin friction law [Eq. (7–13)] was used to obtain the initial value for CF(1) = 0.002665. The initial velocity profiles were formed with the law of the wake for U and $V = 0.0$.

We select MODEL = 1 for the mixing length model and then MODEL = 2 for an eddy viscosity model.

Of course, the relevant sections of the code have to be modified to treat other flow problems. One obvious place is in the loop DO 5 N = 1, NMAX, where the desired inviscid velocity distribution denoted UE(N) and DUEDX(N) must be specified.

All velocities are normalized by UINF inside the code.

Selected output from the two runs with a mixing length and an eddy viscosity model is shown in the following table ("MIXL" means mixing length, and "MUT" means eddy viscosity):

N	X	CF(MIXL)	CF(MUT)
1	5.00	0.00267	0.00267
6	5.05	0.00241	0.00257
11	5.10	0.00248	0.00251
16	5.15	0.00252	0.00252
21	5.20	0.00255	0.00254
26	5.25	0.00256	0.00256
31	5.30	0.00258	0.00257
36	5.35	0.00259	0.00259
41	5.40	0.00259	0.00260
46	5.45	0.00260	0.00261
51	5.50	0.00260	0.00261
56	5.55	0.00261	0.00262
61	5.60	0.00261	0.00263
66	5.65	0.00261	0.00263
71	5.70	0.00261	0.00263
76	5.75	0.00261	0.00264
81	5.80	0.00261	0.00264
86	5.85	0.00261	0.00264
91	5.90	0.00261	0.00265
96	5.95	0.00261	0.00265
101	6.00	0.00261	0.00265

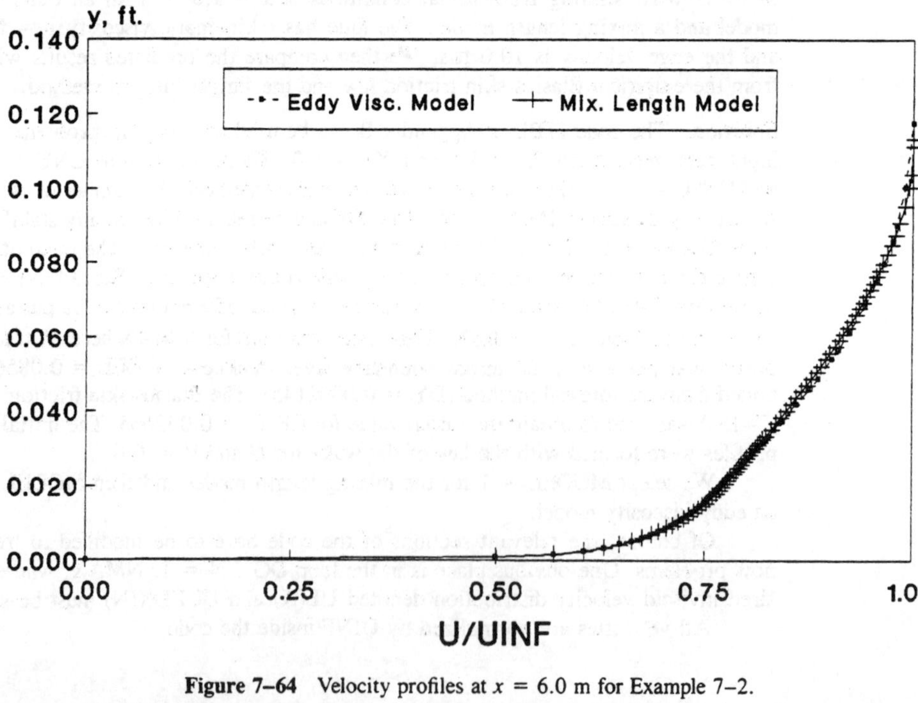

Figure 7–64 Velocity profiles at $x = 6.0$ m for Example 7–2.

The Blasius skin friction law gives $C_f = 0.00257$ at the conditions at $x = 6.0$ ft, and both numerical predictions are in good agreement with that value. The small difference between the two predictions should not be judged significant. Profiles are shown in Fig. 7–64. There is a small difference between the predicted profiles in the outer region. Note that not every numerical point is shown on the graphs, because they would then become too cluttered.

EXAMPLE 7–3. Numerical Solution of a Turbulent Boundary Layer with Stretched Grid

Let us calculate the boundary layer development of a fluid with $\nu = 10^{-6}$ m²/s on a flat plate with $U_\infty = 10.0$ m/s to a length of 1.0 m with an eddy viscosity model. We use a stretched grid to reduce the computational effort. We then compare the result for the skin friction coefficient with the Schoenherr formula. How thick is the laminar sublayer at the end of the plate?

Solution Program ITBLS in Appendix B can handle the requirements of this problem. The conditions of the problem give a high Reynolds number (10^7) at the end of the plate, so we can assume turbulent flow from the leading edge with only a small loss of accuracy. We enter XI = 0.0 and XF = 1.0 first. Then, we let CNU = 0.000001 and UINF = 10.0. With the stretched grid, MMAX = 98 should be adequate. We select NMAX = 100 to give a fine x-grid to go with the small y-spacing near the wall. Finally, $U_e = U_\infty$ (UE(N) = 1.0) for a flat plate. Some output at various stations is listed in the following table:

N	X	THETA	DELTS	CF
4	0.0303	0.00008	0.00012	0.00433
8	0.0707	0.00017	0.00024	0.00388
12	0.1111	0.00025	0.00035	0.00357
16	0.1515	0.00032	0.00044	0.00338
20	0.1919	0.00039	0.00053	0.00325
24	0.2323	0.00045	0.00061	0.00315
28	0.2727	0.00052	0.00070	0.00307
32	0.3131	0.00058	0.00078	0.00301
36	0.3535	0.00064	0.00085	0.00296
40	0.3939	0.00070	0.00093	0.00291
44	0.4343	0.00076	0.00100	0.00286
48	0.4747	0.00081	0.00108	0.00283
52	0.5151	0.00087	0.00115	0.00279
56	0.5556	0.00093	0.00122	0.00276
60	0.5960	0.00098	0.00129	0.00273
64	0.6364	0.00104	0.00136	0.00270
68	0.6768	0.00109	0.00143	0.00268
72	0.7172	0.00115	0.00150	0.00266
76	0.7576	0.00120	0.00157	0.00264
80	0.7980	0.00125	0.00164	0.00262
84	0.8384	0.00131	0.00170	0.00260
88	0.8788	0.00136	0.00177	0.00258
92	0.9192	0.00141	0.00184	0.00256
96	0.9596	0.00146	0.00190	0.00255
100	1.0000	0.00152	0.00197	0.00253

A plot of some of these results with the skin friction coefficient compared to the Schoenherr empirical law is shown in Fig. 7–65(A). The velocity profile at the end of the plate is plotted in Fig. 7–65(B). A semilogarithmic scale was used in the latter plot so that the points close to the wall could be clearly seen. Note that the first grid point above the wall is well within the laminar sublayer with this strongly stretched grid and MMAX = 98. That condition is generally considered necessary for accurate skin friction results; one cannot achieve that condition with a uniform grid and only about 500 points across the layer. From the figure, it is possible to estimate the thickness of the laminar sublayer to be about 0.01 cm.

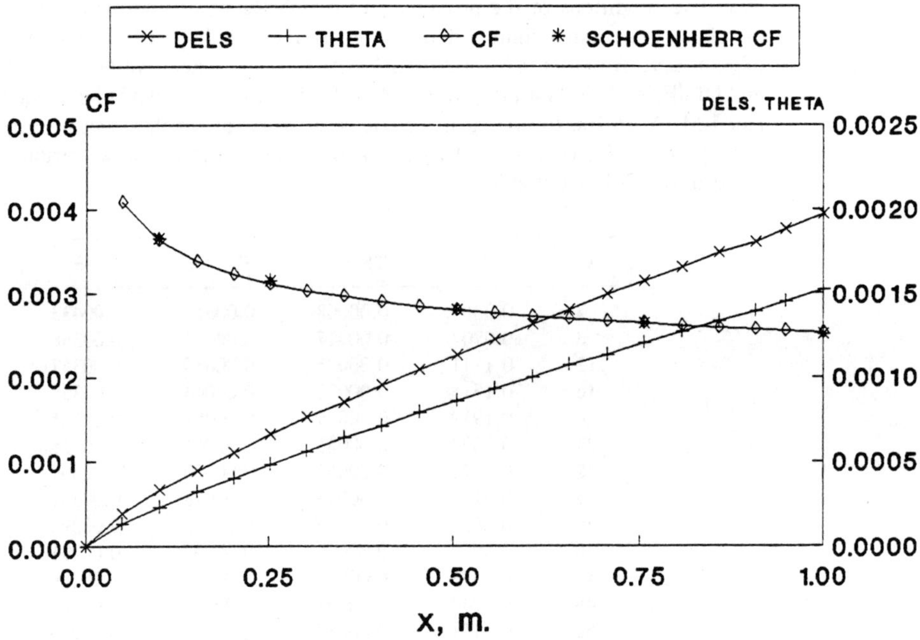

(A) Streamwise variation of integral boundary layer properties

Figure 7–65 Predicted results for Example 7–3.

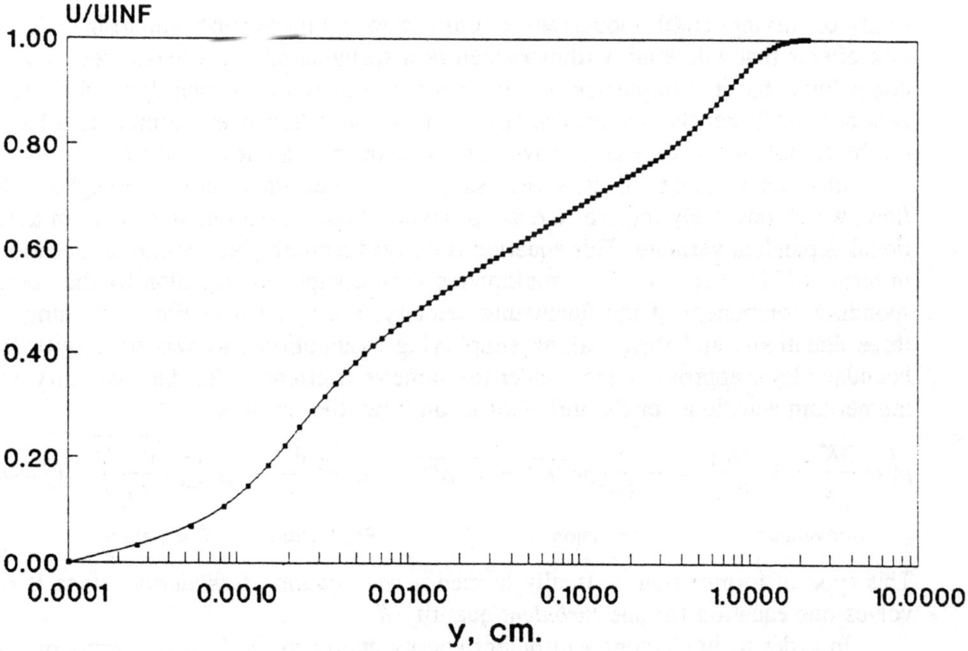

(B) Velocity profiles at x = 1.0 m.

Figure 7–65 *(continued)*

7–10 FORMULATIONS BASED ON TURBULENT KINETIC ENERGY

It is possible to construct turbulent flow formulations at higher levels than those based on the mean properties of the flow alone. If any single quantity is to be selected to represent the fluctuating character of the turbulence, it is the turbulent kinetic energy (TKE) of the fluctuations, defined in Eq. (7–25), and picked by virtually all researchers in the field. There are two general reasons for seeking a formulation that directly involves the fluctuating character of the flow. The first is philosophical: To many people, it just does not seem sensible to analyze a turbulent flow without explicitly treating such a basic feature of turbulence as the fluctuations. Also, one can argue that such a formulation will be at a more *fundamental* level than a mean flow formulation and thus will, presumably, produce better predictions for general flows. The second reason for pursuing this level of formulation is that there are some physical situations for which a mean flow analysis is clearly inadequate. One easily understood example is a boundary layer flow over a surface with a high level of free stream turbulence. The boundary layers on the blades of the latter stages of a multistage turbine represent a practical instance where such free stream turbulence occurs. These flows fall outside of the data base on which the eddy viscosity or mixing length models were constructed, so there is no way to include the new effects that will result within a mean flow formulation. This reason seems quite compelling, but it is important to note that the occurrence of such flows in practice is rather small. For the two general reasons mentioned, however, formulations based on the turbulent kinetic energy have been developed to an advanced state.

In order to utilize K, it is necessary to calculate its variation throughout the flow, which obviously requires the use of an additional equation, since K is an additional dependent variable. This equation is derived from the Navier-Stokes equations in terms of $(U + u')$, etc., by multiplying each component equation by the corresponding component of the fluctuating velocity, averaging over time, summing all three equations, and then making simplifying assumptions, as was done with the boundary layer approximation. Under the same restrictions as for the continuity and momentum equations for the turbulent mean flow, the result is

$$\underbrace{\rho\left(U\frac{\partial K}{\partial x} + V\frac{\partial K}{\partial y}\right)}_{\text{Convection}} = \underbrace{-\frac{\partial}{\partial y}(\rho\overline{v'K'} + \overline{v'p'})}_{\text{Diffusion}} \underbrace{- \rho\overline{u'v'}\frac{\partial U}{\partial y}}_{\text{Production}} \underbrace{- \mu\sum\overline{\left(\frac{\partial u_i'}{\partial x_j}\right)^2}}_{\text{Dissipation}} \tag{7–94}$$

This type of formulation is usually termed a *one-equation formulation,* since it involves one equation for one *turbulent* quantity K.

In order to implement a turbulent energy approach, each of the terms on the right-hand side of Eq. (7–94) (or its equivalent in other geometries) must be *modeled,* in the same sense that $-\rho\overline{u'v'}$ had to be modeled in Eq. (7–34a), via, for example, Eq. (7–36) or Eq. (7–37). That is, these terms must be related to the mean flow variables (U, V), the turbulent shear $-\rho\overline{u'v'}$, and/or the turbulent kinetic energy K. This is so because there are only three equations to determine the three unknowns (U, V, K). Indeed, we see that $-\rho\overline{u'v'}$ must still be related to some combination of U, V, and K.

The modeling of the last term on the right-hand side of Eq. (7–94) is generally viewed as noncontroversial. Since viscous dissipation of turbulent energy takes place predominantly at the smaller eddy sizes, and since these scales have been found to be nearly locally *isotropic* (independent of direction), the exact result for dissipation under those conditions is generally simply carried over directly, that is,

$$\mu\sum\overline{\left(\frac{\partial u_i'}{\partial x_j}\right)^2} \longrightarrow C_D\frac{\rho K^{3/2}}{\ell} \tag{7–95}$$

where ℓ is a length scale similar, but not equal, to the mixing length ℓ_m and C_D is an empirical constant of order $O(1/10)$.

Two different approaches have been proposed for the modeling of the second term on the right-hand side of Eq. (7–94) and the relation of $-\rho\overline{u'v'}$ to K. Prandtl (1945) reinvoked the eddy viscosity concept, but introduced K through

$$\mu_T = \rho\sqrt{K}\,\ell \tag{7–96}$$

This relation follows directly upon observing that viscosity is proportional to *density* × *velocity* × *length* and upon wishing to involve K. The characteristic velocity is now $\sqrt{K}$ rather than u_*. Note that the gradient transport formulation, together with its limitations, remains. Also, one must still model $\ell(y/\delta)$. The second general type of model was introduced by Bradshaw et al. (1967) as

$$\tau_T = -\rho\overline{u'v'} = a_1\rho K \tag{7-97}$$

where $a_1 \approx 0.30$. This equation is known to be approximately true from experiment (see Fig. 7-20), especially in the inner region. Note that the model does avoid the gradient transport formulation. It does appear to be valid mainly for external, wall-bounded flows.

The first term on the right-hand side of Eq. (7-94) is generally looked upon as *diffusion*. The modeling of this term is particularly difficult, since good direct data for terms involving the pressure fluctuations out in the flow are not available. By crude analogy with laminar diffusion or, for that matter, with the gradient transport model for the diffusion of momentum, Eq. (7-36), researchers who use Eq. (7-96) employ

$$-(\rho\overline{v'K'} + \overline{v'p'}) \longrightarrow \frac{\mu_T}{\sigma_K}\frac{\partial K}{\partial y} \tag{7-98}$$

where σ_K is a Prandtl number for turbulent kinetic energy (generally, $\sigma_K \approx 1$). This model is called the *Prandtl energy method*. Bradshaw et al. (1967) proposed that

$$(\rho\overline{v'K'} + \overline{v'p'}) = \tau_T\sqrt{\frac{\tau_{max}}{\rho}}\,G\left(\frac{y}{\delta}\right) \tag{7-98a}$$

where $G(y/\delta)$ is a universal empirical function (see Fig. 7-66) determined by recourse to flat-plate flow data. Note the central role of the maximum shear stress. This model avoids the gradient transport formulation, but it still requires that $\ell(y/\delta)$

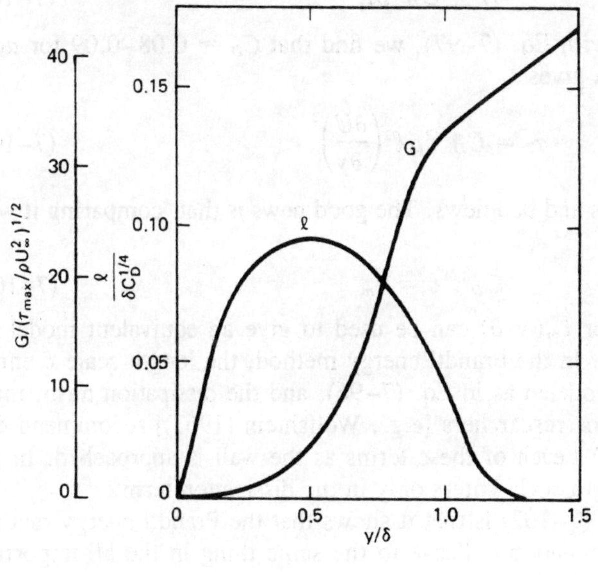

Figure 7-66 Empirical functions used by Bradshaw et al., 1967.

and $G(y/\delta)$ be modeled. A suggested variation for $\ell(y/\delta)$ was also determined from flat-plate flow data, and it is presented as another empirical function also shown in Fig. 7–66.

For the Prandtl energy method, the modeled form of the TKE equation becomes

$$\rho\left(U\frac{\partial K}{\partial x} + V\frac{\partial K}{\partial y}\right) = \frac{\partial}{\partial y}\left(\frac{\mu_T}{\sigma_k}\frac{\partial K}{\partial y}\right) + \mu_T\left(\frac{\partial U}{\partial y}\right)^2 - C_D\frac{\rho K^{3/2}}{\ell} \tag{7–99}$$

With a model for $\ell(y/\delta)$, it is clearly possible to solve Eq. (7–99) by using the numerical methods developed for the momentum equation, since the forms are so similar. However, with Bradshaw's modeling, the TKE equation takes on a *hyperbolic,* not *parabolic,* form, and the usual methods for the boundary layer equations do not apply. Bradshaw developed suitable methods, and the interested reader should refer to the original paper for full details.

Just as for the eddy viscosity or mixing length, the modeling at this level is completed by looking at limiting cases of the equations of motion. Near a rigid wall, *convection* and *diffusion* of K are negligible compared to *production* and *dissipation* (see Fig. 7–24), so Eq. (7–99) becomes

$$\rho\sqrt{K}\,\ell\left(\frac{\partial U}{\partial y}\right)^2 = C_D\frac{\rho K^{3/2}}{\ell} \tag{7–99a}$$

Noting Eqs. (7–36) and (7–96), we can rewrite Eq. (7–99a) as

$$\tau_T^2 = \left(\mu_T\frac{\partial U}{\partial y}\right)^2 = C_D\rho^2K^2 \tag{7–99b}$$

or

$$\tau_T = C_D^{1/2}\rho K \tag{7–100}$$

Comparing this equation with Eq. (7–97), we find that $C_D \approx 0.08–0.09$ for $a_1 \approx 0.30$. A little more algebra gives

$$\tau_T = C_D^{-1/2}\rho\,\ell^2\left(\frac{\partial U}{\partial y}\right)^2 \tag{7–101}$$

This relation has good news and bad news. The good news is that, comparing it with Eq. (7–37), we obtain

$$C_D^{-1/4}\ell = \ell_m \tag{7–102}$$

so that any good model for $\ell_m(y/\delta)$ can be used to give an equivalent model for $\ell(y/\delta)$, which was needed. In the Prandtl energy method, the length scale ℓ enters into the formula for μ_T, modeled as in Eq. (7–96), and the dissipation term, modeled as in Eq. (7–95). Some researchers [e.g., Wolfshtein (1967)] recommend different damping functions for each of these terms as the wall is approached. In the Bradshaw method, the length scale enters only in the dissipation term.

The bad news in Eq. (7–102) is that it shows that the Prandtl energy method and the mixing length formulation collapse to the same thing in the all-important wall region. Note also that the preceding analysis of the wall region, especially that leading to Eq. (7–100), shows that the Bradshaw TKE formulation also is closely related to the Prandtl energy method in the inner region. Thus, the mixing length model, the eddy viscosity model (Van Driest or Reichardt), the Prandtl energy method, and the Bradshaw TKE method are all essentially equivalent in the inner region for flows over smooth, solid surfaces with moderate pressure gradients. One can, therefore, expect that any differences in performance in comparisons with data will be due to greater generality (no explicit restriction to equilibrium pressure gradients) in the treatment of the outer region or better modeling of the flow as separation is approached.

As with the mean flow models, some workers use a *wall function method* for TKE methods. Since the match point is in the wall region, Eq. (7–97) can be used with $\tau_T \approx \tau_w$ to obtain $K = u^{*2}/a_1$ at the match point. If the solution is carried all the way to the wall, then $K = 0$ on the wall, and a laminar diffusion term $\partial/\partial y(\mu\, \partial K/\partial y)$ must be included on the right-hand side of Eq. (7–99).

For flat-plate flow, predictions for the inner region based on the Prandtl energy method are indistinguishable from those based on a Reichardt-Clauser eddy viscosity model, as shown in Fig. 7–67. Comparisons of predictions using the Bradshaw TKE method and three pressure gradient test cases are shown in Figs. 7–68 to 7–70. Generally, better agreement with the data is demonstrated than for any of the methods discussed previously.

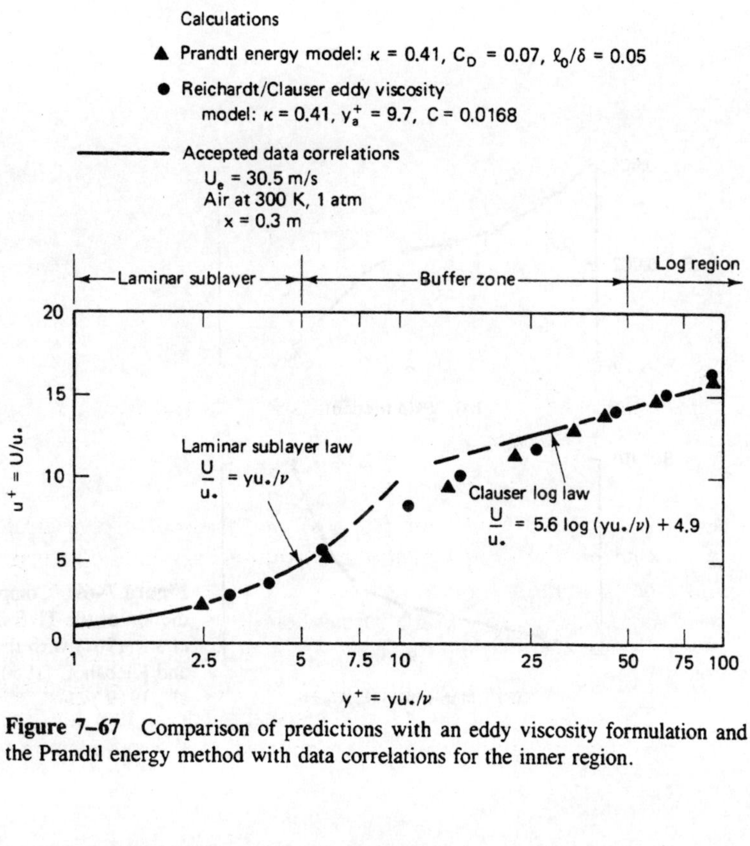

Figure 7–67 Comparison of predictions with an eddy viscosity formulation and the Prandtl energy method with data correlations for the inner region.

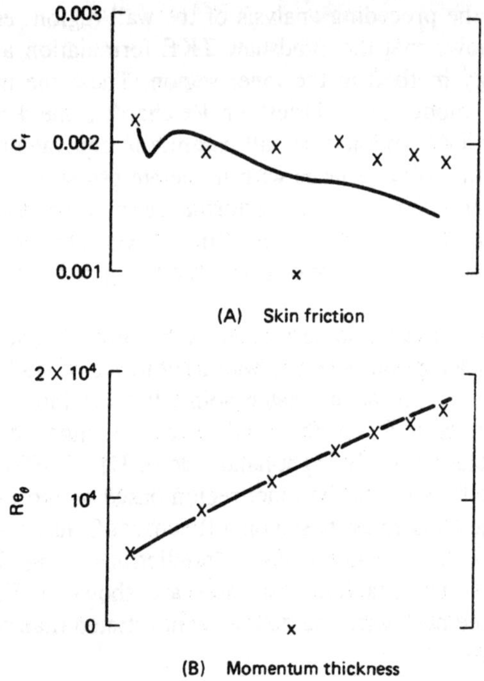

Figure 7–68 Comparison of the prediction of the TKE model of Bradshaw et al. (1967) with the data of Clauser (1956). (From Kline et al., 1969.)

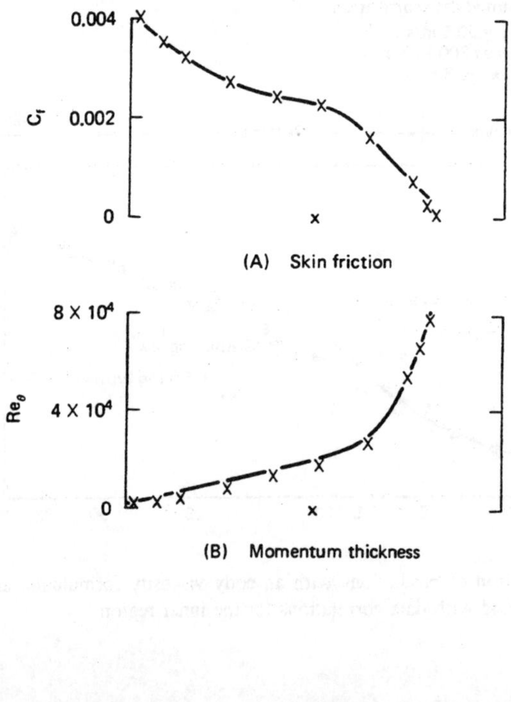

Figure 7–69 Comparison of the prediction of the TKE model of Bradshaw et al. (1967) with the data of Schubauer and Klebanoff (1950). (From Kline et al., 1969.)

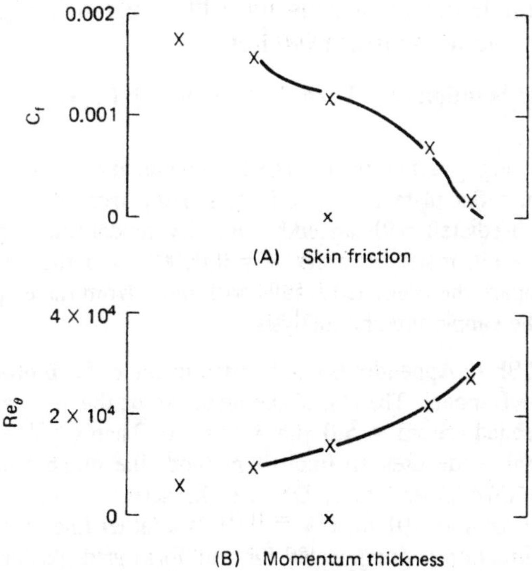

Figure 7–70 Comparison of the prediction of the TKE model of Bradshaw et al. (1967) with the data of Newman (1951). (From Kline et al., 1969.)

The 1981 Stanford conference included two separate TKE calculations by Murphy and Orlandi for the Samuel and Joubert (1974) experiment. Both predictions are shown in Fig. 7–71, compared with the measurements. Recent predictions with the Bradshaw method are also shown. Note the rather large difference in the predictions. Murphy used the Prandtl energy method, as implemented by Glushko (1966). Orlandi also used what is essentially the Prandtl energy method, but he introduced modifications to the dissipation and the pressure-work term. Bradshaw used his original method. Some researchers feel that the poor predictions are a result of neglecting the normal stress terms.

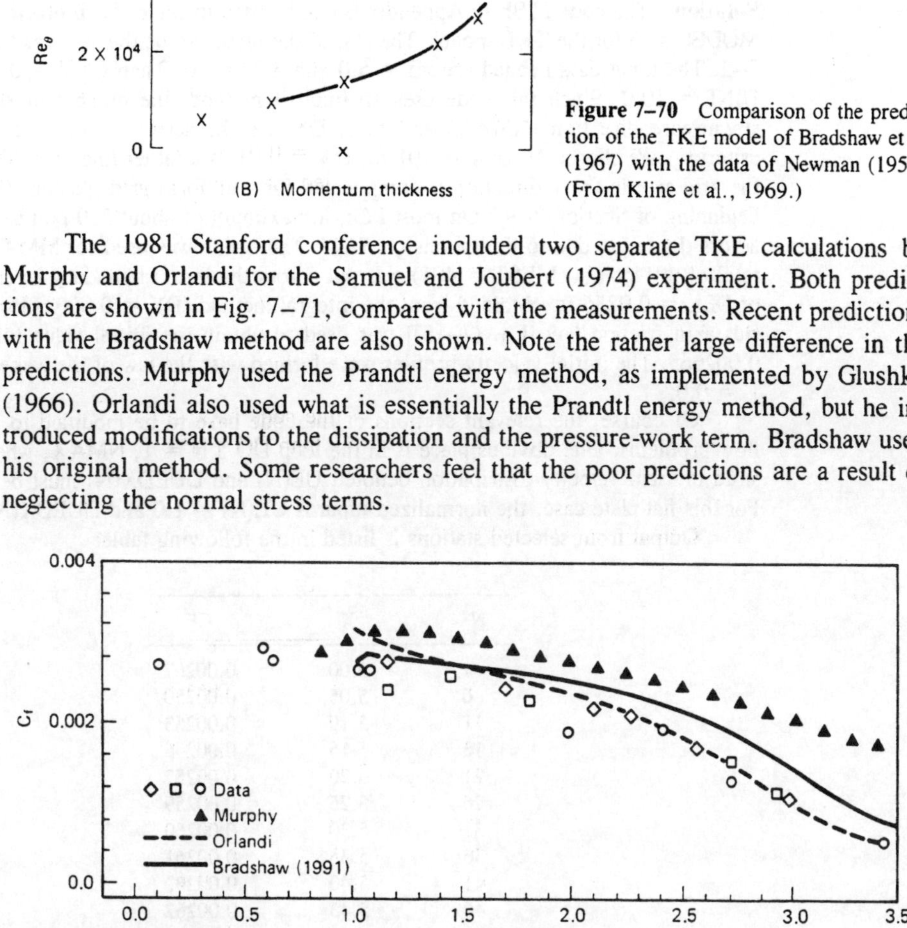

Figure 7–71 Comparison of the predictions of Murphy and Orlandi using turbulent kinetic models with the data of Samuel and Joubert (1974) for an adverse pressure gradient flow (From Kline et al., 1982.)

Appendix B has a simple computer code for a PC, implementing the Prandtl energy method, that is suitable for student exercises.

EXAMPLE 7–4. Numerical Solution of a Turbulent Boundary Layer with a TKE Model

Let us use the Prandtl energy method to compare the predicted development of a turbulent boundary layer on a flat plate to $x = 6.0$ ft, starting from initial conditions at $x = 5.0$ ft, with that predicted with an eddy viscosity model and a mixing length model. The fluid has a kinematic viscosity $\nu = 0.00001$, and the edge velocity is 10.0 ft/s. We then compare the predicted results with those from the empirical Blasius skin friction law and the simple integral analysis.

Solution The code ITBL in Appendix B can be used to solve this problem. We select MODEL = 3 for the TKE model. The rest of the input can be the same as for Example 7–2. The input data needed are XI = 5.0 and XF = 6.0. Then CNU = 0.00001 and UINF = 10.0. Since this code uses an implicit method, the number of steps in the streamwise direction (NMAX) and thus, DX, can be selected free of any stability criterion. We choose NMAX = 101 for DX = 0.01 as a rather fine grid. The issue of the grid size in the y-direction is complicated for a uniform grid spacing. (Reread the beginning of Section 7–9.) On most PCs, a maximum of about 550 points can be put across the layer, due to limitations on *RAM*. That value was used for MMAX here. A total of 401 points (MEST = 401) was put across the initial boundary layer thickness of DEL = 0.0856 ft, obtained from the integral method (DY = 0.0002145). The Blasius skin friction law [Eq. (7–13)] was used to obtain the initial value for CF(1) = 0.002665. The initial velocity profiles were formed with the law of the wake for U and $V = 0.0$.

Of course, the relevant sections of the code have to be modified to treat other flow problems. One obvious place is in the loop DO 5 N = 1, NMAX, where the desired inviscid velocity distribution denoted UE(N) and DUEDX(N) must be specified. For this flat-plate case, the normalized input is UE(N) = 1.0 and DUEDX(N) = 0.0.

Output from selected stations is listed in the following table:

N	X	CF
1	5.00	0.00267
6	5.05	0.00250
11	5.10	0.00253
16	5.15	0.00256
21	5.20	0.00257
26	5.25	0.00259
31	5.30	0.00260
36	5.35	0.00261
41	5.40	0.00262
46	5.45	0.00262
51	5.50	0.00263
56	5.55	0.00263
61	5.60	0.00263
66	5.65	0.00263
71	5.70	0.00263
76	5.75	0.00263

81	5.80	0.00263
86	5.85	0.00263
91	5.90	0.00263
96	5.95	0.00263
101	6.00	0.00263

The Blasius skin friction law gives $C_f = 0.00257$ at the conditions at $x = 6.0$ ft, and this solution gives a prediction about equivalent to those with the mixing length and eddy viscosity models in Example 7–2. That is to be expected for this rather simple flow problem.

The final velocity profile predicted here is compared with those predicted by the mean flow models discussed earlier in Fig. 7–72. Again, good agreement can be seen.

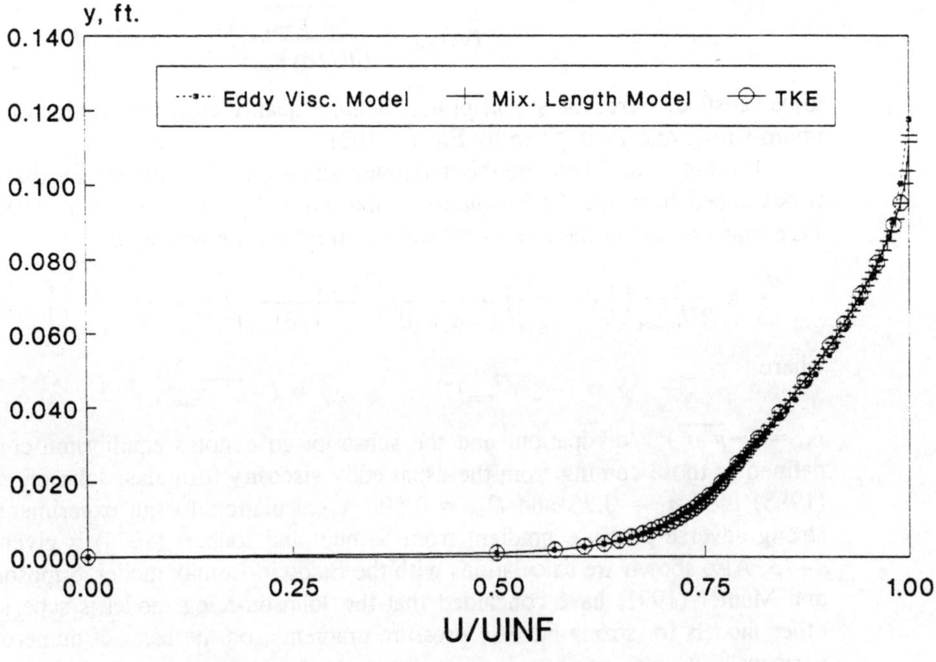

Figure 7–72 Velocity profiles at $x = 6.0$ m for Example 7–4.

Earlier, we observed that, for very strong adverse pressure gradients, the simplest types of turbulence models cannot represent the observed flow physics. That is also true of the basic TKE models discussed. The core of the problem is now felt to be the *upstream history effects* felt by a boundary layer subjected to strong and rapidly changing adverse pressure gradients. A model that includes such effects has been developed by Johnson and King (1985). An ordinary differential equation, derived from the TKE equation, is used to described the streamwise development of the maximum stress, together with an assumed eddy viscosity distribution that has the square root of the maximum stress as the velocity scale, rather than the square root of the TKE, as in the Prandtl energy method. The *kinematic eddy viscosity,*

$\nu_T = \mu_T/\rho$, is represented as

$$\nu_T = \nu_{To}(1 - \exp(-\nu_{Ti}/\nu_{To})) \tag{7-103}$$

where ν_{Ti} and ν_{To} are the kinematic eddy viscosities in the inner and outer regions, respectively. The inner eddy viscosity is modeled as

$$\nu_{Ti} = D^2 \kappa y \sqrt{-\overline{u'v'}_{\max}} \tag{7-104}$$

Here, D is the Van Driest damping factor [the term in brackets in Eq. (7–61)], as adjusted by Cebeci and Smith. Johnson and King also use $A^+ = 15$, rather than the usual value of 26. The outer eddy viscosity ν_{To} is taken to be a constant times the intermittency function of Klebanoff. The value of the constant is determined during the calculation by requiring the following relation

$$(\nu_T)_{\max} = \frac{-\overline{u'v'}_{\max}}{(\partial U/\partial y)_{\max}} \tag{7-105}$$

to be satisfied. Here, the subscript max denotes quantities at the location of the maximum stress, and ν_T is given by Eq. (7–103).

It remains to determine the maximum stress, and an equation for that quantity is developed from the TKE equation, modeled as by Bradshaw et al. (1967). The TKE equation along the line of maximum stress can be written as

$$\frac{dg}{dx} = \frac{a_1}{2U_m L_m}\left(\left(1 - \frac{g}{g_{eq}}\right) + \frac{C_{dif} L_m}{a_1 \delta(0.7 - (y/\delta)_{\max})}\left(1 - \sqrt{\frac{\nu_{To}}{\nu_{To,eq}}}\right)\right) \tag{7-106}$$

where

$$g \equiv (-\overline{u'v'}_{\max})^{-1/2}, \qquad g_{eq} \equiv (-\overline{u'v'}_{\max,eq})^{-1/2} \tag{7-107}$$

$L_m = (-\overline{u'v'})^{3/2}$/dissipation, and the subscript eq denotes equilibrium conditions, defined as those coming from the usual eddy viscosity formulas. Johnson and King (1985) used $a_1 = 0.25$ and $C_{dif} = 0.50$. A calculation for an experiment with a strong adverse pressure gradient from Samuel and Joubert (1974) is given in Fig. 7–73. Also shown are calculations with the Baldwin-Lomax model. Simpson (1987) and Menter (1991) have concluded that the Johnson-King model is superior to all other models for strong adverse pressure gradients, on the basis of numerous comparisons.

The TKE methods also have the practical advantage that they can treat flows for which the turbulence field does not have the same relation to the mean flow field as it does for the usual cases used in the data base for the mean flow formulations. Again, we cite flows with high free stream turbulence as, perhaps, the most common example. The author (see Schetz, et al., 1982) has made calculations using the Prandtl energy method for comparison with the results of Huffman et al. (1972) for the experimental situation of a flat plate for which the free stream turbulence level was varied over a rather wide range. A comparison of predictions with experiment is shown in Fig. 7–74. It appears that the TKE analysis underpredicts somewhat the effects of increasing K_e, although there is considerable uncertainty in the data. Some of the discrepancy can also be traced to the fact that the experiment involved a leading-edge sand strip to trip the flow, for which the analysis cannot account.

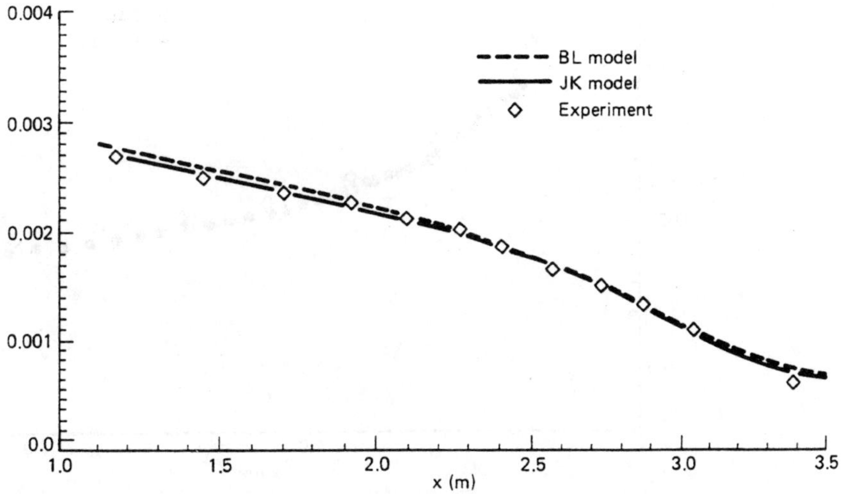

Figure 7–73 Comparison of predictions from the Johnson-King and Baldwin-Lomax models with experiment for the Samuel and Joubert Flow. (From Menter, 1991.)

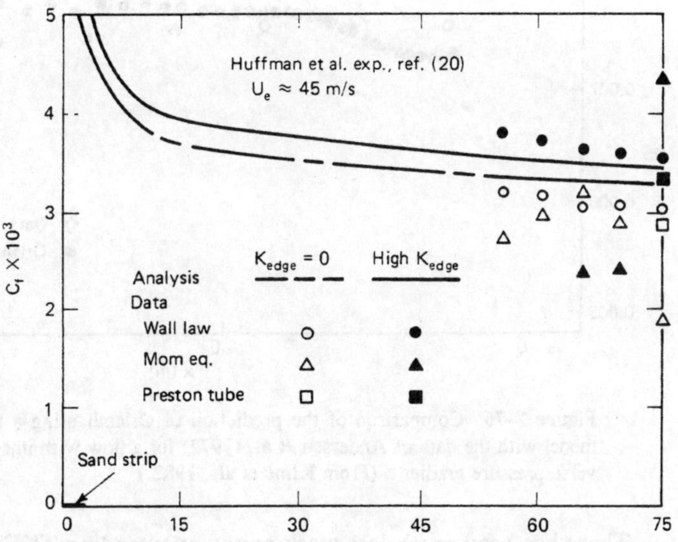

Figure 7–74 Comparison of prediction with the Prandtl energy method and the data of Huffman et al. (1972) for low-speed flow with high free stream turbulence. (From Schetz, Billig and Favin, 1982.)

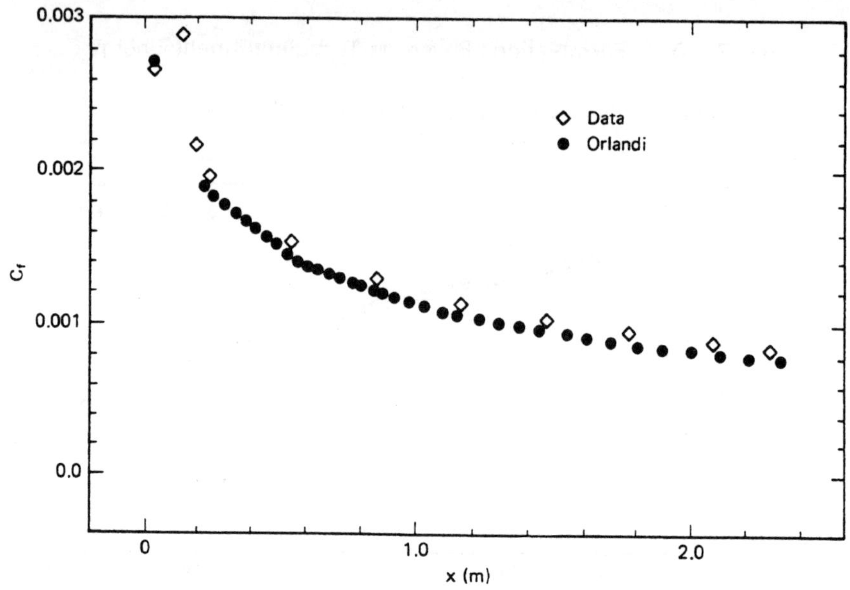

Figure 7–75 Comparison of the prediction of Orlandi using a turbulent kinetic model with the data of Andersen et al. (1972) for a flow with injection. (From Kline et al., 1982.)

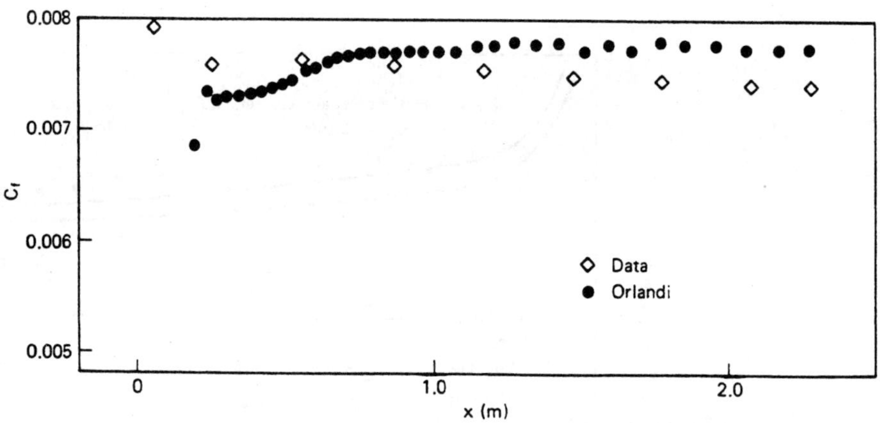

Figure 7–76 Comparison of the prediction of Orlandi using a turbulent kinetic model with the data of Andersen et al. (1972) for a flow with injection and an adverse pressure gradient. (From Kline et al., 1982.)

There has been much less work aimed at extending TKE models to cases with roughness and suction or injection than for the mean flow models. Bradshaw reports such extensions for his model, but no comparisons with data have appeared in the open literature. Orlandi reported calculations for two cases with injection in the 1981 Stanford conference, using his modified Prandtl energy method. The experiments are from the work of Anderson et al. (1972). Results from the first case with no pressure gradient are shown in Fig. 7–75, and results from the second case, which had an adverse pressure gradient, are given in Fig. 7–76. Clearly, the added complication of a pressure gradient degrades the predictions.

7–11 FORMULATIONS BASED ON TURBULENT KINETIC ENERGY AND A LENGTH SCALE

Many of the deficiencies in the performance of the TKE models have been attributed to the manner in which the length scale is modeled. That quantity generally appears directly in the modeled form of the various terms, for example, Eqs. (7–95), (7–96), and (7–98). At the level of the TKE models, the most that can be accommodated is an algebraic formula, usually taken to be $\ell \sim y$ or $\ell \sim \delta$. There are some physical situations in which the simple relation of the length scale to some distance or thickness is not clear. Perhaps the easiest case to understand occurs in a free turbulent flow—the merging of the mixing zones from several parallel, coaxial jets, all exhausting into one cross-sectional plane. Near the point of injection, each jet is independent of the others, and the mixing length can be simply related to the local width of the mixing zone of a single jet. Further downstream, the several mixing zones have grown to the point where they merge, more or less abruptly. The *width* of the mixing zone has now suddenly grown much larger. If there are 50 jets, it becomes 50 times larger. Clearly, the local flow in one of the middle jets cannot respond instantly to the presence of all the other jets in a linear way, as would be implied by taking ℓ to be proportional to the new mixing zone width. We must have something better, and one approach has been to seek an independent equation for ℓ or for a new quantity that is a combination of K and ℓ, say, $Z \equiv K^m \ell^n$. Such an equation can also be found by manipulating the Navier-Stokes equations, and the result, under the same restrictions as before and after some modeling, is

$$\underbrace{\rho\left(U\frac{\partial Z}{\partial x} + V\frac{\partial Z}{\partial y}\right)}_{\text{Convection}} = \underbrace{\frac{\partial}{\partial y}\left(\frac{\rho\sqrt{K}\ell}{\sigma_z}\frac{\partial Z}{\partial y}\right)}_{\text{Diffusion}} + \underbrace{C_{Z1}\frac{\rho Z}{K}\mathcal{P}}_{\text{Production}} - \underbrace{C_{Z2}\frac{\rho Z\sqrt{K}}{\ell}}_{\text{Destruction}} + \underbrace{S_Z}_{\text{Source}} \tag{7–108}$$

Here, σ_Z is a *Prandtl number* for diffusion of Z, S_Z are source terms that appear in some models, C_{Z1} and C_{Z2} are constants, and $\mathcal{P}$ is production of turbulent kinetic energy. This equation is almost identical in form to some of the modeled forms of the TKE equation. One still works with the mean flow equations and the turbulent kinetic energy (TKE) equation, along with the new Z equation, and this approach has been named a *two-equation model*.

The most active proponents of the two-equation models have been Spalding and his co-workers [see Launder and Spalding (1972) and Rodi (1980)]. These researchers have concentrated mainly on the choice $m = 3/2$, $n = -1$, giving $Z = K^{3/2}/\ell$, which is proportional to the dissipation rate $\epsilon = C_D K^{3/2}/\ell$. Also, $\mu_T = C_\mu \rho K^2/\epsilon$. These models are, therefore, often termed *Kε models*. They are not without conceptual problems, however. If one looks closely at the various terms of the equation for Z, one sees that it is really modeling terms that can be described as *dissipation of dissipation*—for example, when $Z = \epsilon$. In spite of such philosophical perplexities, these approaches are aimed at solving a real problem—better and dynamic specification of the length scale—and they do offer some practical advantages.

Each term on the right-hand side of the equation for Z (now ϵ) still has to be modeled. This is accomplished largely by analogy with the modeling of Reynolds stress and the TKE equation. For example, the diffusion of ϵ is written as

$$-\rho\overline{v'\epsilon'} \sim \frac{\mu_T}{\sigma_\epsilon}\frac{\partial \epsilon}{\partial y} \tag{7-109}$$

The value of $C_{\epsilon 2}$ is found by considering the simple case of the decay of turbulence in uniform flow (in the transverse direction) behind a screen. In that situation, one can neglect *production* and *diffusion*, and Eq. (7–99) reduces to

$$\rho U\frac{dK}{dx} = -\rho\epsilon \sim x^{-q} \tag{7-110}$$

where the decay x^{-q} is known from experiment. Equation (7–108), for ϵ, reduces to

$$\rho U\frac{d\epsilon}{dx} = -C_{\epsilon 2}\frac{\rho\epsilon^2}{K} \tag{7-111}$$

Thus,

$$K \propto x^{-1/(C_{\epsilon 2}-1)} \tag{7-112}$$

which implies that $C_{\epsilon 2} \approx 2.0$.

To set the value of $C_{\epsilon 1}$, it is again helpful to consider the logarithmic portion of the wall region, where we have Eq. (7–100), and where

$$\frac{\partial U}{\partial y} = \frac{u_*}{\kappa y} = \frac{\sqrt{\tau/\rho}}{\kappa y} \tag{7-113}$$

which comes from Eq. (7–55) and the assumption $\tau = \tau_w$. In this region, the convection and diffusion of ϵ can be neglected, and Eq. (7–108), for ϵ, reduces to give

$$C_{\epsilon 1} = C_{\epsilon 2} - \frac{\kappa^2}{\sigma_\epsilon C_\mu^{1/2}} \tag{7-114}$$

Thus, for $Z = (K^{3/2}/\ell) \sim \epsilon = C_D K^{3/2}/\ell$, one has $C_{\epsilon 1} \approx 1.45$, $C_{\epsilon 1} \approx 1.92$, $\sigma_\epsilon \approx 1.3$, $C_\mu \approx a_1^2 \approx 0.09$, and $\kappa \approx 0.41$ [see Rodi (1980)].

Ng and Spalding (1970) used a two-equation model with K and $K\ell$ $(m = n = 1)$ as the two variables. They presented calculations for many of the cases from the 1968 Stanford competition and found that the results were almost identical to their earlier results with the mixing length model (Ng et al., 1969), except for strong nonequilibrium pressure gradients. In those cases, the two-equation model perfomed better when compared with experiment.

There have been many studies of and with the $K\epsilon$ model since its formulation. On the negative side, the method seems to be ill posed, and only fine-tuning of the numerous constants involved permits calculations to be made. The analysis of Schneider (1989) proves this conclusively (see also Takemitsu, 1990). On the other hand, many extensions of the method have been proposed that permit the treatment of complex situations. The predictions are not always good, but a more or less unified approach is possible. This unified treatment has made the $K\epsilon$ model popular for general-purpose computer codes. The aero-hydrodynamicist must carefully read any report to discern exactly which $K\epsilon$ model is being used. Some refer to a so-called *standard $K\epsilon$ model* that has the constants just given and the equations

$$U\frac{\partial K}{\partial x} + V\frac{\partial K}{\partial y} = \frac{\partial}{\partial y}\left(\frac{\nu_T}{\sigma_K}\frac{\partial K}{\partial y}\right) + \nu_T\left(\frac{\partial U}{\partial y}\right)^2 - \epsilon \tag{7–115}$$

and

$$U\frac{\partial \epsilon}{\partial x} + V\frac{\partial \epsilon}{\partial y} = \frac{\partial}{\partial y}\left(\frac{\nu_T}{\sigma_\epsilon}\frac{\partial \epsilon}{\partial y}\right) + C_{\epsilon 1}\frac{\epsilon}{K}\nu_T\left(\frac{\partial U}{\partial y}\right)^2 - C_{\epsilon 2}\frac{\epsilon^2}{K} \tag{7–116}$$

In addition, many researchers use an extended version of the *wall function method* (see Sections 7–9 and 7–10) in conjuction with the $K\epsilon$ model. In this case, one needs "boundary" values for K and ϵ at the match point above the physical wall. Since the match point is in the wall region, Eq. (7–97) can be used with $\tau_T \approx \tau_w$ to obtain a value for K at the match point. The corresponding value for ϵ is found by taking *production* $\approx$ *dissipation,* i.e., $\epsilon = u_*^2 \partial U/\partial y$, and using the log law for $\partial U/\partial y$. Thus, the values at the match point become

$$K = \frac{u_*^2}{a_1}; \qquad \epsilon = \frac{u_*^3}{\kappa y} \tag{7–117}$$

If wall functions are not used and one is solving all the way to the wall, some modifications for the low Reynolds numbers near the wall are necessary. The basic work in this area is from Jones and Launder (1972), but there have been numerous other suggestions in the literature. In this situation, laminar terms must be added to the equations for K and ϵ. In particular, $\partial/\partial y(\nu\, \partial K/\partial y)$ must be added to the right-hand side of Eq. (7–115), and $\partial/\partial y(\nu\, \partial\epsilon/\partial y)$ must be added to the right-hand side of Eq. (7–116). In addition, factors f_1 and f_2 are added to the last two terms on the right-hand side of Eq. (7–116). Finally, the eddy viscosity relation is modified with another factor, f_μ, to read $\nu_T = C_\mu f_\mu K^2/\epsilon$. It is these extra factors f_1, f_2, and f_μ that contain the near-wall modifications. The wall boundary conditions are $K = 0$ and $\partial\epsilon/\partial y = 0$.

Some results from the 1981 Stanford conference are included here to show the level of performance that can be expected. Murphy gave results using the Jones and Launder (1973) near-wall modifications without any wall functions, and results for the Samuel and Joubert (1974) experiment are given here in Fig. 7–77. Rodi presented calculations using a later near-wall treatment due to Lam and Bremhorst (1978) without any wall functions, and those results are also plotted in the figure. Generally disappointing results such as these have led to a continual process of developing "new and improved" near-wall treatments for the $K\epsilon$ model. A recent example is the work of Nagano and Tagawa (1990), who developed expressions for f_μ and f_2 based on specified behavior of U, V, K, ϵ, ν_T, and τ_T near the wall. The relations they used were

$$\begin{gathered} f_1 = 1.0, \qquad f_2 = [1 - 0.3\exp(-(R_t/6.5)^2)][1 - \exp(-y^+/6)]^2 \\ f_\mu = [1 - \exp(-y^+/26)]^2[1 + 4.1/R_t^{3/4}] \end{gathered} \tag{7–118}$$

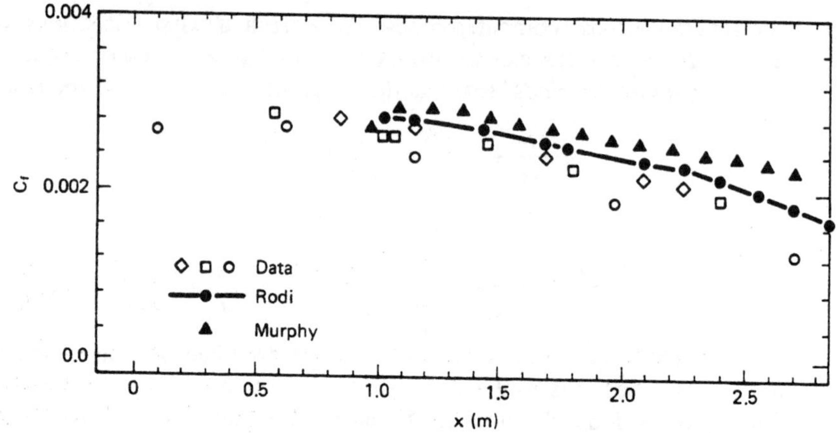

Figure 7–77 Comparison of the predictions of Murphy and Rodi using $K\epsilon$ models with the data of Samuel and Joubert (1974) for an adverse pressure gradient flow. (From Kline et al., 1982.)

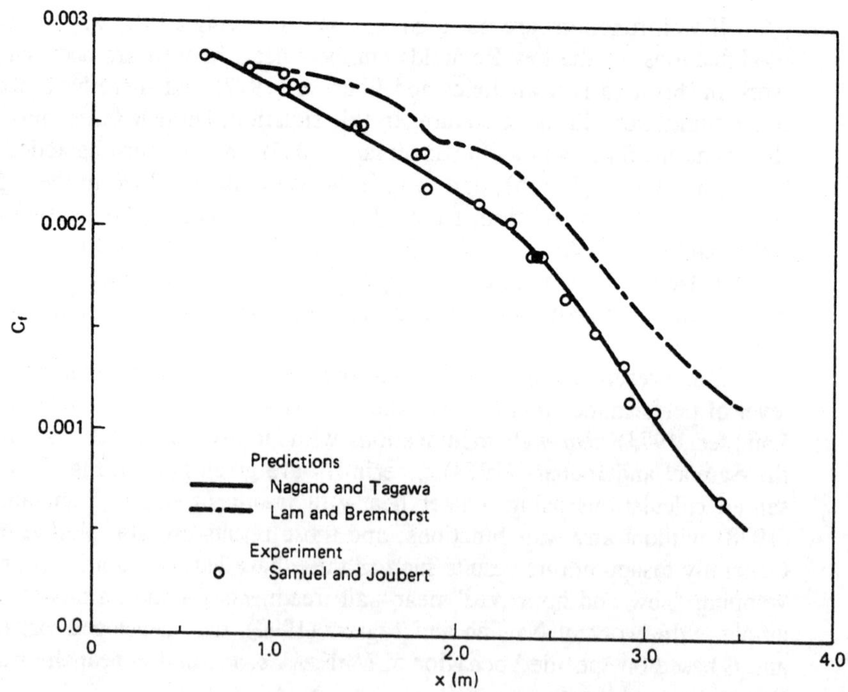

Figure 7–78 Comparison of the prediction of Nagano and Tagawa (1990) using a $K\epsilon$ model with the data of Samuel and Joubert (1974) for an adverse pressure gradient flow. (From Nagano and Tagawa, 1990.)

with $R_t = K^2/\nu\epsilon$, $C_{\epsilon 1} = 1.45$, $C_{\epsilon 2} = 1.90$, $C_\mu = 0.09$, $\sigma_K = 1.4$, and $\sigma_\epsilon = 1.3$. A comparison of their results with those of the Samuel and Joubert (1974) experiment are shown in Fig. 7–78. These predictions are certainly better than those with the Lam and Bremhorst model, but one becomes uneasy with so many adjustable constants and ad hoc modifications that are apparently necessary. Nonetheless, we will see that even more modifications are required for free turbulent flows.

There has been much less work with the $K\epsilon$ model for flows with roughness or suction or injection. For the 1981 Stanford Conference, Rodi produced predictions for the constant-pressure, injection case of Andersen et al. (1972), and a comparison of these predictions with the data is presented in Fig. 7–79. The disappointing results are apparent.

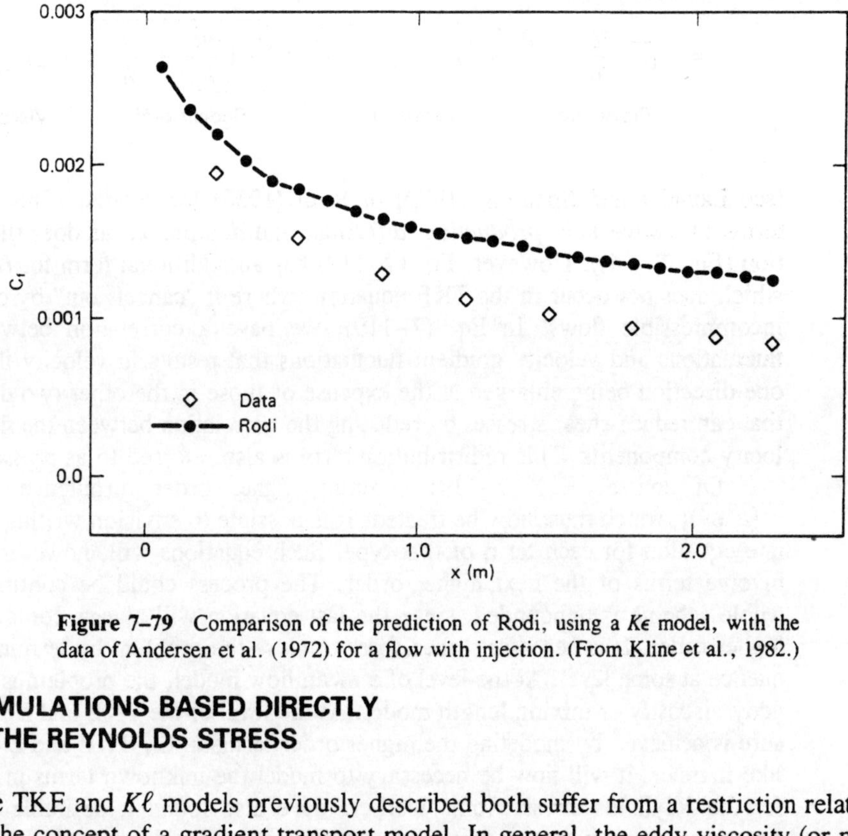

Figure 7–79 Comparison of the prediction of Rodi, using a $K\epsilon$ model, with the data of Andersen et al. (1972) for a flow with injection. (From Kline et al., 1982.)

7–12 FORMULATIONS BASED DIRECTLY ON THE REYNOLDS STRESS

The TKE and $K\ell$ models previously described both suffer from a restriction relating to the concept of a gradient transport model. In general, the eddy viscosity (or mixing length) concept, combined with the gradient transport relationship, imply too direct a connection between the mean flow field and the turbulent stress. This limitation applies as well to models that are more advanced than the mean flow models, even though other facets of the turbulence have been included. One way around the problem is to formulate a transport equation for the Reynolds stress, $-\rho\overline{u'v'}$, itself. One is then directly attacking the term that caused the central problem at the lowest level of analysis, in Eq. (7–34a).

The transport equation for $-\rho\overline{u'v'}$ is obtained by manipulating the Navier-Stokes equations and making the boundary layer assumptions. The time-averaged momentum equation is subtracted from the time-dependent Navier-Stokes equations for both the x- and y-momentum. The resulting equation for the x-component is multiplied by v', and the resulting equation for the y-component is multiplied by u'. Summation of these two equations and averaging over time then yields the desired equation. With our usual restrictions for this volume, the equation for the Reynolds stress is

$$\underbrace{U\frac{\partial\overline{u'v'}}{\partial x} + V\frac{\partial\overline{u'v'}}{\partial y}}_{\text{Convection}}$$

$$= \underbrace{-\overline{v'^2}\frac{\partial U}{\partial y}}_{\text{Production}} \underbrace{- \frac{\partial}{\partial y}\left(\overline{v'u'v'} + \frac{\overline{p'u'}}{\rho}\right)}_{\text{Diffusion}} + \underbrace{\overline{\frac{p'}{\rho}\left(\frac{\partial u'}{\partial y} + \frac{\partial v'}{\partial x}\right)}}_{\text{Redistribution}} \underbrace{- 2\nu\sum\overline{\left(\frac{\partial u'}{\partial x_j}\frac{\partial v'}{\partial x_j}\right)}}_{\text{Viscous Dissipation}} \tag{7-119}$$

[see Launder and Spalding (1972) or Rodi (1980) for details]. This equation has terms for *convection, production, diffusion,* and *dissipation,* as does the TKE equation (Eq. (7–94)). However, Eq. (7–119) has an additional term for *redistribution,* which does not occur in the TKE equation, where it "cancels out" by continuity for incompressible flows. In Eq. (7–119), we have a correlation between pressure fluctuations and velocity gradient fluctuations that results in velocity fluctuations in one direction being enlarged at the expense of those in the other two directions and that can reduce shear stresses by reducing the correlation between the fluctuating velocity components. This redistribution term is also referred to as *pressure strain.*

Of course, Eq. (7–119) contains higher order turbulence terms [e.g., $\overline{v'(u'v')}$], which must now be treated. It is possible to envision writing a new, separate equation for each term of that type. Such equations will, however, necessarily involve terms of the next higher order. The process could be continued, but the whole scheme is unbounded, since the last equation will always contain terms of a higher order. It is necessary, therefore, to *close* the problem by terminating the sequence at some level. At the level of a mean flow model, the problem is closed by an eddy viscosity or mixing length model. At the level of the TKE and $K\ell$ models, closure is achieved by modeling the higher order terms in the TKE and Z equations. In like manner, it will now be necessary to model the unknown terms in Eq. (7–119) for the Reynolds stress. Some workers have also found it desirable to add other equations to the system at this level. Equations for ℓ and/or the normal stresses $\overline{u'^2}$ and so on are common choices.

Although it was not the first model from a chronological point of view, the model of Hanjalic and Launder (1972) follows most simply in logical order from the previous discussion. The *production* term in Eq. (7–119) is modeled by asserting that $\overline{v'^2} \propto K$. The *diffusion* term is modeled through the generalized gradient transport framework, that is,

$$-\frac{\partial}{\partial y}\left(\overline{v'u'v'} + \frac{\overline{p'u'}}{\rho}\right) \sim \frac{\partial}{\partial y}\left(\frac{K^2}{\epsilon}\frac{\partial \overline{u'v'}}{\partial y}\right) \sim \frac{\partial}{\partial y}\left(\sqrt{K}\ell\frac{\partial \overline{u'v'}}{\partial y}\right) \tag{7–120}$$

Note that here a turbulent viscosity ($\nu_T \sim K^{1/2}\ell$) has crept back into the formulation. The *dissipation* term is neglected with respect to the other terms. Finally, the *redistribution* term is modeled via two processes: one with turbulent interactions alone, $(\epsilon/K)\overline{u'v'}$, and one that arises via the mean flow gradient, $\sim K\,\partial U/\partial y$. The forms of the equation for the Reynolds stress and those for K and ϵ become

$$U\frac{\partial \overline{u'v'}}{\partial x} + V\frac{\partial \overline{u'v'}}{\partial y} = C_s\frac{\partial}{\partial y}\left(\frac{K^2}{\epsilon}\frac{\partial \overline{u'v'}}{\partial y}\right) - C_{\phi 1}\left(C_\mu K\frac{\partial U}{\partial y} + \frac{\epsilon}{K}(\overline{u'v'})\right) \tag{7–121a}$$

$$U\frac{\partial K}{\partial x} + V\frac{\partial K}{\partial y} = 0.9C_s\frac{\partial}{\partial y}\left(\frac{K^2}{\epsilon}\frac{\partial K}{\partial y}\right) - \overline{u'v'}\frac{\partial U}{\partial y} - \epsilon \tag{7–121b}$$

$$U\frac{\partial \epsilon}{\partial x} + V\frac{\partial \epsilon}{\partial y} = C_\epsilon\frac{\partial}{\partial y}\left(\frac{K^2}{\epsilon}\frac{\partial \epsilon}{\partial y}\right) - C_{\epsilon 1}\overline{u'v'}\frac{\epsilon}{K}\frac{\partial U}{\partial y} - C_{\epsilon 2}\frac{\epsilon^2}{K} \tag{7–121c}$$

with $C_s \equiv 0.10$, $C_\epsilon = 0.07$, and $C_{\phi 1} \approx 2.8$.

Note that in the wall region, where *convection* and *diffusion* may be neglected, compared with the other terms, Eq. (7–121a) collapses to

$$0 = -C_{\phi 1}\left(C_\mu K\frac{\partial U}{\partial y} + \frac{\epsilon}{K}\overline{u'v'}\right) \tag{7–121d}$$

or

$$\sqrt{K}\ell\frac{\partial U}{\partial y} \sim -\overline{u'v'} \tag{7–121e}$$

which is like the *Prandtl energy method* and which itself was shown (see Section 7–10) to be equivalent to the mixing length model. Thus, we should not expect any large improvement in the predictions from this model for wall-bounded flows of the ordinary type.

The model of Hanjalic and Launder has been extended and refined by Launder et al. (1975), although, for boundary layer flows of the type under discussion here, the basic model will serve.

At a slightly higher level of mathematical complexity, Donaldson (1971) chose an equation for the Reynolds stress, an algebraic relation for ℓ, and separate equations for the normal stresses. Since each normal stress is found separately, an equation for one-half their sum (i.e., K) is not needed. It is worth noting that the concept of a turbulent viscosity is not invoked in any form. The length scale is, however, not determined from a differential equation. Some comparisons of predictions by Rubesin et al. (1977), based on the Donaldson model and experimental results for a flat boundary layer, are shown in Fig. 7–80. The mean flow, C_f, $\overline{u'v'}$, and $\overline{v'^2}$ are well predicted, but $\overline{u'^2}$, and $\overline{w'^2}$ are poorly predicted. The fact that the mean flow and the Reynolds stress can be accurately predicted with a formulation that gives poor predictions for $\overline{u'^2}$, and $\overline{w'^2}$ can be interpreted to indicate that it is perhaps not worthwhile to solve separate equations for $\overline{u'^2}$, $\overline{v'^2}$, and $\overline{w'^2}$.

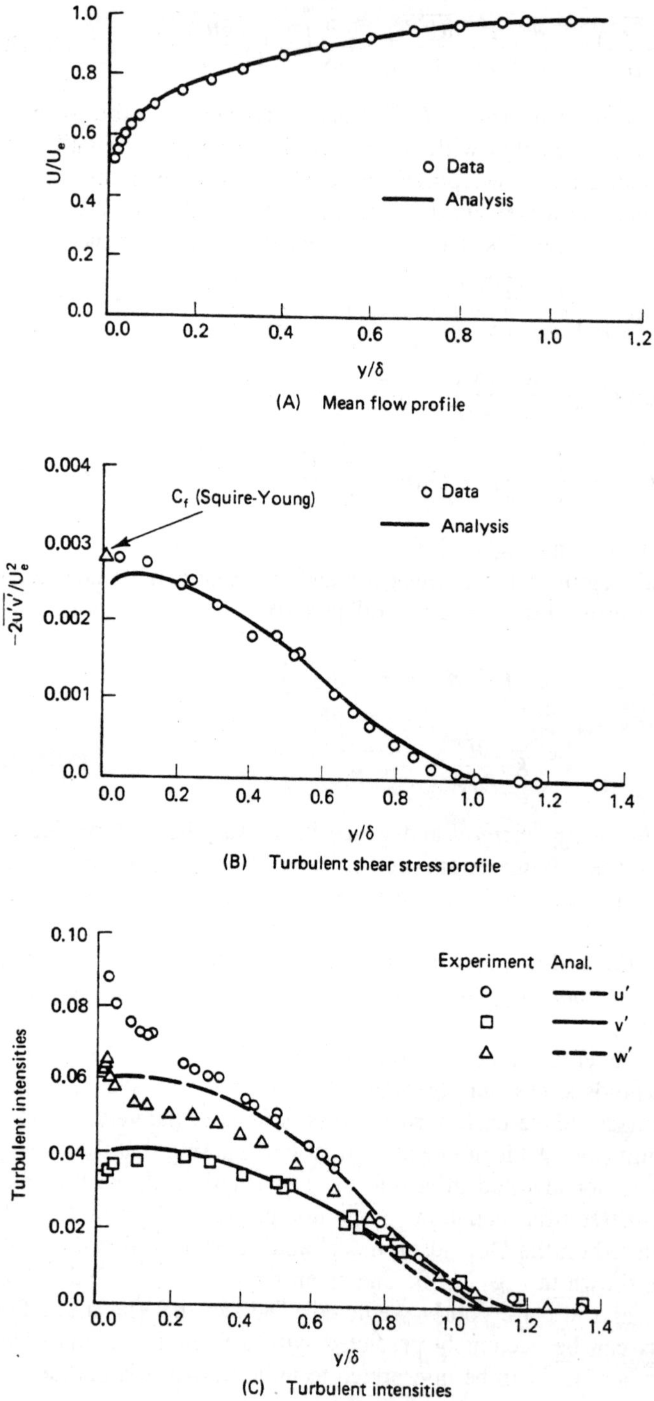

Figure 7–80 Comparison of predictions from the Donaldson Reynolds stress model by Rubesin et al. (1977) with the flat-plate data of Klebanoff (1955).

The model of Daly and Harlow (1970) employs differential equations for the Reynolds stress, each of the normal stresses, and ℓ (via ϵ), so this model entails the solution of five equations for turbulent quantities, in addition to two (or more) mean flow equations. The modeling of the terms in the equations for the Reynolds and normal stresses is similar to that used by Hanjalic and Launder. The Daly and Harlow model is little used at the present time, since it is complex and has been found to produce comparatively poor predictions.

A treatment based on, for example, the Hanjalic and Launder model involves partial differential equations for K, ϵ, and $-\overline{u'v'}$, and the computational load becomes burdensome. One is, therefore, tempted to look for simplifications. Gradients of the dependent variables occur only in the *convection* and *diffusion* terms; if these terms can be modeled or even eliminated, the differential equations will reduce to algebraic equations. This kind of formulation is called an *algebraic stress model.* The simplest such model, which is adequate for most purposes, just neglects all the differential terms. A more general model can be found in Rodi (1976). The simple model gives

$$-\overline{u'v'} = \left[\frac{2}{3}\frac{1-\gamma}{C_1}\frac{C_1 - 1 + \gamma\mathcal{P}/\epsilon}{(C_1 - 1 + \mathcal{P}/\epsilon)^2}\right]\frac{K^2}{\epsilon}\frac{\partial U}{\partial y} \tag{7–122}$$

Here, $\mathcal{P}$ is the stress production of K (see Eq. (7–94)), $C_1 = 1.5$, and $\gamma = 0.6$. Looking back at Section 7–11, we can see that Eq. (7–122) amounts to an eddy viscosity model, with the bracketed term playing the role of a varying C_μ. Flows in local equilibrium have $\mathcal{P}/\epsilon = 1$, which leads to a constant value for C_μ. Thus, an algebraic stress model should be more capable than a simple eddy viscosity model would be of representing flows that are out of equilibrium.

As with all turbulence models, the issue of the treatment of the near-wall region must be discussed for Reynolds stress models of any kind. The wall function method has been extended to Reynolds stress models, and the value of $-\overline{u'v'}/u_*^2$ at the match point in the inner region is obtained from measurements such as those in Fig. 7–21 [Launder et al. (1975)]. If the solution is carried all the way to the wall, the boundary condition on $-\overline{u'v'}$ is zero. Also, low Reynolds number modifications similar to those for the $K\epsilon$ model must be introduced. The suggestions of Chieng and Launder (1980) are widely used.

The 1968 and 1981 Stanford conferences had far fewer calculations with Reynolds stress models than with the less elaborate models. For the 1981 conference, Hanjalic presented calculations with his model using wall functions, and Donaldson presented calculations with his model without wall functions. The predictions are compared with the data from the Samuel and Joubert (1974) experiment in Fig. 7–81. Good agreement was achieved. The same two groups presented calculations for the injection experiment of Andersen et al. (1972), and predictions of these calculations are compared with Andersen's data in Fig. 7–82. Again, good agreement was achieved. No calculations for either of these two cases with an algebraic stress model were presented.

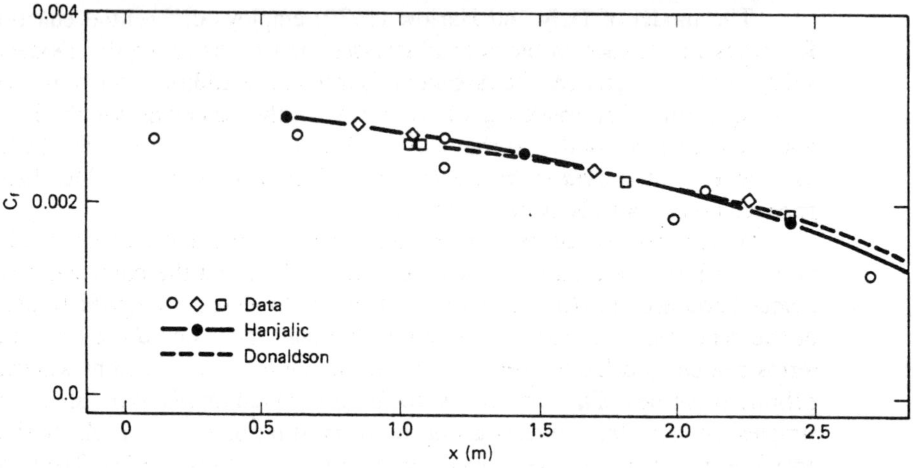

Figure 7–81 Comparison of the predictions of Donaldson and Hanjalic, using Reynolds stress models, with the data of Samuel and Joubert (1974) for an adverse pressure gradient flow. (From Kline et al., 1982.)

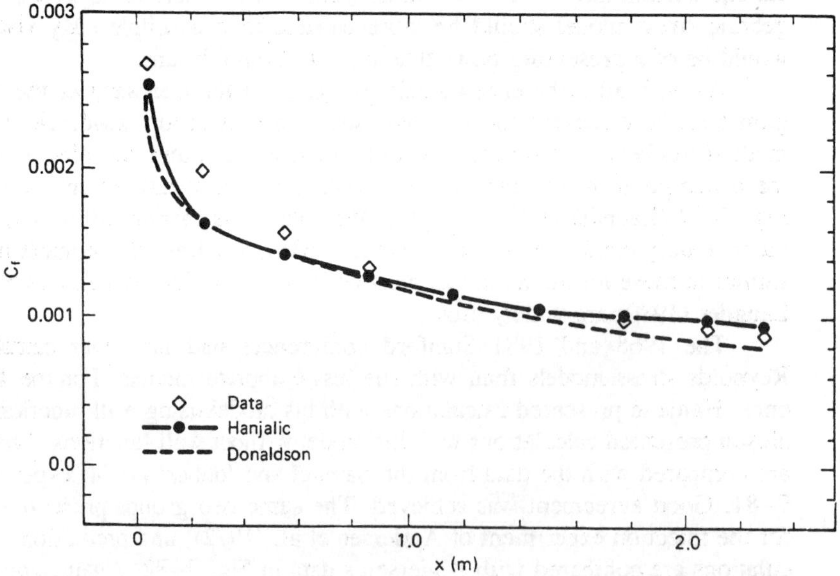

Figure 7–82 Comparison of the predictions of Hanjalic and Donaldson, using Reynolds stress models, with the data of Andersen et al. (1972) for a flow with injection. (From Kline et al., 1982.)

7–13 DIRECT NUMERICAL SIMULATIONS

Up to this point, we have not made any direct connection with the large body of knowledge that has developed on the detailed instantaneous structure of turbulent shear flows. Only recently have even limited connections between research into turbulent structure and the needs of the engineer been made. Here, we shall call such models *direct numerical simulations*. The basic problem with attempting to treat the fluctuating turbulent flow directly is that these flows are characterized by a range of excited scales of motion over several orders of magnitude, as discussed earlier in this chapter (see Fig. 7–22). Even the most optimistic projections for the capacity of future computing machines fall far short of that required to meet this need. Only at low-Reynolds-number conditions does *all scale resolution* of the turbulence seem at all likely in the foreseeable future.

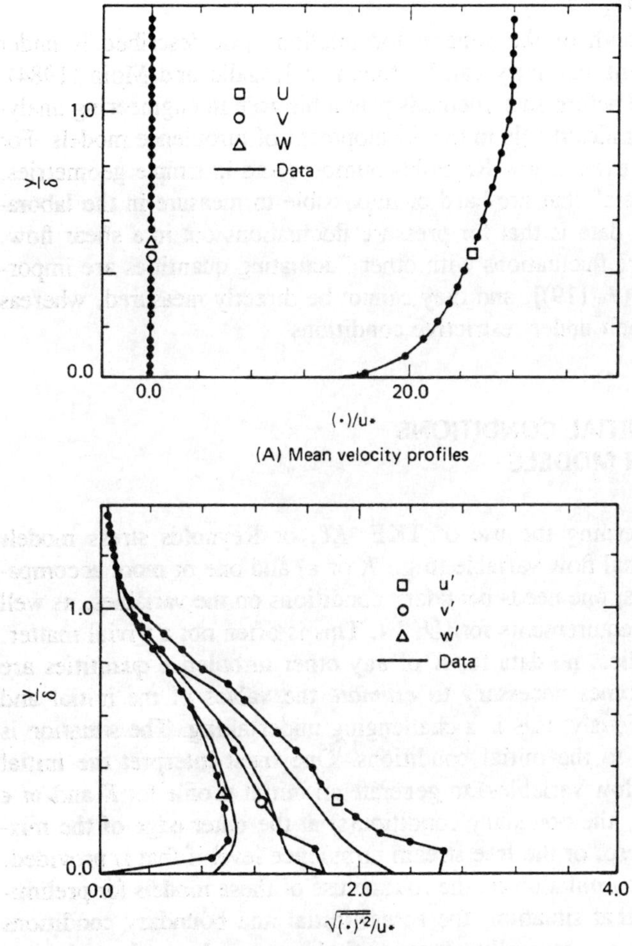

Figure 7–83 Comparison of predictions from a large eddy simulation with experiment for turbulent boundary layer on a flat plate. (From Schmitt et al., 1985.)

One appealing method that has been proposed for alleviating this problem is to limit the attempt at direct, three-dimensional, unsteady treatment of the turbulence to only those scales above a certain size. All the scales that cannot be resolved are then modeled as *subgrid turbulence,* using an eddy viscosity or other transport approximation. The hope underlying this approach is that the smaller scale structure of the turbulence is nearly universal, so that accurate resolution is not required. The large eddy structure that is presumed to contain the part of the turbulence that changes markedly from flow to flow or condition to condition is treated directly. These models are called *large eddy simulations*. Here also, however, the treatment of flows of engineering interest is a ways in the future, due to computer cost and size limitations. Finally, it appears that the subgrid modeling necessary to treat the turbulence has a surprisingly large effect on the results, and predictions superior to other, simpler turbulence models are not easily achieved. Some typical results from a large eddy simulation of turbulent boundary layer flow over a flat plate are compared with experiment in Fig. 7–83.

Important work in both of the general formulations just described is under way, and a relatively recent summary can be found in Rogallo and Moin (1984). While it will be some time before such methods play a big role in engineering analysis, they already play a significant role in the development of turbulence models. For example, full simulations, even at low Reynolds numbers and in simple geometries, can provide turbulence "data" that are hard or impossible to measure in the laboratory. A clear case of such data is that for pressure fluctuations out in a shear flow. Correlations of the pressure fluctuations with other fluctuating quantities are important [see Eqs. (7–94) and (7–119)], and they cannot be directly measured, whereas they can be calculated, albeit under restrictive conditions.

7–14 BOUNDARY AND INITIAL CONDITIONS FOR HIGHER ORDER MODELS

An important point concerning the use of TKE, $K\ell$, or Reynolds stress models emerges. With the additional flow variable (e.g., K or ϵ) and one or more accompanying differential equations, one needs boundary conditions on the variables, as well as dealing with the usual requirements for (U, V). This is often not a trivial matter. In some experimental studies, no data for K or any other turbulence quantities are reported. It therefore becomes necessary to *estimate* the values of the initial and boundary conditions. Obviously, this is a challenging undertaking. The situation is most difficult with regard to the initial conditions. One must interpret the initial profiles for U (i.e., mean flow variables) to generate an initial profile for K and/or ϵ and/or $-\rho\overline{u'v'}$. Generally, the boundary condition(s) at the outer edge of the mixing region is taken to be zero, or the free stream turbulence level if that is provided. This matter is an important limitation on the routine use of these models for preliminary design estimates. In that situation, the actual initial and boundary conditions for the mean flow variables are generally not precisely known and must be estimated themselves. Carrying the process further to produce an accurate initial profile for K, for example, is difficult.

Despite these admonitions, the situation in practice is not as severe as one might fear. Since the equations for K and ϵ are always differential equations, and since the equation for the Reynolds stress usually is also (unless an algebraic stress model is used), the system is *dynamic*, and it adjusts to poor estimates for initial conditions. Thus, if K is estimated, for example, too high, the equations quickly reduce the level to a more appropriate value by high initial dissipation.

7-15 NON-NEWTONIAN TURBULENT FLOWS

Most non-Newtonian fluids of practical interest have such a high effective viscosity that they are usually in a laminar state for conditions of interest. One important exception is the addition of a small amount of a polymer that is a long-chained molecule to a water flow under conditions of high Reynolds number. The amounts of the additive are so small that the viscosity of the fluid is about the same as pure water, but a substantial reduction in drag can occur. Much of the work in this area has been for flow in pipes, so we will reserve the details for discussion in a later chapter on *internal flows*.

Skelland (1966) has extended the simple integral analysis for turbulent flow over a flat plate presented in Section 7–7 to non-Newtonian fluids of the power-law type. The "one-seventh expression" for the velocity profile (see Eq. (7–41)) is extended to

$$\frac{U}{U_e} = \left(\frac{y}{\delta}\right)^{\beta p/(2-\beta(2-p))} \tag{7-123}$$

Here, $\beta(p)$ is a function of the power-law exponent p, given as a graph. It is actually the exponent on the effective Reynolds number in a skin friction law corresponding to Eq. (7–13) for Newtonian fluids. For the Newtonian case, $p = 1$ and $\beta = 0.25$, so the exponent in Eq. (7–123) becomes $1/7$. Using an extension of Eq. (7–13) for power-law fluids, and proceeding as in Section 7–7, gives the following result for the growth of the boundary layer along the plate:

$$\frac{\delta}{x} = \left[\frac{(\beta p + 1)\phi_1(\alpha, \beta)}{\phi_2(\beta)}\left(8^{p-1}\left(\frac{3p+1}{4p}\right)^p\right)^\beta \left(\frac{\rho U_e^{2-p} x^p}{\mu_{PL}}\right)^{-\beta}\right]^{1/(\beta p+1)}$$

$$\phi_1(\alpha, \beta) = \frac{\alpha(0.817)^{2-\beta(2-p)}}{2^{\beta p+1}}$$

$$\phi_2(\beta) = \frac{2-\beta(2-p)}{2(1-\beta+\beta p)} - \frac{2-\beta(2-p)}{2-2\beta+3\beta p} \tag{7-124}$$

In this equation, $\alpha(p)$ is the coefficient in the skin friction law corresponding to Eq. (7–13), which is also provided as a graph. For Newtonian fluids, Eq. (7–124) reduces to Eq. (7–43).

There have been some efforts to apply more elaborate turbulence models to non-Newtonian flows. For example, simple extensions of the $K\epsilon$ model have been proposed, but the data base upon which such models can be constructed and against which they can be tested is very sparse.

PROBLEMS

7.1. Air at standard temperature and pressure (STP) is flowing over a flat plate at 30 m/s. What is the value of C_f at $x = 1$ m? Calculate u_*.

7.2. For the flow in Problem 7.1, what is the thickness of the laminar sublayer? Estimate the distance to the inner and outer edges of the logarithmic region.

7.3. CO_2 at STP flows over a flat plate at 10 m/s. What is the value of C_f for a smooth plate at $x = 1.0$ m? What is the *allowable* roughness size for *smooth*? What roughness size corresponds to *fully* rough?

7.4. Air at standard temperature and pressure is flowing over a flat plate at 5.0 m/s. What is the value of the skin friction coefficient at a distance 30 cm from the leading edge? What about 500 cm? What would happen if the pressure were lowered to 0.05 atm?

7.5. A flat plate is immersed in a 100-m/s air stream (at STP). Plot the Reynolds stress versus y at a distance 1 m from the leading edge.

7.6. To test your knowledge of the first part of Chapter 7, try, *without looking back at the text*, to state the restrictions and limits on regions of applicability, if any, on the following items:

(a) $u^+ = g(y^+)$

(b) $u(x, y, z, t) = U(x, y) + u'(x, y, z, t)$

(c) $\ell_m = \kappa y$

(d) $(U - U_e)/u_* = f(y/\delta)$

(e) $\tau_T = -\rho \overline{u'v'}$

(f) $U/u_* = A \log(yu_*/\nu) + C$

(g) $\Delta U/u_* = F(k^+)$

(h) $\mu_T = 0.018\rho U_e \delta^*$

7.7. Water at 25°C flows over a flat plate at 5 m/s. At 1.0 m from the leading edge, what are the values of δ, δ^*, θ, C_f, and u^*? What is the thickness of the laminar sublayer? What is the value of the velocity at $y^+ = 1.0$, 10.0, and 100.0?

7.8. For the conditions of Problem 7.7, what is the influence on C_f of uniform sand roughness of average height 0.5 mm? You may assume that the boundary layer thickness is changed only negligibly. Also, what is the velocity at $y^+ = 100.0$ now?

7.9. A useful approximation for flat-plate turbulent boundary layers is

$$\frac{U}{U_e} = \left(\frac{y}{\delta}\right)^{1/n}$$

where, for low Re_x, $n \approx 6$, and for very high Re_x, $n \approx 8$. How would the predicted growth rate of the boundary layer thickness be influenced? How do the shapes of the profile differ?

7.10. Consider two airflows over solid surfaces for which the boundary layer thickness δ is the same and U_e is the same, but one is laminar and one is turbulent. (*Note:* This could happen for a Reynolds number near transition and two different levels of background disturbance.) What is the ratio of the shear at $y = \delta/2$ in the two cases? You may neglect pressure gradient effects.

7.11. Suppose we have turbulent flow of a fluid with $\nu = 20 \times 10^{-5}$ ft²/s over a surface. For the first 10.0 ft of the surface, the inviscid velocity is constant at 200.0 ft/s. Starting at $x = 10.0$ ft, the inviscid velocity varies as $U_e(x) = 300 - x^2$ ft/s. Using the Moses integral method, calculate the boundary layer development to $x = 15.0$ ft.

7.12. Apply the Moses integral method to the flow of Problem 2.3 for water with $L = 5.0$ ft and $C = 10.0$ ft/s. Calculate up to $x = 4.0$ ft, and start with the laminar solution at $x = 0.5$ ft.

7.13. Use the Moses integral method to calculate the turbulent portion of the flow described in Problem 2.14. Use the Thwaites-Walz method for the laminar portion and the Michel method to predict the location of transition. Repeat the calculation for $U_\infty =$ 350 ft/s.

7.14. Consider the flow of water over a flat plate at 8 m/s. At a location 1 m from the leading edge, what are δ, δ^*, θ, C_f, and u_*? What is the thickness of the laminar sublayer? What is the eddy viscosity at a distance $\delta/2$ above the plate? What is the mixing length at the same point?

7.15. Integrate Eq. (7–60) with $\kappa = 0.41$ and $24 \le A^+ \le 28$, and determine the value of A^+ that gives a good match to the data in the logarithmic region of Fig. 7–2. Van Driest obtained $A^+ = 26$ with $\kappa = 0.40$.

7.16. Von Karman made a little-used suggestion for the mixing length:

$$\ell_m = \kappa \frac{\partial U/\partial y}{|\partial^2 U/\partial y^2|}$$

What would this equation yield in the logarithmic region? If the outer region profile is described as

$$U/U_e = (y/\delta)^{1/7}$$

what would the von Karman expression yield in that region? Compare these results with the usual results.

7.17. Suppose water flows over a surface at a location where $\delta = 2$ cm and the edge velocity is 10 m/s. The local skin friction coefficient is $C_f = 0.002$. What is the value of the eddy viscosity and the mixing length at $y = \delta/1{,}000$? At $\delta/20$? What is the turbulent shear at the same points? What are the eddy viscosity and the turbulent shear at $y = \delta/2$?

7.18. Show that, for the Reichardt model, Eq. (7–63), the eddy viscosity behaves as y^3 near the wall, $y = 0$.

7.19. Show that for the Van Driest model, Eq. (7–62), the eddy viscosity behaves as y^4 near the wall, $y = 0$.

7.20. Assume an alternative universe in which there is no law of the wall and turbulent velocity profiles are well represented by

$$U/U_e = (y/\delta)^{1/7}$$

everywhere except at $y = 0$. The skin friction coefficient is $C_f = 0.002$. Lastly, turbulent exchange processes are everywhere greater than laminar processes. Derive an eddy viscosity model, a mixing length model, and a Prandtl energy method for $0 < y/\delta \leq 0.1$.

7.21. A small research and development company has designed an efficient undersea remotely piloted vehicle that travels at 10.0 knots with very thin, flat laminar flow wings of 1.5-ft chord and 4.0-ft span, assuming "flight" in a uniform, quiet ocean. If the vehicle encounters unstable, turbulent water, how much will the friction drag increase?

7.22. On a strange planet in a far galaxy, turbulent flow behaves somewhat differently than here on earth. The inhabitants of the planet find a laminar sublayer, but in about the same region as our log law, they find that

$$u^+ = 4.9 + 1.1\sqrt{y^+}$$

Derive an eddy viscosity model and a mixing length model for that region. Compare your results with those here on earth.

7.23. Test the sensitivity of numerical predictions of turbulent boundary layer development to the values of the required modeling constants. Use $0.40 \leq \kappa \leq 0.43$ (the range of values quoted in the literature) in the mixing length model, and calculate the boundary layer on a flat plate for the case in Example 7-2. Compare the skin friction coefficients.

7.24. Test the sensitivity of numerical predictions of turbulent boundary layer development to the values of the required modeling constants. Use 0.018 (the original Clauser value) and 0.0168 (the value suggested by Cebeci and Smith) in the Clauser eddy viscosity model, and calculate the boundary layer on a flat plate for the case in Example 7-2. Compare the velocity profiles in the outer region of the boundary layer at the end of the plate. How do the skin friction coefficients compare with each other?

7.25. Test the sensitivity of numerical predictions of turbulent boundary layer development to the values of the required modeling constants. Use $0.08 \leq C_D \leq 0.10$ (the range of values quoted in the literature) in the Prandtl energy method, and calculate the boundary layer on a flat plate for the case in Example 7-2. Compare the boundary layer thicknesses and the skin friction coefficients.

7.26. Suppose we have turbulent flow of a fluid with $\nu = 10^{-5}$ m^2/s over a surface. For the first 5.0 m of the surface, the inviscid velocity is constant at 10.0 m/s. Starting at $x = 5.0$ m, the inviscid velocity varies as $U_e(x) = 15.0 - x$ m/s. Using the mixing length model and a finite difference method, calculate the boundary layer development to $x = 7.0$ ft. Compare the results with those from the Moses integral method.

7.27. Assume that the flow on the side of the hull of a slender oceangoing vessel can be treated locally as that over a flat plate. We wish to consider two stations along the hull—10.0 ft and 100.0 ft from the bow—as the ship is moving at 3.0 knots. What are the values of the eddy viscosity and mixing length at $\delta/2$ at each station? Can you *infer* the value of K? What about ϵ?

7.28. Suppose we have turbulent flow of a fluid with $\nu = 10^{-5}$ over a surface. For the first 5.0 m of the surface, the inviscid velocity is constant at 10.0 m/s. Starting at $x =$ 5.0 m, the inviscid velocity varies as $U_e(x) = 15.0 - x$ m/s. Using the eddy viscosity model and a finite difference method, calculate the boundary layer development to $x = 7.0$ m. Compare the results with those from the Moses integral method.

7.29. Water at 25°C is flowing over a flat plate at 5 m/s. What is the value of the TKE at $x = 0.5$ m and $y = \delta/2$? What is the value of the length scale ℓ? What is the value of the eddy viscosity?

7.30. Suppose we have turbulent flow of a fluid with $\nu = 10^{-5}$ m²/s over a surface. For the first 5.0 m of the surface, the inviscid velocity is constant at 10.0 m/s. Starting at $x = 5.0$ m, the inviscid velocity varies as $U_e(x) = 15.0 - x$ m/s. Using the Prandtl energy method and a finite difference method, calculate the boundary layer development to $x = 7.0$ m. Compare the results with those from the Moses integral method.

7.31. The effect of "stretching" the grid in the y-direction in numerical solutions of turbulent boundary layer problems can be very important for computer time and cost, but one still needs numerous points. Convince yourself of this fact by considering a trial problem of the flow over a flat plate to $Re_L = 10^7$. Use the code ITBLS in Appendix B with 30–95 points across the region. Plot C_f and δ versus $y^+(1)$.

7.32. Calculate the turbulent portion of the flow in the second part of Prob. 7.13 ($U_\infty = 350$ ft/s.) using a numerical method with an eddy viscosity model. Use the Thwaites-Walz method for the laminar portion and the Michel method to find the location of transition. Construct an initial profile for $u(x_t, y)$ using Eq. (2–23) with (2–27) and relating λ to Λ using the information in Table 2–1.

7.33. Use the Prandtl energy method to calculate the influence of free stream turbulence levels of 5, 10, and 20% on the skin friction on a flat plate for a case as in Example 7–2. Present the results as $C_f/(C_f)_{K_e=0}$.

CHAPTER 8

INTERNAL FLOWS

8–1 INTRODUCTION

In the preceding chapters in this book, and also in later chapters, the emphasis is on *external flows*—that is flows confined by at most one rigid lateral boundary. In this chapter, the flows of interest are confined by rigid boundaries on all sides, and those boundaries are close enough to the viscous zone to constrain the flow in important ways. Ackeret (1967) describes four special features of these *internal flows*:

1. *The continuity equation plays a predominant role*. The flow in a duct of any cross-sectional shape is modified by the growing boundary layer displacement thickness on the walls, which effectively narrows the channel and alters the pressure distribution and any inviscid flow in the central core. This interaction between the inviscid and viscous flows leads to an integral equation that must be solved. In long ducts, there is no clear distinction between the boundary layer and the central core flow. All of these effects are more important for compressible flows.
2. Internal flows routinely have *large deflections*, such as in pipe bends, whereas large changes in angle are generally avoided in external flows. Thus, strong *secondary flows*, centrifugal forces, and even other forces due to turning can be encountered much more frequently in internal than in external flows. Sec-

ondary flows are the flow components in a cross plane of a duct, such as in a pipe bend or the corner of a rectangular straight duct.

3. In internal flows, *separation* is usually followed quickly by *reattachment*, since the duct walls prevent the continued expansion of the free separation streamlines and wakes, unless the cross section of the duct is changing very rapidly. An analytical treatment of these phenomena is beyond the scope of this volume, but the practicing engineer must be aware of these effects.
4. *Three-dimensional boundary layers* occur in internal flows, as well as in external flows (see Chapter 11 for a detailed discussion), but in internal flows, they often occur on *rotating surfaces*, such as disks and blade rotors.

One could add at least two other important characteristics of internal flows: First, great complexity in the flow patterns is routine; second, the occurrence of internal flows in practice is pervasive. For instance, every home in the industrialized world contains a myriad of examples of internal flows, while it may or may not have examples of important external flow devices. This makes the subject of internal flows a very old one. The sketch in Fig. 8–1 from Leonardo da Vinci is a quite accurate representation of the high-Reynolds-number flow in a simple duct tee, showing separation at the corner, reattachment in the lower branch, and confinement of the eddies by the walls in the right-hand branch.

The coverage of this broad and complex subject area in this book follows that adopted for other subject areas—laminar flow, transition, and then turbulent flow, with the discussion in each moving from simple to more complex cases. However, since the emphasis here is on *boundary layer analysis*, some subjects that have practical importance in the general field of internal flow but that defy current methods of analysis, such as the flow in valves and fittings, will be omitted. The engineer must rely on purely empirical information about the flow in such devices. Finally, some coverage of the internal flow of non-Newtonian fluids is included.

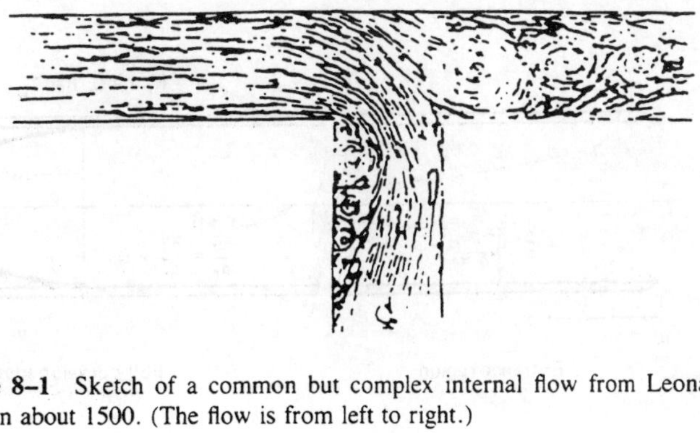

Figure 8–1 Sketch of a common but complex internal flow from Leonardo da Vinci in about 1500. (The flow is from left to right.)

8–2 LAMINAR INTERNAL FLOWS

8–2–1 Fully Developed Flows

As stated, we begin with a simple, but useful, situation. The flow of a viscous fluid in a round tube or pipe is obviously of great engineering interest. The item of primary practical concern is the loss in pressure head, or *pressure drop,* along the pipe, since that must be made up by pumping. It happens that this flow problem is a member of the group of flows known as *parallel flows* (see Section 4–2), for which an exact solution of the equations of motion is rather easy to obtain.

Most applications of pipe flows involve cases in which the length of the pipe is measured in many, many diameters. In that instance, the details of the flow near the beginning of the pipe are not important. If the entrance to the pipe had sharp lips, a boundary layer similar to that on a flat plate would initially form. However, as the boundary layer grew along the wall, it would interact with the inner inviscid flow and then with that on the opposite wall until, finally, the flow would be fully viscous. After the bounday layer fills the pipe, it can no longer grow, and an equilibrium state is reached in which changes in the boundary layer profile along the pipe cease. This condition is called *fully developed* flow.

The flow problem is shown schematically in Fig. 8–2, and we will analyze the velocity field first. The appropriate equations are the continuity equation, Eq. (3–7a), and the momentum equation, Eq. (3–25), less the unsteady term and assuming constant properties. The fully developed flow condition is stated mathematically as $u = u(r) \neq f(x)$. Substituting $\partial u/\partial x = 0$ into the continuity equation and integrating, one sees that $v \equiv 0$ satisfies the equation and the boundary condition, $v(x, R) = 0$. Using $v = 0$ and the previously stated condition on $u(r)$ in the momentum equation produces

$$0 = -\frac{1}{\rho}\frac{dp}{dx} + \frac{\nu}{r}\frac{d}{dr}\left(r\frac{du}{dr}\right) \tag{8-1}$$

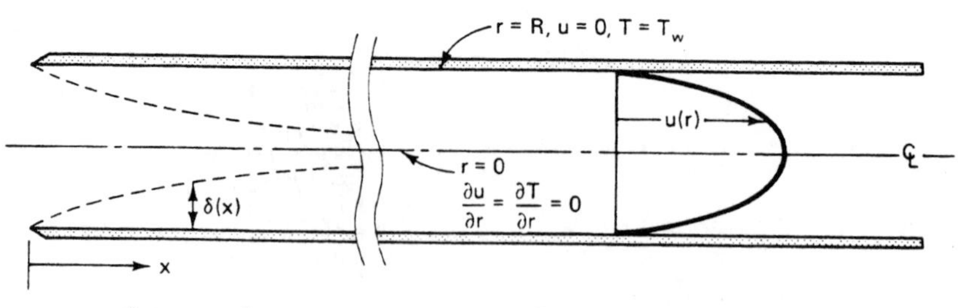

Figure 8–2 Schematic of the flow in a pipe.

or

$$\frac{\nu}{r}\frac{d}{dr}\left(r\frac{du}{dr}\right) = \frac{1}{\rho}\frac{dp}{dx} = \text{constant} \tag{8–2}$$

This result can be seen from the fact that $u \neq f(x)$ and $p \neq g(r)$. Thus, the equations have been rigorously reduced to a single *linear* ordinary differential equation for $u(r)$. This equation can easily be integrated twice to give

$$u(r) = -\frac{1}{4\mu}\frac{dp}{dx}(R^2 - r^2) \tag{8–3}$$

where the boundary conditions are no slip on the wall and symmetry on the axis, that is,

$$u(x, R) = 0, \qquad \frac{du}{dr}(x, 0) = 0 \tag{8–4}$$

The solution is written with the minus sign in front because there will be a pressure loss along the tube (i.e., $dp/dx < 0$). The velocity profile described by Eq. (8–3) is a paraboloid, and the geometry of that shape leads to a value of the average velocity across the pipe of

$$u_{\text{ave}} = \left(-\frac{dp}{dx}\right)\frac{R^2}{8\mu} = \frac{u_c}{2} \tag{8–5}$$

where u_c is the centerline velocity.

The *pressure drop* is caused by the viscous shear on the pipe wall, and that relation can be developed using Eq. (8–3). We obtain

$$\tau_w = -\mu \frac{du}{dr}\bigg|_{r=R} = -\frac{R}{2}\frac{dp}{dx} \tag{8–6}$$

The *pressure drop per unit length* along the pipe is usually expressed in terms of a *resistance coefficient* λ, defined as

$$\lambda \equiv \frac{(-dp/dx)D}{\frac{1}{2}\rho u_{\text{ave}}^2} \tag{8–7}$$

Substituting for dp/dx from Eq. (8–5) yields

$$\lambda = \frac{64}{\rho u_{\text{ave}}(2R)/\mu} = \frac{64}{\text{Re}_D} \tag{8–8}$$

The success of this relation is demonstrated on the far left-hand side of Fig. 8–3 by comparison with data. This result is sometimes presented in terms of a *friction factor* $f \equiv \lambda/4 = 16/\text{Re}_D$.

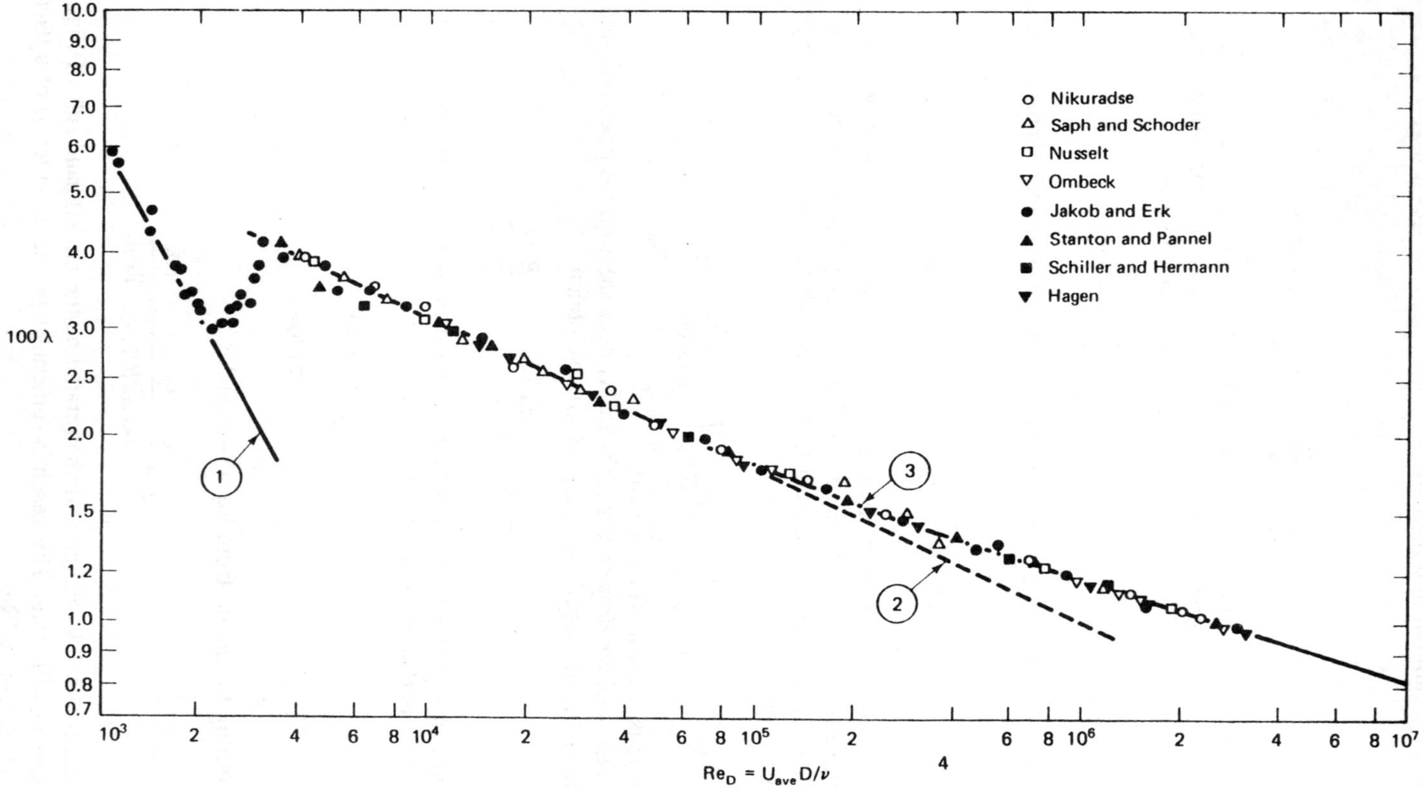

Figure 8–3 Frictional resistance in a smooth pipe. Curve 1: Eq. (8–8); curve 2: Eq. (8–40); curve 3: Eq. (8–41). (Data collected by Schlichting, 1968.)

It is interesting to note that Eq. (8–1) is actually the correct, reduced form of the full equations of motion for this problem, and the boundary layer assumption was not necessary in order to obtain the solution given by Eq. (8–8).

Many devices (e.g., heat exchangers), employ pipes or tubes to promote heat transfer and other cases of pipe flow simply involve heat transfer because a temperature difference exists between the wall and the fluid. If the temperature difference is not too large, the problem can still be analyzed while retaining the incompressible, constant-property assumptions. The appropriate energy equation is the low-speed, steady form of Eq. (3–54).[1] Using the fully developed flow conditions, that equation rigorously reduces to

$$\rho c_p u \frac{\partial T}{\partial x} = \frac{k}{r}\frac{\partial}{\partial r}\left(r\frac{\partial T}{\partial r}\right) \tag{8–9}$$

where $u(r)$ is known from Eq. (8–3).

The boundary conditions on $T(r)$ are

$$\left.\frac{\partial T}{\partial r}\right|_{r=0} = 0$$

$$T(x, R) = T_w(x) \qquad \text{or} \qquad -\left(-k\left.\frac{\partial T}{\partial r}\right|_{r=R}\right) = q_w(x) \tag{8–10}$$

That is, either the wall temperature or the wall heat transfer rate may be specified. A simple case occurs with the use of a specified constant q_w. Consider an energy balance for a fluid flowing through two adjacent stations along the tube under that condition. It is clear that the average value of the temperature must increase linearly with distance x along the tube. For *fully developed* flow, this must also be true for any value of the radius r, so we can say that $\partial T/\partial x$ = constant. Equation (8–9) can easily be integrated for $\partial T/\partial x$ = constant to give

$$T(r) - T_w = \frac{2u_{\text{ave}}R^2}{(k/\rho c_p)}\left(\frac{\partial T}{\partial x}\right)\left[\frac{1}{4}\left(\frac{r}{R}\right)^2 - \frac{1}{16}\left(\frac{r}{R}\right)^4 - \frac{3}{16}\right] \tag{8–11}$$

after the boundary conditions are applied. It is common at this point to introduce the average or *bulk* temperature

$$T_b \equiv \frac{\int_0^R \rho u c_p T\, 2\pi r\, dr}{\int_0^R \rho u c_p 2\pi r\, dr} = T_w - \frac{11}{48}\frac{u_{\text{ave}}R^2}{k/\rho c_p}\left(\frac{\partial T}{\partial x}\right) \tag{8–12}$$

[1] Note that there is no difference between c_p and c_v for the special case of a constant-density, constant-property fluid.

Basing a film coefficient $\hbar$ on the difference between the wall and bulk temperatures, that is,

$$q_w = \hbar(T_w - T_b) \tag{8-13}$$

we get

$$\text{Nu}_D \equiv \frac{\hbar(2R)}{k} = 4.364 \tag{8-14}$$

For cases with constant T_w, the value of Nu_D is 3.66. In that situation, both q_w and $(T_w - T_b)$ tend to zero with increasing distance down the pipe as the bulk temperature is heated (or cooled) to approach the wall temperature. In the situation of constant q_w, both T_b and T_w increase continuously.

It is a simple matter to envision a fully developed flow with mass transfer in a tube or a pipe. The wall might, for example, be covered with a material that is soluble in the fluid flowing in the pipe. That case is analogous to a constant-temperature wall. On the other hand, the walls might be porous with a foreign fluid forced in at a set rate. That case is analogous to a constant wall heat flux. In Chapter 2, we observed a close interrelationship among momentum, heat, and mass transfer. Thus, the preceding results for heat transfer in a pipe can be used for mass transfer by replacing the Nusselt number with the Nusselt number for diffusion and the Prandtl number with the Schmidt number if the transverse velocity at the wall is small and its influence can be neglected. A better procedure would be to begin with results for heat transfer in the presence of injection or suction (with the flow still restricted to one species) and then apply the analogy between heat and mass transfer. However, that is seldom done for pipe flows, and the simple procedure is the most common.

One is sometimes interested in the flow in noncircular ducts that are long enough to have fully developed flow. In such cases, the momentum and energy equations can be written as

$$0 = -\frac{1}{\rho}\frac{dp}{dx} + \nu\left(\frac{\partial^2 u}{\partial y^2} + \frac{\partial^2 u}{\partial z^2}\right) \tag{8-15}$$

$$\rho c_p u \frac{\partial T}{\partial x} = k\left(\frac{\partial^2 T}{\partial y^2} + \frac{\partial^2 T}{\partial z^2}\right) \tag{8-16}$$

These equations are clearly close relatives of Eqs. (8–1) and (8–9), except that now there are two coordinates (y, z) in the cross-sectional plane of the duct to describe the nonaxisymmetric flow field. The pressure across the duct is still presumed constant, and the secondary flow components (v, w) are assumed negligible in laminar flow. (In turbulent flows, that assumption is not viable.) Solutions for a number of geometries have been obtained by Shah and London (1978), and some results for the resistance coefficient and Nusselt number are listed in Table 8–1 in terms of a Reynolds number Re_{D_h} based on an equivalent or *hydraulic* diameter defined as $D_h = 4A/P$, where A is the cross-sectional area and P is the perimeter of the duct. (For a circular duct, $D_h = D$.) Heat transfer results in terms of a Nusselt number for

TABLE 8–1 LAMINAR FRICTION AND HEAT TRANSFER RESULTS IN NONCIRCULAR DUCTS

Shape of duct	$\lambda\, \mathrm{Re}_{D_h}$	$\mathrm{Nu}_{T_w=C}$	$\mathrm{Nu}_{q_w=C}$
Ellipse, 1 : 1	64.00	3.66	4.36
Ellipse, 2 : 1	67.29	3.74	4.56
Ellipse, 4 : 1	72.96	3.79	4.88
Rectangle, 1 : 1	56.91	2.98	3.61
Rectangle, 2 : 1	62.19	3.39	4.12
Rectangle, 4 : 1	72.93	4.44	5.33
Triangle, 1 : 1 : 1	53.35	2.47	3.11
Hexagon	60.22	3.34	4.00

constant wall temperature are in the third column of the table, and those for constant heat transfer rate are in the last column. The results show that the shape of the duct can have a large effect on heat transfer and pressure drop. The heat transfer results can be extended to cases of mass transfer using the analogy between heat and mass transfer described before.

8–2–2 Entrance Region Flows

The flow near the entrance of a duct before the flow field is fully developed is also of interest. Analysis of this flow is much more complicated than that for the fully developed flow region, so we begin with some simplified treatments to develop a basic understanding.

First, consider a force and momentum balance between a station at the entrance of a pipe where the velocity can be taken to be uniform at u_{ave} and another station where the velocity profile is fully developed and is given by Eq. (8–3). Figure 8–2 shows such a situation. The entrance velocity must be uniform at u_{ave}, since the mass flow through any cross section of the pipe is the same in steady flow. Let the inlet pressure be p_i and the wall shear and velocity distribution in the fully developed region be τ_{FD} and u_{FD}, respectively. Then a force and momentum balance from the entrance ($x = 0$) to a station in the fully developed region ($x = x_{\mathrm{FD}}$) can be written

$$(p_i - p_{\mathrm{FD}})\pi R^2 = \tau_{\mathrm{FD}} 2\pi R x_{\mathrm{FD}} + \int_0^R \rho (u_{\mathrm{FD}}^2 - u_{\mathrm{ave}}^2) 2\pi r\, dr + \int_0^{x_{FD}} (\tau - \tau_{\mathrm{FD}}) 2\pi R\, dx \tag{8–17}$$

The pressure drop in the entrance region is higher than in the fully developed region, because the wall friction is higher and the inviscid core is accelerating. This fact can be expressed by an additional term in the equation for the pressure drop in the fully developed region [Eq. (8–7)]. We obtain

$$\frac{\Delta p}{\frac{1}{2}\rho u_{\mathrm{ave}}^2} = \lambda \frac{x_{\mathrm{FD}}}{D} + K \tag{8–18}$$

where

$$K = \frac{2}{3} + \int_0^{x_{\text{FD}}/R} \frac{4(\tau - \tau_{\text{FD}})}{\rho u_{\text{ave}}^2 R}\, dx \tag{8-19}$$

Experiment indicates that $K \approx 1.25$, as suggested by Shah and London (1978).

The length of the entrance region has been correlated with Re_D by Shah and London (1978), who propose that

$$\frac{L_{\text{FD}}}{D} = \frac{0.6}{1 + 0.035\,\text{Re}_D} + 0.056\,\text{Re}_D \tag{8-20}$$

This region is thus quite long. For a typical laminar $\text{Re}_D = 1{,}000$, Eq. (8–20) indicates that $L_{\text{FD}} = 56D$.

An entrance region type of problem that can be solved analytically is the case where the velocity field in a pipe is already fully developed, but the heat transfer problem begins abruptly at some station x_0 by a sudden change in the wall temperature, which was equal to the fluid bulk temperature before that point. The energy equation for this case is again Eq. (8–9), with $u(x, r)$ from Eq. (8–3), but the boundary and initial conditions are not those in Eq. (8–10), but rather

$$\begin{aligned} T(x_0, r) &= T_0 = \text{constant} \\ \left.\frac{\partial T}{\partial r}\right|_{r=0} &= 0 \\ T(x, R) &= T_w = \text{constant} \end{aligned} \tag{8-21}$$

This problem was solved by Nusselt (1910) using *separation of variables,* and the results are illustrated in Fig. 8–4. After about $(x/D)/(\text{Re}_D\,\text{Pr}) \approx 0.05$, the temperature field is also fully developed.

The general entrance region problem where the velocity and temperature fields develop from the pipe entrance requires a numerical solution. The presentation here follows the work of Cebeci and co-workers [Cebeci and Smith (1974), Cebeci and Chang (1978), Cebeci and Bradshaw (1977), and Cebeci and Bradshaw (1984)]. As with external flows, the equations of motion for internal flows can be treated numerically in either physical coordinates or transformed coordinates. The transformed coordinates have some important advantages, especially in controlling the transverse extent of the computational region as the boundary layer grows. In internal flows, however, that advantage is overridden as the boundary layer nears the duct axis. We will consider only the untransformed equations here for clarity, but the equations will be made dimensionless. The continuity and momentum equations can be combined by the introduction of a dimensionless stream function defined as

$$F = \frac{\psi(x, r)}{\sqrt{u_{\text{ave}}\nu R}} \tag{8-22}$$

This leads to a third-order equation

$$[(1 - t)^2 F'']' = \frac{dp^*}{dx} + F'\frac{\partial F'}{\partial x} - F''\frac{\partial F}{\partial x} \tag{8-23}$$

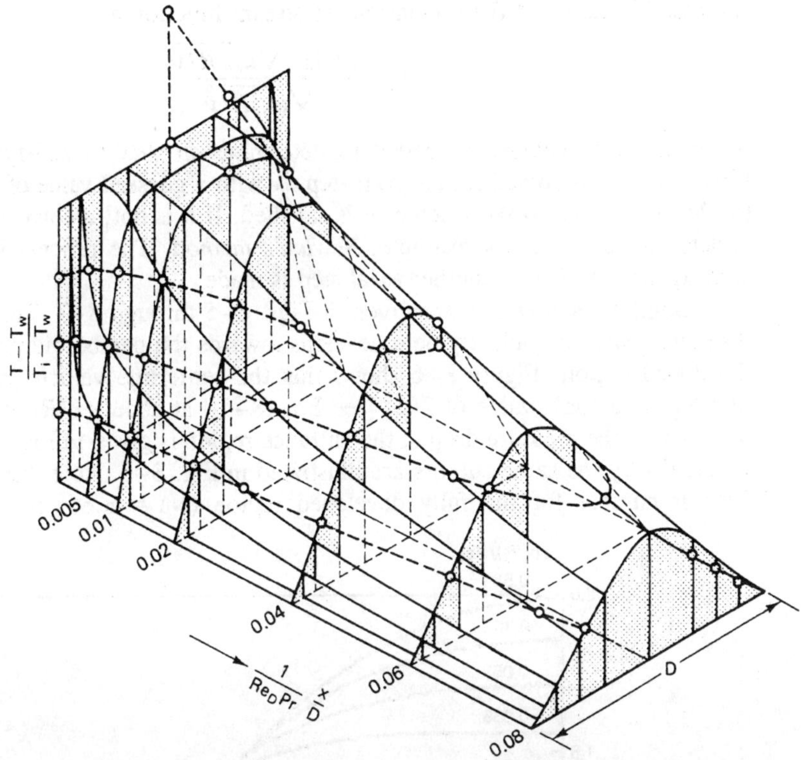

Figure 8–4 Temperature distribution for laminar flow in the entrance region of a pipe with constant wall temperature and fully developed velocity field. (From Nusselt, 1910.)

where $p^* = p/\rho u_{ave}^2$ and t is a term representing the effects of transverse curvature. The boundary conditions on a solid pipe wall are $F = F' = 0$. The usual symmetry condition on the pipe axis is evaluated by examining Eq. (8–23) on the axis, which gives

$$F''_{CL} = -\frac{1}{2}\sqrt{u_{ave}R/\nu}\left[\frac{dp^*}{dx} + \frac{1}{2}\frac{d}{dx}(F')^2\right] \tag{8–24}$$

This formulation resembles that for external flows, except for the important difference that in internal flows the pressure gradient term is unknown and is not specified by an inviscid solution, as it is in external flows. This additional unknown requires an additional equation, which is obtained by requiring a constant flow rate through any section of the pipe for steady flow, i.e.,

$$\rho u_{ave}\pi R^2 = \int_0^R \rho u 2\pi r\, dr \tag{8–25}$$

Since a stream function is directly related to the mass flow rate, Eq. (8–25) can be expressed in terms of the dimensionless stream function as

$$1 = \frac{2F(x, \sqrt{u_{ave}R/\nu})}{\sqrt{u_{ave}R/\nu}} \tag{8-26}$$

An iterative procedure is needed to determine dp^*/dx to satisfy this condition. First, the flow is solved for an axial step dx with a guessed value of dp^*. Then, Eq. (8–26) is checked to see whether it is satisfied. If it is not, a corrected value of dp^* is determined using, for example, *Newton's method.* The process is repeated until convergence, and then another axial step is made.

Some typical results are given in Figs. 8–5 through 8–7. The profiles in Fig. 8–5 show how the velocity field develops towards the paraboloid profile in the fully developed region. Figure 8–6 shows that the centerline velocity reaches the fully developed region's value of $2u_{ave}$ [see Eq. (8–5)] at about $(x/R)/\text{Re}_D \approx 0.15$. The variation in the pressure drop in the entrance region is given in Fig. 8–7. Heat transfer results by the same authors are illustrated in Fig. 8–8. Note that the value of the Nusselt number for the fully developed region (Nu = 3.66) is reached at about

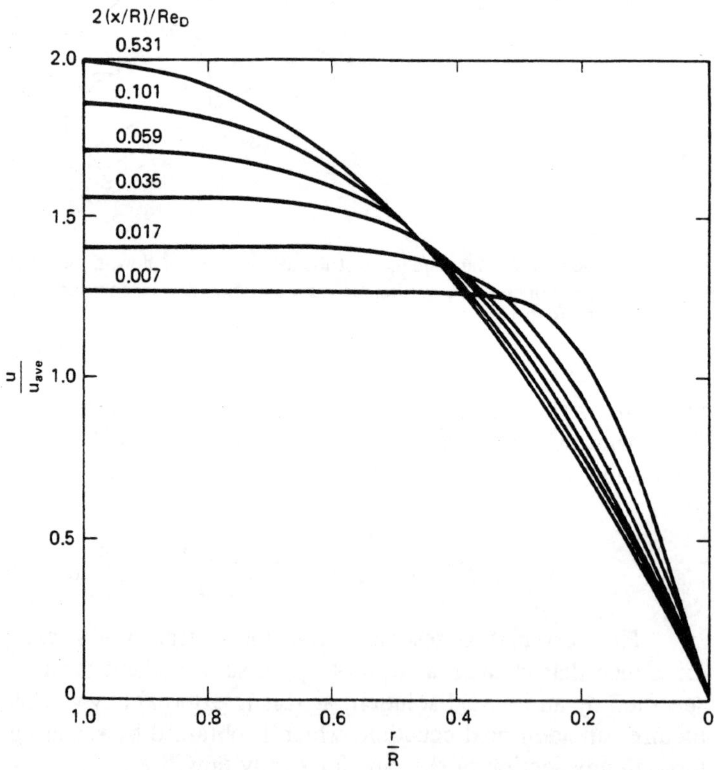

Figure 8–5 Calculated velocity profiles in the entrance region of a pipe. (From Cebeci and Bradshaw, 1977.)

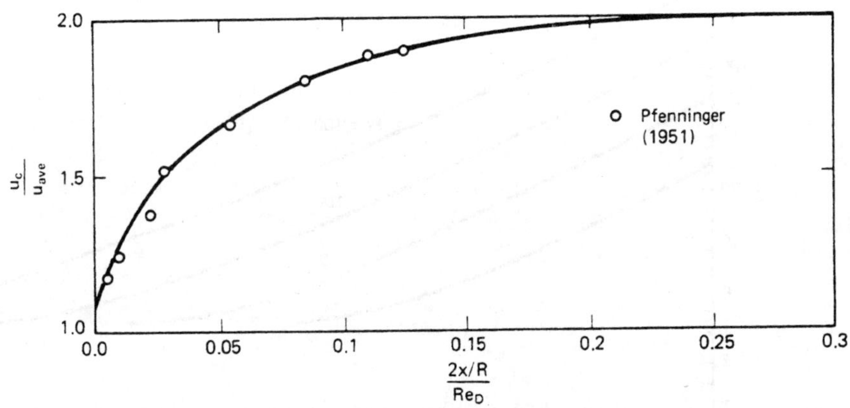

Figure 8–6 Calculated and experimental variation in centerline velocity in the entrance region of a pipe. (From Cebeci and Bradshaw, 1977.)

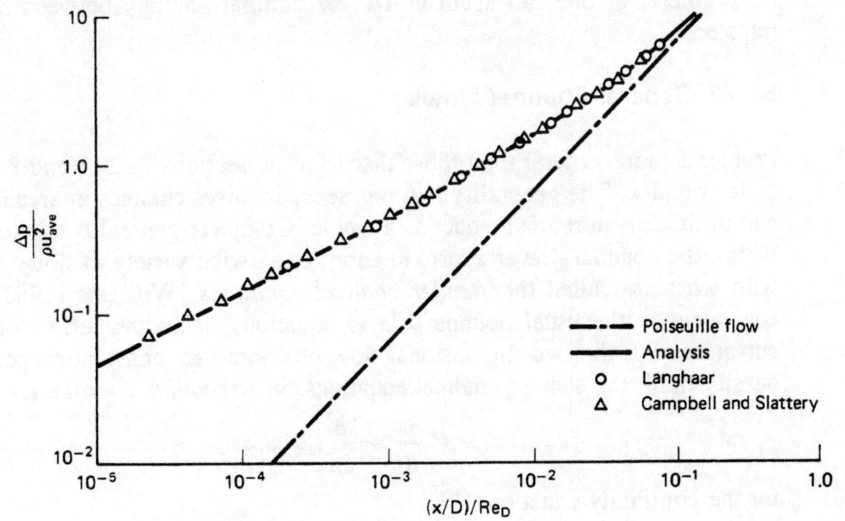

Figure 8–7 Calculated and experimental variation in pressure gradient in the entrance region of a pipe. (From Cebeci and Chang, 1978.)

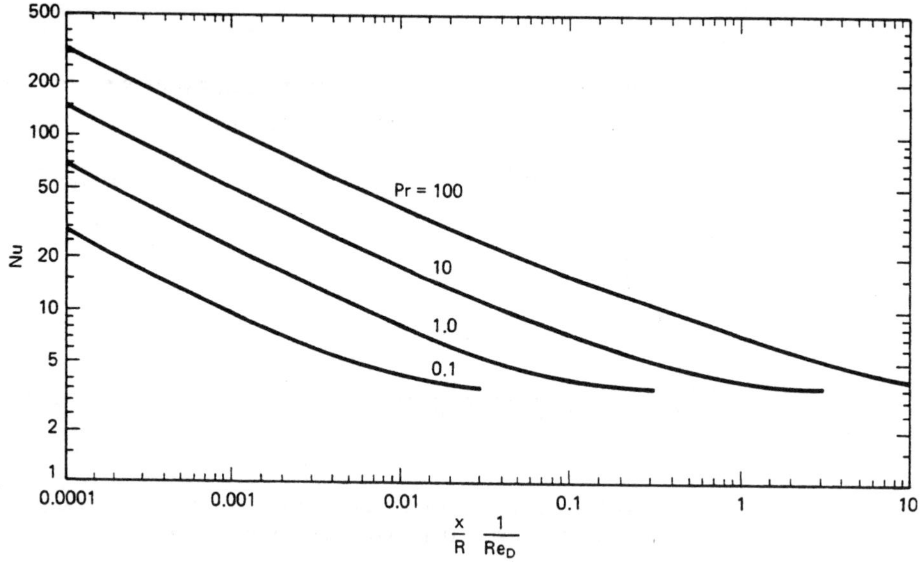

Figure 8–8 Variation in Nusselt number for laminar flow in the entrance region of a pipe with constant wall temperature. (From Cebeci and Bradshaw, 1984.)

$(x/R)/\mathrm{Re}_D \approx 0.25$ for $\mathrm{Pr} = 1$. This length is extended greatly for high Prandtl numbers.

There has been much less work on mass transfer in the entrance region than on momentum and heat transfer. The same methods could be applied to the mass transfer problem, or one can again utilize the familiar analogy between heat and mass transfer.

8–2–3 General Channel Flows

Problems more general than those discussed in Sections 8–2–1 and 8–2–2 become quite complex. The generality that one seeks involves changes in area along the duct and in the curvature of the duct as a whole. Complete generality cannot be achieved within the boundary layer approximation, but a wide variety of flows can be treated with what are called the *slender channel equations* [Williams (1963)]. These are equivalent to the usual boundary layer equations, including terms for longitudinal curvature. For the two-dimensional flow of a laminar, constant-property, constant-density fluid, the slender channel equations, in *streamline coordinates*, are

$$\frac{\partial u}{\partial s} + \frac{\partial}{\partial n}(vh_c) = 0 \tag{8–27}$$

for the continuity equation,

$$u\frac{\partial u}{\partial s} + v\frac{\partial}{\partial n}(uh_c) = -\frac{1}{\rho}\frac{\partial p}{\partial s} + \frac{\nu}{h_c}\frac{\partial}{\partial n}\left[h_c^2\left(\frac{\partial u}{\partial n} - \frac{\kappa u}{h_c}\right)\right] \tag{8–28}$$

for the streamwise momentum equation, and

$$\frac{\partial p}{\partial n} = \frac{\kappa}{h_c} \rho u^2 \tag{8-29}$$

for the transverse momentum equation. The term κ is the reciprocal of the radius of curvature of the channel, and h_c is the metric coefficient which is equal to $1 + \kappa n$.

Blottner (1977) developed numerical procedures for treating these equations in the entrance region of straight and curved two-dimensional channels. A representative result for the case of a curved channel of constant height is shown in Fig. 8-9.

It is easy to envision much more complicated problems in this general class of flows, such as problems with three-dimensional duct cross sections and/or separation, but the analysis of such flows is beyond the scope of this book.

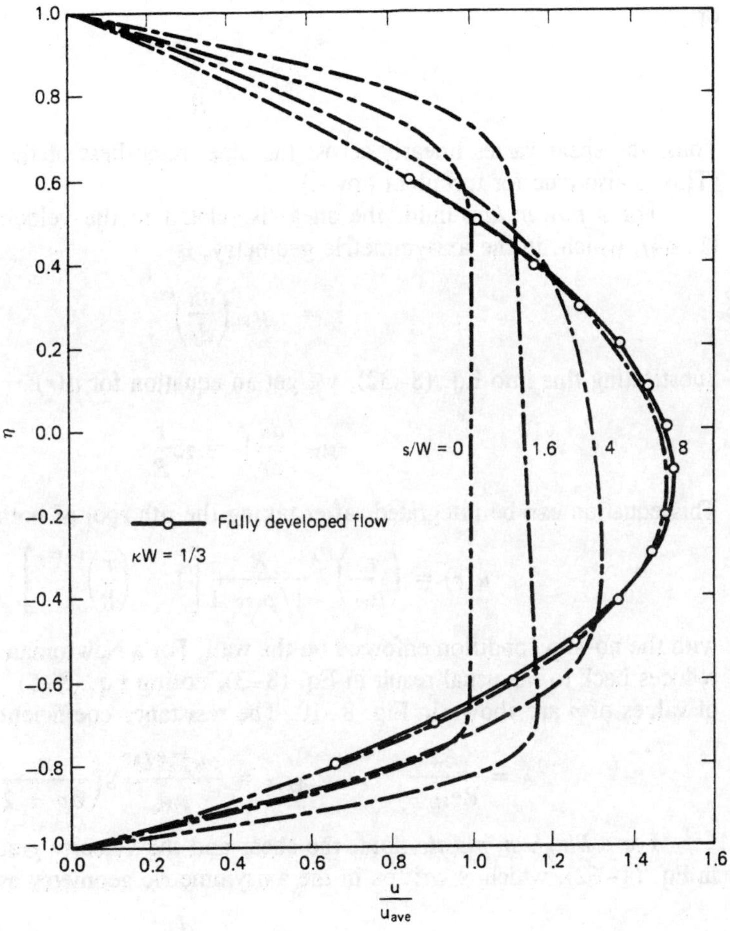

Figure 8-9 Calculated velocity profiles in the entrance region of a curved channel with constant height. (From Blottner, 1977.)

8–2–4 Non-Newtonian Fluids

The effects of non-Newtonian fluid behavior will be shown by the example of fully developed flow in a pipe with *power law* and *Bingham plastic* fluids (see Section 1–7 for definitions of these terms). We begin with

$$0 = -\frac{1}{\rho}\frac{dp}{dx} - \frac{1}{\rho r}\frac{d}{dr}(r\tau) \tag{8–30}$$

which is Eq. (8–1) rewritten in a more general form without specifying fluid behavior. This equation can be integrated once to give

$$\tau(r) = -\frac{r}{2}\frac{dp}{dx} \tag{8–31}$$

or

$$\tau(r) = \tau_w \frac{r}{R} \tag{8–32}$$

Thus, the shear varies linearly across the pipe, regardless of the type of fluid in it. (This is also true for turbulent flows.)

For a *power law* fluid, the shear is related to the velocity gradient by Eq. (1–64), which, in the axisymmetric geometry, is

$$\tau = -\mu_{PL}\left(\frac{du}{dr}\right)^p \tag{8–33}$$

Substituting this into Eq. (8–32), we get an equation for $u(r)$:

$$-\mu_{PL}\left(\frac{du}{dr}\right)^p = \tau_w \frac{r}{R} \tag{8–34}$$

This equation can be integrated, after taking the pth root of both sides, to give

$$u(r) = \left(\frac{\tau_w}{\mu_{PL}}\right)^{1/p} \frac{R}{1/p + 1}\left[1 - \left(\frac{r}{R}\right)^{1+1/p}\right] \tag{8–35}$$

with the no-slip condition enforced on the wall. For a Newtonian fluid ($p = 1$), this reduces back to the usual result in Eq. (8–3), noting Eq. (8–5). Results for a range of values of p are shown in Fig. 8–10. The resistance coefficient becomes

$$\lambda = \frac{64}{\mathrm{Re}_{PL,D}}, \qquad \mathrm{Re}_{PL,D} = \frac{\rho u_{ave}^{2-p} D^p}{\mu_{PL}} 8\left(\frac{p}{6p + 2}\right)^p \tag{8–36}$$

For a *Bingham plastic* fluid, the shear and the velocity gradient are related as in Eq. (1–62), which is written in the axisymmetric geometry as

$$\tau = \tau_0 - \mu_{BP}\frac{du}{dr} \tag{8–37}$$

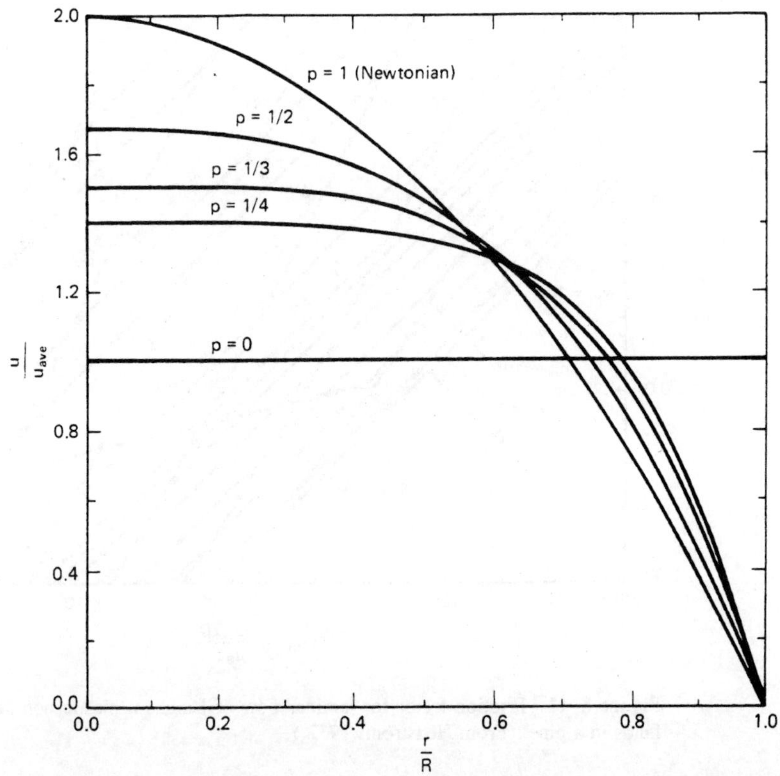

Figure 8–10 Calculated velocity profiles for fully developed flow of power law non-Newtonian fluids in a pipe.

Since the linear variation in the shear expressed in Eq. (8–32) holds in general, the behavior in Eq. (8–37) indicates that there must be a plug of radius r_p of uniform-velocity fluid near the axis of the pipe. This is the region where $\tau \le \tau_0$. After substituting the shear relation, integrating, and applying boundary conditions, the velocity profile emerges as

$$u(r) = \frac{1}{\mu_{BP}}\left[-\frac{1}{4}\frac{dp}{dx}(R^2 - r^2) - \tau_0(R - r)\right] \qquad (8\text{–}38)$$

for $r_p \le r \le R$. For $0 \le r \le r_p$, u_p is given by Eq. (8–38), evaluated at $r = r_p = 2\tau_0/(-dp/dx)$. The results for the friction factor f are presented by Hedstrom (1952) in the chart given as Fig. 8–11. Here, $N_{He} = \rho\tau_0 D^2/\mu_{BP}^2$ and $N_{Pl} = \tau_0 D/\mu_{BP} u_{ave}$. For $\tau_0 = 0$, these results collapse back to the usual Newtonian result.

The flow of a *power law* fluid in the entrance region of a plane channel has been studied numerically by Crochet et al. (1984). The development of the velocity profiles and the length of the entrance region as a function of the power law index are shown in Fig. 8–12. The profiles are fuller for lower values of p.

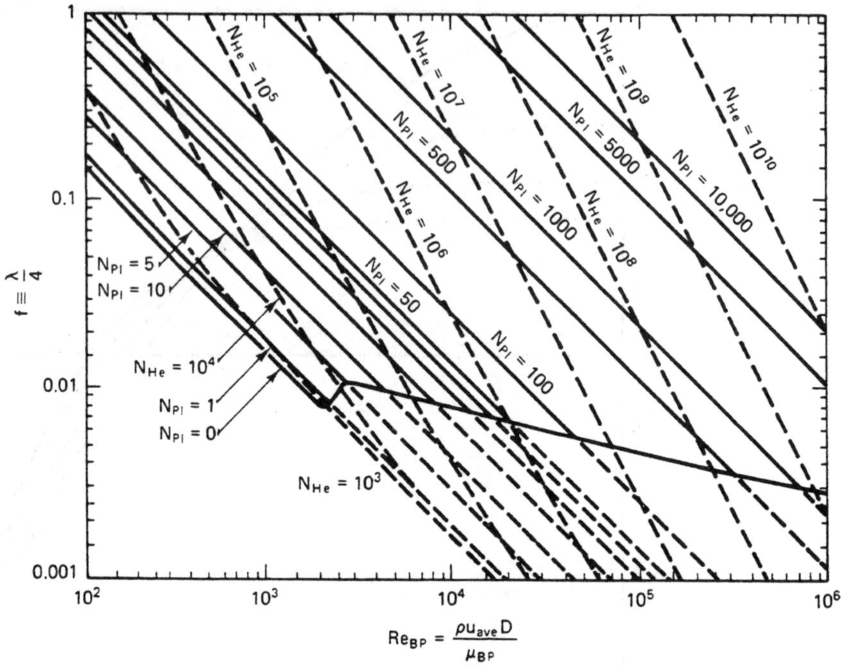

Figure 8–11 Friction factor for laminar flow of Bingham plastic non-Newtonian fluids in a pipe. (From Hedstrom, 1952.)

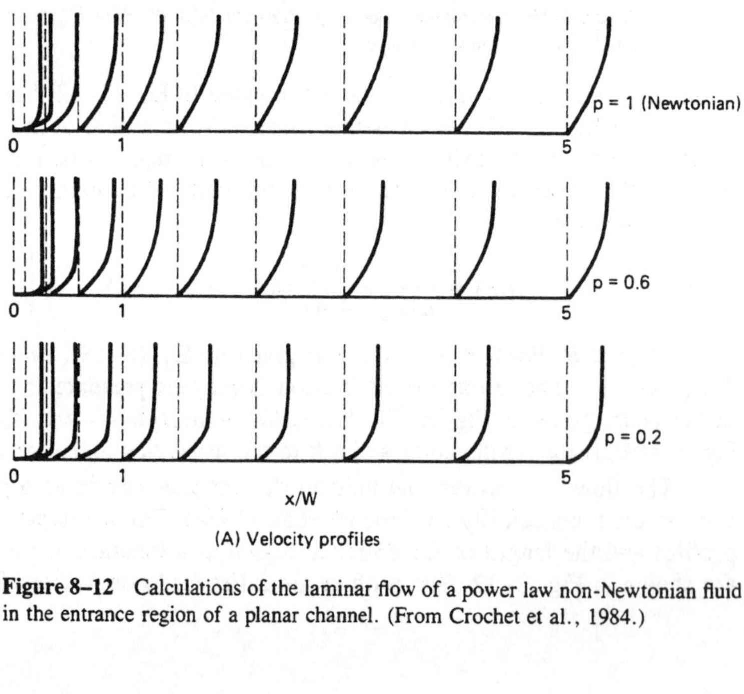

Figure 8–12 Calculations of the laminar flow of a power law non-Newtonian fluid in the entrance region of a planar channel. (From Crochet et al., 1984.)

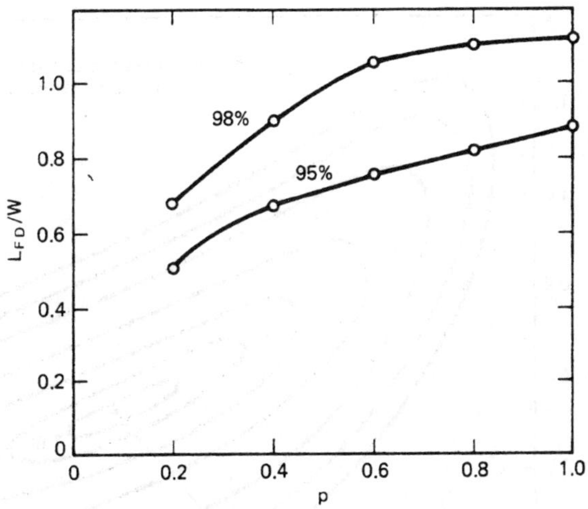

(B) Length to fully developed flow

Figure 8–12 *(continued)*

8–3 TRANSITION TO TURBULENT FLOW

All discussions of transition actually begin with an internal flow—a fully developed flow in a pipe, as studied by Reynolds and introduced in Section 6–1. The *critical Reynolds number* of 2,300 determined by Reynolds for industrial environments is seen to be confirmed by the data in Fig. 8–3. Note that again there is an extensive range over which transition occurs before the flow is fully turbulent. It was emphasized in Chapter 6 that a simple so-called critical value of a Reynolds number is too crude an idea to predict transition satisfactorily. The influence of the background environment is great, and one generally needs a more refined prediction method. The value 2,300 for pipes and other such engineering *rules of thumb* for other cases of flow are really the value below which the flow will remain laminar, even in the presence of large background disturbances. For noncircular pipes, values of the critical Reynolds number based on the *hydraulic diameter* comparable to the value 2,300 for circular pipes have been reported to be 2,100 for a square cross section and 1,600 for a rectangle with sides in the ratio 2.43 : 1.

The *hydrodynamic stability theory* has been applied to internal flows also (see Section 6–3 for an introduction to the subject). Results for a planar, fully developed flow between two plates are shown in Fig. 8–13. Again, as with external flows, the Reynolds number for instability is much lower than that for transition, so these kind of results have limited practical utility. They can be used to show the influence of certain parameters, such as wall temperature ratio, on stability and thus, ultimately, on transition.

Flows in the entrance region or internal flows that never become fully developed can be analyzed in essentially the same manner as are external flows. The e^N *method* (see Section 6–4) is a good candidate for such a treatment.

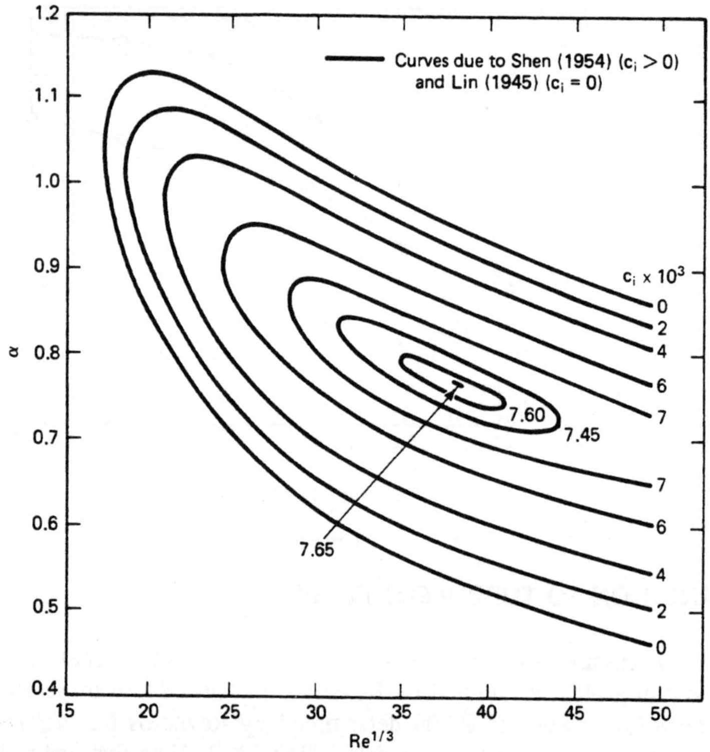

Figure 8–13 Hydrodynamic stability theory predictions for fully developed flow in a planar channel. (From Lin, 1945 and Shen, 1954.)

8–4 TURBULENT FLOWS

As with external flows, most internal flows of engineering interest occur in the turbulent regime. The practice established in Chapter 7 of presenting mean flow data and turbulence data and then introducing turbulence modeling when discussing turbulent flows will be adopted here, except that the discussion is subdivided by type of flow into (1) fully developed flows, (2) entrance region flows, and (3) general channel flows. A final section on non-Newtonian fluids completes the coverage.

8–4–1 Fully Developed Flows

The simplest wall-bounded internal turbulent flow is that in a round tube or pipe. For a sufficiently long pipe (measured in L/D), the flow again becomes *fully developed,* such that the velocity profile does not change with distance along the pipe. Some useful results can be obtained from a simple analysis following the method used in Section 8–2–1 for laminar flows. This analysis again leads to the general result

$$\tau_w = \frac{-R}{2}\frac{dP}{dx} \tag{8–39}$$

Thus, the wall shear can be inferred from the pressure drop along the pipe. Before the development of the skin friction balance, the only reliable turbulent skin friction measurements were obtained in that way. As with laminar flow, the *resistance coefficient* λ [see Eq. (8–7)] is used to present data for the frictional resistance. Data for smooth pipes in the laminar, transitional, and turbulent regimes are shown in Fig. 8–3, together with some empirically based equations from Blasius (1913), namely,

$$\lambda = 0.316(\mathrm{Re}_D)^{-1/4} \tag{8–40}$$

and Prandtl (1935), i.e.,

$$\frac{1}{\sqrt{\lambda}} = 2.0 \log(\mathrm{Re}_D \sqrt{\lambda}) - 0.8 \tag{8–41}$$

The figure shows clearly the important fact that turbulent skin friction is generally higher than the corresponding laminar value, especially for Reynolds numbers of the order of the value for transition.

The influence of uniform roughness on the frictional resistance to flow in pipes is usually given in terms of the well-known *Moody chart,* as shown in Fig. 8–14. Note that in this case k has been made dimensionless simply with the pipe diameter D, rather than by forming a Reynolds number, as is the practice for external flows.

For noncircular tubes, it is again common to use a Reynolds number based on the *hydraulic diameter* D_h. With that number, Eqs. (8–40) and (8–41) can be used.

The empirical heat transfer relation for fully developed pipe flow, from Petukhov (1970), is

$$\mathrm{St} = \frac{\lambda/8}{1.07 + 12.7(\mathrm{Pr}^{2/3} - 1)\sqrt{\lambda/8}} \tag{8–42}$$

This value is accurate to 5–6% for $10^4 \le \mathrm{Re}_D \le 5 \times 10^6$ and $0.5 \le \mathrm{Pr} \le 200$. Equation (8–42) is sometimes approximated further as

$$\mathrm{St} = 0.027(\mathrm{Re}_D)^{-0.2}\mathrm{Pr}^{-2/3} \tag{8–43}$$

This result compares well with measurements, some of which are given in Fig. 8–15 for the whole range of laminar, transitional, and turbulent flow. In the figure, μ_b stands for the laminar viscosity evaluated at the *bulk* temperature [see Eq. (8–12)]. For heat transfer in noncircular tubes, we again apply a Reynolds number based on the *hydraulic diameter*. With that number, Eqs. (8–42) and (8–43) can be used.

Heat transfer results in rough pipes are usually developed directly from resistance coefficients as in Fig. 8–14. Norris (1971) suggests that

$$\frac{\mathrm{Nu}}{\mathrm{Nu}_{\mathrm{smooth}}} = \left(\frac{\lambda}{\lambda_{\mathrm{smooth}}}\right)^n \tag{8–44}$$

where $n = 0.68\ \mathrm{Pr}^{0.215}$.

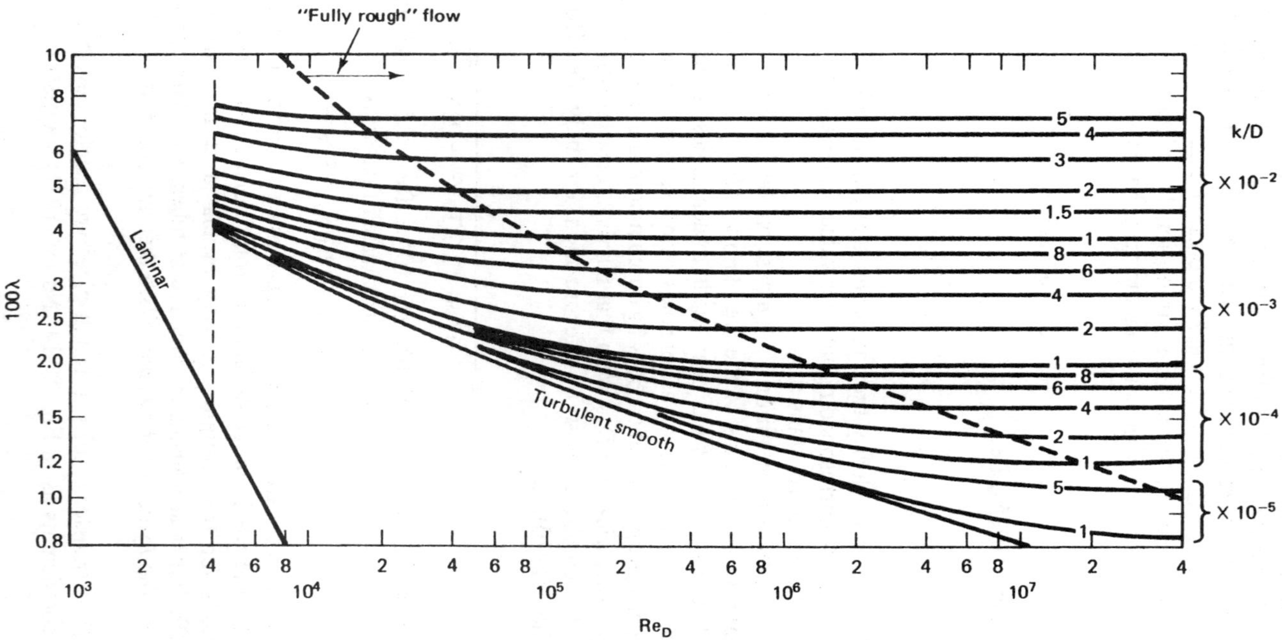

Figure 8–14 Frictional resistance in uniformly roughened pipes. (From Moody, 1944.)

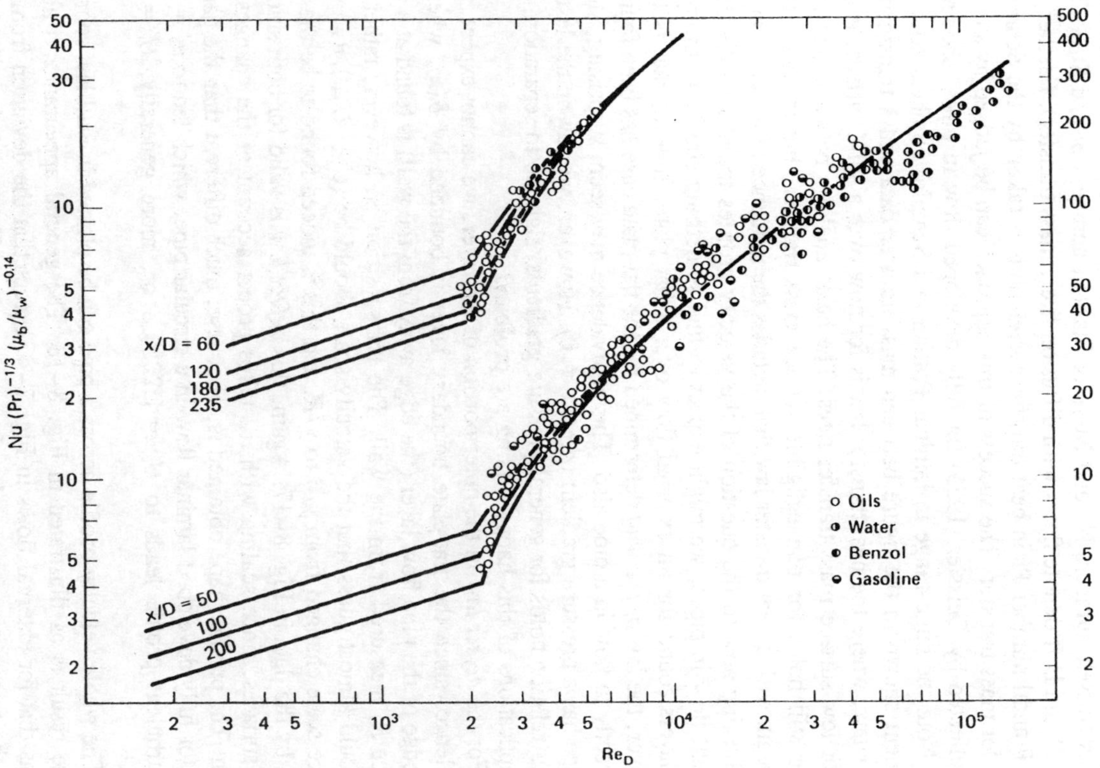

Figure 8–15 Laminar, transitional, and turbulent heat transfer in pipes. (From Sieder and Tate, 1936.)

Most of the consideration of internal flows with surface mass transfer has been in terms of the analogy between heat and mass transfer introduced earlier for laminar flows. Within that framework, one presumes a direct correspondence between a film coefficient for heat, h, and for diffusion, h_D, usually presented in dimensionless terms as Nusselt numbers Nu and Nu_{Diff} or Stanton numbers St and St_{Diff}. The subscripts D and Diff denote flows with diffusion (i.e., mass transfer). Further, the role of the Prandtl number Pr in heat transfer correlations is taken by the Schmidt number Sc for mass transfer. The success of this approach can be judged by studying the data collected by Deissler (1955) for fully developed flow in pipes, shown in Fig. 8–16. Note the huge range in laminar Prandtl and Schmidt numbers covered. The agreement shown in the figure between mass transfer data and a transformed heat transfer law, where the heat transfer law is for flow over a solid surface, is obtained only for low surface mass transfer rates. The basic analogy between heat and mass transfer still holds for nonnegligible surface mass transfer, but in such cases, one must begin with a heat transfer law that includes the influence of injection or suction.

Turning now to the question of the velocity profiles found for fully developed turbulent flow in pipes, we might expect to find something similar to the inner and outer layers found for an external flow over a flat plate. Indeed, that is the case; moreover, the *law of the wall* determined for the flat-plate case has been found to apply directly to flow in a pipe also. This equivalence may seem surprising, since flow in a pipe has a pressure gradient ($dP/dx \neq 0$). However, we saw in the last chapter that this relation holds for general pressure gradients, right up to separation, and that the implications of this happy finding are profound.

For the *defect law* in the outer portion of the layer, we cannot expect complete equivalence, since the flat-plate boundary layer is bounded by $\delta(x)$, whereas the outer edge of the pipe flow layer (the edge away from the wall) is found at a constant distance R, measured from the wall. The transposition is, however, rather simple. One could almost guess that the velocity scale should be $(U - U_{\text{max}})/u_*$, with the distance scale changed from y/δ to y/R, and this is indeed found to be the case, as shown by the data in Fig. 8–17. Again, this *defect law* is valid for both smooth and rough surfaces, since scaling with u_* takes proper account of the changes in wall friction. The profile shape obtained is, of course, much different than the paraboloid found for fully developed laminar flow in a circular pipe, which led to $u_c = 2.0u_{\text{ave}}$. The turbulent profile leads to $U_c \approx 1.22U_{\text{ave}}$, or, more generally, $U_c = U_{\text{ave}} + 4.07u_*$.

The whole velocity profile across a pipe can be plotted in wall law coordinates, and the result is as illustrated in Fig. 8–18. The general appearance is about the same as that for external flows in Fig. 7–4, except that the deviation from the log law in the outer region is much less for the pipe flow. If one translates the *law of the wake*, $W(y/\delta)$ (see Section 7–3–1), to the pipe case as $W(y/R)$, W is seen to be very small in the pipe case.

One more feature of the mean flow in a fully developed pipe flow is worthy of notice. Writing Eq. (8–1) for turbulent flow with the sum of the laminar and turbulent shear stresses implies that

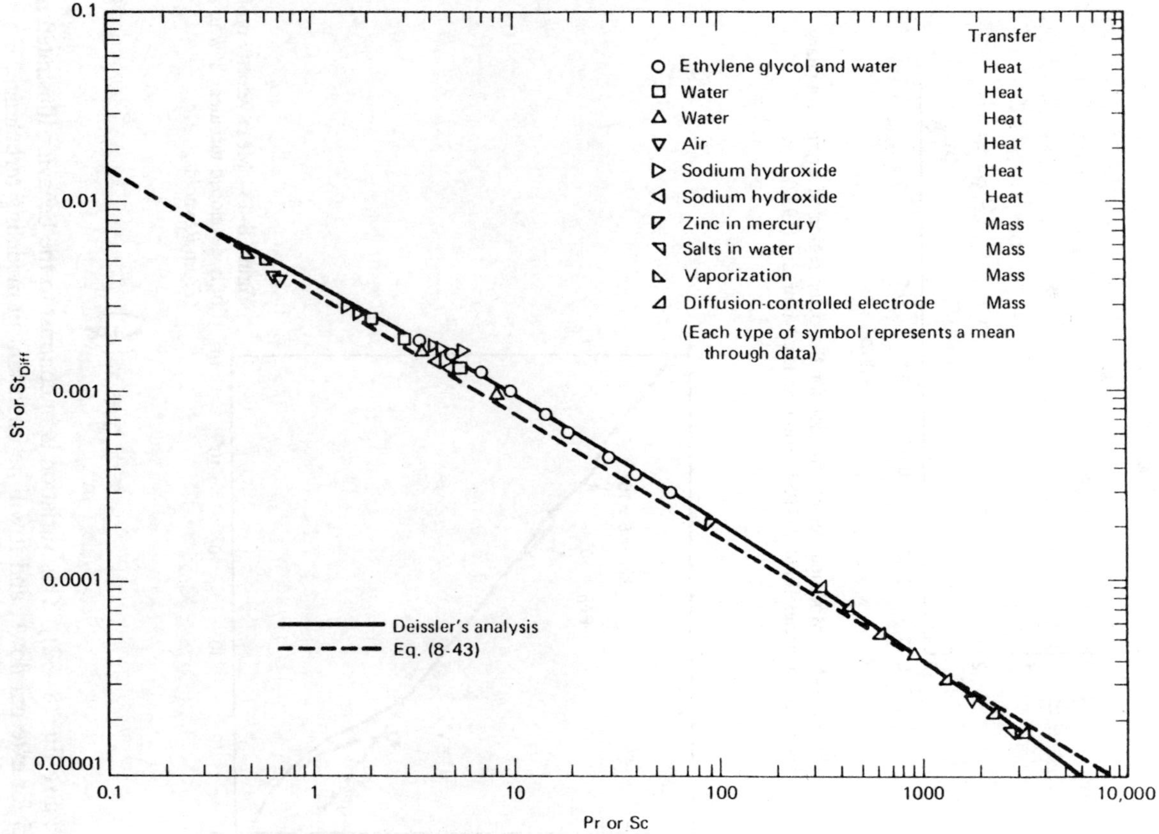

Figure 8–16 Data for heat and mass transfer in pipes (Re_D = 10,000), showing the close relationship between the two processes. (Collected by Deissler, 1955.)

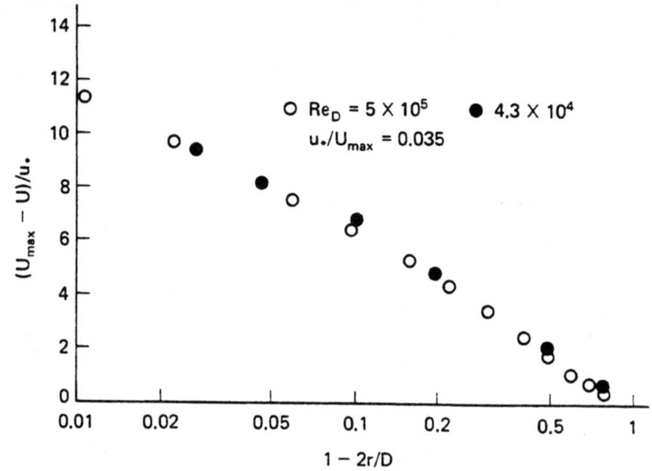

Figure 8–17 Modified defect law plot for the outer region of a pipe boundary layer. Open circles: Laufer (1954); solid circles: Nikuradse (1932).

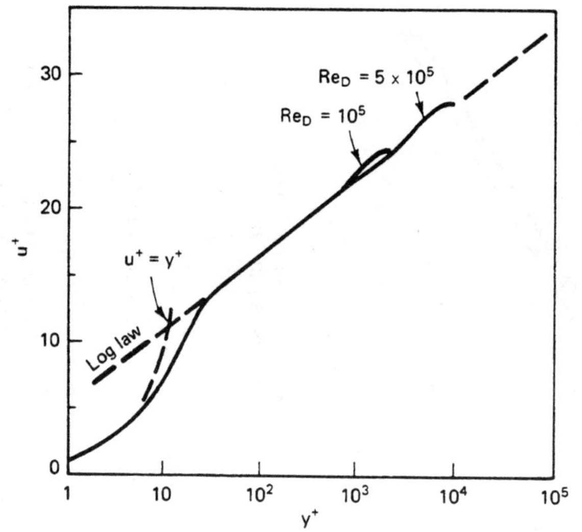

Figure 8–18 Mean velocity profiles for fully developed turbulent flow in a pipe. (From Reynolds, 1974.)

$$\tau + \tau_T = \tau_w\left(\frac{r}{R}\right) \tag{8–45}$$

[see also Eq. (8–32)]. This variation is in contrast to the behavior illustrated in Fig. 7–5 for external flows, and it will be important in modeling turbulence.

In noncircular ducts, we encounter the important phenomenon of *secondary flow*. This is generally taken to be the components of the velocity in the cross plane perpendicular to the main direction of fluid motion. The subject will play a prominent role in Chapter 11, which covers three-dimensional, external boundary layer flows. Here, the emphasis is on a class of secondary flows produced by the turbulence itself. Other important secondary flows that occur in curved ducts are produced by the pressure field, and they will be discussed in a later section in this chapter. Considering a force balance between a unit area on a wall and a plane off the wall, one finds that

$$P + \rho\overline{v'^2} = P_w \tag{8–46}$$

This simple statement has important consequences, since it establishes that the normal stresses can produce changes in the mean static pressure. Unless the flow is symmetrical in all directions, such as in a circular pipe, these pressure changes are not in balance without secondary mean flows. The secondary flows generate cross-plane shear stresses that balance the normal stresses. Such processes also occur in curved ducts, but their effects are dominated by the pressure-induced secondary flows. Some typical results for the secondary flow in a straight, rectangular duct are shown in Fig. 8–19. The secondary flow tends to make the velocity distribution uniform over the duct cross section. This uniformity leads to the useful and simple notion that the shear on the wall of a noncircular duct can be taken to be nearly that in a circular pipe with the same core velocity. This close approximation is the basis of the use of a Reynolds number based on the *hydraulic diameter* introduced earlier.

The extensive data of Laufer (1954) for turbulence quantities in fully developed pipe flow have been selected here to exhibit the phenomena encountered. These data will be contrasted with the corresponding data for flat-plate flow in Section 7–4 to illustrate some differences between internal and external flows. The individual turbulence intensities are plotted in Fig. 8–20(A) and (B), and they can be compared with the flat-plate results in Fig. 7–17. The general behaviors are similar. The axial intensity is the largest, followed by the out-of-plane intensity, and then the vertical intensity. Also, the maximum values are comparable, and they occur at about the same location, $y^+ \approx 15$. The profiles of K and τ_T are shown in Fig. 8–21, and they correspond to the flat-plate data in Fig. 7–20. Again in pipe flow, one can note that, roughly, $\tau_T \sim K$, and the proportionality constant is about the same as for external flows at approximately 0.30. Figure 8–22 displays the variation in the Reynolds stress near the wall, where the close agreement with the flat-plate data in Fig. 7–21 is clear. Spectra of the axial turbulence intensity in a pipe from Lawn (1971) are given in Fig. 8–23. Comparing these with the flat-plate spectra in Fig. 7–22, we can easily observe generally similar behavior. There is again a range of four to five orders of magnitude in the scales of motion containing significant energy. Also, near the wall, there is more energy in the small-scale fluctuations for both situations. A turbulent kinetic energy balance is plotted in Fig. 8–24. [Note that the *diffusion* term has been subdivided into its two components; see Eq. (7–94).] The *convection* is identically zero in fully developed flow. The important result is that *production* and *dissipation* are dominant near the wall, the same as for external flows, as shown in Fig. 7–24.

From all of this, one can conclude that the major features of the mean and turbulent flow for a fully developed pipe flow and an external boundary layer flow are qualitatively essentially the same and quantitatively similar.

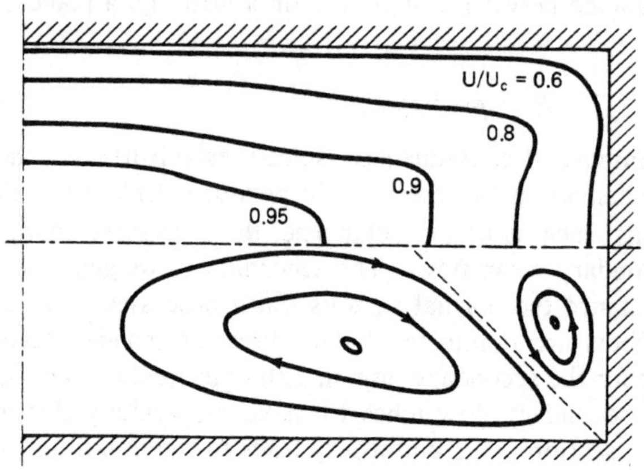

Figure 8–19 Contours of constant axial velocity and secondary flow pattern for fully developed turbulent flow in a rectangular channel with sides in the ratio of 3 : 1 and at $Re_{D_h} \approx 6 \times 10^4$. (From Brundett and Baines, 1964.)

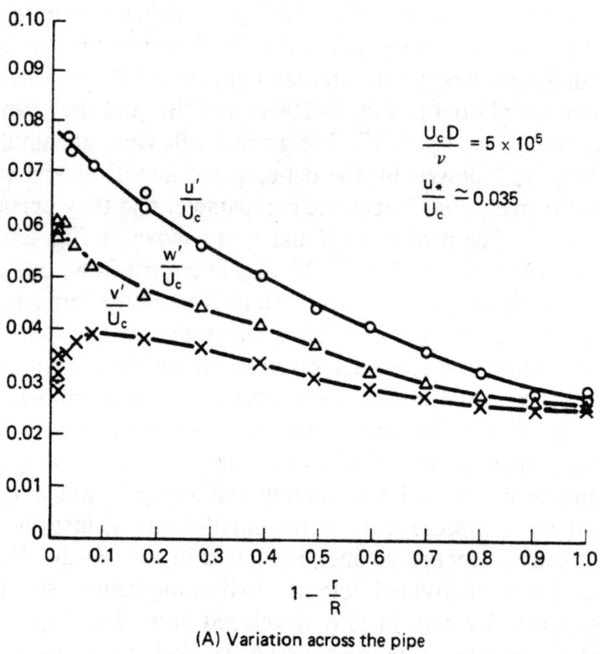

Figure 8–20 Profiles of turbulence intensities for fully developed flow in a pipe. (From Laufer, 1954.)

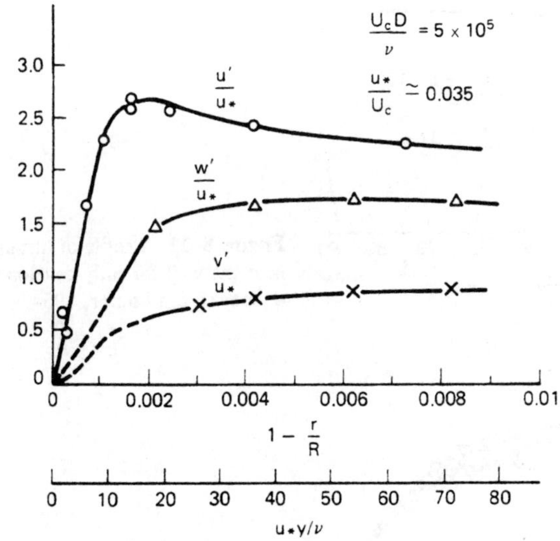

(B) Variation near the pipe wall

Figure 8–20 *(continued)*

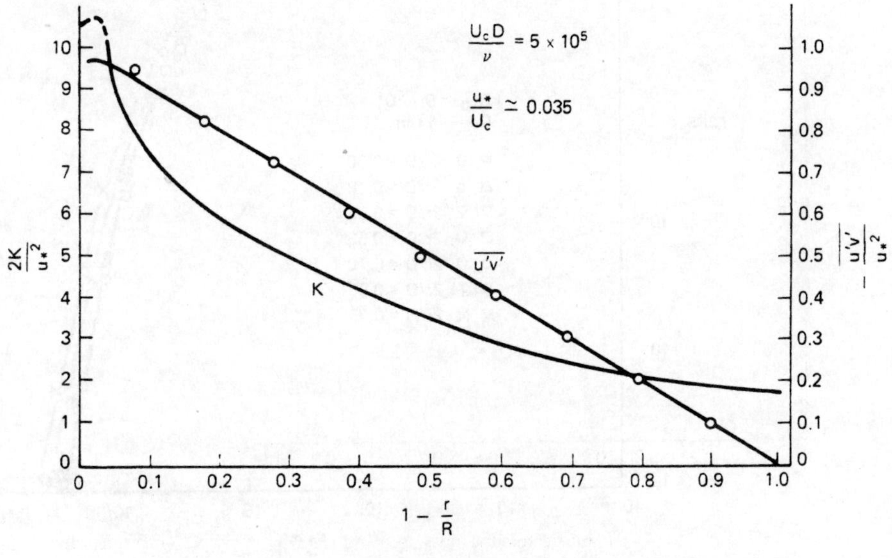

Figure 8–21 Profiles of turbulent kinetic energy and turbulent stress for fully developed flow in a pipe. (From Laufer, 1954.)

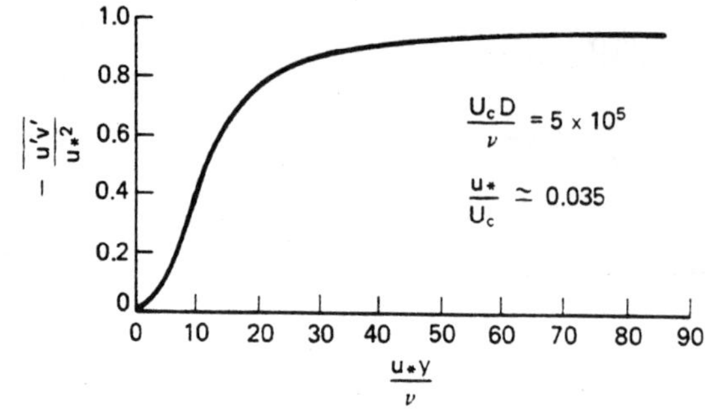

Figure 8–22 Profile of turbulent stress near the wall for fully developed flow in a pipe. (From Laufer, 1954.)

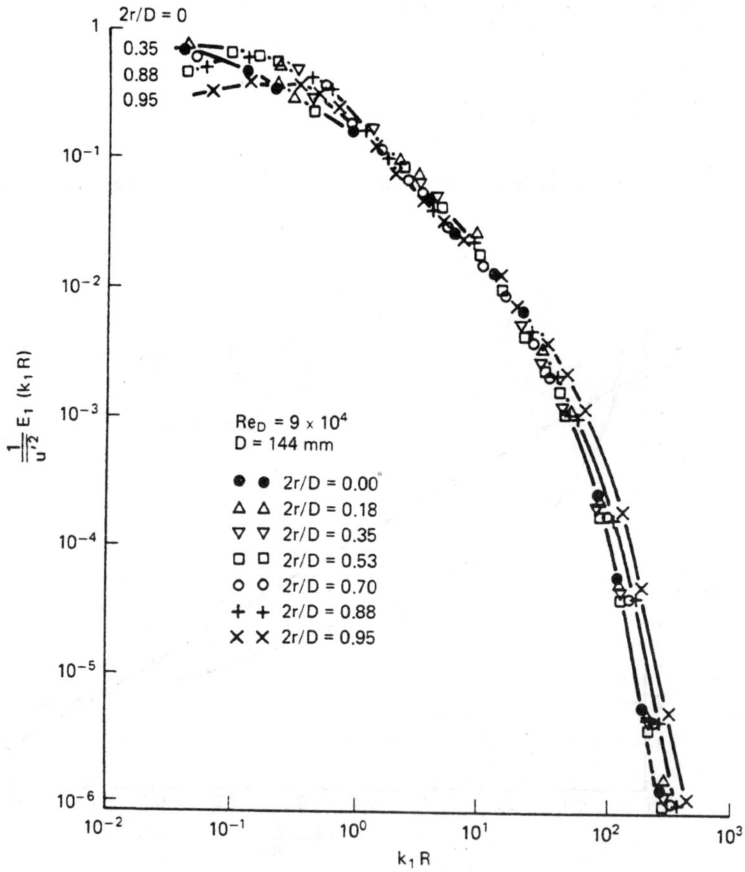

Figure 8–23 Spectra of axial turbulence intensity for fully developed flow in a pipe. (From Lawn, 1971.)

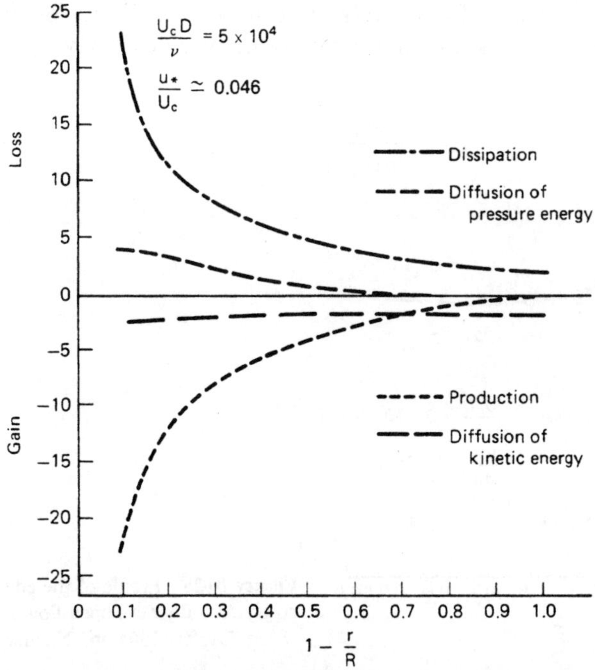

Figure 8–24 Turbulent kinetic energy balance for fully developed flow in a pipe. (From Laufer, 1954.)

With this background, we begin the development of turbulence models for the fully developed regime, starting with *mean flow models*. The development is simplified by the fact that the total shear distribution across the pipe is known exactly from Eq. (8–45), rewritten here in terms of y rather than r:

$$\tau + \tau_T = \tau_w\left(1 - \frac{y}{R}\right) \tag{8–47}$$

This equation permits us to have a direct procedure for finding either a mixing length or an eddy viscosity model, rather than the inverse procedure that applies in general (see Section 7–8). Very near the wall, the Van Driest mixing length model [Eq. (7–61) with $A^+ = 26$] works very well for smooth pipes. Once we get out into the log region, laminar shear can be neglected, and the definition of the mixing length [Eq. (7–37)] can be substituted into Eq. (8–47) without the laminar shear term. The resulting equation can easily be integrated, using the measured velocity distribution, to yield

$$\frac{\ell_m}{R} = 0.14 - 0.08\left(1 - \frac{y}{R}\right)^2 - 0.06\left(1 - \frac{y}{R}\right)^4 \tag{8–48}$$

from Nikuradse (1932). For small y/R, this equation reduces to $\ell_m = \kappa y \approx 0.40y$, which it should, in order to reproduce the log law actually observed (see Section 7–8–1). This same equation also holds for rough pipes in the same way that $\ell_m = 0.09\delta$ for boundary layers over both smooth and rough surfaces. Near the wall, roughness effects must be included.

A parallel analysis can be used for an eddy viscosity model. First, following Laufer, we can use the Reynolds stress and mean velocity data to give μ_T directly from

$$\mu_T = \frac{-\rho \overline{u'v'}}{\partial U/\partial y} \tag{8-49}$$

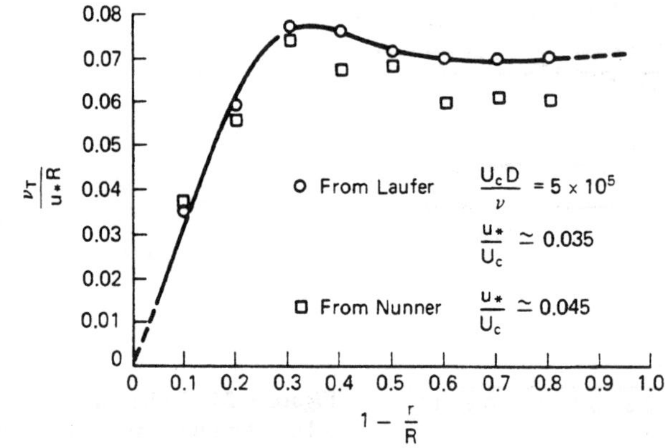

Figure 8–25 Profile of the eddy viscosity for fully developed flow in a pipe. (From Laufer, 1954 and Nunner, 1956.)

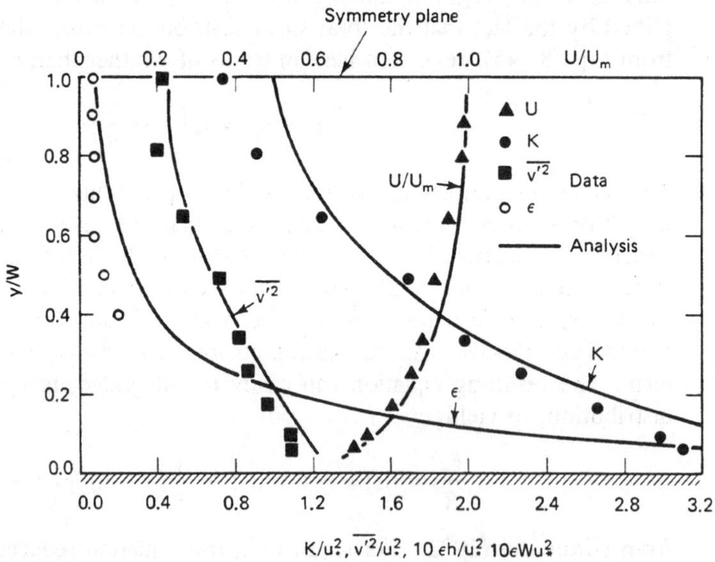

Figure 8–26 Comparison of predictions from a $K\epsilon$ model by Hossain (1979) with the data of Laufer (1951) for fully developed flow in a channel.

Some results are given in Fig. 8–25. The *core* or *outer-region* eddy viscosity is nearly constant, and Reynolds (1974) suggests that

$$\mu_{To} = 0.192\kappa\rho u_* R \tag{8–50}$$

In the log region, the same expression as for external boundary layer flows results, namely,

$$\mu_{T,\log} = \kappa\rho u_* y \tag{8–51}$$

In the inner region, one can again adopt the Reichardt expression in Eq. (7–63) or construct an eddy viscosity model from van Driest's mixing length model using Eq. (7–40). This kind of composite model will not have the small peak in eddy viscosity exhibited by the data shown in Fig. 8–25. However, that deficiency does not seem to degrade predictions based on these models.

There has been little work with TKE models for fully developed duct flows. A $K\epsilon$ model was applied to the planar case by Hossain (1979), and a comparison of predictions with the data of Laufer (1951) is given here in Fig. 8–26. The agreement for this relatively simple flow is quite good.

Cases with noncircular ducts are much more challenging, because predicting the secondary flows is difficult. Indeed, the popular $K\epsilon$ model simply includes no mechanism for the development of stress-driven secondary flows. There has, thus, been more emphasis on *Reynolds stress* and *algebraic stress* models for such cases. Fully developed flow in a square duct has been a subject of much interest. Gosman and Rapley (1978) compared predictions from the Hanjalic and Launder Reynolds stress model (see Section 7–12), two other Reynolds stress models, and an algebraic stress model with the experiments of Launder and Ying (1973), as shown in Fig. 8–27. All of these models predict the axial and secondary mean velocities well, but they do not predict the turbulent quantities accurately.

Fully developed flows represent a class to which the application of *direct numerical simulations* has enjoyed considerable success. The calculations are very expensive, very complex and quite restricted, but they are yielding much basic information about turbulence in shear flows.

8–4–2 Entrance Region Flows

The structure of the turbulent boundary layers on the interior walls near the entrance of a duct is essentially the same as that for external boundary layers described in Chapter 7. The main difference is that the continuity constraint drives the external velocity distribution in the case of internal flow. Data for the pressure drop, variation in centerline velocity, and velocity profiles in the entrance region of a pipe at high Reynolds number are shown in Fig. 8–28. Note how the centerline velocity asymptotically approaches the fully developed flow value of about $1.22U_{\text{ave}}$, as mentioned earlier. Representative heat transfer data are plotted in Fig. 8–29(A) and (B). The solid curves are predictions to be described later.

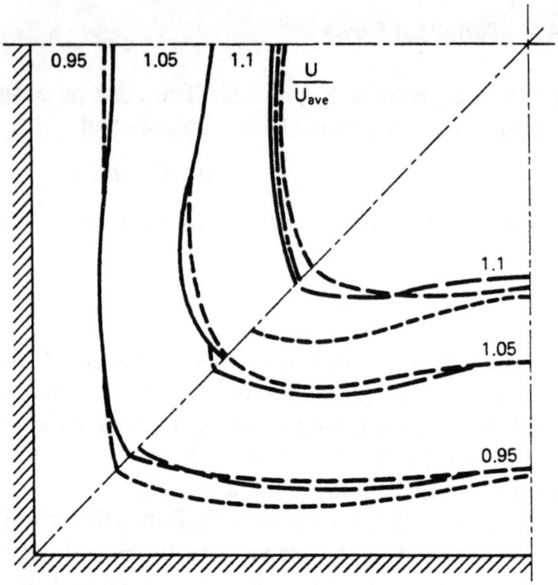

(A) Axial velocity contours: — — Experiment from Launder and Ying, 1972; — · — Prediction from Launder and Ying, 1973; - - Prediction from Noat et al. 1974; — — — Prediction from Reece, 1976; —— Prediction from Gosman and Rapley, 1980.

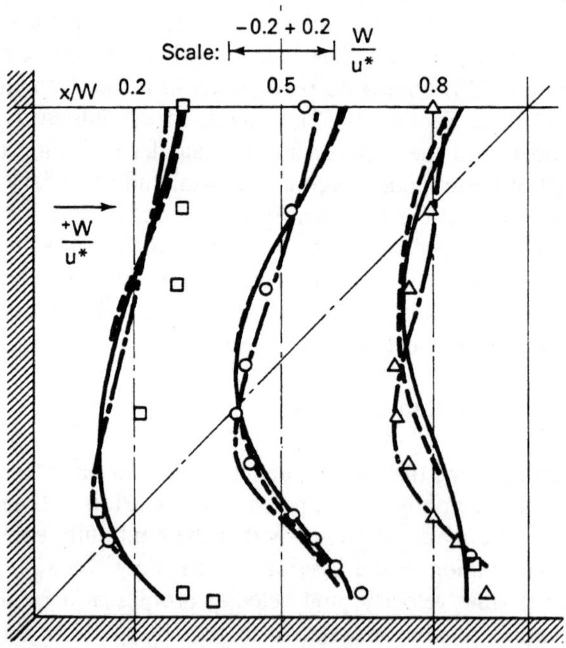

(B) Secondary velocity profiles. O□△ Experiment from Launder and Ying, 1972; - - - Prediction from Launder and Ying, 1973; — - — Prediction from Reece, 1976; —— Prediction from Gosman and Rapley, 1980

Figure 8–27 Comparison of predictions with experiment for fully developed turbulent flow in a square duct.

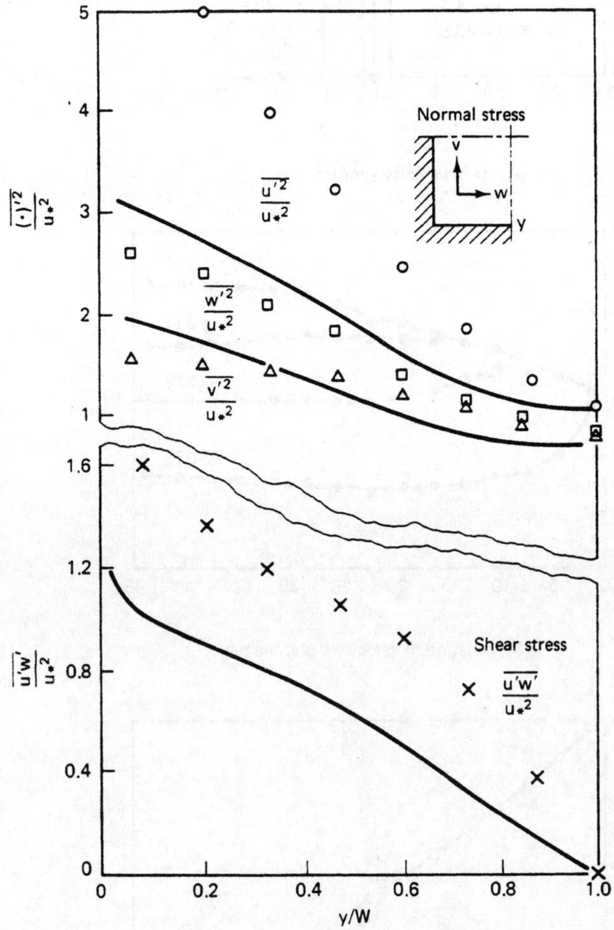

(C) Reynolds stress profiles: Experiment from Brundett and Baines, 1964;
——— Prediction from Gosman and Rapley, 1980

Figure 8–27 *(continued)*

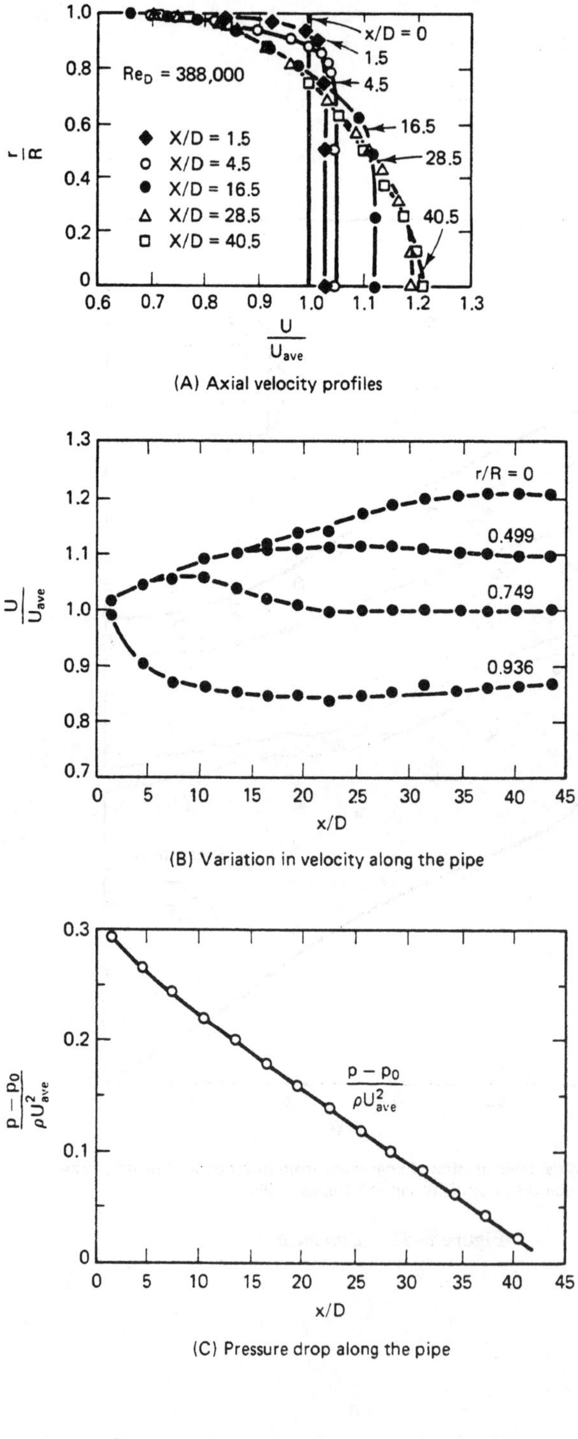

Figure 8–28 Turbulent flow in the entrance region of a pipe. (From Barbin and Jones, 1963.)

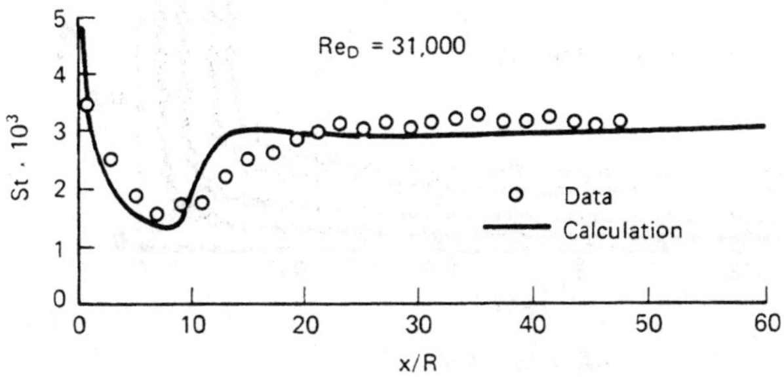

(A) Constant heat flux

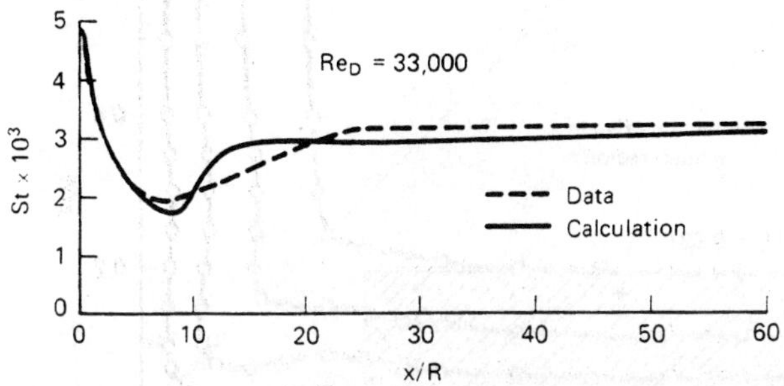

(B) Constant wall temperature

Figure 8–29 Measured and computed heat transfer results in the entrance region of a pipe. Experiments from Hall and Khan (1964). Computations from Cebeci and Chang (1978).

The flow in the entrance region of a square duct is complicated by the secondary flows. The development of the flow in the corner with increasing distance from the entrance can be seen in the isotach profiles in Fig. 8–30. The variation in the axial velocity along the centerline is plotted in Fig. 8–31. The peak value is higher than in the case of a pipe, and it then decreases to the fully developed value. Velocity profiles along a wall bisector and a corner bisector are displayed in Fig. 8–32.

Some turbulence data in the entrance region of a pipe are given in Fig. 8–33. The axial turbulence intensity increases faster along the pipe than does the tangential intensity. This suggests that the turbulent motions first arise as fluctuations in the direction of the main flow and then are transferred to the other directions.

With this brief introduction to the data base for entrance region flows, we are in a position to discuss analysis. The methods developed by Cebeci and his coworkers for laminar entrance region flows (see Section 8–2–2) can be extended to turbulent cases using eddy viscosity models. For heat transfer, the eddy thermal conductivity can be modeled with a constant value of the *turbulent Prandtl number,* defined as

$$\mathrm{Pr}_T \equiv \frac{\mu_T c_p}{k_T} \tag{8–52}$$

Most workers use $\mathrm{Pr}_T \approx 0.8$ for wall-bounded flows. Thus, the modeling question is reduced to specifying a model for the eddy viscosity. There is very little work in the

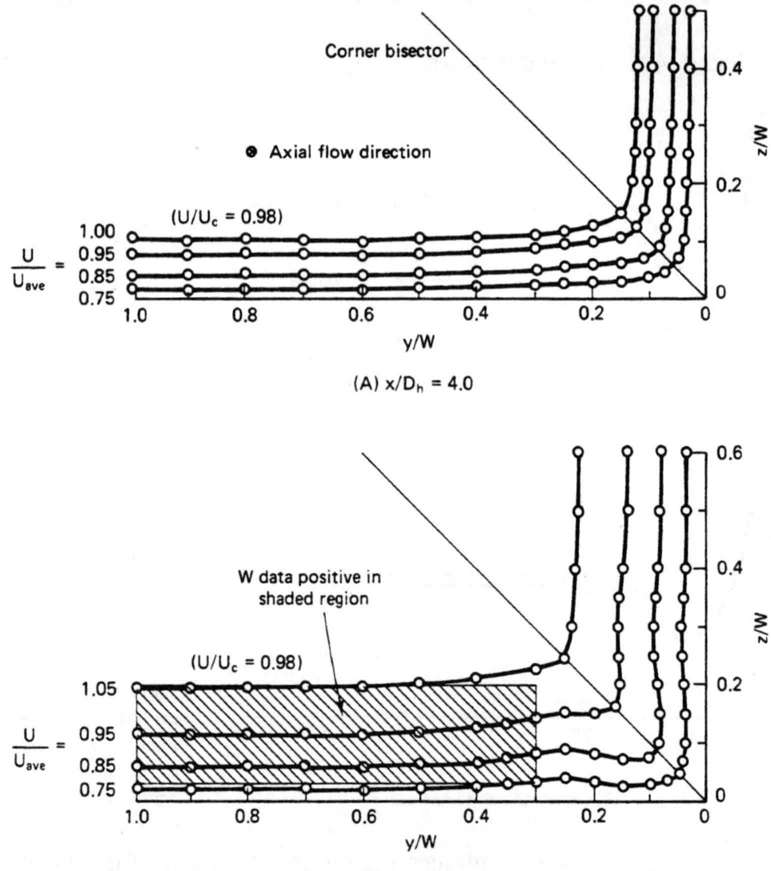

Figure 8–30 Axial velocity contours in the entrance region of a square duct. (From Gessner, 1973.)

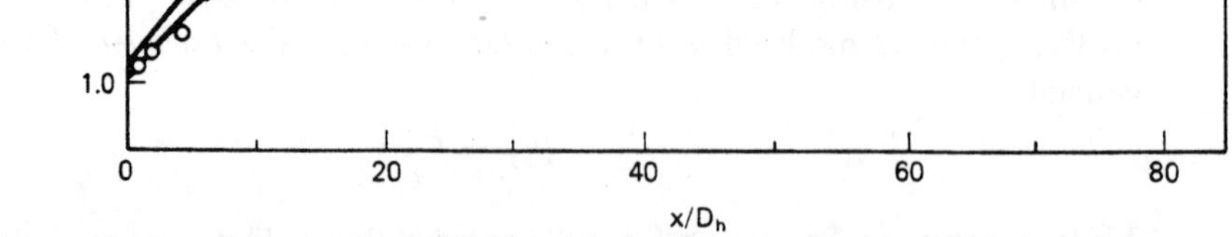

Figure 8–31 Comparisons of experiment with predictions of centerline velocity in the entrance region of a square duct. Experiments from Melling (1975) (+) at Re_D = 42,000 and Gessner and Emery (1981) (o) at Re_D = 250,000. Predictions of Demuren and Rodi (1984) with different wall-damping models.

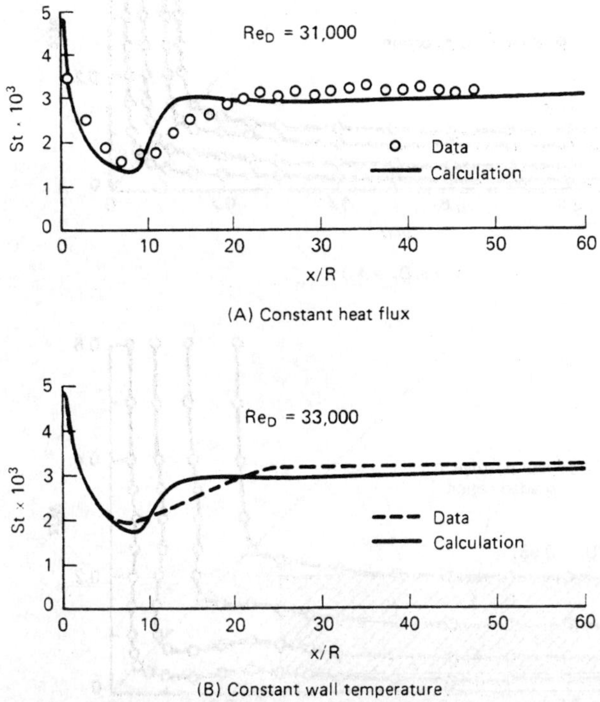

Figure 8–29 Measured and computed heat transfer results in the entrance region of a pipe. Experiments from Hall and Khan (1964). Computations from Cebeci and Chang (1978).

The flow in the entrance region of a square duct is complicated by the secondary flows. The development of the flow in the corner with increasing distance from the entrance can be seen in the isotach profiles in Fig. 8–30. The variation in the axial velocity along the centerline is plotted in Fig. 8–31. The peak value is higher than in the case of a pipe, and it then decreases to the fully developed value. Velocity profiles along a wall bisector and a corner bisector are displayed in Fig. 8–32.

Some turbulence data in the entrance region of a pipe are given in Fig. 8–33. The axial turbulence intensity increases faster along the pipe than does the tangential intensity. This suggests that the turbulent motions first arise as fluctuations in the direction of the main flow and then are transferred to the other directions.

With this brief introduction to the data base for entrance region flows, we are in a position to discuss analysis. The methods developed by Cebeci and his co-workers for laminar entrance region flows (see Section 8–2–2) can be extended to turbulent cases using eddy viscosity models. For heat transfer, the eddy thermal conductivity can be modeled with a constant value of the *turbulent Prandtl number*, defined as

$$\mathrm{Pr}_T \equiv \frac{\mu_T c_p}{k_T} \tag{8–52}$$

Most workers use $\mathrm{Pr}_T \approx 0.8$ for wall-bounded flows. Thus, the modeling question is reduced to specifying a model for the eddy viscosity. There is very little work in the

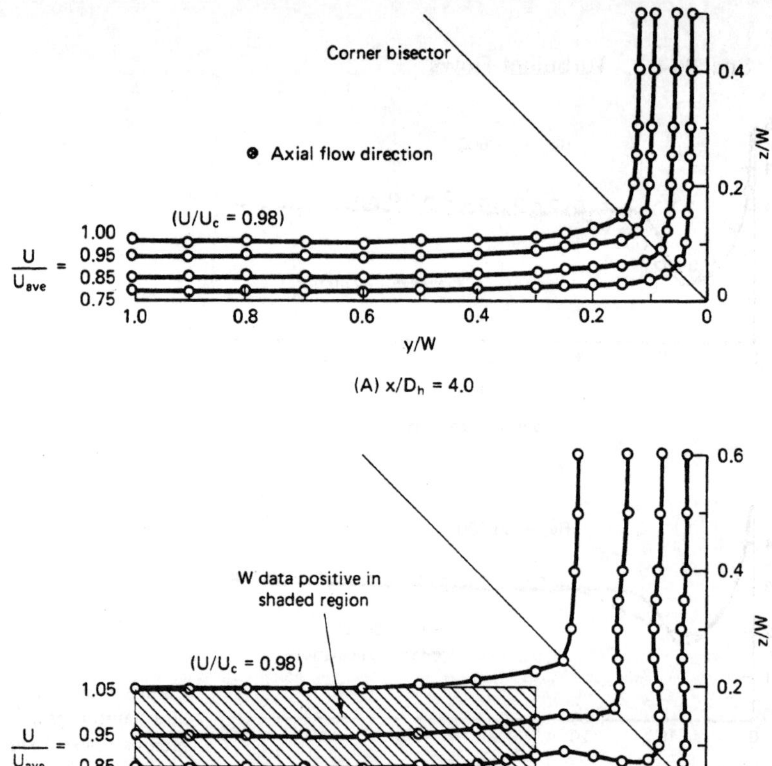

Figure 8–30 Axial velocity contours in the entrance region of a square duct. (From Gessner, 1973.)

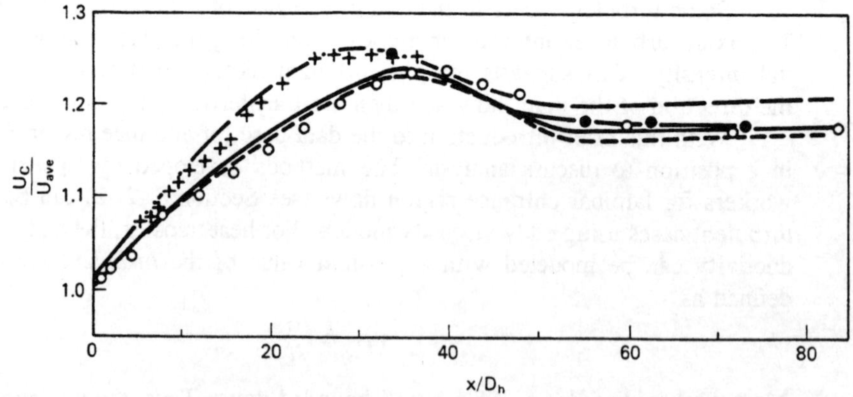

Figure 8–31 Comparisons of experiment with predictions of centerline velocity in the entrance region of a square duct. Experiments from Melling (1975) (+) at Re_D = 42,000 and Gessner and Emery (1981) (o) at Re_D = 250,000. Predictions of Demuren and Rodi (1984) with different wall-damping models.

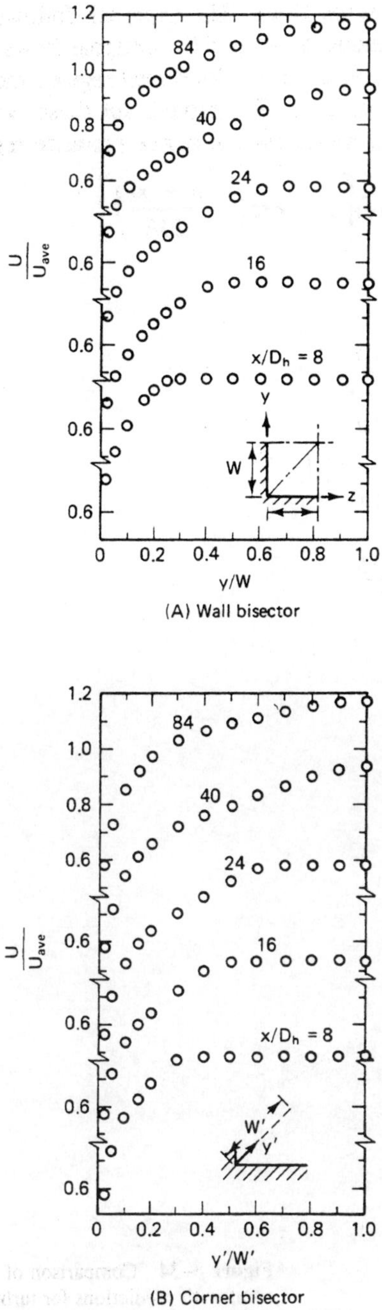

Figure 8–32 Axial velocity profiles in the entrance region of a square duct. (From Gessner and Emery, 1981.)

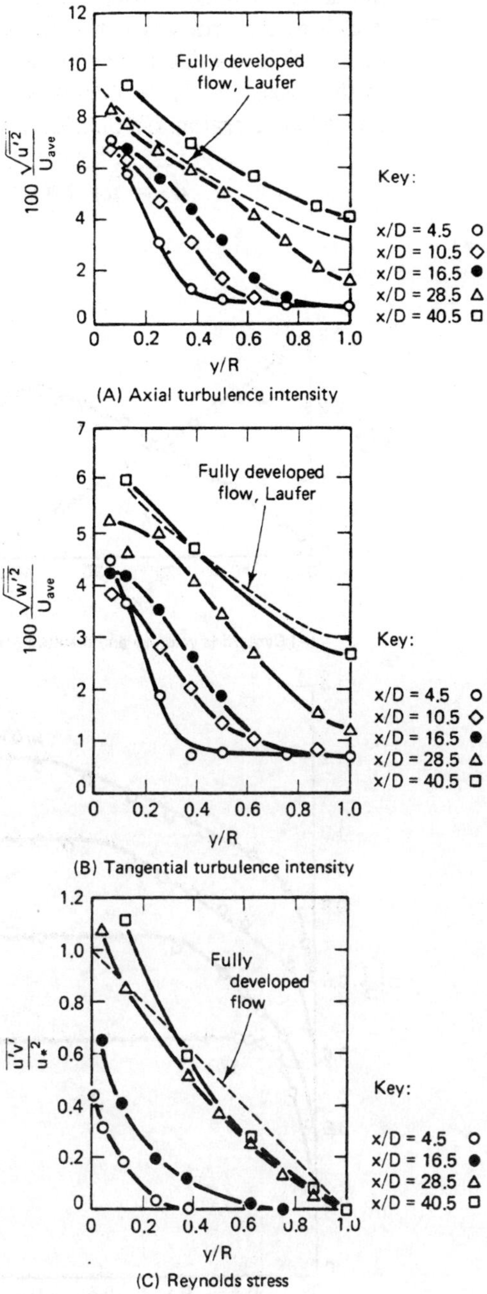

Figure 8–33 Turbulence data in the entrance region of a pipe. (From Barbin and Jones, 1963.)

literature for mass transfer in entrance region flows. The approach followed by Cebeci is to use ordinary eddy viscosity models developed for external flows (see Section 7–8) just inside the entrance region. In the fully developed region, models such as that discussed in Section 8–4–1 can be used. The models for these two ends of the entrance region are blended to give a model for the whole entrance region:

$$\mu_T = \mu_T^I + (\mu_T^{FD} - \mu_T^I)\left[1 - \exp\left(-\frac{x - x_0}{20R}\right)\right] \tag{8–53}$$

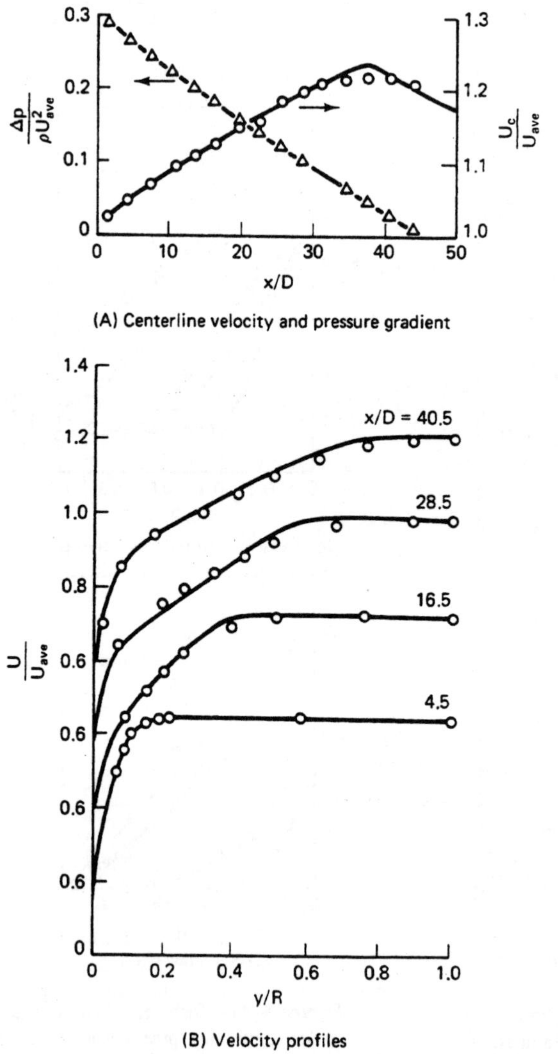

Figure 8–34 Comparison of experiment with predictions for turbulent flow in the entrance region of a pipe. Data of Barbin and Jones (1963). (From Cebeci and Chang, 1978.)

In this equation, x_0 is the point where the wall layers nearly merge, μ_T^i is the model used near the inlet, and μ_T^{FD} is the model used in the fully developed region. This rather arbitrary procedure works quite well. Some comparisons of predictions with experiments can be found in Figs. 8–29, 8–34, and 8–35.

Some work has been done on higher order turbulence models for the case of a square duct. In fact, this was one of the test cases for the 1981 Stanford conference, but only three predictions were presented. The subject has been reviewed recently by Nallasamy (1987). The lowest order model that has the capability to predict the stress-driven secondary flow is a Reynolds stress model, and most workers have adopted an *algebraic stress model*. The most success has been achieved by Demuren and Rodi (1984) using an *extended algebraic stress model* in which the secondary velocity gradient terms are retained and modeled and the pressure-strain term is modified to account for the proximity of the wall. Some results have already been given in Fig. 8–31 for the variation in the centerline velocity. Predictions of secondary velocity profiles are compared with experiment in Fig. 8–36. The secondary velocities are underpredicted at both stations by Demuren and Rodi (1984). All the predictions show decreasing V along the duct, while the data indicate a slight increase. The Demuren and Rodi model does predict the correct behavior of the turbulent normal stresses (not shown in the figure).

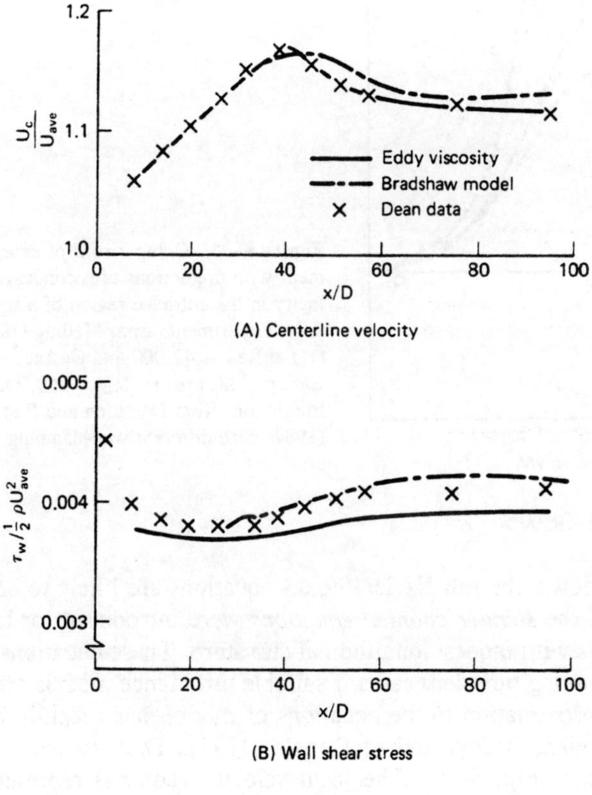

Figure 8–35 Comparison of experiment with predictions for turbulent flow in the entrance region of a channel. Data of Dean (1972). (From Cebeci and Chang, 1978.)

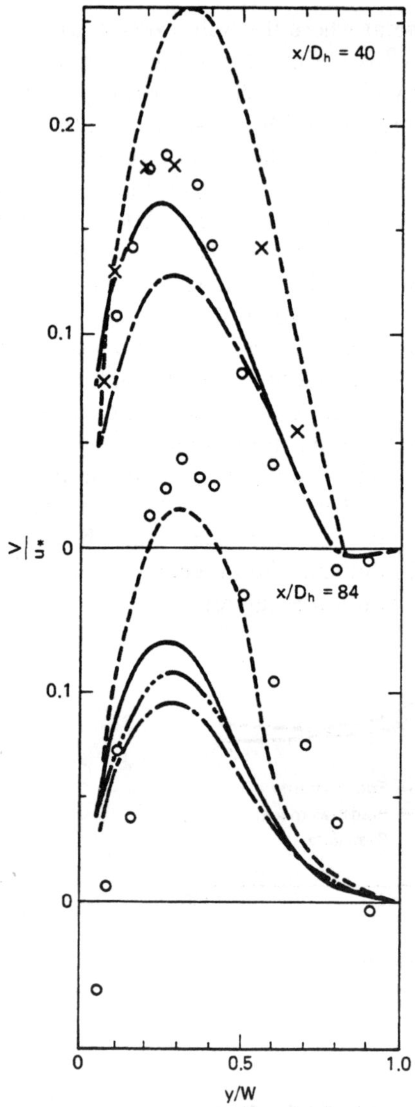

Figure 8–36 Comparisons of experiment with predictions of secondary velocity in the entrance region of a square duct. Experiments from Melling (1975) (×) at Re_D = 42,000 and Gessner and Emery (1981) (o) at Re_D = 250,000. Predictions from Demuren and Rodi (1984) with different wall-damping models.

8–4–3 General Channel Flows

For truly general channel flows, the full Navier-Stokes equations are likely to be required. In Section 8–2–3, the *slender channel equations* were introduced for laminar flow in slender channels with modest longitudinal curvature. These equations can also be applied to corresponding turbulent cases if suitable turbulence models are introduced. This kind of approximation to the equations of motion has recently been extended further by Govindan, Briley, and McDonald (1991). The essence of the approximation can be seen in Fig. 8–37. The local velocity vector is represented

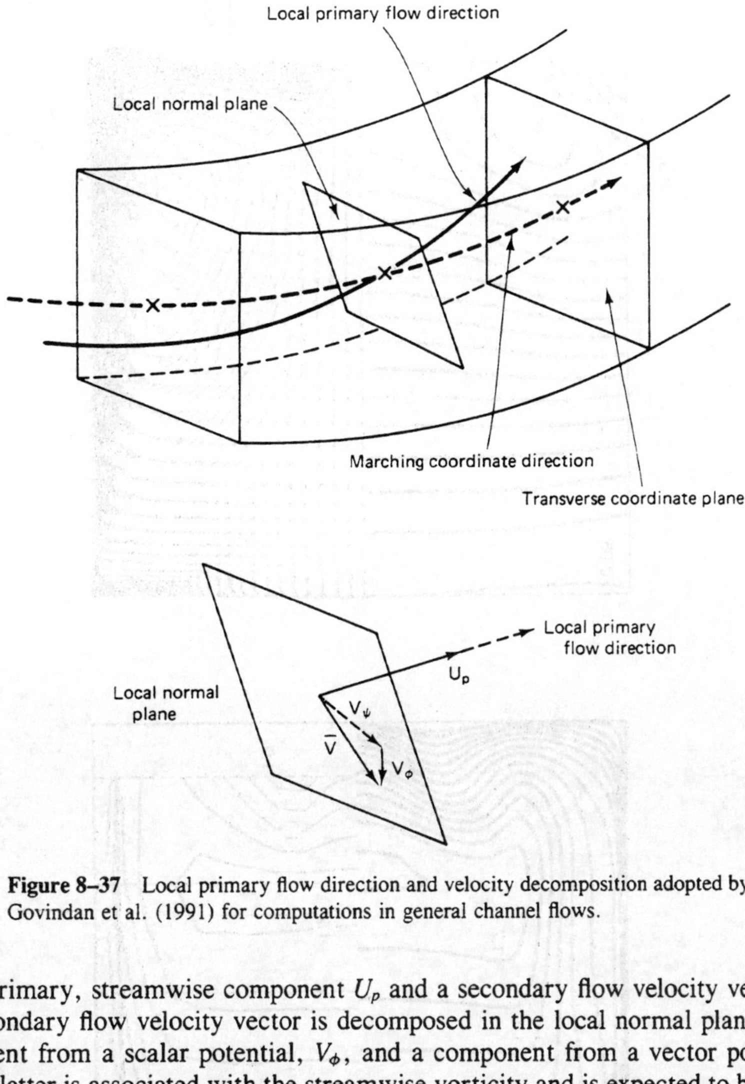

Figure 8–37 Local primary flow direction and velocity decomposition adopted by Govindan et al. (1991) for computations in general channel flows.

with a primary, streamwise component U_p and a secondary flow velocity vector **V**. The secondary flow velocity vector is decomposed in the local normal plane into a component from a scalar potential, V_ϕ, and a component from a vector potential, V_ψ. The latter is associated with the streamwise vorticity and is expected to be large. The former is expected to be small, so it is neglected in the transverse momentum equations, but not in the streamwise momentum or continuity equations. Also, viscous diffusion in the streamwise direction is neglected. The result is a system of equations that can be solved numerically by marching downstream. With further approximations, these equations reduce to the slender channel equations or the two-dimensional boundary layer equations. Govindan et al. (1991) have applied the equations with a mixing length model, calculated with an integrated form of the TKE equation for the turbulence, to the challenging case of flow through a 90° bend in a square duct. Some comparisons of predictions with data are shown in Fig. 8–38, where good agreement can be noted.

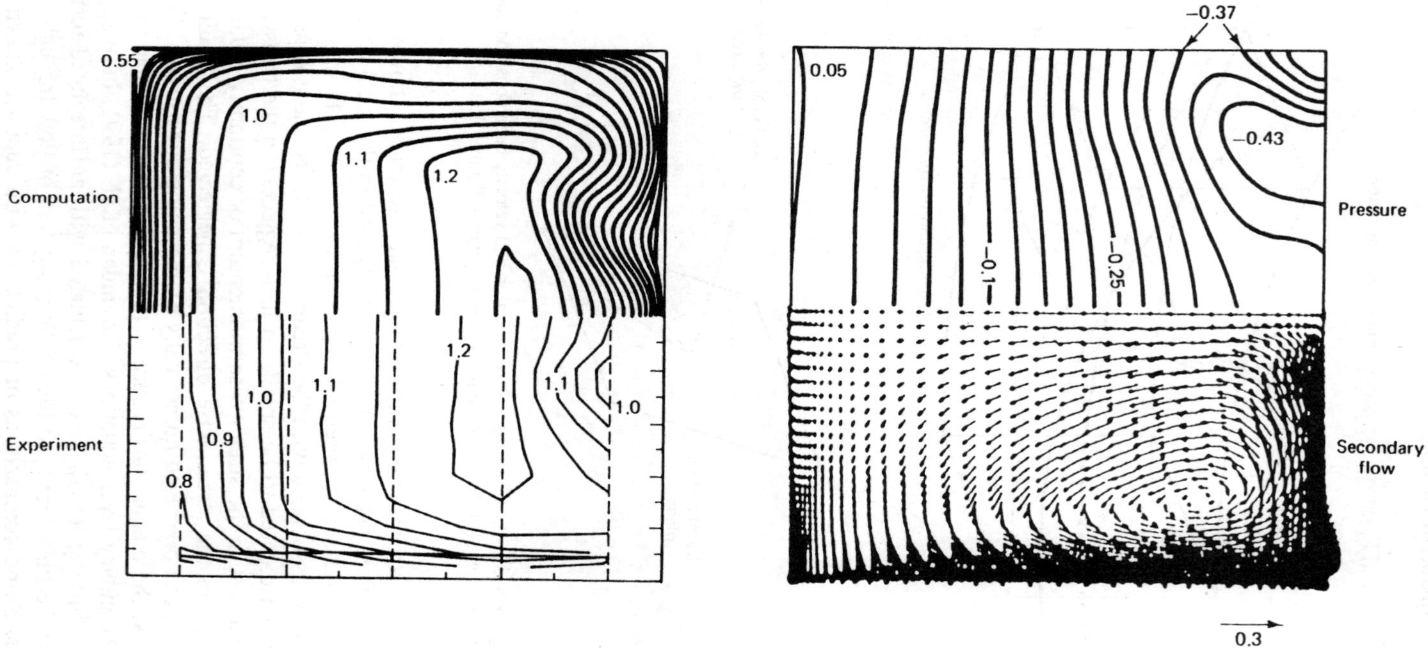

Figure 8–38 Comparison of predictions with experiment for turbulent flow at 60 degrees in a square duct with a 90-degree bend. (From Govindan, Briley, and McDonald, 1991.)

8–4–4 Non-Newtonian Fluids

Most non-Newtonian fluids have very high effective viscosities, so they normally flow in the laminar regime. There is, however, one important exception: The experimental fact that the pressure drop found for dilute solutions of polymers in pipes is dramatically lower than for the same flow rate of the pure solvent has led to great interest in this kind of non-Newtonian flow. This lower pressure drop occurs even though the viscosity of the solution is higher than that of the pure solvent. Clearly, the presence of the polymer must be affecting the turbulence.

The following material comes mainly from a review of the subject by Hoyt (1989). A typical result is illustrated in Fig. 8–39 for polyacrylamide in water in small concentrations. Obviously, the effect is large. Further, the long-chain polymer molecule apparently becomes more extended in seawater, making it even more effective. Other polymers are more or less effective. The criteria are solubility and a linear, extended molecule. The most effective molecules have a linear structure and the maximum extensivity for a given molecular weight.

The inner region is strongly affected by the polymer. A wall law plot for such a

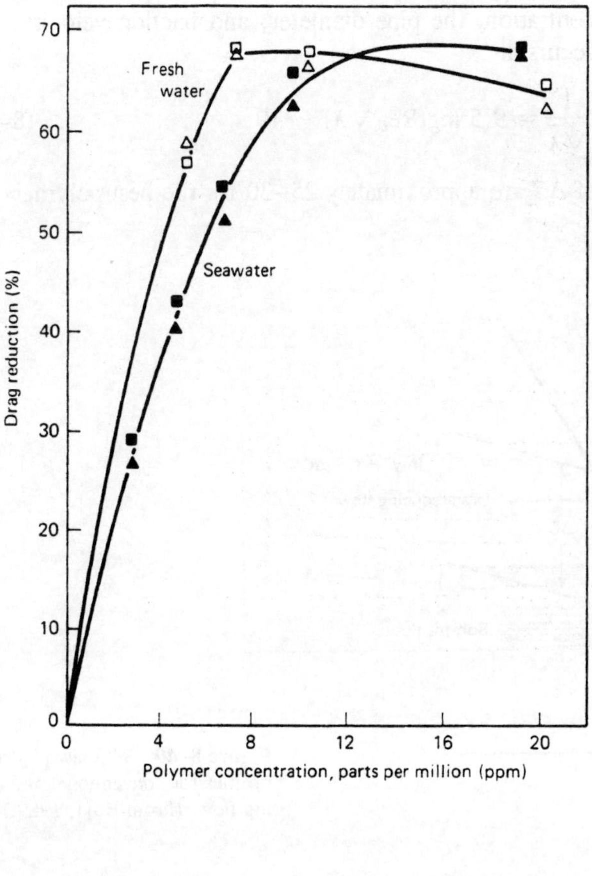

Figure 8–39 Drag reduction tests on two high-molecular-weight polyacrylamides in water in pipes at Re_D = 14,000. (From Hoyt, 1989.)

solution takes the form shown in Fig. 8–40, and the log law becomes

$$u^+ = \frac{1}{\kappa}\ln(y^+) + C + \Delta C \tag{8–54}$$

Here, the shift in the log region is up rather than down, as for roughness (see Section 7–3–2). Instead of intersecting the usual laminar sublayer solution, $u^+ = y^+$, the polymer solution profiles intersect an empirically determined maximum drag reduction line

$$u^+ = 11.7\ln(y^+) - 17 \tag{8–55}$$

The limited turbulence data available suggest that the main effect of the polymers is in reducing the level of v', but u' also seems to be reduced somewhat. In addition, the polymer increases the time between bursts (see Section 7–4).

The drag reduction in a pipe has been correlated as

$$\frac{1}{\sqrt{\lambda}} = 2\log(\mathrm{Re}_D\sqrt{\lambda}) - 0.8 + \frac{\Delta C}{2\sqrt{2}} \tag{8–56}$$

One can see that this is a simple modification of the usual relation, Eq. (8–41). The important parameter ΔC is a complicated function of the type of polymer used, its molecular weight and concentration, the pipe diameter, and friction velocity. The maximum drag reduction occurs at

$$\frac{1}{\sqrt{\lambda}} = 9.5\log(\mathrm{Re}_D\sqrt{\lambda}) - 19 \tag{8–57}$$

and the maximum values of ΔC are approximately 25–30 for the best polymers.

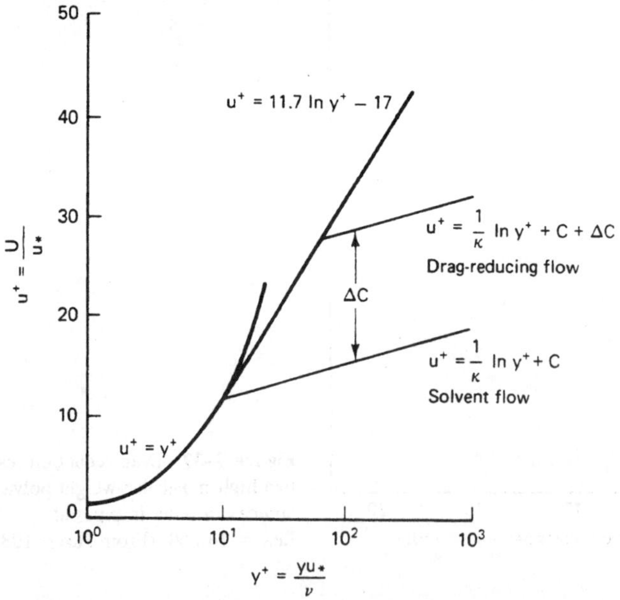

Figure 8–40 Wall law plot of velocity profiles for conventional and drag-reducing flow. (From Hoyt, 1989.)

8–5 HIGH-SPEED FLOWS

High-speed, laminar external flows were discussed in Chapter 5, and high-speed, turbulent external flows will be the subject of Chapter 10. Most high-speed internal flows of practical interest occur in cases where the flow is far from fully developed, so they can often be treated in the same way as external flows. There is, however, one important matter concerning viscous effects on high-speed internal flows that warrants a special discussion.

Consider a one-dimensional flow of an *ideal gas* through a duct with friction on the walls. For the simplest case, a constant-area duct, the equations of motion, together with $f = \tau_w/(1/2\rho u_{\text{ave}}^2)$ and $\gamma \equiv c_p/c_v$, give

$$\frac{1}{M^2}\frac{dM^2}{dx} = \frac{\gamma M^2\left(1 + \dfrac{(\gamma - 1)}{2}M^2\right)}{(1 - M^2)}\frac{4f}{D} \tag{8–58}$$

This equation shows that the effect of wall friction is always to drive the average Mach number in the duct towards unity. If the flow is supersonic, friction decreases the Mach number. If the flow is subsonic, friction increases the Mach number. The locus of states corresponding to these kinds of flows on an (h, s) or *Mollier* diagram are called *Fanno lines,* as illustrated in Fig. 8–41. (The *Rayleigh line* shown in the figure will be explained shortly.) The upper branch of the Fanno line corresponds to subsonic flow, the lower branch corresponds to supersonic flow, and the point of

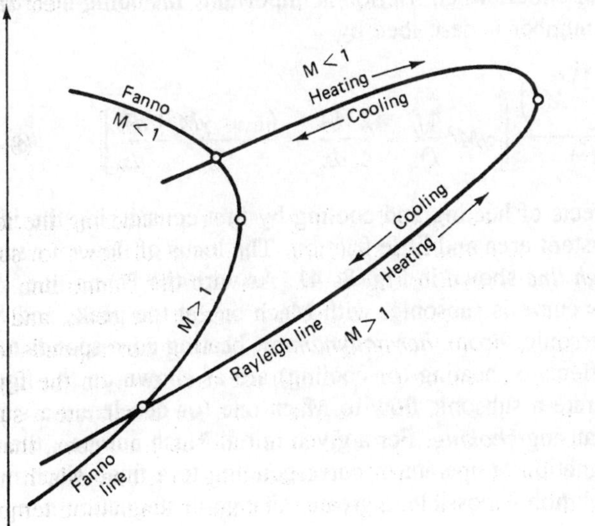

Figure 8–41 *Fanno* and *Rayleigh lines* in compressible internal flows.

maximum entropy, s, corresponds to Mach one. In a constant-area duct, the Mach number cannot pass through unity, and we can speak of a maximum length of duct for a given friction coefficient and given initial conditions, without altering the initial conditions and without introducing inviscid discontinuities. Integrating Eq. (8–58) from the initial Mach number to unity gives

$$L_{max}\frac{4\bar{f}}{D} = \frac{1 - M_i^2}{\gamma M_i^2} + \frac{\gamma + 1}{2\gamma}\ln\left[\frac{(\gamma + 1)M_i^2}{2\left(1 + \frac{\gamma - 1}{2}M_i^2\right)}\right] \tag{8–59}$$

If the duct does not have a constant area, but rather $A = A(x)$, then the equations of motion give

$$\frac{1}{M^2}\frac{dM^2}{dx} = \frac{\left(1 + \frac{(\gamma - 1)}{2}M^2\right)}{(1 - M^2)}\left[\gamma M^2\frac{4f}{D} - \frac{2}{A}\frac{dA}{dx}\right] \tag{8–60}$$

The reader is perhaps more familiar with this equation without the frictional term, where it is used to show that an inviscid flow in a duct can pass from subsonic to supersonic flow only through a minimum-area *throat*, where $dA/dx = 0$. With friction, the sonic point occurs, not at the minimum area, but where the whole term in brackets on the right-hand side of Eq. (8–60) is equal to zero. The effect of friction is still to increase the Mach number of a subsonic flow and to decrease the Mach number of a supersonic flow, but now there is the added effect of the change in area. An increase in area tends to reduce the Mach number in subsonic flow, but to increase the Mach number in supersonic flow.

Since momentum and heat transfer are such analogous processes, we often have significant heat transfer effects when friction is important. Including heat transfer, the variation in Mach number is described by

$$\frac{1}{M^2}\frac{dM^2}{dx} = \frac{\left(1 + \frac{(\gamma - 1)}{2}M^2\right)}{(1 - M^2)}\left[\gamma M^2\frac{4f}{D} - \frac{2}{A}\frac{dA}{dx} + \frac{(1 + \gamma M^2)}{T_t}\frac{dT_t}{dx}\right] \tag{8–61}$$

It is simplest to see the effects of heating and cooling by first considering the idealized limiting case with constant area and little friction. The locus of flows for such a process falls on the *Rayleigh line* shown in Fig. 8–41. As with the Fanno line, flow on the upper branch of this curve is subsonic, with Mach one at the peak, and flow on the lower branch is supersonic. From *thermodynamics,* heating corresponds to increasing entropy, so the effects of heating (or cooling) are as shown on the figure. Indeed, heating can accelerate a subsonic flow to Mach one (or decelerate a supersonic flow to Mach one), causing *choking*. For a given initial Mach number, there is a maximum change in stagnation temperature, corresponding to a final Mach number of unity, for which a solution is possible; a greater change in stagnation temperature will cause a disruption in the incoming flow or unsteady flow. When acting alone or together, friction and heating increase the Mach number for a subsonic flow and decrease the Mach number for a supersonic flow, while an increase in area decreases the Mach number for a subsonic flow and increases the Mach number for a supersonic flow.

Finally, we have stressed throughout this book the close linkage between momentum, heat, and mass transfer. Here, we restrict the discussion to cases where the molecular weight and ratio of specific heats remain fixed, for clarity. [It is not difficult to include variations in these quantities also; see Shapiro (1953), Chapter 8.] The equations of motion now can be written

$$\frac{1}{M^2}\frac{dM^2}{dx} = \frac{\left(1 + \frac{(\gamma - 1)}{2}M^2\right)}{(1 - M^2)} \cdot \left[\gamma M^2 \frac{4f}{D} - \frac{2}{A}\frac{dA}{dx} + \frac{(1 + \gamma M^2)}{T_t}\frac{dT_t}{dx} + \frac{2(1 + \gamma M^2)}{\dot{m}}\frac{d\dot{m}}{dx}\right] \tag{8-62}$$

where it has been assumed that fluid entering the flow by mass transfer does so perpendicularly to the wall. All of this can be summarized by saying that friction, heating, and the addition of mass, acting singly or together, increase the Mach number for a subsonic flow and decrease the Mach number for a supersonic flow, whereas an increase in area decreases the Mach number for a subsonic flow and increases the Mach number for a supersonic flow.

This general subject has many interesting features, including the important matter of the fluid's making a continuous passage through a sonic point. That issue was raised briefly in Section 5-13. Looking at Eq. (8-62), we see that a passage through $M = 1$ from the subsonic side requires that $(1 - M^2)\, dM^2/dx$ be positive until $M = 1$, where it must pass through zero and then become negative on the supersonic side. This means that the terms in the brackets involving friction, change in area, heat transfer, and mass transfer must be balanced such that the numerator goes to zero in a special way as $(1 - M^2)$ in the denominator goes to zero. This type of behavior is called a *saddle point singularity* in *mathematics*. The interested reader should refer to the classic text by Shapiro (1953) for a detailed exposition.

PROBLEMS

8.1. A liquid with the same density as water but twice its viscosity at room temperature is flowing through a long tube 1.0 cm in diameter at a rate of 20.0 cm³/s. What is the pressure drop per meter of length? How would your analysis change if the flow rate were doubled?

8.2. Water flowing through a 3-m section of a long pipe of 1-cm diameter at a rate of 15 cm³/s is cooled from 100°C to 30°C. The inside wall temperature of the pipe is 20°C. What is the heat transfer coefficient?

8.3. The inside of a pipe is covered with a material that highly absorbs CO_2. The inside diameter of the pipe is 4 cm. A mixture of 5% CO_2 in air enters the pipe at 0.5 m/s at standard temperature and pressure. The CO_2 is so strongly absorbed that the mass fraction of CO_2 on the wall may be taken to be zero. What is the *bulk* concentration of CO_2 after 10 m of pipe are passed, assuming fully developed flow over the whole length of the pipe?

8.4. Repeat Problem 8.1 for a square tube and an elliptical tube of eccentricity equal to 2 : 1, each with the same cross-sectional area as the round tube.

8.5. Suppose we wish to estimate the length of the entrance region in a 1.0-in-diameter pipe with air at 0.5 atm and 65°F flowing in the entrance at 5.0 ft/s. Make the crude assumption that the flat-plate boundary layer formulas can be used for this purpose. How long is the entrance region, in diameters?

8.6. Consider water flow in the entrance region of a pipe 0.75 cm. in diameter. The flow rate is 5 cm^3/s. Plot the pressure drop *vs.* distance from the pipe entrance to the fully developed flow station.

8.7. Plot the length of the thermal entrance region *vs.* the value of the Prandtl number for laminar flow in a tube.

8.8. Plot the ratio of the centerline velocity to the average velocity in a pipe for laminar, fully developed flow for *Newtonian* and *non-Newtonian* fluids of the *power law* type as a function of the exponent p.

8.9. For water at 25°C flowing in a pipe 0.5 m in diameter at 2 m^3/s, what is the value of λ for a smooth wall? What is the corresponding value for a concrete pipe? Take ($k \approx$ 0.002 m.)

8.10. A heat exchanger has water flowing in tubes with an inside diameter of 1.0 in at a flow rate of 0.005 ft^3/s. The average water temperature is 100°F. What is the Stanton number for a smooth tube? What will be the percentage increase in the film coefficient if the tubes are roughened to have $k = 0.01$ in?

8.11. Using analytical models, plot the variation in the mixing length and eddy viscosity across a pipe in fully developed flow. Make qualitative comparisons to data.

8.12. Compare the results for the *resistance coefficient* λ or the *friction factor* f for fully developed flow in a pipe *vs.* a square duct as a function of Reynolds number.

8.13. How much reduction in pressure drop can be obtained by adding 10 ppm of polyacrylamide to water flowing at $Re_D = 10{,}000$ in a pipe?

8.14. Suppose air at 2.0 atm and 350° K enters a tube at a Mach number of 0.3. The average *friction factor* f is approximately 0.003. What is the maximum length in diameters of tube where the exit Mach number is unity? What if the initial Mach number is 2.0?

CHAPTER 9

FREE SHEAR FLOWS

9–1 INTRODUCTION

In this chapter, we shall be concerned with those flows where there is no rigid surface present in the region of interest. The three general representatives of this class of free shear flows are wakes, jets, and simple shear layers, as illustrated schematically in Fig. 9–1. Attention is focused on the flow downstream of the body producing the wake or the walls that initially bound the jet or shear layer. The major simplifying feature of such flows for turbulent cases is that the treatment of an inner, wall-dominated region is eliminated. Thus, one is concerned only with the equivalent of the outer region of a wall-bounded turbulent layer.

Both wake and jet flows can be viewed as consisting of two or more regions. The *near wake,* right behind the body, and the *potential core,* near the injection station of the jet, are relatively complex, and they depend on the details of the body shape or the jet nozzle. The near-wake region is complicated and often involves separated flow, so it is outside the scope of this volume. In the idealized case of small boundary layers on the outside and the inside of the jet injector, there is a substantial initial region before the developing shear layers from the edges of the jet merge on the axis. This region is called the *potential core*, since the flow of the jet fluid in it is uniform and inviscid. If the flow coming out of the injector is fully developed, or nearly so, there is little or no potential core. Far behind the body or the jet noz-

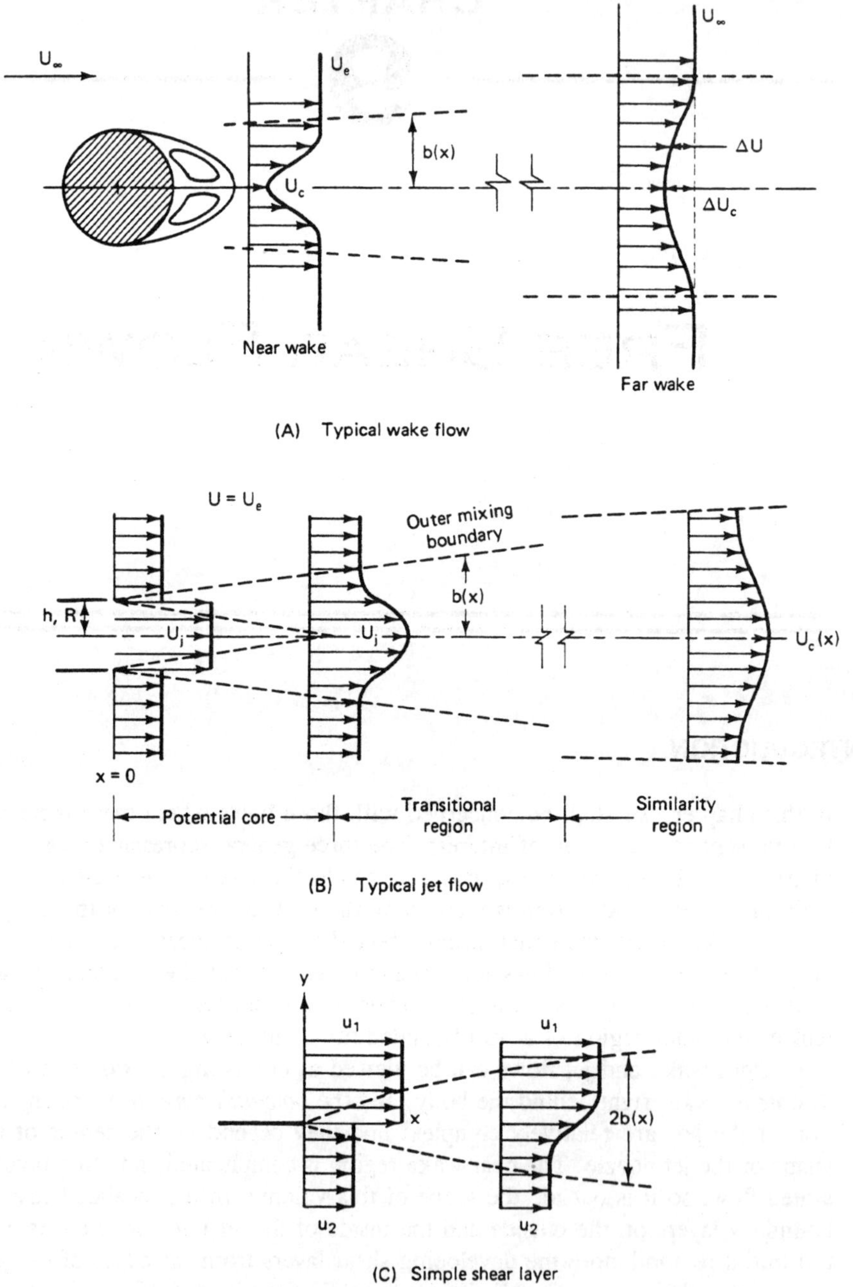

(A) Typical wake flow

(B) Typical jet flow

(C) Simple shear layer

Figure 9–1 Schematic illustrations of free shear flows.

zle, the flows assume a more universal behavior, and analysis is simpler. A transitional region between the near and far fields is often also envisioned. We will concentrate on the far wake and the main mixing region of jets and simple shear layers.

Most free shear flows of engineering interest are turbulent. This is because the velocity profiles generally involve inflection points (see Fig. 9–1), so the flows are quite unstable. Thus, the discussion here will emphasize turbulent flows. The treatment of laminar free shear flows will serve largely as an introduction to the analysis of the turbulent cases.

9–2 LAMINAR FREE SHEAR FLOWS

9–2–1 The Simple Shear Layer

The simplest case of a free mixing flow is that formed by two parallel streams with different velocities that are initially separated by a thin splitter plate. Downstream of the plate, the two streams interact due to viscous shear, and a growing mixing zone develops (see Fig. 9–1(C)). This flow is sometimes called a *mixing layer*, and it can be either laminar or turbulent. The turbulent case will be discussed later.

An exact solution can be developed for the incompressible case when the boundary layers on both sides of the splitter plate can be neglected. The initial profile for the mixing layer at the end of the plate ($x = 0$) is then simply a discontinuous jump at $y = 0$ from the velocity above the plate, U_1, to that below the plate, U_2. The boundary layer equations admit to a *similarity solution* for this case, with

$$\frac{u(x, y)}{U_1} = f'(\eta(x, y))$$

$$\psi(x, y) = U_1 b(x) f(\eta) \tag{9–1}$$

$$\eta = \frac{y}{b(x)}$$

where $b(x)$ is a length that scales the width of the viscous region. Substituting into the boundary layer equations gives

$$f''' + \frac{U_1 b}{\nu}\frac{db}{dx} ff'' = 0 \tag{9–2}$$

For a similarity solution to hold, the coefficient of the second term must be a constant, which can be conveniently taken to be equal to 1/2. This leads to

$$b(x) = \left(\frac{\nu x}{U_1}\right)^{1/2}, \qquad \eta = \frac{y}{x}\left(\frac{U_1 x}{\nu}\right)^{1/2} \tag{9–3}$$

and

$$f''' + 1/2\, ff'' = 0 \tag{9–4}$$

This equation is essentially the same as that for the Blasius solution in Section 4–3–1, but the boundary conditions are different here, namely,

$$f(0) = 0$$
$$\lim_{\eta \to \infty} f'(\eta) = U_1 \qquad (9\text{–}5)$$
$$\lim_{\eta \to -\infty} f'(\eta) = U_2$$

A numerical solution is again required, and some typical results from Lock (1951) are shown in Fig. 9–2. It can be seen that the width of the mixing zone grows as $x^{1/2}$.

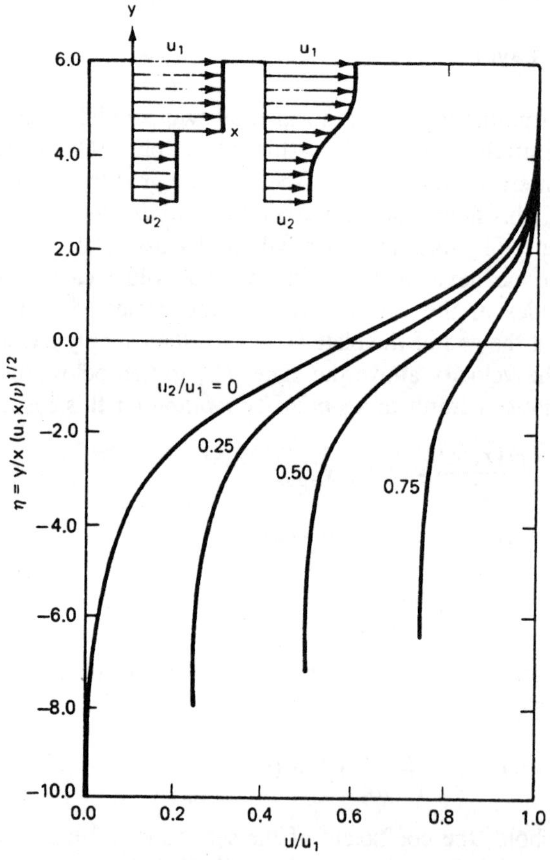

Figure 9–2 Velocity profiles for the laminar simple shear layer. (From Lock, 1951.)

The case of an initial boundary layer on the splitter plate for $U_2 = 0$ has been treated by Kubota and Dewey (1964), with an approximate analysis based on the integral momentum equation approach described in Chapter 2 for wall boundary layers. The variation in the velocity along the dividing streamline from the end of the plate is shown in Fig. 9–3. The exact solution given by Eq. (9–4) for the case with no initial boundary layer gives $0.587U_1$ for all axial stations. In this case, that condi-

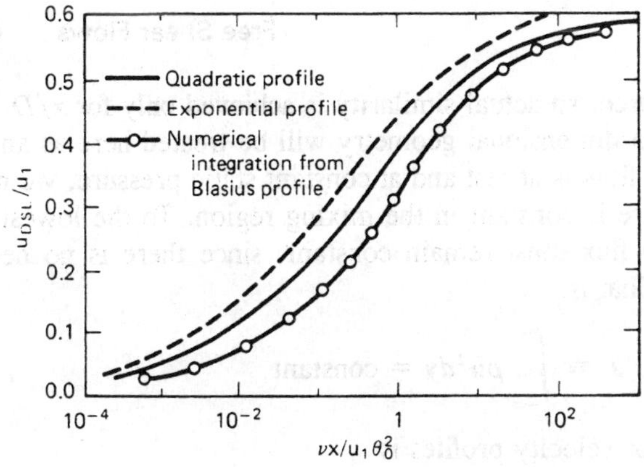

Figure 9–3 Velocity along the dividing streamline for the laminar simple shear layer. (From Kubota and Dewey, 1964.)

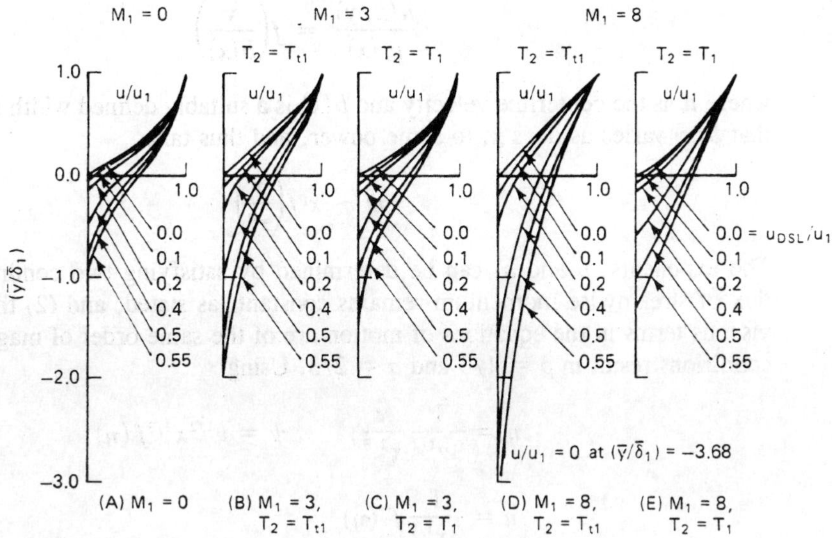

Figure 9–4 Velocity profiles for the laminar simple shear layer with an initial boundary layer. (From Kubota and Deway, 1964.)

tion is achieved only as $x/\theta_0 \to \infty$, where θ_0 is the initial momentum thickness. The profiles develop downstream, as shown in Fig. 9–4(A).

High-speed flows have been considered by a number of different authors. The analysis of Kubota and Dewey (1964) was extended to cases of variable density by the use of compressibility transformations. Some profile results at Mach 3.0 and 8.0 are shown in Figs. 9–4(B)–(E). Note that the vertical coordinate is based on the transformed values.

9–2–2 Jets

The simplest jet flows are laminar, constant-density cases where the surrounding fluid is at rest. These flows also allow *similarity solutions*. One imagines the jet to be from a point or line source, so actual similarity is achieved only for $x/D \gg 1$ in real cases. The planar, two-dimensional geometry will be treated here as an example. Since the surrounding flow is at rest and at constant static pressure, we may assume that the static pressure is constant in the mixing region. To the lowest order, the total axial momentum flux must remain constant, since there is no net shear force acting on the flow. That is,

$$J \equiv \int_{-\infty}^{\infty} \rho u^2 dy = \text{constant} \tag{9–6}$$

The similarity condition for velocity profiles is

$$\frac{u(x, y)}{u_c(x)} = f\left(\frac{y}{b(x)}\right) \tag{9–7}$$

where u_c is the centerline velocity and $b(x)$ is a suitably defined width scale. Assume that $b(x)$ varies as does x, to some power, and thus take

$$\psi \sim x^p f\left(\frac{y}{x^q}\right) \tag{9–8}$$

The exponents, p and q, can be determined by satisfying two conditions: (1) The flux of streamwise momentum remains constant, as stated, and (2) the inertial and viscous terms in the equations of motion are of the same order of magnitude. These conditions result in $p = 1/3$ and $q = 2/3$. Using

$$\eta = \frac{1}{3\nu^{1/2}} \frac{y}{x^{2/3}}, \qquad \psi = \nu^{1/2} x^{1/3} f(\eta)$$

$$u = \frac{1}{3x^{1/3}} f'(\eta) \tag{9–9}$$

$$v = -\frac{1}{3} \nu^{1/2} x^{-2/3} (f - 2\eta f')$$

We see that the equations of motion reduce to the nonlinear equation

$$f''' + ff'' + (f')^2 = 0 \tag{9–10}$$

The boundary conditions are

$$\lim_{\eta \to \infty} f'(\eta) = f(0) = f''(0) = 0 \tag{9–11}$$

and Schlichting (1933) developed the closed form solution

$$u = \frac{2}{3} a^2 x^{-1/3} (1 - \tanh^2(a\eta)) \tag{9–12}$$

to this equation, where

$$a = \left(\frac{9J}{16\rho\nu^{1/2}}\right)^{1/3} \tag{9–13}$$

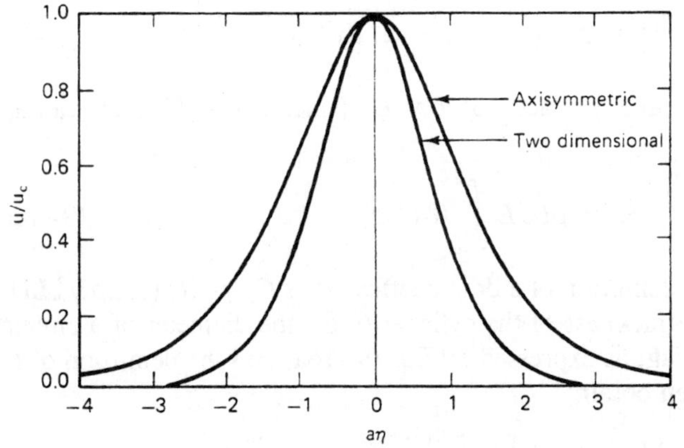

Figure 9–5 Velocity profiles in axisymmetric and two-dimensional laminar jets. (From Schlichting, 1933.)

Velocity profiles for this and the corresponding axisymmetric case are shown in Fig. 9–5. Also, $b \sim x^{2/3}$ and $u_c \sim x^{-1/3}$ for the planar jet, and $b \sim x$ and $u_c \sim x^{-1}$ for the axisymmetric jet.

In recent times, careful analysis (Schneider, 1985) has shown that the momentum flux does not remain constant if there are any walls in the flow field. Thus, if the jet issues from a hole in a wall, the axial momentum flux in the jet actually decreases slowly with distance from the orifice. A detailed discussion of these matters is beyond the scope of this book.

9–2–3 Wakes

The analysis of wakes is simple only in the far wake, where the velocity difference across the wake has decayed to a small value, permitting a linearization of the equations of motion. We will consider the planar, constant-density, constant-property case as an example.

If $\Delta u = U_\infty - u_c \ll 1$, the boundary layer momentum equation can be approximated as

$$U_\infty \frac{\partial u}{\partial x} = \nu \frac{\partial^2 u}{\partial y^2} \tag{9–14}$$

This is a simple, linear partial differential equation, and its solution can be found by standard methods. With the boundary condition of $\lim_{y\to\infty} u(x, y) = U_\infty$ and symmetry on the axis, the solution is

$$u(x, y) = U_\infty - \frac{C}{\sqrt{x}} U_\infty \exp\left(-\frac{U_\infty y^2}{4\nu x}\right) \tag{9–15}$$

The remaining constant C is determined to match the momentum defect in the wake with the drag of the body. If the body is a general two-dimensional (planar) cylinder, we may write the drag $\mathfrak{D}$ for a body length L perpendicular to the plane of the flow as

$$\mathfrak{D} = \rho L \int_{-\infty}^{\infty} u(U_\infty - u)\, dy \tag{9-16}$$

Noting the definition of Δu and the fact that $u \approx U_\infty$ (since $\Delta u \ll U_\infty$), we can approximate Eq. (9–16) as

$$\mathfrak{D} \approx \rho U_\infty L \int_{-\infty}^{\infty} (\Delta u)\, dy \tag{9-16a}$$

With this equation and the definition of a drag coefficient as $C_D = \mathfrak{D}/(1/2\rho U_\infty^2 LD)$, where D is a characteristic thickness of the cylinder (e.g., the diameter of a circular cylinder), we can set the drag, as expressed by Eq. (9–16a) and the definition of C_D equal to each other and then obtain

$$\frac{U_\infty - u(x, y)}{U_\infty} = 2^j C_D \left(\frac{\text{Re}}{16\pi} \frac{D}{x}\right)^{(j+1)/2} \exp\left(-\frac{U_\infty y^2}{4\nu x}\right) \tag{9-17}$$

where $\text{Re} = U_\infty D/\nu$. In this way, C may be determined.

Note that the axisymmetric result obtained by a parallel analysis is included here for $j = 1$; the planar result corresponds to $j = 0$.

9–3 MEAN FLOW AND TURBULENCE DATA FOR A ROUND JET IN A MOVING STREAM

9–3–1 Introduction

As in earlier chapters, we follow the practice here of beginning the discussion of turbulent cases with a presentation of data for representative flows. These flows then form the basis for the critical discussion of turbulence modeling. To limit the scope of this chapter to a manageable level, the case of a round jet in a coflowing external stream has been chosen as a representative flow to illustrate the nature of this class of flows. Some data for simple shear layers will also be included to illustrate one important matter. A more general treatment of the subject can be found in the monograph by Schetz (1980). Finally, the presentation will concentrate on the main mixing region beyond the potential core, since the behavior of the main mixing region is much more universal than that in the potential core. Harsha (1971) has given the following data correlation for the length of the potential core for air-air, axisymmetric jets:

$$\frac{x_c}{D} = 2.13(\text{Re}_D)^{0.097} \tag{9-18}$$

This equation can be used for purposes of estimation.

9–3–2 Constant-Density Flows

Figure 9–6 shows the variation of the centerline velocity, in terms of $U_e/[U_c(x/D) - U_e]$ and a characteristic width $r_{1/2}$ defined by the relation

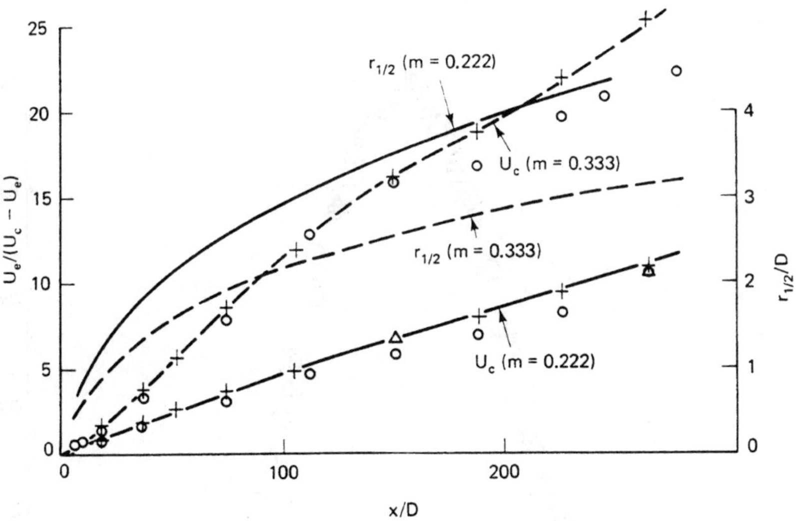

Figure 9–6 Variation of centerline velocity and half-radius for an axisymmetric jet: ○, △, hot wire; + Pitot tube. (From Antonia and Bilger, 1973.)

$$U(r_{1/2}, x) = \frac{U_c(x) + U_e}{2} \tag{9–19}$$

with the axial distance. This latter quantity is frequently called the *half-radius,* and it can be determined accurately more easily than can some ill-defined total width where the mixing zone merges asymptotically into the external stream. First, it can be seen that the influence of the parameter $m(\equiv U_e/U_j)$ on the flow field is quite profound. Second, the total distance to the final decay, where $U_c \rightarrow U_e$, is very long when measured in terms of jet diameters. Nondimensional radial velocity profiles are given in Fig. 9–7 for various x/D. These data indicate that the profiles are apparently *similar* for $x/D \geq 40$, meaning that the profiles, expressed in terms of coordinates such as those in the figure, remain unchanged with x/D. This fact will prove useful for analysis later.

Some data of Antonia and Bilger (1973) for the axial turbulence intensity are given in Figs. 9–8 and 9–9. For free turbulent flows, the intensity of the turbulence is usually normalized with the mean flow velocity difference across the layer $\Delta U_c = U_\infty - U_c$, rather than with U_e or U. Note that the average fluctuations are a substantial fraction of the mean flow velocity difference. The streamwise behavior of the centerline values and the characteristic shape of the transverse profiles can easily be seen. The maximum generally occurs somewhat off the centerline, since the mean velocity gradient and, hence, the production of turbulence, is greater there [see Fig. 9–7 and Eq. (7-94)]. Observe that the profiles of this variable have not attained a similarity condition by $x/D \approx 200$. The data of Gibson (1963) for $U_e = 0$ showed that the relative magnitudes of the axial, transverse, and peripheral turbulence intensities were about the same as for wall-bounded flows.

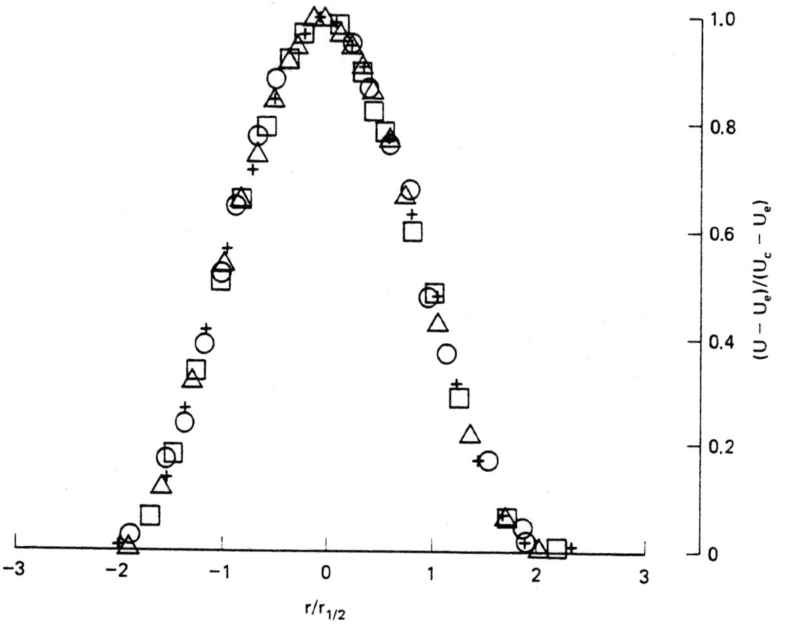

Figure 9–7 Nondimensional radial profiles across an axisymmetric jet: $m = 0.222$; $x/D = 38$ (+), 76 (△), 152 (□), 248 (○). (From Antonia and Bilger, 1973.)

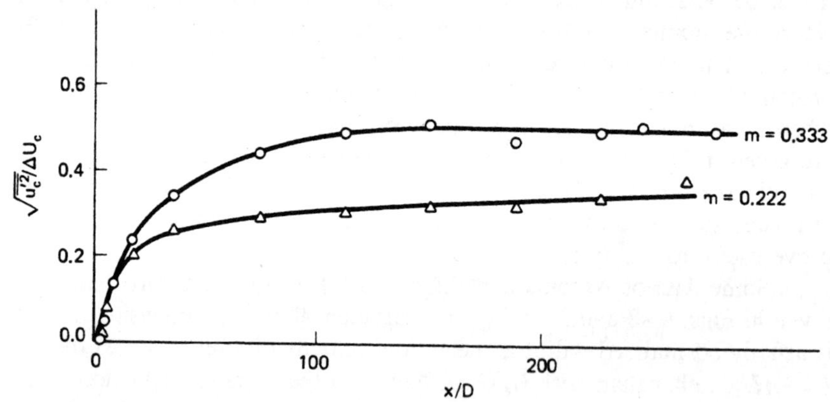

Figure 9–8 Streamwise variation of centerline axial turbulence intensity in axisymmetric jets. (From Antonia and Bilger, 1973.)

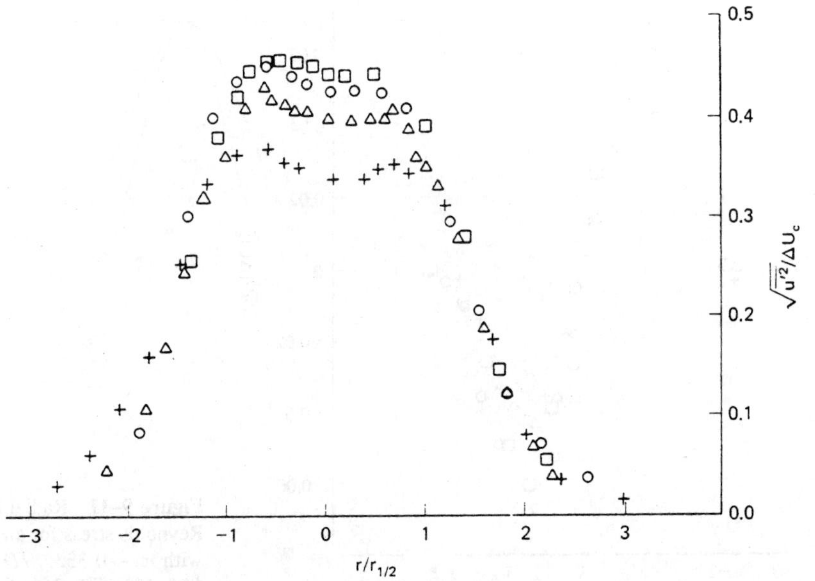

Figure 9–9 Radial profiles of axial turbulence intensity in axisymmetric jets: x/D = 38 (+), 76 (△), 152 (□), 266 (○); m = 0.333. (From Antonia and Bilger, 1973.)

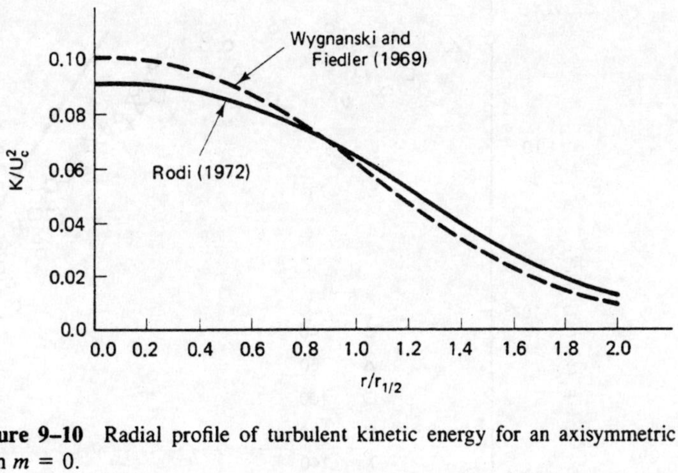

Figure 9–10 Radial profile of turbulent kinetic energy for an axisymmetric jet with m = 0.

The variation in the turbulent kinetic energy across the layer, again for U_e = 0, is shown in Fig. 9–10 from Rodi (1972) and Wygnanski and Fiedler (1969). Information on the Reynolds turbulent shear stress for $U_e \neq 0$ is shown in Fig. 9–11. Note the antisymmetry of this quantity in a symmetrical flow. About a line (or plane) of symmetry, $-\overline{u'v'} = 0$ when $\partial U/dy = 0$.

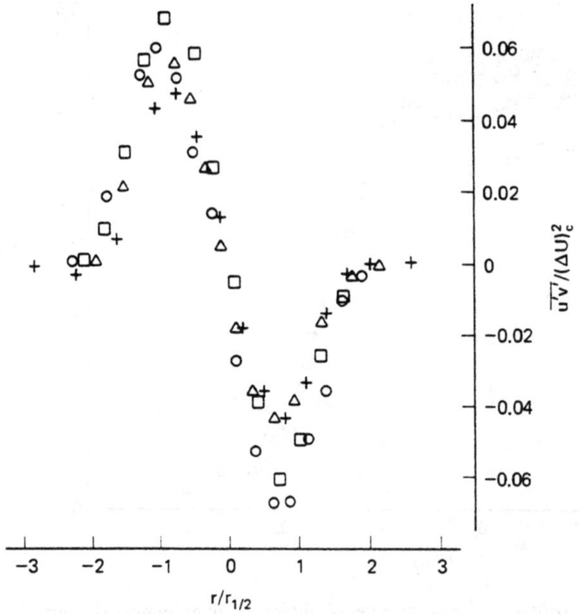

Figure 9–11 Radial profiles of Reynolds stress for an axisymmetric jet with $m = 0.333$; $x/D = 38$ (+), 76 (△), 152 (□), 266 (○). (From Antonia and Bilger, 1973.)

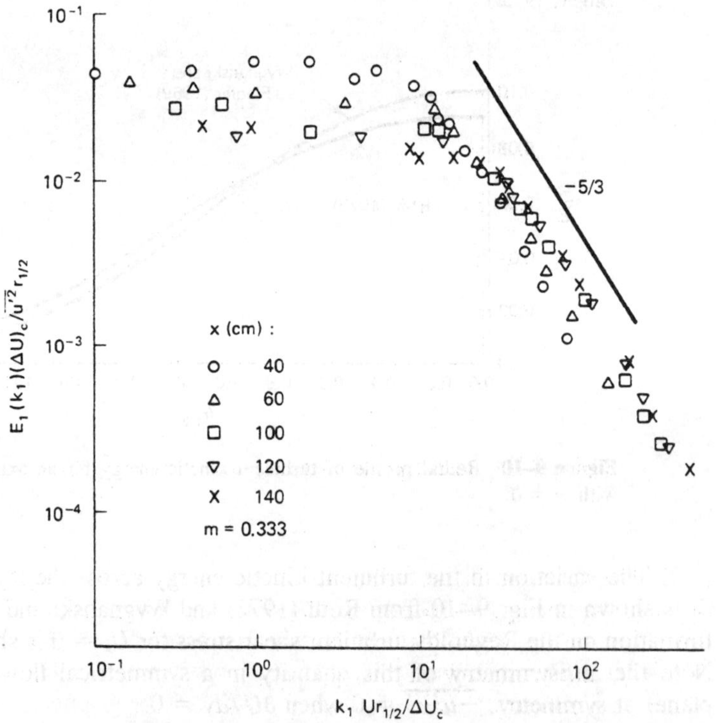

Figure 9–12 Spectra of the axial component for an axisymmetric jet. (From Antonia and Bilger, 1973.)

Spectra of the axial turbulence intensity are shown in Fig. 9–12. Again, as in wall boundary layers, motion of several decades in wave number k_1 can be observed. Finally, Fig. 9–13 shows the energy balance across the flow, where the magnitude of the individual terms in Eq. (7–94) can be seen and studied. The most complete and reliable information available is the basic data of Wygnanski and Fiedler (1969) for $U_e = 0$, as modified by Rodi (1975). This balance is much different than that for a boundary layer (cf. Fig. 7–24). In free shear flows, the energy production by the normal stresses

$$-(\overline{u'^2} - \overline{v'^2})\frac{\partial U}{\partial x} \tag{9–20}$$

in addition to production by shear stress, was included for completeness. It can be seen that the contribution from normal stresses is the smaller of the two, except near the axis.

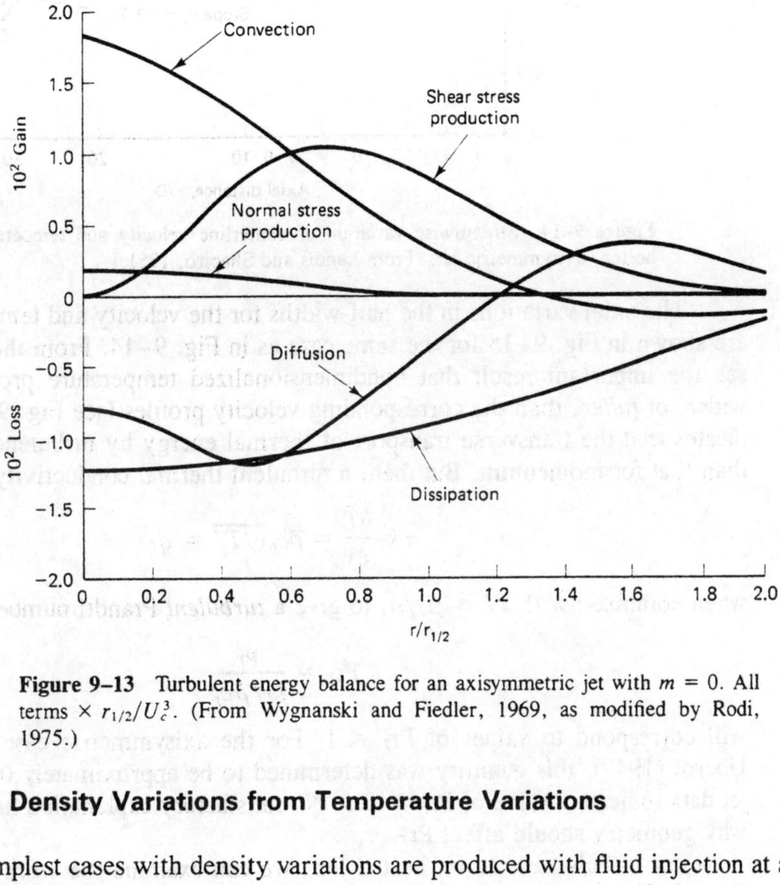

Figure 9–13 Turbulent energy balance for an axisymmetric jet with $m = 0$. All terms × $r_{1/2}/U_c^3$. (From Wygnanski and Fiedler, 1969, as modified by Rodi, 1975.)

9–3–3 Density Variations from Temperature Variations

The simplest cases with density variations are produced with fluid injection at a temperature different from that of the main flow. Experiments of that type in the axisymmetric geometry were reported by Landis and Shapiro (1951). Mean flow results, in terms of the axial variation of the centerline velocity and temperature, are shown in Fig. 9–14 for a case with $T_e/T_j = 0.77$ and $U_e/U_j \equiv m = 0.50$. The density ratio $\rho_e/\rho_j \equiv n$ for such cases is related to the temperature ratio as $T_e/T_j = \rho_j/\rho_e = 1/n$. It can be seen that the power decay law exponent P_i (i.e., in $\Delta\overline{T} \sim xP_T$, $\Delta U \sim xP_v$, etc.) is slightly greater for this case than that for the nearly constant-density case presented by Forstall and Shapiro (1950).

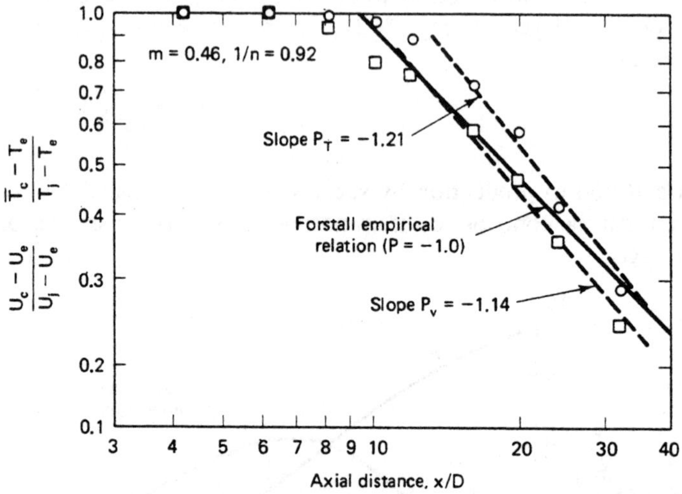

Figure 9–14 Streamwise variation of centerline velocity and temperature in a heated, axisymmetric jet. (From Landis and Shapiro, 1951.)

The axial variations in the half-widths for the velocity and temperature profiles are shown in Fig. 9–15 for the same case as in Fig. 9–14. From these data, we can see the important result that nondimensionalized temperature profiles are always wider, or *fuller*, than the corresponding velocity profiles (see Fig. 9–16), which indicates that the transverse transport of thermal energy by turbulence is more rapid than that for momentum. But then, a turbulent thermal conductivity k_T defined by

$$-k_T\frac{\partial\overline{T}}{\partial y} = \overline{\rho}c_p\overline{v'T'} = q_T \tag{9–21}$$

when combined with $\nu_T \equiv \mu_T/\rho$, to give a *turbulent* Prandtl number

$$\mathrm{Pr}_T = \frac{\nu_T}{k_T/\overline{\rho}c_p} \tag{9–22}$$

will correspond to values of $\mathrm{Pr}_T < 1$. For the axisymmetric case of Corrsin and Uberoi (1949), this quantity was determined to be approximately 0.7, while planar jet data indicate a value of $\mathrm{Pr}_T \approx 0.5$. No satisfactory explanation has been found of why geometry should affect Pr_T.

For turbulence data on heated jets, we can examine the work of Chevray and Tutu (1978) with $U_e = 0$. Velocity and temperature fluctuations at $x/D = 15$ are

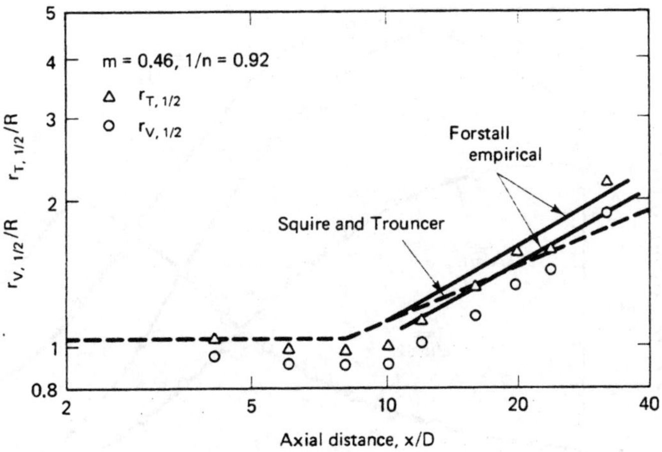

Figure 9–15 Variation in half-radii based on velocity and temperature in a heated, axisymmetric jet. (From Landis and Shapiro, 1951.)

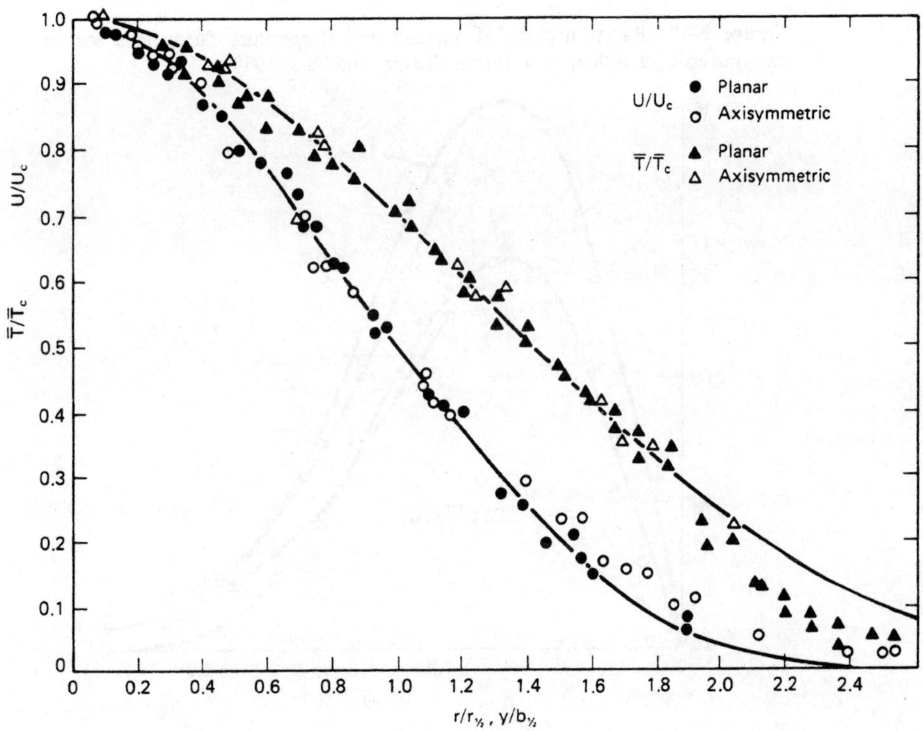

Figure 9–16 Transverse velocity and temperature profiles for planar and axisymmetric jets with $m = 0$. (From Ginevskii, 1966.)

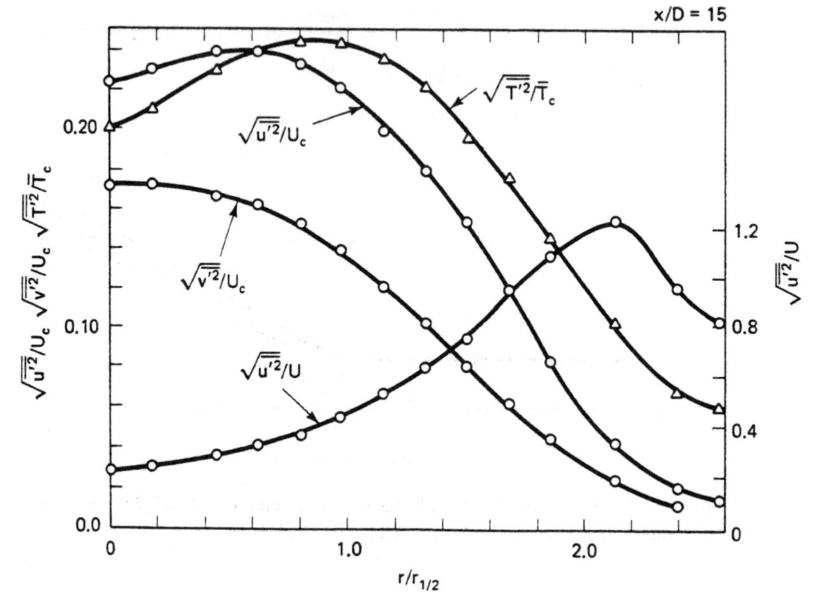

Figure 9–17 Radial profiles of velocity and temperature fluctuations for an axisymmetric jet with $m = 0$. (From Chevray and Tutu, 1978.)

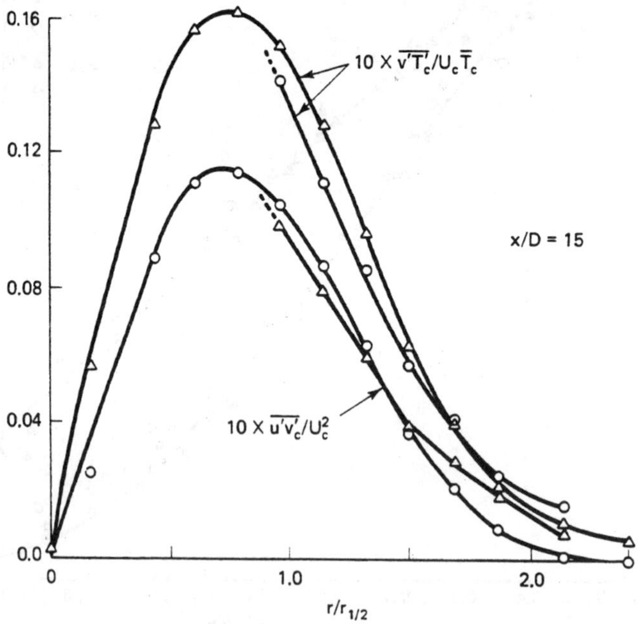

Figure 9–18 Radial profiles of turbulent shear and heat flux for an axisymmetric jet with $m = 0$. (From Chevray and Tutu, 1978.)

given in Fig. 9–17. Radial variations in turbulent shear and heat transfer are shown in Fig. 9–18. These profiles have similar shapes, and their peaks occur at roughly the same radial station, which is not, however, at the location of the maximum value of either $\partial U/\partial y$ or $\partial \overline{T}/\partial y$. These facts suggest that the transport mechanisms for momentum and heat are quite similar and that gradient transport models are suspect, even for this simple flow problem.

Finally, high-speed, free turbulent flows, such as the wake behind a supersonic body, produce substantial variations in temperature. The new effects that can occur under those circumstances will be illustrated with a discussion of the simple shear layer in supersonic flow later in this chapter.

9–3–4 Density Variations from Composition Variations

Many important practical applications involve injection of one fluid into the surroundings of a different fluid. In this book, the discussion is restricted to cases of one phase, either gas or liquid, and binary mixtures. A discussion of more general cases can be found in Schetz (1980).

An experimental correlation for the length of the *potential core*, in terms of the concentration of the injected species, was developed by Chriss (1968), who found the relation to be

$$\frac{x_c}{D} = 6.5\sqrt{\frac{\rho_j U_j}{\rho_e U_e}} \tag{9–23}$$

For data on the influence of variations in composition on the axial decay of centerline flow variables, the tests of Chriss (1968) can be used again. Figure 9–19 shows the obviously strong influence of density and velocity ratios on the centerline velocity decay. The axial variation in the centerline composition of the injectant is presented in Fig. 9–20.

When the axial decay of the centerline values is plotted on logarithmic paper, a power-law decay rate again appears. Many workers have attempted to determine the values of the relevant exponents for each type of variable: velocity, temperature, and composition. Schetz (1969) looked at a large group of data from various investigators to develop a correlation for P_c in terms of the single parameter $\rho_j U_j/\rho_e U_e$, and the result is shown in Fig. 9–21.

As with density changes produced by temperature variations alone, the half-width of concentration profiles is larger than that for velocity profiles and the concentration profiles themselves are consequently *fuller*, as shown in Fig. 9–22. Indeed, nondimensional temperature and concentration profiles have been found to be virtually identical.

Introducing a turbulent mass diffusion coefficient D_T through the equation

$$-\overline{\rho} D_T \frac{\partial C_i}{\partial y} = \overline{\rho}\,\overline{v' c_i'} \tag{9–24}$$

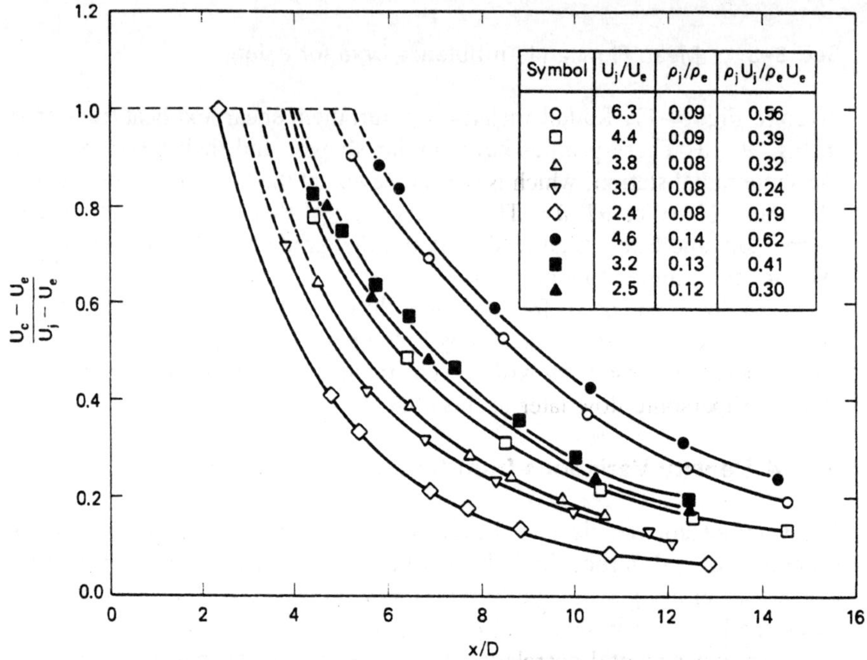

Figure 9–19 Streamwise variation in centerline velocity for axisymmetric H_2 jets into air. (From Chriss, 1968.)

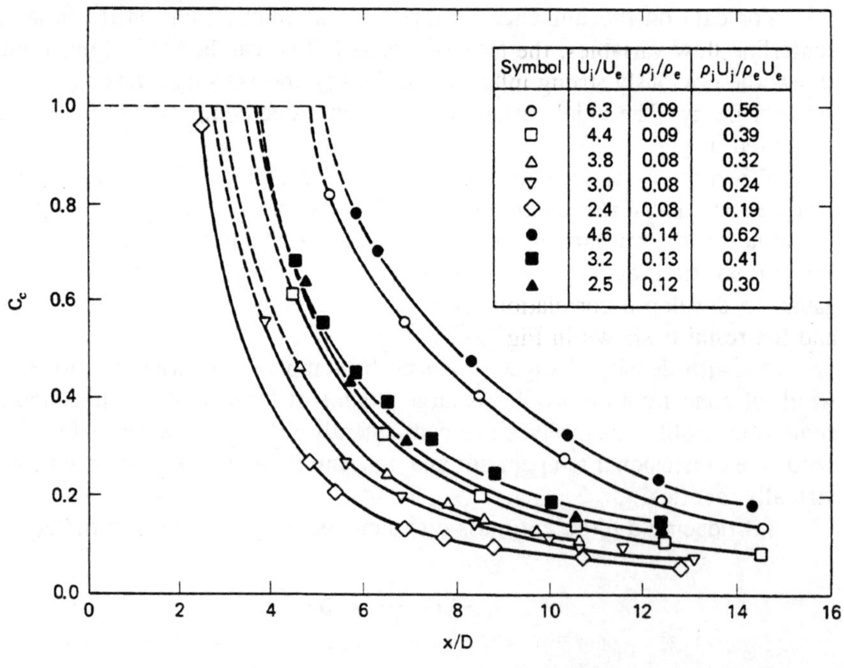

Figure 9–20 Streamwise decay of centerline concentration for an axisymmetric H_2 jet into air. (From Chriss, 1968.)

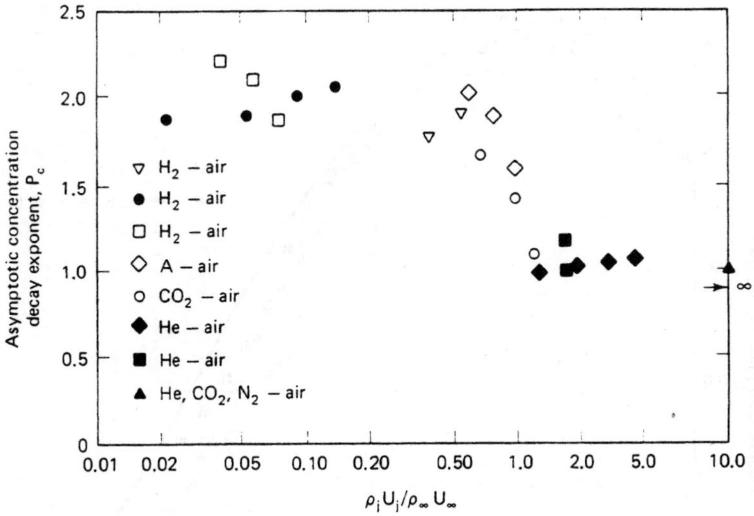

Figure 9–21 Experimental determination of asymptotic concentration decay exponent. (From Schetz, 1969.)

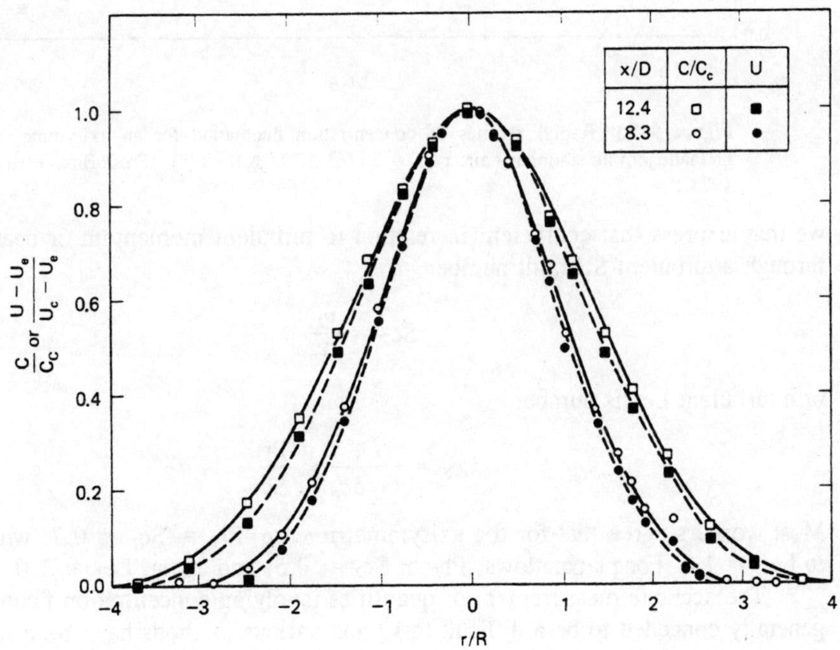

Figure 9–22 Radial profiles of velocity and concentration for an axisymmetric H_2 jet into air. (From Chriss, 1968.)

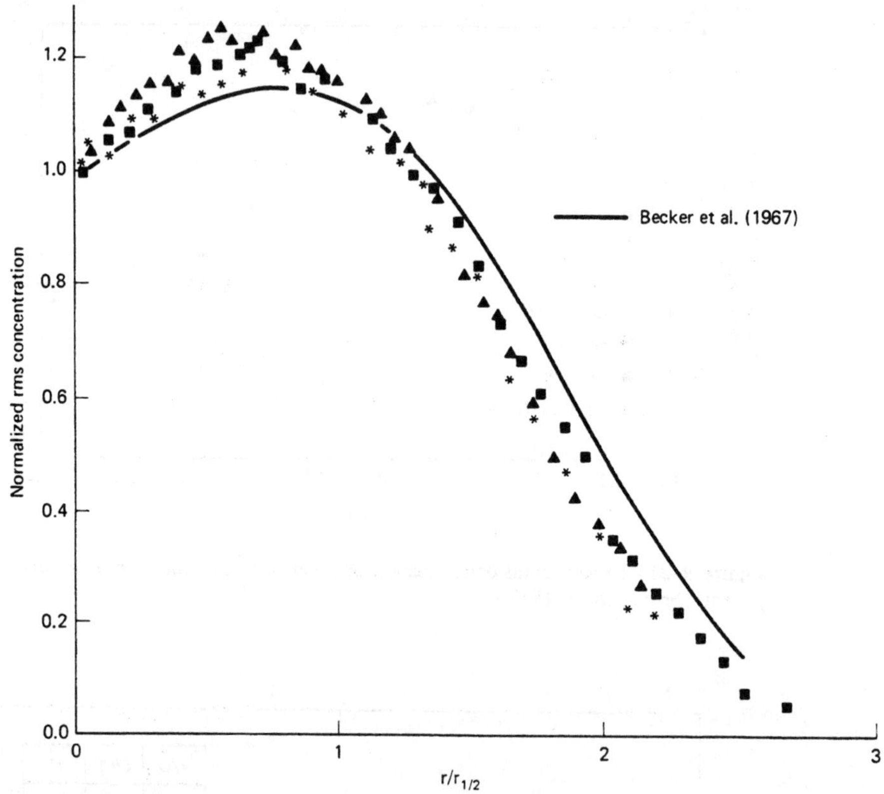

Figure 9–23 Radial profiles of concentration fluctuation for an axisymmetric methane jet into stationary air: x/D = 20 (■), 30 (▲), 40 (*). (From Birch et al., 1978.)

we may express that coefficient in relation to turbulent momentum or heat transfer through a turbulent Schmidt number

$$Sc_T = \frac{\nu_T}{D_T} \tag{9–25}$$

or a turbulent Lewis number

$$Le_T = \frac{D_T}{k_T/\rho c_p} = \frac{Pr_T}{Sc_T} \tag{9–26}$$

Most workers agree that for the axisymmetric case, $Pr_T \approx Sc_T \approx 0.7$, which leads to $Le_T \approx 1.0$. For planar flows, $Pr_T \approx Sc_T \approx 0.5$, and again, $Le_T \approx 1.0$.

The accurate measurement of quantities involving concentration fluctuations is generally conceded to be a difficult task, and various methods have been tried. The experiments of Becker et al. (1967) were for the axisymmetric geometry with $U_e = 0$, and the variable composition was provided by oil-generated smoke mixed with the air in the injectant. The measurements were made using a light-scattering technique. In another experiment, conducted by Birch et al. (1978), Raman scattering of laser light was used to obtain detailed mean and turbulent measurements in a round natural gas (95% methane) jet exhausting into air. Radial profiles of the root mean square of the normalized concentration fluctuations from both groups at various x/D are shown in Fig. 9–23. These profiles are in close agreement with the corresponding measurements for temperature by Antonia et al. (1975). The axial variation in the intensity of the concentration fluctuations is shown in Fig. 9–24, together with corresponding results for the velocity and temperature fields.

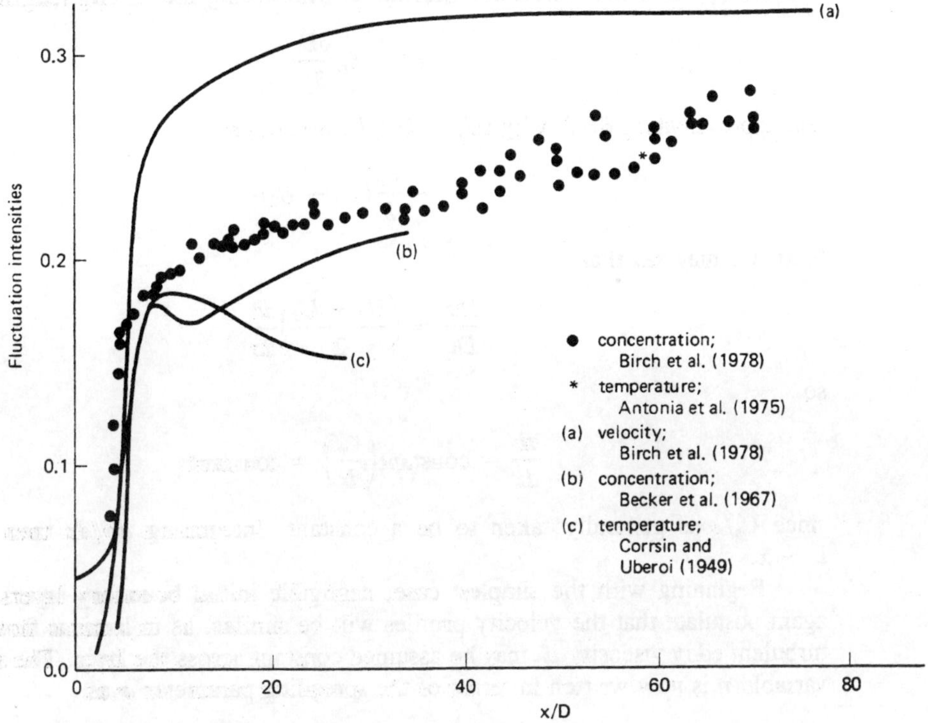

Figure 9–24 Streamwise variation of centerline fluctuation intensities for an axisymmetric jet with $m = 0$.

9–4 MEAN FLOW TURBULENT ANALYSES

9–4–1 Similarity Analyses

As with laminar free shear flows, the easiest turbulent case to analyze is the simple shear layer in Fig. 9–1(C). For all turbulent flows, the mathematical specification of the flow is not complete with the usual equations of motion, and a turbulent transport model must also be given. For the simple shear layer problem, this information is usually expressed in terms of a *spreading parameter* denoted as σ. This quantity reflects the rate at which the shear layer grows laterally with axial distance.

The dependence of the increase in width $b(x)$ of the shear layer on the axial distance can be developed from a simple physical argument. We postulate that the rate of growth of the turbulent region is directly proportional to the value of the transverse velocity fluctuations, i.e.,

$$\frac{Db}{Dt} \sim v' \tag{9–27}$$

where $D(\ )/Dt$ is the convective derivative. Introducing the mixing length

$$v' \sim \ell_m \frac{\partial U}{\partial y} \tag{9–28}$$

and approximating $\partial U/\partial y$ by $(U_1 - U_2)/b$, we obtain

$$v' \sim \frac{\ell_m}{b}(U_1 - U_2) \tag{9–29}$$

Next, we may say that

$$\frac{Db}{Dt} \approx \left(\frac{U_1 + U_2}{2}\right)\frac{db}{dx} \tag{9–30}$$

so

$$\frac{db}{dx} \approx \text{constant}\left(\frac{\ell_m}{b}\right) = \text{constant} \tag{9–31}$$

since ℓ_m/b is generally taken to be a constant. Integrating db/dx then leads to $b \sim x$.

Beginning with the simplest case, negligible initial boundary layers, we can again postulate that the velocity profiles will be similar, as in laminar flows, if the turbulent eddy viscosity μ_τ may be assumed constant across the layer. The similarity variable η is now written in terms of the spreading parameter σ as

$$\eta \equiv \frac{\sigma y}{x} \tag{9–32}$$

Since $b \sim x$, we write

$$\begin{aligned} \psi &\sim x \\ &= x\left(\frac{U_1 + U_2}{2}\right) f(\eta) \end{aligned} \tag{9–33}$$

Then, since $U = \partial\psi/\partial y$,

$$U = \left(\frac{U_1 + U_2}{2}\right)\sigma f'(\eta) \tag{9–34}$$

Görtler (1942) used these relations with the Prandtl (1942) eddy viscosity model

$$\mu_T \sim \rho b(U_1 - U_2) \sim \rho x(U_1 - U_2) \tag{9–35}$$

Prandtl simply reasoned that $\mu_T \sim$ *density* $\times$ *velocity* $\times$ *length* and then took the length scale as the width and the velocity scale as the velocity difference across the layer. (Recall that $b \sim x$.) This line of reasoning reduces the equations of motion to

$$f''' + 2\sigma^2 ff'' = 0 \tag{9–36}$$

The solutions for various values of σ were compared with the experiments of Reichardt (1942) with $U_2 = 0$, and the value $\sigma_0 = 13.5$ was selected. The results are shown in Fig. 9–25. Other workers have proposed $9.0 \leq \sigma_0 \leq 13.5$; a value of $\sigma_0 = 11$ is now generally accepted, based on more recent data. The influence of the parameter U_2/U_1 on σ has also been studied by several groups. A survey by Birch and Eggers (1971) suggests that the best agreement with the data is given by

$$\frac{\sigma_0}{\sigma} = \frac{U_1 - U_2}{U_1 + U_2} \tag{9–37}$$

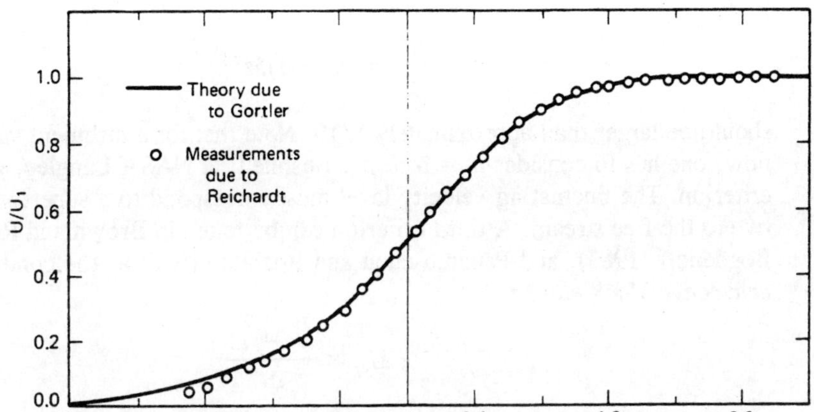

Figure 9–25 Velocity profiles in a turbulent simple shear layer. (From Görtler, 1942.)

The analysis of the initial boundary layer effects for laminar flows in Kubota and Dewey (1964) was extended to turbulent flows with $U_2 = 0$ by Alber and Lees (1968). They chose to work with a new spreading parameter σ_θ based on the momentum thickness θ. The new parameter is closely related to the usual spreading parameter σ, and Alber and Lees chose a value $\sigma_\theta = 9.42$. The velocity on the dividing streamline is then as given by Kubota and Dewey (1964) [see Fig. 9–3], except that the independent variable involving axial distance, x, becomes $x/\sigma_\theta\theta_0$. Alber and Lees (1968) give the simple algebraic formula

$$\frac{x}{\sigma_\theta \theta_0} = \frac{19.7 u_{DSL}^3}{2 - u_{DSL} - 4u_{DSL}^2} \tag{9–38}$$

for $u_{DSL}(x/\sigma_\theta \theta_0)$.

The effect of variable density on the value of the spreading parameter has been the subject of considerable research and debate. Studies by Alber and Lees (1968) suggest that

$$\frac{(\sigma_\theta)_{\text{variable}\,\rho}}{(\sigma_\theta)_{\text{constant}\,\rho}} = \left(\frac{\rho_1}{\rho_t}\right)^2 \tag{9–39}$$

The data in Fig. 9–26, from Birch and Eggers, illustrates the scope of the problem. The data indicate a reduction in mixing rate at high Mach number (a higher σ means slower spreading) if the experiments are free of the influences of the initial boundary layers on the splitter plate. This is generally viewed as an effect of *compressibility*, not just density variation, and it is now possible to estimate with reasonable certainty when such effects might be expected. Following Morkovin and Bradshaw in Bradshaw (1977), one criterion is that a representative density fluctuation level

$$\frac{\rho'}{\overline{\rho}} \approx (\gamma - 1)M^2 \frac{\Delta U}{U} \tag{9–40}$$

should be larger than approximately 1/10. Note that for a turbulent variable-density flow, one has to consider $\rho = \overline{\rho} + \rho'$. Bushnell, at NASA Langley, gives a second criterion: The fluctuating velocity level must correspond to a supersonic speed relative to the free stream. A third criterion can be found in Brown and Roshko (1974), Bogdanoff (1983), and Papamoschou and Roshko (1986) as the conditions where a *convective Mach number*

$$M_{c1} \equiv \frac{U_1 - u_c}{a_1} \tag{9–41}$$

approaches unity. Here, u_c is the convective speed of large-scale turbulent structures. The influence of M_{c1} on the width growth rate can be presented as

$$\frac{d\delta}{dx} = C_3 f(M_{c1}) \frac{(1 - U_2/U_1)(1 + \sqrt{\rho_2/\rho_1})}{\left(1 + \sqrt{\dfrac{\rho_2 U_2^2}{\rho_1 U_1^2}}\right)} \tag{9–42}$$

with $C_3 \approx 0.17$ and $f(M_{c1})$ a function equal to unity up to about $M_{c1} = 0.5$ and then decreasing to a value of about 0.3 by $M_{c1} \approx 1.0$, to mimic the data shown in Fig. 9–27.

The oldest and simplest analyses that exist for jet flows are based on the observed similarity of profiles, as was displayed in Figs. 9–7, 9–16, and 9–22. Taking the case of a constant-density round jet with $U_e = 0$ as an extremely easy example, one proceeds by asserting that the integral of the streamwise momentum remains

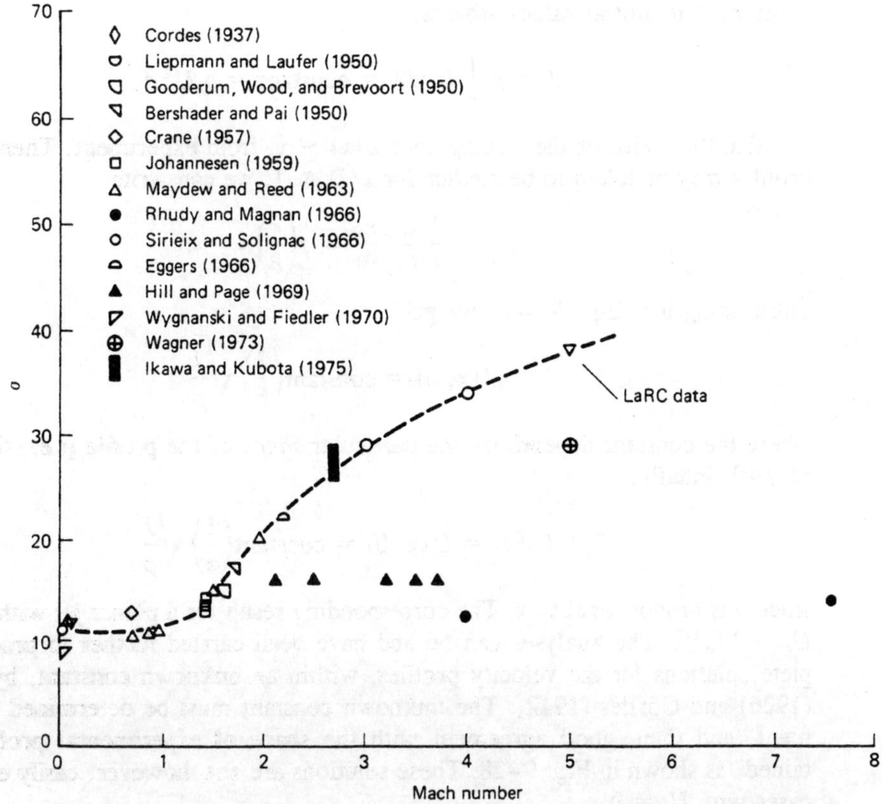

Figure 9–26 Spreading parameter in the turbulent simple shear layer as a function of Mach number. (From Birch and Eggers, 1973, with additional data from Wagner, 1973 and Ikawa and Kubota, 1975.)

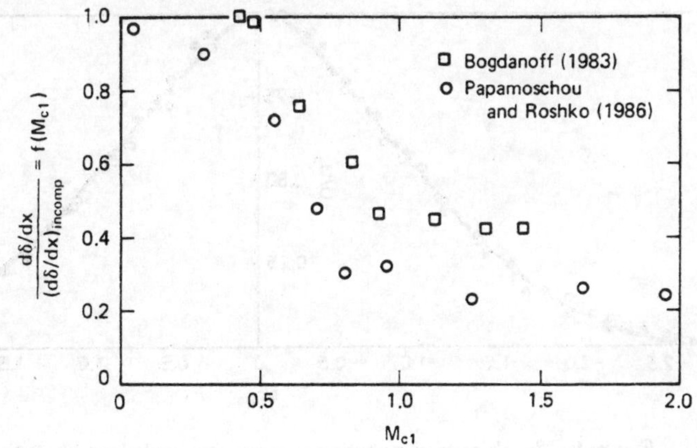

Figure 9–27 The effect of convective Mach number on the spreading rate in the turbulent simple shear layer.

constant at its initial value—that is,

$$J = \rho \int U^2 dA = \text{constant} = \rho_j U_j^2 A_j \tag{9–43}$$

and that the width of the mixing zone $b(x) \sim x$, from experiment. Then, since the profiles may be taken to be *similar* for $x/D \gg 1$, we can write

$$\frac{U(x, r)}{U(x, 0)} = f\left(\frac{r}{b}\right) \tag{9–44}$$

Substituting into Eq. (9–43), we get

$$U(x, 0) = \text{constant}\left(\frac{1}{b}\right)\sqrt{\frac{J}{\rho}} \tag{9–45}$$

where the constant depends on the particular *shape* of the profile [i.e., the form of $f(r/b)$]. Finally,

$$U_c(x) = U(x, 0) = \text{constant}\left(\frac{1}{x}\right)\sqrt{\frac{J}{\rho}} \tag{9–45a}$$

since b is proportional to x. The corresponding result for a planar jet with $U_e = 0$ is $U_c \sim 1/x^{1/2}$. The analyses can be and have been carried further to produce complete solutions for the velocity profiles, within an unknown constant, by Tollmien (1926) and Görtler (1942). The unknown constant must be determined by experiment, and then, good agreement with the shape of experimental profiles is obtained, as shown in Fig. 9–28. These solutions are not, however, easily extended to cases with $U_e \neq 0$.

Again, as with the laminar case, more recent analysis (Schneider, 1985) shows that the total momentum flux does not really stay constant in these flows. Schneider

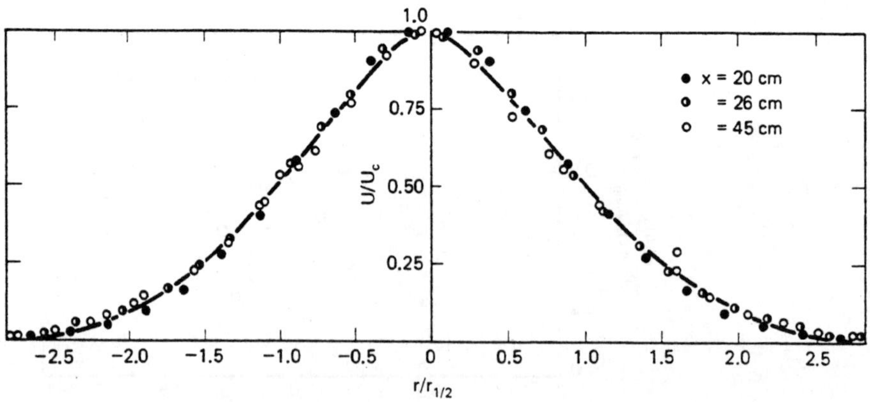

Figure 9–28 Velocity distribution in a circular, turbulent jet. (Measurements due to Reichardt, 1942; analysis due to Tollmien, 1926.)

derived the formula

$$\frac{dJ}{dx} = -\frac{C}{2}Jx^{-1} \tag{9–46}$$

with $C \approx 0.013$ for axisymmetric jets. Using these results, we find that the axial velocity in the jet decays slightly more rapidly than predicted by the classical analysis, and the induced velocities and induced pressure are slightly smaller than they are for constant J. For planar jets, the value of C can be deduced from the measured entrainment rate; it is $C \approx 0.085$. With that result, the predictions of Eq. (9–46) are in good agreement with careful experiments.

For the wake behind a body, the profiles become *similar* only at a distance of many characteristic body dimensions downstream, where the velocity defect $\Delta U \equiv (U_\infty - U)$ has become small compared to U_∞. Also, at such a location, the static pressure will return to the free stream value ahead of the body. The *momentum theorem* can then be applied to a control volume that contains the body and whose downstream surface is in the *similar* profile region. The situation is like that for the laminar far wake discussed in Section 9–2–3, so Eq. (9–16a) holds here also if one reads U for u. The value of the integral will be proportional to $(\Delta U_c)(b(x))$, with the proportionality factor depending on the details of the profile shape $\Delta U(y)$. (Here, ΔU_c is the centerline value of ΔU.) With this information and the definition of the drag coefficient C_D, we can set the two expressions for the drag equal to each other and obtain

$$\frac{\Delta U_c}{U_\infty} \sim \frac{C_D D}{2b} \tag{9–47}$$

The growth of the width $b(x)$ is again determined on the basis of the assumption that the time rate of increase of the width is proportional to the transverse velocity fluctuation v'; that is,

$$\frac{Db}{Dt} \sim v' \tag{9–48}$$

with

$$\frac{Db}{Dt} \sim U_\infty \frac{db}{dx} \tag{9–49}$$

and

$$v' \sim \ell_m \frac{\partial U}{\partial y} \approx \ell_m \frac{\Delta U_c}{b} \tag{9–50}$$

As usual, for a flow not restricted by a rigid wall, we may take ℓ_m to be proportional to the width of the layer (i.e., $\ell_m = K_2 b$). We obtain

$$\frac{db}{dx} \sim K_2 \frac{\Delta U_c}{U_\infty} \tag{9–51}$$

Now, substituting for ΔU_c from Eq. (9–47), we get

$$2b\,\frac{db}{dx} \sim K_2 C_D D \tag{9–52}$$

Integration then gives

$$b \sim (K_2 x C_D D)^{1/2} \tag{9–53}$$

Using this relation in Eq. (9–47) yields

$$\frac{\Delta U_c}{U_\infty} \sim \left(\frac{C_D D}{K_2 x}\right)^{1/2} \tag{9–54}$$

Comparison with experiment gives the values of the proportionality constants and leads to

$$b = 0.57(x C_D D)^{1/2} \tag{9–53a}$$

and

$$\frac{\Delta U_c}{U_\infty} = 0.98\left(\frac{C_D D}{x}\right)^{1/2} \tag{9–54a}$$

The corresponding results for a circular wake are

$$b \sim (K_2 C_D A x)^{1/3} \tag{9–55}$$

and

$$\frac{\Delta U_c}{U_\infty} \sim \left(\frac{C_D A}{K_2^2 x^2}\right)^{1/3} \tag{9–56}$$

where A is the projected area of the body.

9–4–2 Analyses Based on Eddy Viscosity or Mixing Length Models

The various approximate methods for treating jet mixing problems with $U_e \neq 0$ and the far wake that have been presented were generally concerned primarily with overcoming the difficulties of solving the equations of motion. The widespread availability of large digital computers that permit *numerically exact* solutions has rendered these methods obsolete now, so we will not consider them any further.

Since the equations of motion can be solved numerically, within the limits of mean flow turbulence models, the discussion reduces to a choice of an eddy viscosity or mixing length model. The major models that have been used are listed in Table 9–1. A computer code for a PC using a mixing length model is in Appendix B.

Example 9–1. Application of Mixing Length Model to Jet Mixing Problem

Let us predict the decay of the centerline velocity for a two-dimensional turbulent jet in a 10.0-m/s flow with an initial velocity ratio $U_j/U_e = 4.0$ in a fluid with $\nu = 20 \times$

10^{-6} m²/s with a nozzle half-height of 2.0 cm. We use a mixing length model. Consider the region to $x = 1.0$ m. What is the length of the *potential core*?

Solution This flow problem can be solved with the code JETWAKE in Appendix B. We input first XI = 0.0 and XF = 1.0 and then CNU = 0.00002 and UINF = 10.0. Next, we pick 500 points across the mixing zone (MMAX = 500), 50 points across the nozzle half height (DY = 0.0004 and MEST = 51), and 100 steps in the streamwise direction (NMAX = 101) for a fine grid to resolve the rapid changes near the jet exit. We specify UJ = 4.0 and UE = 1.0. Finally, we select MODEL = 1 for a mixing length model.

It is easy to see how the code is to be modified for other problems with, for example, a different initial velocity ratio.

The output for the centerline velocity is listed in the following table and plotted in Fig. 9–29. From these results, one can pick off the length of the *potential core* as the value of x where the centerline velocity begins to fall below the initial value, say, $x \approx 0.2$ m. A logarithmic plot is also presented to show how the flow tends towards a similarity decay, i.e., $U_c \sim x^{-n}$ (a straight line on a logarithmic plot) for large x.

N	X	UC	BW
1	0.000	4.000	0.0004
11	0.100	4.000	0.0160
21	0.200	4.000	0.0388
31	0.300	3.643	0.0644
41	0.400	3.244	0.0852
51	0.500	2.974	0.1068
61	0.600	2.778	0.1244
71	0.700	2.631	0.1420
81	0.800	2.515	0.1592
91	0.900	2.421	0.1740
101	1.000	2.342	0.1897

Prandtl's third model (which is the most widely used of the first five for jets) introduced the concept of velocity difference for $\nu_T \sim \Delta U$. Note, however, that $\delta^* \sim \Delta U$, so the Clauser outer region boundary layer model contains that dependence also. It is important to observe that the three entries following Prandtl's third model can be shown to be equivalent to it and to each other.

Consider first the *extended* Clauser model,

$$\nu_T = \frac{\mu_T}{\rho} = CU_e\delta^* \tag{9–57}$$

where δ^* must be interpreted to be based on the absolute value of $[1 - (U/U_e)]$. Compare this to the Prandtl model, using the half-width $b_{1/2}(x)$:

$$\nu_T = 0.037 b_{1/2}|U_{\max} - U_{\min}| \tag{9–58}$$

For profile shapes that are assumed to be simple, such as a rectangular, triangular,

TABLE 9–1 KINEMATIC EDDY VISCOSITY MODELS FOR MAIN MIXING REGION OF JETS AND WAKES

Author	Planar	Axisym-metric	Variable density	Expression	Remarks
Prandtl (1925)	x	x		$\ell_m^2\left(\dfrac{\partial U}{\partial y}\right)$	ℓ_m proportional to the width of the mixing region
Von Kármán (1930)	x	x		$\kappa^2 \dfrac{(\partial U/\partial y)^3}{(\partial^2 U/\partial y^2)^2}$	
Taylor (1932)	x	x		$\frac{1}{2}\ell_w^2\left(\dfrac{\partial U}{\partial y}\right)$	$\ell_w = \sqrt{2}\ell_m$
Prandtl (1942)	x	x		$\ell_m^2\sqrt{\left(\dfrac{\partial U}{\partial y}\right)^2 + \ell_1^2\left(\dfrac{\partial^2 U}{\partial y^2}\right)^2}$	Requires two mixing lengths
Prandtl (1942)	x	x		$\kappa_1 b(U_{max} - U_{min})$	Introduced concept of velocity difference with b taken as $b_{1/2}$, $\kappa_1 = 0.037$ in planar jets; Schlichting extended this to axisymmetric jets with $r_{1/2}$ and $\kappa_1 = 0.025$
Schlichting (1942)	x			$0.0222 U_e C_D D$	Wake of a cylinder of arbitrary cross section
Clauser (1956)	x			$CU_e\delta^* = C\int_0^\infty \lvert U_e - U \rvert \, dy$	Applied to wakelike outer region of a boundary layer $0.0168 < C < 0.018$

Author	Planar	Axisym-metric	Variable density	Expression	Remarks
Hinze (1959)	x			$0.016U_eD$	Wake of a circular cylinder
Ting-Libby (1960)		x	x	$\rho^2\nu_T = \frac{\rho_c^2\nu_{T_0}}{r^2}\int_0^r 2\frac{\rho}{\rho_c}r'\,dr'$	ν_{T_0} is the constant-density eddy viscosity, and ρ_c is the centerline density
Ting-Libby (1960)	x		x	$\rho^2\nu_T = \rho_c^2\nu_{T_0}$	
Ferri et al. (1962)		x	x	$\rho\nu_T = 0.025r_{1/2}[(\rho U)_{\max} - (\rho U)_{\min}]$	Extended Prandtl's third model to variable density; introduced concept of mass flow difference
Schetz (1964)	x		x	$\rho^2\nu_T = 0.037b_{1/2}\rho_c[(\rho U)_{\max} - (\rho U)_{\min}]$	Application of mass flow difference to planar flows
Schetz (1968)	x	x	x	"Turbulent viscosity proportional to mass flow defect (or excess) in the mixing region"	
Baldwin and Lomax (1978)	x		x	$\rho\nu_T = \frac{0.027\rho y_{\max}[U_{\max} - U_{\min}]^2}{(y\lvert\partial U/\partial y\rvert)_{\max}}$	Equivalent to Prandtl's third model

*x = yes

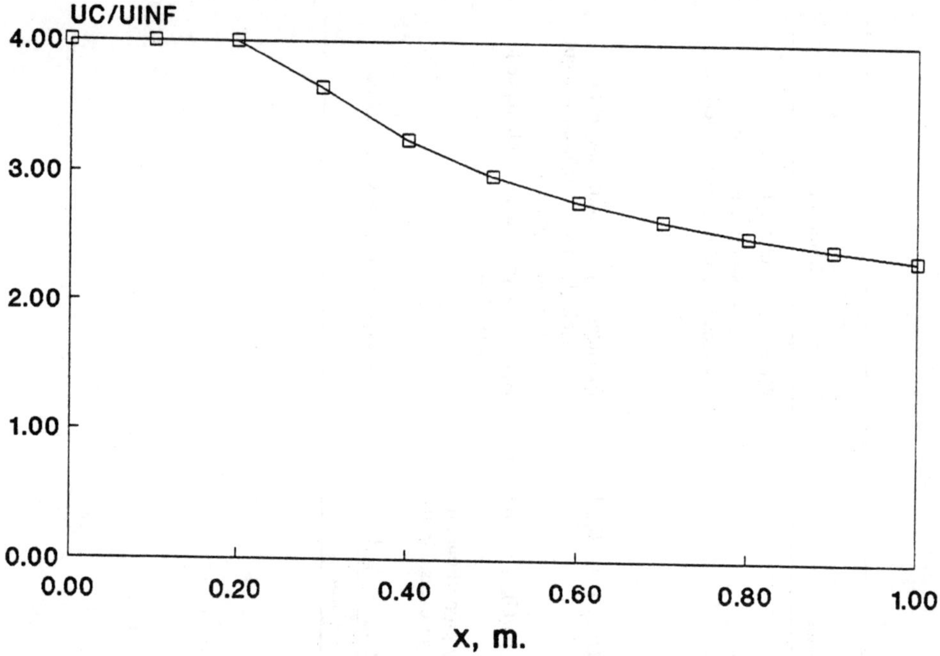

(A) Centerline velocity decay

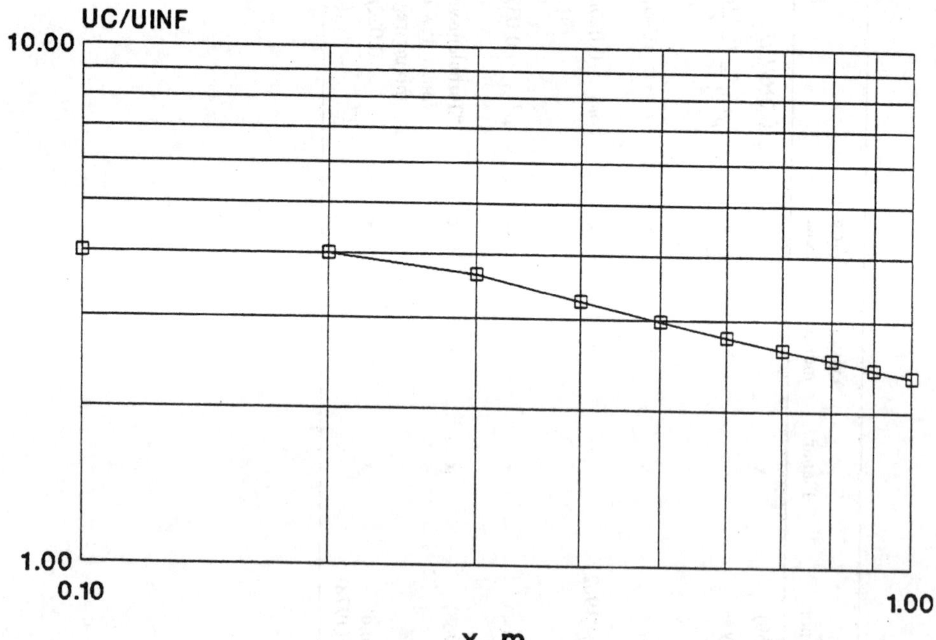

(B) Logarithmic plot of centerline velocity decay

Figure 9–29 Predicted results for Example 9–1.

or cosine velocity defect (or excess), the two expressions agree exactly in form and even to the extent that the proportionality constant equals 0.036 versus 0.037. The wake models of Schlichting and Hinze can be reduced to the same form as the extended Clauser model. Observing that

$$C_D D = 2\Theta\left(\equiv 2 \int_{-\infty}^{\infty} \frac{U}{U_e}\left(1 - \frac{U}{U_e}\right) dy\right) \tag{9–59}$$

and taking a crude representative value of $C_D \approx 1.20$ for a circular cylinder, we find that these expressions become, respectively,

$$\nu_T = 0.044 U_e \Theta \tag{9–60}$$

and

$$\nu_T = 0.027 U_e \Theta \tag{9–61}$$

In the treatment of wake problems, it is common to neglect the factor U/U_e in the definition of Θ, since $U/U_e \approx 1$ (see Section 9-2-3). Accordingly, we obtain

$$\Theta \approx \Delta^*\left(\equiv \int_{-\infty}^{\infty} \left|1 - \frac{U}{U_e}\right| dy\right) \tag{9–62}$$

so that Eqs. (9–60) and (9–61) could as well be written

$$\nu_T = 0.044 U_e \Delta^* \tag{9–60a}$$

and

$$\nu_T = 0.027 U_e \Delta^* \tag{9–61a}$$

Written in these terms, the extended Clauser model is

$$\nu_T = 0.018 U_e \Delta^* \tag{9–57a}$$

where we have taken the displacement thickness appropriate to a *two-sided*, planar free-mixing problem, rather than the *one-sided* boundary layer case considered by Clauser. Figures 9–30 and 9–31 present some comparisons of experimental data with predictions based on Eq. (9–57a) for a constant-density, planar jet flow from Schetz (1971). The good agreement is evident. It should be reemphasized here that the Prandtl model and the extended Clauser model are essentially equivalent in the planar case, so that the same level of agreement could be obtained with the Prandtl model. The question arises as to why the proportionality constant for wakes ($U_c/U_e < 1$) is so much larger than that for jets ($U_c/U_e > 1$). (cf. Eq. (9–60a) or (9–61a) with Eq. (9–57a).) Abramovich (1960) notes this effect and attributes it to increased turbulence caused by the separated base flow in the wake case. For our purposes, however, the important result is that these free-mixing eddy viscosity models are all equivalent in functional form. We will return to the matter of the value of the constants again shortly.

The models listed in Table 9–1 following the wake models are attempts to extend the basic Prandtl model to problems involving significant density variations. The Ting-Libby model results from an attempt to apply transformation theory to tur-

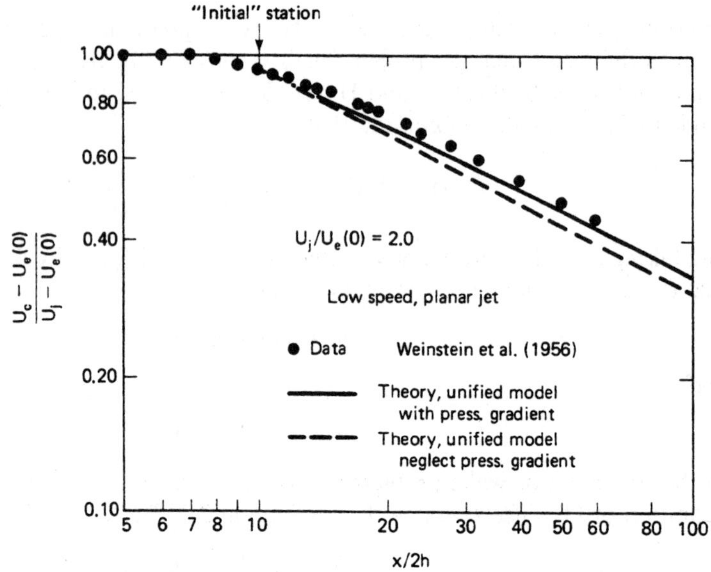

Figure 9–30 Comparison of prediction with experiment for centerline velocity variation of a planar jet. (From Schetz, 1971.)

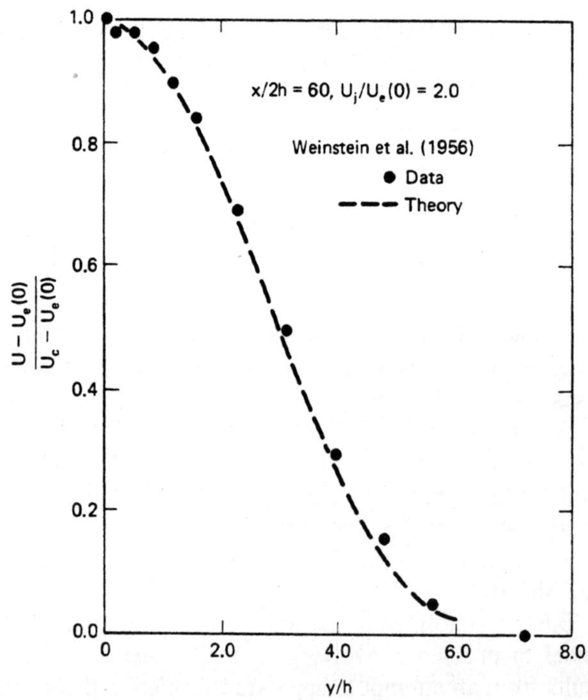

Figure 9–31 Comparison of prediction with experiment for velocity profile in a planar jet. (From Schetz, 1971.)

bulent free mixing; it has been shown to be unreliable. Ferri's suggestion of using a mass flow difference to replace the velocity difference in the Prandtl model has provided predictions of unreliable accuracy for the axisymmetric case. However, when the mass flow difference was applied to the planar, variable-density case, a good prediction was achieved (Schetz, 1964).

Since it was possible to demonstrate some unity in the models for planar, constant-density cases, it was instructive to examine new means for extending these models to the compressible case. As these models are all equivalent, one could begin with any one. Rather than starting with the Prandtl model, as was customary, Schetz (1968) chose to begin with the extended Clauser model. It was simple, at least formally, to extend this model to varying density, that is,

$$\nu_T = 0.018 U_e \Delta^* = 0.018 U_e \int_{-\infty}^{\infty} \left| 1 - \frac{\overline{\rho} U}{\rho_e U_e} \right| dy \tag{9–57b}$$

and to show that, for simple assumed profile shapes, this expression is equivalent to the planar mass flow difference model given in Table 9–1, which had been shown to produce good predictions for planar, variable-density cases [see Schetz (1964)].

This still left the axisymmetric case without a satisfactory eddy viscosity model, especially for variable-density cases. Schetz (1968) generalized the extended Clauser model to the axisymmetric geometry. The line of reasoning was as follows. We rewrite Eq. (9–57b) as

$$\mu_T = \overline{\rho} \nu_T = 0.018 \overline{\rho} U_e \Delta^* \tag{9–57c}$$

This equation can be read to say: "The turbulent viscosity is proportional to the mass flow defect (or excess) per unit width in the mixing region." We can carry the statement over into axisymmetric flow by defining a new displacement thickness δ_r^*, given by

$$\pi \rho_e U_e (\delta_r^*)^2 \equiv \int_0^{\infty} |\rho_e U_e - \overline{\rho} U| 2\pi r \, dr \tag{9–63}$$

This relation follows directly from the logic used for the ordinary displacement thickness (see Section. 2–2) and the geometry of the situation. Using Eq. (9–63) in the statement quoted gives

$$\mu_T = \frac{C_3 \rho_e U_e \pi (\delta_r^*)^2}{R} \tag{9–64}$$

where R is needed to make the dimensions correct. The proportionality constant C_3 had to be determined by a comparison between theory and experiment for one case, as is done with all eddy viscosity models, and the experiments of Forstall and Shapiro (1950) were employed. The results are shown in Fig. 9–32; the value of $C_3 \pi$ determined is 0.018. This constant is held fixed at that value for all further calculations. The model gives excellent qualitative as well as quantitative agreement with the data. Schlichting's extension of the Prandtl model to axisymmetric jets produces poor predictions for the rate of centerline velocity decay.

To consider variable-density jets or wakes, it is necessary to add energy and/or

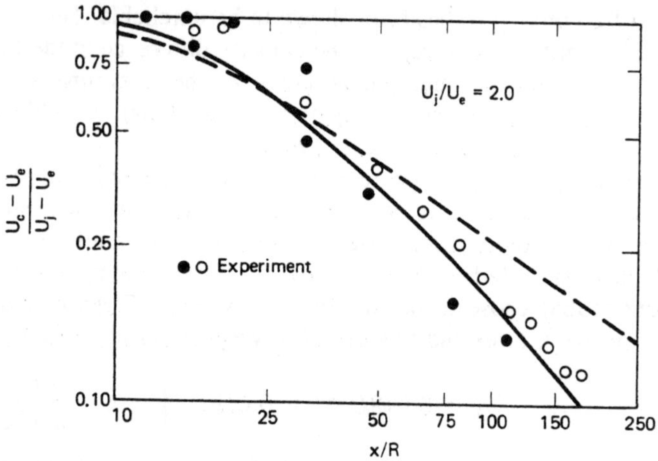

Figure 9–32 Comparison of prediction with experiment for centerline velocity variation of an axisymmetric jet. Dashed line: analysis with Schlichting axisymmetric version of Prandtl model; solid line: analysis with Eq. (9–64). (From Schetz, 1968; data from Forstall and Shapiro, 1950.)

species conservation equations to the system. Heat transfer is, in general, simply modeled using an eddy viscosity or mixing length model and a constant value of the *turbulent* Prandtl number [see Eqs. (9–21) and (9–22)]. Mass transfer is correspondingly modeled using a constant value of the *turbulent* Schmidt or Lewis number [see Eqs. (9–24) to (9–26)]. The case of a jet of hydrogen injected into an airstream provides a stringent test of the theory, since there is a very large density gradient across the mixing zone. The data of Chriss (1968) were used to make the comparisons of theory with experiment, and the Ferri model and Eq. (9–64) were used with constant *turbulent* Prandtl and Schmidt numbers of 0.75 (Schetz, 1968). Results for the case with $\rho_i U_j/\rho_\infty U_\infty = 0.56$, $U_j/U_\infty = 6.3$, are shown in Fig. 9–33. The calculation was started with experimental profiles at $x/R = 5.9$. The model of Eq. (9–64) produced a superior prediction.

In addition to predicting the behavior of centerline values, predicting transverse profiles is also of interest. A comparison of prediction with experiment is given in Fig. 9–34. Again, good agreement is obtained using the eddy viscosity model of Eq. (9–64).

The Baldwin-Lomax eddy viscosity model introduced in Section 7–8–2 also has a proposed form for free-mixing flows (see Baldwin and Lomax, 1984). In such cases, these researchers propose that

$$\mu_T = \frac{0.018(1.60)\rho y_{\max}(\Delta U_c)^2}{F_{\max}} \tag{9–65}$$

where F is as defined in Eq. (7–86) and $y_{\max}$ is y at the location of $F_{\max}$. For simple profile shapes, Eq. (9–65) reduces to the Prandtl eddy viscosity model. Indeed, it was derived to do just that. This means that the Baldwin-Lomax model will perform well whenever the Prandtl model performs well. Similarly, for variable-density free

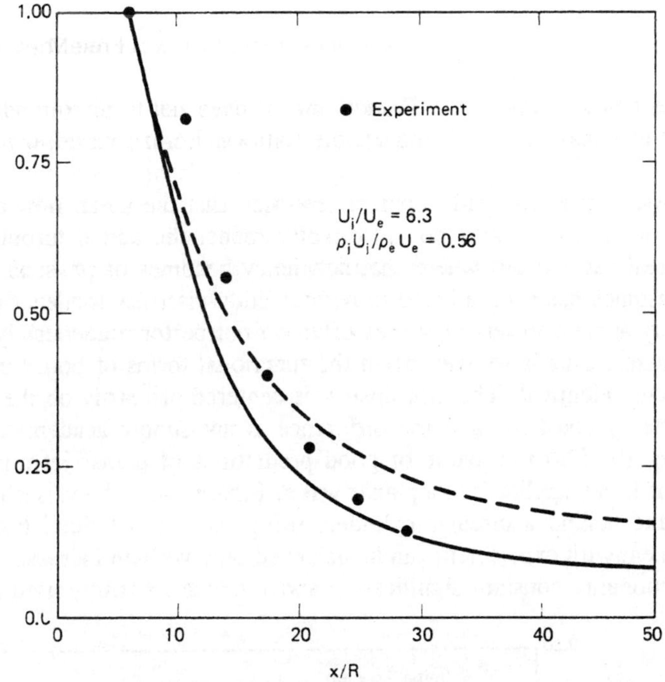

Figure 9–33 Comparison of prediction with experiment for centerline concentration decay for an axisymmetric H_2 jet into air. Dashed line: analysis with Ferri model; solid line: analysis with Eq. (9–64). (From Schetz, 1968; data from Chriss, 1968.)

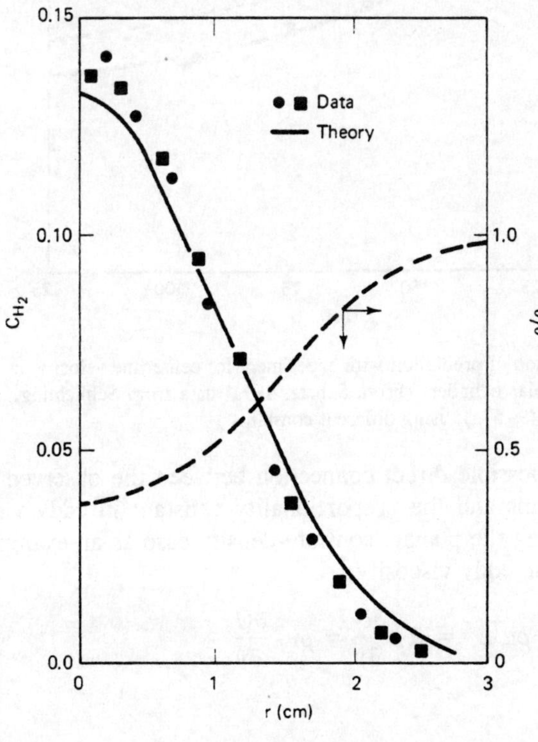

Figure 9–34 Comparison of prediction with experiment for a radial concentration profile in an axisymmetric H_2 jet into air. (From Schetz, 1968; data from Chriss, 1968; analysis with Eq. (9–64).)

turbulent flows, where the Prandtl model does not perform well (unless it is extended to a mass flow difference), the Baldwin-Lomax model will not perform well either.

Aside from the philosophical objection that the mean flow models (eddy viscosity or mixing length) do not directly reflect the actual turbulent nature of the flow, real cases occur where that deficiency becomes of physical importance. One such instance has been alluded to before: Eddy viscosity models that have been successfully applied to jets or wakes often do not perform adequately when applied to other cases. This is so even when the functional forms of both kinds of models are essentially identical. The discrepancy is centered primarily on the value of the proportionality constant, and the difference is not simply academic. We have shown that Eq. (9–57a) is capable of good predictions of planar jet mixing flows. What happens if one applies it to a planar wake? Figure 9–35 shows such a comparison for the wake behind a circular cylinder, using the data of Schlichting (1930). Good agreement with experiment can be achieved only with an increase in the value of the proportionality constant significantly above that successfully used for jet problems.

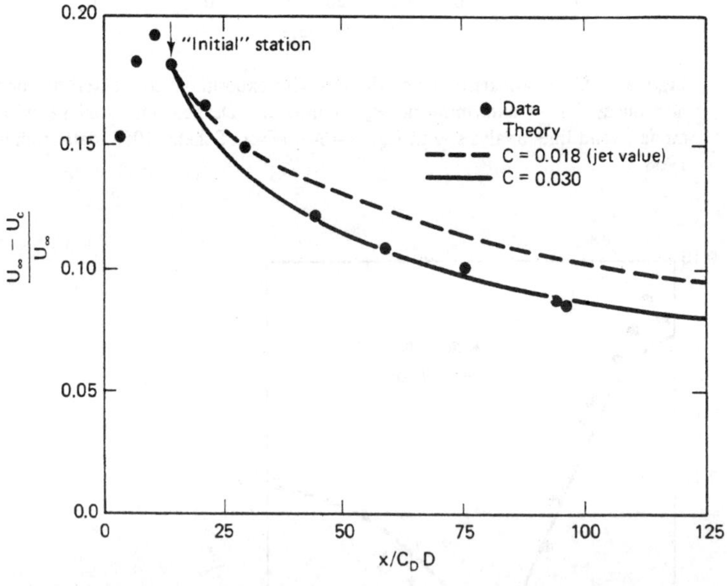

Figure 9–35 Comparison of prediction with experiment for centerline velocity in the wake behind a circular cylinder. (From Schetz, 1971; data from Schlichting, 1930; analysis with Eq. (9–57a), using different constants.)

In Schetz (1971), the possible direct connection between the observed turbulence in various flow problems and the proportionality constant in eddy viscosity models was investigated. Using the planar, constant-density case as an example, we start with the definition of the eddy viscosity,

$$-\rho\overline{u'v'} = \mu_T \frac{\partial U}{\partial y} = \rho\nu_T \frac{\partial U}{\partial y} \tag{7-36}$$

and approximate the velocity gradient crudely as $\partial U/\partial y \approx \Delta U_c/b$. Then, working with the extended Clauser model [Eq. (9–57a)] and replacing the numerical value of the now unknown proportionality constant by C_4, we have, using the approximate velocity gradient

$$-\rho\overline{u'v'} = C_4\rho\int_{-\infty}^{\infty}|U_e - U|\,dy\left(\frac{\Delta U_c}{b}\right) \tag{9–66}$$

The displacement thickness integral can be adequately represented by a constant C_5 times the product of the velocity difference ΔU_c and the width b. The actual value of C_5 depends on the specific profile shape, and typical shapes yield a value of roughly one-half. Thus,

$$-\overline{u'v'} \approx C_4C_5(\Delta U_c)^2 \tag{9–67}$$

In general, we may write

$$-\overline{u'v'} \approx C_6\sqrt{\overline{u'^2}}\sqrt{\overline{v'^2}} \tag{9–68}$$

where C_6 is also about one-half. Further, in shear flows,

$$u' \approx -v' \tag{9–69}$$

so the final result emerges as

$$C_4 \approx \left(\frac{C_6}{C_5}\right)\frac{\overline{(u')^2}}{(\Delta U_c)^2} \tag{9–70}$$

where C_6/C_5 is expected to be near unity. Equation (9–70) achieves the desired result of relating the proportionality constant in an eddy viscosity model to a characteristic of the turbulence. Some data for the turbulence and mean flow that appear in Eq. (9–70) are presented in Fig. 9–36. The fact that the quantity in Eq. (9–70) is

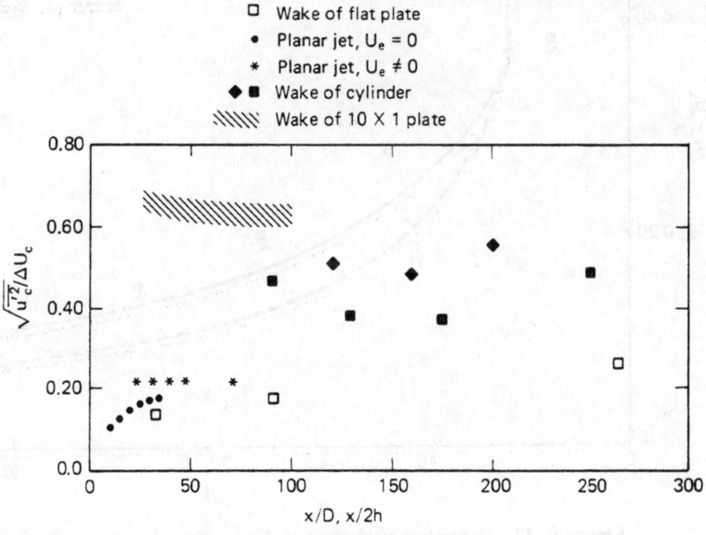

Figure 9–36 Axial turbulence intensities for planar mixing flows collected by Schetz (1971).

more or less constant for a given flow problem, except in the *near wake* or *potential core*, is comforting in light of Eqs. (9–57a) and (9–70), since C_4 is presumed constant. Comparing the results for the planar jet with $U_e \neq 0$ with those for the wake behind a circular cylinder in the figure, we see that the increase in C_4 by a factor of 5/3, required in going from a jet to a wake behind a circular cylinder (see Fig. 9–35), is roughly predicted by Eq. (9–70). For wakes behind streamlined bodies, $\overline{u_c'^2}/(\Delta U_c)^2$ is quite close to the values found in jets, so the appropriate proportionality constant is correspondingly the same as for jets. This correspondence is demonstrated in Fig. 9–37, which shows comparisons of predictions with data for the wake behind a slender, axisymmetric body. No increase in the "jet" value of the proportionality constant is necessary for this wake case. In summary, it is not the case that all jets are different from all wakes, but rather that wakes behind bodies of different shapes have different turbulence levels in relation to the mean flow velocity difference. (It is interesting to note that the Chevray wake experiment was a test case for the 1981 Stanford conference, and no researcher chose to treat it.)

With the foregoing results, the first-order influence of turbulence can be crudely incorporated into an eddy viscosity (or mixing length) model, as long as information of the type in Fig. 9–36 is available for the flow problem under consideration. Results for axisymmetric cases can be found in Schetz (1971).

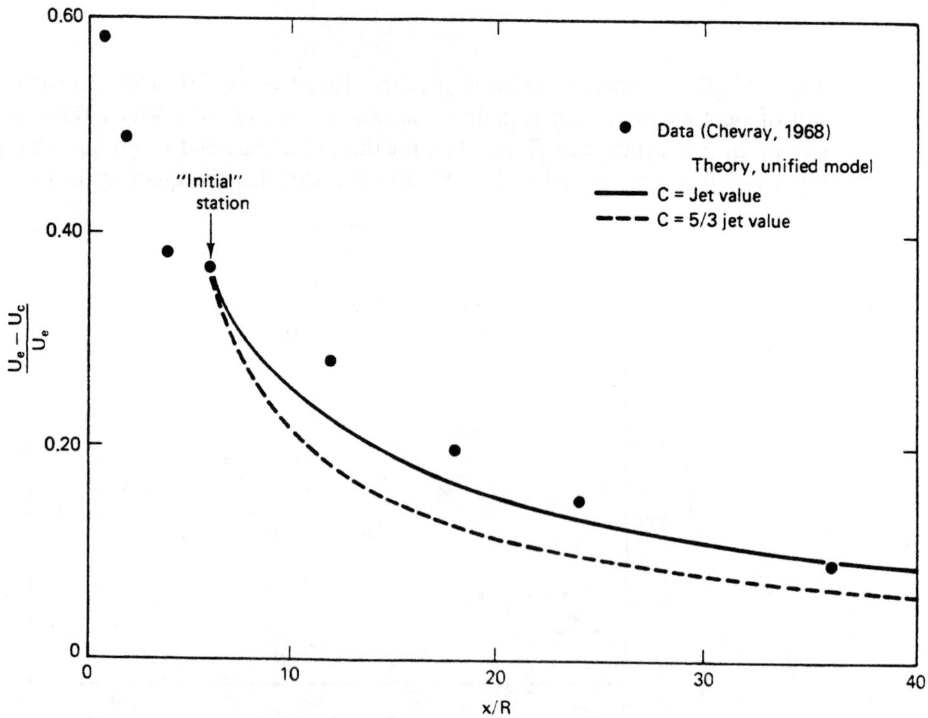

Figure 9–37 Centerline velocity distribution for a low-speed axisymmetric wake. (From Schetz, 1971; data from Chevray, 1968.)

9–5 ANALYSES BASED ON TURBULENT KINETIC ENERGY

The development of the two basic TKE models in use for wall-bounded flows—the Prandtl energy method and the Bradshaw model—was described in Section 7–10. For free-mixing flows, these models are used in the same forms, except that it is not necessary to modify ℓ for wall effects. A simple computer code for a PC based on the Prandtl energy method for incompressible jets and wakes is in Appendix B.

Example 9–2. Application of TKE Model to Jet Mixing

Let us use the Prandtl energy method to predict the decay of the centerline velocity for a two-dimensional turbulent jet in a 10.0-m/s flow with an initial velocity ratio $U_j/U_e = 4.0$ in a fluid with $\nu = 20 \times 10^{-6}$ m²/s with a nozzle half-height of 2.0 cm. Consider the region up to $x = 1.0$ m. What is the length of the *potential core*? How do these results compare with those with the mixing length model?

Solution This flow problem can also be solved with the code JETWAKE in Appendix B. First, we select MODEL = 2, for the TKE model. All the other input data can be the same as for Example 9–1. Thus, we set XI = 0.0 and XF = 1.0 and then CNU = 0.00002 and UINF = 10.0. Next, we pick 500 points across the mixing zone (MMAX = 500), 50 points across the nozzle half-height (DY = 0.0004 and MEST = 51), and 100 steps in the streamwise direction (NMAX = 101) for a fine grid to resolve the rapid changes near the jet exit. Finally, we specify UJ = 4.0 and UE = 1.0.

Predictions for the centerline velocity are listed in the following table and plotted in Fig. 9–38. From these results, we can pick off the length of the *potential core* as the value of x where the centerline velocity begins to fall below the initial value, say, $x \approx 0.2$ m. A logarithmic plot is also presented to show how the flow tends towards a similarity decay, i.e., $U_c \sim x^{-n}$ (a straight line on a logarithmic plot) for large x. Plainly, the results agree closely with those from the mixing length model toward the end of the computational region. The rate of decay is a little faster, however, with the TKE model.

N	X	UC	BW
1	0.000	4.000	0.0004
11	0.100	4.000	0.0156
21	0.200	3.992	0.0360
31	0.300	3.757	0.0528
41	0.400	3.403	0.0628
51	0.500	3.123	0.0740
61	0.600	2.896	0.0856
71	0.700	2.704	0.0968
81	0.800	2.545	0.1080
91	0.900	2.413	0.1192
101	1.000	2.304	0.1304

The Imperial College group contributed a comprehensive survey paper (Launder et al., 1973) to the 1972 NASA Conference on Free Turbulent Shear Flows

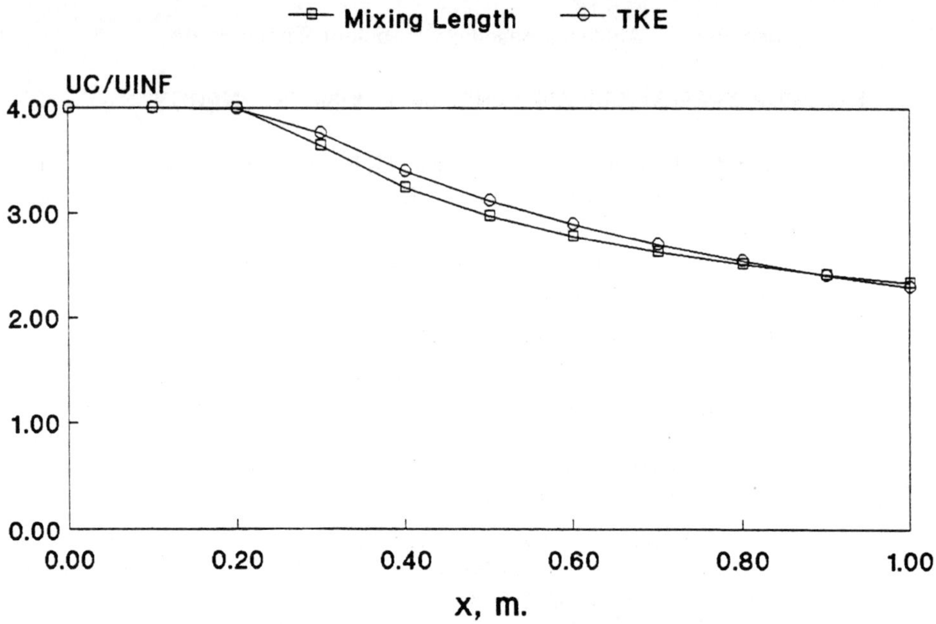

(A) Centerline velocity decay

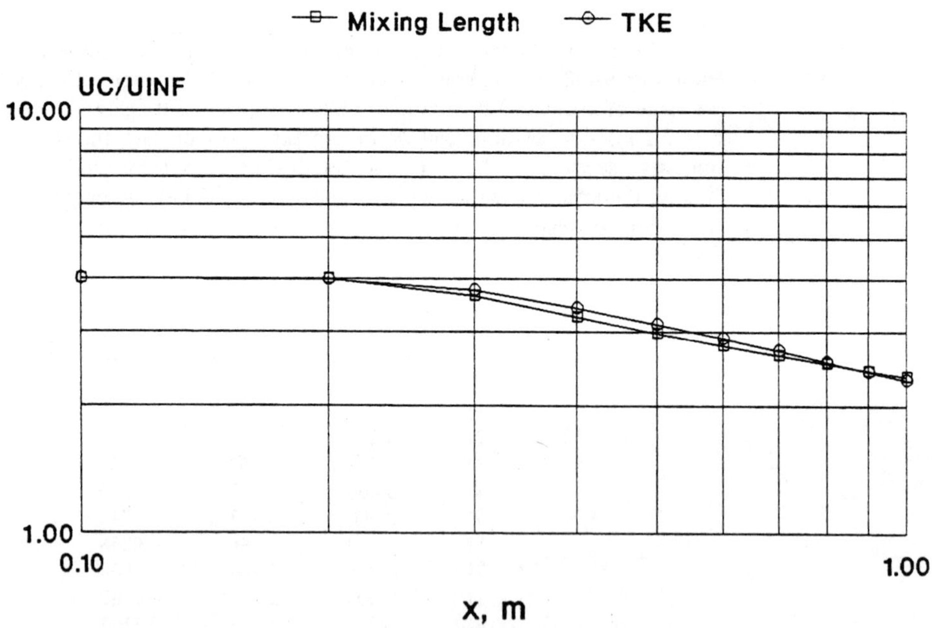

(B) Logarithmic plot of centerline velocity decay

Figure 9–38 Predicted results for Example 9–2.

which compared the adequacy of several turbulence models with free-mixing data, all on the same basis. At the TKE level, they used the Prandtl energy method [Eq. (7–99)]. Harsha (1973) extended the Bradshaw TKE approach to free shear flows.

To compare predictions with experiment, we begin with the case of $U_j/U_\infty = 4.0$, from Forstall and Shapiro (1950), selected for the 1972 NASA conference. In Fig. 9–39, the centerline velocity predictions based on the eddy viscosity model [Eq. (9–64)] of Schetz (1971), the mixing length model, the TKE model based on Eq. (7–99) from Launder et al. (1973), and the extended Bradshaw TKE model from Harsha (1973) are shown, compared with experiment. The predictions are all in reasonable agreement with the data. Moreover, it appears that the various models would give predictions that were even closer together if they were started at the end of the potential core ($x/D \approx 7$), rather than at the station specified by the NASA conference organizers, which was still in the potential core, a complex region.

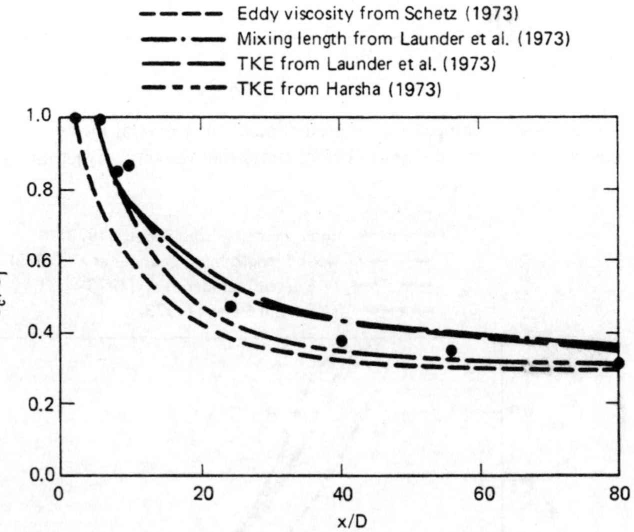

Figure 9–39 Comparison of predictions, using several models, with the axisymmetric jet experiments of Forstall and Shapiro (1950).

The H_2-air experiments of Chriss (1968) provide test cases with very large differences in density across the mixing layer. Figures 9–40 and 9–41 show the variations in velocity and concentration along the centerline, as compared with predicted data, using the same models as in Fig. 9–39. In this case again, it is hard to detect any clear advantage of either TKE model over the mean flow models when one looks at the whole range of x/D covered.

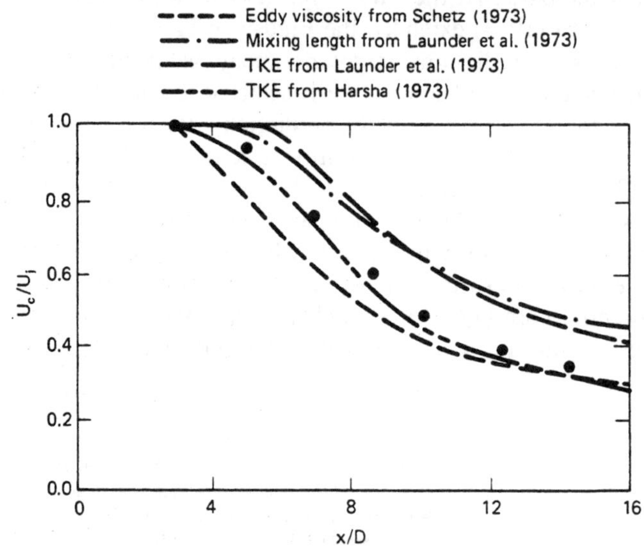

Figure 9–40 Comparison of predictions, using several models, with the axisymmetric experiment of Chriss (1968): centerline velocity, H_2 jet into air.

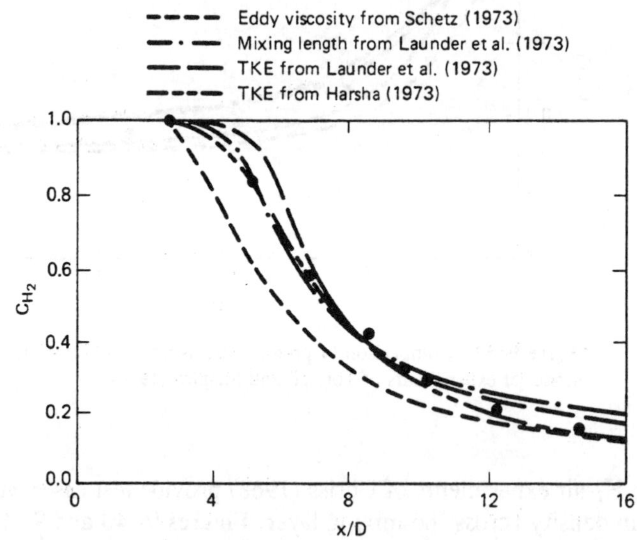

Figure 9–41 Comparison of predictions, using several models, with the axisymmetric experiment of Chriss (1968): centerline concentration, H_2 jet into air.

9–6 ANALYSES BASED ON EQUATIONS FOR TURBULENT KINETIC ENERGY AND LENGTH SCALE

The basic material for analyses based on equations for TKE and length scale was developed in Section 7–11, but again, it is not necessary in the present case to treat a special wall-dominated region. Launder et al. (1973) used $Z \equiv \epsilon$ for their calculations for the 1972 NASA conference, and the final modeled form of the equation for ϵ was taken to be

$$\rho\left(U\frac{\partial \epsilon}{\partial x} + V\frac{\partial \epsilon}{\partial y}\right) = \frac{1}{y^j}\frac{\partial}{\partial y}\left(y^j\frac{\mu_T}{\sigma_\epsilon}\frac{\partial \epsilon}{\partial y}\right) + C_{\epsilon 1}\frac{\rho\epsilon}{K}\nu_T\left(\frac{\partial U}{\partial y}\right)^2 - C_{\epsilon 2}\frac{\rho\epsilon^2}{K} \tag{9–71}$$

with $\sigma_\epsilon = 1.3$, $C_{\epsilon 1} = 1.43$, and $C_{\epsilon 2} = 1.92$ for planar flows. (*Note:* $j \equiv 0$ for planar flows, and $j \equiv 1$ for axisymmetric flows.) For axisymmetric flows,

$$C_{\epsilon 2} = 1.92 - 0.0667F \tag{9–72}$$

where

$$F = \left[\frac{b}{2\Delta U_c}\left(\frac{dU_c}{dx} - \left|\frac{dU_c}{dx}\right|\right)\right]^{1/5} \tag{9–73}$$

The TKE equation [Eq. (7–99)] is retained with $\sigma_K = 1.0$ and

$$\nu_T = \frac{C_\mu K^2}{\epsilon} = \frac{(0.09 - 0.04F)K^2}{\epsilon} \tag{9–74}$$

This model is called $K\epsilon 1$. For weak shear flows (i.e., for cases where the velocity defect or excess, ΔU_c, is a small fraction of U_e), an extended version called $K\epsilon 2$ was presented. The extended version used the same equations as its progenitor, but with $C_{\epsilon 1} = 1.40$, $C_{\epsilon 2} = 1.94$, and

$$C_\mu = 0.09g(\overline{\mathcal{P}/\epsilon}); \qquad C_\mu = 0.09g(\overline{\mathcal{P}/\epsilon}) - 0.0534F \tag{9–75}$$

for planar and axisymmetric flows, respectively. In this equation,

$$\overline{\mathcal{P}/\epsilon} = \frac{\int_0^b \tau_T\left(\frac{\mathcal{P}}{\epsilon}\right)y^j\,dy}{\int_0^b \tau_T y^j\,dy} \tag{9–76}$$

where $\mathcal{P}$ is production of K, and $g(\overline{\mathcal{P}/\epsilon})$ is given as a graph in Fig. 9–42.

Now consider the low-speed, axisymmetric jet case of Forstall and Shapiro (1950), which was discussed before; comparison of predictions with data are given in Fig. 9–43. It can be seen that the $K\epsilon 1$ and $K\epsilon 2$ models perform very well. For axisymmetric jets of H_2 mixing with air, the situation is more confused. For the tests of Eggers (1971), the TKE, $K\epsilon 1$, and $K\epsilon 2$ models perform essentially equally well. However, for the data of Chriss (1968), the relative performance of both $K\epsilon$ models comes out poorer, as shown in Fig. 9–44.

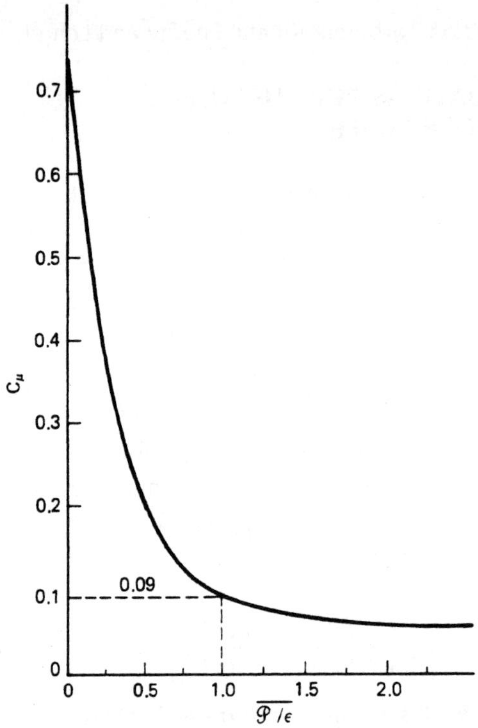

Figure 9–42 Empirical function $C_\mu = f(\overline{\mathcal{P}/\epsilon})$ for the $K\epsilon 2$ model. (From Launder et al., 1973.)

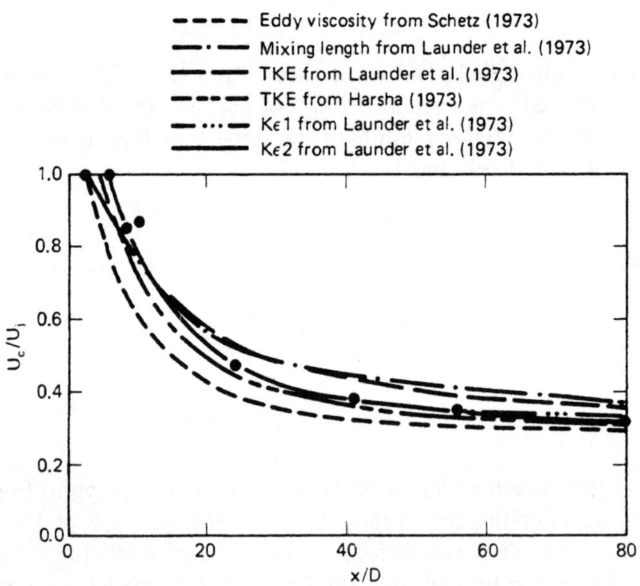

Figure 9–43 Comparison of predictions, using mean flow, one-equation, and two-equation models, with the axisymmetric jet experiment of Forstall and Shapiro, 1950.

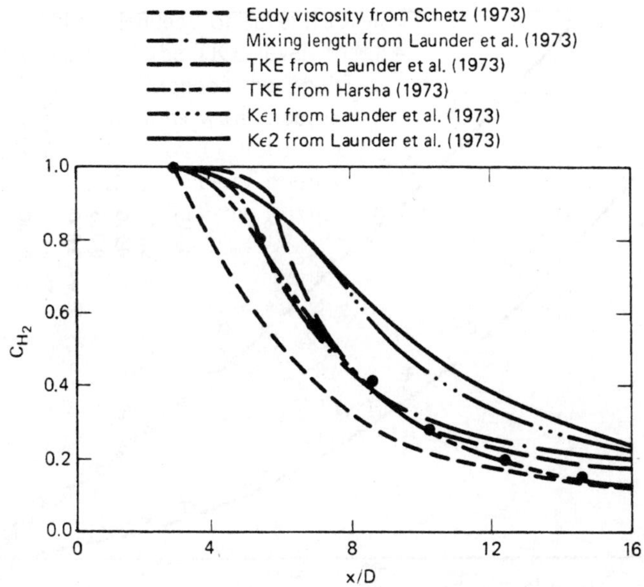

Figure 9–44 Comparison of predictions, using mean flow, one-equation, and two-equation models, with the axisymmetric experiment of Chriss (1968): centerline concentration, H_2 jet into air.

Rodi has conducted a continuing, thorough, and objective study of $K\epsilon$ models and their performance. Some results from those studies for planar and axisymmetric jets and an axisymmetric wake are given in Figs. 9–45 and 9–46. Plotting the results in the manner shown exaggerates any differences between prediction and measurement. The one-equation model mentioned in the figure is the Prandtl energy method.

The 1981 Stanford conference had the low-speed simple shear layer as a test case. Few researchers treated it, but a prediction with a $K\epsilon$ model from Rodi is shown, as compared with several sets of data, in Fig. 9–47.

For supersonic flow, the simple shear layer can show an effect of compressibility, as discussed in Section 9–4–1. The $K\epsilon 2$ model will not predict this effect, as demonstrated in Fig. 9–48. In order to capture the effect, it is necessary to introduce yet another correction to the $K\epsilon$ model. Dash et al. (1975) developed an ad hoc correction factor $K(M_\tau)$ to the eddy viscosity relation [see the left-hand subequation of Eq. (9–74)]. Here, $M_\tau = (K_{\max})^{1/2}/a$, where a is the local speed of sound. The correction factor was determined simply to reproduce the variation of σ found experimentally. The result is given in Fig. 9–49, and the behavior of the new model, called $K\epsilon 2$, cc ("cc" denotes *compressibility corrected*), is shown in Fig. 9–48.

In general, looking at all of the various comparisons of prediction with experiment that are available, many, but not all, observers see an advantage to the two-equation models for free shear flows.

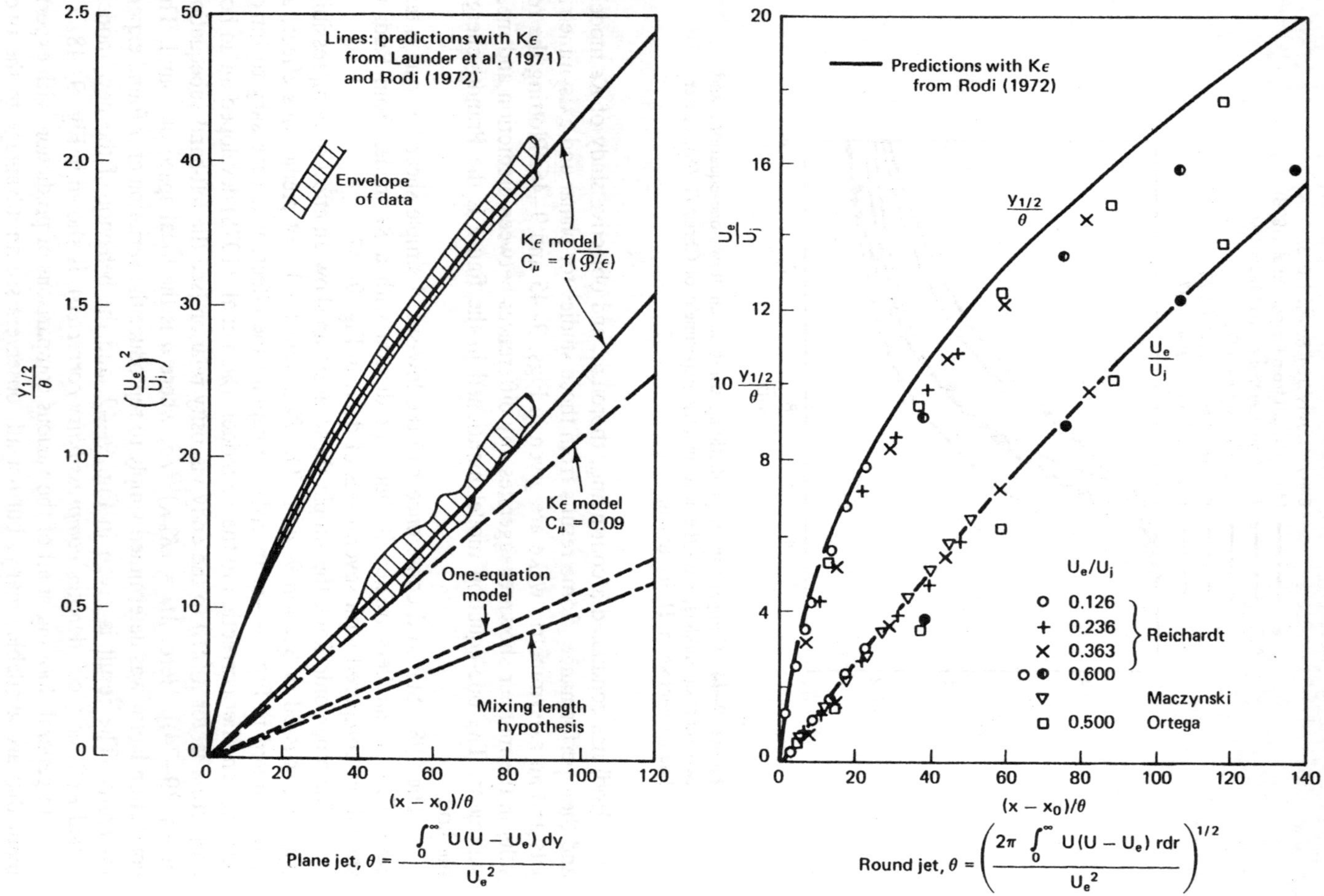

Figure 9–45 Development of maximum excess velocity and half-width for turbulent jets. (From Rodi, 1980.)

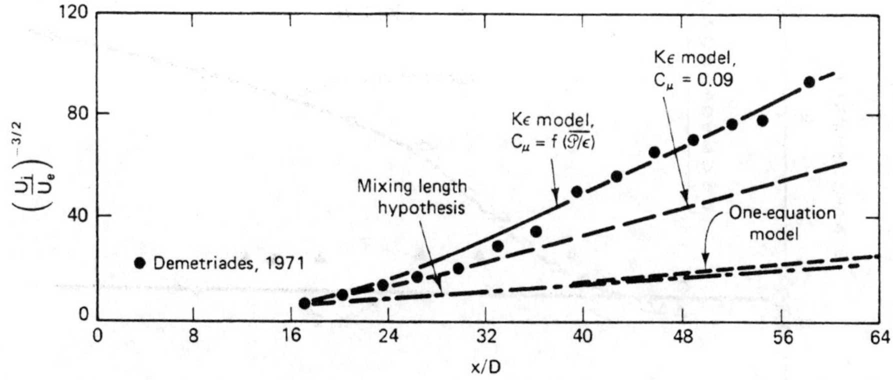

Figure 9–46 Development of maximum velocity defect in a compressible, axisymmetric wake. (From Rodi, 1980.)

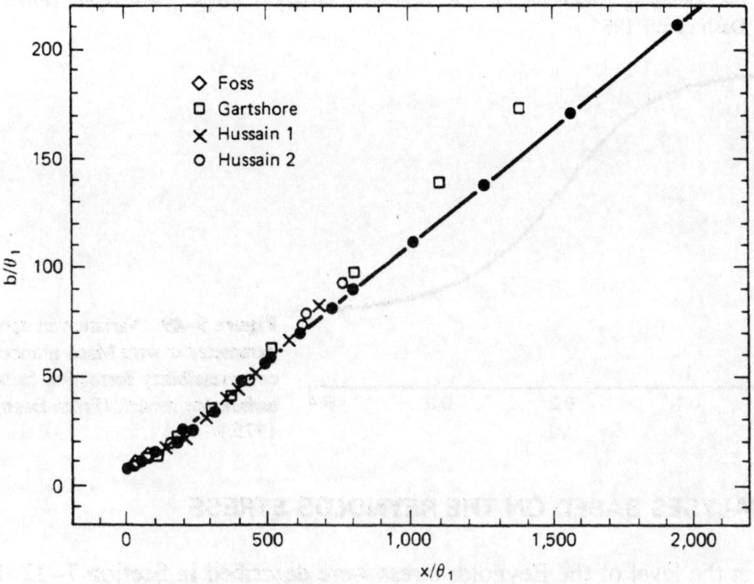

Figure 9–47 Comparison of prediction by Rodi, using a $K\epsilon$ model, with experiment for the low-speed turbulent simple shear layer. (From Kline et al., 1982.)

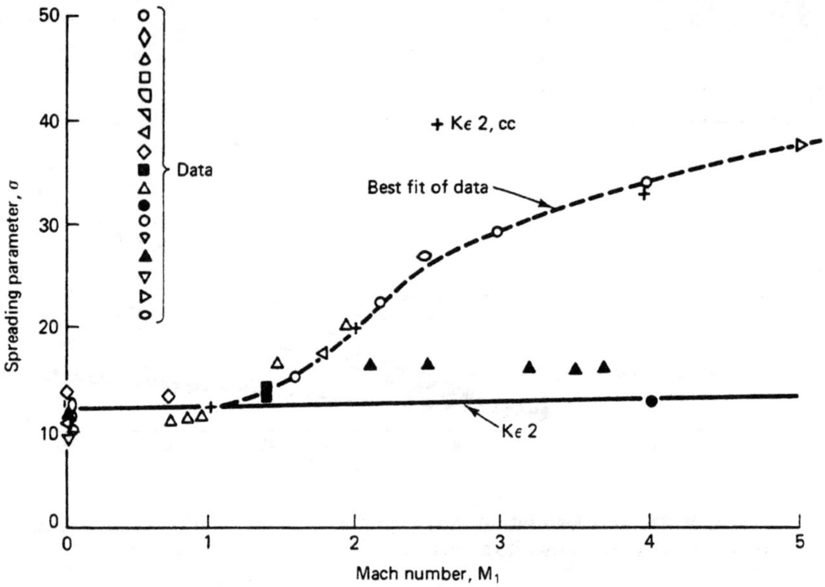

Figure 9–48 Comparison of predictions, using $K\epsilon$ models, with experiment for the spreading parameter for the supersonic turbulent simple shear layer. (From Dash et al., 1985.)

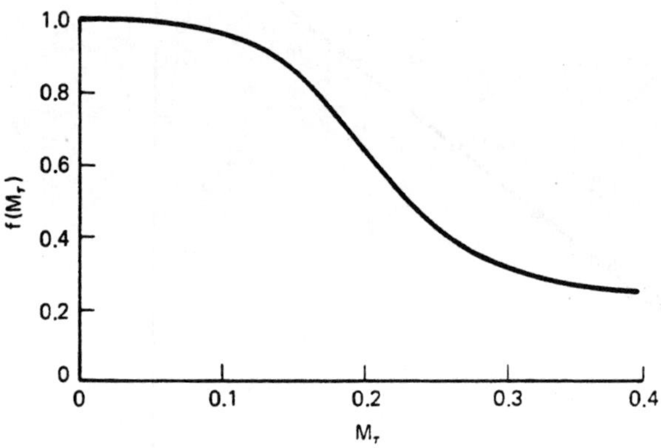

Figure 9–49 Variation in spreading parameter σ with Mach number and compressibility correction factor for $K\epsilon$ 2 turbulence model. (From Dash et al., 1975.)

9–7 ANALYSES BASED ON THE REYNOLDS STRESS

Models at the level of the Reynolds stress were described in Section 7–12. Here we will present only some comparisons of predictions with experiment.

The predictions of the Hanjalic-Launder model are compared with data for a planar jet with $U_e = 0$ in Fig. 9–50. From that comparison and only the one other given by Launder et al. (1973), no clear advantage over the two-equation models is apparent.

The results of the Donaldson (1971) model for Reynolds and normal stresses

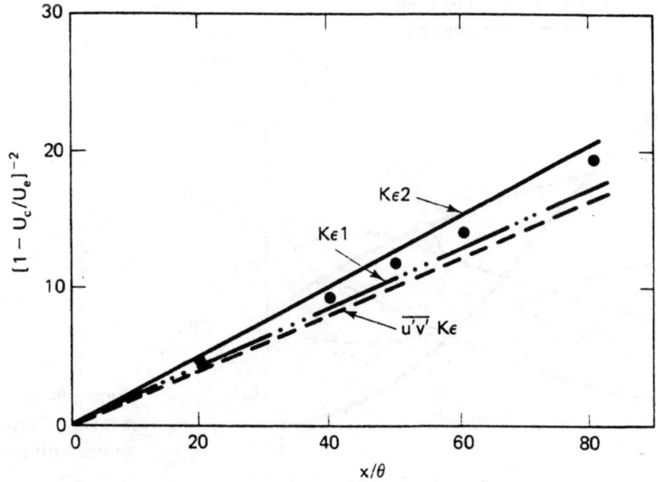

Figure 9–50 Comparison of predictions from two-equation and Reynolds stress models with the planar jet experiment of Everitt and Robins (1978); calculations from Launder et al. (1973).

are compared with experimental data for a round jet with $U_e = 0$ in Fig. 9–51. Some adjustments to the many constants involved were made to obtain the results shown, and the reader is referred to the original paper for all the details. Nonetheless, good results are obviously achievable with this model.

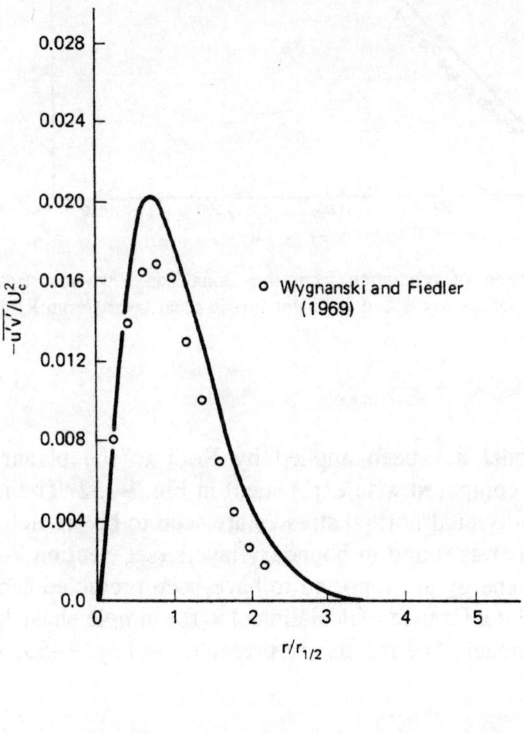

Figure 9–51 Comparison of a prediction from a Reynolds stress model with an experiment for an axisymmetric jet with $U_e = 0$ for Reynolds stress. (From Donaldson, 1971.)

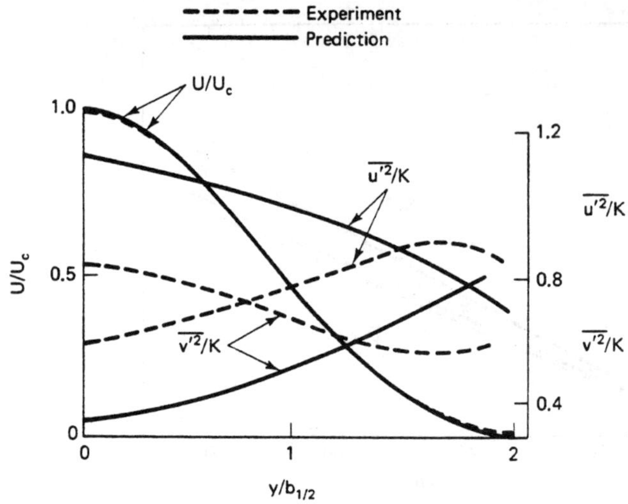

Figure 9–52 Comparison of prediction from the Daly-Harlow Reynolds stress model with experiment for a planar jet after Rodi.

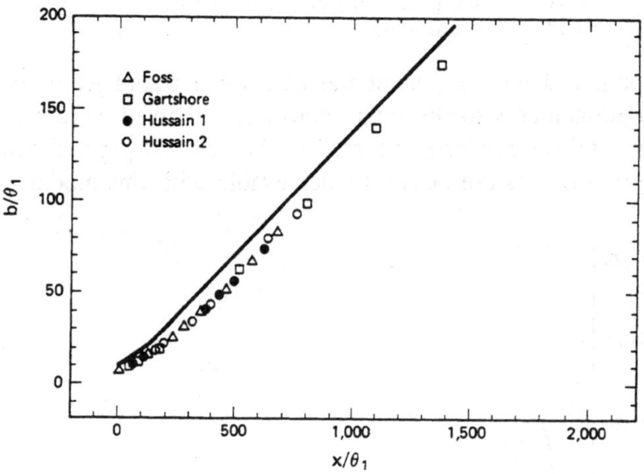

Figure 9–53 Comparison of prediction from the Donaldson Reynolds stress model with experiment for the low-speed turbulent simple shear layer (From Kline et al., 1982.)

The Daly-Harlow model has been applied by Rodi to the planar jet with $U_e = 0$, and the results are compared with experiment in Fig. 9–52. The mean flow is predicted well, but the individual normal stresses are seen to be predicted poorly. This is the same situation as was found in boundary layers (see Section 7–12). The shear stress and turbulence energy are reported to have been predicted accurately.

For the 1981 Stanford conference, calculations for the simple shear layer were made with the Donaldson model. The results are presented in Fig. 9–53, where the fair agreement can be seen.

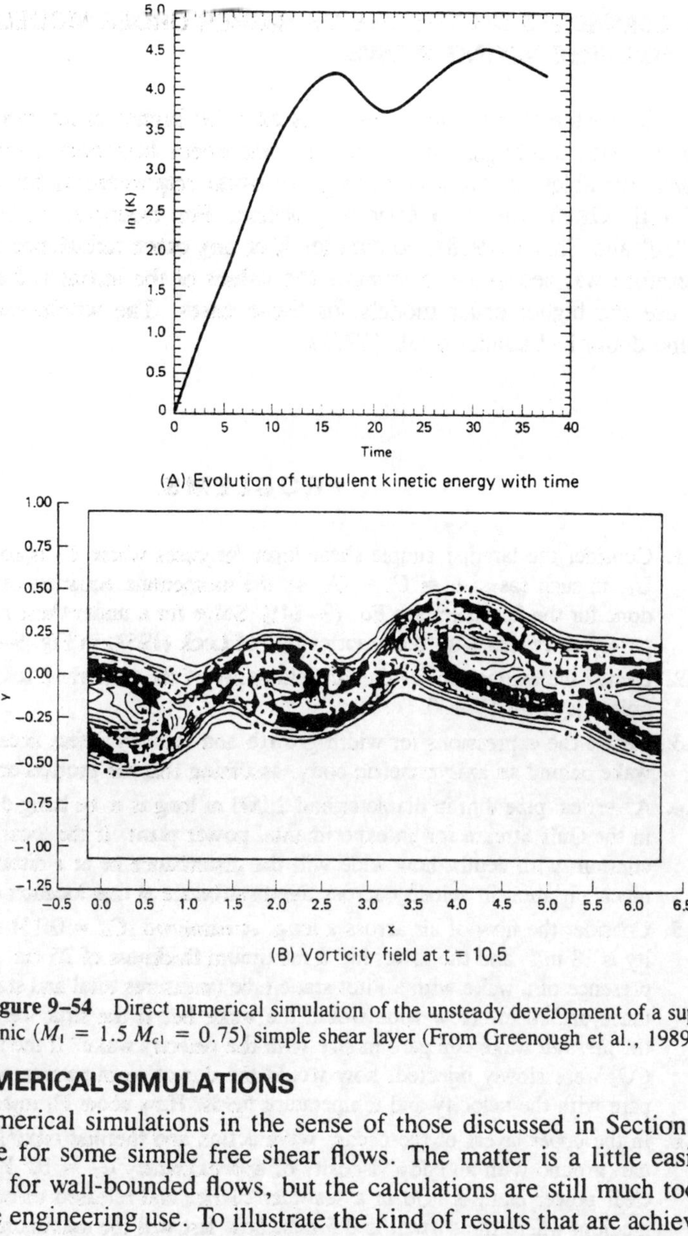

Figure 9–54 Direct numerical simulation of the unsteady development of a supersonic ($M_1 = 1.5\ M_{c1} = 0.75$) simple shear layer (From Greenough et al., 1989.)

9–8 DIRECT NUMERICAL SIMULATIONS

Direct numerical simulations in the sense of those discussed in Section 7–13 have been made for some simple free shear flows. The matter is a little easier for such cases than for wall-bounded flows, but the calculations are still much too expensive for routine engineering use. To illustrate the kind of results that are achievable, some calculations by Greenough et al. (1989) are given in Fig. 9–54. These calculations use the inviscid equations of motion and start with a perturbation of a simple shear layer in the range where linear stability theory predicts instability. The case with $M_1 = 1.5$ and $M_{c1} = 0.75$ in the figure has an initial time rate of change of the turbulent kinetic energy (see Fig. 9–54(A)) that agrees closely with the linear theory. It levels off at a dimensionless time of about 8.0. The vorticity field in Fig. 9–54(B) shows not only the excited fundamental mode, but also the third harmonic.

9-9 COMMENTS ON THE USE OF HIGHER ORDER MODELS FOR FREE MIXING FLOWS

The discussion in Section 7–14 holds also for higher order models for free-mixing flows. With the higher order models, one needs boundary conditions on the additional variables, as well as having the usual requirements for (U, V) [and maybe $(\bar{T}, C)$]. Often, this is a serious problem. For example, in Forstall and Shapiro (1950) and Chriss (1968), no data for K or any other turbulence data are reported. It therefore was necessary to *estimate* the values of the initial and boundary conditions to use the higher order models for those cases. The whole matter is discussed in some detail in Launder et al. (1971).

PROBLEMS

9.1. Consider the laminar simple shear layer for cases where U_2 is not very different from U_1. In such cases, $u \approx U_1 \approx U_2$, so the momentum equation can be linearized, as is done for the far wake [see Eq. (9–14)]. Solve for u under these restrictions, and compare your results with the exact results of Lock (1951) in Fig. 9–2.

9.2. Derive the expression for the far-field decay of the centerline velocity for a planar turbulent jet with $U_e = 0$.

9.3. Derive the expressions for width growth and velocity defect decay in the turbulent far wake behind an axisymmetric body, assuming that the profiles are *similar*.

9.4. A vertical pipe 8 m in diameter and 2,000 m long is to be hung down from the surface in the Gulf Stream for an experimental power plant. If the local current is 1 m/s and constant with depth, how wide will the disturbance be at a distance of 0.5 km? How much variation in velocity across the disturbance at that location can be expected?

9.5. Consider the flow of air across a long, streamlined ($C_D = 0.15$) tower. The air velocity is 18 m/s, and the tower has a maximum thickness of 25 cm. Could you detect the presence of a wake with a Pitot-static tube (measures total and static pressure) at a distance of 500 m? How wide would the wake be? If the strut were heated, how would the *thermal* wake compare in size with the *velocity* wake? If the strut were porous and CO_2 were slowly injected, how would the downstream pattern of concentration compare with the velocity and temperature fields? How about H_2 injection?

9.6. In the upper layers of the ocean, wave action and thermal mixing produce a turbulent background with an eddy viscosity of approximately $\mu_T \approx 10^2$ P. If waste from a nuclear power plant is held in a seawater carrier and released through a slit into such a medium where the current is 0.5 m/s, how fast will the maximum concentration decay with downstream distance?

9.7. Air at 20.0 mph is flowing over a round rod with a 2.0-in diameter. How far behind the rod will the maximum velocity difference across the wake drop to 3%? How wide will the wake be at that location? If a streamlined shield with $C_D = 0.10$ is fitted to the rod, how will these values change?

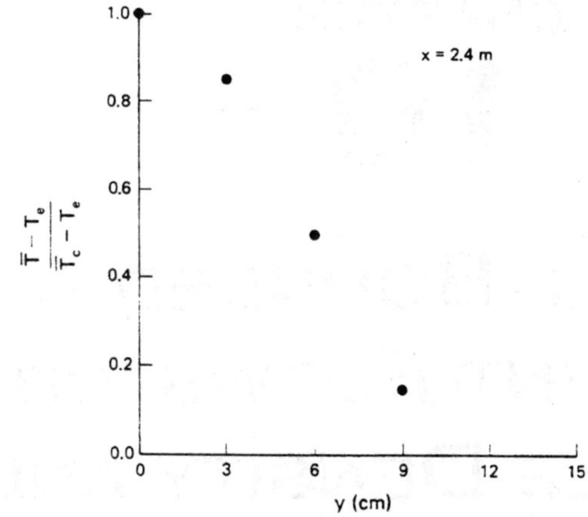

Figure P9–8

9.8. Suppose an experiment and an analysis are to be compared to determine a turbulent Prandtl number. Hot air (50°C) is injected at 3 m/s through a two-dimensional jet (total height = 3 cm) into a room at 25°C. The analysis gives the result

$$\frac{\overline{T}(x, y) - T_e}{\overline{T}_c(x) - T_e} = \frac{1 + \cos(\pi y/5\alpha_T x)}{2}$$

where α_T ($\equiv k_T/\overline{\rho}c_p$) is the turbulent diffusivity, with units m²/s, and $\overline{T}_c(x) \sim 1/(x)$. You may assume that the kinematic eddy viscosity is constant throughout the flow at a value of 0.005 m²/s. Using the *data* shown in the accompanying figure, what is the Prandtl number?

9.9. A chemical plant has a very tall stack that is 1.0 m in diameter. A new zoning code says that the plant must reduce the maximum variation in the mean velocity in the wake by 25% and the maximum rms velocity fluctuation in the wake by 30% at a fence line 500 m from the stack when the prevailing wind blows at 15.0 mph. How might the plant comply with the code at low cost?

9.10. Use a numerical method and a mixing length model to calculate the development of a jet mixing problem with air at STP. There is an external stream at 50.0 ft/s, and the jet exit velocity is 100.0 ft/s from a planar nozzle 0.5 in high. Compare your prediction for the centerline velocity decay with the data from Weinstein (1956) shown in Fig. 9–30.

9.11. Modify the code JETWAKE in Appendix B to make it capable of treating laminar jet and wake problems in the planar geometry.

9.12. Use a numerical treatment and the Prandtl energy method to calculate the development of a jet mixing problem with air at STP. There is an external stream at 50.0 ft/s, and the jet exit velocity is 100.0 ft/s from a planar nozzle 0.5 in high. Compare your prediction for the centerline velocity decay with the data from Weinstein (1956) shown in Fig. 9–30.

9.13. Modify the code JETWAKE in Appendix B to make it capable of employing a $K\epsilon$ turbulence model for planar jet and wake problems. To do this, you must add Eq. (9–71) to the system.

CHAPTER

10

WALL-BOUNDED TURBULENT FLOWS WITH VARIABLE DENSITY AND HEAT AND MASS TRANSFER

10-1 INTRODUCTION

This chapter is concerned with cases in which there are differences between the wall and free stream temperatures or injection of a foreign fluid, so as to produce heat and/or mass transfer. We also allow substantial variations in fluid density and laminar, thermophysical properties, such as those encountered in high-speed flight. The laminar thermophysical properties remain important in wall-bounded turbulent flows because of the *laminar sublayer*. The organization of the chapter will follow that used in earlier sections dealing with turbulent flows: Experimental data for mean and fluctuating flow quantities will be presented first, followed by transport modeling at various levels, and then comparisons of predictions with experiment.

10-2 EXPERIMENTAL INFORMATION

10-2-1 Mean Flow Data

For mean flow profiles, one obviously has to deal with at least one new variable, the mean temperature $\bar{T}$. With the injection of a foreign fluid, it is also necessary to treat the mean concentration C. A few of these cases will still permit the assumption of constant density and laminar properties. In those cases, it is reasonable to ask if there exist inner and outer region scaling laws for mean temperature profiles, such as were found for the mean velocity profiles in Section 7–3–1. For the inner, wall-dominated regions, an effort of that type has been pursued for low-speed, constant-property flows by a number of researchers, with the most comprehensive treatment being that by Kader (1981). The key idea is that of a *heat transfer temperature*

$$T_* \equiv \frac{q_w}{\rho c_p u_*} \tag{10–1}$$

which corresponds to the *friction velocity* u_*. This concept leads to a dimensionless temperature

$$T^+ \equiv \frac{T_w - \bar{T}}{T_*} \tag{10–2}$$

which corresponds to u^+. This equation in turn will permit the development of a *temperature law of the wall,*

$$T^+ = g_T(y^+, \mathrm{Pr}) \tag{10–3}$$

where Pr is the laminar Prandtl number. For the flow in the *laminar sublayer*, the same kind of development as used for the velocity field gives

$$T^+ = y^+\mathrm{Pr} \tag{10–4}$$

One assumes laminar flow and a thin layer, so that the heat flux can be taken to be constant at its wall value. There is also a logarithmic region described by

$$T^+ = \frac{1}{\kappa_T}\ln(y^+) + C_T(\mathrm{Pr}) \tag{10–5}$$

with

$$C_T(\mathrm{Pr}) = (3.85\mathrm{Pr}^{1/3} - 1.3)^2 + 2.12\ln(\mathrm{Pr}) \tag{10–6}$$

This region is analogous to the logarithmic region for the velocity profile. Kader (1981) further developed functions to join these two regions smoothly and also for the outer region corresponding to the velocity defect law. His final, complete equation is

$$T^+ = \mathrm{Pr}\cdot y^+\exp(-\Gamma) + \left[2.12\ln\left((1 + y^+)\frac{2.5(2 - y/\delta)}{1 + 4(1 - (y/\delta))^2}\right) + C_T(\mathrm{Pr})\right]\exp\left(-\frac{1}{\Gamma}\right) \tag{10–7}$$

where

$$\Gamma \equiv \frac{0.01(y^+\cdot\mathrm{Pr})^4}{1 + 5y^+\cdot\mathrm{Pr}^3} \tag{10–8}$$

The success of this representation for fluids with a wide range of laminar Prandtl numbers is shown in Fig. 10–1. In addition, Kader (1981) gives a corresponding law for pipe and channel flows.

A useful result can be deduced with the aid of the logarithmic laws for both temperature and velocity [Eqs. (10–5) and (7–7c)], extended out to the edge of the boundary layers. We have

$$T_e^+ = \frac{1}{\kappa_T}\ln(\delta_T^+) + C_T(\text{Pr}) \tag{10–9}$$

and

$$u_e^+ = \frac{1}{\kappa}\ln(\delta^+) + C \tag{10–10}$$

for $\text{Pr} = 0(1)$, $\delta_T \approx \delta$, and $\kappa_T \approx \kappa$, so we can subtract Eq. (10–10) from Eq. (10–9) to eliminate the logarithmic terms and get

$$T_e^+ - u_e^+ = C_T(\text{Pr}) - C \tag{10–11}$$

Using the definitions of T^+ and u^+, we see that this becomes

$$\frac{\rho c_p u_*(T_w - T_e)}{q_w} - \frac{U_e}{u_*} = C_T(\text{Pr}) - C \tag{10–12}$$

Noting that $\hbar = q_w/(T_w - T_e)$, $\text{St} = \hbar/(\rho U_e c_p)$, and $C_f = 2(u_*/U_e)^2$, we obtain

$$\sqrt{\frac{2}{C_f}} + C_T(\text{Pr}) - C = \text{St}^{-1}\sqrt{\frac{C_f}{2}} \tag{10–12a}$$

This result is not tidy, so many people use the equivalent approximation of von Karman (1939), i.e.,

$$\text{St} = \frac{C_f}{2\text{Pr}^{0.4}(T_w/T_e)^{0.4}} \tag{10–13}$$

which is good for $0.7 \leq \text{Pr} \leq 10.0$. This expression can be seen to be closely related to the *Reynolds analogy* [see Eq. (2–61)]. The prediction of Eq. (10–13) is compared with data in Fig. 10–2.

Most of the consideration of flows with surface mass transfer has been in terms of the analogy between heat and mass transfer introduced earlier for laminar flows. Within that framework, one presumes a direct correspondence between a film coefficient for heat, $\hbar$, and for diffusion, $\hbar_D$, usually presented in dimensionless terms as the Nusselt numbers Nu and Nu_{Diff} or the Stanton numbers St and St_{Diff}. The subscripts D and Diff denote flows with diffusion (i.e., mass transfer). Further, the role of the Prandtl number Pr in heat transfer correlations is taken by the Schmidt number Sc in mass transfer correlations. The correspondence between mass transfer data and a transformed heat transfer law ($\text{St} \rightarrow \text{St}_{\text{Diff}}$ with $\text{Pr} \rightarrow \text{Sc}$), where the heat transfer law is for flow over a solid surface, is obtained only for very low surface

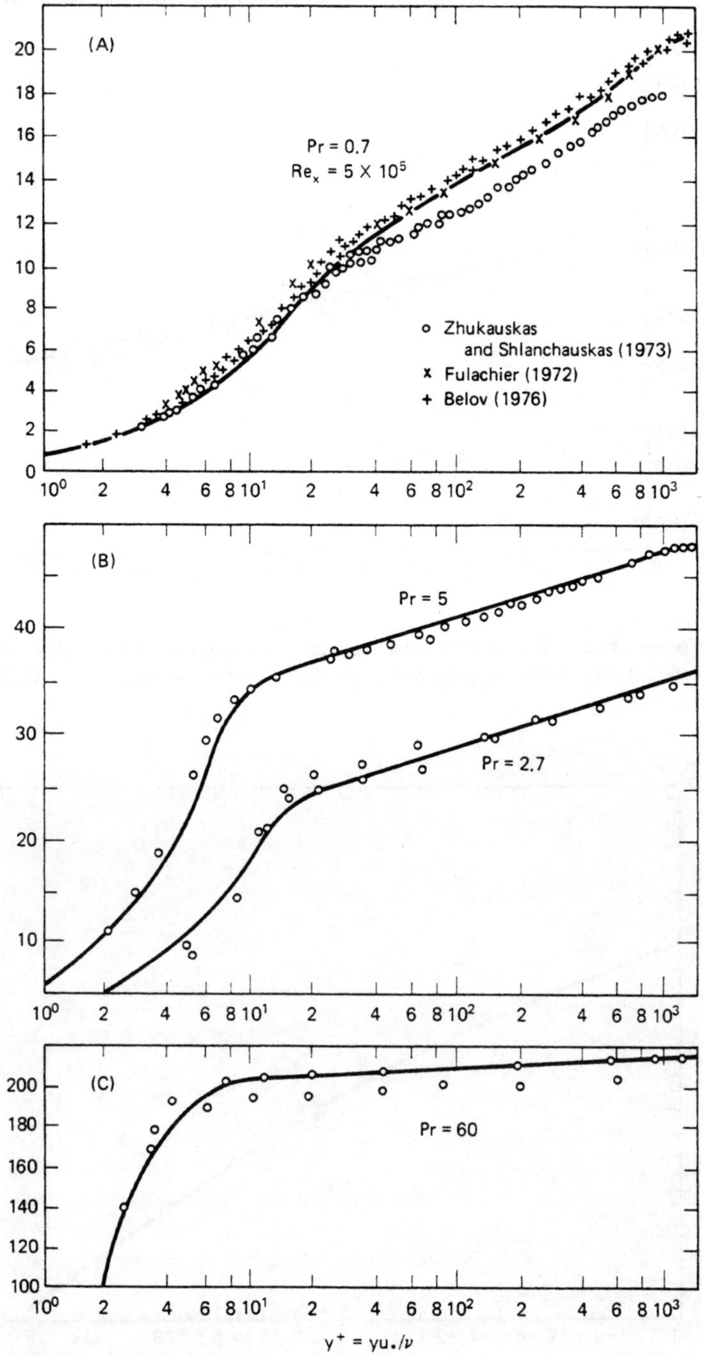

Figure 10–1 Comparison of Eq. (10–7) with data for the temperature law of the wall. (From Kader, 1981.)

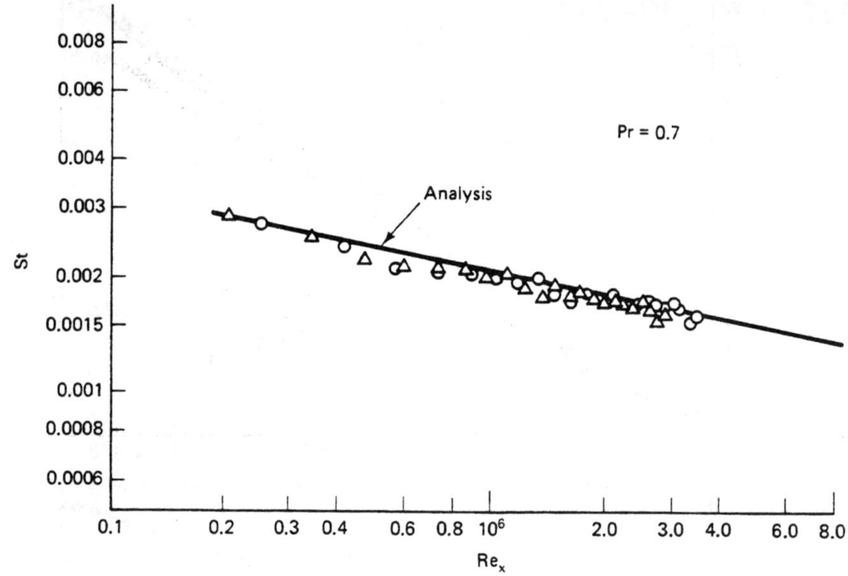

Figure 10–2 Dimensionless heat transfer coefficient for turbulent flow over a constant-temperature flat plate in low-speed flow. (From Reynolds et al., 1958.)

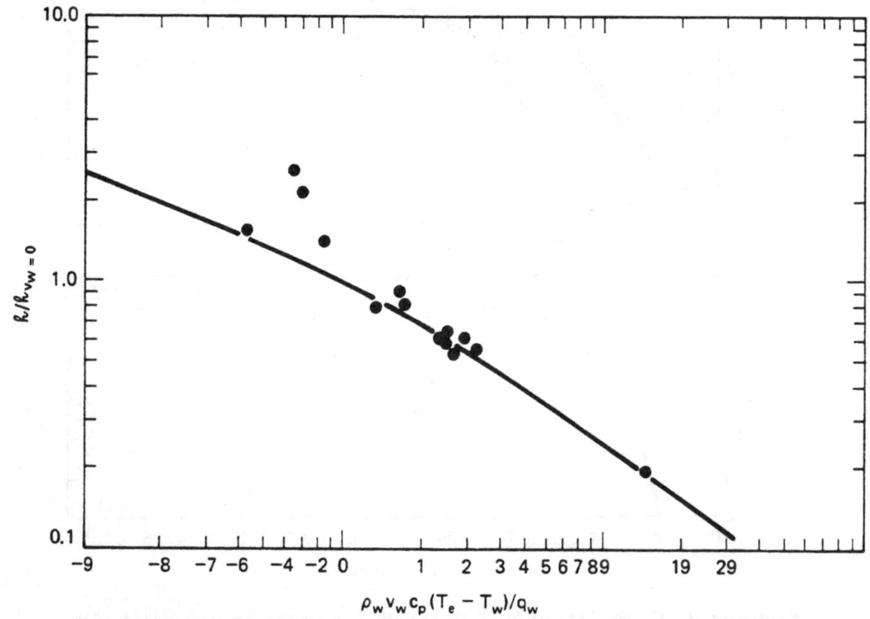

Figure 10–3 Influence of injection or suction on the film coefficient for flow over a flat plate. (From Mickley et al., 1954.)

mass transfer rates. The basic analogy between heat and mass transfer still holds for nonnegligible surface mass transfer, but in such cases, one must begin with a heat transfer law for flow over a porous surface with injection or suction. The influence of injection or suction on heat transfer is substantial as was shown earlier for laminar flow (see Sections 4–5–1 and 5–10–2). Figure 10–3 shows this for turbulent flow. This kind of information has been generalized considerably by relating the ratio of skin friction coefficients without and with injection or suction to the ratio of the Stanton numbers without and with injection or suction through two parameters, $-B_f = (\rho_w v_w/\rho_e U_e)/(C_f/2)$ and $B_h = (\rho_w v_w/\rho_e U_e)/\text{St}$. The simple correlation formula

$$\frac{\text{St}}{\text{St}_{v_w=0}} = \frac{C_f}{C_{f_{v_w=0}}} = \left[\frac{\ln(1 + B_{f,h})}{B_{f,h}}\right]^{5/4}(1 + B_{f,h})^{1/4} \tag{10–14}$$

works very well over a considerable range of injection or suction rates and different pressure gradients, as illustrated in Fig. 10–4. Here, Re_θ for the two flows must match. This information is useful not only for mass transfer work, but also for momentum and/or heat transfer in single-species flows.

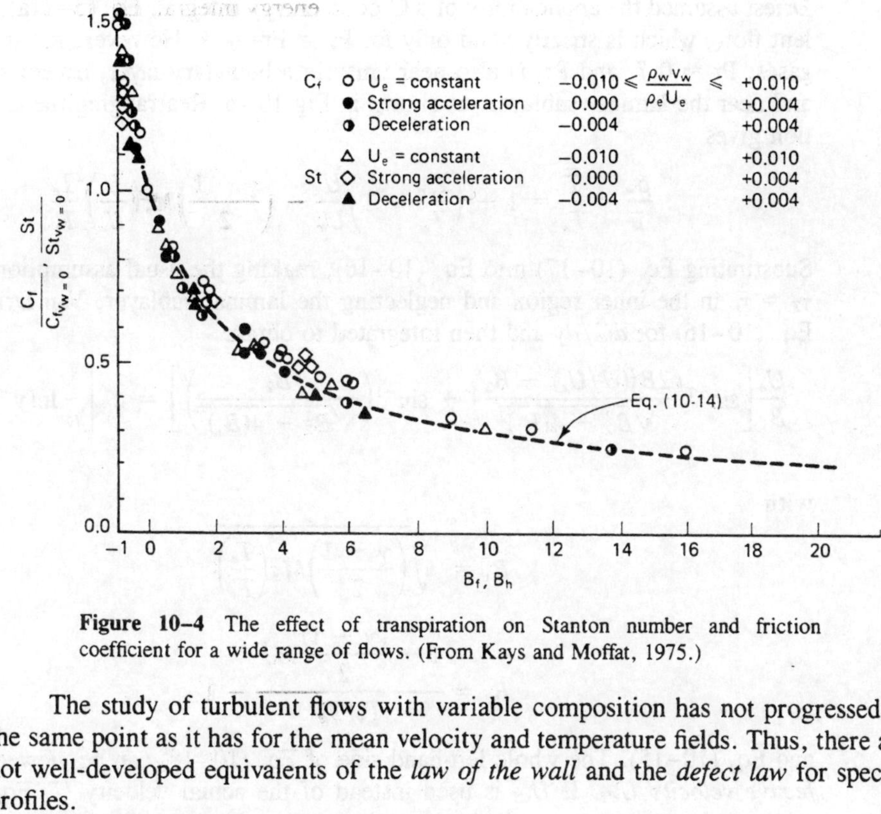

Figure 10–4 The effect of transpiration on Stanton number and friction coefficient for a wide range of flows. (From Kays and Moffat, 1975.)

The study of turbulent flows with variable composition has not progressed to the same point as it has for the mean velocity and temperature fields. Thus, there are not well-developed equivalents of the *law of the wall* and the *defect law* for species profiles.

For variable-density and variable-property flows produced directly by high free stream velocities, the *law of the wall* for velocity developed for low-speed, constant-density, constant-property flows no longer holds, as shown in Fig. 10–5 from Lee et al. (1969). In this plot, the simple choices

$$y^+ \equiv \frac{\rho_w y u_*}{\mu_w}, \qquad u_* \equiv \sqrt{\frac{\tau_w}{\rho_w}} \tag{10–15}$$

were made, since density and properties vary with y. There have been a number of attempts to develop a law suitable for high-speed cases, and those usually referred to as *Van Driest I* and *II* from Van Driest (1951) and (1956b), respectively, are generally viewed as the most successful.

For *Van Driest I,* one takes the mixing length formulation, Eq. (7–37), but now with the fluid density as a variable:

$$\tau_T = \overline{\rho}\ell_m^2\left|\frac{\partial U}{\partial y}\right|\frac{\partial U}{\partial y} \tag{10–16}$$

The assumption $\ell_m = \kappa y$ is retained. To obtain a simple relation for the mean density $\overline{\rho}$, one needs a relation for $\overline{T}$, since P is constant across the boundary layer. Van Driest assumed the applicability of a Crocco energy integral, Eq. (5–21a), for turbulent flow, which is strictly valid only for $\mathrm{Pr} = \mathrm{Pr}_T = 1$. However, for air and most gases, $\mathrm{Pr} \approx 0.7$, and Pr_T is also near unity in a boundary layer, except at the edge and near the laminar sublayer, as shown in Fig. 10–6. Rearranging the Crocco relation gives

$$\frac{\rho_w}{\overline{\rho}} = \frac{\overline{T}}{T_w} = 1 + \left(\frac{T_{aw}}{T_w} - 1\right)\frac{U}{U_e} - \left(\frac{\gamma - 1}{2}\right)M_e^2\left(\frac{U}{U_e}\right)^2\frac{T_e}{T_w} \tag{10–17}$$

Substituting Eq. (10–17) into Eq. (10–16), making the usual assumption that $\tau + \tau_T = \tau_w$ in the inner region and neglecting the laminar sublayer, Van Driest solved Eq. (10–16) for dU/dy and then integrated to obtain

$$\frac{U_e}{B_1}\left[\sin^{-1}\left(\frac{2B_1^2(U/U_e) - B_2}{\sqrt{B_2^2 + 4(B_1)^2}}\right) + \sin^{-1}\left(\frac{B_2}{\sqrt{B_2^2 + 4(B_1)^2}}\right)\right] = u_*\left[\frac{1}{\kappa}\ln(y^+) + C\right] \tag{10–18}$$

with

$$B_1 = \sqrt{\left(\frac{\gamma - 1}{2}\right)M_e^2\left(\frac{T_e}{T_w}\right)}$$
$$B_2 = \frac{1 + \dfrac{(\gamma - 1)}{2}M_e^2}{T_w/T_e} - 1 \tag{10–19}$$

and Eq. (10–15). The whole left-hand side of Eq. (10–18) can be viewed as an *effective velocity* U_{eff}. If U_{eff} is used instead of the actual velocity U, Eq. (10–18) takes on the same form as the low-speed, constant-density, constant-property *law of*

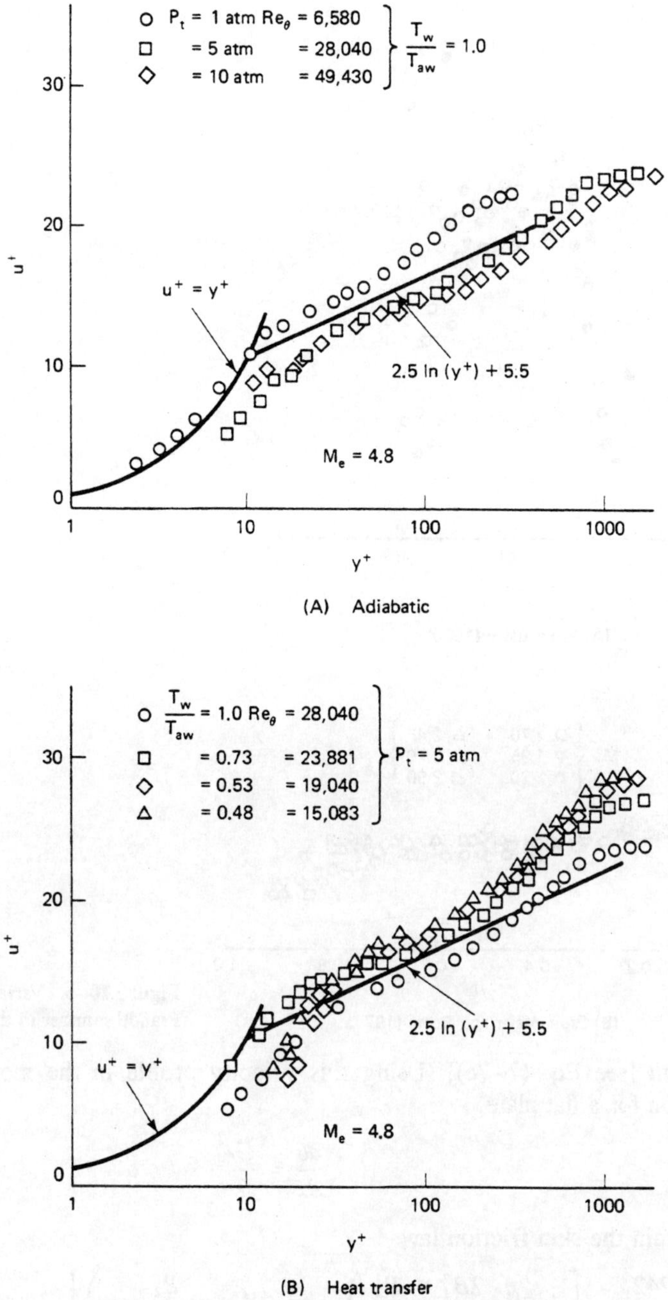

Figure 10–5 Law-of-the-wall plot of high-speed data and low-speed equations. (From Lee et al., 1969.)

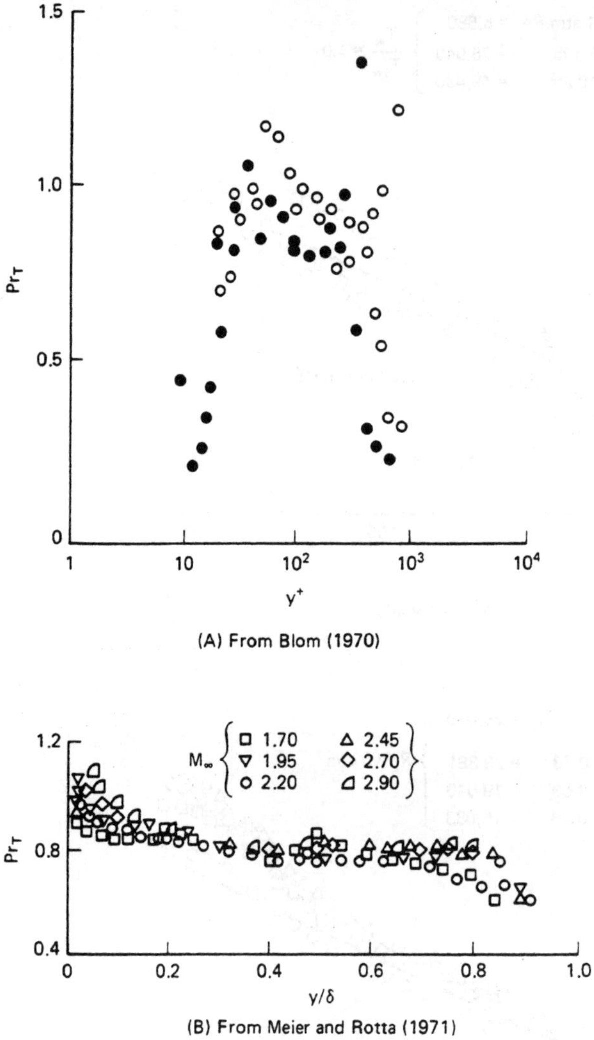

Figure 10–6 Variation of the turbulent Prandtl number in the boundary layer.

the wall [see Eq. (7–7c)]. Using this velocity profile in the momentum integral equation for a flat plate,

$$\frac{d\theta}{dx} = \frac{C_f}{2}$$

we obtain the skin friction law

$$\frac{0.242}{B_1\sqrt{C_f(T_w/T_e)}}\left[\sin^{-1}\left(\frac{2B_1^2 - B_2}{\sqrt{B_2^2 + 4B_1^2}}\right) + \sin^{-1}\left(\frac{B_2}{\sqrt{B_2^2 + 4B_1^2}}\right)\right]$$
$$= 0.41 + \log(\text{Re}_x C_f) - \left(\frac{1}{2} + \omega\right)\log\left(\frac{T_w}{T_e}\right) \tag{10–20}$$

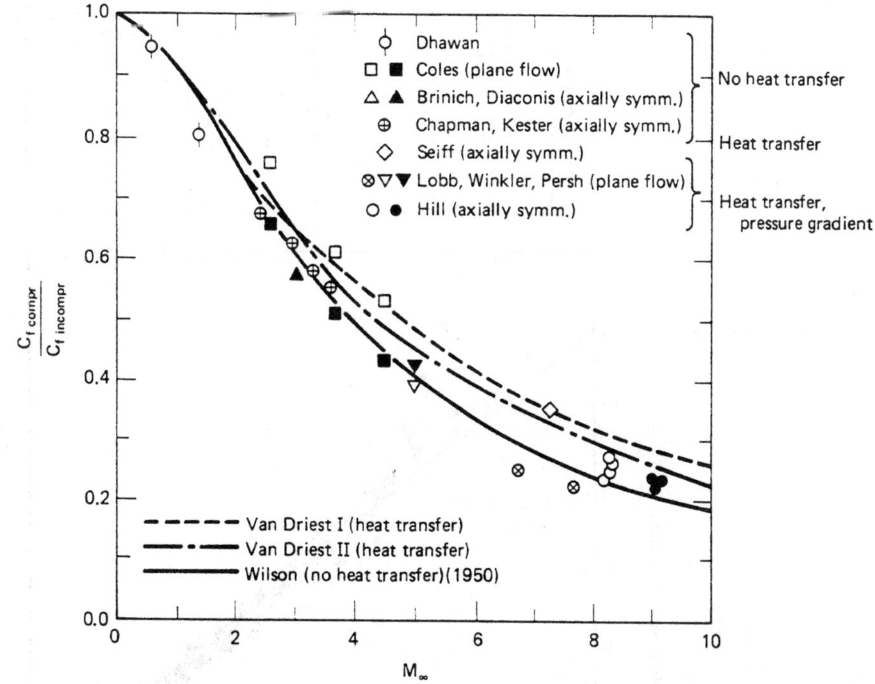

Figure 10–7 Variation of skin friction coefficient for turbulent flow over a flat plate as a function of Mach number at $Re_x = 10^7$. (From Hill, 1956.)

where ω enters from the assumption that $\mu \sim T^\omega$. This result is compared with data as $C_f/(C_f)_{M\approx 0}$ versus M_∞ in Fig. 10–7.

Since the profile in Eq. (10–18) is based on $\ell_m = \kappa y$, it does not have a wake component in the sense of Coles's *law of the wake* (see Section 7–3–1). Maise and McDonald (1968) considered whether the Van Driest profile could be extended to have a wake component by simple analogy with the constant-density case [see Eq. (7–9)], using U_{eff} for U. The result is shown in Fig. 10–8, in comparison with some supersonic profile measurements over an adiabatic wall. Generally, good agreement was found for this case, but results for heat transfer cases were not good.

For *Van Driest II*, the Von Karman (1931) mixing length model

$$\ell_m = \kappa \left| \frac{\partial U/\partial y}{\partial^2 U/\partial y^2} \right| \tag{10–21}$$

was used instead of $\ell_m = \kappa y$. That use is a bit surprising, since the Von Karman model has never gained wide acceptance. In addition, the temperature relationship is a *modified* Crocco integral, to account for cases in which the recovery factor r is not unity. Using those assumptions again with the integral momentum equation, a different skin friction law is developed, viz.,

	M_e	Re_θ	
▽	1.47	4.56×10^3	Seddon
△	2.669	7.02×10^5	Moore and Harkness
□	2.91	5.4×10^4	Moore and Harkness
○	4.93	5.53×10^3	Lobb et al.
■	1.85	3.09×10^3	Michel
▼	2.10	2.78×10^3	Michel
▲	2.57	3.24×10^3	Michel
●	2.96	2.64×10^3	Michel
●	1.966	2.98×10^3	Coles
■	4.512	3.47×10^3	Coles

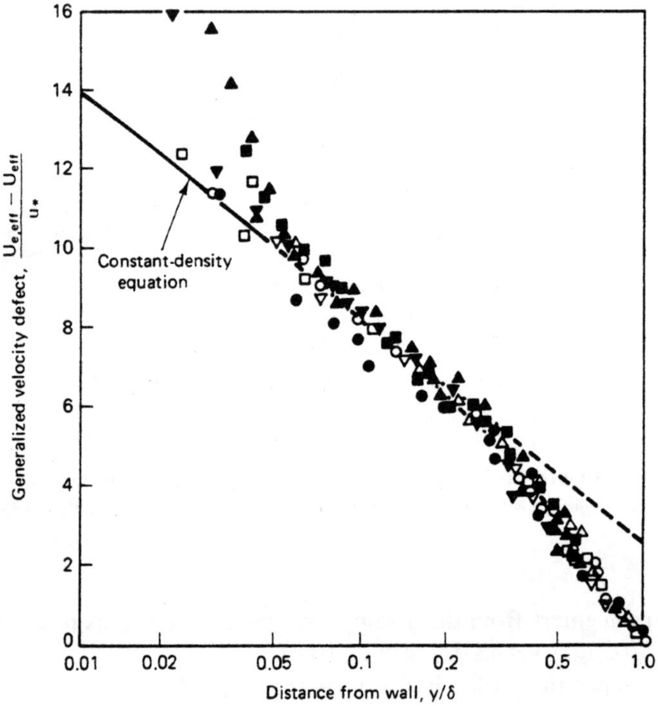

Figure 10–8 Generalized velocity profiles for compressible, adiabatic flat-plate boundary layers. (From Maise and McDonald, 1968.)

$$\frac{0.242}{B_1'\sqrt{C_f(T_w/T_e)}}\left[\sin^{-1}\left(\frac{2(B_1')^2 - B_2'}{\sqrt{(B_2')^2 + 4(B_1')^2}}\right) + \sin^{-1}\left(\frac{B_2'}{\sqrt{(B_2')^2 + 4(B_1')^2}}\right)\right] = 0.41 + \log(\mathrm{Re}_x C_f) - \omega \log\left(\frac{T_w}{T_e}\right) \tag{10–22}$$

where

$$B_1' = \sqrt{\frac{((\gamma - 1)/2)M_e^2 r}{T_w/T_e}} \qquad B_2' = \frac{1 + \dfrac{(\gamma - 1)}{2}M_e^2 r}{T_w/T_e} - 1 \tag{10–23}$$

The improved agreement of this prediction with experiment can be seen in Fig. 10–7.

When U_{eff} [the left-hand side of Eq. (10–18)] is defined using B_1' and B_2', a good correlation of profiles in both the inner and outer regions is obtained, as shown in Fig. 10–9 from Hopkins et al. (1972). This correlation has been tested to very

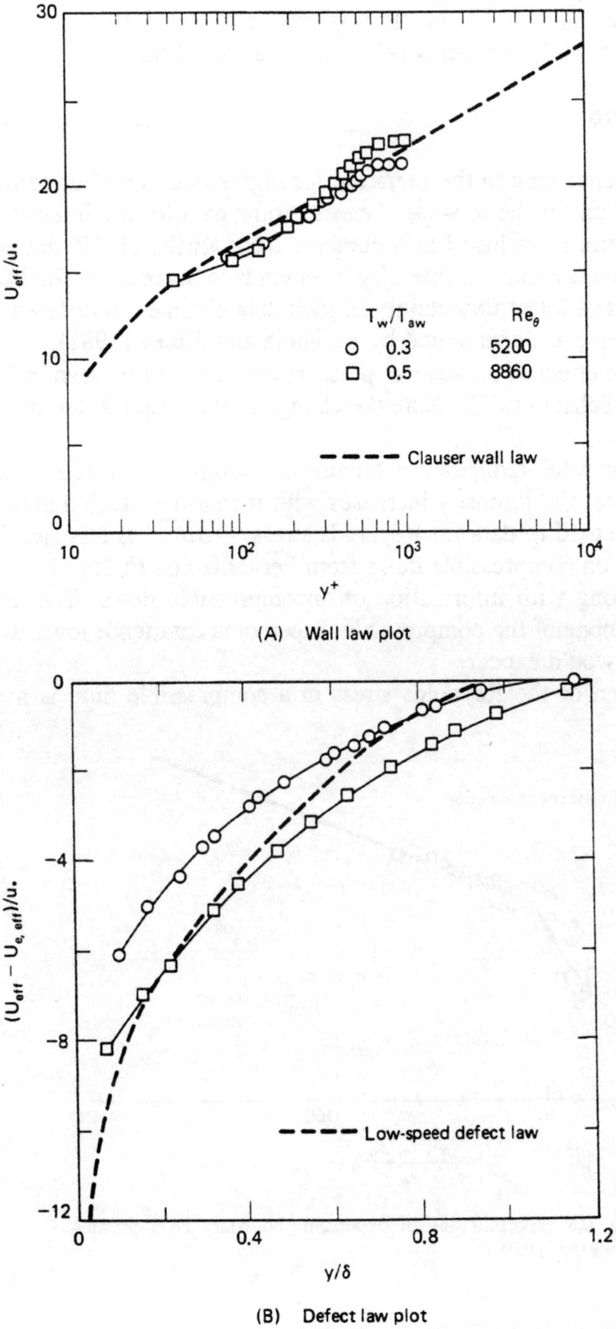

Figure 10–9 Comparison of correlation of velocity profiles by Van Driest II with Mach 7 data. (From Hopkins et al., 1972.)

high Mach numbers, with experiments in helium wind tunnels up to Mach 47, and the success of the correlation is illustrated in Fig. 10–10.

For heat transfer calculations in high-speed flow, one needs to calculate the adiabatic wall temperature, and that requires information on the recovery factor r. A sampling of data are given in Fig. 10–11. The approximation $r \approx \mathrm{Pr}^{1/3}$ for turbulent flow is often used. This information is restricted to attached flows.

10–2–2 Turbulence Data

There is much less turbulence data in the literature for high-speed, variable-density, variable-property cases than for the low-speed cases. Some data for the intensity of the axial velocity turbulence at various Mach numbers from Kistler (1959) are plotted in Fig. 10–12. One can see that the intensity level tends to decrease as the Mach number increases. There is a lot of uncertainty in such data obtained in different facilities by different workers, as documented by Fernholz and Finley (1981).

Some measure of the effect of an adverse pressure gradient can be seen in Fig. 10–13 from Waltrup and Schetz (1973). Note the change in the shape of the profile near the wall.

Next, some data for total temperature fluctuations from Kistler (1959) are shown in Fig. 10–14. Here, the intensity increases with increasing Mach number.

The situation with regard to data for Reynolds stress, $-\bar{\rho}\overline{u'v'}$ is less satisfactory. A collection of data on compressible flows from Fernholz and Finley (1981) is shown in Fig. 10–15, along with information on incompressible flows. The large scatter is apparent. Also, none of the compressible, supersonic data tends towards τ_w as y goes to zero, as one would expect.

Finally, the definition of the Reynolds stress in a compressible fluid is more

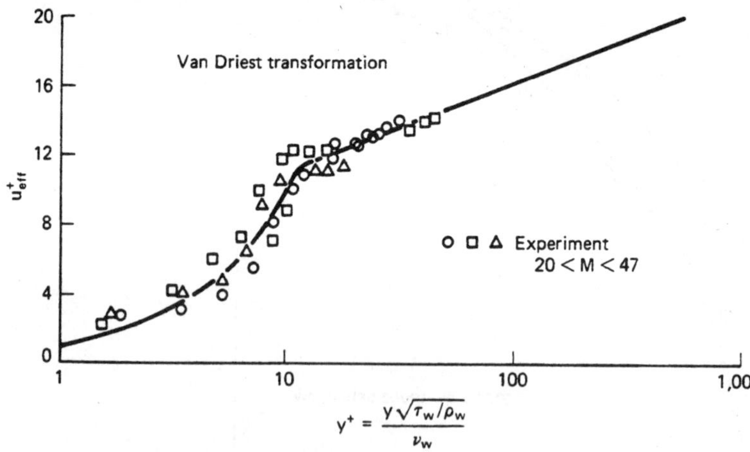

Figure 10–10 A log law representation of hypersonic boundary layer profiles. (From Marvin and Coakley, 1989.)

	Wind tunnel	M_∞	Type of cone		Wind tunnel	M_∞	Type of cone
●	Ames 1 × 3 ft No. 1	1.97	10° hollow; steel	⊙	Aberdeen	2.18	10° wood
□	Lewis 8 × 6 ft	1.98	10° hollow; steel	X	GALCIT 5 × 5 in.	6.0	20° ceramic
△	Lewis 18 × 18 in.	1.94	10° hollow; steel	○	Ames 1 × 3 ft No. 1	2.0	20° hollow; steel
▽	Ames 6 × 6 ft	1.9	10° hollow; steel	▲	JPL 18 × 20 in.	4.50	5° fiberglass
◇	Ames 10 × 14 in.	4.48	10° hollow; steel	⊡	JPL 18 × 20 in.	1.63	13° Lucite
▼	Ames 2 × 2 ft	1.21	10° hollow; steel	+	JPL 18 × 20 in.	4.50	13° Lucite
⊕	Lewis 2 × 2 ft	3.93	10° hollow; steel	◬	JPL 12 × 12 in.	1.63	13° Lucite
◆	Ames 1 × 3 ft No. 2	3.00	10° hollow; steel	*	JPL 12 × 12 in.	2.54	13° Lucite

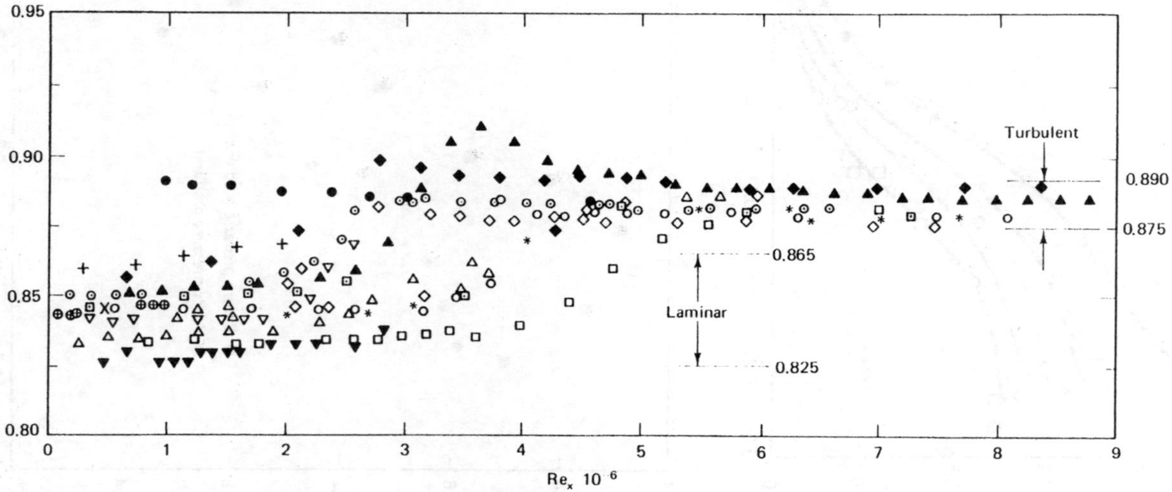

Figure 10–11 Recovery factors on cones from Mach 1.2 to Mach 6.0. (From Mack, 1954.)

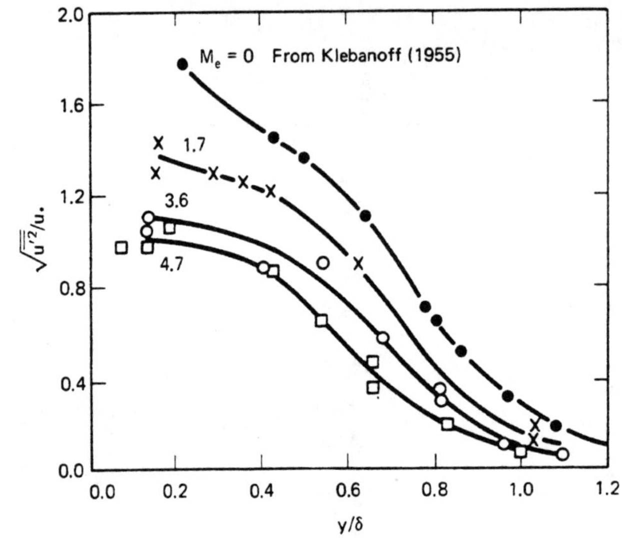

Figure 10–12 Data for axial turbulent intensity distribution in the boundary layer at various Mach numbers. (From Kistler, 1959.)

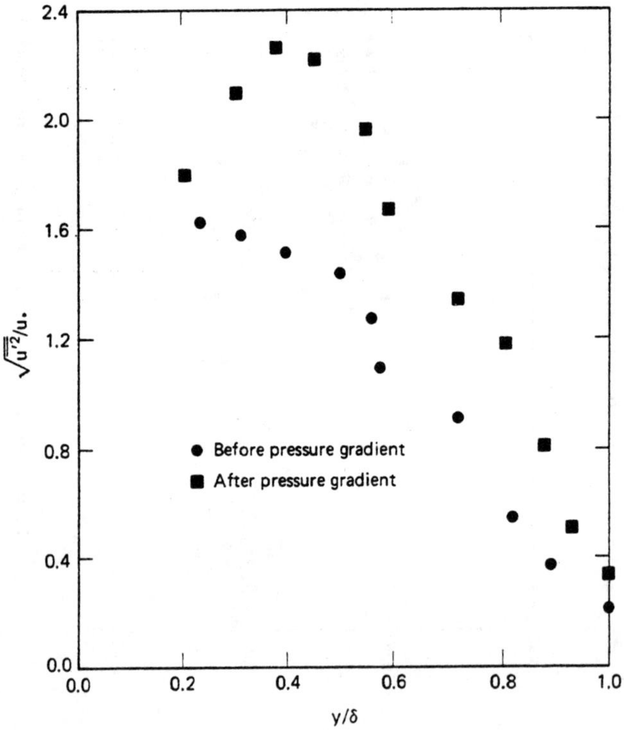

Figure 10–13 Effect of pressure gradient on the distribution of axial turbulence intensity in the boundary layer at supersonic speed. (From Waltrup and Schetz, 1973.)

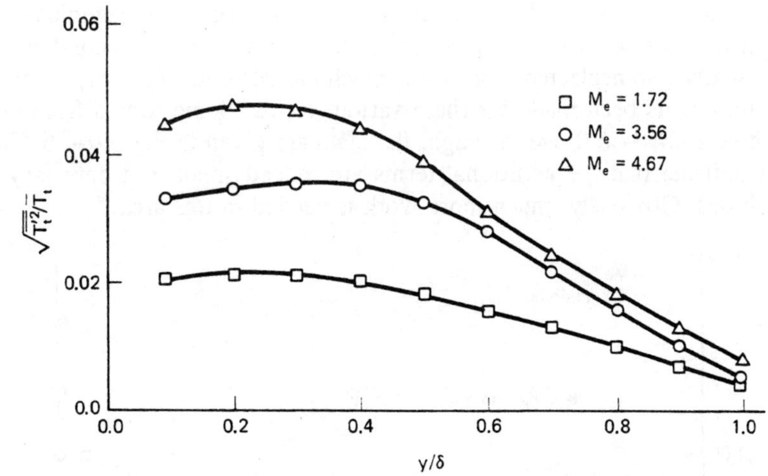

Figure 10–14 Distribution of total temperature fluctuations in supersonic boundary layers. (From Kistler, 1959.)

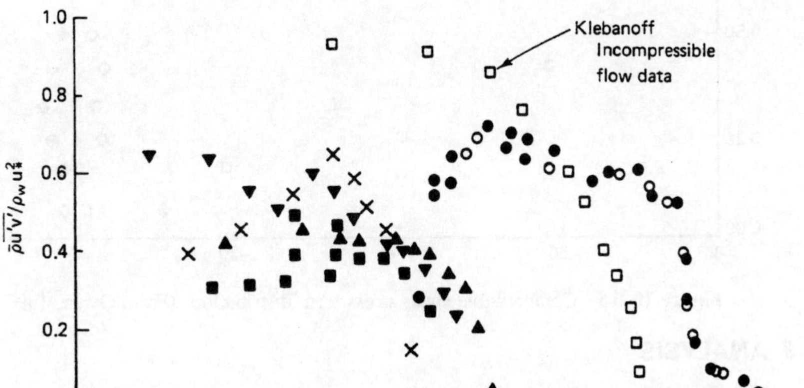

Figure 10–15 Reynolds shear stress distribution in compressible turbulent boundary layers. (From Fernholz and Finley, 1981.)

complicated than it might seem from the discussion. In a boundary layer, the largest stress component has four parts, i.e.,

$$\tau_T \equiv -\bar{\rho}\overline{u'v'} - \overline{U}\overline{\rho'v'} - \overline{V}\overline{\rho'u'} - \overline{\rho'u'v'} \tag{10–24}$$

One generally argues that the last term is smallest and negligible. That seems justified, since it involves a triple correlation. However, the second and third terms are usually also neglected, and that is much harder to justify. Only recently have any measurements been made for these various terms. Some results from Owen (1990) for flow at $M_e = 6.0$ over a rough, flat plate are given in Fig. 10–16. These limited data indicate that the additional terms are indeed small in a boundary layer up to Mach 6.0. Obviously, much more work is needed in this area.

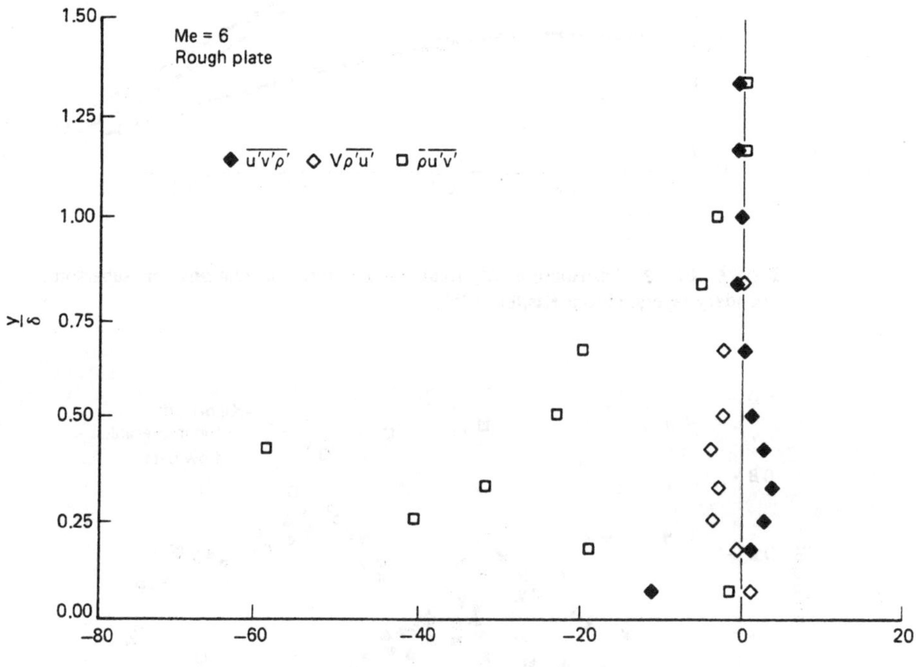

Figure 10–16 Compressible shear stress term distribution. (From Owen, 1990.)

10–3 ANALYSIS

10–3–1 Boundary Layer Equations of Motion

The matter of splitting a variable into its mean and fluctuating parts, substituting into the equations of motion, and then taking the time mean of the equations to produce equations governing the mean flow is more complicated for a variable-density fluid than for a fluid of constant density. If one follows the procedure of Section 7–5 for constant-density flows using

$$U(x, y) = \frac{1}{T_0}\int_0^{T_0} u(x, y, z, t)\, dt$$

additional terms such as $\overline{\rho' u'}$ appear, since the varying density now has a mean $\overline{\rho}$ and fluctuating ρ' part. It has sometimes proven more convenient, therefore, to follow the lead of Van Driest (1951) and use a *mass-weighted averaging* procedure. A *mass-weighted mean* velocity is defined as

$$\overline{\overline{u}} = \frac{\overline{\rho u}}{\overline{\rho}} \tag{10–25}$$

and then

$$u(x, y, z, t) = \overline{\overline{u}}(x, y) + u''(x, y, z, t) \tag{10–26}$$

instead of $u = U + u'$. Also,

$$\rho u = (\overline{\rho} + \rho')(\overline{\overline{u}} + u'') = \overline{\rho}\,\overline{\overline{u}} + \rho'\overline{\overline{u}} + \rho' u'' + \overline{\rho} u'' \tag{10–27}$$

Taking the time average gives

$$\overline{\rho u} = \overline{\rho}\,\overline{\overline{u}} + \overline{\rho u''} \tag{10–28}$$

Comparing this equation with Eq. (10–25), we see that $\overline{\rho u''} = 0$. The important differences between the two methods of time averaging can be summarized as

$$\overline{u'} \equiv 0, \qquad \overline{\rho u'} \not\equiv 0$$

versus

$$\overline{u''} \not\equiv 0, \qquad \overline{\rho u''} \equiv 0 \tag{10–29}$$

Also, it can be shown that

$$\frac{\overline{u''}}{U} = 1 - \frac{\overline{\rho u}}{\overline{\rho} U} \tag{10–30}$$

There are few experimental results in the literature for this kind of *mass-averaged turbulence intensity*. Some recent measurements by Hyde et al. (1990) are shown in Fig. 10–17(A) for heated and unheated slot injection into a Mach 3.0 flow. For the unheated case, where the density variations are small, the mass-averaged intensity is small compared to the corresponding conventional intensity shown in Fig. 10–17(B).

These distinctions are important in principle, but under the assumptions usually invoked for boundary layer flows, current thinking implies that

$$\overline{\overline{u}} \approx U, \qquad \overline{\overline{v}} \approx V + \frac{\overline{\rho' v'}}{\overline{\rho}} \tag{10–31}$$

(see Schubauer and Tchen, 1961). But the second term is not always small compared to the first, which is itself small. Also,

$$\overline{\rho}\,\overline{\overline{v}} = \overline{\rho} V + \overline{\rho' v'} = \overline{\rho v} \tag{10–31a}$$

With all this information, the boundary layer equations of motion for steady, planar flow can then be written as

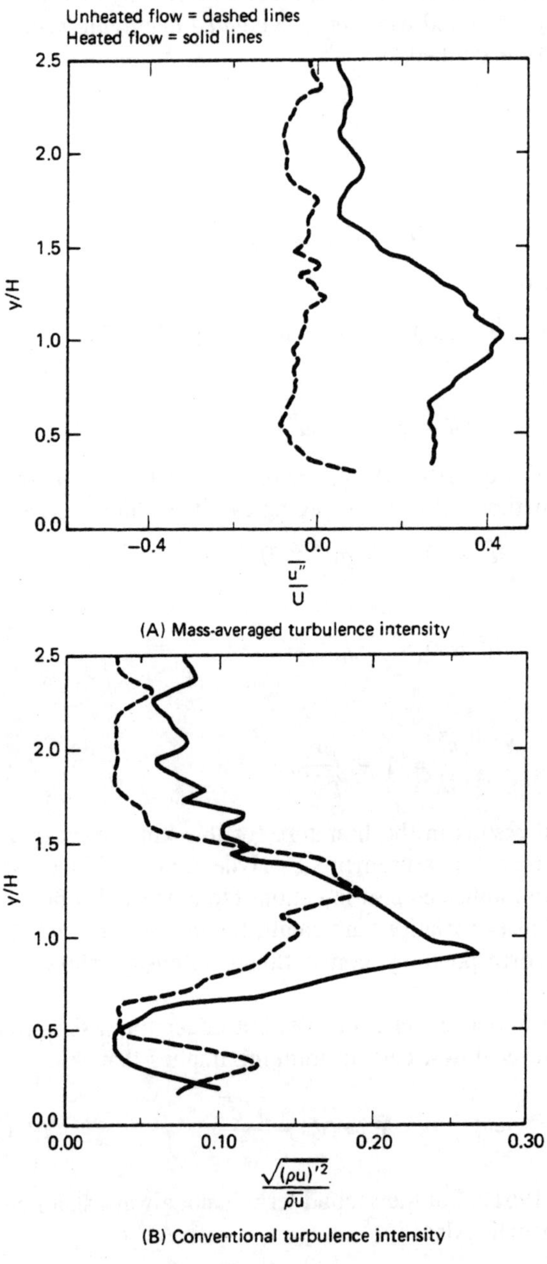

Figure 10–17 Turbulence data in a supersonic slot injection flow. (From Hyde, Smith, Schetz, and Walker, 1990.)

$$\frac{\partial}{\partial x}(\bar{\rho}U) + \frac{\partial}{\partial y}(\overline{\rho v}) = 0 \tag{10-32}$$

for the continuity equation,

$$\bar{\rho}U\frac{\partial U}{\partial x} + \overline{\rho v}\frac{\partial U}{\partial y} = -\frac{dP}{dx} + \frac{\partial}{\partial y}\left(\mu\frac{\partial U}{\partial y} - \bar{\rho}\overline{u'v'}\right) \tag{10-33}$$

for the momentum equation, and

$$\bar{\rho}U\frac{\partial \bar{h}}{\partial x} + \overline{\rho v}\frac{\partial \bar{h}}{\partial y} = \frac{\partial}{\partial y}\left(\frac{\mu}{\Pr}\frac{\partial \bar{h}}{\partial y} - \bar{\rho}\overline{v'h'}\right) + \left(\mu\frac{\partial U}{\partial y} - \bar{\rho}\overline{u'v'}\right)\frac{\partial U}{\partial y} + U\frac{dP}{dx} \tag{10-34}$$

for the energy equation in terms of the mean static enthalpy. If there is diffusion, the second term in the first set of parentheses on the right-hand side becomes

$$-\sum_i \bar{\rho}\, C_i\overline{v'h_i'} \tag{10-34a}$$

and two additional terms for heat transport via mass transfer must be added:

$$\frac{\partial}{\partial y}\left(\sum_i \bar{h}_i\left(\bar{\rho}D_{12}\frac{\partial C_i}{\partial y} - \bar{\rho}\overline{v'c_i'}\right)\right) \tag{10-34b}$$

Also, a species conservation equation is then required:

$$\bar{\rho}U\frac{\partial C_i}{\partial x} + \overline{\rho v}\frac{\partial C_i}{\partial y} = \frac{\partial}{\partial y}\left(\bar{\rho}D_{12}\frac{\partial C_i}{\partial y} - \bar{\rho}\overline{v'c_i'}\right) \tag{10-35}$$

In Eqs. (10–33) through (10–35), it is implied that $\bar{\mu} = \mu = \mu(\bar{T})$, $\bar{k} = k = k(\bar{T})$, and $\bar{D}_{12} = D_{12} = D_{12}(\bar{T})$. Also, only the first term in Eq. (10–24) and the corresponding terms for turbulent heat and mass transfer have been included.

10-3-2 Mean Flow Models for the Eddy Viscosity and Mixing Length

From Eqs. (10–33) through (10–35), it is clear that it is again necessary to *model* the turbulent shear, $-\bar{\rho}\overline{u'v'}$, heat transfer $\bar{\rho}\overline{v'h'}$, and mass transfer $\bar{\rho}\overline{v'c'_i}$. With an *eddy viscosity formulation,* one writes

$$-\bar{\rho}\overline{u'v'} = \mu_T\frac{\partial U}{\partial y} \tag{10-36}$$

and with a *mixing length formulation*

$$-\bar{\rho}\overline{u'v'} = \bar{\rho}\ell_m^2\left|\frac{\partial U}{\partial y}\right|\frac{\partial U}{\partial y} \tag{10-37}$$

Turbulent heat transfer is almost universally treated via a constant *turbulent*

Prandtl number $\Pr_T = \mu_T c_p / k_T$ [see Eq. (9–21)] with a value of approximately 0.8. Similarly, turbulent mass transfer calculations are usually based on a constant *turbulent* Schmidt number $\mathrm{Sc}_T = \mu_T/(\rho D_T)$ [see Eq. (9–24)] of about 0.8. Thus, $\mathrm{Le}_T \approx 1.0$.

One might have expected that the derivation of mean flow models for the inner and outer regions for variable-density flows would follow the procedure used for constant-density flows. Recall, in that case, that the procedure started with the *law of the wall* and a *defect law* as *given,* and models were found that, when used in the equations of motion, would reproduce those laws (see Section 7–8). An equivalent development for variable-density flows would start with a *law of the wall* and a *defect law* for variable-density boundary layers and then proceed along the same lines as before. Perhaps surprisingly, that was not the procedure followed. Indeed, the models currently in use come about as ad hoc extensions of the constant-density models.

For the inner region, Patankar and Spalding (1967) and Cebeci (1971) simply used the Van Driest model [Eq. (7–61)], with the density and viscosity evaluated at the local conditions as a function of y in the inner region. Anderson and Lewis (1971) did the same thing with the Reichardt model [Eq. (7–63)]. Corresponding mixing length models can be found with Eq. (7–40). No soundly based analysis has been presented to justify these assumptions. Instead, the justification for adopting them is the pragmatic argument that the models seem to "work." By that, one means that predictions obtained by using the subject model in a numerical solution procedure agree, within an acceptable tolerance, with data for a few specific cases.

For the outer region, Maise and McDonald (1968) concluded that the constant-density mixing length model $\ell_m = 0.09\delta$ [see Eq. (7–73)] could be applied to variable-density cases up to about Mach 5.0, as seen in Fig. 10–18. This conclusion agrees with the criterion in Eq. (9–40), which indicates that there are no effects of compressibility in boundary layers up to Mach 5.0. That assumption was used in Patankar and Spalding (1967). There is no agreement as to how to proceed above Mach 5.0.

The matter is more complicated for eddy viscosity models in the outer region. Some researchers who favor the Clauser model for the eddy viscosity in the outer region of constant-density flows, Eq. (7–85), have tried to extend the model to variable-density flows in the form

$$\begin{aligned}\mu_T &= C\bar{\rho}U_e\delta_k^* \\ &= C\bar{\rho}U_e\int_0^\delta\left(1-\frac{U}{U_e}\right)dy\end{aligned} \tag{10–38}$$

where δ_k^*, a so-called *kinematic displacement thickness* is used in place of the true displacement thickness for variable-density flows,

$$\delta^* \equiv \int_0^\delta\left(1-\frac{\bar{\rho}U}{\rho_e U_e}\right)dy \tag{10–39}$$

It is important to note that Eq. (10–38) is not rederived for variable-density flows using an appropriate *defect law,* following the development of Clauser. The

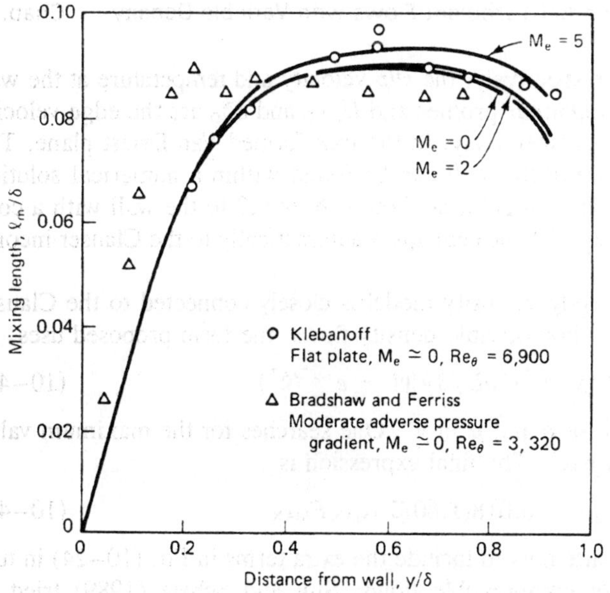

Figure 10–18 Calculated mixing length variation in compressible, flat-plate boundary layers, $Re_\theta = 10^4$. (From Maise and McDonald, 1968.)

justification is, again, that it "seems to work." The form adopted in Eq. (10–38) is, however, contrary to that found appropriate for free-mixing flows with large variations in density (see Section 9–4–2). For free-mixing flows, the Clauser model extended on the basis of the true δ^*, not δ_k^*, was successful. This apparent contradiction can be partially resolved by noting that many turbulent boundary layers that have been considered for comparisons between prediction and experiment have much smaller variations in density than do the free-mixing flows that have been considered. (Recall the H_2-air jet mixing cases discussed in Section 9–4–2, where $\rho_j/\rho_e \approx 1/15$.) Thus, the distinction between δ_k^* and δ^* is often less important in boundary layer flows than in free-mixing flows.

Kiss and Schetz (1992) have rationally extended the Clauser model to compressible cases by following the logical path of the original Clauser derivation (see Section 7–8–2). To repeat the development for the compressible case, it was necessary first to use the Howarth-Dorodnitzin transformation (see Section 5–6–1) to obtain the required compressible *pseudolaminar* profiles from the incompressible profiles used by Clauser. Next, the Van Driest transformation discussed earlier in this chapter was applied to the experimental incompressible *velocity defect law* to obtain the corresponding compressible law. Finally, the compressible *pseudolaminar* profiles could be rescaled to make them match the compressible *defect law*, just as Clauser did for the incompressible equivalents. The resulting compressible, outer region model is

$$\mu_T = \frac{1}{(7.40 - 0.40M_e)\sqrt{1 - \dfrac{\dfrac{(\gamma-1)}{2}M_e^2}{1+\dfrac{(\gamma-1)}{2}M_e^2}\dfrac{U(0)^2}{U_e^2}}}\frac{\bar{T}}{\bar{T}(0)}\overline{\rho_w}U_{e,\mathrm{VD}}\delta^*_{\mathrm{VD}} \quad (10\text{–}40)$$

where $U(0)$ and $T(0)$ are, respectively, the *slip* velocity and temperature at the wall corresponding to the *pseudolaminar* profiles and $U_{e,\text{VD}}$ and δ^*_{VD} are the edge velocity and displacement thickness, respectively, in the transformed Van Driest plane. The *slip* velocity and temperature at the wall can be found within a numerical solution procedure by extrapolating from a point at about $y/\delta \approx 0.2$ to the wall with a constant slope. The model in Eq. (10–40) collapses automatically to the Clauser incompressible result as $M_e \to 0$.

The Baldwin-Lomax eddy viscosity model is closely connected to the Clauser model (see Section 7–8–2). For variable-density flows, the form proposed uses

$$F(y) = y|\partial U/\partial y|(1 - e^{-y^+/A^+}) \tag{10–41}$$

where wall values are used for ρ and μ in y^+. One searches for the maximum value of F, and y at that point is y_{MAX}. The final expression is

$$\mu_T = 0.018(1.60)\overline{\rho}\, y_{\text{MAX}} F_{\text{MAX}} \tag{10–42}$$

There have been few attempts to include the extra terms in Eq. (10–24) in turbulent transport models for compressible flows. Situ and Schetz (1989) tried to model the second term with an extended mixing length approach. One takes $u' \sim \ell_m \partial U/\partial y$ and $v' \sim \ell_m \partial U/\partial y$ as before, but now also $\rho' \sim \ell_\rho \partial\overline{\rho}/\partial y$. With these, the complete model becomes

$$\tau_T = \ell_m^2 \left| \overline{\rho}\frac{\partial U}{\partial y} + \frac{U}{S}\frac{\partial \overline{\rho}}{\partial y} \right| \frac{\partial U}{\partial y} \tag{10–43}$$

The quantity S involves the ratio of ℓ_m to ℓ_ρ, and it is reasoned to be approximately equivalent to Pr_T. Clearly, when $\partial\overline{\rho}/\partial y$ is small, this model reduces to the conventional mixing length model, Eq. (10-37). One can obtain a model for μ_T from Eq. (10–43) as $\tau_T/(\partial U/\partial y)$.

We are now in a position to discuss comparisons between experiment and predictions based on the various mean-flow models for variable-density boundary layer flows. The numerical solutions are obtained from Eqs. (10–32) through (10–34), with an eddy viscosity or mixing length model for the turbulent transport terms, transformed into Levy-Lees coordinates, as described for laminar, compressible cases in Section 5–11. With a mean-flow turbulence model, the equations are *pseudolaminar* with an augmented viscosity due to turbulence. Thus, the numerical procedures to be employed are the same in the laminar and turbulent cases. Only the stretching of the grid in the transverse direction (see Section 7–9) need be different. A comparison of prediction with experiment was already presented in Fig. 5–16 from Anderson and Lewis (1971), who used the extended Reichardt model with Eq. (10–38). Cebeci and co-workers (see Cebeci, 1971) have produced a number of comparisons using the extended Van Driest model with Eq. (10–38), with generally good results. Herring and Mellor (1968) used their own inner region model, which is essentially a mixing length model, for both constant- and variable-density cases, together with Eq. (10–38) and, again, obtained generally good results. Some comparisons of predictions with experiments from the work of Cebeci are given in Figs. 10–19 and 10–20 to show the type of results that can be expected. These represent

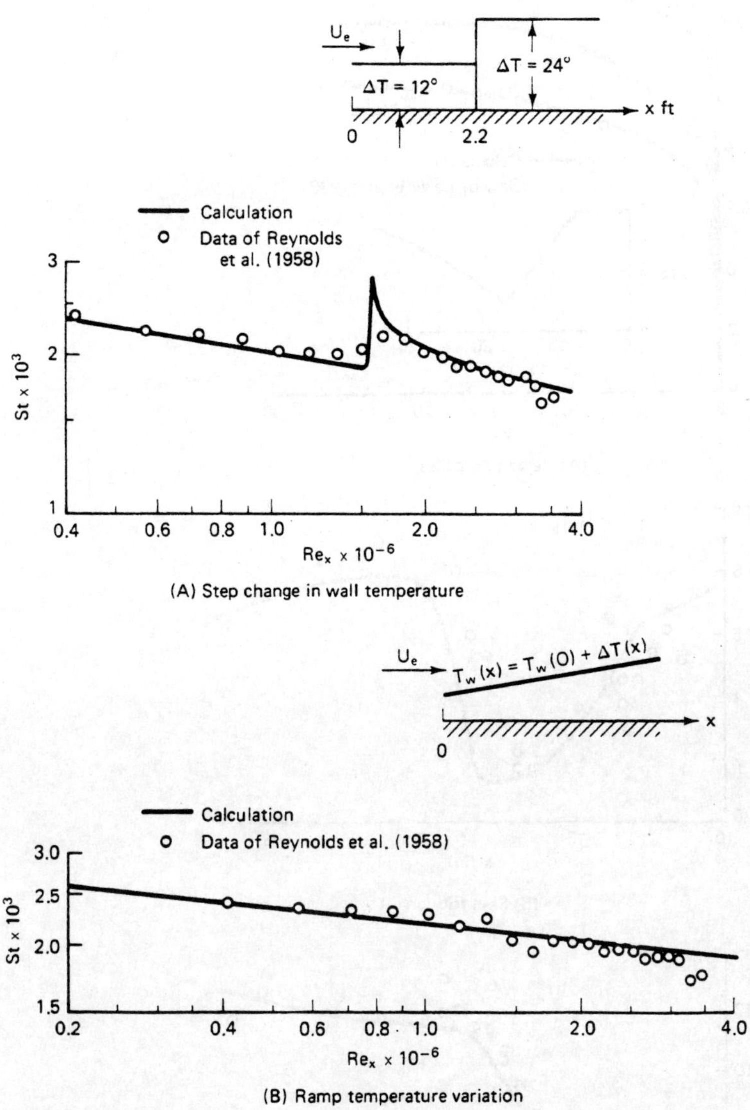

Figure 10–19 Comparison of calculated with experimental Stanton numbers for the flat-plate boundary layer measured by Reynolds et al. (1958). (From Cebeci and Smith, 1974.)

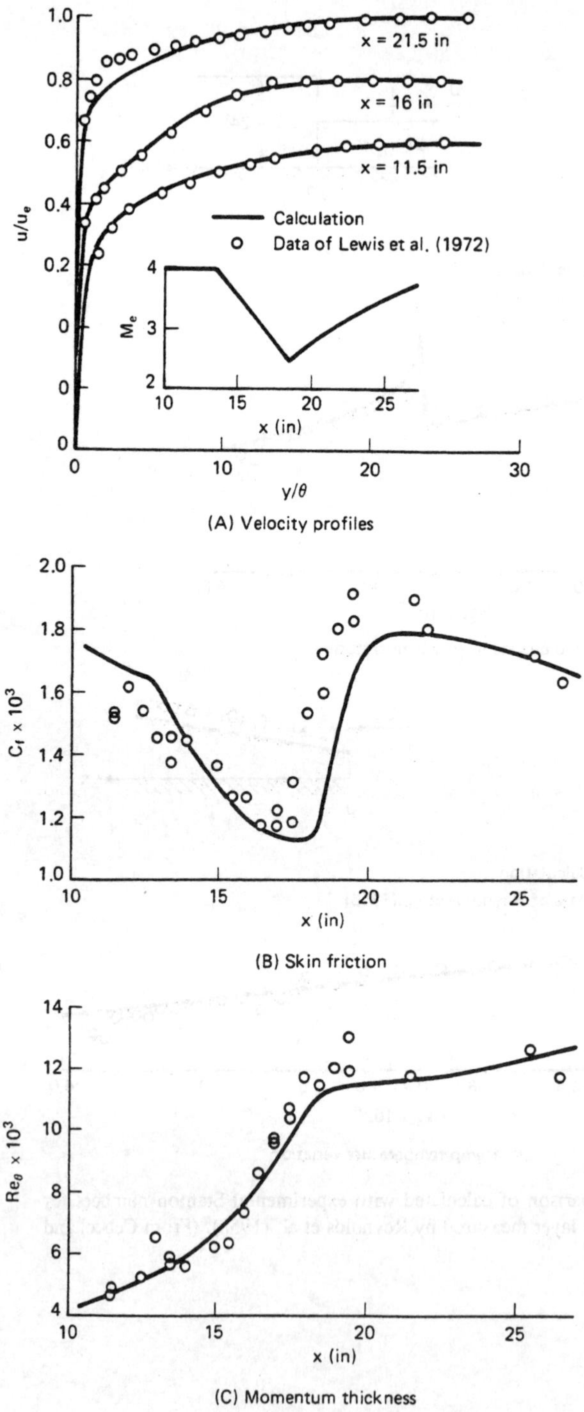

Figure 10–20 Comparison of calculated and experimental results for an adiabatic compressible turbulent boundary layer with an adverse and a favorable pressure gradient. (From Cebeci and Smith, 1974.)

rather complicated flow problems, and the good agreement with experimental data can be noted.

Schetz and Favin (1971) used the extended Reichardt model for the inner region and compared predictions from the Clauser model based on both δ^* and δ_k^* with experiment. Some results are given in Figs. 10–21 and 10–22 for supersonic flows at a moderate Mach number without and with wall heat transfer. Clearly, based on these results, it is hard to argue convincingly for δ_k^* over δ^* (or vice versa).

Flows with light gas injection such that sizable variations in density are produced, even at low-speed conditions, were also treated in Schetz and Favin (1971). The procedure used the Reichardt model extended to injection, Eq. (7–70), with local $\overline{\rho}$ and μ for the inner region and the Clauser model with both δ^* and δ_k^* for the outer region. In Fig. 10–23(A), the predictions are compared with the data of Scott et al. (1964) for helium injection in terms of the skin friction coefficient and the integral boundary layer thicknesses. Both models give predictions that are low for the momentum thickness and C_f, with somewhat better agreement for the δ^* model, and both agree well with the data on displacement thickness. The velocity profile measured at the farthest downstream station is compared with the theory in Fig. 10–23(B); the agreement is quite good for both models, but is better for the δ^* model near the wall and at the outer edge of the layer. The measurements were interpreted to provide the turbulent eddy viscosity distribution across the layer. A comparison with the eddy viscosity distributions obtained numerically using δ^* and δ_k^* in the outer region is shown in Fig. 10–23(C). In the outer region, the δ^* model is better. However, at large y/δ, the theoretical models deviate considerably, suggesting that the Klebanoff *intermittency factor* used falls off too rapidly toward the outer edge of the layer.

The Baldwin-Lomax model can produce results that are equivalent to those with other mean-flow models. Predictions of the variation in skin friction coefficient with Reynolds number at two supersonic Mach numbers are compared to the Hopkins-Inouye data correlation in Fig. 10–24, where the generally good agreement is plain.

As the Mach number increases, the variation in density across the layer increases, and one can expect that the difference between δ^* and δ_k^* will increase markedly. In that case, both of the ad hoc models using δ^* and δ_k^* should be discarded in favor of the more soundly based model, Eq. (10–40). This assertion is demonstrated by the comparisons of predictions with data for the Mach 3.7 experiment of Coles (1952) shown in Fig. 10–25. The prediction with the rational extension of the Clauser model to compressible flows is clearly superior in the outer region, where it applies.

For cases with strong compressibility effects, the Situ-Schetz mixing length model might be expected to give improved predictions. That this is so is shown in Fig. 10–26 for high Mach number flow over flat plates. Figure 10–26(A) has $M_e = 4.67$, which is below the generally accepted value of $M_e \approx 5$ for important compressibility effects. Predictions with the ordinary Prandtl mixing length model

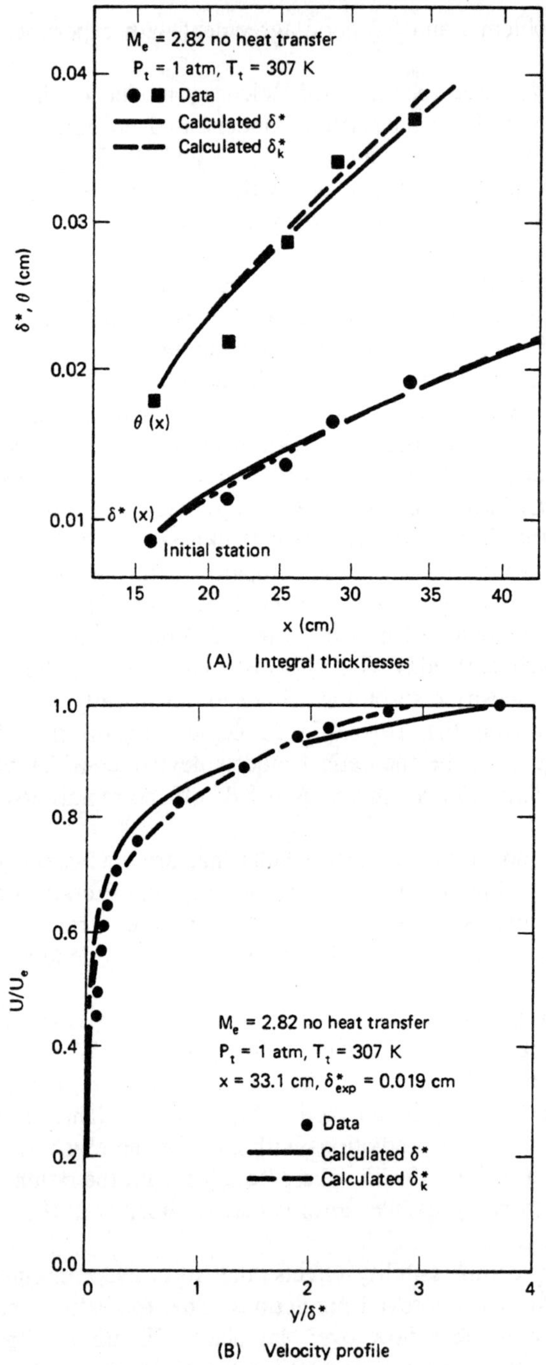

Figure 10–21 Comparison of predictions from mean flow models with the data of Monaghan and Cooke (1952) for adiabatic flow. (From Schetz and Favin, 1971.)

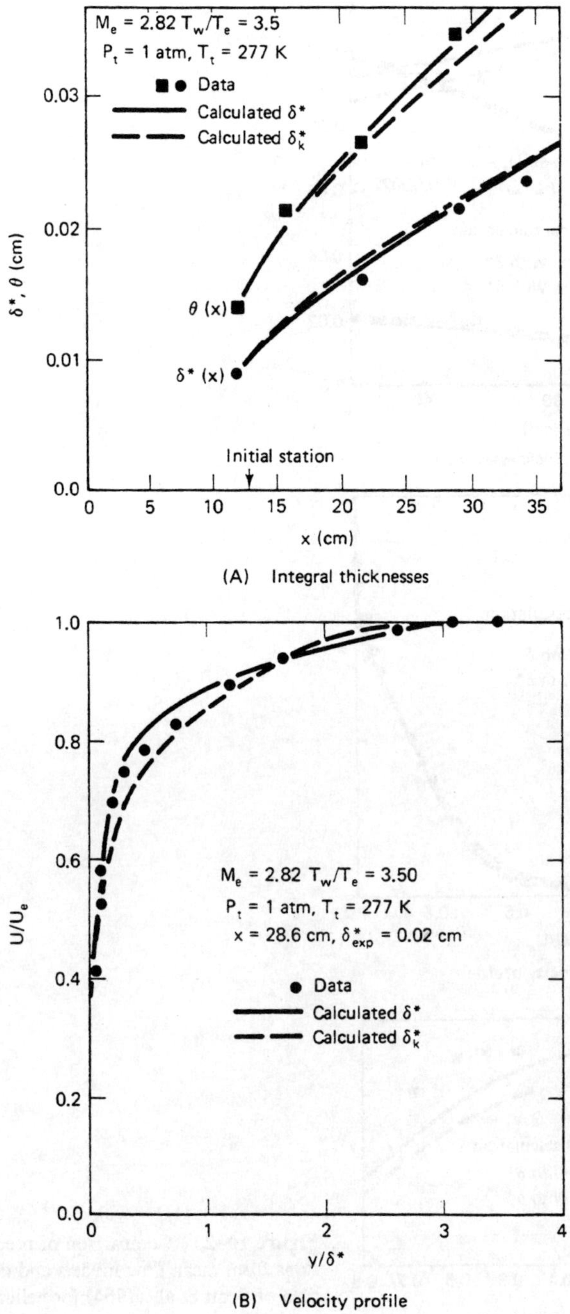

Figure 10–22 Comparison of predictions from mean flow models with the data of Monaghan and Cooke (1952) for heat transfer cases. (From Schetz and Favin, 1971.)

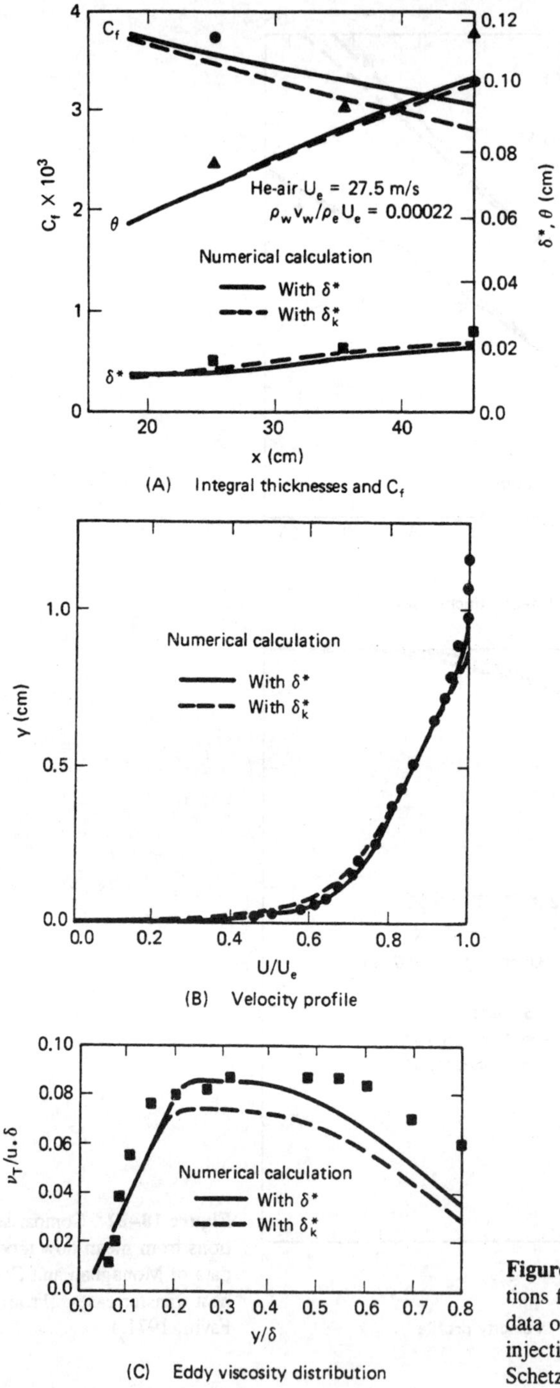

Figure 10–23 Comparison of predictions from mean flow models and the data of Scott et al. (1964) for helium injection into the boundary layer. (From Schetz and Favin, 1971.)

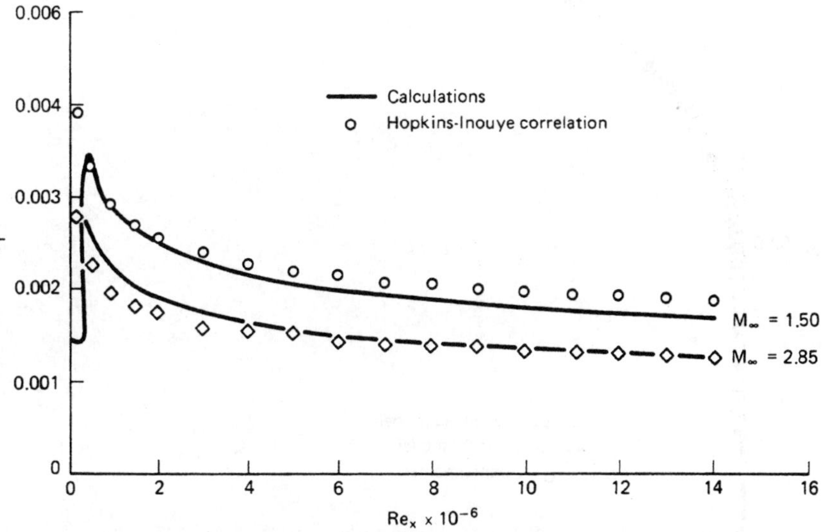

Figure 10–24 Comparison of calculated skin friction on a flat plate with Hopkins-Inouye correlation. (From Baldwin and Lomax, 1978.)

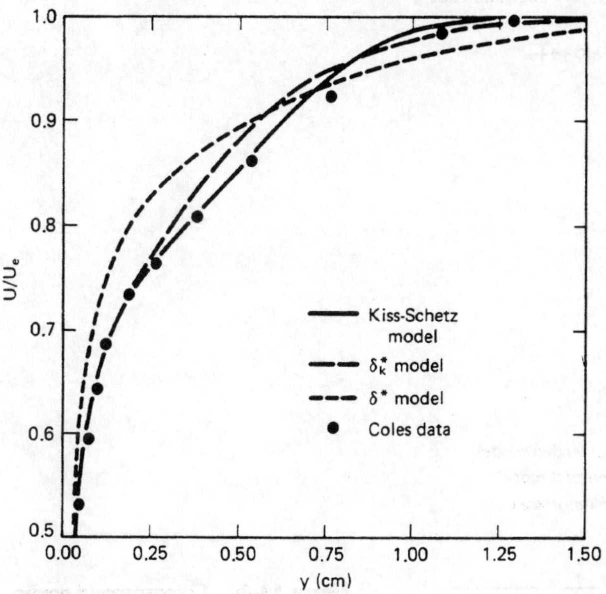

Figure 10–25 Comparisons of predictions from various outer region eddy viscosity models and the Mach 3.7 flat-plate measurements of Coles (1952). (From Kiss and Schetz, 1992.)

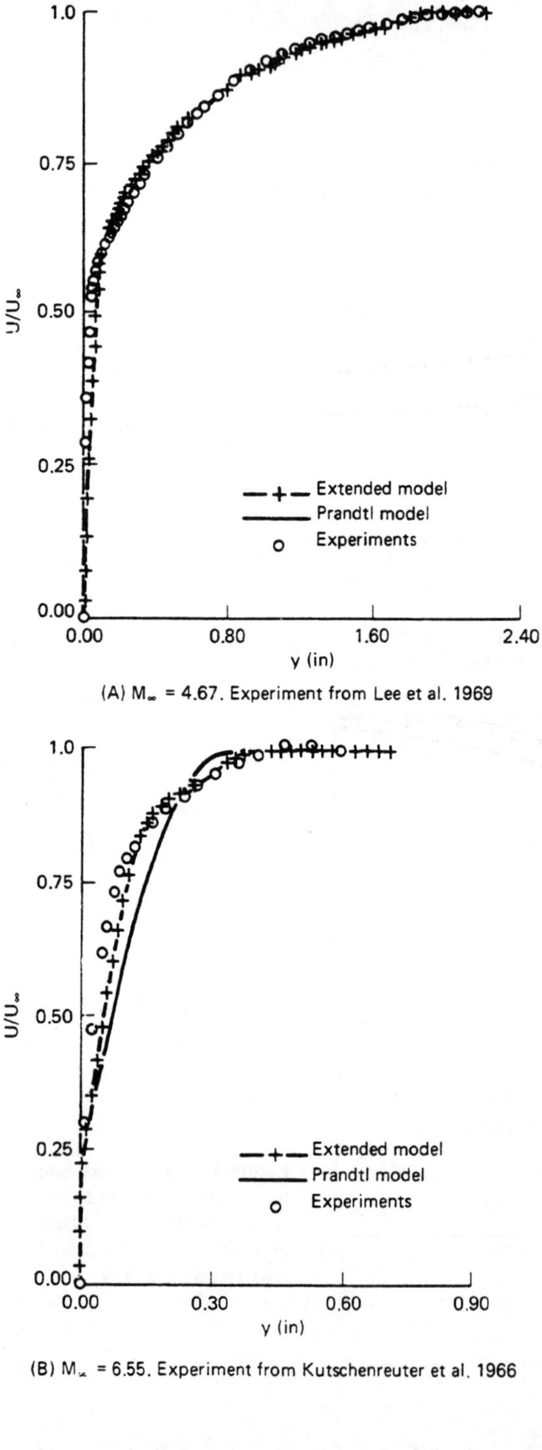

Figure 10–26 Comparison of predictions with experiments for hypersonic constant-pressure boundary layer flows. (From Situ and Schetz, 1991.)

and the extended mixing length model are essentially the same, as they should be, and both are in good agreement with experiment. Figure 10–26(B) has $M_e = 6.55$, which is above the critical value, so it should have compressibility effects. The predictions from the extended model are in much better agreement with the data than the predictions from the simple Prandtl model.

10–3–3 Methods Based on Models for the Turbulent Kinetic Energy

Variable-density versions of the two approaches to TKE models (the Prandtl energy method and the Bradshaw model) have been developed. The modeling of the various terms in the TKE equation is as described in Section 7–10. Mass-averaged versions of the TKE equation exist, but for boundary layer flows, the distinction is not considered important.

Schetz et al. (1982) developed a computer code based on the Prandtl energy method. The required model for the length scale ℓ [see Eqs. (7–95) and (7–96)] was obtained from the Van Driest model for ℓ_m, Eq. (7–61), with local $\overline{\rho}$ and μ for the inner region and $\ell_m = 0.09\delta$ for the outer region, both with $\ell = C_D^{1/4}\ell_m$. Some comparisons of prediction with experiment for the supersonic flat-plate case of Coles (1954) are shown in Fig. 10–27. Coles reported $C_f = 0.00162$ at this station, and the calculation predicted $C_f = 0.00177$. These results clearly agree well with experiment, but it is worth noting that the eddy viscosity methods discussed in the preceding section do as well (see Anderson and Lewis, 1971).

Bradshaw developed a code incorporating a variable-density extension to his method. Sivasegaram and Whitelaw (1971) made numerous comparisons between predictions and data for adiabatic wall cases. They also used the mixing length method of Patankar and Spalding (1967). Two comparisons are shown in Figs. 10–28 and 10–29. A couple of points are clear from all these results: There is no significant difference between the results of the mixing length and TKE models, and the poorest accuracy is found for adverse pressure gradients.

The Johnson-King model (see Section 7–10) has also been used in variable-density flows with strong pressure gradients. A comparison of predictions from that model and predictions from the Cebeci-Smith approach with data is presented in Fig. 10–30. For this very strong pressure gradient flow, the Johnson-King model provides a better prediction.

10–3–4 Methods Based on Models for Turbulent Kinetic Energy and a Length Scale

Of this general class of models, the $K\epsilon$ model has enjoyed the greatest popularity. It has been used for variable-density flows by a number of authors. The basic model of this variety was described in Sections 7–11 and 9–6. The so-called *compressibility correction* described in Section 9–6 for free shear flows has not found favor in wall-bounded flows.

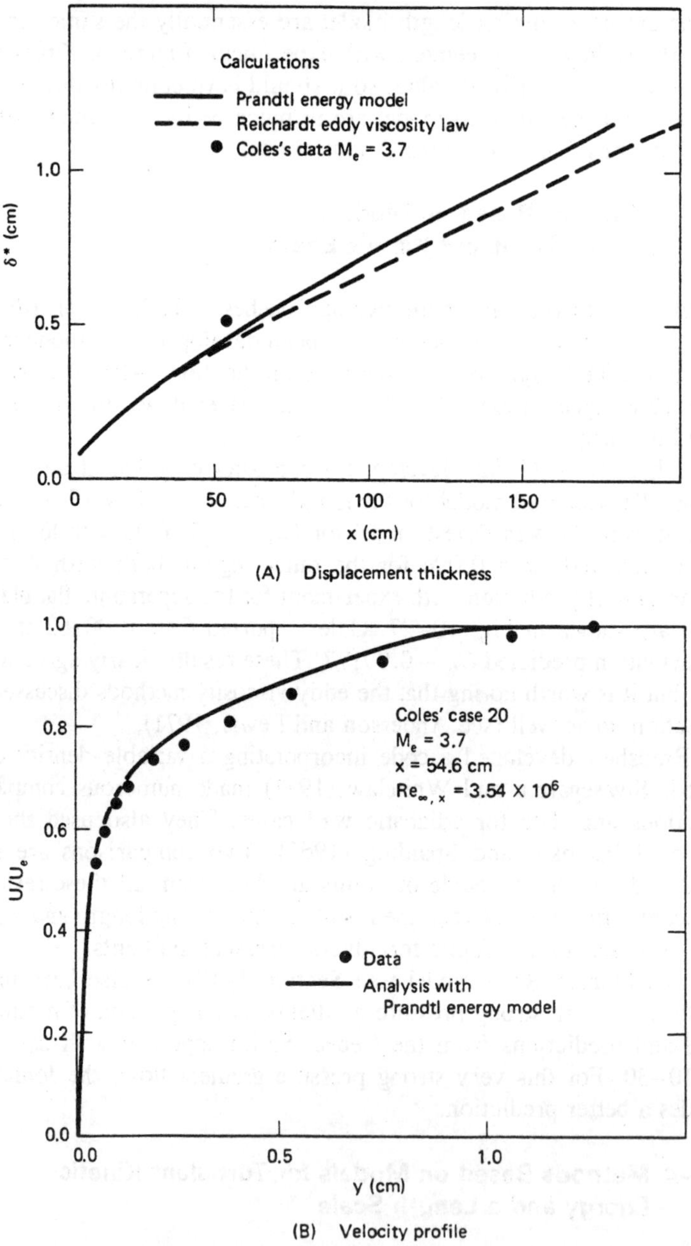

Figure 10–27 Comparison of predictions from the Prandtl energy method with the data of Coles (1954) for a supersonic flow over a flat-plate (From Schetz et al., 1982.)

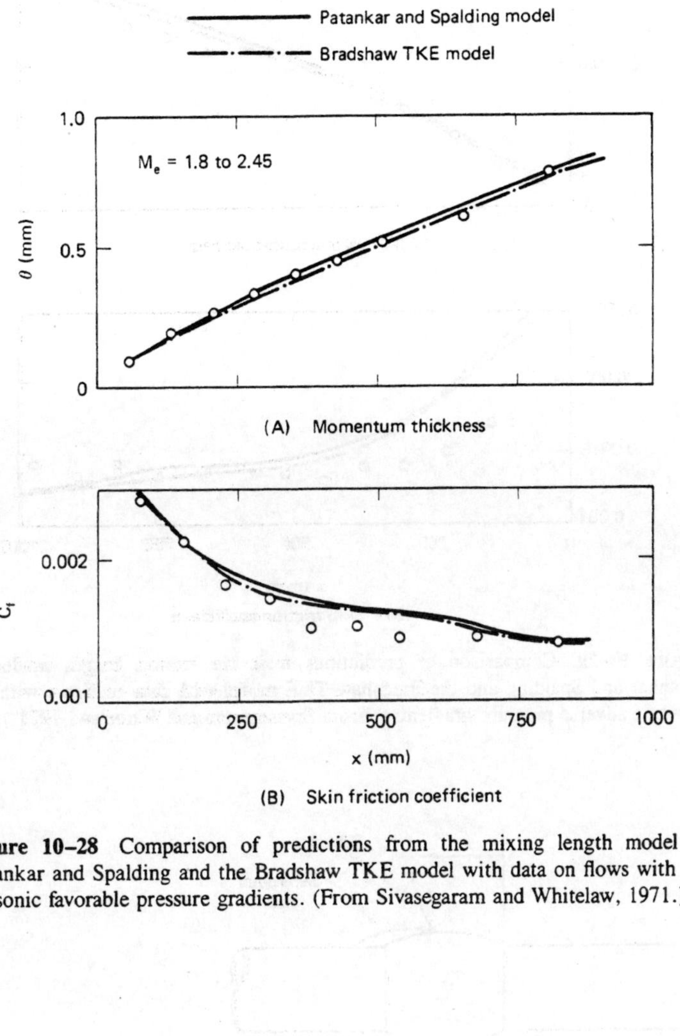

Figure 10–28 Comparison of predictions from the mixing length model of Patankar and Spalding and the Bradshaw TKE model with data on flows with supersonic favorable pressure gradients. (From Sivasegaram and Whitelaw, 1971.)

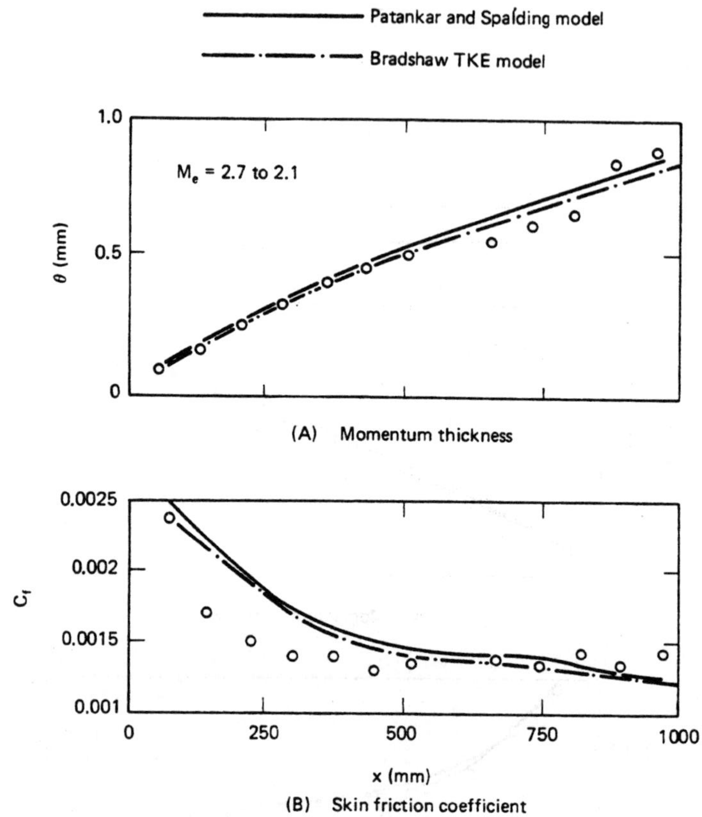

Figure 10–29 Comparison of predictions from the mixing length model of Patankar and Spalding and the Bradshaw TKE model with data on flows with supersonic adverse pressure gradients. (From Sivasegaram and Whitelaw, 1971.)

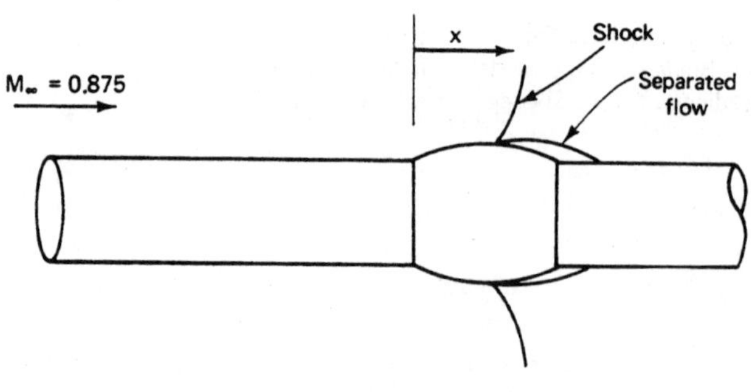

Figure 10–30 Comparison of predictions with experiment for a transonic boundary layer with strong pressure gradient. (From Johnson and King, 1985.)

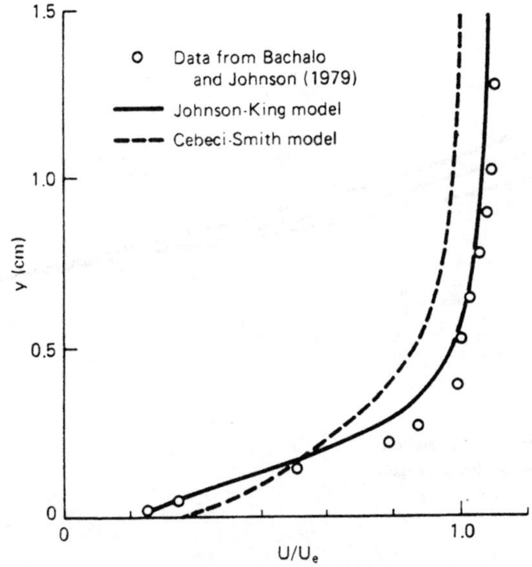

(B) Profiles at x/c = 0.75

Figure 10–30 *(continued)*

The 1981 Stanford conference [see Kline et al. (1982)] had some high-speed test cases, and a few predictions with $K\epsilon$ models were presented. One test case was for the influence of T_w/T_{aw} on the skin friction coefficient at Mach 5.0. Ha presented the $K\epsilon$ predictions shown in Fig. 10–31. Also shown are the accepted data correlation and predictions from Viegas using an eddy viscosity model for comparison.

The use of *wall functions* in supersonic boundary layer flows has been the subject of considerable study and debate. A detailed discussion can be found in Rubesin and Viegas (1985).

10-3-5 Methods Based Directly on the Reynolds Stress

The model of Donaldson (1971) (see Sections 7–12 and 9–7) has been applied by Rubesin et al. (1977) to various cases, including the subsonic, compressible (M_∞ = 0.6) experiments of Acharya (1976). Some comparisons are shown in Fig. 10–32. As with the constant-density cases (see Fig. 7–80), the mean flow, the turbulent shear, and v' are predicted most accurately. However, v' is not predicted as well for the subsonic, compressible flow as for the constant-density flow.

Again, we can look to the results of the 1981 Stanford conference for a few comparisons of predictions with data. Reynolds stress model results from Donaldson for the same Mach 5.0 case with the effects of wall temperature ratio as that considered earlier are shown in Fig. 10–33. Another test case was for the influence of edge Mach number on skin friction, and comparisons of predictions with data correlation are given in Fig. 10–34. No clear advantage for the Reynolds stress model, or for that matter for the $K\epsilon$ model, are apparent from these comparisons.

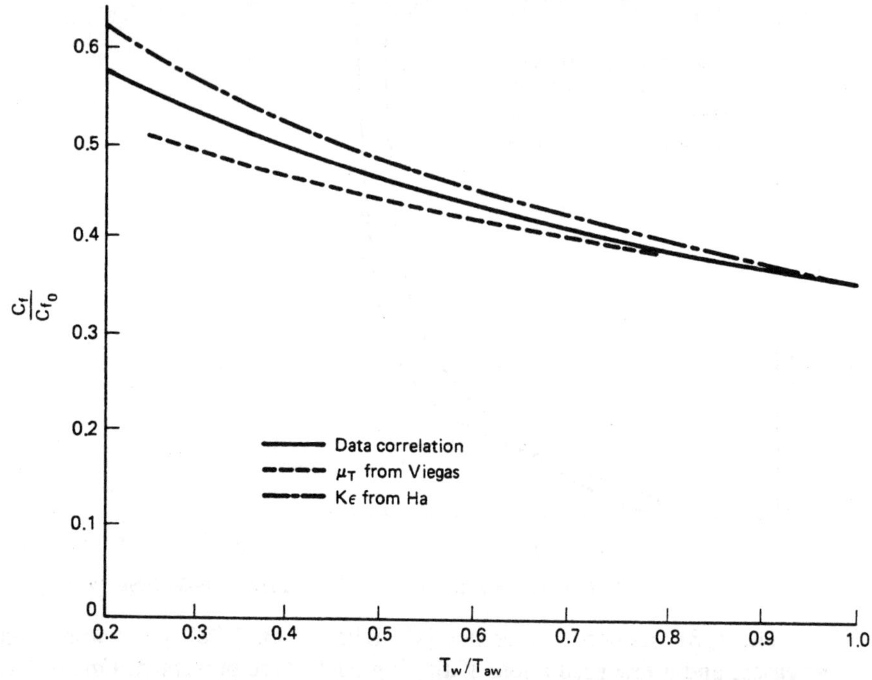

Figure 10–31 Comparison of predictions from eddy viscosity and $K\epsilon$ models with data correlation for skin friction as a function of wall temperature ratio at Mach 5.0. (From Kline et al., 1982.)

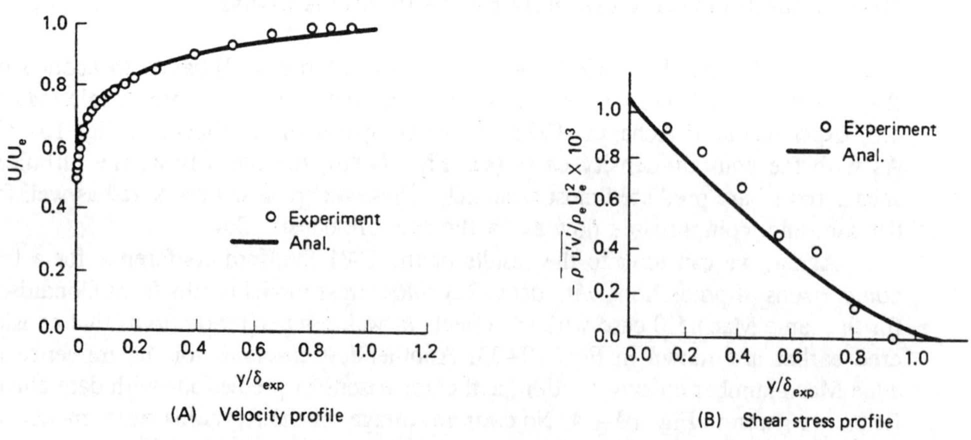

Figure 10–32 Comparison of predictions from the Donaldson Reynolds stress model with the Mach 0.6 data of Acharya (1976). (From Rubesin et al., 1977.)

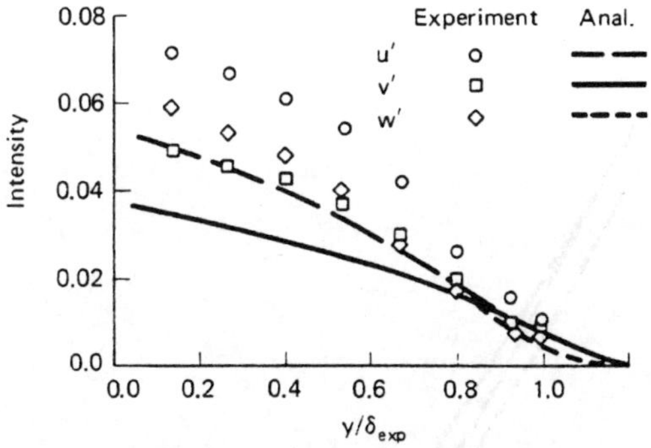

(C) Turbulent intensity profiles

Figure 10–32 *(continued)*

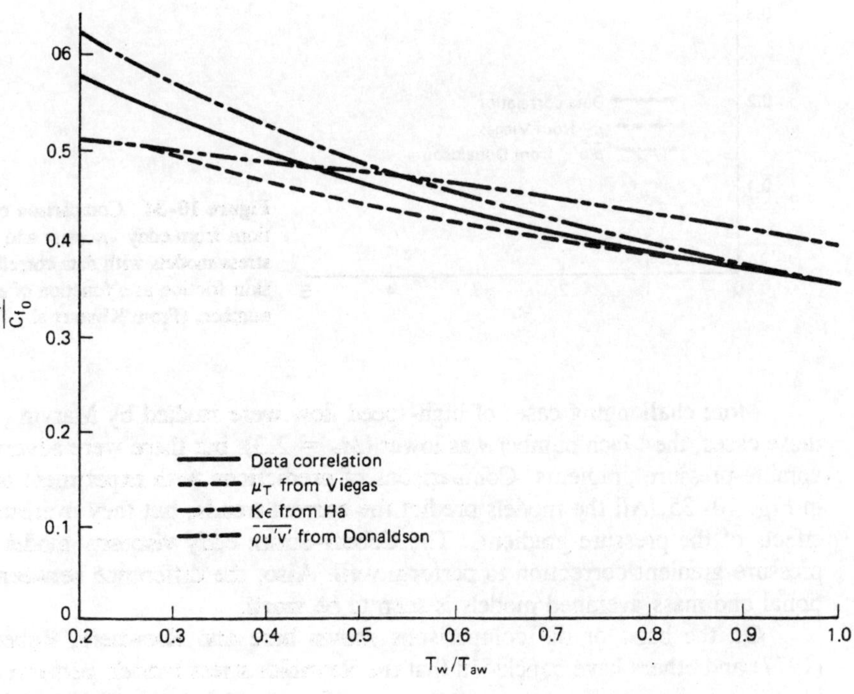

Figure 10–33 Comparison of predictions from eddy viscosity, $K\epsilon$, and Reynolds stress models with data correlation for skin friction as a function of wall temperature ratio at Mach 5.0. (From Kline et al., 1982.)

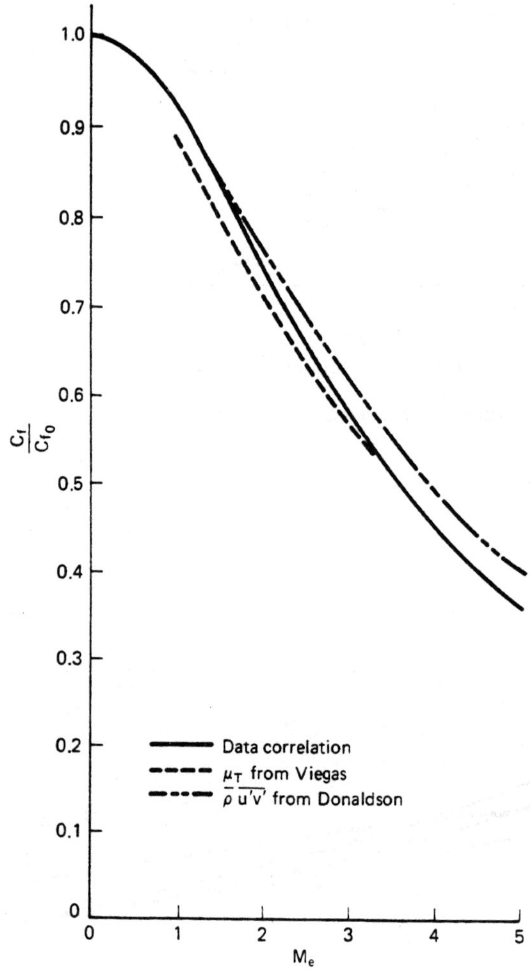

Figure 10–34 Comparison of predictions from eddy viscosity and Reynolds stress models with data correlation for skin friction as a function of edge Mach number. (From Kline et al., 1982.)

More challenging cases of high-speed flow were studied by Marvin (1982). In those cases, the Mach number was lower ($M_\infty = 2.3$), but there were adverse and favorable pressure gradients. Comparisons of predictions with experiment are shown in Fig. 10–35. All the models predict the correct trends, but they overestimate the effects of the pressure gradients. The Cebeci-Smith eddy viscosity model needs its pressure gradient correction to perform well. Also, the difference between conventional and mass-averaged models is seen to be small.

On the basis of the comparisons shown here and elsewhere, Rubesin et al. (1977) and others have concluded that the Reynolds stress models perform comparably to and sometimes better than the mean flow models in predicting skin friction for attached boundary layers.

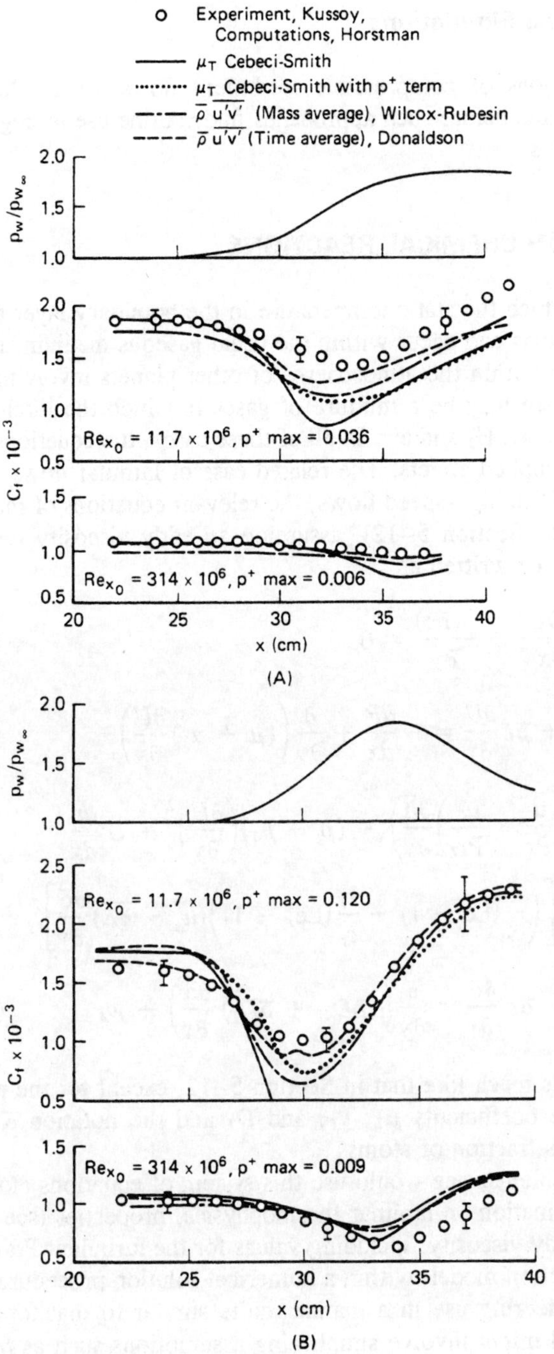

Figure 10–35 Effects of pressure gradient and Reynolds number on a compressible turbulent boundary layer with $M_\infty = 2.3$. (A) adverse pressure gradient to a plateau; (B) adverse pressure gradient followed by favorable pressure gradient (From Marvin, 1983.)

10–3–6 Direct Numerical Simulations

Direct numerical simulations of compressible, turbulent flows are in their early stages. It will be a long time before such approaches find routine use in engineering work with high-speed flows.

10–4 HIGH-SPEED FLOWS WITH CHEMICAL REACTIONS

For high-speed flows in which the static temperature in the boundary layer becomes very high, chemical reactions can occur within the usual gaseous medium, i.e., air. Similar things can happen within the atmospheres of other planets involving different gaseous media. Or there may be a mixture of gases in which the likelihood of reaction is more obvious, e.g., H_2 and air. In all of these cases, the equations of motion must include all the implied effects. The related case of laminar flows was discussed in Section 5–12. With high-speed flows, the relevant equations of motion for a dissociation reaction (see Section 5–12), assuming an eddy viscosity turbulence model as an example, can be written as

$$\frac{\partial(\bar{\rho}U)}{\partial x} + \frac{\partial(\overline{\rho v})}{\partial y} = 0 \tag{10–44}$$

$$\bar{\rho}U\frac{\partial U}{\partial x} + \overline{\rho v}\frac{\partial U}{\partial y} = -\frac{dP}{dx} + \frac{\partial}{\partial y}\left((\mu + \mu_T)\frac{\partial U}{\partial y}\right) \tag{10–45}$$

$$\begin{aligned}\bar{\rho}U\frac{\partial \bar{h}}{\partial y} + \overline{\rho v}\frac{\partial \bar{h}}{\partial y} &= \frac{\partial}{\partial y}\left(\left(\frac{\mu}{\mathrm{Pr}} + \frac{\mu_T}{\mathrm{Pr}_T}\right)\frac{\partial \bar{h}}{\partial y}\right) + (\mu + \mu_T)\left(\frac{\partial U}{\partial y}\right)^2 + U\frac{dp}{dx}\\ &+ \frac{\partial}{\partial y}\left[\left(\frac{k}{c_p}(\mathrm{Le} - 1) + \frac{k_T}{c_p}(\mathrm{Le}_T - 1)\right)(\overline{h_A} - \overline{h_{A_2}})\frac{\partial \bar{\alpha}}{\partial y}\right]\end{aligned} \tag{10–46}$$

$$\bar{\rho}U\frac{\partial \bar{\alpha}}{\partial x} + \overline{\rho v}\frac{\partial \bar{\alpha}}{\partial y} = \frac{\partial}{\partial y}\left((\bar{\rho}D_{12} + \bar{\rho}D_T)\frac{\partial \bar{\alpha}}{\partial y}\right) + w_A \tag{10–47}$$

This system of equations is much like that in Section 5–12, except for the presence of the turbulent exchange coefficients μ_T, k_T, and D_T and the notation $\bar{\alpha}$ for the *mean* value of α, the mass fraction of atoms.

For general flow problems, one would use this system of equations along with an *equation of state*, information on laminar thermophysical properties (see Section 5–12), a model for the eddy viscosity, including values for the turbulent Prandtl and Lewis numbers, and a reaction model, within a numerical solution procedure to produce predictions for engineering use in a manner quite similar to that for laminar flows. The reaction model might involve simplifying assumptions such as *frozen* or *equilibrium* flow or a complex finite reaction rate. Some solutions of this type were already shown in Fig. 5–16. As noted in Chapter 5, however, the difference between results for a perfect gas and for a real gas in that experiment was not large. Nonetheless, the agreement with experiment was quite good.

This general approach can be extended to treat cases with more complex mixtures and chemistry. Calculations with a TKE model for the boundary layer on the wall of a supersonic combustor with H_2 fuel burning in air can be found in Schetz et al. (1982).

10–4–1 Approximate Solutions for Special Cases

Since calculations of the type just described are complex and relatively expensive, there is a useful role for approximate analyses, especially for preliminary design purposes. The basis of the approximate analysis is the assumption that $\mathrm{Pr} = \mathrm{Pr}_T = \mathrm{Le} = \mathrm{Le}_T = 1$, which, as we have seen, results in considerable simplifications (see Section 5–10–1). The effects of the real situation with nonunity Prandtl and Lewis numbers are then introduced as a "correction."

The direct effect of unity Lewis numbers on the energy equation for a reacting mixture can be seen clearly in Eq. (10–46). It is more instructive, however, to look directly at the heat flux component normal to a surface. That flux can be written as

$$q = -\left(k\frac{\partial \bar{T}}{\partial y} + \bar{\rho}D_{12}\sum \bar{h}_i \frac{\partial C_i}{\partial y}\right) - \left(k_T\frac{\partial \bar{T}}{\partial y} + \bar{\rho}D_T\sum \bar{h}_i \frac{\partial C_i}{\partial y}\right) \tag{10–48}$$

Recall that

$$h_i = \int_0^T c_{pi}\,dT + h_i^0\,, \qquad h = \sum h_i C_i \tag{10–49}$$

and substitute into Eq. (10–48) to get, after a little manipulation,

$$\begin{aligned} q = &-\frac{k}{c_p}\left[\left(\frac{\partial \bar{h}}{\partial y} - \sum \bar{h}_i \frac{\partial C_i}{\partial y}\right) + \frac{\bar{\rho}D_{12}c_p}{k}\sum \bar{h}_i \frac{\partial C_i}{\partial y}\right] \\ &- \frac{k_T}{c_p}\left[\left(\frac{\partial \bar{h}}{\partial y} - \sum \bar{h}_i \frac{\partial C_i}{\partial y}\right) + \frac{\bar{\rho}D_T c_p}{k_T}\sum \bar{h}_i \frac{\partial C_i}{\partial y}\right] \end{aligned} \tag{10–50}$$

Note that for $\bar{\rho}D_{12}c_p/k = \mathrm{Le} = 1 = \mathrm{Le}_T = \bar{\rho}D_T c_p/k_T$, the last two terms in each set of large brackets cancel each other, and one can write

$$q = -\frac{(k + k_T)}{c_p}\frac{\partial \bar{h}}{\partial y} \tag{10–51}$$

independently of the chemical reactions in the mixture.

The next step in the development is to introduce the *Reynolds analogy* [see Eq. (2–61)], rewritten as

$$\mathrm{St} = \left(\frac{C_f}{C_{fi}}\right)\frac{C_{fi}}{2\mathrm{Pr}^{-2/3}} \tag{10–52}$$

where C_{fi} is the incompressible skin friction coefficient. For a zero pressure gradient region, one can use a simple formula such as Eq. (7–13) for C_{fi}. Rose et al. (1958) used $C_{fi} = 0.058\ \mathrm{Re}_x^{-0.2}$ and compared the predictions of Eq. (10–52) first with $C_f/C_{fi} = 1$ with their shock tube data for dissociated air over a hemispheric cylinder, as shown in Fig. 10–36. This simple calculation is represented by the solid line, and one can see that the predictions and data disagree at the higher dissociation levels. They next estimated C_f/C_{fi} using Eckert's *reference temperature method* (see Section 5–3) in terms of enthalpy. The resulting prediction is shown as the dashed curve in the figure, and the predictions are consistently high. The final analysis was developed by introducing a nonunity Lewis number correction by analogy with laminar flow. In that case, we can write, for the dissociation reaction,

$$q = -\frac{k}{c_p}\frac{\partial(h + h^0 C_A)}{\partial y} + \frac{k}{c_p}(\mathrm{Le} - 1)\frac{\partial(h^0 C_A)}{\partial y} \tag{10–53}$$

assuming that the specific heats of the molecules and atoms are equal. Assuming further that the profiles across the boundary layer are approximately linear for highly cooled walls, this equation can be used to deduce the relation

$$\frac{q_{Le\neq 1}}{q_{Le=1}} = 1 + (\mathrm{Le}^{\beta} - 1)\left(\frac{h_{\mathrm{De}}}{H_e}\right) \tag{10–54}$$

where the exponent $\beta \approx 0.5–1.0$ was introduced on a semiempirical basis and h_{De} is the energy in dissociation at the boundary layer edge. The results are in good agreement with the data in Fig. 10–36 for $\beta \approx 1.0$ using a value of Le ≈ 1.4.

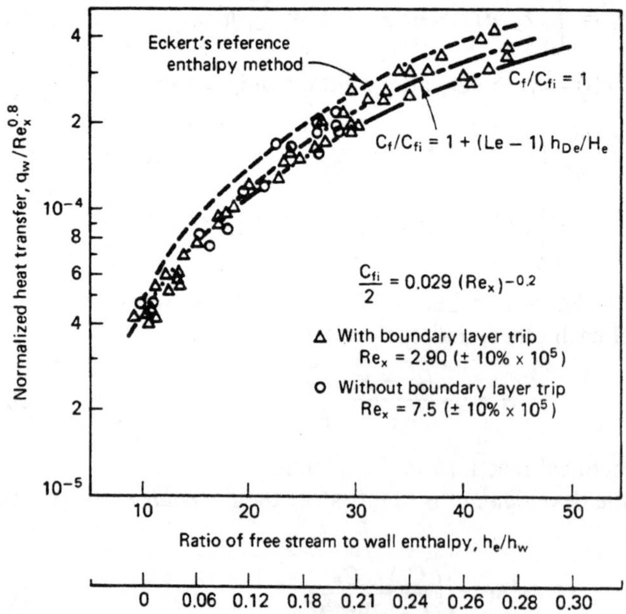

Figure 10–36 Comparison of data with predictions for effects of dissociation on heat transfer in region of zero pressure gradient. (From Rose et al., 1958.)

10–5 VISCOUS-INVISCID INTERACTIONS

As discussed in Section 5–13, pressure gradient effects in supersonic flow can be quite different than in subsonic flows. The main difference is the presence of shock waves, leading to shock–boundary layer interactions. Also, other types of viscous-inviscid interactions can be important.

The flow pattern accompanying a shock–boundary layer interaction in a turbulent flow has some important differences, compared with the pattern accompanying a laminar flow. Typical experimental observations are given in Fig. 10–37. The schlieren photograph in Fig. 10–37(A) can be compared with that for the laminar case in Fig. 5–27(A). The most obvious difference between the two is that the length of the interaction has been dramatically shortened in the turbulent case. This effect can also be seen in the pressure distributions in Fig. 10–37(B). The length of the interaction is $O(100\delta)$ in the laminar case and $O(10\delta)$ in the turbulent case. This difference is explained by the fact that the shapes of the velocity profiles are very different, such that the distance above the wall to the point where $M \approx 1$ in the boundary layer is much larger for a laminar flow, given the same total boundary layer thickness in the two cases. The subsonic region is the channel through which the upstream influence is communicated.

A turbulent boundary layer can tolerate a much stronger adverse pressure gradient than a laminar boundary layer can, and the very simple indicator from Mager (1958), i.e.,

$$\frac{M_2^2}{M_1^2} \approx 0.55 \tag{10–55}$$

gives a surprisingly good estimate of how much deceleration (increased pressure) a supersonic turbulent boundary layer can tolerate before separation. This close correlation is demonstrated in Fig. 10–38, where the sensitivity to the value of the constant selected is also indicated.

Clearly, a boundary layer analysis cannot handle a flow past the separation point. Indeed, a boundary layer analysis will have difficulty in the whole upstream interaction region, since it presumes prior knowledge of the inviscid flow. But the inviscid flow prediction of the region upstream of shock impingement would simply be constant pressure, which does not agree with experiment [see Fig. 10–37(B)]. Thus, either a Navier-Stokes formulation is required, or one can adopt some simple model of the inviscid-viscous interaction, combined with a boundary layer approach. Since Navier-Stokes formulations are outside the scope of this text, the latter approximate approach will be illustrated with a slot injection case. For the interested reader, the 1981 Stanford conference [see Kline et al. (1982)] had some comparisons of predictions from Navier-Stokes formulations with experiment for flows with strong interactions. The review of Marvin and Coakley (1989) also has some material of that type.

The Crocco-Lees mixing theory was introduced in Section 5–13 for laminar

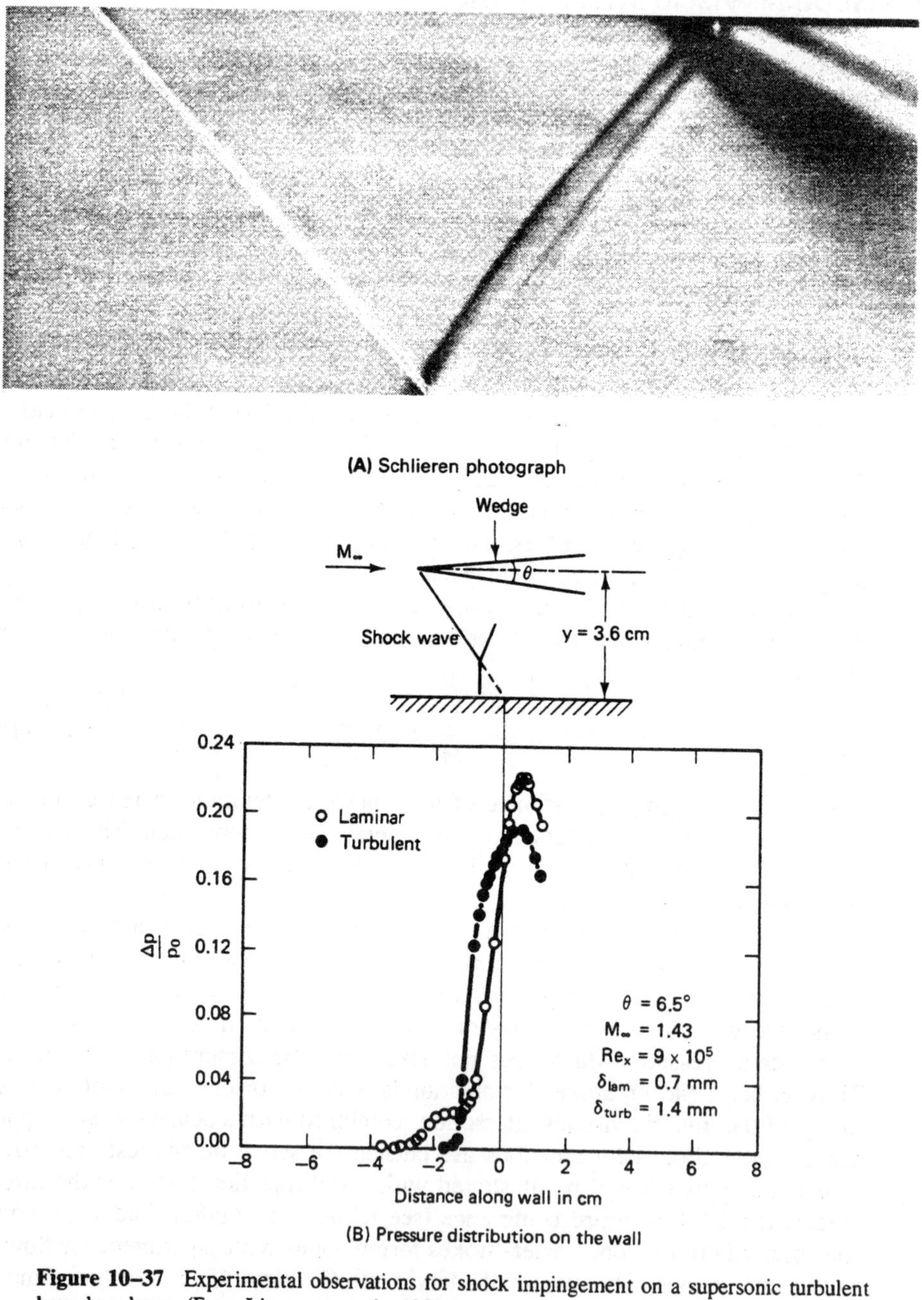

Figure 10–37 Experimental observations for shock impingement on a supersonic turbulent boundary layer. (From Liepmann et al., 1952.)

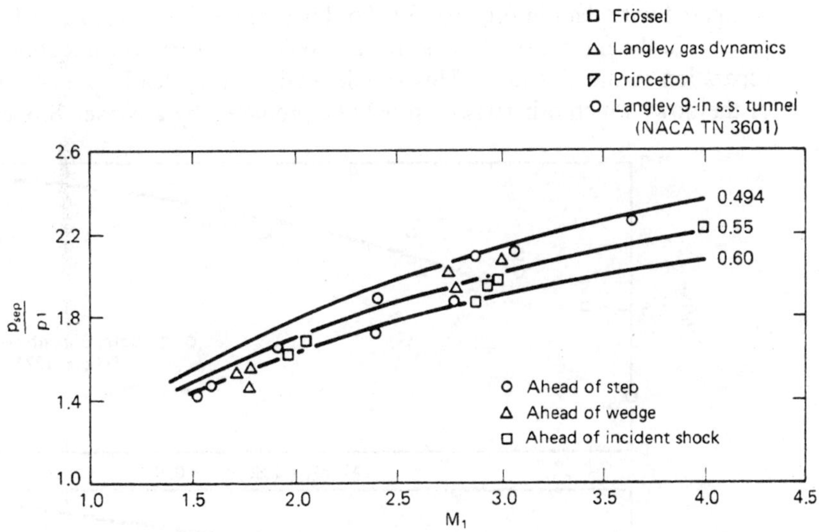

Figure 10–38 Test of empirical rule [Eq. (10–55)] for pressure rise to separation for supersonic turbulent boundary layers. (From Mager, 1958.)

flows, and the turbulent counterpart uses the same key elements [see Fig. 5–28 and Eqs. (5–94) through (5–97)]. This approach has been applied to supersonic slot injection problems and base injection problems by Schetz et al. (1976) and to similar configurations with hypersonic Mach numbers and combustion of H_2 in more recent work [see Schetz et al. (1991)]. For turbulent flow, the constant in Eq. (5–96) must be increased above that for laminar flow. Crocco and Lees (1952) suggest a value of 0.01–0.03 for turbulent flows, with the higher figure for separated flows. For slot and base injection problems, Eq. (5–96) was extended to

$$\frac{d\dot{m}}{dx} = k_m \rho_e U_e \left| 1 - \frac{\overline{\rho U}}{\rho_e U_e} \right| \tag{10–56}$$

where $\overline{\rho U}$ signifies the average value in the inner viscous region. This model seeks to reproduce the minimal mixing found when $(\rho U)_{\text{inj}} \approx \rho_e U_e$. Recent data on rates of high-speed shear layer growth [see Eq. (9–42)] have also been used to deduce an entrainment law, but the results were essentially the same as Eq. (10–56). Using Eq. (5–97) and a corresponding energy equation with Eqs. (5–94), (5–95), and (10–56), we can produce complete flow field solutions for the variation in the static pressure, the height of the viscous region, and flow variables averaged across the inner viscous layer. An interesting result is that one cannot arbitrarily set both the injection rate and the pressure in the injectant for subsonic injection into a supersonic external stream. That pressure is fixed by the inviscid-viscous interaction and the downstream *viscous throat*, where the average Mach number in the inner stream is accelerated to unity by the supersonic external stream (see Section 5–13) and the prediction of the pressure is an important test of any analysis. Some predictions are compared with data in Fig. 10–39. For higher injection rates, $k_m \approx 0.01$ gives good results. In the figure, the S on some data points for very low injection rates indicates separation in the slot flow. This simple analysis can clearly predict critical features of the flow, which otherwise can only be predicted by a Navier-Stokes formulation.

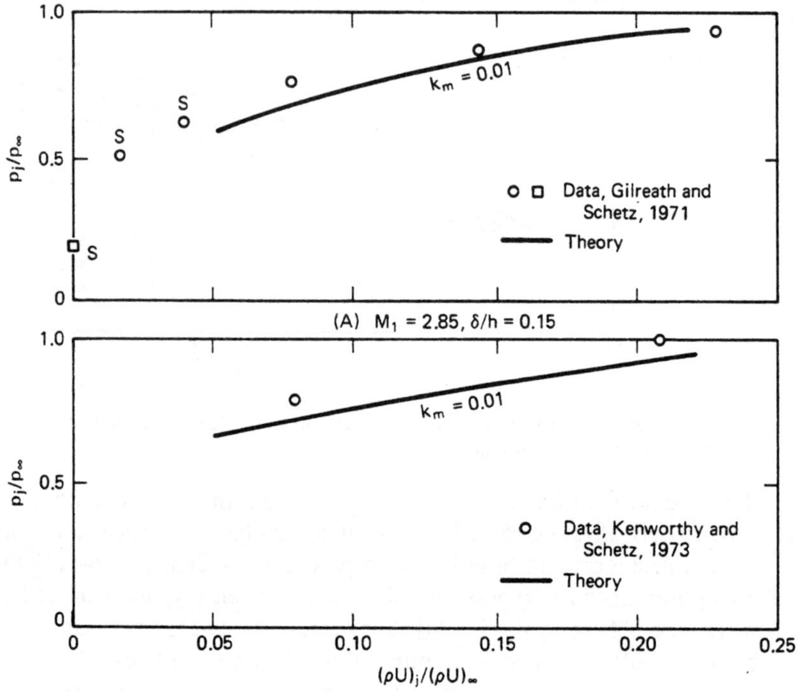

(B) M_1 = 2.36, δ/h = 1.06

Figure 10–39 Comparison of data with predictions of pressure at injection for subsonic slot injection into supersonic streams. (From Schetz et al., 1976.)

PROBLEMS

10.1. Compare the thickness of the thermal laminar sublayer for a flow of air, water, and H_2 at 300°K, all at the same Reynolds number and U_e.

10.2. Consider a flow of water at 25°C at 10 m/s over a flat plate. What are C_f and St at L = 1.0 m? For a wall temperature of 20°C, what are u_* and T_*?

10.3. Suppose we have air at 1.0 atm and 70°F flowing at 100 ft/s over a flat plate with a wall temperature of 100°F. Plot the variation in the film coefficient and the heat transfer rate along the length of the plate until x = 10.0 ft. Use any method you choose to determine the location of transition.

10.4. A designer wishes to reduce the heat transfer to a surface in a gas flow by 25% by injection. At Re = 10^7, how much injection is required? Express your answer as a fraction of the mass flux in the free stream.

10.5. Water at 25°C is flowing at 3.0 m/s over a square plate 3.0 m on a side. The wall temperature is 20°C. What is the total heat transfer?

10.6. For flow over a 5° half-angle wedge of length 1.0 m, compare the wall shear τ_W at $M_\infty = 2.0$ and 4.0 at sea level.

10.7. Air at $M_e = 2.0$ flows over an insulated flat plate. Assuming that $U/U_e = (y/\delta)^{1/7}$, and using the Crocco integral, compare δ^*/δ with δ_k^*/δ.

10.8. Use the extended *law of the wake*, Eq. (7–9), with U_{eff} from Eq. (10–18) for U, for a velocity profile and the Crocco integral for a temperature profile to compare the predictions of the Baldwin-Lomax outer-region eddy viscosity model with the Clauser outer-region model extended to variable density with both δ^* and δ_k^* for $2.0 \le M_e \le 5.0$ on an adiabatic wall.

10.9. Estimate the pressure rise P_2/P_1 for separation vs. M_1 for $1.0 \le M_1 \le 10.0$, using the Mager rule in Eq. (10–55).

10.10. Compare the pressure rise for separation for laminar and turbulent flows at $M_1 = 2.0$ and 4.0. Assume representative Reynolds numbers.

10.11. Use the extended *law of the wake*, Eq. (7-9) with U_{eff} from Eq. (10–18) for U, for a velocity profile and the Crocco integral for a temperature profile to compare the predictions of the Clauser outer-region eddy viscosity model extended to variable density with δ^* and δ_k^* and the Kiss-Schetz model, Eq. (10–40), for $2.0 \le M_e \le 6.0$ on an adiabatic wall.

10.12. Use the extended *law of the wake*, Eq. (7–9), with U_{eff} from Eq. (10–18) for U, for a velocity profile and the Crocco integral for a temperature profile to evaluate the importance of the added term in the Situ-Schetz mixing length model, Eq. (10-43), as a function of Mach number at $M_e = 2.0$, 4.0, 6.0, and 8.0 on an adiabatic wall.

CHAPTER 11

THREE-DIMENSIONAL EXTERNAL BOUNDARY LAYER FLOWS

11–1 INTRODUCTION

The basic ideas of boundary layer theory as discussed to this point for two-dimensional flows carry over directly to the three-dimensional situation. However, the increased role of surface geometry and the curving of the external inviscid flow make the physical and mathematical description of the three-dimensional cases more complex. In this section, the differences between the two situations will be emphasized and some new concepts introduced.

A two-dimensional (2D) boundary layer can be formally defined as that which forms on a plane surface of infinite lateral extent when the projections of the external, inviscid streamlines on that surface are straight and perpendicular to the leading edge. Any other case is three dimensional (3D). By this definition, the axisymmetric flow over a cone, where the external streamlines are straight but the surface has lateral curvature is strictly 3D, even though it can be mathematically described by two coordinates, x and r. The flow is 3D from a physical viewpoint, because the boundary layer is stretched and thinned as it flows along the surface of the cone, which has a growing radius and thus increasing surface area. Individual streamlines thus must diverge.

The most important feature of general 3D boundary layer flows is the existence of so-called *secondary flows,* in addition to the *primary flow*. The primary flow in the boundary layer is usually taken to be in the direction of the external, inviscid flow, and the secondary flow is then in the direction orthogonal to that and parallel to the surface. There are at least two general classes of secondary flows, but one is found only in turbulent cases, so it will be discussed later in the appropriate section. The most significant type of secondary flow is referred to as *pressure-driven secondary flow*. This general phenomenon can be understood by considering the special case of circular external streamlines above a flat surface. In that case, the pressure gradient is balanced by the centrifugal force corresponding to the velocity and curvature of the streamline. By the boundary layer approximation, the external flow pressure field is imposed on the viscous flow in the boundary layer beneath it. But the velocity in the boundary layer is decreased below the inviscid edge value by viscosity, down to zero at the surface. Thus, the pressure gradient cannot be balanced by a centrifugal force corresponding to the reduced local velocity and the same streamline curvature as in the external flow. Indeed, the streamline curvature in the boundary layer must be considerably greater, leading to a streamline pattern on any surface in the boundary layer that is quite different than that in the external flow. This *streamline skewing* is a key feature of 3D viscous flows. One therefore commonly talks of a *streamwise velocity component*, aligned with the local edge streamline, and a *cross-flow velocity component*, orthogonal to the streamwise velocity component and parallel to the surface. The cross-flow component comprises the secondary flow.

Another difference between 2D and 3D boundary layer flows is their response to an adverse pressure gradient. In a 2D flow, the flow is constrained to *climb the pressure gradient hill* if it is physically possible to do so. If it is not, the flow separates from the surface. In the 3D case, the flow can take the path of least resistance by moving in the lateral direction. Thus, a strong adverse pressure gradient does not always lead to separation in the 3D case. The result can be a large change in direction of the flow. Separation in 3D is also more complex, and we shall return to that matter shortly.

In 2D flows, we find it convenient to introduce some integral quantities, such as the displacement and momentum thicknesses, that have simple physical interpretations. That simplicity is lost in the 3D case. Since the direction of the total velocity vector varies through the boundary layer, we must have two velocity components to describe the velocity profile. That leads to two displacement thicknesses and four momentum thicknesses. (These are discussed more fully in Section 11-2-1.)

Three-dimensional boundary layer flows can be divided into two general classes. The first are thin 3D layers over surfaces. These correspond closely to 2D boundary layers and are sometimes called *boundary sheets*. If x is the main streamwise direction, e.g., along the centerline of an aircraft fuselage, y is locally normal to the surface, and z is in the transverse direction, then the direct 3D analog of the 2D boundary layer will have v small compared with u and w and the pressure constant in the y-direction. This implies that changes in *both* the x- and z-directions are slow, compared with changes in the y-direction. Such a flow occurs, for example, on finite wings and propeller blades, excluding the regions near the root and tip. The flow in those regions is called *slender 3D flow,* and changes in the y- and z-direction are rapid, compared with changes in the x-direction. Some slender 3D flows develop discrete longitudinal vortices in the boundary layer. An example is the *horseshoe vortex* that forms in a wing-body junction. Most of the analytical and numerical work in the literature has dealt with boundary sheet flows.

The boundary layer approximation in 3D flow also implies some new mathematical properties of the equation system. The 3D Navier-Stokes equations are *elliptic,* which means that information from a given point in the flow can be propagated by convection, viscous diffusion, or pressure perturbations. In a 3D boundary layer, propagation by the pressure has been eliminated by taking $\partial p/\partial y \approx 0$. Also, viscous diffusion has been eliminated in the x- and z-directions. Thus, information is propagated in the cross-flow plane only by convection at an angle $\tan^{-1}(w/u)$ in the plan view. Taken all together, these properties imply that the influence of the solution at one point is transferred to other points first by viscous diffusion along a line across the boundary sheet through that point in the y-direction and then by convection downstream along all streamlines through that vertical line. This forms a *zone of influence* bounded by the body surface at the bottom, the boundary sheet edge at the top, and the two characteristic surfaces normal to the surface containing the external, inviscid and wall streamlines. This configuration is illustrated in Fig. 11–1. On the other hand, the solution at a given point depends only on the solution in another zone called the *zone of dependence.* This zone is upstream of the point in question and is formed by the body surface, the boundary sheet edge, and the two characteristic surfaces passing through the point. It is closed on the upstream end by an initial-value surface that cannot coincide with a characteristic surface. The latter will sound familiar to those who have studied the *method of characteristics* for inviscid, supersonic flow. The zone of dependence is also shown in Fig. 11–1. These matters are very important in developing appropriate integral and differential methods for solving 3D boundary layer flows. Before we begin a discussion of those techniques, one more important difference between 2D and 3D boundary layer flows needs to be addressed, at least briefly.

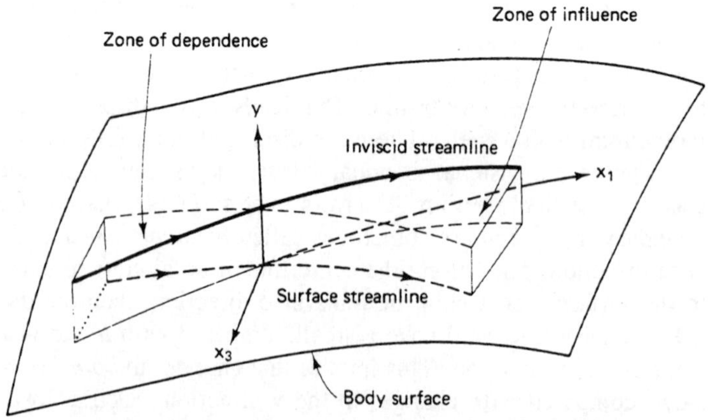

Figure 11–1 Schematic of the regions of influence and dependence in a 3D boundary layer.

11-1-1 Separation of Three-Dimensional Boundary Layers

It was mentioned earlier that a 3D flow in an adverse pressure gradient can turn laterally and avoid separation. This possibility leads to significant changes and complexity in separation patterns. In 2D flows, one simply looks for the location of any points where $(\partial u/\partial y)_w = 0$, and the flow pattern is as shown in Fig. 1-10. In 3D flows, the mathematical condition can be generalized to say that separation corresponds to $(\partial q/\partial y)_w = 0$, where q is the component of velocity perpendicular to the separation line. This is much less clear from a physical point of view. Maskell (1955) distinguishes two broad types of 3D separations, as illustrated in Fig. 11-2. The case in Fig. 11-2(A) is said to form a *separation bubble* that encloses fluid that is not part of the main stream, but rather is carried along with the body. Formation of the bubble entails one singular saddle point, denoted by S on Fig. 11-2(A). The behavior of the flow at this point, and this point alone, resembles that in a 2D sepa-

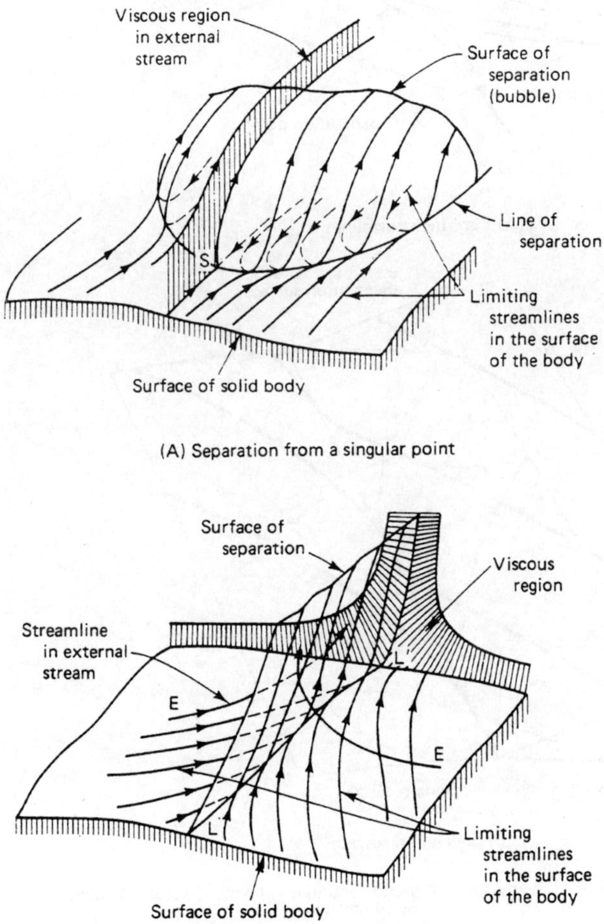

Figure 11-2 Separation in a three-dimensional boundary layer. (After Maskell, 1955.)

ration. In the second case, shown in Fig. 11–2(B), the space on either side of the surface of separation is filled with fluid from the main stream. Maskell suggested that a separation surface be viewed as consisting of the separating limiting streamlines coming from upstream and downstream (reversed-flow) regions. He argued that separation occurs when two such streamlines meet. They then merge and leave the surface as a single separation streamline. This is shown schematically in Fig. 11–3. In the ordinary case, the separation streamline first is tangent to the body surface and then lifts off, as drawn in Fig. 11–3(B). For a singular separation, such as that at the point S in Fig. 11–2(A), the separation streamline inclines to the body surface immediately.

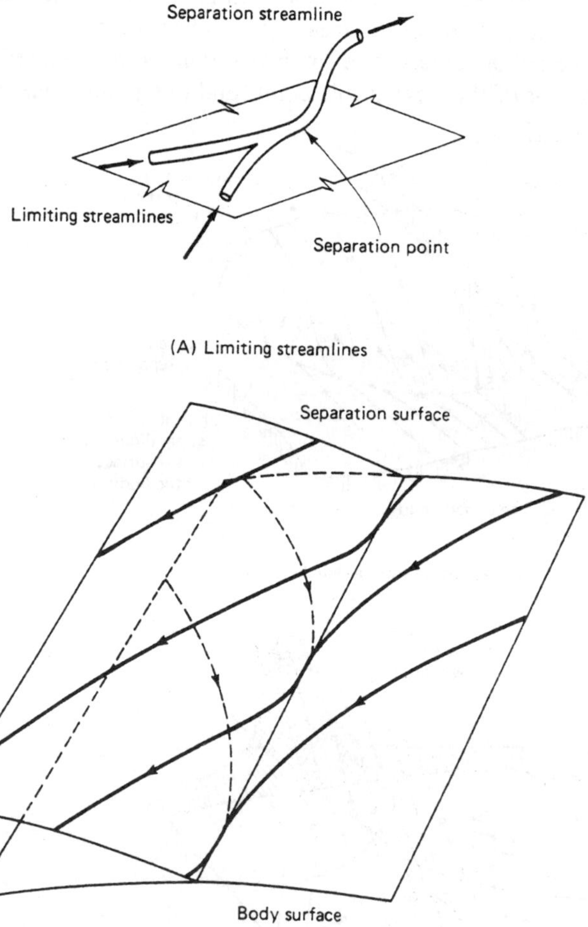

(B) Separation surface

Figure 11–3 Separation streamlines in a 3D boundary layer. (After Maskell, 1955.)

11–2 LAMINAR THREE-DIMENSIONAL BOUNDARY LAYERS

11–2–1 Boundary Layer Equations in Three Dimensions

To maintain a reasonable level of generality, we adopt an orthogonal coordinate system (x_1, y, x_3) on the surface over which the boundary layer flows, where y is normal to the surface and $y = 0$ on the surface. Lines of constant x_1 and x_3 form an orthogonal system of coordinates on the surface. It is common to take x_1 to be the direction of the external inviscid flow. The square of an element of arc length $d\ell$ on the surface may be written

$$d\ell^2 = h_1^2 dx_1^2 + h_3^2 dx_3^2 \tag{11–1}$$

where h_1 and h_3 are *metric coefficients*. With the boundary layer approximation, the boundary layer is thin, and the metric coefficient in the y-direction can be taken to be unity. Thus, the square of a general element of arc length ds within the boundary layer becomes

$$ds^2 = h_1^2 dx_1^2 + dy^2 + h_3^2 dx_3^2 \tag{11–2}$$

It is assumed here that the radii of curvature of the surface are large compared to the boundary layer thickness.

From Eq. (11–2), the steady form of the convective derivative in the given coordinate system is

$$\frac{D(\cdot)}{Dt} = \frac{u}{h_1}\frac{\partial(\cdot)}{\partial x_1} + v\frac{\partial(\cdot)}{\partial y} + \frac{w}{h_3}\frac{\partial(\cdot)}{\partial x_3} \tag{11–3}$$

where (u, v, w) are the velocity components in the directions of the (x_1, y, x_3) axes. Assuming a perfect gas, a single species, and constant c_p and Pr, we can write the boundary layer equations. We have

$$\frac{1}{h_1 h_3}\left[\frac{\partial(h_3\rho u)}{\partial x_1} + \frac{\partial(h_1 h_3 \rho v)}{\partial y} + \frac{\partial(h_1 \rho w)}{\partial x_3}\right] = 0 \tag{11–4}$$

for the continuity equation,

$$\frac{Du}{Dt} + uwK_1 - w^2K_3 = -\frac{1}{\rho h_1}\frac{\partial p}{\partial x_1} + \frac{1}{\rho}\frac{\partial}{\partial y}\left(\mu\frac{\partial u}{\partial y}\right) \tag{11–5}$$

for the x_1-momentum equation,

$$\frac{Dw}{Dt} - u^2K_1 + uwK_3 = -\frac{1}{\rho h_3}\frac{\partial p}{\partial x_3} + \frac{1}{\rho}\frac{\partial}{\partial y}\left(\mu\frac{\partial w}{\partial y}\right) \tag{11–6}$$

for the x_3-momentum equation, and finally,

$$\rho c_p \frac{DT}{Dt} = \frac{Dp}{Dt} + \frac{\partial}{\partial y}\left(\frac{\mu c_p}{Pr}\frac{\partial T}{\partial y}\right) + \mu\left(\frac{\partial u}{\partial y}\right)^2 + \mu\left(\frac{\partial w}{\partial y}\right)^2 \tag{11–7}$$

for the energy equation. Of course, the y-momentum equation is simply $\partial p/\partial y \approx 0$, as in the 2D case. The pressure gradient terms in the two momentum equations are partial derivatives because, in 3D flows, $p = (x_1, x_3)$ from the inviscid flow. We have also introduced the *geodesic curvatures* of the surface coordinate lines, namely,

$$K_1 = \frac{1}{h_1 h_3}\frac{\partial h_1}{\partial x_3} \quad \text{for } x_3 = \text{constant} \tag{11–8a}$$

$$K_3 = \frac{1}{h_1 h_3}\frac{\partial h_3}{\partial x_1} \quad \text{for } x_1 = \text{constant} \tag{11–8b}$$

The system is completed with an equation of state, e.g., for a perfect gas,

$$p = \rho RT \tag{11–9}$$

and $\mu(T)$.

The boundary and initial conditions also need to be discussed. For flow over an impermeable surface, the velocity boundary conditions on the surface, $y = 0$, are just $u = v = w = 0$. For a spinning, axisymmetric body, one would have a specified $w \neq 0$ on the surface. Of course, with suction or injection, $v \neq 0$ on the surface. The other surface boundary condition will be a specified wall temperature or heat transfer rate.

The boundary sheet "edge" boundary conditions are direct extensions of those used for 2D flows, namely,

$$\begin{aligned} \lim_{y\to\infty} u &= U_e(x_1, x_3) \\ \lim_{y\to\infty} w &= W_e(x_1, x_3) \\ \lim_{y\to\infty} T &= T_e(x_1, x_3) \end{aligned} \tag{11–10}$$

The edge conditions themselves are found from the inviscid equations on the surface, which can be written

$$\frac{U_e}{h_1}\frac{\partial U_e}{\partial x_1} + \frac{W_e}{h_3}\frac{\partial U_e}{\partial x_3} + U_e W_e K_1 - W_e^2 K_3 = -\frac{1}{\rho_e h_1}\frac{\partial p}{\partial x_1} \tag{11–11}$$

$$\frac{U_e}{h_1}\frac{\partial W_e}{\partial x_1} + \frac{W_e}{h_3}\frac{\partial W_e}{\partial x_3} - U_e^2 K_1 + U_e W_e K_3 = -\frac{1}{\rho_e h_3}\frac{\partial p}{\partial x_3} \tag{11–12}$$

$$\rho_e c_p\left(\frac{U_e}{h_1}\frac{\partial T_e}{\partial x_1} + \frac{W_e}{h_3}\frac{\partial T_e}{\partial x_3}\right) = \frac{U_e}{h_1}\frac{\partial p}{\partial x_1} + \frac{W_e}{h_3}\frac{\partial p}{\partial x_3} \tag{11–13}$$

With the pressure available from a separate inviscid solution, as in 2D flows, Eqs. (11–11) to (11–13) can be used to find U_e, W_e, and T_e.

The last conditions required are the upstream initial conditions. We need sufficient conditions on an upstream plane perpendicular to the surface. Then, the boundary layer equations can be solved in a region downstream of that plane if the zones of influence and dependence introduced in Section 11-1 are properly taken into account.

In the case where the surface that the boundary layer flows over is *developable*, the metric and geodesic terms can be made to disappear from the boundary layer equations. A developable surface is a surface that can be rolled out into a plane without being stretched, but allowing suitable cuts. The equations can then be written in a Cartesian (x, y, z) form as follows:

$$\frac{\partial(\rho u)}{\partial x} + \frac{\partial(\rho v)}{\partial y} + \frac{\partial(\rho w)}{\partial z} = 0 \tag{11–14}$$

$$u\frac{\partial u}{\partial x} + v\frac{\partial u}{\partial y} + w\frac{\partial u}{\partial z} = -\frac{1}{\rho}\frac{\partial p}{\partial x} + \frac{1}{\rho}\frac{\partial}{\partial y}\left(\mu\frac{\partial u}{\partial y}\right) \tag{11–15}$$

$$u\frac{\partial w}{\partial x} + v\frac{\partial w}{\partial y} + w\frac{\partial w}{\partial z} = -\frac{1}{\rho}\frac{\partial p}{\partial z} + \frac{1}{\rho}\frac{\partial}{\partial y}\left(\mu\frac{\partial w}{\partial y}\right) \tag{11–16}$$

$$\rho c_p\left(u\frac{\partial T}{\partial x} + v\frac{\partial T}{\partial y} + w\frac{\partial T}{\partial z}\right) = u\frac{\partial p}{\partial x} + w\frac{\partial p}{\partial z} + \frac{\partial}{\partial y}\left(\frac{\mu c_p}{Pr}\frac{\partial T}{\partial y}\right) + \mu\left(\frac{\partial u}{\partial y}\right)^2 + \mu\left(\frac{\partial w}{\partial y}\right)^2 \tag{11–17}$$

Integral equations corresponding to those introduced for 2D flows in Chapter 2 can also be derived. For 3D constant-density and constant-property flows with the x_1 axis taken to be along the inviscid streamlines, these equations may be written

$$\frac{1}{h_1}\frac{\partial}{\partial x_1}(U_e^2\theta_{11}) + U_e\delta_1^*\frac{1}{h_1}\frac{\partial U_e}{\partial x_1} + \frac{1}{h_3}\frac{\partial}{\partial x_3}(U_e^2\theta_{12}) + U_e\delta_2^*\frac{\partial U_e}{\partial x_3} + K_1U_e^2(\theta_{12} + \theta_{21}) + K_3U_e^2(\theta_{11} - \theta_{22}) = \frac{\tau_{w1}}{\rho} \tag{11–18}$$

and

$$\frac{1}{h_1}\frac{\partial}{\partial x_1}(U_e^2\theta_{21}) + \frac{1}{h_3}\frac{\partial}{\partial x_3}(U_e^2\theta_{22}) + 2K_3U_e^2\theta_{21} + K_1U_e^2(\theta_{22} - \theta_{11} - \delta_1^*) = \frac{\tau_{w3}}{\rho} \tag{11–19}$$

with the various integral thicknesses defined as

$$\delta_1^* \equiv \int_0^\infty\left(1 - \frac{u}{U_e}\right)dy, \qquad \delta_2^* \equiv -\int_0^\infty \frac{w}{U_e}\,dy$$

$$\theta_{11} \equiv \int_0^\infty \frac{u}{U_e}\left(1 - \frac{u}{U_e}\right)dy, \qquad \theta_{12} \equiv \int_0^\infty \frac{w}{U_e}\left(1 - \frac{u}{U_e}\right)dy \tag{11–20}$$

$$\theta_{21} \equiv -\int_0^\infty \frac{u}{U_e}\frac{w}{U_e}\,dy, \qquad \theta_{22} \equiv -\int_0^\infty\left(\frac{w}{U_e}\right)^2 dy$$

11–2–2 Special Cases

As in 2D flows, there are some special cases of 3D flows that admit to a relatively easy solution. The first is constant-density, constant-property flow over a flat plate, where the external, inviscid flow has streamlines in planes parallel to the plate, but forming a system of parabolic translates whose common axis is parallel to the leading edge. This configuration is illustrated in Fig. 11–4. For such a flow, it is more convenient to align the coordinate system with the plate, rather than the external flow. The edge velocity is

$$U_e = \text{constant}, \qquad W_e = U_e(\chi - Cx) \tag{11–21}$$

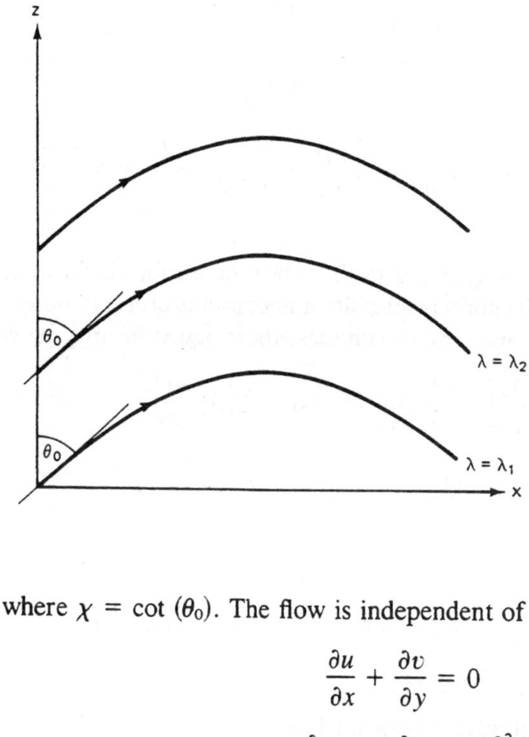

Figure 11–4 Parabolic streamlines of a curved external flow over a flat plate.

where $\chi = \cot(\theta_0)$. The flow is independent of z, and the equations reduce to

$$\begin{aligned} \frac{\partial u}{\partial x} + \frac{\partial v}{\partial y} &= 0 \\ u\frac{\partial u}{\partial x} + v\frac{\partial u}{\partial y} &= \nu\frac{\partial^2 u}{\partial y^2} \\ u\frac{\partial w}{\partial x} + v\frac{\partial w}{\partial y} &= -CU_e^2 + \nu\frac{\partial^2 w}{\partial y^2} \end{aligned} \tag{11–22}$$

If $C = 0$, the problem collapses down to the Blasius problem considered in Section 4–3–1, and the solution for $C \neq 0$ can also be obtained by similarity methods. Sowerby (1954) introduced a similarity transformation using

$$\eta_1 = \frac{y}{2}\sqrt{\frac{U_e}{\nu x}}, \qquad u = \frac{1}{2}U_e f'$$
$$v = \frac{1}{2}(\eta_1 f' - f)\sqrt{\frac{U_e \nu}{x}} \tag{11-23}$$
$$w = U_e[\chi g(\eta_1) - Cxh(\eta_1)]$$

whereupon the equations become

$$f''' + ff'' = 0$$
$$g'' + fg' = 0 \tag{11-24}$$
$$h'' + fh' - 2f'h + 4 = 0$$

with boundary conditions

$$f(0) = f'(0) = g(0) = h(0) = 0$$
$$f' \rightarrow 2, g \rightarrow 1, h \rightarrow 1 \text{ as } \eta_1 \rightarrow \infty \tag{11-25}$$

Clearly, the first equation for f is related to the Blasius solution, and his results can be adapted for use here. Comparing the second equation for g with the first equation for f and the boundary conditions for each, we see that $g = f'/2$. The last equation is linear in h, and it was solved numerically by Sowerby (1954). The solutions for a particular case are plotted in Fig. 11–5 in terms of a velocity component in the direction of the inviscid streamlines, called u_s in this coordinate system, and another component perpendicular to the inviscid streamlines, called u_n. The component u_n is what we have called the secondary flow, and it is directed towards the center of curvature of the inviscid streamlines, reaching a value about 64 percent of the edge velocity. The component u_s has an *overshoot*—i.e., $u_s > U_e$—something we did not see in 2D flows. The streamline skewing in the boundary layer that was mentioned earlier is quite substantial here. The results in Fig. 11–5(B) show a 38° difference between the streamline angle on the wall and the inviscid streamline at the edge.

The second special case that will be discussed is the flow at a 3D stagnation point, such as that on a blunt-nosed body that is neither planar nor axisymmetric. The problem was solved by Howarth (1951), who formulated it as shown in Fig. 11–6, where the coordinate system he chose is depicted. If the flow in the external flow is irrotational, it can be written as $U_e = Ax$ and $W_e = Bz$ near the stagnation point, where A and B depend on the specific geometry of the body. This situation can be seen to be closely akin to that for a 2D body discussed in Section 4–3–2, where we had $U_e = Ax$ at a 2D stagnation point. Howarth introduced

$$\eta_2 \equiv y\sqrt{\frac{U_e}{\nu x}}, \qquad u = U_e f'(\eta_2), \qquad w = W_e g'(\eta_2) \tag{11-26}$$

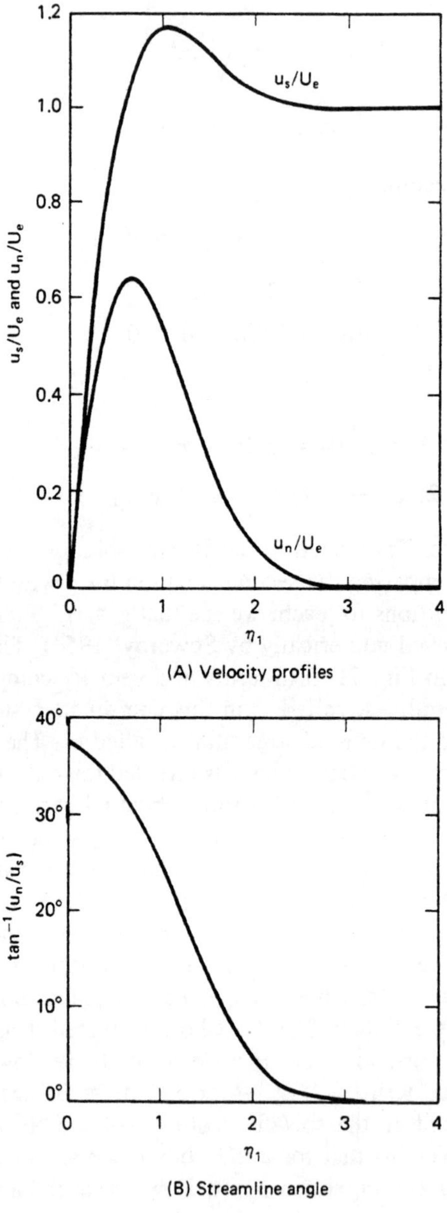

Figure 11–5 Solution of 3D laminar boundary layer with parabolic external flow, $\chi = 1$, $Cx = 2$. (From Sowerby, 1954.)

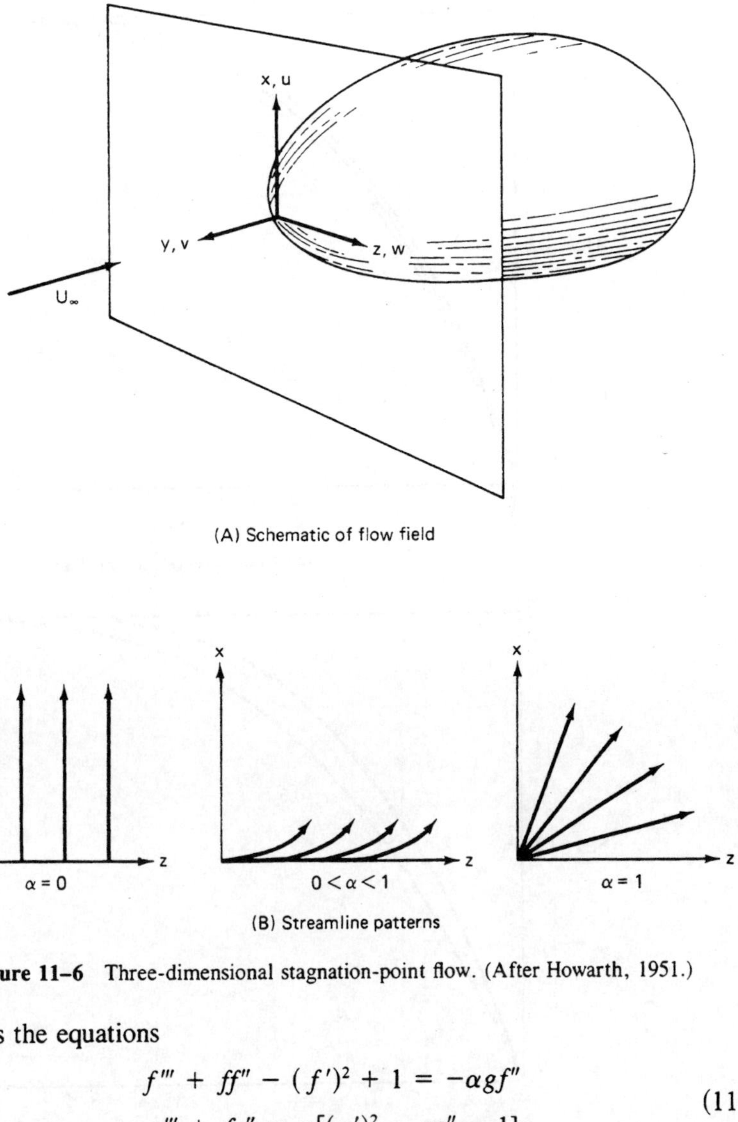

Figure 11–6 Three-dimensional stagnation-point flow. (After Howarth, 1951.)

which yields the equations

$$\begin{aligned} f''' + ff'' - (f')^2 + 1 &= -\alpha g f'' \\ g''' + fg'' &= \alpha[(g')^2 - gg'' - 1] \end{aligned} \tag{11–27}$$

where $\alpha = B/A$. When $\alpha = 0$, the flow reduces to 2D, and when $\alpha = 1$, the flow reduces to axisymmetric. The shapes of the streamlines in the external flow are shown schematically in Fig. 11–6(B). Howarth (1951) obtained numerical solutions for various values of α, and some velocity profiles are shown in Fig. 11–7, where it can be observed that the value of α has a small effect on the profiles in the direction of the major axis, but a large effect on the profiles in the direction of the minor axis. There is also a strong effect on the angle of the surface streamlines, compared with those in the external flow. The results show differences of 14°, 6°, 3°, and 0° for

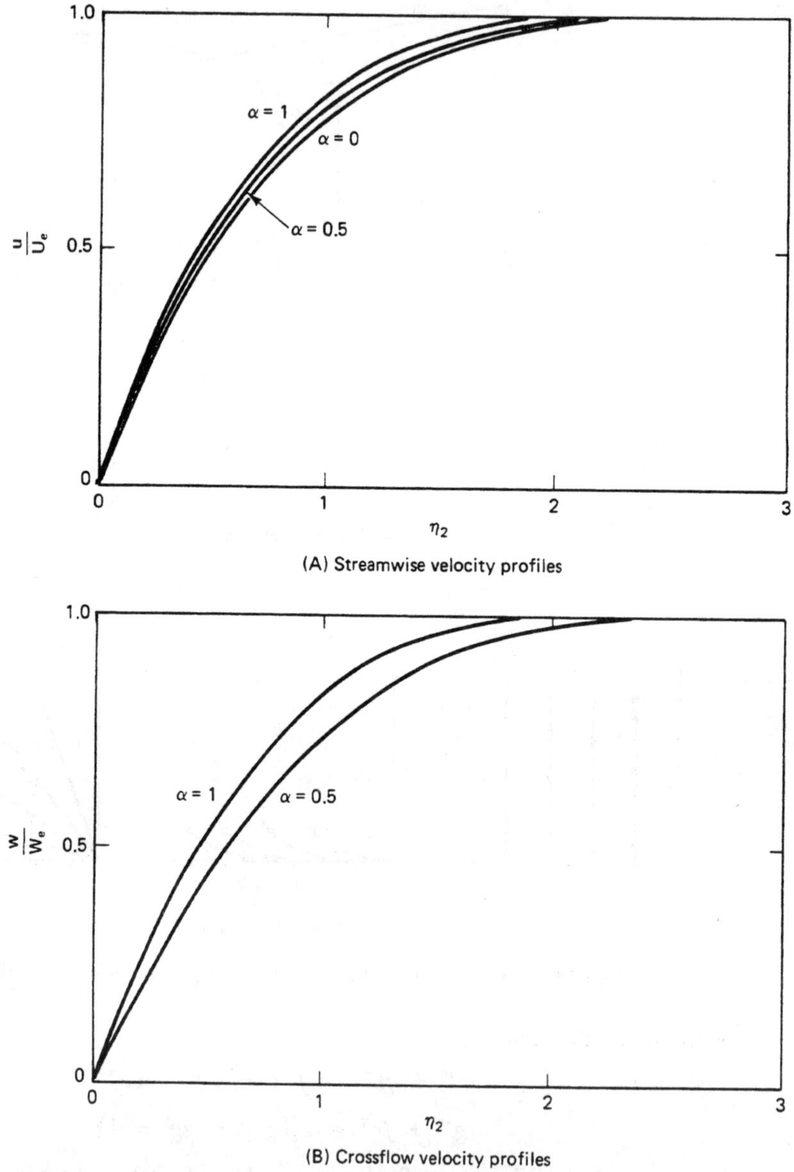

Figure 11–7 Velocity profiles near a general stagnation point; $\alpha = 0$ is 2D, $\alpha = 1$ is axisymmetric. (From Howarth, 1951.)

$\alpha = 1/4$, $1/2$, $3/4$, and 1, respectively. These solutions are interesting in their own right, but in modern times they are most useful as starting profiles for general 3D problems treated with numerical methods.

The last special case to be considered is the *infinite swept wing*, or, more generally, infinite yawed cylinder. Flows over finite swept wings are of interest in aerodynamic applications, and the case of an infinite version allows great simplification to the equations of motion. The inviscid, external flow over an infinite swept cylinder can be found by the superposition of two flows, one lying in planes normal to the cylinder axis and the second parallel to the generators of the cylinder. Since a cylinder is a developable surface, one can use the boundary layer equations in Cartesian coordinates. Here, we treat only the constant-density, constant-property case, so the equations reduce from Eqs. (11–14) through (11–16) to

$$\frac{\partial u}{\partial x} + \frac{\partial v}{\partial y} = 0$$

$$u\frac{\partial u}{\partial x} + v\frac{\partial u}{\partial y} = U_e\frac{dU_e}{dx} + \nu\frac{\partial^2 u}{\partial y^2} \qquad (11\text{–}28)$$

$$u\frac{\partial w}{\partial x} + v\frac{\partial w}{\partial y} = \nu\frac{\partial^2 w}{\partial y^2}$$

where x is measured normal to the leading edge, y is measured normal to the surface, and z is in the spanwise direction. The flow is independent of z, since the cylinder is of infinite extent and the approach flow is taken to be uniform. The boundary conditions are

$$u(x, 0) = v(x, 0) = w(x, 0) = 0$$
$$u \rightarrow U_e, w \rightarrow W_e \text{ as } y \rightarrow \infty \qquad (11\text{–}29)$$

Interestingly, the solution for u and v can be found from the first two equations independently of w. Then, the last equation is linear, and it can be solved for w using the known u and v as coefficients. This is reminiscent of the case for the low-speed energy equation in Section 4–4. Separation can occur only through the effects of an adverse pressure gradient formed by $U_e(x)$, not W_e. This whole special behavior was called the *independence principle* by Prandtl (1946).

The simplest example is, as usual, a flat plate, in this case with the leading edge yawed to the uniform approach flow. Thus, both U_e and W_e are constant. The solution for u and v is just the Blasius 2D solution, and the equations can be seen to admit of the simple solution $w/u = W_e/U_e$. At each point across the boundary layer, the resultant of u and w forms a profile that is itself a Blasius profile. The flow is then independent of yawing of the leading edge, and there is no streamline skewing. This kind of analysis has been extended to Cooke's (1950) yawed equivalent of the Falkner-Skan *wedge flows* discussed in the 2D case in Section. 4–3–2. There really is no great practical utility to this very restrictive, special class of flows.

11–2–3 Integral Methods

As with all integral methods, the main issue in 3D integral methods is the selection of the shape(s) of the velocity profile(s). Experience with 2D cases is helpful in selecting the shape of the profile of the streamwise velocity component. However, the choice of shape for the profile of the crossflow velocity component presents new and difficult problems. If the streamline skewing is all in one direction for a given problem, the choice is not so trying. But, if, as in the general case, the skewing first is in one direction and then changes direction under the influence of the inviscid pressure gradient, the crossflow profile can have a complicated S-shape. The crossflow velocity in one part of the layer is in one direction, and that in another part of the layer is in the opposite direction. The development or relaxation of this behavior is obviously important in predicting the flow, and it is hard to "build in" the correct behavior in the choice of profile. Most workers have chosen to employ two boundary layer thicknesses to describe the two velocity profiles. Wild (1949) proposed

$$\frac{u}{U_e} = F\left(\frac{y}{\delta}\right) + \frac{\delta^2}{\nu}\frac{dU_e}{dx}G\left(\frac{y}{\delta}\right), \qquad \frac{w}{W_e} = F\left(\frac{y}{\Delta}\right)$$
$$F = \frac{y}{\delta}\left[2 - 2\left(\frac{y}{\delta}\right)^2 + \left(\frac{y}{\delta}\right)^3\right], \qquad G = \frac{1}{6}\frac{y}{\delta}\left(1 - \frac{y}{\delta}\right)^2 \tag{11-30}$$

for his treatment of the case of the infinite yawed wing. In that case, δ is known from the equivalent 2D solution, and Δ is the remaining unknown. The latter is found from the remaining momentum equation. With the assumptions in Eq. (11–30), the w-profile is independent of the u-profile. In other methods, especially for turbulent flows, it is common to express the shape of the w-profile in terms of the shape of the u-profile. For the case of an infinite swept wing, Rott and Crabtree (1952) have extended the Thwaites-Walz 2D method described in Section 2–3–2.

All in all, the serious limitations of integral methods for 3D laminar flows has resulted in their being little used in modern times. The use of integral methods for 3D turbulent cases is more common, and we shall consider that subject later in the chapter. If one needs accurate solutions for 3D laminar flows under fairly general conditions, the only choice is numerical solutions.

11–2–4 Numerical Solution Techniques

The general methods discussed in Chapter 4, 5, 7, and 10 for 2D flows can be extended to 3D flows. The new items that need special attention are the complexities of 3D geometries and the zones of influence and dependence introduced in Section 11–1. The zone of dependence influences numerical stability, and it must be reckoned in determining the "marching" direction. Also, obtaining the required 3D inviscid flow solution is not trivial, but that subject will not be treated in any depth here.

The thorough review of numerical methods for 3D boundary layer flows by Blottner (1975) and other reviews by Van den Berg et al. (1988) and Humphreys and Lindhout (1988) will be used as the basis of the discussion here. The boundary layer equations in physical coordinates, as described in Section 11–2–1, or in the various transformed forms that will be described later, all have first derivatives in the three coordinate directions and second derivatives in the normal, y-direction. A general grid system and notation are shown in Fig. 11–8 for reference in discussing the several finite difference methods that are in use. In the figure, the y-direction is up out

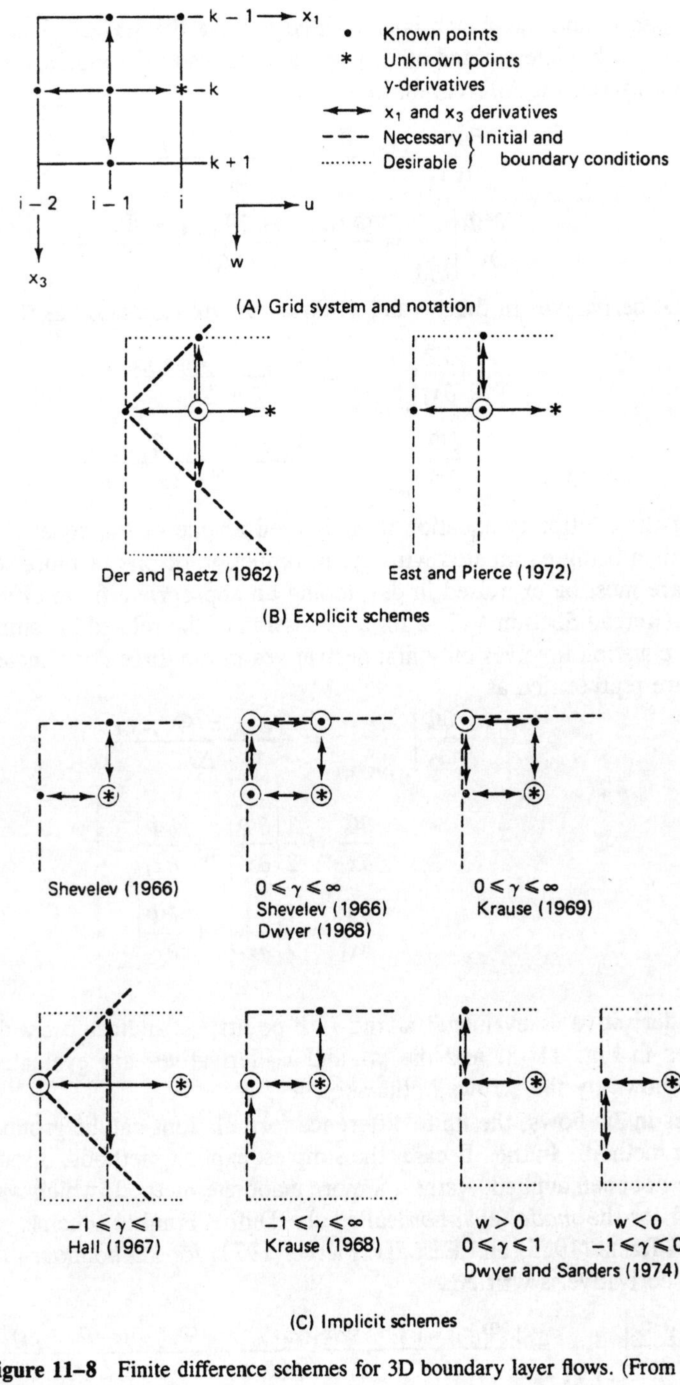

Figure 11–8 Finite difference schemes for 3D boundary layer flows. (From Blottner, 1975.)

of the paper, and the sketch is at a plane j above the surface. Unknowns are at the grid point i, k. Unless noted otherwise, derivatives in the y-direction are always represented by central differences; i.e.,

$$\left.\frac{\partial \Phi}{\partial y}\right|_{i,j,k} = \frac{\Phi_{i,j+1,k} - \Phi_{i,j-1,k}}{2 \cdot \Delta y} \tag{11-31}$$

$$\left.\frac{\partial^2 \Phi}{\partial y^2}\right|_{i,j,k} = \frac{\Phi_{i,j+1,k} - 2\Phi_{i,j,k} - \Phi_{i,j-1,k}}{(\Delta y)^2} \tag{11-32}$$

The first derivatives in the x_1- and x_3-direction are expressed as the differences

$$\left.\frac{\partial \Phi}{\partial x_1}\right|_{i-1,j,k} = \frac{\Phi_{i,j,k} - \Phi_{i-2,j,k}}{2 \cdot \Delta x_1} \tag{11-33}$$

$$\left.\frac{\partial \Phi}{\partial x_3}\right|_{i-1,j,k} = \frac{\Phi_{i-1,j,k+1} - \Phi_{i-1,j,k-1}}{2 \cdot \Delta x_3} \tag{11-34}$$

If the continuity equation itself is used as one of the equations in the system, rather than being eliminated with the introduction of one or more stream functions, then care must be exercised in developing an appropriate finite difference representation. (Reread Section 4–7–2 for a discussion of the related 2D situation.) The continuity equation involves only first derivatives in the three coordinate directions, and these are represented as

$$\left.\frac{\partial \Phi}{\partial y}\right|_{i,j-1/2,k} = \frac{\Phi_{i,j,k} - \Phi_{i,j-1,k}}{\Delta y} \tag{11-35a}$$

$$\frac{\partial \Phi}{\partial x_1} = \frac{1}{2}\left|\left.\frac{\partial \Phi}{\partial x_1}\right|_j + \left.\frac{\partial \Phi}{\partial x_1}\right|_{j-1}\right| \tag{11-35b}$$

$$\frac{\partial \Phi}{\partial x_3} = \frac{1}{2}\left|\left.\frac{\partial \Phi}{\partial x_3}\right|_j + \left.\frac{\partial \Phi}{\partial x_3}\right|_{j-1}\right| \tag{11-35c}$$

The y-derivative is evaluated at the i, k points, as indicated by the circle in the sketches in Fig. 11–8, and the x_1- and x_3-derivatives are evaluated with the i, k points shown by the arrows in the sketches.

As in 2D flows, the finite difference formulations can be grouped as *explicit* or *implicit* methods. In the 3D case, the simplest explicit methods, discussed in Chapter 2, have not been applied. Rather, a more elaborate method, which was originally developed for the *model equation* called the Dufort-Frankel scheme, was adopted by Der and Raetz (1962) and East and Pierce (1972) for 3D boundary layer flows. The second derivative is written

$$\left.\frac{\partial^2 \Phi}{\partial y^2}\right|_{i-1,j,k} = \frac{|(\Phi_{i-1,j+1,k} + \Phi_{i-1,j-1,k}) - (\Phi_{i,j,k} + \Phi_{i-2,j,k})|}{(\Delta y)^2} \tag{11-36}$$

The two variants of this method are illustrated in Fig. 11–8(B), which shows how the initial data can be used. Heavy and light dashed lines represent the options. For the options given as light dashed lines, information at two stations, $(i-1)$ and $(i-2)$, is required, which means that some special starting procedure is needed at the very first station. Stability for 3D schemes is described in terms of the parameter

$$\gamma \equiv \frac{h_1 w \Delta x_1}{h_3 u \Delta x_3} \tag{11-37}$$

and these schemes are rated as stable for $-1 \leq \gamma \leq 1$ with the Der and Raetz scheme and for $0 \leq \gamma \leq \infty$ with the Pierce and East scheme. Thus, the Der and Raetz scheme can be used for reverse crossflow. The only unknown in these schemes is at (i, j, k), so an explicit relation for that unknown in terms of known quantities at that and other points can be written.

There are several implicit methods in use for 3D boundary layer calculations that are related to the *simple implicit* and *Crank-Nicolson* schemes in Chapter 2. As in the 2D case, these schemes result in equations for unknowns at (i, j, k), $(i, j+1, k)$, and $(i, j-1, k)$ in terms of knowns at those points and others. However, the implicit 3D schemes are not all unconditionally stable, as was the case for 2D flows. The stability limits, where available, are shown along with schematics of the various schemes in Fig. 11–8(C). The Krause (1968) *zigzag* scheme requires a special treatment of the x_3-derivative:

$$\left.\frac{\partial \Phi}{\partial x_3}\right|_{i-1/2,j,k} = \frac{\left|(\Phi_{j-1,j,k+1} - \Phi_{i-1,j,k}) + (\Phi_{i,j,k} - \Phi_{i,j,k-1})\right|}{2\Delta x_3} \tag{11-38}$$

Only the schemes in the lower part of Fig. 11–8(C) are stable for reverse crossflow. The Dwyer-Sanders (1974) scheme is the only one that can be used to calculate the solution downstream from a plane of initial data. The finite-difference equations for all these methods result in a system of algebraic equations that are linearized, as in the 2D case, giving a system that can be solved with the *Thomas algorithm*. The continuity equation can again be written in finite-difference form to give an explicit relation for v as the only unknown.

One can distinguish three broad classes of 3D boundary layer flows for numerical treatment:

1. *Tip, stagnation point, or leading edge flows*. The equations are functions of y and x_3 alone. A 2D finite-difference scheme is needed, and the solution is started where the crossflow is zero.
2. *Plane-of-symmetry flows*. The equations are functions of y and x_1 alone. A 2D finite-difference scheme is needed, and the solution is started at the tip, stagnation point, or leading edge.
3. *General 3D flows*. The equations are functions of x_1, y, x_3. One of the 3D finite-difference schemes is needed.

Finally, just as in the 2D case, it has proven convenient to introduce transformations that approximately scale the computational region with the growth of the boundary layer. Numerous suggestions have been made. To illustrate, we pick one suggestion from Dwyer and McCroskey (1971) for constant-density, constant-property flow and Cartesian coordinates:

$$\xi = x_1$$
$$\eta = y\sqrt{\frac{U_e}{2\nu\xi}} \tag{11-39}$$
$$\omega = x_3$$

For the dependent variables, they chose

$$F = \frac{u}{U_e}$$
$$G = \frac{w}{U_e} \tag{11-40}$$
$$V = v\sqrt{\frac{x}{2\nu U_e}} + \frac{1}{2}F(\beta_\xi - 1) + \frac{1}{2}G\eta\beta_\omega$$

where

$$\beta_\xi = \frac{\xi}{U_e}\frac{\partial U_e}{\partial \xi}$$
$$\beta_\omega = \frac{\xi}{U_e}\frac{\partial U_e}{\partial \omega} \tag{11-41}$$

These choices are closely related to the variables in Section 4–7–4 for 2D flows. Other suggestions that include the effects of compressibility and nondevelopable surfaces can be found in Blottner (1975).

11–3 TRANSITION TO TURBULENT FLOW

There is much less data for transition in 3D boundary layer flows than for the corresponding 2D cases. Combined with the fact that the parameter space for 3D cases is much larger than for 2D cases, this leads to the result that the data base for transition in 3D flows is exceedingly sparse.

One can make some general observations about transition in 3D versus 2D boundary layer flows. In the 2D case, we saw in Chapter 6 that velocity profiles with an inflection point were much less stable and thus were much more prone to transition. In a 3D boundary layer, there is necessarily at least one inflection point in the crossflow velocity profile, since the crossflow velocity must, even in the simplest cases, vary from zero at the wall up to a maximum absolute value in the layer and then decay to zero at the boundary layer edge. This leads to the growth of longitudinal vortices with their axes near the point of inflection, and the most noticeable are those that are aligned with axes that have profiles with zero net crossflow velocity at the point of inflection. This behavior is shown as the circles in Fig. 11–9. Such longitudinal vortices are prominent features in visualizations of 3D transition in the flow over swept wings, for example. Computational stability studies have shown that this kind of instability overwhelms the Tollmein-Schlichting mode examined in Chapter 6 for 2D flows. Also, attempts to extend the e^N method to 3D flows have indicated that one needs $N \approx 40$ in 3D flows, compared with $N \approx 10$ in 2D flows (see Section 6–4). Owen and Randall (1952) suggested transition for $w_{max}\delta_w/\nu > 200$, where δ_w is y at $w/w_{max} = 0.10$. Pate (1980) proposed the same criterion up to Mach 8, but more recent work indicates an increase in the crossflow Reynolds number for transition as the speed increases.

For strongly swept wings, the major mechanism for transition is *spanwise contamination,* a condition wherein flow along the swept leading edge transports and propagates turbulence from the fuselage boundary layer and the wing/fuselage junction out along the span of the wing, resulting in rapid transition on the wing. This mechanism is much stronger than the other mechanisms of transition, which become

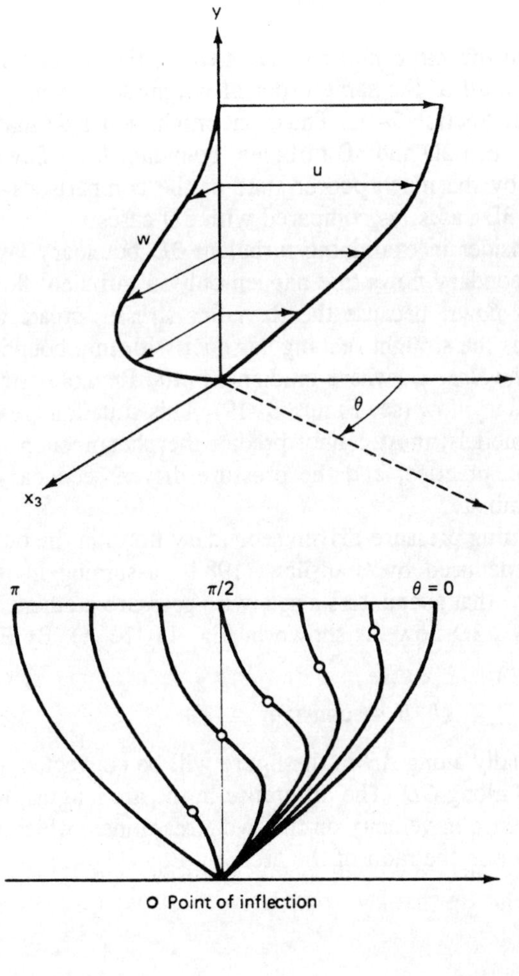

Figure 11–9 Effect of axis orientation on profiles in 3D boundary layers.

irrelevant in relation to it. Bradshaw (1987) reports a critical value of $Re_\theta \geq 100$ as a simple criterion for spanwise contamination in the leading edge boundary layer to occur. The spanwise spread of transition can also occur from streamwise corners, longitudinal vortices, or other localized disturbances. This was mentioned in Chapter 6. The angle of the spread is about the same as that from turbulent spots, i.e., about 11° in low-speed flows.

Supersonic flows are strongly unstable on swept wings. Scott-Wilson and Capps (1955) suggest that compressibility has a destabilizing effect on all 3D flows.

Suction has been found to stabilize a laminar 3D flow perhaps more strongly than in 2D flows. This is because the suction reduces the magnitude of the crossflow velocity, as well as the boundary layer thickness.

11–4 TURBULENT THREE-DIMENSIONAL BOUNDARY LAYERS

11–4–1 Introduction

Recall that *turbulence is always three-dimensional*; that is, the fluctuating parts of the velocity components are all of the same order of magnitude, even if the mean flow is two dimensional (see Section 7–4). Thus, one might conclude that there are no essential differences between 2D and 3D turbulent boundary layer flows. But that is incorrect, as evidenced by the much poorer state of the comparisons of predictions with experiments for 3D cases, as compared with 2D cases.

The first thing to consider in examining turbulent 3D boundary layers is the occurrence of a class of secondary flows that happen only in turbulent flows. These are *shear-driven secondary flows,* because the Reynolds stresses create them. The simplest practical example is the straight rectangular duct with thin boundary layers (i.e., the flow is not *fully developed*), where gradients in the Reynolds stresses near the corners drive the secondary flow (see Figure 8–19). This situation creates a very severe test for turbulence models; most cannot predict the phenomenon. However, ducts are seldom straight in practice, and the pressure-driven secondary flows in curved ducts usually predominate.

A useful result concerning pressure-driven secondary flows in the outer part of a 3D boundary layer was deduced by Bradshaw (1987), assuming inviscid flow. Consider an initially 2D flow that encounters a spanwise pressure gradient leading to a turn to the right and thus a 3D flow, as shown in Fig. 11–10(A). By Bernoulli's equation, the velocity field is

$$U \cdot r = \text{constant} \qquad (11\text{–}42)$$

Thus, a fluid element originally along AB in the figure will be convected and turned after a time δt to be located along CD. The difference in the arc lengths AC and BD is simply δt times the difference in velocity on the two streamlines, which is proportional to minus the difference in the radii of the arcs, by Eq. (11–42). So the angle

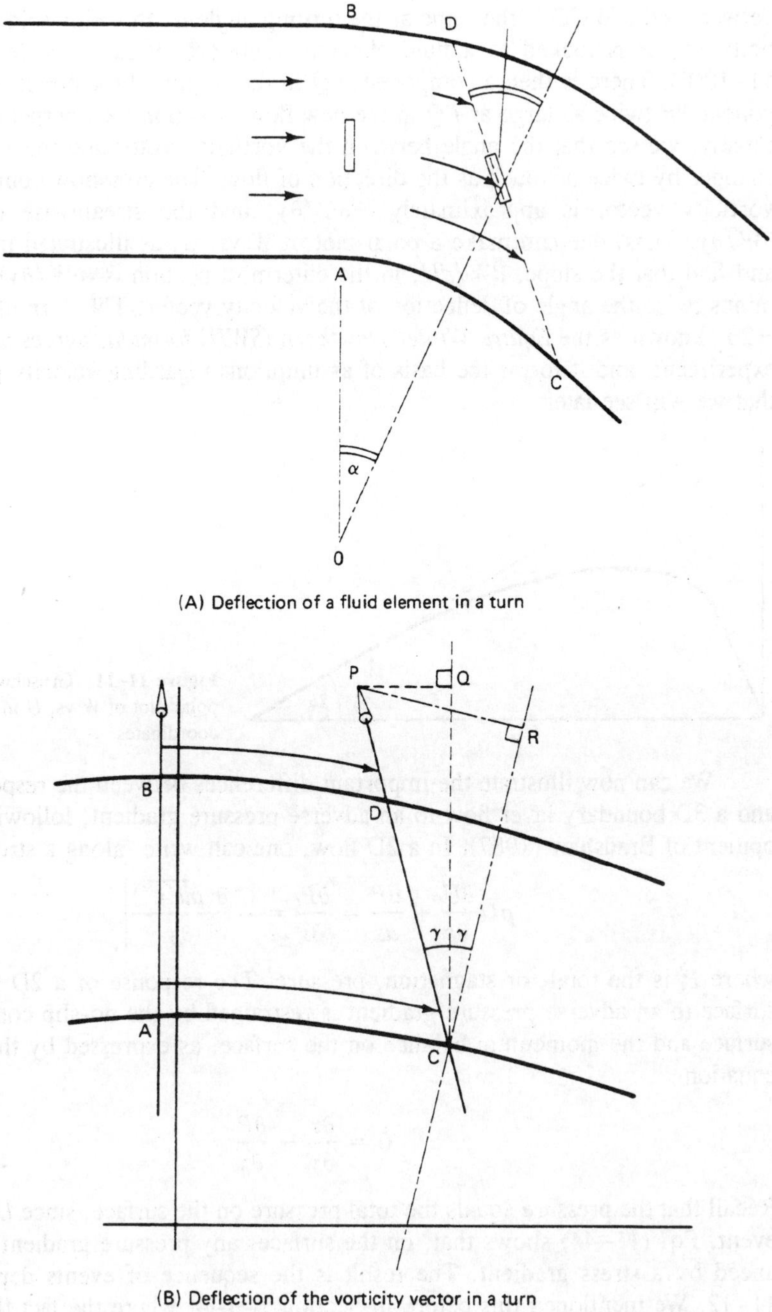

Figure 11–10 Illustration of the Squire-Winter-Hawthorn secondary flow equation. (From Bradshaw, 1987.)

between *AB* and *CD* is the same as the turning angle α. In an inviscid flow, the vorticity vector is locked to a fluid element; therefore, it turns as depicted in Fig. 11–10(B). There is then a component *PQ* in the original flow direction and a component *PR* twice as large as *PQ* in the new flow direction i.e., perpendicular to *CR*. Finally, we see that the angle between the vorticity vector and the velocity vector changes by twice as much as the direction of flow. The crossflow component of the vorticity vector is approximately $-\partial U/\partial y$, and the streamwise component is $\partial W/\partial y$. Thus, one can make a polar plot of W vs. U, as illustrated in Fig. 11–11, and find that the slope, dW/dU, in the outermost portion is $(\partial W/\partial y)/(\partial U/\partial y)$, or minus twice the angle of deflection of the velocity vector. This formula, $dW/dU = -2\alpha$, known as the *Squire-Winter-Hawthorn (SWH) formula,* agrees very well with experiment, and it forms the basis of assumptions regarding velocity profile shapes that we will see later.

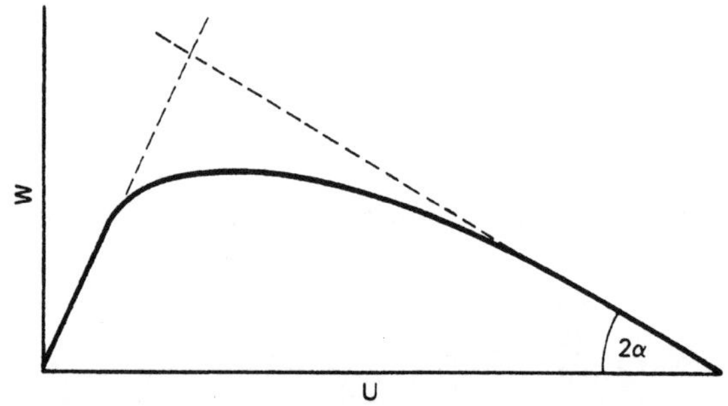

Figure 11–11 Gruschwitz-Johnston polar plot of W vs. U in streamline coordinates.

We can now illustrate the important differences between the responses of a 2D and a 3D boundary layer flow to an adverse pressure gradient, following the development of Bradshaw (1987). In a 2D flow, one can write, along a streamline,

$$\rho U \frac{\partial U}{\partial s} + \frac{dP}{ds} = \frac{\partial P_t}{\partial s} = -\frac{\partial(\rho \overline{u'v'})}{\partial y}\bigg|_s \tag{11–43}$$

where P_t is the total, or stagnation, pressure. The response of a 2D flow near the surface to an adverse pressure gradient is restrained by the no-slip condition on the surface and the momentum balance on the surface, as expressed by the momentum equation

$$0 = \frac{\partial \tau}{\partial y} - \frac{\partial P}{\partial x} \tag{11–44}$$

Recall that the pressure equals the total pressure on the surface, since $U = 0$. In any event, Eq. (11–44) shows that, on the surface, any pressure gradient must be balanced by a stress gradient. The result is the sequence of events depicted in Fig. 11–12. We mentioned this before in Section 7–3–4, where the fact that the maximum stress moves off the wall in a strong adverse pressure gradient was noted. The effects of an adverse pressure gradient on the total pressure in a 3D flow are qualitatively the same in a 2D flow, but the effects on the turbulent stresses are quite differ-

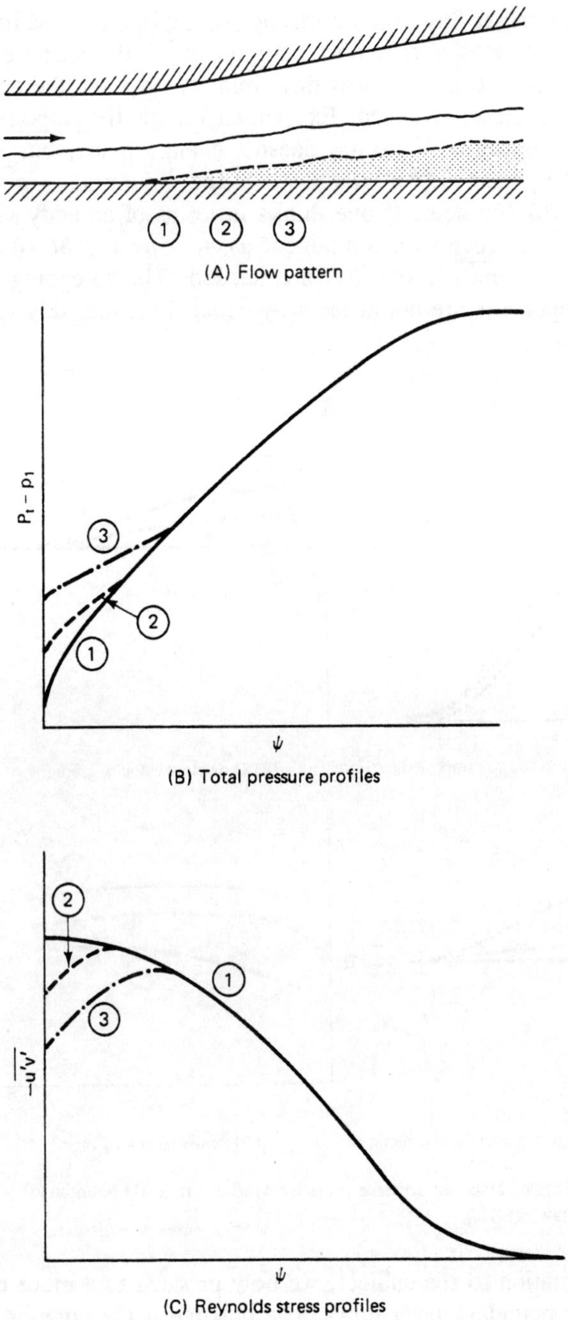

Figure 11–12 Effects of a strong adverse pressure gradient in a 2D boundary layer. (From Bradshaw, 1987.)

ent in the two cases. In a laminar flow, the occurrence of a crossflow and its gradient $\partial W/\partial y$ immediately produces a crossflow component of the shear equal to $\mu(\partial W/\partial y)$. In a turbulent flow, the mean crossflow velocity field and the crossflow turbulent stresses are not so closely coupled. Experiments show the response illustrated in Fig. 11–13. At first, the U profile doesn't change much, but $-\rho\overline{u'v'}$ changes significantly in the presence of a crossflow. At the same time, $-\rho\overline{w'v'}$ increases more slowly than $\partial W/\partial y$ does. If one thinks in terms of an eddy viscosity concept in 3D, one sees that a streamwise component $\mu_{Tx} = -\rho\overline{u'v'}/(\partial U/\partial y)$ and a crossflow component $\mu_{Tz} = -\rho\overline{w'v'}/(\partial W/\partial y)$ are needed. The foregoing discussion shows that these components are not necessarily equal, i.e., the eddy viscosity is not, in general, *isotropic*.

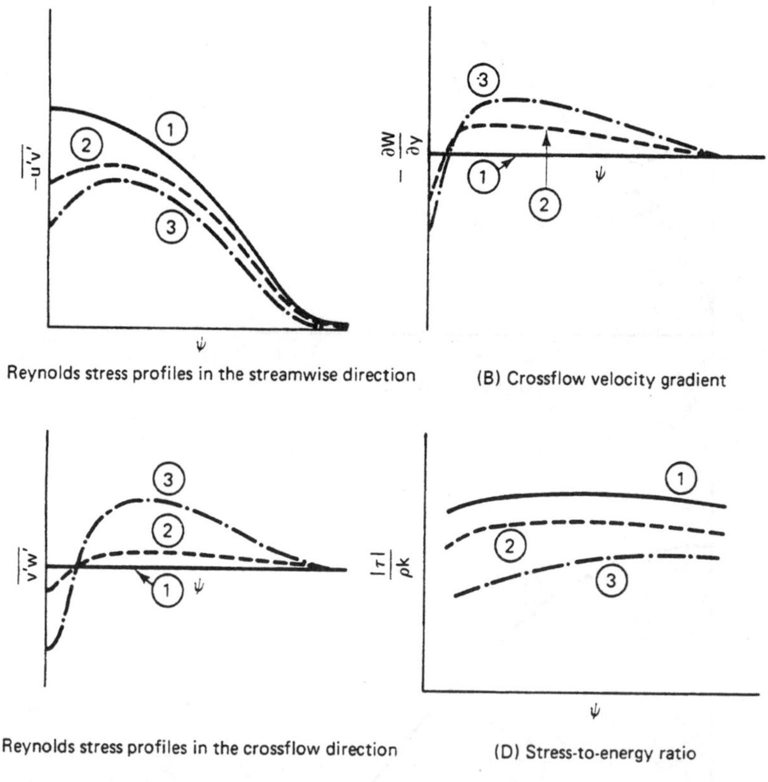

Figure 11–13 Effects of a strong adverse pressure gradient in a 3D boundary layer. (From Bradshaw, 1987.)

With this brief orientation to the subject, we now proceed to a more general discussion of 3D turbulent boundary layer flows. The practice is the same as earlier in this book. First, representative mean flow data are presented. This is followed by a presentation of turbulence data. Next, we have a description of turbulence models for 3D flows. Finally, comparisons of predictions with experiments are given, including both integral and differential methods. A great portion of the experimental data and computational results available are for the case of the so-called *infinite swept wing* (see Section 11–2–2), so much of the discussion is restricted to that special case.

11–4–2 Mean Flow Observations

A number of researchers have studied the infinite swept wing in low-speed flow by simulating the situation in a wind tunnel with curved walls. Data from a set of such experiments by Bradshaw and Pontikos (1985) will be used here and in the next section as representative of this kind of flow and of 3D boundary layer flows in general.

Mean velocity profiles in terms of the magnitude of the total velocity, $Q = (U^2 + V^2)^{1/2}$, regardless of flow angle, across the boundary layer are presented in Fig. 11–14. There is nothing noteworthy about such profiles plotted in this way; they look much like 2D profiles proceeding into an adverse pressure gradient. The

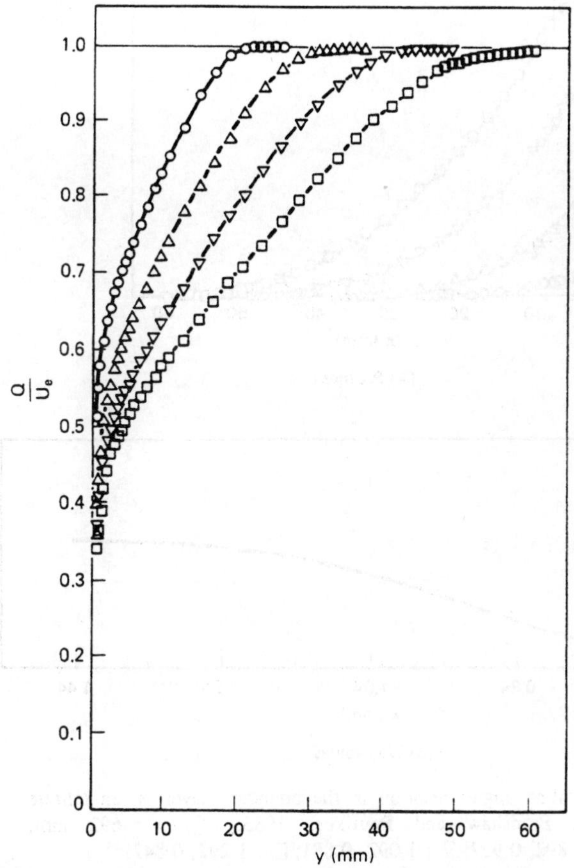

Figure 11–14 Mean resultant velocity profiles in the boundary layer on an infinite swept wing. (From Bradshaw and Pontikos, 1985.) ○ : x = 692 mm, $U_e/U_{e,\text{ref}}$ = 1.00; △ : 892, 0.932; ▽ : 1,092, 0.881; □ : 1,292, 0.847.

variation in the flow angle across the layer is given in Fig. 11–15. In this case, the maximum difference between the direction of the edge velocity vector and that on the surface (actually the wall shear vector) is about 15°. A polar plot such as that sketched in Fig. 11–11 for these data is shown in Fig. 11–16. Such plots have been suggested by Gruschwitz (1935) and Johnston (1960). The straight lines in the outer

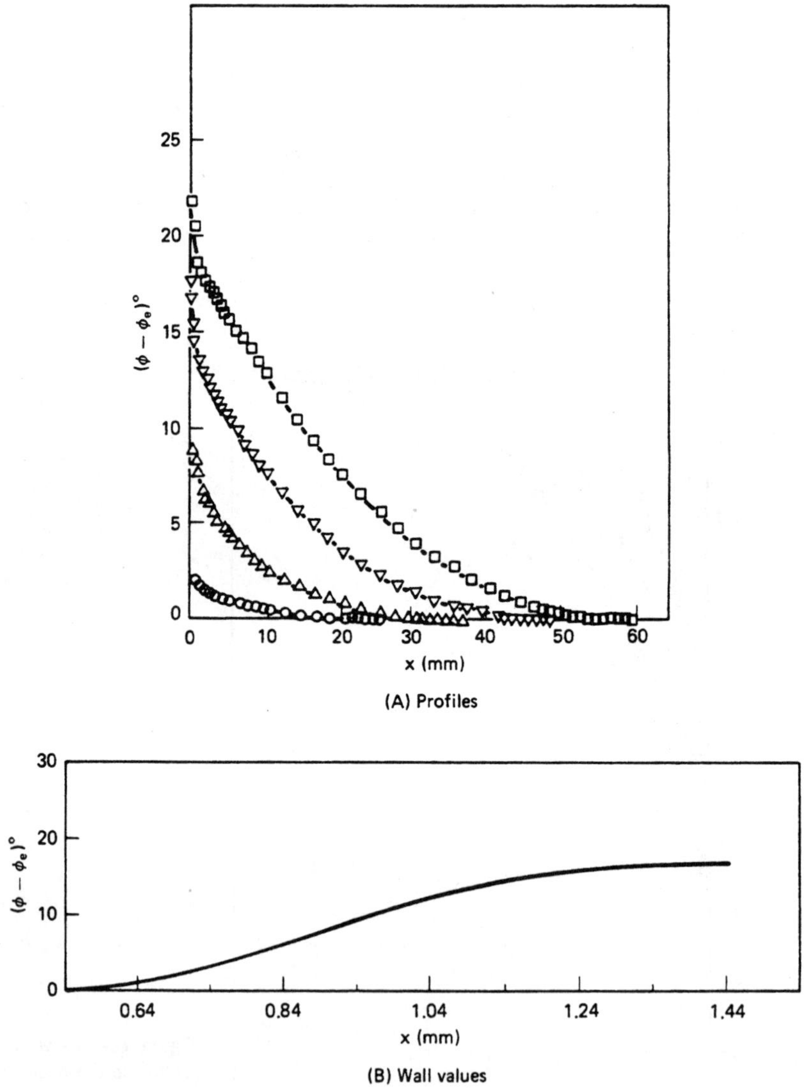

Figure 11–15 Streamline angle variation in the boundary layer on an infinite swept wing. (From Bradshaw and Pontikos, 1985.) ○ : x = 692 mm, $U_e/U_{e,\text{ref}}$ = 1.00; △ : 892, 0.932; ▽ : 1,092, 0.881; □ : 1,292, 0.847.

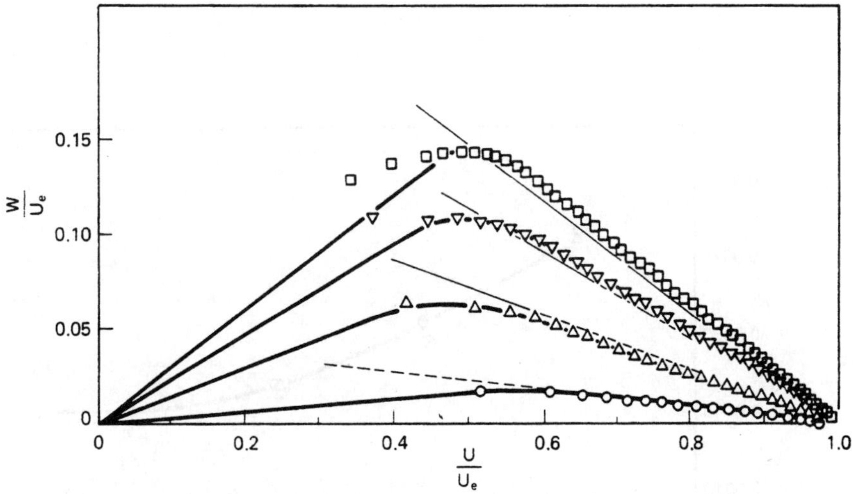

Figure 11–16 Polar plot of mean velocity profiles in the boundary layer on an infinite swept wing. (From Bradshaw and Pontikos, 1985.) ○ : x = 692 mm, $U_e/U_{e,\text{ref}}$ = 1.00; △ : 892, 0.932; ▽ : 1,092, 0.881; □ : 1,292, 0.847.

part of the layer are predictions of the SWH formula, and they represent the data quite well. The skin friction data are given in Fig. 11–17.

The matter of correlating the preceding and other 3D mean flow data has been considered by Ölcmen and Simpson (1989) and others in some depth. Ölcmen and Simpson conclude, based on comparisons of eight sets of data with nine different suggestions for correlations in the near-wall region, that the correlation model of Johnston (1960) provides the best results overall. Johnston assumed that a fictitious velocity component defined as $U/\cos(\beta_w)$ obeyed the 2D law of the wall of Clauser, giving, for 3D flow,

$$\frac{U}{q^* \cos(\beta_w)} = \frac{1}{\kappa}\ln\left(\frac{yq^*}{\nu}\right) + C \tag{11–45}$$

Here, β_w is the difference between the directions of the velocity vector at the boundary layer edge and the velocity vector on the wall (actually the direction of the skin friction), and q^* is a friction velocity formed with the total wall shear. The performance of this correlation scheme is shown in Fig. 11–18. Himeno and Tanaka (1975) have suggested an extended *law of the wake* (see Section 7–3–1) for 3D flows, viz.,

$$Q = u_* g\left(\frac{yu_*}{\nu}\right) + U_e \cdot W\left(\frac{y}{\delta}\right) \tag{11–46}$$

In this relation, $g(\,)$ is a function to express the law of the wall, and $W(\,)$ is a function to express the extended wake function. There have been a number of proposals for representing the W velocity profile, and Ölcmen and Simpson (1990) found that

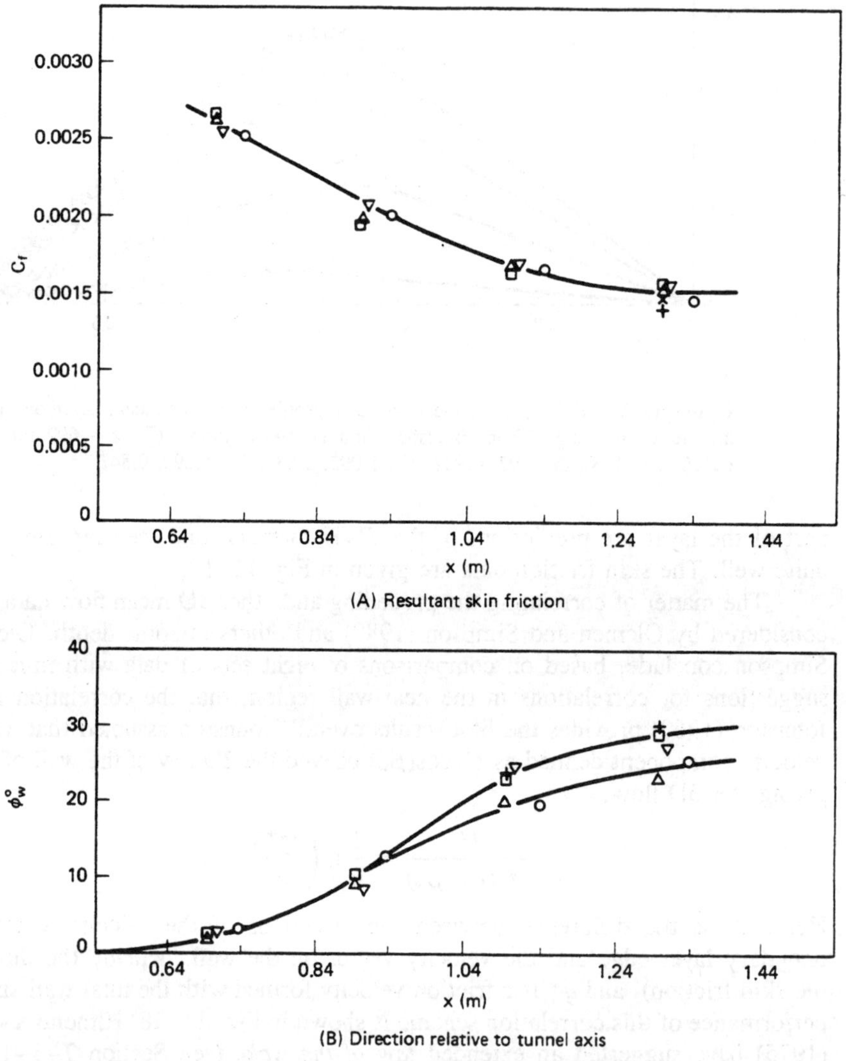

Figure 11–17 Skin friction on an infinite swept wing. (From Bradshaw and Pontikos, 1985.)

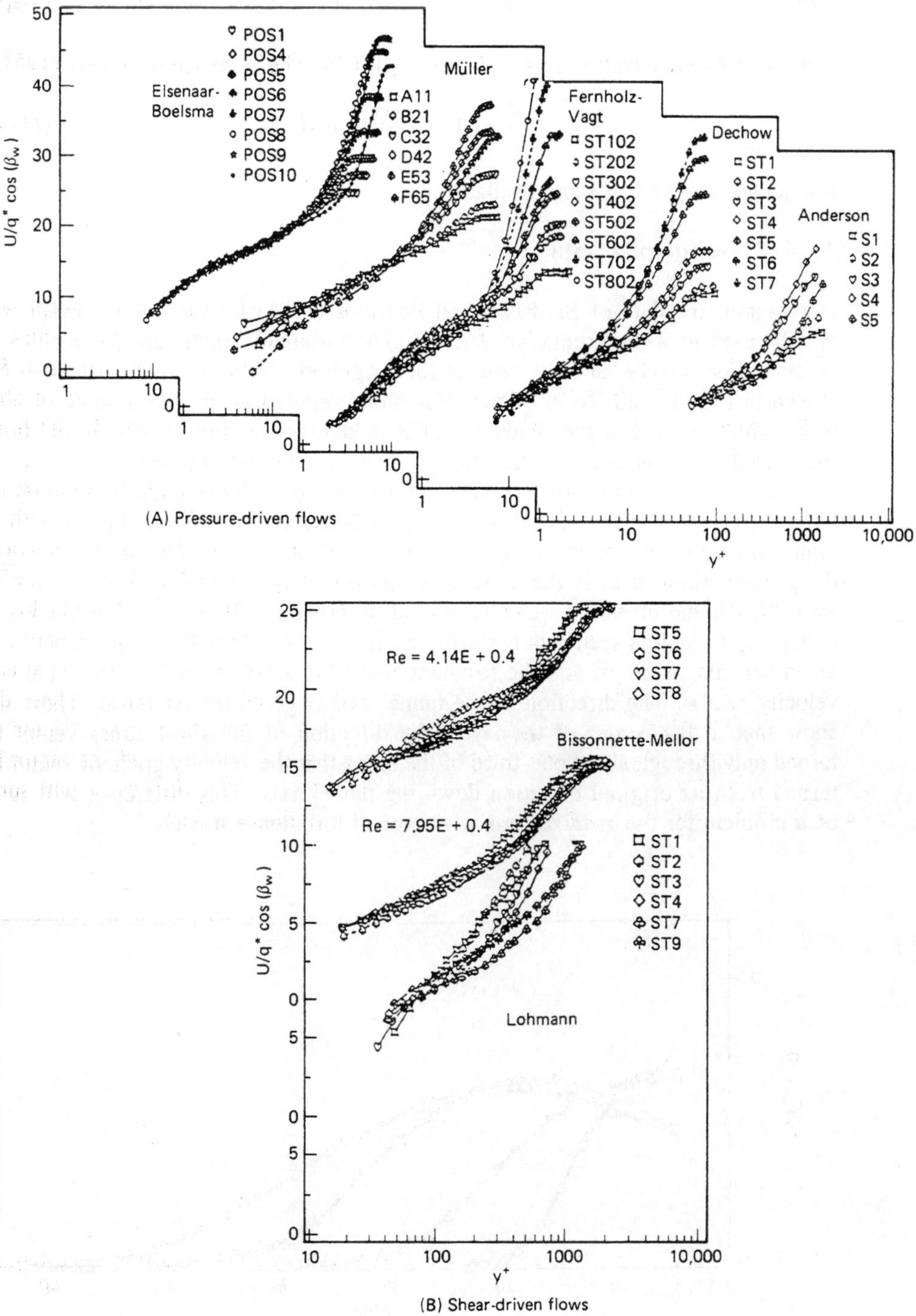

Figure 11–18 Three-dimensional boundary layer velocity profiles in the law-of-the-wall coordinates of Johnston. (From Olcmen and Simpson, 1989.)

none worked very well over a wide range of data. The proposal of Mager (1951),

$$\frac{W}{U} = \left(1 - \frac{y}{\delta}\right)^2 \tan(\beta_w) \tag{11-47}$$

was judged the best of those available.

11-4-3 Turbulence Data

Here again, the data of Bradshaw and Pontikos (1985) for the infinite swept wing will be used as a representative data set. The turbulent kinetic energy profiles are given in Fig. 11–19, and the ratio of the magnitude of the Reynolds stress to K is shown in Fig. 11–20. In 2D flows, this ratio, denoted as a_1, has a value of about 0.30, which is used in the Bradshaw TKE model (see Section 7–10). In 3D flows, the ratio drops rapidly as the crossflow velocity component increases.

In a 3D flow, the choice of axes is important in displaying information on turbulent stress. In Fig. 11–21, Reynolds stress data referred to axes aligned with the tunnel axis, and not the local edge velocity vector, are given. The most noteworthy thing about these data is the slow development of the crossflow stress, $-\rho\overline{w'v'}$, even though the crossflow velocity W and its gradient $\partial W/\partial y$ develop quickly. A better way to look at such data is shown in Fig. 11–22, where the mean velocity, the mean velocity gradient, and the resultant shear stress are referred to the local edge velocity vector. The direction of the tunnel axis is given for reference. These data show that in the center of the layer, the direction of the shear stress vector has turned only through about one-third of the angle that the velocity gradient vector has turned from its original direction down the tunnel axis. This difference will surely be a problem for the *gradient transport* type of turbulence models.

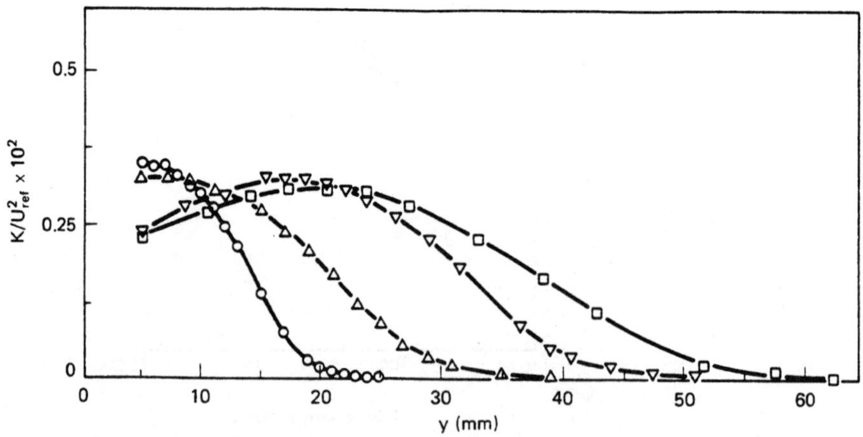

Figure 11–19 Turbulent kinetic energy profiles in the boundary layer on an infinite swept wing. (From Bradshaw and Pontikos, 1985.) ○ : x = 692 mm, $U_e/U_{e,\text{ref}}$ = 1.00; △ : 892, 0.932; ▽ : 1,092, 0.881; □ : 1,292, 0.847.

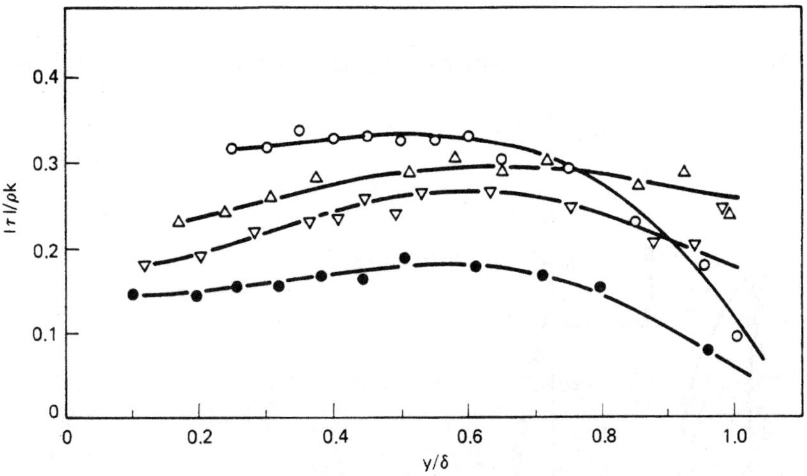

Figure 11–20 Ratio of resultant Reynolds stress to K in the boundary layer on an infinite swept wing. (From Bradshaw and Pontikos, 1985.) ○ : x = 692 mm, $U_e/U_{e,\text{ref}}$ = 1.00; △ : 892, 0.932; ▽ : 1,092, 0.881; ● : 1,292, 0.847.

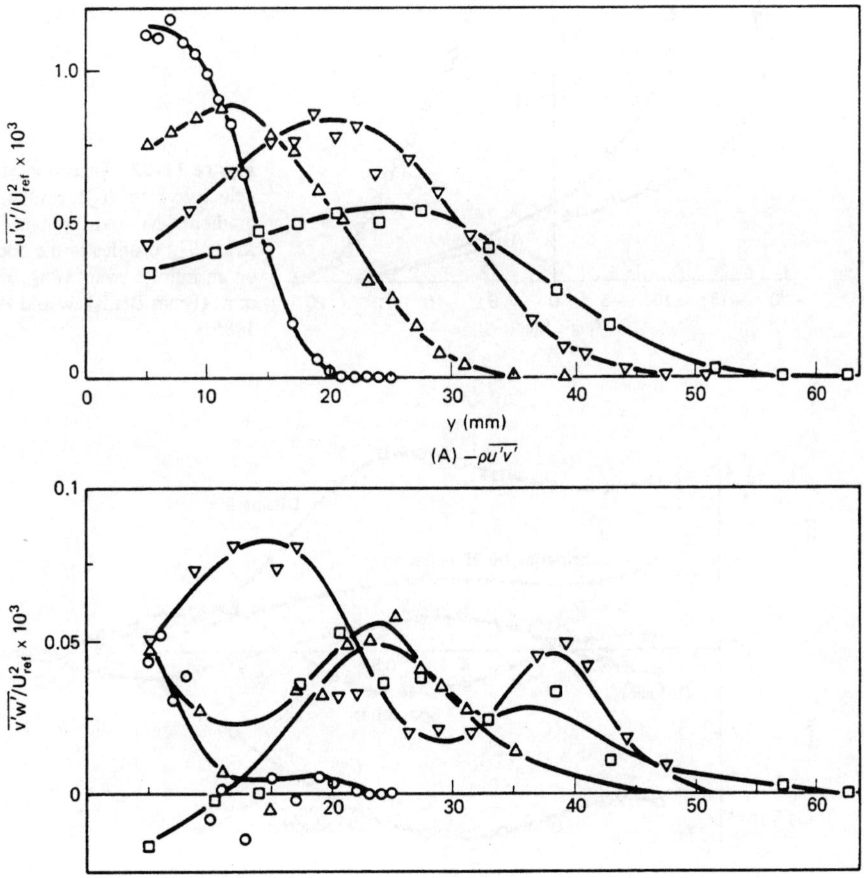

Figure 11–21 Reynolds stress profiles (in coordinates aligned with tunnel axis) in the boundary layer on an infinite swept wing. (From Bradshaw and Pontikos, 1985.) ○ : x = 692 mm, $U_e/U_{e,\text{ref}}$ = 1.00; △ : 892, 0.932; ▽ : 1,092, 0.881; □ : 1,292, 0.847.

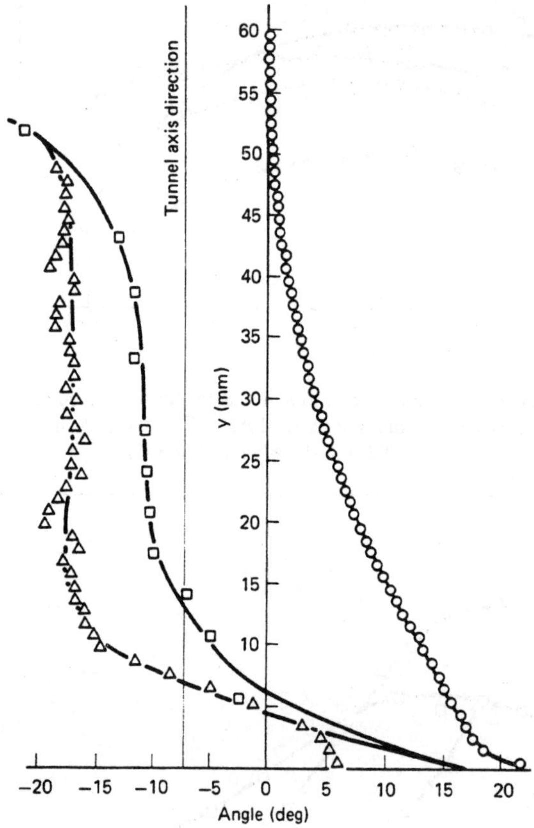

Figure 11–22 Direction of resultant velocity vector (○), resultant velocity gradient (△), and resultant Reynolds stress (□) profiles in the boundary layer on an infinite swept wing, x = 1,292 mm. (From Bradshaw and Pontikos, 1985.)

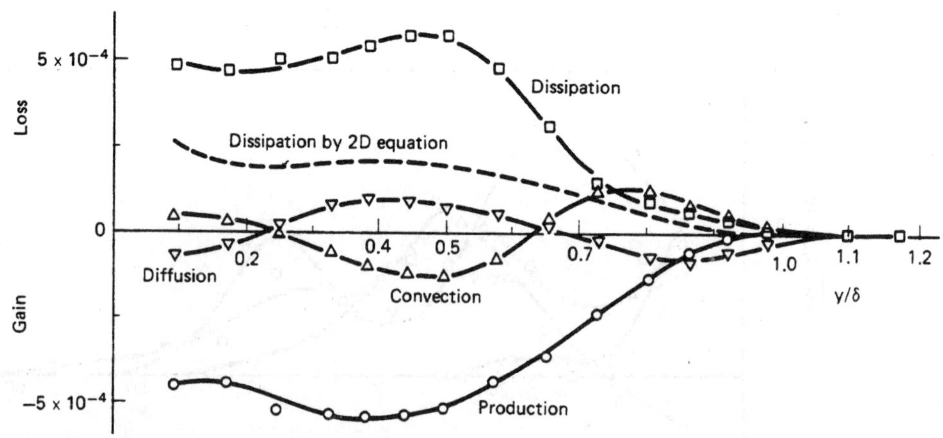

Figure 11–23 Turbulent kinetic energy balance in the boundary layer on an infinite swept wing, x = 1,292 mm. (From Bradshaw and Pontikos, 1985.)

A *TKE balance* is shown in Fig. 11–23. The general appearance of this balance is similar to that for 2D flows in an adverse pressure gradient, except that the viscous dissipation is larger than would be expected based on the 2D equation

$$\epsilon \approx \frac{(|\tau|/\rho)^{3/2}}{\delta/10} \tag{11-48}$$

Finally, the response of the intermittency to the developing 3D character of the flow is plotted in Fig. 11–24. It is clear that the effect is small and in the direction of an increase in intermittency in the outermost part of the layer.

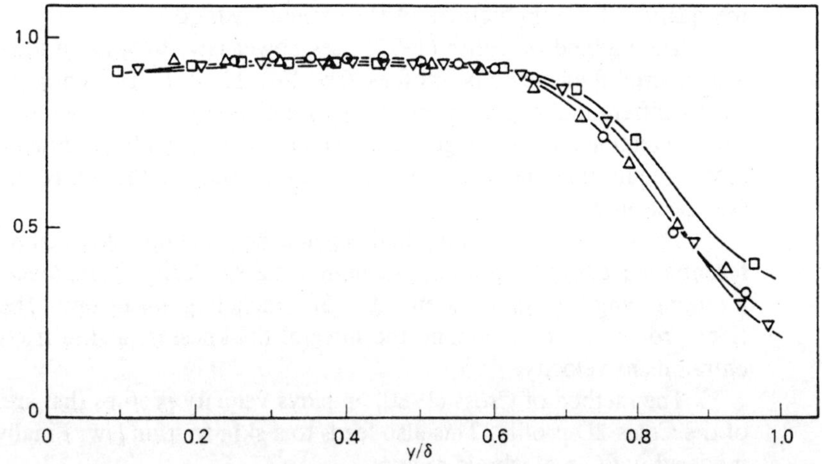

Figure 11–24 Intermittency profiles in the boundary layer on an infinite swept wing. (From Bradshaw and Pontikos, 1985.) ○ : x = 692 mm, $U_e/U_{e,\text{ref}}$ = 1.00; △ : 892, 0.932; ▽ : 1,092, 0.881; □ : 1,292, 0.847.

In summing up their turbulence results, Bradshaw and Pontikos (1985) conclude that turbulent activity decreases as three-dimensionality increases, leading to reductions in turbulent transport of momentum, turbulent energy, and turbulent shear stress across the layer. The hypothesis is that the large eddies in the initially 2D flow are "tilted" sideways by the crossflow velocity gradient, and this tilting reduces their efficiency as transporters of those three quantities in the transverse direction.

11-4-4 Integral Analyses

Integral analyses have remained competitive for turbulent 3D flows up until the present, while for 2D flows they are now little used. The situation for turbulent 2D cases is a little hard to understand, since modern integral methods perform about as well as the best differential methods in 2D at a fraction of the cost. The reason for the continued emphasis on integral methods in turbulent 3D cases may well be that the cost of differential methods for these cases remains high.

In a Eurovisc Workshop in 1982, calculations were compared with experiment for a number of test cases, and several integral methods were considered. The results are discussed in Van den Berg et al. (1988). Recently, the integral method of Moses (see Section 7–7) has been extended to 3D flows by Caille and Schetz (1992). Two of the test cases studied at the Eurovisc Workshop are experiments from Van den Berg and Elsenaar (1972) and Müller and Krause (1979). The first of these is a close simulation of an *infinite swept wing*, so the flow does not vary in the spanwise direction. That should be a comparatively easy case. The second configuration is superficially similar, but the flow does vary in the spanwise direction, thus providing a more stringent test of analyses. These two test cases will be used here to illustrate the quality of the predictions of the various methods.

The method of Smith (1972) uses power law streamwise velocity profiles and Mager crossflow velocity profiles (see Eq. 11–47). The entrainment is modeled with a differential equation accounting for changes in the structure of the turbulence from divergence or convergence of the external streamlines. Finally, a skin friction law is required as input, as with all integral methods that utilize only a single strip (see Section 7–7).

The Cousteix (1974) method depends on functions obtained by solving the differential equations of motion, assuming local similarity. Turbulence is modeled with a mixing length formula extended to 3D, including anisotropy. The similarity solutions provide relations among the integral thicknesses, a skin friction law, and the entrainment velocity.

The method of Cross (1980) employs velocity profiles that are a generalization of the Coles 2D profile. This also leads to a skin friction law. Finally, entrainment is modeled with an algebraic relation.

The integral method of Caille and Schetz (1992) aims to eliminate the need for a skin friction law as input, in the same way that Moses accomplished that goal for 2D cases. Integral equations are written for two control volumes, one encompassing the whole boundary layer, as usual, and the second covering the region from the wall up through a fraction (about 35 percent) of the boundary layer thickness. With this method, one needs velocity profile shapes and a model for turbulent shear at the outer boundary of the inner strip, i.e., $y \approx 0.35\delta$. The velocity profiles are constructed using a generalization of the Moses 2D expression for the streamwise velocity,

$$\begin{aligned} \frac{U}{U_e} &= A\cos(\beta_w)\ln(C_1 A\,\mathrm{Re}_\delta \eta) - B\eta^2(3 - 2\eta) \\ A &= \frac{1}{\kappa}\sqrt{\frac{C_f}{2}} \\ B &= A\cos(\beta_w)\ln(C_1 A\,\mathrm{Re}_\delta) - 1 \end{aligned} \tag{11–49}$$

and the Johnston model for the crossflow velocity,

$$\frac{W}{U_e} = \frac{U}{U_e}\tan(\beta_w), \qquad \eta < \eta_{\mathrm{mc}}$$

$$\frac{W}{U_e} = C_{mc}\left(1 - \frac{U}{U_e}\right), \qquad \eta > \eta_{mc} \tag{11–50}$$

$$C_{mc} = \frac{\dfrac{W_{mc}}{U_e}}{\left(1 - \dfrac{U_{mc}}{U_e}\right)}$$

The constant η_{mc} denotes the location of the maximum crossflow velocity W_{mc}, and U_{mc} is the streamwise velocity at that point. In all of this, $\kappa = 0.41$, $C_1 = 3.057$, and $\eta = y/\delta$.

Unlike the other integral methods, the Caille-Schetz method (and any two-strip method) utilizes a turbulence model to represent the shear at the outer boundary of the inner strip. This is an advantage of the method, since turbulent transport is modeled at a much more fundamental level than the level of methods that require a skin friction law or an entrainment model as input. Caille and Schetz chose to use an eddy viscosity formulation, and the model used was an extension of the Clauser 2D outer-region model [see Eq. (7–85)], written as

$$\mu_{Tx} = C\rho U_e \delta_1^* \tag{11–51}$$

for the streamwise eddy viscosity with $C \approx 0.0168$. This is simply the usual Clauser 2D model written using the streamwise displacement thickness. The ratio μ_{Tx}/μ_{Tz} is generally taken as approximately unity, since there is no agreement in the literature as to a universal value. Some workers report values less than unity and others values greater than unity. Only an outer-region model is required, since the outer edge of the inner strip is in the outer region ($y/\delta \approx 0.35$).

All of the foregoing methods do a good job of predicting the skin friction in the Van den Berg and Elsenaar flow, except for the Cousteix method near the last axial station, as shown in Fig. 11–25. The crossflow angle on the wall is predicted less

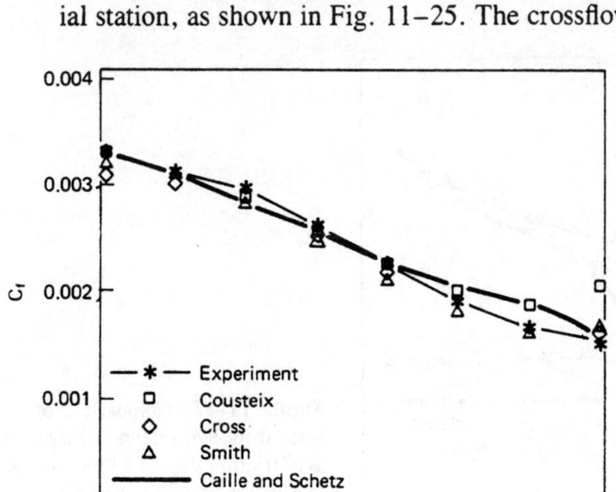

Figure 11–25 Comparison of predictions from several integral methods with the skin friction data of Van den Berg and Elsenaar (1972) for flow over an infinite swept wing. (From Caille and Schetz, 1992.)

well, as demonstrated in Fig. 11–26. The Cross method appears to do the best job, but it also has the most empirical information for input.

The Müller and Krause flow is more complex, and only Cross submitted predictions for the Eurovisc Workshop. Caille and Schetz have also treated that case. Some comparisons of predictions with data are given in Figs. 11–27 and 11–28 for skin friction and wall streamline angle. Generally good agreement can be seen, with Caille and Schetz's results being perhaps a little better than Cross's.

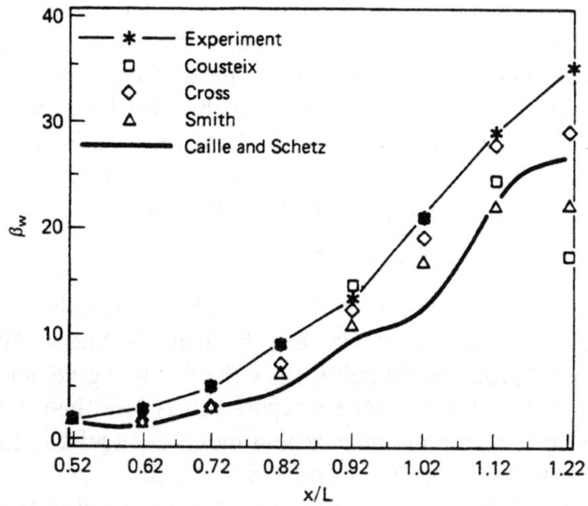

Figure 11–26 Comparison of predictions from several integral methods with data on the wall crossflow angle from Van den Berg and Elsenaar (1972) for flow over an infinite swept wing. (From Caille and Schetz, 1992.)

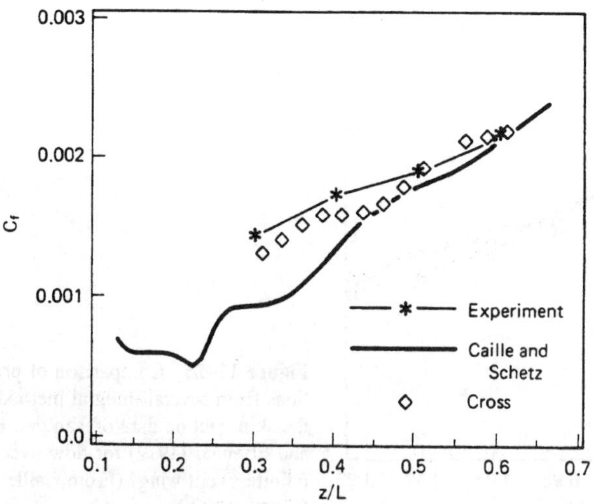

Figure 11–27 Comparison of predictions from two integral methods with the skin friction data of Müller and Krause (1979) for flow over a swept wing at $x/L = 0.6$. (From Caille and Schetz, 1992.)

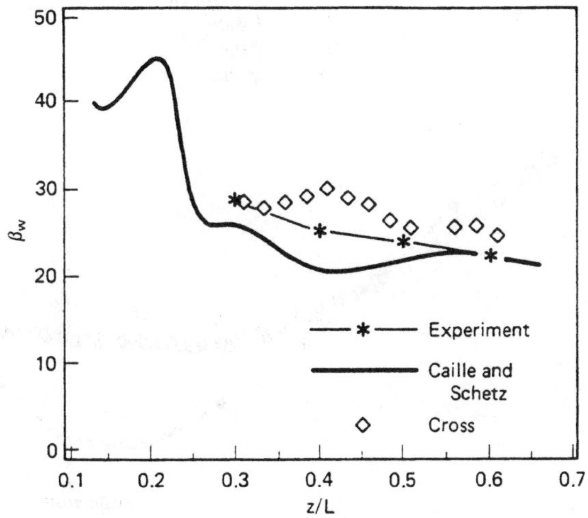

Figure 11–28 Comparison of predictions from two integral methods with data on the wall crossflow angle from Müller and Krause (1979) for flow over a swept wing at $x/L = 0.6$. (From Caille and Schetz, 1992.)

11–4–5 Differential Methods

As with differential methods for other turbulent flow problems, it is logical here to divide the methods by the level of the turbulence model employed, starting with mean flow models. Cebeci (1974) extended the 2D Van Driest–Clauser eddy viscosity model to 3D, using

$$\mu_{Ti} = \rho \ell_m^2 \left[\left(\frac{\partial U}{\partial y} \right)^2 + \left(\frac{\partial W}{\partial y} \right)^2 \right]^{1/2}$$
$$\mu_{To} = 0.0168\rho \left| \int_0^\infty [(U_e^2 + W_e^2)^{1/2} - (U^2 + W^2)^{1/2}]\, dy \right| \tag{11–52}$$

with ℓ_m from the Van Driest form [see Eq. (7–61)] using q^* instead of u^* and A^+ modified for pressure gradients, as in his 2D approach [see Eq. (7–93)]. There is no distinction between μ_{Tx} and μ_{Tz}. Chang and Patel (1975) use the same turbulence model with a different numerical method. Müller (1982) uses a mixing length model for the whole layer, following the suggestion of Michel et al. (1968), with a Van Driest damping function where the pressure gradient correction of Cebeci is employed on A^+.

These three mean flow methods were used for some of the cases in the Eurovisc Workshop. Results from there and elsewhere show the performance of such methods. Figures 11–29 and 11–30 show comparisons of predictions with experiment for the case of the infinite swept wing of Van den Berg and Elsenaar (1972). Similar results for the more complex Müller and Krause (1979) flow are shown in Fig. 11–31 and 11–32. These comparisons are disappointing, since one certainly cannot see any clear advantage these approaches have over the better integral methods (see Figs. 11–25 through 11–28). Recall, however, that much the same thing happened for 2D flows, as illustrated in Sections 7–7 and 7–9.

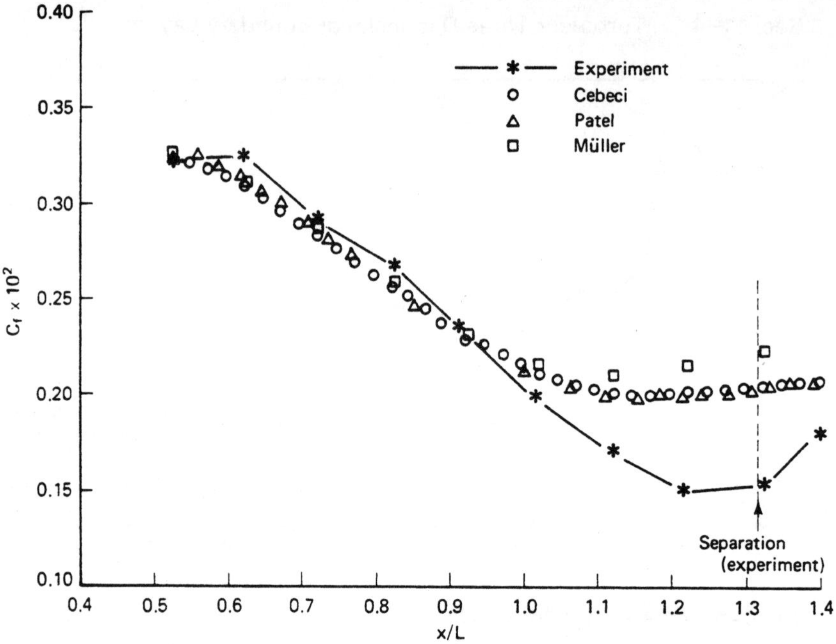

Figure 11–29 Comparison of predictions from three differential methods with the skin friction data of Van den Berg and Elsenaar (1972) for flow over an infinite swept wing. (From Van den Berg et al., 1988.)

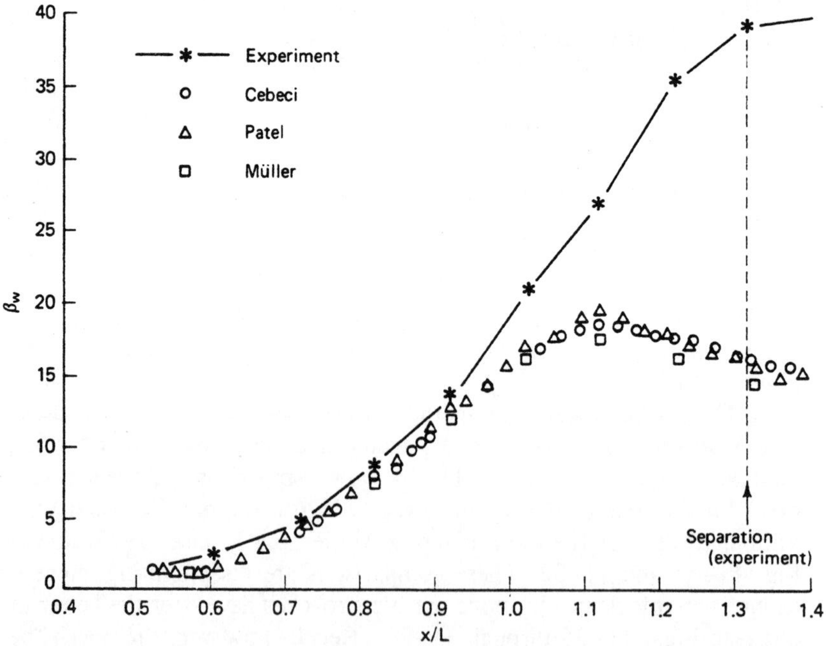

Figure 11–30 Comparison of predictions from three differential methods with data on the wall crossflow angle from Van den Berg and Elsenaar (1972) for flow over an infinite swept wing. (From Van den Berg et al., 1988.)

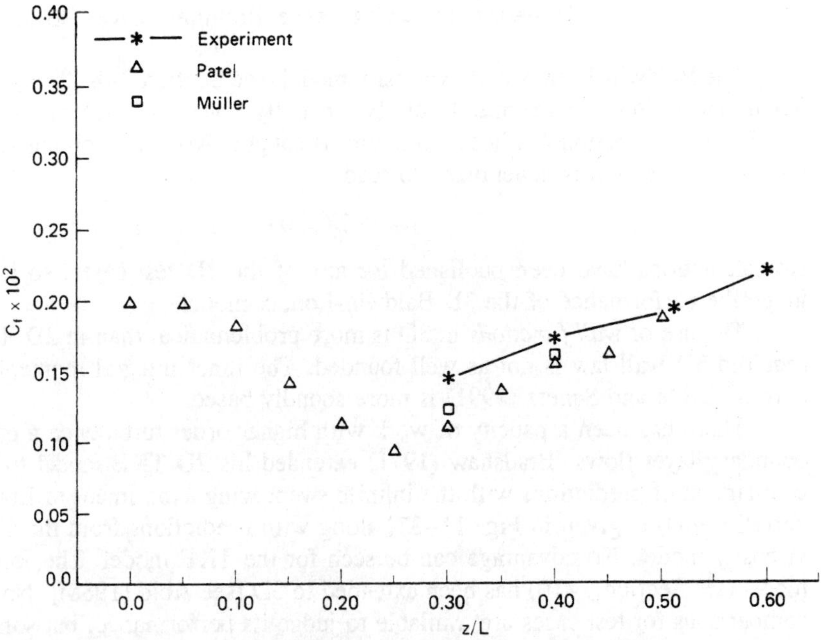

Figure 11–31 Comparison of predictions from two differential methods with the skin friction data of Müller and Krause (1979) for flow over a swept wing at $x/L = 0.6$. (From Van den Berg et al., 1988.)

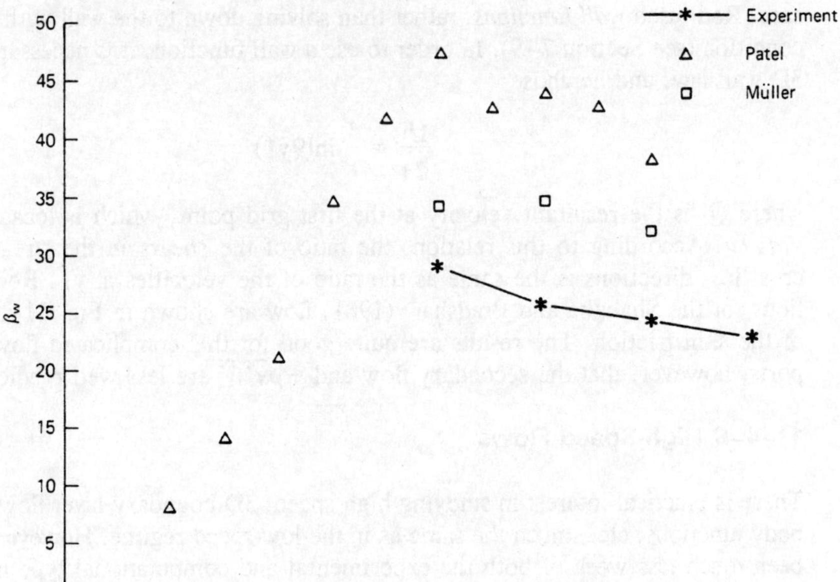

Figure 11–32 Comparison of predictions from two differential methods with data on the wall crossflow angle from Müller and Krause (1979) for flow over a swept wing at $x/L = 0.6$. (From Van den Berg et al., 1988.)

The Baldwin-Lomax eddy viscosity model (see Section 7–8–2) is applicable in 3D if one writes the magnitude of the vorticity, $|\omega|$, in place of the derivative $\partial U/\partial y$ in the function F whose maximum is sought. Also, the mixing length–eddy viscosity expression is generalized to read

$$\mu_{Ti} = \overline{\rho}\ell_m^2|\omega| \tag{11–53}$$

No calculations have been published for any of the 3D test cases, so it is hard to judge the performance of the 3D Baldwin-Lomax model.

The use of *wall functions* in 3D is more problematical than in 2D, because the required 3D wall law is not as well founded. The inner integral momentum procedure of Caille and Schetz (1991) is more soundly based.

There has been a paucity of work with higher order turbulence models for 3D boundary layer flows. Bradshaw (1971) extended his 2D TKE model to 3D, and a comparison of predictions with the infinite swept wing experiment of Bradshaw and Terrell (1969) is given in Fig. 11–33, along with predictions from the Cebeci eddy viscosity model. No advantage can be seen for the TKE model. The Johnson-King model (see Section 7–10) has been extended to 3D [see Abid (1988)]. No systematic comparisons for test cases are available to judge its performance, but some workers report better results than the Baldwin-Lomax and Cebeci-Smith models for finite wings with strong pressure gradients.

Rodi presented calculations from a $K\epsilon$ model for a low-speed, 3D wing-body junction flow [Shabaka and Bradshaw (1981)] that was one of the test cases in the 1981 Stanford conference. It is surprising how few researchers tried to treat that case. Rodi used *wall functions,* rather than solving down to the wall with the no-slip condition (see Section 7–9). In order to use a wall function, it is necessary to have a 3D wall law, and he chose

$$\frac{Q_1}{q_*} = \frac{1}{\kappa}\ln(9y_1^+) \tag{11–54}$$

where Q_1 is the resultant velocity at the first grid point, which is located at $y_1^+ = y_1 q_*/\nu$. According to this relation, the ratio of the shears in the streamwise and crossflow directions is the same as the ratio of the velocities at y_1^+. Rodi's predictions for the Shabaka and Bradshaw (1981) flow are shown in Fig. 11–34 in terms of the skin friction. The results are quite good for this complicated flow. Rodi reports, however, that the secondary flow and $-\rho\overline{u'v'}$ are less well predicted.

11–4–6 High-Speed Flows

There is practical interest in studying high-speed, 3D boundary layer flows for wing-body junctions, etc., much the same as in the low-speed regime. However, there has been much less work of both the experimental and computational type in the high-speed area.

The 1981 Stanford conference had a number of high-speed, 3D test cases, but few researchers tried them. Also, the more recent reviews of 3D boundary layer

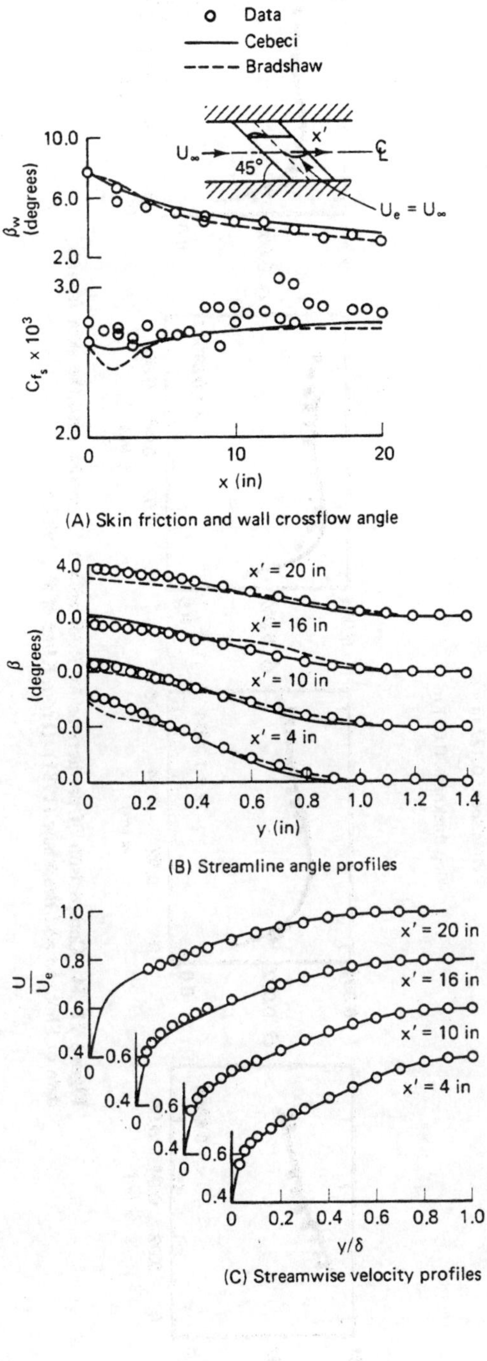

(A) Skin friction and wall crossflow angle

(B) Streamline angle profiles

(C) Streamwise velocity profiles

Figure 11-33 Comparison of predictions from an eddy viscosity model by Cebeci (1974) and a TKE model by Bradshaw (1971) with the data of Bradshaw and Terrell (1969) for flow over an infinite swept wing.

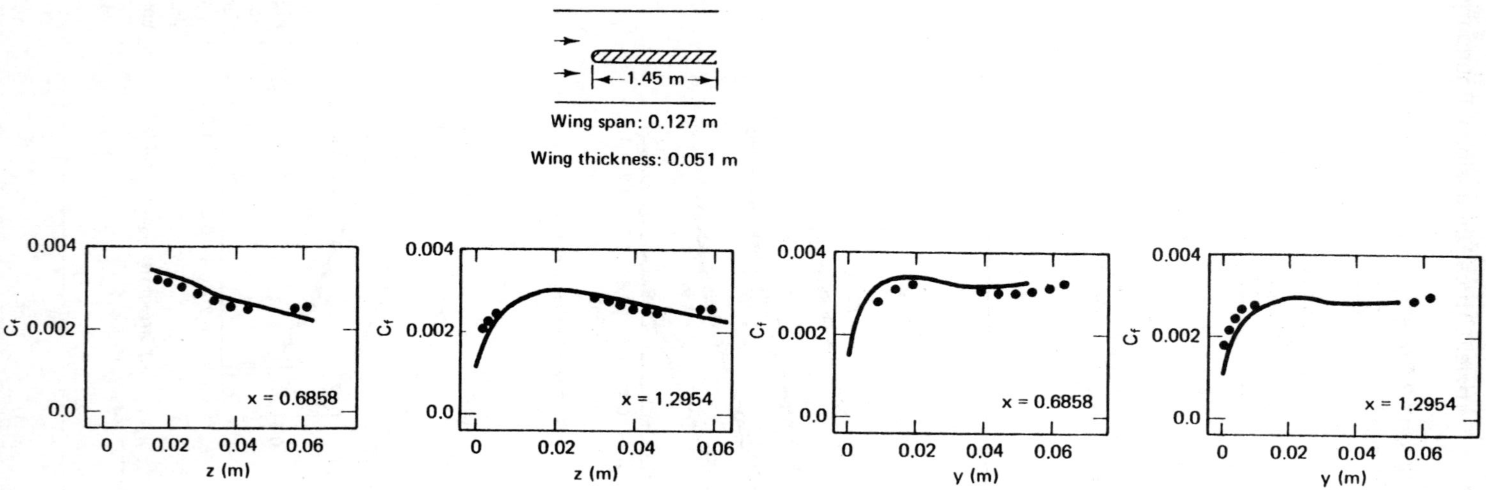

Figure 11–34 Comparison of predictions by Rodi with a $K\epsilon$ model with the skin friction data of Shabaka and Bradshaw (1981). (From Kline et al., 1982.)

flows that have been published have tended to ignore the high-speed cases. Rakich presented predictions of the 3D shock-impingement flow shown in Fig. 11–35 for the 1981 Stanford conference, using an eddy viscosity model. The eddy viscosity model adopted was the Van Driest model for the inner region (see Section 10–3–2) and the Clauser model for the outer region using δ_k^* [see Eq. (10–38)], both unchanged from their 2D form in this 3D application. The comparison of predictions with data in Fig. 11–35 shows fair agreement. No TKE, $K\epsilon$, or Reynolds stress model calculations were presented for this case.

The Situ-Schetz mixing length model for high-speed flows [see Eq. (10–42)] can be extended to 3D cases in the form

$$\mu_T = \ell_m^2 \left| \bar{\rho}\frac{U}{Q}\frac{\partial U}{\partial y} + \bar{\rho}\frac{W}{Q}\frac{\partial W}{\partial y} + \frac{Q}{S}\frac{\partial \bar{\rho}}{\partial y} \right| \tag{11–55}$$

No comparisons of predictions with experiment have been made for 3D cases at high enough Mach numbers ($M_e > 6$) for compressibility effects to be important.

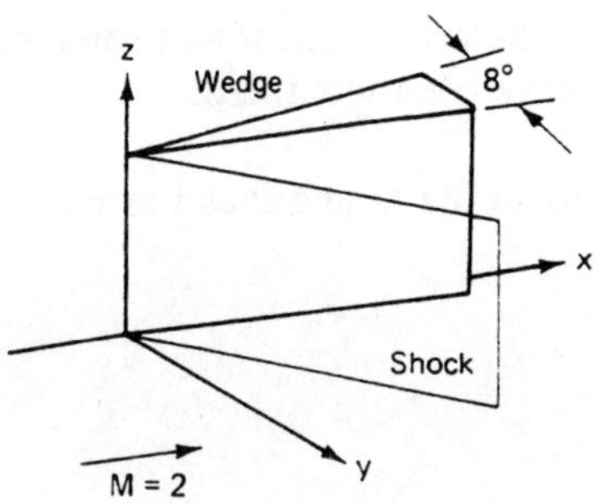

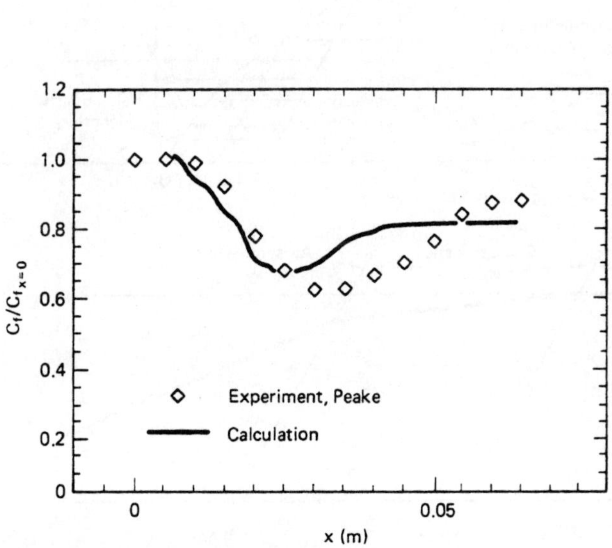

Figure 11–35 Comparison of predictions by Rakich from an eddy viscosity model with the skin friction data of Peake (1976). (From Kline et al., 1982.)

11-5 THREE-DIMENSIONAL TURBULENT JETS

11-5-1 Introduction

The practical interest in three-dimensional jets stems from their frequent application as a *mixing augmentation scheme*. In the simplest view, mixing is increased because the peripheral area around a 3D jet is increased, compared with a 2D (e.g., round) jet with the same fluid exhaust rate. But there is more to it than that, as the turbulent structure in the jet can also be altered. This alteration has been the subject of a number of recent studies, with those of Schadow and his co-workers [e.g., Ho and Gutmark (1987)] being perhaps the best known.

11-5-2 Experimental Data

For purposes of illustration, consider a turbulent, rectangular jet characterized by $e = d/\ell_1$, the ratio of the minor axis to the major axis of the orifice, as a representative example of this class of flows. The arrangement and some notation are shown in Fig. 11-36. Three main regions have been identified:

1. *Potential core (PC)*. The mixing at the boundaries has not yet reached the inviscid core.

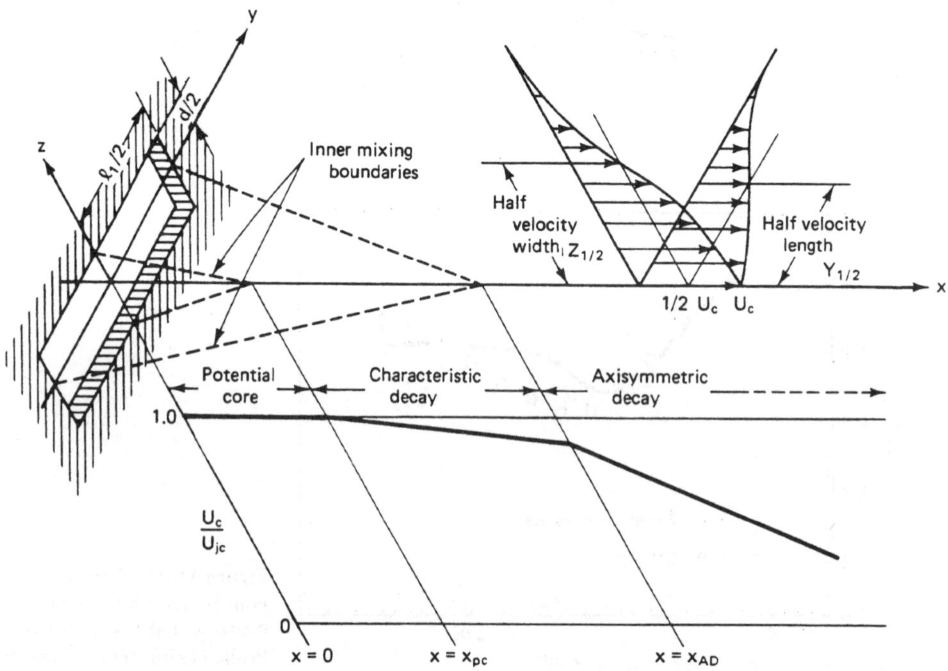

Figure 11-36 Schematic of three-dimensional jets showing the three regions.

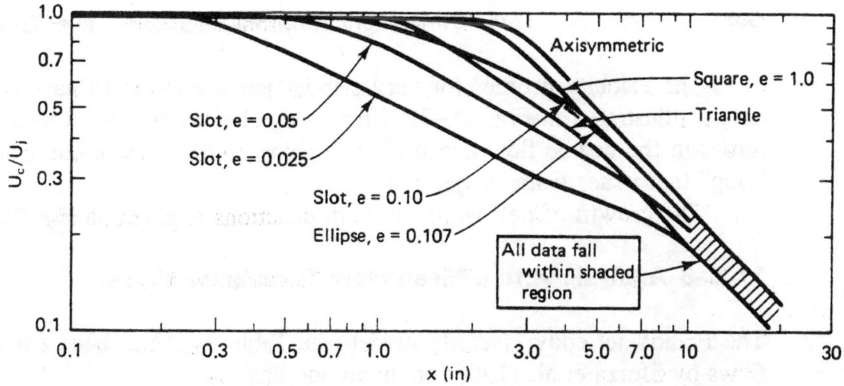

Figure 11–37 Streamwise variation in axis velocity for three-dimensional turbulent jets with $U_e = 0$. (From Sforza et al., 1966.)

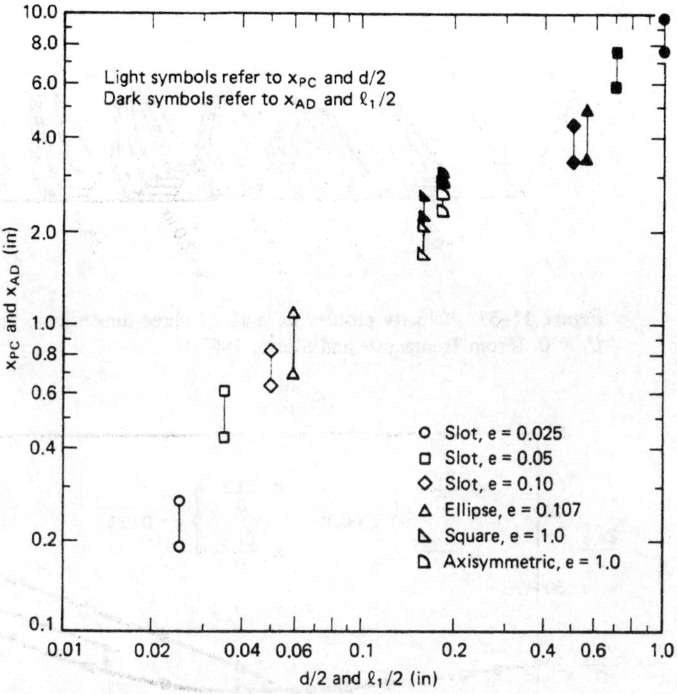

Figure 11–38 Length of the two main regions for three-dimensional turbulent jets with $U_e = 0$. (From Trentacoste and Sforza, 1967.)

2. *Characteristic decay (CD) region.* The decay along the axis is dependent on the geometry of the orifice, and the velocity profiles in the plane of the minor axis are *similar*, while those in the plane of the major axis are not.
3. *Axisymmetric decay (AD) region.* The decay of the axis velocity is axisymmetric in nature, and the whole flow approaches axisymmetry.

The decay of the axis velocity is shown in Fig. 11–37, and the extent of some of these regions is given in Fig. 11–38.

The velocity profiles for very slender jets are found to have some irregularities, as illustrated in Fig. 11–39. These irregularities are not a result of nonuniformities in the nozzle flow ahead of the orifice; rather, they come from a 3D vortex "ring" that issues from the jet exit.

The growth of half-widths in both directions is given in Fig. 11–40.

11–5–3 Analysis with a Mean Flow Turbulence Model

The Prandtl jet eddy viscosity model (see Table 9–1) has been extended to 3D jet flows by Sforza et al. (1966) on an ad hoc basis as

$$\mu_T = 0.037\bar{\rho}[Y^a_{1/2}Z^a_{1/2}(Y^a_{1/2} + Z^a_{1/2})]^{1/a}U_c \tag{11–56}$$

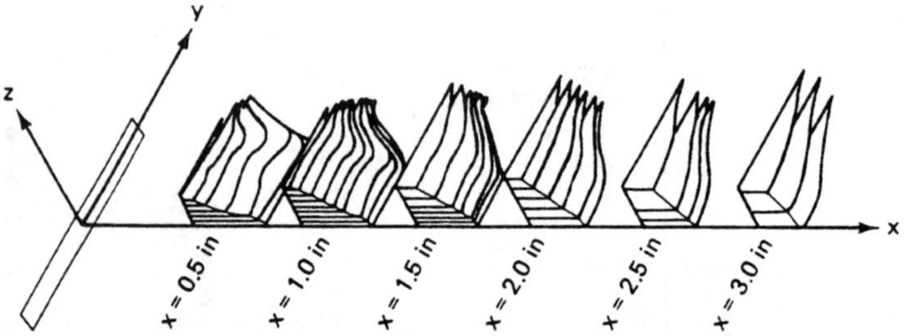

Figure 11–39 Velocity profiles for a 10 : 1 three-dimensional turbulent jet with U_e = 0. (From Trentacoste and Sforza, 1967.)

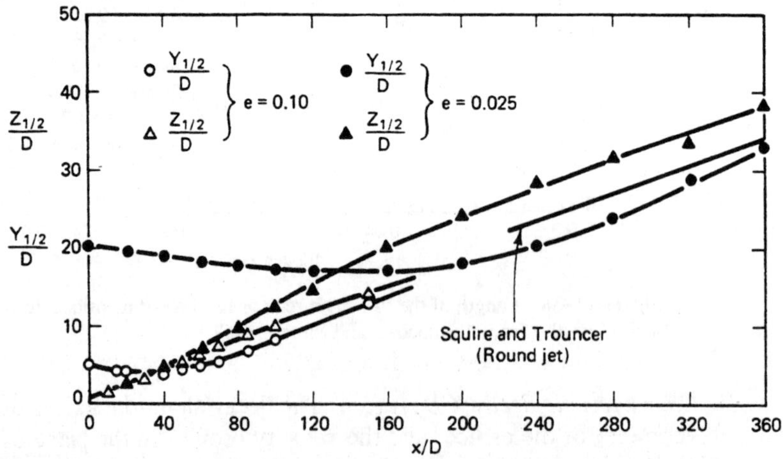

Figure 11–40 Variation in half-height and half-width for three-dimensional turbulent jets with U_e = 0. (From Trentacoste and Sforza, 1967.)

with $a = 0.53$. For a 2D planar jet, this equation collapses to the usual Prandtl formula, and for a round jet, it collapses to the Schlichting round-jet extension of the Prandtl model with the correct constant (see Table 9–1). A comparison of prediction with experiment for a 10 : 1 jet is shown in Fig. 11–41.

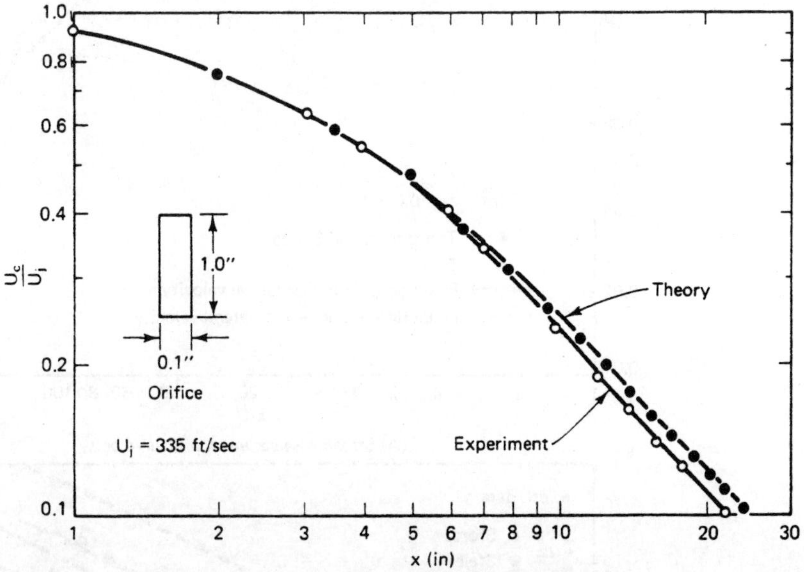

Figure 11–41 Comparison of prediction from an eddy viscosity model with experiment for the streamwise variation in axis velocity for a 10 : 1 three-dimensional turbulent jet with $U_e = 0$. (From Sforza et al., 1966.)

11–5–4 Analysis with a $K\epsilon$ Turbulence Model

A $K\epsilon$ model was applied by McGuirk and Rodi (1979) to rectangular jet flows. It was found necessary to relate $C_{\epsilon 1}$ to the centerline velocity gradient. The result was

$$C_{\epsilon 1} = 1.14 - 5.31 \frac{Y_{1/2}}{U_c} \frac{dU_c}{dx} \tag{11–57}$$

This expression was developed, and is only recommended, for jets in stagnant surroundings. Predictions from it are compared with data in Fig. 11–42. The agreement is generally good, but it was necessary to make estimates of the initial transverse (lateral) velocities in the exit flow to get such agreement. The complex features observed in the velocity profiles in the experiments (see Fig. 11–39) are not predicted by the model.

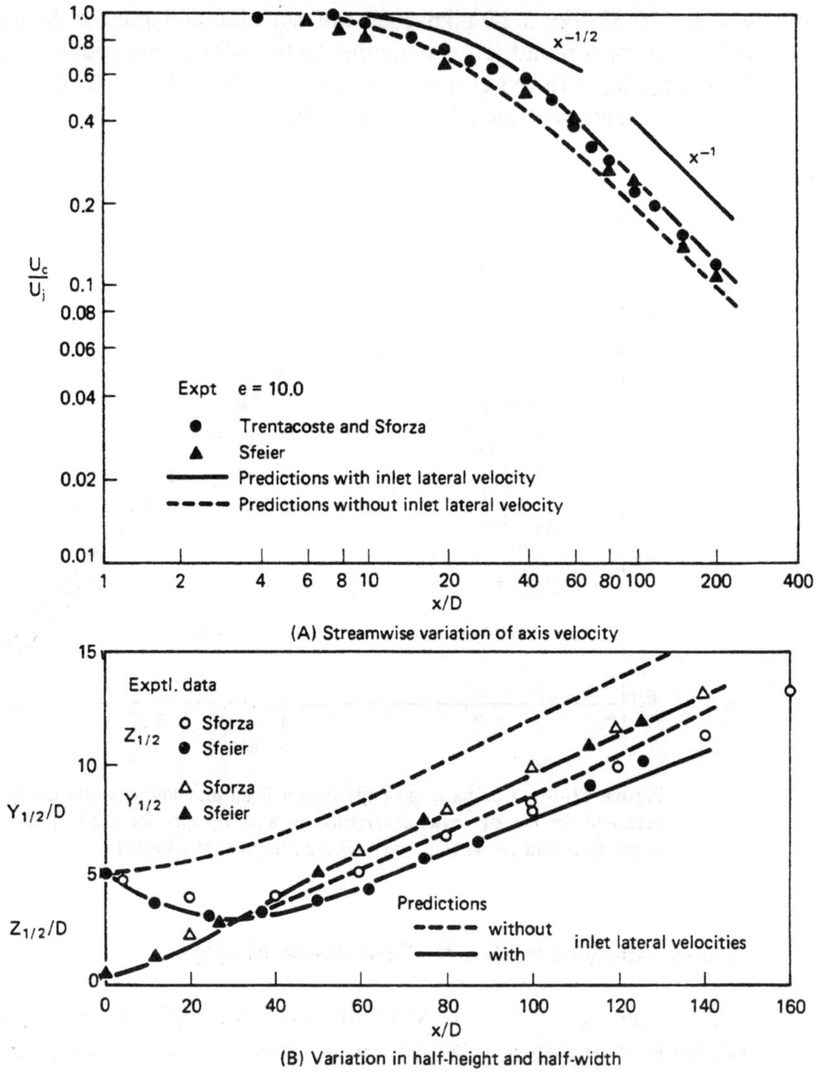

Figure 11–42 Comparison of prediction from a $K\epsilon$ model with experiment for a 10 : 1 three-dimensional turbulent jet with $U_e = 0$. (From McGuirk and Rodi, 1979.)

PROBLEMS

11.1. Write and run a computer code to solve the third equation listed under Eq. (11–24) for $\chi = 0.8$, 1.0, and 1.2, with $Cx = 2.0$. Plot the velocity profiles and discuss the results.

11.2. Consider the boundary layer profiles near a stagnation point. Compare the values obtained for the total velocity divided by the edge velocity at $y = \delta/2$ for a planar flow, an axisymmetric flow, and a 3D case, where $U_e = Ax$ and $W_e = Az/2$.

11.3. The so-called *laminar supersonic cone rule* states that, for the same Re_x and T_w/T_e,

$$C_{f,\text{cone}} = \sqrt{3}\, C_{f,\text{plate}}$$

Starting with the Levy-Lees transformation, Eq. (5–38), applied to a cone and then a flat plate, derive this relation. Next, evaluate the skin friction coefficient for the two cases in the form used for the similarity solutions in Section 5–9 for $\bar{\beta} = 0$. [*Hint:* You really need only the definition of C_f and $d\bar{\eta}/dy$] Now compare the two relations.

APPENDIX A

LAMINAR THERMOPHYSICAL PROPERTIES OF SELECTED FLUIDS

This appendix contains information on the laminar thermophysical properties of a few selected fluids. It is designed as an aid to the reader in working the problems at the end of each chapter. The information included was obtained from two main sources. For gases, the computer code developed by Svehla and McBride (1973) was used. For liquid water, the data compiled by the Thermophysical Properties Research Center (Touloukian and Ho, 1970, 1975) were used. Both are excellent, convenient sources of information on a wide range of fluids. A brief listing of conversion factors for viscosity and thermal conductivity is included, since data from different sources are often given in a variety of units.

Only Newtonian fluids are included in this appendix. There are simply so many different types of non-Newtonian fluids, that a short, comprehensive listing is not possible.

TABLE A–1 CONVERSION FACTORS FOR VISCOSITY

Multiply by ↓ to obtain →	$N \cdot s/m^2$	$Pa \cdot s$	Poise	$lb_f \cdot s/ft^2$
$N \cdot s/m^2$	1	1	10	2.0885×10^{-2}
$Pa \cdot s$	1	1	10	2.0885×10^{-2}
Poise	10^{-1}	10^{-1}	1	2.0885×10^{-3}
$lb_f \cdot s/ft^2$	4.7880×10^{1}	4.7880×10^{1}	4.7880×10^{2}	1

TABLE A–2 CONVERSION FACTORS FOR THERMAL CONDUCTIVITY

Multiply by ↓ to obtain→	BTU/h · ft · R	cal/s · cm · K	J/s · cm · K	W/cm · K
BTU/h · ft · R	1	4.1338×10^{-3}	1.7296×10^{-2}	1.7296×10^{-2}
cal/s · cm · K	2.4191×10^{2}	1	4.184	4.184
J/s · cm · K	5.7818×10^{1}	2.3901×10^{-1}	1	1
W/cm · K	5.7818×10^{1}	2.3901×10^{-1}	1	1

TABLE A–3 AIR AT 1.0 ATM

Temperature K	Viscosity Poise	Conductivity cal/s · cm · K	Specific heat cal/gm · K	Prandtl number	Density kg/m^3
300	184×10^{-6}	63×10^{-6}	0.240	0.706	1.177
400	227	78	0.242	0.706	0.883
500	265	92	0.246	0.706	0.705
600	299	106	0.251	0.706	0.588
700	331	121	0.257	0.706	0.503
800	362	135	0.263	0.706	0.441
900	391	148	0.268	0.706	0.393
1,000	419	162	0.273	0.706	0.352

TABLE A–4 HYDROGEN AT 1.0 ATM

Temperature K	Viscosity Poise	Conductivity cal/s · cm · K	Specific Heat cal/gm · K	Prandtl number	Density kg/m^3
300	89×10^{-6}	439×10^{-6}	3.421	0.695	0.0819
400	108	539	3.458	0.694	0.0614
500	126	629	3.471	0.693	0.0492
600	142	713	3.477	0.692	0.0409
700	157	794	3.489	0.692	0.0349
800	172	874	3.512	0.692	0.0306
900	186	954	3.544	0.691	0.0272
1,000	200	1035	3.581	0.691	0.0245

TABLE A–5 CARBON DIOXIDE AT 1.0 ATM

Temperature K	Viscosity Poise	Conductivity cal/s · cm · K	Specific heat cal/gm · K	Prandtl number	Density kg/m³
300	150×10^{-6}	40×10^{-6}	0.208	0.729	1.787
400	194	60	0.224	0.732	1.340
500	234	77	0.242	0.734	1.072
600	271	95	0.257	0.736	0.894
700	305	112	0.269	0.736	0.766
800	335	127	0.279	0.737	0.670
900	363	142	0.288	0.737	0.596
1,000	389	156	0.295	0.736	0.536

TABLE A–6 LIQUID WATER

Temperature K	Viscosity Poise	Conductivity cal/s · cm · K	Specific heat cal/gm · K	Prandtl number	Density kg/m³
273.15	1.753×10^{-2}	1.349×10^{-3}	1.003	13.03	1,000
300	0.823	1.454	0.998	5.65	997
350	0.360	1.595	1.002	2.26	974
400	0.215	1.641	1.013	1.33	-
450	0.151	1.608	1.055	0.99	-
500	0.117	1.518	1.110	0.86	-
600	0.097	1.365	1.259	0.90	-
647*	0.042	-	-	-	-

*Critical temperature

TABLE A–7 SEAWATER

Temperature K	Viscosity Poise	Kinematic Viscosity m²/s	Density kg/m³
280	1.48×10^{-2}	1.49×10^{-6}	1,027
285	1.29	1.30	1,026
290	1.13	1.14	1,025
295	1.02	1.02	1,024
300	0.92	0.91	1,023

TABLE A-8 BINARY DIFFUSION COEFFICIENTS, cm^2/s

Pair	300 K	400 K	500 K	600 K	700 K	800 K	900 K	1,000 K
CO_2—CO_2	0.110	0.190	0.287	0.399	0.524	0.658	0.803	0.958
CO_2—H_2	0.553	0.932	1.384	1.899	2.477	3.107	4.094	4.892
CO_2—H_2O	0.108	0.211	0.347	0.517	0.717	0.942	1.194	1.469
CO_2—N_2	0.174	0.293	0.436	0.600	0.782	0.982	1.034	1.233
CO_2—O_2	0.161	0.277	0.418	0.580	0.762	0.965	1.050	1.256
H_2—H_2	1.466	2.386	3.482	4.760	6.150	7.696	9.390	11.237
H_2—H_2O	0.886	1.502	2.237	3.080	4.024	5.057	6.179	7.386
H_2—N_2	0.622	1.013	1.473	1.994	2.576	3.217	3.912	4.660
H_2—O_2	0.914	1.490	2.170	2.944	3.800	4.745	5.772	6.876
H_2O—H_2O	0.088	0.202	0.373	0.603	0.891	1.235	1.636	2.086
H_2O—N_2	0.258	0.420	0.610	0.826	1.067	1.331	1.618	1.926
H_2O—O_2	0.258	0.442	0.665	0.922	1.210	1.527	1.871	2.241
N_2—N_2	0.204	0.335	0.490	0.665	0.861	1.075	1.309	1.561
N_2—O_2	0.206	0.340	0.498	0.679	0.884	1.113	1.368	1.643
O_2—O_2	0.206	0.341	0.502	0.686	0.894	1.126	1.381	1.656

APPENDIX B

Computer Codes for Students

This appendix contains simple FORTRAN computer codes that are intended for use in solving the problems at the end of each chapter in the text and other similar problems on a PC of the IBM PS/2 Model 30 class (with 8087 math coprocessor). They are specifically not intended as general-purpose codes for use by working professionals in the field. The goal has been to keep their formulation, logic, and programming as simple as possible, so that the student can easily grasp the flow of the calculations. Thus, primitive variables (u, v); (x, y) are employed with no transformations. Last, only limited information on physical properties is included. The *input* information is called for in the code directly, so one must recompile and then run each new problem. The codes have all been compiled and tested with Microsoft FORTRAN 77.

The simple codes presented are meant only to relieve the student of the burden of writing and debugging numerous codes during a one- or two-semester course. The hope is to thereby leave sufficient time and energy for the working of actual boundary layer problems of reasonable complexity. The student is not relieved of the burden to think. You should always estimate the answer you expect the computer solution to yield. For example, you could use a simple integral solution to estimate the level of the skin friction expected. Do not fall into the trap of assuming that, because the computer produces nice, neat output, it must be correct. You, and not the computer, are the analyst, and you are responsible for getting the correct answer.

There are eight separate codes for the various methods covered: (1) Thwaites-Walz incompressible laminar integral method; (2) incompressible laminar boundary layers by an explicit numerical method; (3) incompressible laminar boundary layers by an implicit numerical method; (4) compressible laminar boundary layers by an implicit numerical method; (5) Moses's incompressible turbulent integral method; (6) incompressible turbulent boundary layers by an implicit numerical method; (7) incompressible turbulent boundary layers by an implicit numerical method with a stretched grid; and (8) turbulent jets and wakes by an implicit numerical method. A brief description of each code precedes the listings given.

For any boundary layer problem, one must first specify the fluid through ρ and μ, or perhaps just ν. In variable-density, variable-property cases, an equation of state and information on the variation of properties must also be given. The next information needed would be the free stream velocity U_∞ and the inviscid edge velocity distribution $U_e(x)$ (and sometimes, the edge velocity gradient dU_e/dx). For high-speed flows, M_∞ and $M_e(x)$ are needed. With heat transfer or compressible flow, the wall temperature $T_w(x)$ or the wall heat transfer distribution, $q_w(x)$, and T_∞ and $T_e(x)$ are required. Mass transfer cases are not included in the codes given here.

Next is the matter of initial conditions. The integral methods require only initial values of quantities such as $\theta(0)$, $\delta(0)$, and $C_f(0)$. The differential methods need complete, initial profiles $u(0, y)$, $v(0, y)$. For turbulent cases, one must select the turbulence model from those available in the code, e.g., the mixing length model, eddy viscosity model, or TKE model with the Prandtl energy method. The TKE model requires an initial profile and boundary conditions for K. There are also numerous so-called *modeling constants*, such as κ, whose values the user may wish to change from the values that are in the code. Also, a transition location must be specified.

Finally, the computational region must be specified. With an integral method, this is simply the length that is of interest. The differential methods require the height and the length of the computational region. Some judgment must be exercised in making the choices. When the equations are not transformed, the computational region should grow with streamwise distance to accommodate the growth in height of the boundary layer (see Section 4–7–4). The programs are based on the untransformed equations to keep the codes simple, so the user must pick a computational region high enough at the initial station to accommodate the estimated growth of the boundary layer by the downstream end of the computational region. If not, the calculation must be stopped in the middle and then restarted with a higher region. The last matter concerns the choice of step size(s). The integral methods only need a choice for dx, while the differential methods need dx and dy. These choices are normally accomplished by picking the number of points across the height and length of the region. For simplicity, most of the codes have fixed step sizes in the x- and y-directions. For a laminar flow, about 20–25 points across the boundary layer are sufficient. Thus, knowing the thickness of the boundary layer at the initial station permits a reasonable choice for dy. With the boundary layer approximation, we can take $dx \gg dy$, and a value of $dx \approx \delta_i/2$ is usually adequate. Of course, for explicit methods, the step size must satisfy the stability criterion (see Section 4–7–2). For a turbulent flow, a stretched grid in the y-direction is much more efficient (see Section 7–9), but that was implemented in only one of the codes, in the interests of simplicity. Therefore, a large number of grid points is necessary—at least 400 across the boundary layer.

The student should reread the relevant sections cited preceding each program before attempting to use any of the codes. Also, he or she should look at the code and see how the required input information described is to be provided.

Program WALZ

This program implements the Thwaites-**WALZ** integral method for incompressible laminar boundary layer flows (see Section 2–3–2). One must start from $x = 0$, at either a sharp or a blunt nose. If the local Λ is at or below -0.09, the program will print, SEPARATION AT OR BEFORE S = XX and then stop.

```
      PROGRAM WALZ
C
C
C      THIS CODE IMPLEMENTS THE THWAITES-WALZ METHOD FOR
C      CONSTANT-PROPERTY, CONSTANT-DENSITY LAMINAR FLOWS.
C      THE USER MUST SPECIFY THE PROBLEM WITH:
C       INITIAL AND FINAL X, XI, AND XF (DIMENSIONS L);
C       KINEMATIC VISCOSITY, CNU (DIMENSIONS L**2/T);
C       FREE STREAM VELOCITY, UINF (L/T);
C       NUMBER OF STATIONS, NMAX;
C       EDGE VELOCITY DISTRIBUTION, UE(N) (L/T), AND
C       THE BODY SHAPE DISTRIBUTION, Y(N) (L);
C       FOR BLUNT-NOSED BODIES, THE NOSE RADIUS, RO (L);
C       "KASE" IS FLOW INDEX:
C         0 FOR 2D FLOW,
C         1 FOR 2D FLOW WITH INITIAL STAGNATION POINT, AND
C         2 FOR AXISYMMETRIC FLOW.
C       FOR EACH NEW CASE, YOU MUST RECOMPILE TO EXECUTE.
C
C      THE OUTPUT CONTAINS A LISTING OF THE INPUT:
C       X(N) (L UNITS), Y(N) (L), UE(N) (L/T), FOR N=1 TO NMAX
C      AND THEN THE RESULTS:
C       SURFACE DISTANCE, S (L UNITS), DISPLACEMENT THICKNESS, DELS (L),
C       MOMENTUM THICKNESS, THETA (L), SHAPE FACTOR, H, AND
C       SKIN FRICTION COEFFICIENT, CF.
C
C     WITHIN THE DO-LOOP, "DO 10 N=1,NMAX" MUST BE GIVEN THE FORMULAS
C     OR DATA FOR THE INVISCID FLOW, UE(N), AND BODY SHAPE, Y(N), FOR
C     EACH NTH STEP. THESE ARE TO BE CHANGED AS NEEDED, BEFORE COMPILING.
C
C     SET THE FOLLOWING DIMENSIONS TO NMAX OR GREATER.
```

```
      DIMENSION X(121),Y(121),S(121),UE(121)
C
C
C  SAMPLE INPUT DATA
C
      NMAX=41
      CNU=2.0E-4
      UINF=10.0
      KASE=0
C
      WRITE(*,*) ' THWAITES-WALZ METHOD'
      WRITE(*,*) ' INPUT: NMAX, CNU, UINF, KASE'
      WRITE(*,*)  NMAX, CNU, UINF, KASE
C
C  AS A SAMPLE, A FLAT PLATE TO X=1.0 PLUS A RAMP TO GIVE AN
C  ADVERSE PRESSURE GRADIENT IS USED HERE.
C
      XI=0.0
      XF=2.0
      DELX=(XF-XI)/(NMAX-1)
C
      WRITE(*,9)
    9 FORMAT(5X,'N',7X,'X(N)',12X,'Y(N)',11X,'UE(N)')
      DO 10 N=1,NMAX
      X(N)=XI+(N-1)*DELX
C     DESCRIBE THE INVISCID FLOW, UE(X),  AND BODY SHAPE, Y(X).
      IF(KASE.EQ.0) Y(N)=0.0
          IF(X(N).LT. 1.0) THEN
      UE(N)=10.0
          ELSE
      UE(N)=10.5-X(N)/2.0
          ENDIF
      WRITE(*,200) N,X(N),Y(N),UE(N)
  200 FORMAT(3X,I3,3X,F7.4,10X,F7.4,10X,F7.4)
   10 CONTINUE
C
C CALCULATION OF DISTANCE ALONG THE SURFACE
C
C
      S(1)=0.0
      DO 20 N=2,NMAX
   20 S(N)=S(N-1)+SQRT((X(N)-X(N-1))**2+(Y(N)-Y(N-1))**2)
C
C
C SOLUTION PROCEDURE
C
      WRITE(*,92)
   92 FORMAT(1H0,2X,'N',5X,'S',10X,'DELTS',6X,'THETA',6X,'H',
     110X, 'CF')
```

```
      CF=0.0
      VSUM=0.0
      F1=0.0
      F2=0.0
      DO 100 N=1,NMAX
      Y2=1.0
      IF(KASE.EQ.2) Y2=Y(N)*Y(N)
      F2=Y2*UE(N)**5
      IF(N .EQ. 1) GO TO 48
      IF(N.EQ.2.AND.KASE.GE.1) GO TO 48
      VSUM=VSUM+0.5*(F1+F2)*(S(N)-S(N-1))
      IF (UE(N).NE.0.0)
      DUDS=(UE(N)-UE(N-1))/(S(N)-S(N-1))
      CONST=0.45*CNU/(F2*UE(N))
      STHETA=CONST*VSUM
      GO TO 95
   48 DUDS=(UE(2)-UE(1))/S(2)
      STHETA=0.0
      IF(KASE.EQ.1)DUDS=2.0*UINF/RO
      IF(KASE.EQ.1) STHETA=0.075*CNU/DUDS
      IF(KASE.EQ.2)DUDS=3.0*UINF/2.0/RO
      IF(KASE.EQ.2) STHETA=0.056*CNU/DUDS
C
   95 THETA=SQRT(STHETA)
      BLAN=STHETA*DUDS/CNU
      IF(BLAN.LT.0.0)GO TO 96
      H1=2.61-3.75*BLAN+5.24*BLAN*BLAN
      S1=0.22+1.57*BLAN-1.8*BLAN*BLAN
      GO TO 97
   96 H1=2.088+0.0731/(0.14+BLAN)
      S1=0.22+1.402*BLAN+0.018*BLAN/(0.107+BLAN)
   97 DELS=THETA*H1
      IF(N.GT.1.AND.UE(N).NE.0.0) CF=2.0*CNU*S1/(UE(N)*THETA)
      IF(BLAN.LE.-0.09)THEN
      WRITE (*,*)'...SEPARATION AT OR BEFORE  S=',S(N)
      WRITE(*,93) N, S(N),DELS,THETA,H1,CF
      F1=F2
  100 CONTINUE
   93 FORMAT(' ',I3,5(1X,F10.6))
      STOP
      END
```

Program ILBLE

This program implements a numerical method for 2D, **I**ncompressible, **L**aminar **B**oundary **L**ayer flows with an **E**xplicit approach (see Section 4–7–2). The step size in the x-direction is controlled by the stability criterion. When the skin friction coefficient falls to a very low value (taken here to be 0.0001), the program will print NEAR SEPARATION AT X =( ) and stop.

```
      PROGRAM ILBLE
C
C 2D BOUNDARY LAYER COMPUTATION, INCOMPRESSIBLE LAMINAR,
C 1ST ORDER, EXPLICIT
C
C EQUATIONS ARE DIMENSIONLESS, USING FREE STREAM VELOCITY, UINF,
C VISCOSITY, MUINF, DENSITY, RHOINF, AND LENGTH, L;
C X/L, Y/L, U/UINF; ALSO, RE=RHOINF*UINF*L/MUINF.
C PICK L = 1.0.
C
C SAMPLE PROBLEM OF A FLAT PLATE WITH UE=UINF=10.0, FOLLOWED BY AN
C ADVERSE PRESSURE GRADIENT. START AT END OF FLAT PLATE X=1.0.
C TAKE NU=2E-4. THEN REX=5.0E4.
C USE BLASIUS SOLN. TO GET INITIAL VALUES.
C CAN BE MODIFIED BY USER TO SOLVE OTHER FLOW CASES.
C
C
C
      COMMON/CONST/ RE, DX, DY
      COMMON/RST/ UE, DUEDX
C
C
      DIMENSION U(100),U0(100),V(100),V0(100)
C
C
      WRITE(*,*) ' PROGRAM ILBLE'
      WRITE(*,*) ' 2-D BOUNDARY LAYER, INCOMPRESSIBLE LAMINAR'
      WRITE(*,*) ' 1ST ORDER'
      WRITE(*,*) ' EXPLICIT METHOD'
C
C INPUT DATA
C
C YOU MUST GIVE INITIAL X (XI), FINAL X (XF), (CNU), (UNINF),
C (MMAX) AND (DY).
C
C PICK MMAX AND DY BASED ON INITIAL BOUNDARY LAYER THICKNESS. USE ABOUT
C 20 POINTS ACROSS THE INITIAL BOUNDARY LAYER, AND ADD ABOUT 50-75 POINTS
C ABOVE THAT. THE X STEP SIZE, DX, WILL BE CALCULATED IN THE CODE
C USING THE STABILITY CRITERION.
C
      XI=1.0
      XF=2.0
      CNU=0.0002
      UINF=10.0
      MMAX=100
      DY=0.00112
C
      RE=UINF*1.0/CNU
C
```

```
C SET INITIAL VALUES.
      X=XI
      DELT=0.0224
      DELST=0.00771
      THETA=0.00297
      CF=0.00297
C FLAT PLATE PLUS RAMP AS AN EXAMPLE, USER MUST GIVE UE AND DUEDX.
      UE=1.0
      DUEDX=-0.05
C
      WRITE(*,*) ' INPUT:  MMAX,  RE, DY'
      WRITE(*,*)          MMAX, RE, DY
C
C
      WRITE(*,*) ' FLAT PLATE PLUS RAMP FLOW'
    6 FORMAT(I5,1X,3F10.5)
C
C INITIAL CUBIC PROFILES CHOSEN HERE. USER CAN CHANGE SHAPE.
C MEST IS THE M INDEX FOR THE INITIAL DELTA.
C
      DO 30 M=1,MMAX
      U(M)=UE
      U0(M)=UE
      V(M)=0.0
   30 V0(M)=0.0
      U(1)=0.0
      U0(1)=0.0
C
C
      MEST=21
      FM1=MEST-1
      DO 40 M=2,MEST
      YOD=(M-1)/FM1
      U0(M)=UE*(1.5*YOD-0.5*YOD**3)
   40 CONTINUE
C
  322 FORMAT(//,6X,'X',10X,'DELT',7X,'DELTS',6X,'THETA',8X,'CF',8X,'UE',
     18X,'DUEDX'/)
      WRITE(*,322)
      WRITE(*,320) X,DELT,DELST,THETA,CF,UE,DUEDX
 1000 CONTINUE
C
C USE STABILITY CRITERION FOR ALLOWABLE DX.
C
      DX=0.4*U0(2)*DY*DY*RE
C
      IF(X .GE. XF) GO TO 99
      IF(X+DX .GT. XF) THEN
```

```
          DX=XF-X
          ENDIF
        X=X+DX
C  FLAT PLATE PLUS RAMP AS AN EXAMPLE, USER MUST GIVE UE AND DUEDX.
        UE=1.05-X/20.0
        DUEDX=-0.05
C
        U0(MMAX)=UE
C
C
        CALL EXPLIC(MMAX,U,U0,V,V0,DELST,THETA)
C
        DELST=DELST*DY
        THETA=THETA*DY
        DO 80 M=1,MMAX
        IF(U(M).GT.0.99*UE) THEN
         MEST=M
         GO TO 120
         ENDIF
   80 CONTINUE
  120 CONTINUE
        DELT=(MEST-1)*DY
        CF=(4.0*U(2)-U(3))/(UE**2*DY*RE)
C CHECK TO SEE IF NEAR SEPARATION.
        IF(CF .LE. 0.0001) THEN
        WRITE(*,*) ' '
        WRITE(*,*) '   NEAR SEPARATION AT X = ',X
        GO TO 99
        ENDIF
C
  300 WRITE(*,320) X,DELT,DELST,THETA,CF,UE,DUEDX
        DO 100 M=2,MMAX
        U0(M)=U(M)
  100 V0(M)=V(M)
  200 CONTINUE
        GO TO 1000
C
  320 FORMAT(1X,7(1X,F10.5))
   99 CONTINUE
        CALL OUTPUT(MEST+25,U,V)
        STOP
        END
C
        SUBROUTINE EXPLIC(MMAX,U,U0,V,V0,ST,TA)
C
C EXPLICIT SCHEME
C
        COMMON/CONST/ RE, DX, DY
```

```
      COMMON/RST/ UE, DUEDX
C
      DIMENSION U(100),U0(100),V(100),V0(100)
C
      ST=0.5
      TA=0.0
C
      DO 100 M=2,MMAX-1
      A=1.0/RE-V0(M)*DY/2.0
      B=1.0/RE+V0(M)*DY/2.0
      C=UE*DUEDX*DY**2-2.0*U0(M)/RE
      H1=0.5*DY/DX
      H=DX/(U0(M)*DY*DY)
      U(M)=U0(M)+H*(A*U0(M+1)+B*U0(M-1)+C)
      V(M)=V(M-1)-H1*(U(M)-U0(M)+U(M-1)-U0(M-1))
C
      F1=1.0-U(M)/UE
      F2=F1*U(M)/UE
      ST=ST+F1
      TA=TA+F2
  100 CONTINUE
      RETURN
      END
C
C
      SUBROUTINE OUTPUT(MMAX,U,V)
      COMMON/CONST/ RE, DX, DY
      DIMENSION U(100),V(100),Y(100)
C
C
      DO 15 M=1,MMAX
   15 Y(M)=(M-1)*DY
      WRITE(*,1)  (M,Y(M),U(M),V(M),M=1,MMAX,1)
    1 FORMAT(//,' M',6X,'Y',10X,'U',10X,'V',/,(1X,I3,3(1X,F10.4)))
      RETURN
      END
```

Program ILBLI

This code implements a numerical method for 2D, Incompressible, Laminar Boundary Layer flows with an Implicit technique (see Section 4–7–3). Here, the user can select DX, since it is not necessary to obey a stability limitation. When the skin friction coefficient falls to a very low value (taken here to be 0.0001), the program will print NEAR SEPARATION AT X =( ) and stop.

```
     PROGRAM ILBLI
C
C 2D BOUNDARY LAYER COMPUTATION,  INCOMPRESSIBLE LAMINAR,
```

```
C 1ST ORDER, IMPLICIT METHOD
C
C EQUATIONS ARE DIMENSIONLESS, USING FREE STREAM VELOCITY, UINF,
C VISCOSITY, MUINF, DENSITY, RHOINF, AND A REFERENCE
C LENGTH, L; X/L, Y/L, U/UINF;  ALSO, RE=RHOINF*UINF*L/MUINF.
C PICK L=1.0.

C
C SAMPLE PROBLEM OF A FLAT PLATE WITH UE=UINF=10.0 FOLLOWED BY AN
C ADVERSE PRESSURE GRADIENT. START AT END OF FLAT PLATE, X=1.0.
C TAKE NU=2E-4. THEN REX=5.0E4.
C USE BLASIUS SOLN. TO GET INITIAL VALUES.
C CAN BE MODIFIED BY USER TO SOLVE OTHER FLOW CASES.
C
C
C
      COMMON/CONST/ RE, DX, DY
      COMMON/RST/ UE, DUEDX
C
C
      DIMENSION U(100),U0(100),V(100),V0(100),X(100),XX(100)
      DIMENSION UE(150),DUEDX(150)
      DIMENSION DELT(150),DELST(150),THETA(150),CF(150)
C
C
      WRITE(*,*) ' PROGRAM ILBLI'
      WRITE(*,*) ' 2-D BOUNDARY LAYER, INCOMPRESSIBLE LAMINAR'
      WRITE(*,*) ' 1ST ORDER'
      WRITE(*,*) ' IMPLICIT METHOD'
C
C
C INPUT DATA
C
C YOU MUST GIVE INITIAL X (XI), FINAL X (XF), (CNU), (UINF), (NMAX)
C (MMAX) AND (DY).
C
C PICK MMAX AND DY BASED ON INITIAL BOUNDARY LAYER THICKNESS. USE ABOUT
C 20 POINTS ACROSS THE INITIAL BOUNDARY LAYER, AND ADD ABOUT 50-100 POINTS
C ABOVE THAT. PICK NMAX BASED ON THE LENGTH OF REGION AND DX DESIRED.
C DX CAN BE ABOUT INITIAL DELTA/TWO.
C
      XI=1.0
      XF=2.0
      CNU=0.0002
      UINF=10.0
      MMAX=100
      NMAX=41
      DY=0.00112
      DX=(XF-XI)/(NMAX-1)
```

```
C
      RE=UINF*1.0/CNU
C
C SET INITIAL VALUES.
      X(1)=XI
      DELT(1)=0.0224
      DELST(1)=0.00771
      THETA(1)=0.00297
      CF(1)=0.00297

      WRITE(*,*) ' INPUT:  MMAX, NMAX, RE, DY'
      WRITE(*,*)          MMAX, NMAX, RE, DY
C
C FLAT PLATE PLUS RAMP AS AN EXAMPLE, USER MUST GIVE UE(N) AND DUEDX(N).
C FOR N = 1, NMAX
C
      DO 5 N=1,NMAX
      XX(N)=X(1)+(N-1)*DX
      UE(N)=1.05-XX(N)/20.0
      DUEDX(N)=-0.05
    5 CONTINUE
      WRITE(*,*) ' FLAT PLATE PLUS RAMP FLOW: N, X/L, UE/UINF AND DUEDX'
      WRITE(*,6)   (N,XX(N), UE(N), DUEDX(N),N=1,NMAX,4)
    6 FORMAT(I5,1X,3F10.5)
C
C INITIAL CUBIC PROFILES CHOSEN HERE. USER CAN CHANGE SHAPE.
C MEST IS THE M INDEX FOR THE INITIAL DELTA.
C
      DO 30 M=1,MMAX
      U(M)=UE(1)
      U0(M)=UE(1)
      V(M)=0.0
   30 V0(M)=0.0
      U(1)=0.0
      U0(1)=0.0
C
C
      MEST=21
      FM1=MEST-1
      DO 40 M=2,MEST
      YOD=(M-1)/FM1
      U0(M)=UE(1)*(1.5*YOD-0.5*YOD**3)
   40 CONTINUE
C
      NMAXP=NMAX
      DO 200 N=2,NMAX
      NNX=N
      X(N)=X(N-1)+DX
```

```
      U0(MMAX)=UE(N)
C
C
      CALL IMPLIC(NNX,MMAX,U,U0,V,V0,DELST,THETA)
C
      DELST(N)=DELST(N)*DY
      THETA(N)=THETA(N)*DY
      DO 80 M=1,MMAX
      IF(U(M).GT.0.99*UE(N)) THEN
      MEST=M
      GO TO 120
      ENDIF
   80 CONTINUE
  120 CONTINUE
      DELT(N)=(MEST-1)*DY
      CF(N)=(4.0*U(2)-U(3))/(UE(N)**2*DY*RE)
C CHECK TO SEE IF NEAR SEPARATION
      IF(CF(N) .LE. 0.0001) THEN
      WRITE(*,*) ' '
      WRITE(*,*) '  NEAR SEPARATION AT X = ',X(N)
      NMAXP=N
      GO TO 300
      ENDIF
C
      DO 100 M=2,MMAX
      U0(M)=U(M)
  100 V0(M)=V(M)
  200 CONTINUE
C
  300 WRITE(*,320) (N,X(N),DELT(N),DELST(N),THETA(N),CF(N),
     1N=1,NMAXP,1)
  320 FORMAT(//,'  N',6X,'X',10X,'DELT',7X,'DELTS',6X,'THETA',6X,'CF',
     1/,(1X,I3,5(1X,F10.5)))
      CALL OUTPUT(MEST+25,U,V)
      STOP
      END
C
C
      SUBROUTINE IMPLIC(NNX,MMAX,U,U0,V,V0,ST,TA)
C
C IMPLICIT SCHEME
C
      COMMON/CONST/ RE, DX, DY
      COMMON/RST/ UE, DUEDX
C
      DIMENSION U(100),U0(100),V(100),V0(100)
      DIMENSION UE(150),DUEDX(150)
      DIMENSION ST(150), TA(150)
```

```
      DIMENSION A(100),B(100),C(100),R(100),RP(100),GAM(100)
C
      MMA1=MMAX-1
      ST(NNX)=0.5
      TA(NNX)=0.0
C
      B(1)=1.0
      C(1)=0.0
      R(1)=0.0
      A(MMAX)=0.0
      B(MMAX)=1.0
      R(MMAX)=UE(NNX)
C
      DENO=RE*DY**2
      DO 20 M=2,MMAX-1
      A(M)=-(0.5*V0(M)/DY+1.0/DENO)
      B(M)=U0(M)/DX+2.0/DENO
      C(M)=-(1.0/DENO-0.5*V0(M)/DY)
      R(M)=UE(NNX)*DUEDX(NNX)+U0(M)**2/DX
   20 CONTINUE
C
      GAM(1)=C(1)/B(1)
      RP(1)=R(1)/B(1)
C
      DO 40 M=2,MMAX
      DENO=B(M)-A(M)*GAM(M-1)
      GAM(M)=C(M)/DENO
      RP(M)=(R(M)-A(M)*RP(M-1))/DENO
   40 CONTINUE
C
      U(MMAX)=RP(MMAX)
      DO 60 M=1,MMA1
      U(MMAX-M)=RP(MMAX-M)-GAM(MMAX-M)*U(MMAX-M+1)
   60 CONTINUE
C
      DO 90 M=2,MMAX
      V(M)=V(M-1)-(0.5*DY/DX)*(U(M)-U0(M)+U(M-1)-U0(M-1))
      F1=1.0-U(M)/UE(NNX)
      F2=F1*U(M)/UE(NNX)
      ST(NNX)=ST(NNX)+F1
      TA(NNX)=TA(NNX)+F2
   90 CONTINUE
      RETURN
      END
C
      SUBROUTINE OUTPUT(MMAX,U,V)
      COMMON/CONST/ RE, DX, DY
```

```
      DIMENSION U(100),V(100),Y(100)
C
C
      DO 15 M=1,MMAX
   15 Y(M)=(M-1)*DY
      WRITE(*,1)  (M,Y(M),U(M),V(M),M=1,MMAX,1)
    1 FORMAT(//,' M',6X,'Y',10X,'U',10X,'V',/,(1X,I3,3(1X,F10.4)))
      RETURN
      END
```

Program CLBL

Here, we treat 2D, **C**ompressible (air), **L**aminar **B**oundary **L**ayer flows with an implicit numerical method (see Section 5–11). The fluid is assumed to be air as a perfect gas, although the ratio of specific heats is allowed to vary with temperature. The viscosity is computed with the Sutherland law [Eq. (1–12)]. Also, the Prandtl number is taken as constant.

```
      PROGRAM CLBL
C
C 2D COMPRESSIBLE AIR BOUNDARY LAYER COMPUTATION.
C 1ST ORDER, IMPLICIT, LAGGED COEFFICIENT,
C LAMINAR WITH SUTHERLAND VISCOSITY LAW.
C
C EQUATIONS ARE DIMENSIONLESS, USING FREE STREAM VELOCITY, UINF,
C VISCOSITY, CMUINF, DENSITY, RHOINF, LENGTH, L, GAMMA,
C MACH NUMBER, XMIF, AND TEMPERATURE, TINF:
C X/L, Y/L, U/UINF, V/UINF, P/RHOINF/UINF**2=P/GAMMA/XMIF**2/PINF.
C T/TINF. ALSO, RE=RHOINF*UINF*L/CMUINF. TAKE L=1.0.
C
C FLAT PLATE WITH A CONSTANT TW (TW/TE=1) AS A SAMPLE.
C FREE STREAM MACH NUMBER, XMIF=4.0,  TINF=300 K.
C START AT LEADING EDGE WITH UNIFORM PROFILES.
C
      COMMON/CONST/ RE,DX,DY,NCON
      COMMON/CON/ XMIF,TINF,XM1
      COMMON/RST/ UE,DUEDX,PR
C
      DIMENSION U(101),U0(101),V(101),V0(101),H(101),H0(101)
      DIMENSION A(101),B(101),C(101),R(101)
      DIMENSION CMU(101),RHO(101),RHN(101),CF(101),ST(101)
      DIMENSION GAMMA(101),TEMP(101)
C
      GAMMA(1)=1.4
C
      WRITE(*,*)  ' PROGRAM CLBL'
```

```
      WRITE(*,*) ' 2-D LAMINAR COMPRESSIBLE AIR BOUNDARY LAYER '
      WRITE(*,*) ' 1ST ORDER IMPLICIT, LAGGED COEFF.'
C
C INPUT DATA
C
C YOU MUST GIVE INITIAL X (XI), FINAL X (XF), (TINF), (UINF), (PINF),
C (XMIF), (CNUINF), (NMAX), (MMAX), (DY) AND TW/TE.
C
C PICK MMAX AND DY BASED ON INITIAL BOUNDARY LAYER THICKNESS.
C FOR LAMINAR FLOW, USE ABOUT 20 POINTS ACROSS THE INITIAL BOUNDARY
C LAYER, AND ADD ABOUT 50-100 POINTS ABOVE THAT. PICK NMAX BASED ON THE
C LENGTH OF REGION AND DX DESIRED. DX CAN BE ABOUT INITIAL DELTA/TWO.
C
      XI=0.0
      XF=1.0
      TINF=300.0
      UINF=1390.0
      XMIF=4.0
      NMAX=101
C
      MMAX=100
      PINF=0.01
      CNUINF=0.00157
      DY=0.0002
      DX=(XF-XI)/(NMAX-1)
      RE=UINF*1.0/CNUINF
      WRITE(*,*) ' INPUT: MMAX, NMAX, RE, DY,  DX'
      WRITE(*,*)          MMAX, NMAX, RE, DY, DX
      WRITE(*,*) '        XMIF, TINF'
      WRITE(*,*)          XMIF, TINF
      PR=0.71
C FOR SUTHERLAND VISCOSITY LAW
      T1=110.33/TINF
C
C GAMMA IS A FUNCTION OF TEMPERATURE. GAMMA(1)=1.4 IS
C THE FREE STREAM RATIO OF SPECIFICS HEATS.
C
      XM1=(GAMMA(1)-1.0)*XMIF**2
C
C SAMPLE CASE OF FLAT-PLATE FLOWS. OTHER CASES WILL
C NEED UE(N) AND DUEDX(N).
C
      UE=1.0
      DUEDX=0.0
      WRITE(*,*) ' FLAT PLATE FLOWS: UE AND DUEDX'
      WRITE(*,*) UE, DUEDX
C
C ASSUME TW = CONSTANT. OTHER CASES NEED TW(N). FOR VARYING
```

```
C UE, YOU MUST GIVE TE(N), RHOE(N), AND PE(N).
      TW=1.0
      TE=1.0
      HE=TE/XM1
      RHOE=1.0
      PE=1.0/(GAMMA(1)*XMIF**2)
C
      DO 10 M=1,MMAX
      U(M)=UE
      U0(M)=UE
      V(M)=0.0
      V0(M)=0.0
      H(M)=HE
      H0(M)=HE
      TREF=XM1*H0(M)
      TEMP(M)=TREF
      GAMMA(M)=GAMMA(1)
      CMU(M)=(1.0+T1)*TREF**1.5/(TREF+T1)
      RHO(M)=1.0/(XM1*H(M))
      RHN(M)=RHO(M)
   10 CONTINUE
      U(1)=0.0
      U0(1)=0.0
      H(1)=TW/XM1
      H0(1)=H(1)
      TREF=XM1*H0(1)
      TEMP(1)=TREF
      CMU(1)=(1.0+T1)*TREF**1.5/(TREF+T1)
      RHO(1)=1.0/(XM1*H(1))
      RHN(1)=RHO(1)
C
C ASSUME UNIFORM INITIAL PROFILES OF U AND V.
C CAN BE CHANGED BY THE USER.
C
      MEST=2
      MMM=MEST+3
      DO 20 M=2,MEST
      U0(M)=UE
      V0(M)=0.0
      H0(M)=HE
   20 RHO(M)=1.0/(XM1*H0(M))
C
C     WRITE(*,*) '  ____________________'
C     WRITE(*,*) ' INITIAL PROFILES OF U AND V'
C     CALL OUTPUT(MMAX,U0,V0,RHO)
      CF(1)=(4.0*U0(2)-U0(3))*CMU(1)/(UE*UE*DY*RE*RHOE)
C
      RR=PR**0.5
```

```
      HAW=HE*(1.0+0.195*RR*XMIF**2)
      ST(1)=(3.0*H0(1)-4.0*H0(2)+H0(3))*CMU(1)
     1/(2.0*PR*UE*DY*RE*RHOE*(H(1)-HAW))
C
      DO 200 N=2,NMAX
      NNX=N
      U0(MMAX)=UE
      H0(MMAX)=HE
      RHO(MMAX)=1.0/(XM1*H0(MMAX))
C
      B(1)=1.0
      C(1)=0.0
      R(1)=0.0
      A(MMAX)=0.0
      B(MMAX)=1.0
      R(MMAX)=UE
      DENO=RE*DY*DY
C
      DO 50 M=2,MMAX-1
      A(M)=-0.5*RHO(M)*V0(M)/DY - CMU(M-1)/DENO
      B(M)=RHO(M)*U0(M)/DX + 0.5*(CMU(M-1)+CMU(M))*2.0
     1/DENO
      C(M)=0.5*RHO(M)*V0(M)/DY - CMU(M)/DENO
      R(M)=RHOE*UE*DUEDX + RHO(M)*U0(M)*U0(M)/DX
   50 CONTINUE
C
      CALL TRID(MMAX, A, B, C, R, U)
C
      B(1)=1.0
      C(1)=0.0
      R(1)=H0(1)
      A(MMAX)=0.0
      B(MMAX)=1.0
      R(MMAX)=HE
      DENO=PR*RE*DY*DY
C
      DO 60 M=2,MMAX-1
      A(M)=-0.5*RHO(M)*V0(M)/DY - CMU(M-1)/DENO
      B(M)=RHO(M)*U0(M)/DX + 0.5*(CMU(M-1)+CMU(M))*2.0
     1/DENO
      C(M)=0.5*RHO(M)*V0(M)/DY - CMU(M)/DENO
      R(M)=-RHOE*UE*DUEDX*U0(M) + RHO(M)*U0(M)*H0(M)/DX
     1+CMU(M)*(U(M+1)-U(M-1))*(U(M+1)-U(M-1))
     2/(4.0*RE*DY*DY)
   60 CONTINUE
C
      CALL TRID(MMAX, A, B, C, R, H)
C
      MT=MEST-8
```

```
      IF(MT.LT.1) MT=1
      DO 90 M=MT,MMAX
      IF(U(M) .GT. 0.99*UE) THEN
      MEST=M
      MMM=MEST+3
      GO TO 120
      ENDIF
   90 CONTINUE
  120 CONTINUE
C
      CALL INTERP(NNX,MMM,GAMMA,TEMP,H)
C
      DO 65 M=2,MMAX-1
      CMU(M)=(1.0+T1)*(0.5*(TEMP(M)+TEMP(M-1)))**1.5
     1/(0.5*(TEMP(M)+TEMP(M-1))+T1)
   65 RHN(M)=PE*GAMMA(M)/(0.5*(H(M-1)+H(M))*(GAMMA(M)-1.0))
      DO 80 M=2,MMAX-1
      RHOUA=RHO(M)*U0(M)
      RHOUB=RHO(M-1)*U0(M-1)
      R1=RHN(M-1)/RHN(M)
      V(M)=R1*V(M-1)-(0.5*DY/DX)*(U(M)+R1*U(M-1)-(RHOUA+RHOUB)/RHN(M))
      IF(U(M).EQ.U(M-1).AND.U0(M).EQ.U0(M-1).AND.V(M).EQ.V(M-1))
     1V(M)=0.0
   80 CONTINUE
      DO 110 M=2,MMAX
      U0(M)=U(M)
      V0(M)=V(M)
      H0(M)=H(M)
  110 RHO(M)=RHN(M)
      CF(N)=(4.0*U0(2)-U0(3))*CMU(1)/(UE*UE*DY*RE*RHOE)
      ST(N)=(3.0*H0(1)-4.0*H0(2)+H0(3))*CMU(1)
     1/(2.0*PR*UE*DY*RE*RHOE*(H(1)-HAW))
  200 CONTINUE
      WRITE(*,*) ' FINAL PROFILES OF U AND V'
      CALL OUTPUT(MMAX,U,V,RHO)
      WRITE(*,*) ' N    CF        ST'
      WRITE(*,320) (N,CF(N),ST(N),N=1,NMAX,5)
C
C
  320 FORMAT(1X,I3,1X,F10.5,1X,F10.5)
      STOP
      END
C
C
C
      SUBROUTINE INTERP(NNX,MM,GAMMA,TEMP,H)
C
C THIS SUBROUTINE PERFORMS LAGRANGIAN INTERPOLATION FOR THE
C RATIO OF SPECIFIC HEATS AND ALSO OBTAINS THE TEMPERATURES.
```

```
C
      COMMON/CON/ XMIF,TINF,XM1
      DIMENSION XT(8),  YGAM(8)
      DIMENSION H(101),GAMMA(101),TEMP(101)
C
C GAMMA IS A FUNCTION OF TEMPERATURE FOR DRY AIR.
C
      DATA XT/273.0,373.0,473.0,573.0,673.0,773.0,873.0,973.0/
C
      DATA YGAM/1.401,1.397,1.390,1.378,1.368,1.357,1.346,1.338/
C
C     WRITE(*,*) XT(8),YGAM(8)
C
C
C
      DO 100 M=2,MM+10
      J=0
      TEMP(M)=XM1*H(M)*TINF
   80 J=J+1
      XINT=TEMP(M)
      IF(J.GT.160) GO TO 70
      YOUT=0.0
      DO 20 K=1,8
      TERM=YGAM(K)
      DO 10 L=1,8
      IF(K.NE.L) TERM=TERM*(XINT-XT(L))/(XT(K)-XT(L))
   10 CONTINUE
      YOUT=YOUT+TERM
   20 CONTINUE
      TEMP(M)=(1.4/YOUT)*(YOUT-1.0)*XMIF**2*H(M)*TINF
      IF(ABS(TEMP(M)-XINT).LE.0.005) GO TO 70
      GO TO 80
   70 CONTINUE
      GAMMA(M)=YOUT
      TEMP(M)=TEMP(M)/TINF
C
  100 CONTINUE
      RETURN
      END
C
C
      SUBROUTINE TRID(MM,A,B,C,R,S)
      DIMENSION A(101), B(101),C(101), R(101)
      DIMENSION S(101),RP(101),GAM(101)
C
C
      GAM(1)=C(1)/B(1)
      RP(1)=R(1)/B(1)
C
      DO 40 M=2,MM
```

```
      DENO=B(M)-A(M)*GAM(M-1)
      GAM(M)=C(M)/DENO
      RP(M)=(R(M)-A(M)*RP(M-1))/DENO
   40 CONTINUE
C
      S(MM)=RP(MM)
      DO 60 M=1,MM-1
      S(MM-M)=RP(MM-M)-GAM(MM-M)*S(MM-M+1)
   60 CONTINUE
      RETURN
      END
C
      SUBROUTINE OUTPUT(MMAX,U,V,RHO)
      COMMON/CONST/ RE, DX, DY, NCON
      DIMENSION U(101),V(101),RHO(101),Y(101)
C
C
      DO 15 M=1,MMAX
   15 Y(M)=(M-1)*DY
      WRITE(*,1)
      WRITE(*,2)  (M,Y(M),U(M),V(M),RHO(M),M=1,MMAX,2)
    1 FORMAT(//,' M',6X,'Y',10X,'U',10X,'V',10X,'RHO',/)
    2 FORMAT(1X,I3,1X,F10.5,1X,F10.5,1X,F10.5,1X,F10.5)
      RETURN
      END
```

Program MOSES

The **MOSES** integral method for 2D, incompressible turbulent flows (see Section 7–7) is the basis for this program.

```
      PROGRAM MOSES
C
C     THIS CODE IMPLEMENTS THE MOSES METHOD FOR CONSTANT-
C     PROPERTY, CONSTANT-DENSITY TURBULENT FLOWS.
C     THE USER MUST SPECIFY THE PROBLEM WITH:
C      KINEMATIC VISCOSITY, CNU (DIMENSIONS L**2/T);
C      NUMBER OF STATIONS, NMAX;
C      EDGE VELOCITY DISTRIBUTION, UE(N) (L/T); AND EDGE VELOCITY
C       GRADIENT, DUE(N) (1/T);
C      INITIAL AND FINAL VALUE OF X, XI, AND XF (L);
C      INITIAL VALUE FOR RE, BASED ON THETA, RTHETA;
C      INITIAL VALUE FOR THE SKIN FRICTION COEFFICIENT, CF;
C      FOR EACH NEW CASE, ONE MUST COMPILE TO EXECUTE.
C
C     THE OUTPUT CONTAINS A LISTING OF THE INPUT, THEN
C      XX(N) (L UNITS), UE(N) (L/T), DUEDX(N) FOR N=1 TO NMAX,
C     AND THE RESULTS:
C      SURFACE DISTANCE, XX (L UNITS),  EDGE VELOCITY, UE (L/T),
```

```
C     SKIN FRICTION COEFFICIENT, CF,  BOUNDARY LAYER THICKNESS,
C     DELTA (L), MOMENTUM THICKNESS, THETA (L),  RE, BASED ON
C     THETA, RTHETA, AND SHAPE FACTOR, H.
C
C
      COMMON/CON/ CNU,CF
C     SET THE FOLLOWING DIMENSION TO NMAX:
      DIMENSION XX(21),UE(21),DUE(21)
C
C INPUT DATA:  THE USER CAN CHANGE ACCORDING TO THE
C        REQUIRED PROBLEM
C SAMPLE PROBLEM OF A FLAT PLATE WITH UE = 10.0, FOLLOWED BY AN
C ADVERSE PRESSURE GRADIENT. START AT END OF FLAT PLATE, X = 5.0.
C USE SIMPLE INTEGRAL SOLUTION TO GET INITIAL VALUES.
      CNU=1.0E-05
      NMAX=21
      XI=5.0
      XF=7.0
      CF=0.002665
      RTHETA=8336.3
      UE(1)=10.0
      DUE(1)=-1.0
C
C
      WRITE(*,*)  ' MOSES METHOD FOR TURBULENT BOUNDARY LAYER'
      WRITE(*,*)  ' INPUT: NMAX, CNU, CF, RTHETA'
      WRITE(*,*)   NMAX, CNU, CF, RTHETA
C
C
      WRITE(*,9)
    9 FORMAT(10X,'N',9X,'XX(N)',11X,'UE(N)',11X,'DUE(N)')
C    SPECIFY EDGE VELOCITY AND GRADIENT.

      DELX=(XF-XI)/(NMAX-1)
      XX(1)=XI
      DO 10 N=2,NMAX
      XX(N)=XI+(N-1)*DELX
      UE(N)=15.0-XX(N)
      DUE(N)=-1.0
      WRITE(*,*) N,XX(N),UE(N),DUE(N)
   10 CONTINUE
      WRITE(*,92)
   92 FORMAT(//,3X,'X',8X,'UE',8X,'CF',8X,'DELTA',
     15X,'THETA',5X,'RTHETA',5X,'H')
C
      CALL INIT(A,RDELTA,RTHETA)
C
      X=XX(1)
      U=UE(1)
```

```
      CALL PLOT(X,U,A,RDELTA,RTHETA,DELTA,THETA,H)
C
      DO 200 N=2,NMAX
      DX=XX(N)-XX(N-1)
      U=UE(N-1)
      DU=DUE(N-1)
C
      CALL DERIV(U,DU,DX,A,RDELTA,DA,DRDELTA)
C
      DA1=DA
      DR1=DRDELTA
      AI=A
      RI=RDELTA
      A=A+DA
      RDELTA=RDELTA+DRDELTA
      U=UE(N)
      DU=DUE(N)
      Z=A*RDELTA
      IF(Z.LE.10.0) GO TO 201
      CALL DERIV(U,DU,DX,A,RDELTA,DA,DRDELTA)
C
      A=AI+(DA1+DA)/2.0
      RDELTA=RI+(DR1+DRDELTA)/2.0
      X=XX(N)
      Z=A*RDELTA
      IF(Z.LE.10.0) GO TO 201
C
      CALL PLOT(X,U,A,RDELTA,RTHETA,DELTA,THETA,H)
C
  200 CONTINUE
  201 CONTINUE
      WRITE(*,95) X,H
   95 FORMAT(//,' SEPARAT AT X = ',F8.4,4X,'H =',F8.4)
      STOP
      END
C
      SUBROUTINE INIT(A,RDTA,RTTA)
      COMMON/CON/ CNU,CF
C
      A=SQRT(CF/2.0)/0.41
      T1=0.0
      DEL=-0.1
      DO 20 J=1,7
      DEL=-0.1*DEL
   80 T1=T1+DEL
      RDTA=RTTA/T1
      B=A*(LOG(A*RDTA)+1.1584)-1.0
      T2=A-0.5*B+1.5833*A*B-2.0*A*A-0.37143*B*B
     1+123.4*A/RDTA-26.65/RDTA
```

```
      Z1=(T1-T2)*DEL/ABS(DEL)
      IF(ABS(Z1).LT.0.00001) GO TO 45
      IF(Z1.LT.0.0) GO TO 80
   20 CONTINUE
C
      WRITE(*,94)
   94 FORMAT(//,3X,'FAILED TO FIND ROOT')
   45 CONTINUE
      RETURN
      END
C
      SUBROUTINE PLOT(X,U,A,RDTA,RTTA,DTA,TTA,H)
      COMMON/CON/ CNU,CF
C
      CF=2.0*(A*0.41)**2
      B=A*(LOG(A*RDTA)+1.1584)-1.0
      D1=A-0.5*B+26.15/RDTA
      T1=A-0.5*B-2.0*A*A+1.5833*A*B-0.37143*B*B
     1-26.15/RDTA+123.4*A/RDTA
      H=D1/T1
      RTTA=RDTA*T1
      DTA=RDTA*CNU/U
      TTA=DTA*T1
      WRITE(*,25) X,U,CF,DTA,TTA,RTTA,H
   25 FORMAT(2F9.4,1X,3(1X,F8.5),2X,F8.2,3X,F6.4)
      RETURN
      END
C
      SUBROUTINE DERIV(U,DU,DX,A,RDTA,DA,DRT)
      COMMON/CON/ CNU,CF
C
      B=A*(LOG(A*RDTA)+1.1584)-1.0
      F1=(-0.0792-0.58320*A-0.02026*B)*RDTA+123.4
      F1=F1+(-0.5+1.58325*A-0.74286*B)*RDTA*LOG(A*RDTA)
      F2=0.5*A-0.5*B-0.41667*A*A+0.84020*A*B-0.37143*B*B
      F3=(0.06321-0.73792*A+0.091565*B)*RDTA+123.4
      F3=F3+(-0.27705+0.91760*A-0.29708*B)*RDTA*LOG(A*RDTA)
      F4=0.02295*A-0.04185*B-0.043606*A*A+0.10214*A*B-0.03934*B*B
      G1=(-2.0*A+B+2.0*A*A-1.58333*A*B+0.3714*B*B)*RDTA-123.4*A
      G1=G1*DU*DX/U+0.41*A*0.41*A*U*DX/CNU
      R1=A-B/2.0-2.0*A*A+1.58333*A*B-0.37143*B*B-26.65/RDTA+123.4
     1*A/RDTA
      C1=((A/0.3)-1.26*B)*(0.0225+125.0/RDTA)*R1
      G2=-1.32238*A+0.5541*B+1.75725*A*A-1.25115*A*B+0.25714*B*B
      G2=(G2*RDTA-123.4*A)*DX*DU/U+((0.41*A)**2-C1)*U*DX/CNU
      DENO=F1*F4-F3*F2
      DA=(F4*G1-F2*G2)/DENO
      DRT=(F1*G2-F3*G1)/DENO
      RETURN
      END
```

Program ITBL

This program uses an implicit numerical method for **I**ncompressible, **T**urbulent **B**oundary **L**ayers, with three options for the turbulence model: (1) mixing length model, (2) eddy viscosity model, and (3) TKE model as the Prandtl energy method. For simplicity, a uniformly spaced grid is used. As discussed in Chapter 7, that is not efficient. With a uniform grid, you must use about 500–700 points across the layer to get accurate values for the skin friction. That may lead to problems with the 640K limit on many PC's with older operating systems. In any event, never use fewer than 400 points across the layer.

```
      PROGRAM ITBL
C 2D BOUNDARY LAYER COMPUTATION, INCOMPRESSIBLE, TURBULENT,
C 1ST ORDER, IMPLICIT
C
C MIXING LENGTH MODEL OR EDDY VISCOSITY MODEL OR TKE MODEL
C
C EQUATIONS ARE DIMENSIONLESS, USING FREE STREAM VELOCITY, UINF,
C VISCOSITY, MUINF, DENSITY, RHOINF, AND A REFERENCE
C LENGTH, L; X/L, Y/L, U/UINF; ALSO, RE=RHOINF*UINF*L/MUINF.
C PICK L=1.0.
C
C SAMPLE PROBLEM OF A FLAT PLATE WITH UINF=10.0. START AT X=5.0.
C GO TO X=6.0. TAKE NU(=MUINF/RHOINF)=1.0E-5. REX=5.0E6.
C
C USE SIMPLE INTEGRAL SOLUTION TO GET INITIAL VALUES. DELTA=0.0856.
C CF=0.002665. OTHER FLOWS CAN BE SET BY USER.
C
      COMMON/CONST/ RE, DX, DY, NCON
      COMMON/RST/ UE, DUEDX, RKAP, YPA
      INTEGER MODEL
C
C
      DIMENSION U(550),UO(550),V(550),VO(550),CF(150)
      DIMENSION A(550), B(550),C(550), R(550)
      DIMENSION TKEO(550),TMU(550),UE(150),DUEDX(150)
C
      RKAP=0.41
      YPA=9.7
C
      WRITE(*,*) ' PROGRAM ITBL'
      WRITE(*,*) ' 2-D BOUNDARY LAYER, INCOMPRESSIBLE TURBULENT'
      WRITE(*,*) ' 1ST ORDER IMPLICIT'
C
C INPUT DATA
C
C YOU MUST GIVE INITIAL X (XI), FINAL X (XF), (CNU), (UNINF), (NMAX),
C (MMAX), AND (DY).
```

```
C
C
C PICK MMAX BASED ON INITIAL BOUNDARY LAYER THICKNESS AND
C NUMBER OF POINTS ACROSS THE LAYER.  USE AT LEAST 400 ACROSS
C DELTA.  ADD AT LEAST 100 POINTS ABOVE DELTA.
C
C PICK NMAX BASED ON LENGTH OF REGION AND DX DESIRED.  DX CAN
C BE OF THE ORDER OF INITIAL DELTA/FIVE. TAKE L=1.0.
C
C
      XI=5.0
      XF=6.0
      CNU=0.00001
      UINF=10.0
      MMAX=550
      NMAX=101
      DY=0.0002145
      DX=(XF-XI)/(NMAX-1)
      CF(1)=0.002665
      DEL=0.0856
C
      RE=UINF*1.0/CNU
C
      WRITE(*,*) ' INPUT:  MMAX, NMAX, RE, DY,  DX'
      WRITE(*,*)         MMAX, NMAX, RE, DY, DX
C
C
      DO 5 N=1,NMAX
      XX=XI+(N-1)*DX
      UE(N)=1.0
      DUEDX(N)=0.0
    5 CONTINUE
C
      WRITE(*,*) ' FLAT PLATE FLOW'
      WRITE(*,*) ' N X UE DUEDX'
      WRITE(*,6)  (N, XX(N),UE(N), DUEDX(N),N=1,NMAX,4)
    6 FORMAT(I5,1X,2F10.5)
C
C CHOOSE TURBULENCE MODEL.
      MODEL=1
      IF(MODEL.EQ.1) WRITE(*,*) ' MIXING LENGTH MODEL'
      IF(MODEL.EQ.2) WRITE(*,*) ' EDDY-VISCOSITY MODEL'
      IF(MODEL.EQ.3) WRITE(*,*) ' TKE MODEL'
C
C
      DO 10 M=1,MMAX
      U(M)=UE(1)
      U0(M)=UE(1)
```

```
      V(M)=0.0
      V0(M)=0.0
   10 TKE0(M)=0.0
      U(1)=0.0
      U0(1)=0.0
C
C THE INITIAL PROFILES OF U AND V CAN BE CHANGED BY THE USER.
C MEST IS THE M INDEX FOR THE INITIAL DELTA.
C ASSUME A COLES WAKE LAW INITIAL VELOCITY PROFILE.
C
      USUE=SQRT(CF(1)/2.0)
      RED=RE*DEL*USUE*UE(1)
      MEST=401
      FM1=MEST-1
      MMM=MEST+2
      DO 20 M=2,MEST
      Y0D=(M-1)/FM1
      U0(M)=USUE*UE(1)*(1.0/RKAP*ALOG(Y0D*RED)+4.90+0.51/RKAP
     1*2.0*(SIN(Y0D*1.5708))**2)
   20 V0(M)=0.0
      NMAXP=NMAX
      MMAXP=MEST+100
      WRITE(*,*) ' '
      WRITE(*,*) ' INITIAL PROFILES OF U AND V'
      CALL OUTPUT(MMAXP,U0,V0)
      NCON=1
C
      DO 200 N=2,NMAX
      NNX=N
      U0(MMAX)=UE(N)
      V0(MMAX)=0.0
C
C
      IF(MODEL.EQ.1) CALL MIXING(NNX,MMAX,MEST,U0,TMU,CF)
      IF(MODEL.EQ.2) CALL EDDY(NNX,MMAX,MEST,U0,TMU,CF)
      IF(MODEL.EQ.3) CALL TKES(NNX,MMM,MEST,U0,V0,TKE0,TMU,CF)
      NCON=2
C
      B(1)=1.0
      C(1)=0.0
      R(1)=0.0
      A(MMAX)=0.0
      B(MMAX)=1.0
      C(MMAX)=0.0
      R(MMAX)=UE(N)
      DENO=RE*DY*DY
C
      DO 50 M=2,MMAX-1
```

```
      A(M)=-0.5*V0(M)/DY-(1.0+TMU(M-1))/DENO
      B(M)=U0(M)/DX+(2.0+TMU(M-1)+TMU(M))/DENO
      C(M)=0.5*V0(M)/DY-(1.0+TMU(M))/DENO
      R(M)=UE(N)*DUEDX(N)+U0(M)*U0(M)/DX
   50 CONTINUE
C
      CALL TRID(MMAX, A, B, C, R, U)
C
      DO 80 M=2,MMAX-1
   80 V(M)=V(M-1)-(0.5*DY/DX)*(U(M)-U0(M)+U(M-1)-U0(M-1))
      MT=MEST-10
      DO 90 M=MT,MMAX
      IF(U(M) .GT. 0.99*UE(N)) THEN
      MEST=M
      MMM=MEST+3
      MMAXP=MEST+100
      GO TO 120
      ENDIF
   90 CONTINUE
  120 CONTINUE
      DO 110 M=2,MMAX
      U0(M)=U(M)
  110 V0(M)=V(M)
      CF(N)=(4.0*U0(2)-U0(3))/
     1(UE(N)**2*DY*RE)
C        CHECK IF NEAR SEPARATION
      IF(CF(N) .LE. 0.0001) THEN
      WRITE(*,*) ' '
      WRITE(*,*) ' NEAR SEPARATION'
      NMAXP=N
      GO TO 300
      ENDIF
  200 CONTINUE
C
  300 CONTINUE
      WRITE(*,*) ' '
      WRITE(*,*) '  N     CF'
      WRITE(*,320) (N,CF(N),N=1,NMAXP,1)
      WRITE(*,*) ' '
      WRITE(*,*) ' FINAL PROFILES OF U AND V'
      IF(MMAXP .GT. MMAX) MMAXP=MMAX
      CALL OUTPUT(MMAXP,U,V)
C
C
  320 FORMAT(1X,I3,1X,F10.5)
      STOP
      END
C
```

```
      SUBROUTINE MIXING(NNX,MMAX,MEST,U0,T,CF)
C
C MIXING LENGTH MODEL
C
C
      COMMON/CONST/ RE, DX, DY, NCON
      COMMON/RST/ UE, DUEDX, RKAP, YPA
      DIMENSION U0(550),T(550),CF(150),UE(150),DUEDX(150)
C
      XLMO=0.09*(MEST-1)*DY
C
      T(1)=0.0
      DO 100 M=2,MMAX-1
      Y=(M-1)*DY
      YP=Y*UE(NNX)*RE*SQRT(0.5*CF(NNX-1))
      XLMM=RKAP*(1.0-EXP(-YP/26.0))*Y
      IF(XLMM.GT.XLMO) XLMM=XLMO
      T(M)=RE*XLMM**2*ABS(U0(M+1)-U0(M-1))/(2.0*DY)
  100 CONTINUE
      T(MMAX)=T(MMAX-1)
      RETURN
      END
C
C
C
      SUBROUTINE  EDDY(NNX,MMAX,MEST,U0,T,CF)
C
C EDDY VISCOSITY MODEL
C
C
      COMMON/CONST/ RE, DX, DY, NCON
      COMMON/RST/ UE, DUEDX, RKAP, YPA
      DIMENSION U0(550),T(550),CF(150),UE(150),DUEDX(150)
C
      DELST=0.0
      DO 10 M=2,MEST
   10 DELST=DELST+DY*(1.0-0.5*(U0(M-1)+U0(M))/UE(NNX))
C
      RMUT=0.018*RE*UE(NNX)*DELST
C
      DO 100 M=1,MMAX
      Y=(M-1)*DY
      YP=Y*UE(NNX)*RE*SQRT(0.5*CF(NNX-1))
      T(M)=RKAP*(YP-YPA*TANH(YP/YPA))
      IF(T(M).GT.RMUT) T(M)=RMUT
  100 CONTINUE
      RETURN
      END
```

```
C
C
C
      SUBROUTINE TKES(NNX,MMM,MEST,U0,V0,TKE0,T,CF)
C
C TKE TURBULENT MODEL
C
      COMMON/CONST/ RE, DX, DY, NCON
      COMMON/RST/ UE, DUEDX, RKAP, YPA
C
      DIMENSION U0(550),V0(550),T(550),CF(150),UE(150),DUEDX(150)
      DIMENSION A(550), B(550),C(550), R(550)
      DIMENSION TKE(550),TKE0(550),TLM(550)
C
      SIGK=1.0
      TKE(MMM)=0.0
      IF(NCON.GT.1) GO TO 300
C
      USS=0.5*UE(NNX)*UE(NNX)*CF(NNX-1)
C
C ASSUME INITIAL PROFILE OF K FROM FIG. 7-20
C
      FM1=MEST-1
      DO 100 M=1,MEST
      Y0D=(M-1)/FM1
      IF(Y0D .LT. 0.001) THEN
      TKE0(M)=5500.0*Y0D*USS
      ELSE IF(Y0D .LE. 0.12) THEN
      TKE0(M)=0.5*USS*(7.0-3.0/0.099*(Y0D-0.11))
      ELSE
      TKE0(M)=0.5*USS*(-7.0/0.90*(Y0D-1.0))
      ENDIF
  100 CONTINUE
C
C
  300 CONTINUE
      CD=0.09
      CDF=CD**0.25
      XLMO=0.09*(MEST-1)*DY
C
      DO 150 M=1,MMM
      Y=(M-1)*DY
      YP=Y*UE(NNX)*RE*SQRT(0.5*CF(NNX-1))
      XLMM=RKAP*(1.0-EXP(-YP/26.0))*Y
      IF(XLMM.GT.XLMO) XLMM=XLMO
      TLM(M)=XLMM*CDF
  150 CONTINUE
```

```
C
      STKM=0.0
      DENO=SIGK*DY*DY
      B(1)=1.0
      C(1)=0.0
      R(1)=0.0
      A(MMM)=0.0
      B(MMM)=1.0
      R(MMM)=TKE(MMM)
C
      DO 40 M=2,MMM-1
      STKM1=STKM
      STKM=SQRT(TKE0(M))
      A(M)=-0.5*V0(M)/DY-STKM1*TLM(M-1)/DENO
      B(M)=U0(M)/DX+CD*STKM/TLM(M)+(STKM1*TLM(M-1)+STKM*
     1TLM(M))/DENO
      C(M)=-STKM*TLM(M)/DENO+0.5*V0(M)/DY
      R(M)=STKM*TLM(M)*(U0(M+1)-U0(M-1))*(U0(M+1)-U0(M-1))/
     1(4.0*DY*DY)+U0(M)*TKE0(M)/DX
   40 CONTINUE
C
      CALL TRID(MMM, A, B, C, R, TKE)
C
      T(1)=0.0
      DO 200 M=2,MMM

      T(M)=RE*TLM(M)*SQRT(TKE0(M))
  200 CONTINUE
C
      DO 250 M=1,MMM
      TKE0(M)=TKE(M)
  250 CONTINUE
      TKE0(1)=0.0
      TKE0(MMM)=0.0
      RETURN
      END
C
      SUBROUTINE TRID(MM,A,B,C,R,S)
      DIMENSION A(550), B(550),C(550), R(550)
      DIMENSION S(550),RP(550),GAM(550)
C
C
      GAM(1)=C(1)/B(1)
      RP(1)=R(1)/B(1)
C
      DO 40 M=2,MM
      DENO=B(M)-A(M)*GAM(M-1)
```

```
      GAM(M)=C(M)/DENO
      RP(M)=(R(M)-A(M)*RP(M-1))/DENO
   40 CONTINUE
C
      S(MM)=RP(MM)
      DO 60 M=1,MM-1
      S(MM-M)=RP(MM-M)-GAM(MM-M)*S(MM-M+1)
   60 CONTINUE
      RETURN
      END
C
      SUBROUTINE OUTPUT(MMAX,U,V)
      COMMON/CONST/ RE, DX, DY, NCON
      DIMENSION U(550),V(550),Y(550)
C
C
      DO 15 M=1,MMAX
   15 Y(M)=(M-1)*DY
      WRITE(*,1)  (M,Y(M),U(M),V(M),M=1,MMAX,4)
    1 FORMAT(//,'M',6X,'Y',10X,'U',10X,'V'/(1X,I3,3(1X,F10.4)))
      RETURN
      END
```

Program ITBLS

The program ITBLS implements an implicit numerical treatment of 2D, **I**ncompressible **T**urbulent **B**oundary **L**ayers with **S**tretched normal grid spacing. The stretching used is very strong, and it is controlled by two parameters YMAX and BETA. Use the values recommended for these parameters in the code. With a strongly stretched grid, one can obtain accurate solutions with 50–100 points across the layer. This is in contrast to the situation with ITBL, where at least 400 points are required to obtain reasonable accuracy. The only turbulence model available is a Reichardt-Clauser eddy viscosity combination. ITBLS is essentially the code first described in Grossman and Schetz (1985).

```
      PROGRAM ITBLS
C
C INCOMPRESSIBLE TURBULENT BOUNDARY LAYER
C REICHARDT-CLAUSER EDDY VISCOSITY MODEL
C IMPLICIT  - COEFFICIENTS LAGGED, 1ST ORDER,
C ***STRETCHED***
C
C
C EQUATIONS ARE DIMENSIONLESS, USING REYNOLDS NUMBER BASED
C ON FREE STREAM VELOCITY, UINF, VISCOSITY, MUINF,
C DENSITY, RHOINF, AND LENGTH, L; RE=RHOINF*UINF*L/MUINF.
```

```
C PICK L = 1.0.
C
C STRETCHING PARAMETERS ARE YMAX AND BETA.
C
C SAMPLE PROBLEM OF A FLAT PLATE WITH UINF = 10.0 M/S.
C START AT LEADING EDGE OF PLATE, X = 0.0, WITH UNIFORM
C PROFILES. TAKE KINEMATIC VISCOSITY AT 1.0E-6 M**2/S.
C REL = 1.0E7. OTHER FLOWS CAN BE SET BY USER.
C
C
      REAL MUTPR, MUT1, MUT1I
C
      DIMENSION Z(101),Y(101),U(101),U0(101),V(101),V0(101),GAM(101)
      DIMENSION YY(101),VV(101),DDEL(101), UE(101), DUEDX(101)
      DIMENSION X(101),DEL(101),DELST(101),THETA(101),CF(101)
C
C
      RKAP=0.41
      YAPL=9.7
C
      WRITE(*,*) ' PROGRAM ITBLS`
      WRITE(*,*) ' TURBULENT BOUNDARY LAYER FLATE PLATE '
      WRITE(*,*) ' EDDY-VISCOSITY MODEL, IMPLICIT METHOD, STRETCHED'
C
C INPUT DATA
C
C YOU MUST GIVE INITIAL X (XI), FINAL X (XF), (CNU), (UINF), (NMAX),
C AND (MMAX).
C
C PICK MMAX BASED ON DESIRED NUMBER OF POINTS ACROSS THE LAYER.
C MMAX OF 98 RECOMMENDED.
C
C PICK NMAX BASED ON LENGTH OF REGION AND DX DESIRED. DX CAN
C BE OF THE ORDER OF INITIAL DELTA/FIVE.
C
C
C
      XI=0.0
      XF=1.0
      CNU=0.000001
      UINF=10.0
      MMAX=98
      NMAX=100
      DZ=1.0/(MMAX-1)
      DX=(XF-XI)/(NMAX-1)
C     STRETCHING PARAMETERS YMAX AND BETA=BE1
      YMAX=100
      BE1=1.001
```

```
      WRITE(*,*) ' '
      WRITE(*,*) ' YMAX AND BETA ARE ',YMAX,BE1
C
      RE=UINF*1.0/CNU
C
      BE2=(BE1+1.0)/(BE1-1.0)
      BE3=2.0*BE1/(YMAX*ALOG(BE2))
      BE4=BE1**2
      ZY0=BE3/(BE4-1.0)
      SREL=SQRT(RE)

      WRITE(*,*) ' '
      WRITE(*,*) ' INPUT:  MMAX, NMAX, RE, DX,  DZ'
      WRITE(*,1)          MMAX, NMAX, RE, DX, DZ
    1 FORMAT(8X,2I6,1P3E15.3)
C
C
      DEL(1)=0
      DELST(1)=0
      THETA(1)=0
      CF(1)=0
      Z(1)=0

      X(1)=0
      UE(1)=1.0
      DUEDX(1)=0
      DO 44 N=2,NMAX
      X(N)=X(N-1)+DX
C FLAT-PLATE FLOW ASSUMED. CAN BE CHANGED BY USER.
      UE(N)=1.0
      DUEDX(N)=0.0
   44 CONTINUE
C
C ASSUME UNIFORM INITIAL PROFILES. CAN BE CHANGED.
C
      DO 5 M=2,MMAX
      Z(M)=(M-1)*DZ
      CC=BE2**(1.0-Z(M))
      Y(M)=YMAX*(BE1-1.0)*(BE2-CC)/(1.0+CC)
      YY(M)=Y(M)/SREL
      U0(M)=UE(1)
      U(M)=UE(1)
      V0(M)=0
      V(M)=0
    5 CONTINUE
      U0(1)=0.0
      U(1)=0.0
      V0(1)=0.0
      V(1)=0.0
```

```
C
      WRITE(*,*) ' '
      WRITE(*,*) ' FLAT PLATE FLOW'
      WRITE(*,*) ' N      X         UE        DUEDX'
      WRITE(*,6)  (N,X(N),UE(N),DUEDX(N),N=1,NMAX,4)
    6 FORMAT(I5,1X,3F10.5)
C
      WRITE(*,*) ' '
      WRITE(*,*) ' INITIAL PROFILES OF U AND V'
      CALL OUTPUT(MMAX,YY,U0,V0)
      WRITE(*,*) ' '
      WRITE(*,*) ' '
      WRITE(*,*) '  N    X    UE    THETA    DELTS  CF
     1     DELT'
C
      DO 200 N=2,NMAX
      U(MMAX)=UE(N)
      THETA(N)=0.0
      DELST(N)=0.5
C
      DO 50 MM=2,MMAX-1
      M=MMAX-MM+1
      MUTPR=0
      MUT1=1.0+0.018*UE(N)*DELST(N-1)*SREL
      CFF=SQRT(CF(N-1)*SREL*0.5)
      YPL=Y(M)*CFF
      RYY=YPL/YAPL
      THYPL=1.0
      IF(RYY .LE. 10.0) THEN
      EXYPL=EXP(RYY)
      THYPL=(EXYPL**2-1.0)/(EXYPL**2+1.0)
      ENDIF
C
      MUT1I=1.0+RKAP*(YPL-YAPL*THYPL)
      IF(MUT1I.LE.MUT1) THEN
      MUT1=MUT1I
      MUTPR=CFF*RKAP*THYPL**2
      ENDIF
      ZY=BE3/(BE4-(1-Y(M)/YMAX)**2)
      ZYY=-2.0*ZY*ZY*(1-Y(M)/YMAX)/(YMAX*BE3)
      A2= ZY*ZY*MUT1
      A1=-ZY*(V0(M)-MUTPR)+MUT1*ZYY
      B1=A2+0.5*DZ*A1
      B3=A2-0.5*DZ*A1
      B2=-2*A2
      HH=DX/(U0(M)*DZ**2)
      DD=1.0/(1.0-HH*B2)
      CC1=B1*HH*DD
      CC2=B3*HH*DD
```

```
      CC3=(HH*DZ**2*UE(N)*DUEDX(N)+U0(M))*DD
      IF(M.EQ.MMAX-1) THEN
      GAM(M) = CC2
      DDEL(M)=CC3+CC1*UE(N)
      ELSE
      DUM=1.0/(1.0-CC1*GAM(M+1))
      GAM(M)=CC2*DUM
      DDEL(M)=(CC3+CC1*DDEL(M+1))*DUM
      ENDIF
   50 CONTINUE
C
      DO 80 M=2,MMAX-1
      ZY=BE3/(BE4-(1-Y(M)/YMAX)**2)
      H1=DZ/(2.0*DX*ZY)
      U(M)=GAM(M)*U(M-1)+DDEL(M)
      V(M)=V(M-1)-H1*(U(M)-U0(M)+U(M-1)-U0(M-1))
   80 VV(M)=V(M)/SREL
      DO 90 M=2,MMAX
      IF(U(M) .GT. 0.99*UE(N)) GO TO 120
   90 CONTINUE
  120 CONTINUE
      ZBL=Z(M-1)+DZ*((.99*UE(N)-U(M-1))/(U(M)-U(M-1)))
      CC=BE2**(1.0-ZBL)
      DEL(N)=YMAX*(BE1-1.0)*(BE2-CC)/(1.0+CC)
      DO 110 M=2,MMAX-1
      ZY=BE3/(BE4-(1-Y(M)/YMAX)**2)
      F1=(1.0-U(M)/UE(N))/ZY
      F2=F1*U(M)/UE(N)
      DELST(N)=DELST(N)+F1
      THETA(N)=THETA(N)+F2
      U0(M)=U(M)
  110 V0(M)=V(M)
      THETA(N)=THETA(N)*DZ
      DELST(N)=DELST(N)*DZ
      CF(N)=ZY0*(4.0*U(2)-U(3))/(UE(N)**2*DZ)
C
      XDEL=DEL(N)/SREL
      XDELST=DELST(N)/SREL
      XTHETA=THETA(N)/SREL
      XCF=CF(N)/SREL
C
      WRITE(*,320) N, X(N), UE(N), XTHETA, XDELST, XCF, XDEL
  200 CONTINUE

C
  300 CONTINUE
      WRITE(*,*) ' '
      WRITE(*,*) ' FINAL PROFILES OF U AND V'
      CALL OUTPUT(MMAX,YY,U,VV)
```

```
C
C
C
  320 FORMAT(1X,I3,1X,6F10.5)
  321 FORMAT(1X,I3,1X,2F10.5)
      STOP
      END
C
      SUBROUTINE OUTPUT(MMAX,X,U,V)
      DIMENSION X(MMAX), U(MMAX), V(MMAX)
C
C
      WRITE(*,1)  (M,X(M),U(M),V(M),M=1,MMAX,1)
    1 FORMAT(/,'   M',6X,'   Y',10X,'U',10X,'V'/(1X,I3,3(1X,F10.7)))
      RETURN
      END
```

Program JETWAKE

Program JETWAKE uses an implicit numerical method for the solution of turbulent free-mixing flows, i.e., **JET**s and **WAKE**s (see Chapter 9), with either a mixing length model or a TKE model as the Prandtl energy method for modeling turbulence.

```
      PROGRAM JETWAKE
C
C 2D JET COMPUTATION,  INCOMPRESSIBLE TURBULENT JETS AND WAKES,
C 1ST ORDER, IMPLICIT
C
C MIXING LENGTH MODEL  OR  TKE MODEL
C
C EQUATIONS ARE DIMENSIONLESS, USING FREESTREAM VELOCITY, UINF,
C VISCOSITY, MUINF, DENSITY, RHOINF, AND A REFERENCE
C LENGTH, L; X/L, Y/L, U/UINF;  ALSO, RE=RHOINF*UINF*L/MUINF.
C PICK L=1.0.
C
C TAKE UE=UINF=CONST. WITH UNIFORM JET AT UJ AS A SAMPLE PROBLEM.
C
      COMMON/CONST/ RE,DX,DY,UJ,UE,NCON
      INTEGER MODEL
C
C
      DIMENSION U(500),U0(500),V(500),V0(500)
      DIMENSION A(500), B(500),C(500), R(500)
      DIMENSION TKE0(500),TMU(500),BW(150), X(150), UC(150)
C
      WRITE(*,*) ' PROGRAM JETWAKE'
      WRITE(*,*) ' 2-D JET, INCOMPRESSIBLE TURBULENT'
```

```
      WRITE(*,*) ' 1ST ORDER, IMPLICIT'
C
C INPUT DATA
C
C YOU MUST GIVE INITIAL X (XI), FINAL X (XF), (CNU), (UINF), (NMAX),
C (MMAX), AND (DY). ALSO, YOU MUST GIVE INITIAL VELOCITY PROFILE.
C
C PICK MMAX BASED ON INITIAL VISCOUS LAYER THICKNESS. USE ABOUT
C 50 POINTS ACROSS THE INITIAL LAYER THICKNESS, AND ADD ABOUT 300-400
C POINTS ABOVE THAT. PICK NMAX BASED ON THE LENGTH OF REGION AND DX
C DESIRED. DX CAN BE ABOUT ONE-HALF INITIAL VISCOUS LAYER WIDTH.
C
      XI=0.0
      XF=1.0
      CNU=0.00002
      UINF=10.0
      MMAX=500
      NMAX=101
      DY=0.0004
      DX=(XF-XI)/(NMAX-1)
C
      RE=UINF*1.0/CNU
      X(1)=XI
C
      WRITE(*,*) ' INPUT:  MMAX, NMAX, RE, DY,  DX'
      WRITE(*,1)          MMAX, NMAX, RE, DY, DX
    1 FORMAT(2I6,F10.1,1P2E15.4)
C
C
      UJ=4.0
      UE=1.0
      DUEDX=0.0
      WRITE(*,*) ' UJ, UE AND DUEDX'
      WRITE(*,*)  UJ, UE, DUEDX
      WRITE(*,*) ' '
C
C CHOOSE TURBULENCE MODEL, MODEL=1 FOR MIXING LENGTH, MODEL=2 FOR TKE.
      MODEL=2
      IF(MODEL.EQ.1) WRITE(*,*) ' MIXING LENGTH MODEL'
      IF(MODEL.EQ.2) WRITE(*,*) ' TKE MODEL'
C
C
      DO 10 M=1,MMAX
      U(M)=UE
      U0(M)=UE
      V(M)=0.0
      V0(M)=0.0
      TMU(M)=1.0E-6
   10 TKE0(M)=(0.005*UE)**2
```

```
C
C THE INITIAL PROFILES OF U CAN BE CHANGED BY THE USER.
C
      MEST=51
      MEST1=MEST-1
      MMM=MEST+5
      DO 20 M=1,MEST
      U(M)=UJ
   20 U0(M)=UJ
      BW(1)=DY*(MEST-MEST1)
      WRITE(*,*) ' '
      WRITE(*,*) ' INITIAL PROFILES OF U AND V'
      CALL OUTPUT(MMAX,U0,V0)
C
      NCON=1
      UC(1)=UJ
      DO 200 N=2,NMAX
      X(N)=(N-1)*DX+X(1)
      NNX=N
      U0(MMAX)=UE
      V0(MMAX)=0.0
C
C
      IF(MODEL.EQ.1) CALL MIXING(NNX,MMM,U0,BW,TMU)
      IF(MODEL.EQ.2) CALL TKES(NNX,MMM,MEST,U0,V0,BW,TKE0,TMU)
      NCON=2
C
      B(1)=1.0
      C(1)=-1.0
      R(1)=0.0
      A(MMAX)=0.0
      B(MMAX)=1.0
      R(MMAX)=UE
      DENO=RE*DY*DY
C
      DO 50 M=2,MMAX-1
      A(M)=-0.5*V0(M)/DY-(1.0+TMU(M-1))/DENO
      B(M)=U0(M)/DX+(2.0+TMU(M-1)+TMU(M))/DENO
      C(M)=0.5*V0(M)/DY-(1.0+TMU(M))/DENO
      R(M)=UE*DUEDX+U0(M)*U0(M)/DX
   50 CONTINUE
C
      CALL TRID(MMAX, A, B, C, R, U)
      UC(N)=U(1)
C
      DO 80 M=2,MMAX-1
   80 V(M)=V(M-1)-(0.5*DY/DX)*(U(M)-U0(M)+U(M-1)-U0(M-1))
      MT=MEST-10
      DO 90 M=MT,MMAX
```

```
      IF(U(M) .LT. 1.01*UE  ) THEN
      MEST=M
      MMM=MEST+5
      GO TO 120
      ENDIF
   90 CONTINUE
  120 CONTINUE
      IF(U(2).LT.0.99*UJ) GO TO 122
      DO 95 M=2,MEST
      IF(U(M).LT.0.99*UJ) THEN
      MEST1=M
      GO TO 125
      ENDIF
   95 CONTINUE
  122 MEST1=2
  125 CONTINUE
      IF(MMM .GE. MMAX) THEN
      WRITE(*,*) ' '
      WRITE(*,*) '  JET NEAR TOP BOUNDARY'
      GO TO 201
      ENDIF
      BW(N)=DY*(MEST-MEST1)
      DO 110 M=2,MMAX
      U0(M)=U(M)
  110 V0(M)=V(M)
      U0(1)=U0(2)
      V0(1)=V0(2)
  200 CONTINUE
C
  201 WRITE(*,*) ' FINAL PROFILES OF U AND V '
      CALL OUTPUT(MMAX,U,V)
      WRITE(*,*) ' '
      WRITE(*,*) '  N     X       UC          BW'
      WRITE(*,320) (N,X(N),UC(N),BW(N),N=1,NMAX,10)
C
C
  320 FORMAT(1X,I3,1X,3F10.5)
      STOP
      END
C
    SUBROUTINE MIXING(NNX,MMM,U0,BW,T)
C
C MIXING LENGTH MODEL
C
C
      COMMON/CONST/ RE,DX,DY,UJ,UE,NCON
      DIMENSION U0(500),T(500),BW(150)
```

```
C
      BW1=BW(NNX-1)
      DO 80 M=2,MMM
      T(M)=(0.15*BW1)**2*RE*ABS((U0(M+1)-U0(M-1))/(2.0*DY))
   80 CONTINUE
      T(1)=T(2)
      RETURN
      END
C
      SUBROUTINE TKES(NNX,MMM,MEST,U0,V0,BW,TKE0,T)
C
C TKE TURBULENT MODEL
C
      COMMON/CONST/ RE,DX,DY,UJ,UE,NCON
C
      DIMENSION U0(500),V0(500),T(500),BW(150)
      DIMENSION A(500),B(500),C(500),R(500)
      DIMENSION TKE(500),TKE0(500)
C
      SIGK=1.0
      IF(NCON.GT.1) GO TO 300
C
C INITIAL PROFILE OF K CAN BE CHANGED BY THE USER.
C
      DO 20 M=1,MEST
      TKE0(M)=0.09*UJ**2
   20 CONTINUE
C
C
  300 CONTINUE
      CD=0.09
      CDF=CD**0.25
      BW1=CDF*0.15*BW(NNX-1)
C
      STKM=SQRT(TKE0(1))
      DENO=SIGK*DY*DY
      B(1)=1.0
      C(1)=-1.0
      R(1)=0.0
      A(MMM)=0.0
      B(MMM)=1.0
      R(MMM)=(0.005*UE)**2
C
C     WRITE(*,*) ' DENO=',DENO
      DO 40 M=2,MMM-1
      STKM1=STKM
      STKM=SQRT(TKE0(M))
```

```
      A(M)=-0.5*V0(M)/DY-STKM1*BW1/DENO
      B(M)=U0(M)/DX+CD*STKM/BW1+(STKM1+STKM)*BW1/DENO
      C(M)=-STKM*BW1/DENO+0.5*V0(M)/DY
      R(M)=STKM*BW1*(U0(M+1)-U0(M-1))*(U0(M+1)-U0(M-1))/
     1(4.0*DY*DY)+U0(M)*TKE0(M)/DX
   40 CONTINUE
C
      CALL TRID(MMM, A, B, C, R, TKE)
C
      DO 200 M=2,MMM
      T(M)=RE*BW1*SQRT((TKE0(M-1)+TKE0(M))/2.0)
  200 CONTINUE
      T(1)=T(2)
C
      DO 250 M=2,MMM
      IF(TKE(M).LT.0.0) TKE(M)=0.0
      TKE0(M)=TKE(M)
  250 CONTINUE
      TKE0(MMM)=0.0
      TKE0(1)=TKE0(2)
      RETURN
      END
C
      SUBROUTINE TRID(MM,A,B,C,R,S)
      DIMENSION A(500), B(500),C(500), R(500)
      DIMENSION S(500),RP(500),GAM(500)
C
      GAM(1)=C(1)/B(1)
      RP(1)=R(1)/B(1)
C
      DO 40 M=2,MM
      DENO=B(M)-A(M)*GAM(M-1)
      GAM(M)=C(M)/DENO
      RP(M)=(R(M)-A(M)*RP(M-1))/DENO
   40 CONTINUE
C
      S(MM)=RP(MM)
      DO 60 M=1,MM-1
      S(MM-M)=RP(MM-M)-GAM(MM-M)*S(MM-M+1)
   60 CONTINUE
      RETURN
      END
C
      SUBROUTINE OUTPUT(MMAX,U,V)
      COMMON/CONST/ RE,DX,DY,UJ,UE,NCON
      DIMENSION U(500),V(500),Y(500)
C
      DO 15 M=1,MMAX
```

```
15 Y(M)=(M-1)*DY
   WRITE(*,1)  (M,Y(M),U(M),V(M),M=1,MMAX,8)
 1 FORMAT(//,' M',6X,'Y',10X,'U',10X,'V',/,(1X,I3,3(1X,F10.5)))
   RETURN
   END
```

REFERENCES

The following abbreviations are used in the listing of the references:

Aeronaut. Eng. Rev.	*Aeronautical Engineering Review*
Aeronaut. Q.	*Aeronautical Quarterly*
J. Aerosp. Sci.	*Journal of the Aerospace Sciences*
J. Aero. Sci.	*Journal of the Aeronautical Sciences*
AIAA J.	*AIAA Journal*
AIChE J.	*AIChE Journal*
ARC	*Aeronautical Research Council*
J. Aircraft	*Journal of Aircraft*
Ann. Rev. Fluid Mech.	*Annual Reviews of Fluid Mechanics*
Q. Appl. Math.	*Quarterly of Applied Mathematics*
J. Appl. Mech.	*Journal of Applied Mechanics*
J. Appl. Phys.	*Journal of Applied Physics*
Astronaut. Acta	*Astronautica Acta*
Ergeb. AVA Göttingen	*Ergebnisse Aerodynamische Versuchsanstalt, Göttingen*
J. Basic Eng.	*Journal of Basic Engineering*
J. Fluids Eng.	*Journal of Fluids Engineering*
J. Fluid Mech.	*Journal of Fluid Mechanics*

Forsch. Ing.-Wes.	*Forschung auf dem Gebiete des Ingenieur-Wesens*
J. Heat Trans.	*Journal of Heat Transfer*
Ind. Eng. Chem.	*Industrial and Engineering Chemistry*
Ing.-Arch.	*Ingenieur-Archiv*
Inst. Aero. Sci.	*Institute of Aerospace Sciences*
Int. J. Heat Mass Trans.	*International Journal of Heat and Mass Transfer*
Jet Prop.	*Jet Propulsion*
Luftfahrt-Forsch.	*Luftfahrt Forschung*
J. Mech. Eng. Sci.	*Journal of Mechanical Engineering Science*
NACA	*National Advisory Committee for Aeronautics*
NASA	*National Aeronautics and Space Administration*
Num. Heat Trans.	*Numerical Heat Transfer*
Phil. Mag.	*Philosophical Magazine*
Philos. Trans. R. Soc. Lond.	*Philosophical Transactions of the Royal Society of London*
Phys. Fluids	*Physics of Fluids*
Proc. Cambridge Phil. Soc.	*Proceedings of the Cambridge Philosophical Society*
Proc. R. Soc. Lond.	*Proceedings of the Royal Society of London*
Prog. Aero. Sci.	*Progress in Aerospace Sciences*
J. Propulsion Power	*Journal of Propulsion and Power*
Trans. ASME	*Transactions of the ASME*
Trans. Camb. Phil. Soc.	*Transactions of the Cambridge Philosophical Society*
Trans. Soc. Nav. Arch. Mar. Eng.	*Transactions of the Society of Naval Architects and Marine Engineers*
VDI Forschungsh.	*Verein Deutsche Ingenieure Forschungsheft*
ZAMM	*Zeitschrift für angewandete Mathematik und Mechanik*
ZAMP	*Zeitschrift für angewandete Mathematik und Physik*
Z. Ver. Deut. Ingr.	*Zeitschrift Verein Deutsche Ingenieure*

Abid, R., "Extension of the Johnson-King Turbulence Model to the 3-D Flows," AIAA-88-0223 (1988).

Abramovich, G. N., *The Theory of Turbulent Jets*, MIT Press, Cambridge, MA, 1960 (English edition).

ACHARYA, M., "Effects of Compressibility on Boundary-Layer Turbulence," AIAA Paper 76-334 (1976).

ACKERET, J., "Aspects of Internal Flow," in G. Sovran (ed.), *Fluid Mechanics of Internal Flow*, Elsevier, Amsterdam (1967).

ACRIVOS, A., SHAH, M.J., and PETERSEN, E. E., "Momentum and Heat Transfer in Laminar Boundary-Layer Flows of Non-Newtonian Fluids Past External Surfaces," *AIChE J.*, Vol. 6, pp. 312–317 (1960).

ADAMSON, T. C., and MESSITER, A. F., "Analysis of Two-Dimensional Interactions between Shock Waves and Boundary Layers," *Ann. Rev. Fluid Mech.*, Vol. 12, pp. 103–138 (1980).

ALBER, I. and LEES, L., "Integral Theory for Supersonic Turbulent Base Flows," *AIAA J.*, Vol. 6, pp. 1343–1351 (1968).

ANDERSEN, P. S., KAYS, W. M., and MOFFAT, R. J., "The Turbulent Boundary Layer on a Porous Plate: An Experimental Study of the Fluid Mechanics for Adverse Pressure Gradients," Tech. Rept. No. 15, Dept. Mech. Eng., Stanford Univ. (1972).

ANDERSON, E. C., and LEWIS, C. H., "Laminar or Turbulent Boundary-Layer Flows of Perfect Gases or Reacting Gas Mixtures in Chemical Equilibrium," NASA CR–1893 (1971).

ANTONIA, R. A., and BILGER, R. W., "An Experimental Investigation of an Axisymmetric Jet in a Co-flowing Air Stream," *J. Fluid Mech.*, Vol. 61, pp. 805–822 (1973).

ANTONIA, R. A., PRABHU, A., and STEPHENSON, S. F., "Conditionally Sampled Measurements in a Heated Turbulent Jet," *J. Fluid Mech.*, Vol. 72, pp. 455–480 (1975).

BACHALO, W. D., and JOHNSON, D. A., "An Investigation of Transonic Turbulent Boundary Layer Separation Generated on an Axisymmetric Flow Model," AIAA 79-1479 (1979).

BACK, L. H., CUFFEL, R. F., and MESSIER, P. F., "Laminarization of a Turbulent Boundary Layer in Nozzle Flow," *AIAA J.*, Vol. 7, pp. 730–733 (1969).

BALDWIN, B. and LOMAX, H., "Thin-Layer Approximation and Algebraic Model for Separated Turbulent Flows," AIAA Paper No. 78-257 (1978).

BARBIN, A. R., and JONES, J. B., "Turbulent Flow in the Inlet Region of a Smooth Pipe," *J. Basic Eng.*, Vol. 85, pp. 29–34 (1963).

BECKER, H. A., HOTTEL, H. C., and WILLIAMS, G. C., "The Nozzle Fluid Concentration Field of the Round Turbulent Free Jet," *J. Fluid Mech.*, Vol. 30, pp. 285–303 (1967).

BELOV, V. M., "Experimental Investigation of Heat Transfer in a Turbulent Boundary Layer with a Step-like Change in Thermal Boundary Conditions on the Wall," Thesis (Cand. Sci.), The Bauman Higher Technical College, 1976.

BERG, B. VAN DEN, and ELSENAAR, A., "Measurements in a Three-Dimensional Incompressible Turbulent Boundary Layer under Infinite Swept Wing Conditions," NLR TR 72092 U (1972).

BERG, B. VAN DEN, HUMPHREYS, D. A., KRAUSE, E., and LINDHOUT, J. P .F., "Three-Dimensional Turbulent Boundary Layers—Calculations and Experiments," in *Notes on Numerical Fluid Mechanics*, Vol. 19, Vieweg, Braunschweig (1988).

BIRCH, A. D., BROWN, D. R., DODSON, M. G., and THOMAS, J. R., "The Turbulent Concentration Field of a Methane Jet," *J. Fluid Mech.*, Vol. 88, pp. 431–449 (1978).

BIRCH, S. F., and EGGERS, J. M., "A Critical Review of the Experimental Data for Developed Free Turbulent Shear Layers," *Free Turbulent Shear Flows*, NASA SP-321 (1973).

BLACKWELDER, R. F., and KAPLAN, R. E., "On the Wall Structure of the Turbulent Boundary Layer," *J. Fluid Mech.*, Vol. 76, pp. 89–112 (1976).

BLASIUS, H., "Das Ähnlichkeitsgesetz bei Reibungsvorgangen in Flüssigkeiten," [The Similarity Law for Frictional Processes in Fluids] *Forsch. Ing-Wes.*, No. 131, Berlin (1913).

BLASIUS, H., "Grenzschichten in Flüssigkeiten mit kleiner Reibung," [Boundary Layers in Fluids with Small Viscosity], *ZAMP.*, Vol. 56, No. 1, pp. 1–37 (1908). [Available in translation as NACA TM 1256 (1950).]

BLOM, J., "An Experimental Determination of the Turbulent Prandtl Number in a Developing Temperature Boundary Layer," Technische Hogeschool, Eindhoven, 1970.

BLOTTNER, F. G., "Finite Difference Methods of Solution of the Boundary-Layer Equations," *AIAA J.*, Vol. 8, pp. 193–205 (1970).

BLOTTNER, F. G., "Computational Techniques for Boundary Layers," in E. Krause, *Computational Methods for Inviscid and Viscous Two- and Three-Dimensional Flow Fields*, AGARD Lecture Series 73, von Karman Inst., Belgium (1975).

BLOTTNER, F. G., "Entry Flow in Straight and Curved Channels with Slender Channel Approximations," *J. Fluids Eng.*, Vol. 99, pp. 666–674 (1977).

BOGDANOFF, D. W., "Compressibility Effects in Turbulent Shear Layers," *AIAA J.*, Vol. 21, pp. 926–927 (1983).

BOUSSINESQ, J., "Théorie de l'écoulement tourbillant" [Theory of Turbulent Flow] [Memoires presentes par diverse savants à l'academie des sciences de l'institut de France] Vol. 23, p. 46 (1877).

BRADSHAW, P., "Calculation of Three-Dimensional Turbulent Boundary Layers," *J. Fluid Mech.*, Vol. 46, pp. 417–445 (1971).

BRADSHAW, P., "Compressible Turbulent Shear Layers," *Ann. Rev. Fluid Mech.*, Annual Reviews, Inc., Palo Alto, CA (1977).

BRADSHAW, P., "Physics and Modelling of Three-Dimensional Boundary Layers," in J. Cousteix, *Computation of Three-Dimensional Boundary Layers Including Separation*, AGARD Rept. No. 741 (1987).

BRADSHAW, P., FERRISS, D. H., and ATWELL, N. P., "Calculation of Boundary Layer Development Using the Turbulence Energy Equation," in S. J. Kline, M. V. Morkovin, G. S. Sovran and D. J. Cockrell, *Computation of Turbulent Boundary Layers—1968 AFOSR-IFP-Stanford Conference*, Stanford University Press, Stanford, CA, 1969. (Original paper published in *J. Fluid Mech.*, Vol. 28, pp. 539–616 (1967).)

BRADSHAW, P., and PONTIKOS, N. S., "Measurements in the Turbulent Boundary Layer on an "Infinite" Swept Wing," *J. Fluid Mech.*, Vol. 159, pp. 105–130 (1985).

BRADSHAW, P., and TERRELL, M. G., "The Response of a Turbulent Boundary Layer on an Infinite Swept Wing to the Sudden Removal of Pressure Gradient," NPL Aer. Rept. 1305, ARC 31514 (1969).

BRINICH, P. F., "Boundary Layer Transition at Mach 3.12 with and without Single Roughness Element," NACA TN 3267 (1954).

BRINICH, P. F., and DIACONIS, N. S., "Boundary Layer Development and Skin Friction at Mach-Number 3.05," NACA TN 2742 (1952).

BROWN, G. L., and ROSHKO, A. A., "On Density Effects and Large Structure in Turbulent Shear Layers," *J. Fluid Mech.*, Vol. 64, pp. 775–816 (1974).

BRUNDRETT, E., and BAINES, W. D., "The Production and Diffusion of Vorticity in Duct Flow," *J. Fluid Mech.*, Vol. 19, pp. 375–394 (1964).

BUDDENBERG, J. W., and WILKE, C. R., "Calculation of Gas Mixture Viscosities," *Ind. Eng. Chem.*, Vol. 41, pp. 1345–1347 (1949).

BUSEMANN, A., "Gasdynamik" [Gas Dynamics], *Handbuch der Experimentalphysik*, Vol. IV, Pt. 1, Leipzig, 1931.

BUSHNELL, D., and BECKWITH, I., "Calculation of Non-equilibrium Hypersonic Turbulent Boundary Layers and Comparisons with Experimental Data," *AIAA J.*, Vol. 8, pp. 1462–1469 (1970).

CAILLE, J., and SCHETZ, J. A., "Three-Dimensional Strip-Integral Method for Incompressible Turbulent Boundary Layers," *AIAA J.*, Vol. 30, pp. 1207–1213 (1992).

CAILLE, J., and SCHETZ, J. A., "New Wall Treatment for Numerical Navier-Stokes Solution of 2D and 3D Incompressible Turbulent Flows," AIAA 91-1736 (1991).

CARSLAW, H. S., and JAEGER, J. C., *Conduction of Heat in Solids*, Clarendon Press, Oxford, 1959.

CEBECI, T., "Calculation of Compressible Turbulent Boundary Layers with Heat and Mass Transfer," *AIAA J.*, Vol. 9, pp. 1091–1097 (1971).

CEBECI, T., "Calculation of Three-Dimensional Boundary Layers, Pt. 1, Swept Infinite Cylinders and Small Cross Flow," *AIAA J.*, Vol. 12, pp. 779–786 (1974).

CEBECI, T., "Calculation of Three-Dimensional Boundary Layers, Pt. 2, Three-Dimensional Flows in Cartesian Coordinates," *AIAA J.*, Vol. 13, pp. 1056–1064 (1975).

CEBECI, T. and BRADSHAW, P., *Momentum Transfer in Boundary Layers*, Hemisphere Publ. Corp., Washington (1977).

CEBECI, T., and BRADSHAW, P., *Physical and Computational Aspects of Convective Heat Transfer*, Springer-Verlag, New York (1984).

CEBECI, T., and CHANG, K. C., "A General Method for Calculating Momentum and Heat Transfer in Laminar and Turbulent Duct Flows," *Num. Heat Transfer*, Vol. 1, pp. 39–68 (1978).

CEBECI, T., and MOSINSKIS, G. J., "Prediction of Turbulent Boundary Layers with Mass Transfer, Including Highly Accelerating Flows," *J. Heat Trans.*, Vol. 93, pp. 271–79 (1971).

CEBECI, T., and SMITH, A. M. O., "A Finite-difference Solution of the Incompressible Turbulent Boundary Layer Equations by an Eddy Viscosity Concept," in S. J. Kline, M. V. Morkovin, G. Sovran, and D. J. Cockerell, *Computation of Turbulent Boundary Layers—1968 AFOSR-IFP-Stanford Conference*, Stanford University Press, Stanford, CA, 1969.

CEBECI, T., and SMITH, A. M. O., *Analysis of Turbulent Boundary Layers*, Academic Press, New York (1974).

CHANG, K. C., and PATEL, V. C., "Calculation of Three-Dimensional Boundary Layers on Ship Forms," Iowa Inst. Hydraulic Res. Rept. No. 178 (1975).

CHAPMAN, D. R., and KESTER, R. H., "Measurements of Turbulent Skin Friction on Cylinders in Axial Flow at Subsonic and Supersonic Velocities," *J. Aerosp. Sci.*, Vol. 20, pp. 441–448 (1953).

CHAPMAN, D. R., KUEHN, D. M., and LARSON, H. K., "Investigation of Separating Flows in Supersonic and Subsonic Streams with Emphasis on the Effects of Transition," NACA Rept. 1356 (1958).

CHEN, F-J, MALIK, M. R., and BECKWITH, I. E., "Boundary-Layer Transition on a Cone and Flat Plate at Mach 3.5," *AIAA J.*, Vol. 27, pp. 687–693 (1989).

CHEVRAY, R., "The Turbulent Wake of a Body of Revolution," *J. Basic Eng.*, Vol. 90, pp. 275–284 (1968).

CHEVRAY, R., and TUTU, N. K., "Intermittency and Preferential Transport of Heat in a Round Jet," *J. Fluid Mech.*, Vol. 88, pp. 133–160 (1978).

CHIENG, C. C., and LAUNDER, B. E., "On the Calculation of Turbulent Heat Transport Downstream from an Abrupt Pipe Expansion," *Num. Heat Trans.*, Vol. 3, pp. 189–207 (1980).

CHRISS, D. E., "Experimental Study of Turbulent Mixing of Subsonic Axisymmetric Gas Streams," Arnold Engineering Development Center, AEDC-TR-68-133 (1968).

CHRISTIAN, W. J., and KEZIOS, S. P., "Sublimation from Sharp-Edged Cylinders in Axisymmetric Flow, Including Influence of Surface Curvature," *AIChE J.*, Vol. 5, pp. 61–68 (1959).

CLAUSER, F. H., "Turbulent Boundary Layers in Adverse Pressure Gradients," *J. Aerosp. Sci.*, Vol. 21, pp. 91–108 (1954).

CLAUSER, F. H., "The Turbulent Boundary Layer," in *Advances in Applied Mechanics*, Vol. IV, Academic Press, New York, 1956.

COHEN, C. B., and RESHOTKO, E., 'The Compressible Laminar Boundary Layer with Heat Transfer and Arbitrary Pressure Gradient," NACA Rept. 1294 (1956a).

COHEN, C. B., and RESHOTKO, E., "Similar Solutions for the Compressible Laminar Boundary Layer with Heat Transfer and Pressure Gradient," NACA Rept. 1293 (1956b).

COLES, D., "Measurements of Turbulent Friction on a Smooth Flat Plate in Supersonic Flows," *J. Aerosp. Sci.*, Vol. 21, pp. 433–448 (1954).

COLES, D., "The Law of the Wall in Turbulent Shear Flow," in *50 Jahre Grenzschichtforschung*, F. Vieweg & Sohn, Braunschweig, 1955.

COLES, D., "The Law of the Wake in the Turbulent Boundary Layer," *J. Fluid Mech.*, Vol. 1, pp. 191–226 (1956).

COLES, D., "The Turbulent Boundary Layer in a Compressible Fluid," The Rand Corp., Rep. R-403-PR (1962).

COLLIER, F. S., and SCHETZ, J. A., "Injection into a Turbulent Boundary Layer through Porous Walls with Different Surface Geometries," AIAA Paper No. 83-0295 (1983).

COOKE, J. C., "The Boundary Layer of a Class of Infinite Yawed Cylinders," *Proc. Cambridge Phil. Soc.*, Vol. 46, pp. 645–648 (1950).

CORRSIN, S., and KISTLER, A. L., "The Free-Stream Boundaries of Turbulent Flows," NACA TN 3133 (1954).

CORRSIN, S., and UBEROI, M., "Further Experiments on the Flow and Heat Transfer in a Heated Turbulent Air Jet," NACA TN 1895 (1949).

CROCCO, L., "Sulla transmissione del calore da una lamina piana a un fluido scorrente ad alta velocita" [On the Heat Transfer from a Flat Plate to a High Speed Flow], *L'Aerotechnica*, Vol. 12, fasc. 2, pp. 181–197 (1932). [Available in translation as NACA TM 690 (1932).]

CROCCO, L., *Lo strato limite laminare nei gas*, [The Laminar Boundary Layer in Gases], Monografie Scientifiche di Aeronautica No. 3, Ministero della Difesa-Aeronautics, Roma, 1946. [Trans. in North American Aviation Aerophysics Lab. Rep. AL-684 (1948).]

CROCCO, L. and LEES, L., "A Mixing Theory for the Interaction between Dissipative Flows and Nearly Isentropic Streams," *J. Aero. Sci.*, Vol. 19, pp. 649–676 (1952).

CROCHET, M. J., DAVIES, A. R., and WALTERS, K., *Numerical Simulation of Non-Newtonian Flow*, Elsevier, Amsterdam (1984).

CROSS, A. G. T., "Calculation of Compressible Three-Dimensional Turbulent Boundary Layers with Particular Reference to Wings and Bodies," British Aerospace (Brough) Note YAD 3379 (1979).

DALY, B. J., and HARLOW, F. H., "Transport Equations in Turbulence," *Phys. Fluids*, Vol. 13, p. 2634 (1970).

DASH, S. M., WEILERSTEIN, G., and VAGLIO-LAURIN, R., "Compressibility Effects in Free Turbulent Shear Flows," AFOSR-TR-75-1436, (1975).

DASH, S. M., WOLF, D. E., and SEINER, J. M., "Analysis of Turbulent Underexpanded Jets, Part I: Parabolized Navier-Stokes Model, SCIPVIS," *AIAA J.*, Vol. 23, pp. 505–514 (1985).

DEAN, R. B., "Interaction of Turbulent Shear Layers in Duct Flow," Ph.D. thesis, Univ. London (1972).

DEISSLER, R. G., "Analysis of Turbulent Heat Transfer, Mass Transfer and Friction in Smooth Tubes at High Prandtl and Schmidt Numbers," NACA Rep. 1210 (1955).

DEMETRIADES, A., "Computation of Numerical data on the Mean Flow from Compressible Turbulent Wake Experiments," Publ. No. U-4970, Aero. Div., Philco-Ford Corp. (1971).

DEMUREN, A. O., and RODI, W., "Calculation of Turbulence-Driven Secondary Motion in Non-Circular Ducts," *J. Fluid Mech.*, Vol. 140, pp. 189–222 (1984).

DER, J., and RAETZ, G. S., "Solution of General Three-Dimensional Laminar Boundary-Layer Problems by an Exact Numerical Method," Inst. Aero. Sci. Paper No. 62-70 (1962).

DEY, J., and NARASHIMA, R., "Integral Method for the Calculation of Incompressible Two-Dimensional Transitional Boundary Layers," *J. Aircraft*, Vol. 27, pp. 859–865 (1990).

DHAWAN, S., "Direct Measurements of Skin Friction," NACA Rep. 1121 (1953).

DONALDSON, C. D., "A Progress Report on an Attempt to Construct an Invariant Model of Turbulent Shear Flows," *Turbulent Shear Flows*, AGARD CP-93 (1971).

DORODNITSYN, A. A., "Boundary Layer in a Compressible Gas," *Prikladnaya Matematika Mekhanika*, Vol. 6 (1942).

DRYDEN, H. L., "Airflow in the Boundary Layer Near a Plate," NACA Rep. 562 (1936).

DRYDEN, H. L., "Review of Published Data on the Effect of Roughness on Transition from Laminar to Turbulent Flow," *J. Aerosp. Sci.*, Vol. 20, pp. 477–482 (1953).

DWYER, H. A., "Solution of a Three-Dimensional Boundary Layer Flow with Separation," *AIAA J.*, Vol. 6, pp. 1336–1342 (1968).

DWYER, H. A., and MCCROSKEY, W. J., "Crossflow and Unsteady Boundary-Layer Effects on Rotating Blades," *AIAA J.*, Vol. 9, pp. 1498–1505 (1971).

DWYER, H. A., and SANDERS, B. R., "A Physically Optimum Difference Scheme for Three-Dimensional Boundary Layers," *Proc. 4th Int. Conf. Num. Meth. Fluid Dyn.*, Boulder, CO (1975).

EAST, J. L., and PIERCE, F. J., "Explicit Numerical Solution of the Three-Dimensional Incompressible Boundary-Layer Equations," *AIAA J.*, Vol. 10, pp. 1216–1223 (1972).

ECKERT, E. R. G., "Die Berechnung des Wärmeübergangs in der laminaren Grenzschicht" [The Calculation of Heat Transfer in Laminar Boundary Layers] *VDI-Forschungsh.*, No. 416 (1942).

ECKERT, E. R. G., "Engineering Relations for Heat Transfer and Friction in High-Velocity Laminar and Turbulent Flow over Surfaces with Constant Pressure and Temperature," *Trans. ASME*, Vol. 78, pp. 1273–1283 (1956).

ECKERT, E. R. G., and DRAKE, R. M., JR., *Heat and Mass Transfer*, McGraw-Hill, New York, 1959.

ECKERT, E. R. G., and DREWITZ, O., "Der Wärmeübergang an eine mit grosser, Geschwindigkeit längsamgeströmte Platte" [The Heat Transfer to a Plate in High Speed Flow], *Forsch. Ing.-wes.*, Vol. 11, pp. 116–124 (1940).

ECKERT, E. R. G., HARTNETT, J. P., and BIRKEBAK, R., "Simplified Equations for Calculating Local and Total Heat Flux to Non-isothermal Surface," *J. Aerosp. Sci.*, Vol. 24, pp. 549–551 (1957).

ECKERT, E. R. G., SCHNEIDER, P. J., HAYDAY, A. A., and LARSON, R. M., "Mass-Transfer Cooling of a Laminar Boundary by Injection of a Light-Weight Foreign Gas," *Jet Prop.*, Vol. 3, pp. 34–39 (1958).

ECKERT, E. R. G., and TEWFIK, O. E., "Use of Reference Enthalpy in Specifying the Laminar Heat-Transfer Distribution around Blunt Bodies in Dissociated Air," *J. Aerosp. Sci.*, Vol. 27, pp. 464–466 (1960).

EGGERS, J. M., "Turbulent Mixing of Coaxial Compressible Hydrogen-Air Jets," NASA TN D-6487 (1971).

EVERITT, K. W., and ROBINS, A. G., "The Development and Structure of Turbulent Plane Jets," *J. Fluid Mech.*, pp. 563–583 (1978).

FAGE, A., and FALKNER, V. M., "Relation between Heat Transfer and Surface Friction for Laminar Flow," ARC R&M, No. 1408 (1931).

FALKNER, V. M., and SKAN, S. W., "Some Approximate Solutions of the Boundary Layer Equations," ARC R&M, No. 1314 (1930).

FAY, J. A., and RIDDELL, F. R., "Theory of Stagnation Point Heat Transfer in Dissociated Air," *J. Aero. Sci.*, Vol. 25, pp. 73–85 (1958).

FEINDT, E. G., "Untersuchungen über die Abhängigkeit des Umschlages Laminar-Turbulent von der Oberflächenraughigkeit und der Druckverteilung" [Investigations of the Dependence of Laminar to Turbulent Transition on Surface Roughness and Pressure Gradient], Diss. Braunschweig, 1956; *Jb. 1956 Schiffbautechn. Ges.*, Vol. 50, pp. 180–203 (1957).

FERNHOLZ, H. H., and FINLEY, P. J., "A Further Compilation of Compressible Boundary Layer Data with a Survey of Turbulence Data," AGARD AG-263 (1981).

FERRI, A., LIBBY, P. A., and ZAKKAY, V., "Theoretical and Experimental Investigation of Supersonic Combustion," *3rd ICAS Conf.*, Stockholm, 1962.

FLÜGGE-LOTZ, I., and BLOTTNER, F. G., "Computation of the Compressible Laminar Boundary-Layer Flow Including Displacement Thickness Interaction Using Finite Difference Methods," Stanford Univ. Tech. Rep. 1313 (1962).

FORSTALL, W., JR., and SHAPIRO, A. H., "Momentum and Mass Transfer in Coaxial Gas Jets," *J. Appl. Mech.*, Vol. 72, pp. 339–408 (1950).

FULACHIER, L., "Contribution à l'étude des analogies des champs dynamique et thermique dans une couche limite turbulent, effet de l'aspiration" [Contribution to the Study of Analogies in the Dynamic and Thermal Fields in a Turbulent Boundary Layer, the Effect of Suction.], Thèse (Doc. Sci.), Phys. Univ. Provence, Marseille, 1972.

GESSNER, F. B., "The Origin of Secondary Turbulent Flow along a Corner," *J. Fluid Mech.*, Vol. 58, pp. 1–25 (1973).

GESSNER, F. B., and EMERY, A. F., "The Numerical Prediction of Developing Turbulent Flow in Rectangular Ducts," *J. Fluids Eng.*, Vol. 103, pp. 445–455 (1981).

GIBSON, M. M., "Spectra of Turbulence in a Round Jet," *J. Fluid Mech.*, Vol. 15, pp. 161–173 (1963).

GILREATH, H. E., and SCHETZ, J. A., "Transition and Mixing in the Shear Layer Produced by Tangential Injection in Supersonic Flow," *J. Basic Eng.*, Vol. 93, Ser. D, pp. 610–618 (1971).

GINEVSKII, A. S., "Turbulent Nonisothermal Jets of a Compressible Gas of Variable Composition," *Promyshlennaya Aerodinamika*, No. 27, pp. 31–54 (1966).

GLUSHKO, G. S., "Turbulent Boundary Layer on a Flat Plate in an Incompressible Flow," NASA TT F-10080 (1966).

GÖRTLER, H., "Berechnung von Aufgaben der freien Turbulenz auf Grund eines neuen Näherungsansatzes" [Calculation of the Problem of Free Turbulence on the Basis of a New Approximation]. *ZAMM.*, Vol. 22, pp. 244–254 (1942).

GOSMAN, A. D., and RAPLEY, C. N., "Fully Developed Flow in Passages of Arbitrary Cross-Section," Chap. 11 in *Rec. Adv. Num. Meth. Fluids,* Vol. 1, C. Taylor and K. Morgan (ed.), Pineridge Press, Swansea (1980).

GOVINDAN, T. R., BRILEY, W. R., and MCDONALD, H., "General Three-Dimensional Viscous Primary/Secondary Flow Analysis," *AIAA J.*, Vol. 29, pp. 361–370 (1991).

GRANVILLE, P. S., "A Near-Wall Eddy Viscosity Formula for Turbulent Boundary Layers in Pressure Gradients Suitable for Momentum, Heat or Mass Transfer," in W. W. Bower and M. J. Morris (eds.), *Forum on Turbulent Flows—1989*, FED-Vol. 76, ASME, New York (1989).

GREENOUGH, J., RILEY, J., SOERTRISNO, M., and EBERHARDT, D., "The Effects of Walls on a Compressible Mixing Layer," AIAA 89-0372 (1989).

GROSSMAN, B., and SCHETZ, J. A., "Teaching Boundary Layer Methods in Fluid Mechanics Using Personal Computers," *Int. Conf. Education, Practice and Promotion Comput. Meth. Engrg. Using Small Computers,* Macao (1985).

GRUSCHWITZ, E., "Turbulente Reibungsschichten mit Sekundärströmungen" [Turbulent Shear Layers with Secondary Flows], *Ing.-Arch.*, Vol. 6, pp. 355–365 (1935).

HALL, A. A., and HISLOP, G. S., "Experiments on the Transition of the Laminar Boundary Layer on a Flat Plate," ARC R&M, No. 1843 (1938).

HALL, M. G., "Numerical Method for Calculating Steady Three-Dimensional Laminar Boundary Layers," RAE Tech. Rept. 67145 (1967).

HALL, W. B., and KHAN, S. A., "Experimental Investigation into the Effect of the Thermal Boundary Condition on Heat Transfer in the Entrance Region of a Pipe," *J. Mech. Eng. Sci.*, Vol. 6, pp. 250–255 (1964).

HAMA, F. R., "Boundary Layer Characteristics for Smooth and Rough Surfaces," *Trans. Soc. Nav. Arch. Mar. Eng.*, Vol. 62, pp. 333–358 (1954).

HANJALIC, K., and LAUNDER, B. E., "A Reynolds Stress Model of Turbulence and Its Application to Asymmetric Shear Flows," *J. Fluid Mech.*, Vol. 52, p. 609 (1972).

HARSHA, P. T., "Free Turbulent Mixing: A Critical Evaluation of Theory and Experiment," *Turbulent Shear Flows*, AGARD CP-93 (1971).

HARSHA, P. T., "Prediction of Free Turbulent Mixing Using a Turbulent Kinetic Energy Method," *Free Turbulent Shear Flows*, NASA SP-321 (1973).

HARTNETT, J. P., and ECKERT, E. R. G., "Mass Transfer Cooling in a Laminar Boundary Layer with Constant Fluid Properties," *Trans. ASME*, Vol. 79, pp. 247–254 (1957).

HARTREE, D. R., "On an Equation Occurring in Falkner and Skan's Approximate Treatment of the Equations of the Boundary Layer," *Proc. Camb. Phil. Soc.*, Vol. 33, p. 223 (1937).

HAYES, W. D, and PROBSTEIN, R. F., *Hypersonic Flow Theory*, Academic Press, New York, 1959.

HEDSTROM, B. O. A., "Flow of Plastic Material in Pipes," *Ind. Eng. Chem.*, Vol. 44, pp. 651–656 (1952).

HERRING, H. J., and MELLOR, G. L., "A Method of Calculating Compressible Turbulent Boundary Layers," NASA CR-114 (1968).

HILL, F. K., "Boundary-Layer Measurements in Hypersonic Flow," *J. Aerosp. Sci.*, Vol. 23, pp. 35–42 (1956).

HINZE, J. L., *Turbulence*, McGraw-Hill, New York, 1959.

HO, C. M., and GUTMARK, E., "Vortex Induction and Mass Entrainment in a Small-Aspect-Ratio Elliptic Jet," *J. Fluid Mech.*, Vol. 179, pp. 383–405 (1987).

HOPKINS, E. J., KEENER, E. R., POLEK, T. E., and DWYER, H. A., "Hypersonic Turbulent Skin-Friction and Boundary-Layer Profiles on Nonadiabatic Flat Plates," *AIAA J.*, Vol. 10, pp. 40–48 (1972).

HORSTMANN, K. H., QUAST, A., and REDEKER, G., "Flight and Wind-Tunnel Investigations on Boundary-Layer Transition," *J. Aircraft*, Vol. 27, pp. 146–150 (1990).

HOSSAIN, M. S., "Mathematische Modellierung von turbulenten Auftriebsströmungen," [Mathematical Modeling of Turbulent Buoyant Flows], Ph.D. thesis, University of Karlsruhe, 1979.

HOWARTH, L., "Concerning the Effect of Compressibility on Laminar Boundary Layers and Their Separation," *Proc. R. Soc. Lond.*, Vol. A194, p. 16 (1948).

HOWARTH, L., "The Boundary Layer in Three-Dimensional Flow, Pt. II: The Flow near a Stagnation Point," *Phil. Mag.*, Vol. 42, pp. 1433–1440 (1951).

HOYT, J. W., "Drag Reduction by Polymers and Surfactants," in D. M. Bushnell and J. Hefner (eds.), *Viscous Drag Reduction in Boundary Layers*, AIAA, New York (1989).

HUFFMAN, G. D., ZIMMERMAN, D. R., and BENNETT, W. A., "The Effect of Free-Stream Turbulence Level on Boundary Layer Behavior," in C. Hirsch, *Boundary Layer Effects in Turbomachines*, AGARDograph 164 (1972).

HUMPHREYS, D. A., and LINDHOUT, J. P. F., "Calculation Methods for Three-Dimensional Turbulent Boundary Layers," *Prog. Aero. Sci.*, Vol. 25, pp. 107–129 (1988).

HYDE, C. R., SMITH, B. R., SCHETZ, J. A., and WALKER, D. A., "Turbulence Measurements for Heated Gas Slot Injection in Supersonic Flow," *AIAA J.*, Vol. 28, pp. 1605–1614 (1990).

HYTOPOULOS, E., SCHETZ, J. A., and GUNZBURGER, M., "Numerical Solution of the Compressible Boundary Layer Equations Using the Finite Element Method," AIAA Paper 92-0666 (1992).

IKAWA, H., and KUBOTA, T., "Investigation of Supersonic Turbulent Mixing Layer with Zero Pressure Gradient," *AIAA J.*, Vol. 13, pp. 566–572 (1975).

JACK, J. R., and DIACONIS, W. S., "Variation of Boundary-Layer Transition with Heat Transfer at Mach Number 3.12," NACA TN 3562 (1955).

JAFFE, N. A., OKAMURA, T. T., and SMITH, A. M. O., "Determination of Spatial Amplification Factors and Their Application to Predicting Transition," *AIAA J.*, Vol. 8, pp. 301–308 (1970).

JOHNSON, D. A. and KING, L. S., "A Mathematically Simple Turbulence Closure Model for

Attached and Separated Turbulent Boundary Layers," *AIAA J.*, Vol. 23, pp. 1684–1692 (1985).

JOHNSTON, J. P., "On the Three-Dimensional Turbulent Boundary Layer Generated by Secondary Flow," *J. Basic Eng.*, Vol. 82, ASME Series D, pp. 233–248 (1960).

JONES, W. P., and LAUNDER, B. E., "The Prediction of Laminarization with a Two-Equation Model of Turbulence," *Int. J. Heat Mass Trans.*, Vol. 15, pp. 301–314 (1972).

JULIEN, H., "The Turbulent Boundary Layer on a Porous Plate: Experimental Study of the Effects of a Favorable Pressure Gradient," Ph.D. thesis, Stanford University, 1969.

KADER, B. A., "Temperature and Concentration Profiles in Fully Turbulent Boundary Layers," *Int. J. Heat Mass Trans.*, Vol. 24, pp. 1541–1544 (1981).

KARPLUS, W. J., "An Electric Circuit Theory Approach to Finite Difference Stability," *Trans. AIEE*, Vol. 77, pp. 210–213 (1958).

KAYS, W. M., and MOFFAT, R. J., "The Behavior of Transpired Turbulent Boundary Layers," in B. E. Launder, *Studies in Convection*, Vol. 1, Academic Press, London, pp. 223–319 (1975).

KEMP, N. H., "Vorticity Interaction at an Axisymmetric Stagnation Point in a Viscous Incompressible Fluid," *J. Aero. Sci.*, Vol. 26, pp. 543–544 (1959).

KEMP, N. H., ROSE, P. H., and DETRA, R. W., "Laminar Heat Transfer around Blunt Bodies in Dissociated Air," *J. Aero. Sci.*, Vol. 26, pp. 421–430 (1959).

KENWORTHY, M. A., and SCHETZ, J. A., "Experimental Study of Slot Injection into a Supersonic Stream," *AIAA J.*, Vol. 11, pp. 585–586 (1973).

KISS, T., and SCHETZ, J. A., "Rational Extension of the Clauser Eddy Viscosity Model to Compressible Boundary Layer Flow," to be published in *AIAA J* (1992).

KISTLER, A. L., "Fluctuation Measurements in a Supersonic Turbulent Boundary Layer," *Phys. Fluids*, Vol. 2, pp. 290–296 (1959).

KLEBANOFF, P. S., "Characteristics of Turbulence in a Boundary Layer with Zero Pressure Gradient," NACA Rep. 1247 (1955).

KLEBANOFF, P. S., and DIEHL, F. W., "Some Features of Artificially Thickened Fully Developed Turbulent Boundary Layers with Zero Pressure Gradient," NACA TN 2475 (1951).

KLINE, S. J., REYNOLDS, W. C., SCHRAUB, F. A., and RUNSTADLER, P. W., "The Structure of Turbulent Boundary Layers," *J. Fluid Mech.*, Vol. 30, pp. 741–773 (1967).

KLINE, S. J., MORKOVIN, M. V., SOVRAN, G., and COCKRELL, D. J., *Computation of Turbulent Boundary Layers—1968 AFOSR-IFP-Stanford Conference*, Vol. I, Stanford University Press, Stanford, CA, 1969.

KLINE, S. J., CANTWELL, B. J., and LILLEY, G. M., *The 1980–81 AFOSR-HTTM-Stanford Conference on Complex Turbulent Flows: Comparison of Computation and Experiment*, Stanford Univ., Stanford, CA, 1982.

KLINE, S. J., and ROBINSON, S. K., "Turbulent Boundary Layer Structure: Progress, Status, and Challenges," in A. Gyr (ed.), *Structure of Turbulence and Drag Reduction*, Springer-Verlag, Berlin (1990).

KOPPENWALLNER, G., "Fundamentals of Hypersonics: Aerodynamics and Heat Transfer," in J. F. Wendt, *Hypersonic Aerothermodynamics*, Short Course Notes, VKI, Belgium (1984).

KONG, F., and SCHETZ, J. A., "Turbulent Boundary Layer over Solid and Porous Surfaces with Small Roughness," AIAA Paper 81-0418 (1981).

KONG, F., and SCHETZ, J. A., "Turbulent Boundary Layer over Porous Surfaces with Different Surface Geometries," AIAA Paper 82-0030 (1982).

KORKEGI, R. H., "Transition Studies and Skin-Friction Measurements on an Insulated Flat Plate at a Mach-Number of 5.8," *J. Aerosp. Sci.*, Vol. 23, pp. 97–102 (1956).

KOVASZNY, L. S. G., "Structure of the Turbulent Boundary Layer," *Phys. Fluids*, Vol. 10, Suppl., pp. S25–S30 (1967).

KRAUSE, E., "Comment on 'Solution of a Three-Dimensional Boundary-Layer Flow with Separation,'" *AIAA J.*, Vol. 7, pp. 575–576 (1969).

KRAUSE, E., HIRSCHEL, E. H., and BOTHMAN, T., "Numerical Stability of Three-Dimensional Boundary Layer Solutions," *ZAMM*, Vol. 48, pp. T205–T208 (1968).

KUBOTA, T., and DEWEY, C. F., JR., "Momentum Integral Methods for the Laminar Free Shear Layer," *AIAA J.*, Vol. 2, pp. 625–629 (1964).

KUTSCHENTEUTER, P. H., JR., BROWN, D. L., and HOELMER, W., "Investigation of Hypersonic Inlet Shock-Wave Boundary Layer Interaction, Part II—Continuous Flow Test and Analysis," AFFDL-TR-65-36, Flight Dynamics Lab., Dayton, OH (1966).

LAM, C. K. G., and BREMHORST, K. A., "Modified Form of the K-ϵ Model for Predicting Wall Turbulence," Univ. Queensland, Dept. Mech. Eng., Res. Rept. 3/78 (1978).

LANDIS, F., and SHAPIRO, A. H., "The Turbulent Mixing of Co-axial Gas Jets," Heat Transfer and Fluid Mechanics Inst., Reprints and Papers, Stanford University Press, Stanford, CA, 1951.

LAUFER, J., "Investigations of Turbulent Flow in a Two-Dimensional Channel," NACA TR 1053 (1951).

LAUFER, J., "The Structure of Turbulence in Fully Developed Pipe Flow," NACA TR 1174 (1954).

LAUNDER, B. E., and SPALDING, D. B., *Mathematical Models of Turbulence*, Academic Press, New York, 1972.

LAUNDER, B. E., and YING, W. M., "Secondary Flows in Ducts of Square Cross Section," *J. Fluid Mech.*, Vol. 54, pp. 289–295 (1972).

LAUNDER, B. E., and YING, W. M., "Prediction of Flow and Heat Transfer in Ducts of Square Cross Section," *Heat Fluid Flow*, Vol. 3, pp. 115–121 (1973).

LAUNDER, B., MORSE, A., RODI, W., and SPALDING, D. B., "Prediction of Free Shear Flows—a Comparison of the Performance of Six Turbulence Models," *Free Turbulent Shear Flows*, NASA SP-321 (1973).

LAUNDER, B. E., REECE, G. J., and RODI, W., "Progress in the Development of a Reynolds Stress Closure Model," *J. Fluid Mech.*, Vol. 68, pp. 537–566 (1975).

LAWN, C. J., "The Determination of the Rate of Dissipation in Turbulent Pipe Flow," *J. Fluid Mech.*, Vol. 48, pp. 477–505 (1971).

LEE, R. E., "Additional Comparisons of Approximations of the Thermodynamic and Transport Properties of Air," NASP Contractor Report 1010 (1988).

LEE, R. E., YANTA, W. J., and LEONAS, A. C., "Velocity Profile, Skin-Friction Balance and Heat-Transfer Measurements of the Turbulent Boundary Layer at Mach 5 and Zero-Pressure Gradient," Nav. Ord. Lab. Rep. TR-69-106 (1969).

LEES, L., "Laminar Heat Transfer over Blunt-Nosed Bodies at Hypersonic Flight Speeds," *Jet Prop.*, Vol. 26, pp. 259–268 (1956).

LEES, L., and REEVES, B. L., "Supersonic Separated and Reattaching Laminar Flows: I. General Theory and Application to Adiabatic Boundary-Layer/Shock-Wave Interactions," *AIAA J.*, Vol. 2, pp. 1907–1920 (1964).

LEVY, S., "Heat Transfer to Constant-Property Laminar Boundary-Layer Flows with Power-

Function Free-Stream Velocity and Wall-Temperature Variation," *J. Aerosp. Sci.*, Vol. 19, p. 341 (1952).

LEWIS, J. E., GRAN, R. L., and KUBOTA, T., "An Experiment in the Adiabatic Compressible Turbulent Boundary Layer in Adverse and Favorable Pressure Gradients," *J. Fluid Mech.*, Vol. 51, pp. 657–672 (1972).

LI, T-Y., and NAGAMATSU, H. T., "Shock-wave Effects on the Laminar Skin Friction of an Insulated Flat Plate at Hypersonic Speeds," *J. Aero. Sci.*, Vol. 20, pp. 345–355 (1953).

LIEPMANN, H. W., ROSHKO, A., and DHAWAN, S. "On Reflection of Shock Waves from Boundary Layers," NACA Rept. 1100 (1952).

LIGHTHILL, M. J., "Dynamics of a Dissociating Gas," *J. Fluid Mech.*, Vol. 2, pp. 1–32 (1957).

LIN, C. C., "On the Stability of Two-Dimensional Parallel Flows," *Q. Appl. Math.*, Vol. 3, pp. 277–301 (1945).

LIN, C. C., *Turbulent Flows and Heat Transfer*, Princeton Univ. Press, Princeton (1959).

LOBB, R. K., WINKLER, E. M., and PERSH, J., "Experimental Investigation of Turbulent Boundary Layers in Hypersonic Flow," NAVORD Rep. 3880 (1955).

LOCK, R. C., "The Velocity Distribution in the Laminar Boundary Layer between Parallel Streams," *Quart. J. Mech.*, Vol. 4, pp. 42–63 (1951).

LOOS, H. G., "A Simple Laminar Boundary Layer with Secondary Flow," *J. Aero. Sci.*, Vol. 22, pp. 35–40 (1955).

LOW, G. M., "Cooling Requirement for Stability of Laminar Boundary Layer with Small Pressure Gradient at Supersonic Speeds," NACA TN 3103 (1954); see also *J. Aerosp. Sci.*, Vol. 22, pp. 329–336 (1955).

LUDWIEG, H., and TILLMANN, W., "Investigation of the Wall Shearing Stress in Turbulent Boundary Layers," NACA TM 1285 (1950).

MACK, L. M., "An Experimental Investigation of the Temperature Recovery-Factor," Jet Propulsion Lab., Rep. 20-80, California Institute of Technology, Pasadena (1954).

MAGER, A., "Transformation of the Compressible Turbulent Boundary Layer," *J. Aero. Sci.*, Vol. 25, pp. 305–311 (1958).

MAISE, G., and MCDONALD, H., "Mixing Length and Kinematic Eddy Viscosity in a Compressible Boundary Layer," *AIAA J.*, Vol. 6, pp. 73–80 (1968).

MALIK, M. R., "Stability Theory for Laminar Flow Control Design," in D. M. Bushnell (ed.), *Viscous Drag Reduction in Boundary Layers*, AIAA, New York (1990).

MANGLER, W., "Boundary Layers on Bodies of Revolution in Symmetrical Flow," *Berichte Aerodynamische Versuchsanstalt, Göttingen*, 45/A/17 (1945).

MARVIN, J. G., "Turbulence Modeling for Computational Aerodynamics," *AIAA J.*, Vol. 21, pp. 941–955 (1983).

MARVIN, J. G., and COAKLEY, T. J., "Turbulence Modeling for Hypersonic Flows," NASP Tech. Memo. 1087, NASA (1989).

MASKELL, E. C., "Flow Separation in Three Dimensions," Rep. Aero. Res. Council London No. 18063 (1955).

MCDONALD. H., "The Effect of Pressure Gradient on the Law of the Wall in Turbulent Flow," *J. Fluid Mech.*, Vol. 35, pp. 311–336 (1969).

MCGUIRK, J. J., and RODI, W., "The Calculation of Three-Dimensional Turbulent Jets," in

F. Durst, B. E. Launder, F. W. Schmidt and J. H. Whitelaw (eds.), *Turbulent Shear Flows I*, Springer-Verlag, Berlin (1979).

MEIER, H. U., and ROTTA, J. C., "Temperature Distributions in Supersonic Turbulent Boundary Layers," *AIAA J.*, Vol. 9, pp. 2149–2156 (1971).

MELLING, A., "Investigation of Flow in Noncircular Ducts and Other Configurations by Laser Doppler Anemometry," Ph.D. thesis, Univ. London (1975).

MENTER, F. R., "Performance of Popular Turbulence Models for Attached and Separated Adverse Pressure Gradient Flows," AIAA-91-1784 (1991).

MICHEL, R., "Étude de la transition sur les profiles d'aile; établissement d'un critère de détermination de point de transition et calcul de la trainée de profile incompressible" [Study of Transition on Wing Sections; Establishment of a Criterion for the Determination of Point of Transition and Calculation of the Wake of an Incompressible Profile], ONERA Rep. 1/1578A (1952).

MICHEL, R., QUEMARD, C., and DURANT, R., Hypotheses on the Mixing Length and Applications to the Calculation of the Turbulent Boundary Layers," in S. J. Kline, M. V. Morkovin, G. Sovran, and D. J. Cockrell, *Computation of Turbulent Boundary Layers—1968 AFOSR-IFP-Stanford Conference*, Stanford University Press, Stanford, CA, 1969.

MICKLEY, H. S., ROSS, R. C., SQUYERS, A. L., and STEWART, W. E., "Heat, Mass and Momentum Transfer for Flow over a Flat Plate with Blowing and Suction," NACA TN 3208 (1954).

MINER, E. W., ANDERSON, E. C., and LEWIS, C. H., "A Computer Program for Two-Dimensional and Axisymmetric Nonreacting Perfect Gas and Equilibrium Chemically Reacting Laminar, Transitional and-or Turbulent Boundary Layer Flows," NASA CR-132601 (1975).

MONAGHAN, R. J., and COOKE, J. R., "The Measurement of Heat Transfer and Skin Friction at Supersonic Speeds—Part IV, Test on a Flat Plate at M = 2.82," RAE Tech. Note Aero. 2171 (1952).

MOODY, L. F., "Friction Factors for Pipe Flow," *Trans. ASME.*, Vol. 66, pp. 671–678 (1944).

MOSES, H. L., "A Strip-Integral Method for Predicting the Behavior of Turbulent Boundary Layers," in S. J. Kline, M. V. Morkovin, G. Sovran, and D. J. Cockrell, *Computation of Turbulent Boundary Layers—1968 AFOSR-IFP-Stanford Conference*, Stanford University Press, Stanford, CA, 1969.

MÜLLER, U. R., "Computation of Incompressible Three-Dimensional Turbulent Boundary Layers and Comparison with Experiment," NASA TM 84230 (1982).

MÜLLER, U. R., and KRAUSE, E., "Measurements of Mean Velocities and Reynolds-Stresses in an Incompressible Three-Dimensional Turbulent Boundary Layer," *Second Sym. Turbulent Shear Flows*, London (1979).

NAGANO, Y., and TAGAWA, M., "An Improved K-ϵ Model for Boundary Layer Flows," *J. Fluids Eng.*, Vol. 112, pp. 33–39 (1990).

NALLASAMY, M., "Turbulence Models and Their Applications to the Prediction of Internal Flows: A Review," *Computers & Fluids*, Vol. 15, pp. 151–194 (1987).

NAVIER, M., "Mémoire sur les lois du mouvement des fluides" [Note on the Laws of Fluid Motion], *Mem. Acad. Sci.*, Vol. 6, pp. 389–416 (1823).

NEWMAN, B. G., "Some Contributions to the Study of the Turbulent Boundary Layer near

Separation," Australia Dept. Supply Rep. ACA-53 (1951).

NG, K. H., PATANKAR, S. V., and SPALDING, D. B., "The Hydrodynamic Boundary Layer on a Smooth Wall Calculated by a Finite-Difference Method," in S. J. Kline, M. V. Morkovin, G. Sovran, and D. J. Cockrell, *Computation of Turbulent Boundary Layers—1968 AFOSR-IFP-Stanford Conference*, Stanford University Press, Stanford, CA, 1969.

NG, K. H., and SPALDING, D. B., "Predictions of Two-Dimensional Boundary Layers on Smooth Walls with a Two-Equation Model of Turbulence," Rep. cw/16, Imperial College, London (1970).

NIKURADSE, J., "Gesetzmässigkeit der turbulenten Strömung in glatten Rohren" [Similarity for Turbulent Flow in Smooth Pipes], *VDI-Forschungsh.*, No. 356 (1932).

NIKURADSE, J., *Laminare Reibungsschichten an der längsamgeströmten Platte* [Laminar Shear Layers on a Plate], Monograph, Zentrale für Wissenschaft, Berichtwesen, Berlin, 1942.

NOAT, D., SHAVIT, A., and WOLFSHTEIN, M., "Numerical Calculation of Reynolds Stresses in a Square Duct with Secondary Flow," *Wärme-und-Stoffübertragung*, Vol. 7, pp. 151–161 (1974).

NORRIS, R. H., *Augmentation of Convection Heat and Mass Transfer,* ASME, New York (1971).

NUNNER, W., "Heat Transfer and Pressure Drop in Rough Tubes," *VDI-Forschungsh. No. 455* (1956).

NUSSELT, W., "Die Abhängigkeit der Wärmeübergangszahl von der Rohrlänge" [The Dependence of Heat Transfer Rate on Pipe Length], *Z. Ver. Deut. Ingr.*, Vol. 54, pp. 1154–1158 (1910).

O'BRIEN, G. G., HYMAN, M. S., and KAPLAN, S., "A Study of the Numerical Solution of Partial Differential Equations," *J. Math. Phys.*, Vol. 29, pp. 223–251 (1952).

ÖLCMEN, S., and SIMPSON, R. L., "Some Near Wall Features of Three-Dimensional Turbulent Boundary Layers," *Fourth Symp. Num. Phys. Aspects Aerodynamic Flows*, Cal. St. Univ., Long Beach, CA (1989).

OWEN, F. K., "Transition and Turbulence Measurements in Hypersonic Flow," AIAA-90-5231 (1990).

OWEN, P. R., and RANDALL, D. D., "Boundary Layer Transition on a Swept Back Wing," RAE TM Aero 277, May (1952).

PAPAMOSCHOU, D., and ROSHKO, A. A., "Observations of Supersonic Free Shear Layers," AIAA 86-0162 (1986).

PARR, W., "Laminar Boundary Layer Calculations by Finite Differences," Nav. Ord. Lab. Rep. TR-63-261 (1963).

PATANKAR, S. V., and SPALDING, D. B., *Heat and Mass Transfer in Boundary Layers*, Intertext Books, London, 1967.

PATE, S. R., "Effects of Wind Tunnel Disturbances on Boundary Layer Transition with Emphasis on Radiated Noise: A Review," AIAA- 80-0431 (1980).

PEAKE, D. J., "Three-Dimensional Swept Shock/Turbulent Boundary-Layer Separations with Control by Air Injection," Aer. Rept. LR-592, Nat. Res. Coun., Canada (1976).

PEARSON, J. R. A., "Instability in Non-Newtonian Flow," *Ann. Rev. Fluid Mech.*, Annual Reviews, Inc., Palo Alto, CA (1976).

PERRY, A. E., and SCHOFIELD, W. H., "Mean Velocity and Shear Stress Distributions in Turbulent Boundary Layers," *Phys. Fluids*, Vol. 16, No. 12, pp. 2068–2074 (1973).

PETUKHOV, B. S., "Heat Transfer and Friction in Turbulent Pipe Flow with Variable Fluid Properties," J. P. Hartnett and T. F. Irvine (eds.), *Advances in Heat Transfer*, Vol. 6, pp. 504–564, Academic Press, New York (1970).

POHLHAUSEN, K., "Der Wärmeaustausch zwischen festen Körpern und Flüssigkeiten mit kleiner Reibung und kleiner Wärmeleitung" [The Heat Transfer between Solid Bodies and Fluids with Small Viscosity and Small Thermal Conductivity], *ZAMM.*, Vol. 1, p. 115 (1921a).

POHLHAUSEN, K., "Zur näherungsweisen Integration der Differentialgleichungen der laminaren Reibungsschicht" [On the Approximate Integration of the Differential Equations of Laminar Shear Layers], *ZAMM.*, Vol. 1, pp. 252–268 (1921b).

POTTER, J. L., and WHITFIELD, J. D., "Effects of Slight Nose Bluntness and Roughness on Boundary-Layer Transition in Supersonic Flows," *J. Fluid Mech.*, Vol. 12, pp. 501–535 (1962).

PRANDTL, L., "Über Flüssigkeitsbewegung bei sehr kleiner Reibung" [On Fluid Motion with very Small Friction], *Proc. 3rd Int. Math. Congr.*, Heidelberg, 1904; see also L. Prandtl, "Gesammelte Abhandlungen zur angewandten Mechanik" [Collected Essays on Applied Mechanics] in W. Tollmien, H. Schlichting, and H. Görtler (eds.), *Hydro- und Aerodynamik* (collected works), Vol. 2, Springer-Verlag, Berlin, 1961.

PRANDTL, L., "Über die ausgebildete Turbulenz" [On Fully Developed Turbulence], *ZAMM.*, Vol. 5, pp. 136–139 (1925).

PRANDTL, L., *Ergeb. AVA Goettingen*, Ser. III, pp. 1–5 (1927).

PRANDTL, L., "The Mechanics of Viscous Fluids," in W. F. Durand, *Aerodynamic Theory*, III, 1935; see also summary by L. Prandtl, "Neuere Ergebnisse der Turbulenzforschung" [New Results in Turbulence Research], *VDIZ.*, Vol. 77, pp. 105–114 (1933); see also *Collected Works*, Vol. 2, pp. 819–845.

PRANDTL, L., "Bermerkungen zur Theorie der freien Turbulenz" [Observations on the Theory of Free Turbulence], *ZAMM.*, Vol. 22, pp. 241–243 (1942).

PRANDTL, L., "Über eine neues Formelsystem für die ausgebildete Turbulenz" [On a New Formulation for Fully Developed Turbulence], *Nachr. Akad. Wiss., Göttingen, Math. Phys. Klasse.*, pp. 6–19 (1945).

PRANDTL, L., "On Boundary layers in Three-Dimensional Flow," *Min. Aircraft Production (Völkenrode) Repts. and Trans.* 64 (1946).

PRANDTL, L., and REICHARDT, H., "Einfluss von Wärmeschichtung auf die Eigenschaften einer turbulenten Stromung" [The Influence of Thermal Layers on the Characteristics of Turbulent Flows] *Deutche Forschung.*, No. 21, pp. 110–121 (1934); see also *Collected Works*, Vol. 2, pp. 846–854.

RAO, K. N., NARASHIMA, R., and BADRI NARAYANAN, M. A., "The Bursting Phenomena in a Turbulent Boundary Layer," *J. Fluid Mech.*, Vol. 48, pp. 339–352 (1971).

REECE, G. A., "A Generalized Reynolds Stress Model of Turbulence," Ph.D. thesis, Univ. London (1976).

REICHARDT, H., "Gesetzmassigkeitender freien Turbulenz" [Similarity in Free Turbulence], *VDI-Forschungsh.*, No. 414 (1942); 2nd ed. (1951).

REICHARDT, H., "Vollstandige Darstellung der turbulenten Geschwindigkeitsverteilung in glatten Leitungen" [Complete Description of Turbulent Velocity Profiles in Smooth Ducts], *ZAMM.*, Vol. 31, pp. 208–219 (1951).

REYNOLDS, A. J., *Turbulent Flows in Engineering*, John Wiley & Sons, London (1974).

REYNOLDS, O., "An Experimental Investigation of the Circumstances which Determine whether the Motion of Water Shall be Direct or Sinuous, and of the Law of Resistance in Parallel Channels," *Philos. Trans. R. Soc. Lond.*, Vol. 174, pp. 935–982 (1883).

REYNOLDS, W. C., KAYS, W. M., and KLINE, S. J., "Heat Transfer in a Turbulent Incompressible Boundary Layer," NASA Memo 12-4-58W (1958).

REYNOLDS, W. C., KAYS, W. M., and KLINE, S. J., "Heat Transfer in the Turbulent Incompressible Boundary Layer,—III Arbitrary Wall Temperature and Heat Flux," NASA Memo 12-3-58W (1958).

RODI, W., "The Prediction of Free Turbulent Boundary Layers by Use of a Two-Equation Model of Turbulence," Ph.D. thesis, University of London, 1972.

RODI, W., "A Review of Experimental Data of Uniform Density Free Turbulent Boundary Layers," in *Studies in Convection*, Vol. 1, B. E. Launder (ed.), Academic Press, London, 1975.

RODI, W., "A New Algebraic Relation for Calculating the Reynolds Stresses," *ZAMM.*, Vol. 56, pp. T219-T221 (1976).

RODI, W., *Turbulence Models and Their Application in Hydraulics*, Int. Assoc. Hydraulics Res., Delft, Netherlands, 1980.

ROGALLO, R. S., and MOIN, P., "Numerical Simulation of Turbulent Flows," in *Ann. Rev. Fluid Mech.*, Vol. 16, Ann. Reviews, Inc., Palo Alto, CA (1984).

ROSE, P. H., PROBSTEIN, R. F., and ADAMS, M. C., "Turbulent Heat Transfer through a Highly Cooled, Partially Dissociated Boundary Layer," *J. Aero. Sci.*, Vol. 25, pp. 751–760 (1958).

ROTT, N., and CRABTREE, L. F., "Simplified Laminar Boundary-Layer Calculations for Bodies of Revolution and Yawed Wings," *J. Aero. Sci.*, Vol. 19, pp. 553–565 (1952).

RUBESIN, M., "An Analytical Investigation of the Heat Transfer between a Fluid and a Flat Plate Parallel to the Direction of Flow Having a Discontinuous Temperature Distribution," master's thesis, University of California, Berkeley, 1947.

RUBESIN, M. W., CRISALLI, A. J., LANFRANCO, M. J., and ACHARYA, M., "A Critical Evaluation of Invariant Second-Order Closure Models for Subsonic Boundary Layers," *Proceedings of the Symposium on Turbulent Shear Flows* (Pennsylvania State University, 1977), Spring-Verlag, Berlin, 1979.

RUBESIN, M. W., and VIEGAS, J. R., "A Critical Examination of the Use of Wall Functions as Boundary Conditions in Aerodynamic Calculations," *Third Symposium on Numerical and Physical Aspects of Aerodynamic Flows,* California St. Univ. at Long Beach (1985).

SAMUEL, A. E., and JOUBERT, P. N., "A Boundary Layer Developing in an Increasingly Adverse Pressure Gradient," *J. Fluid Mech.*, Vol. 66, pp. 481–505 (1974).

SCHETZ, J. A., "On the Approximate Solution of Viscous Flow Problems," *J. Appl. Mech.*, Vol. 30, pp. 263–268 (1963).

SCHETZ, J. A., "Supersonic Diffusion Flames," *Supersonic Flow, Chemical Processes and Radiative Transfer*, Pergamon Press, London (1964).

SCHETZ, J. A., "Analytic Approximations of Boundary Layer Problems," *J. Appl. Mech.*, Vol. 33, pp. 425–428 (1966).

SCHETZ, J. A., "Turbulent Mixing of a Jet in a Coflowing Stream," *AIAA J.*, Vol. 6, pp. 2008–2010 (1968).

SCHETZ, J. A., "Analysis of the Mixing and Combustion of Gaseous and Particle Laden Jets in an Airstream," AIAA Paper 69-33 (1969).

SCHETZ, J. A., "Some Studies of the Turbulent Wake Problem," *Astronaut. Acta.*, Vol. 16, pp. 107–117 (1971).

SCHETZ, J. A., "Free Turbulent Mixing in a Co-flowing Stream," *Free Turbulent Shear Flows*, NASA SP-321 (1973).

SCHETZ, J. A., *Injection and Mixing in Turbulent Flow*, AIAA, New York, 1980.

SCHETZ, J. A., BILLIG, F. S., and FAVIN, S., "Simplified Analysis of Supersonic Base Flows Including Injection and Combustion," *AIAA J.*, Vol. 14, pp. 7–8 (1976).

SCHETZ, J. A., BILLIG, F. S., and FAVIN, S., "Flowfield Analysis of a Scramjet Combustor with a Coaxial Fuel Jet," *AIAA J.*, Vol. 20, pp. 1268–1274 (1982).

SCHETZ, J. A., BILLIG, F. S., and FAVIN, S., "Analysis of Slot Injection in Hypersonic Flow," *J. Propulsion Power.*, Vol. 7, pp. 115–122 (1991).

SCHETZ, J. A., and FAVIN, S., "Numerical Calculation of Turbulent Boundary Layer with Suction or Injection and Binary Diffusion," *Astro. Acta*, Vol. 16, pp. 339–352 (1971).

SCHETZ, J. A., and JANNONE, J., "A Study of Linearized Approximations to the Boundary Layer Equations," *J. Appl. Mech.*, Vol. 32, pp. 757–764 (1965).

SCHETZ, J. A., HYTOPOULOS, E., and GUNZBURGER, M., "Numerical Solution of the Boundary Layer Equations Using the Finite Element Method," *ASME Forum on Finite Element Analysis in Fluid Dynamics*, Atlanta (1991).

SCHETZ, J. A., and NERNEY, B., "The Turbulent Boundary Layer with Injection and Surface Roughness," *AIAA J.*, Vol. 15, pp. 1288–1294 (1977).

SCHLICHTING, H., "Über das ebene Windschattenproblem" [On the Two-dimensional Wake Problem], Diss., Göttingen, 1930; *Ing.-Arch.*, Vol. 1, pp. 533–571 (1930).

SCHLICHTING, H., "Laminare Strahlausbreitung" [Laminar Jet Spreading], *ZAMM.*, Vol. 13, pp. 260–263 (1933).

SCHLICHTING, H., "Turbulenz bei Wärmeschichtung" [Turbulence in Thermal Layers], *ZAMM.*, Vol. 15, pp. 313–338 (1935); also *Proc. 4th Int. Congr. Appl. Mech.*, Cambridge, 1935, p. 245.

SCHLICHTING, H., *Boundary Layer Theory*, McGraw-Hill, New York, 1942; 6th ed., 1968.

SCHNEIDER, W., "Decay of Momentum Flux in Submerged Jets," *J. Fluid Mech.*, Vol. 154, pp. 91–110 (1985).

SCHNEIDER, W., and MÖRWALD, K., "Asymptotic Structure of Turbulent Free Shear Layers and Implications for Turbulence Modelling, *ZAMM.*, Vol. 69, pp. T 626–627 (1989).

SCHOENHERR, K. E., "Resistance of Flat Plates Moving through a Fluid," *Trans. Soc. Nav. Arch. Mar. Eng.*, Vol. 40, pp. 279–313 (1932).

SCHUBAUER, G. B., "Turbulent Process as Observed in Boundary Layer and Pipe," *J. Appl. Phys.*, Vol. 25, pp. 188–196 (1954).

SCHUBAUER, G. B., and KLEBANOFF, P. S., "Investigation of Separation of the Turbulent Boundary Layer," NACA TN 2133 (1950).

SCHUBAUER, G. B., and SKRAMSTAD, H. K., "Laminar Boundary-Layer Oscillations and Stability of Laminar Flow," *J. Aerosp. Sci.*, Vol. 14, pp. 69–78 (1947).

SCHUBAUER, G. B., and TCHEN, C. M., *Turbulent Flow*, Princeton University Press, Princeton, N.J., 1961.

SCHULTZ-GRUNOW, F., "A New Resistance Law for Smooth Plates," *Luftfahrt-Forsch.*, Vol. 17, pp. 239–246 (1940).

SCHMITT, L., RICHTER, K., and FRIEDRICH, R., "Large-Eddy Simulation of Turbulent Boundary Layer and Channel Flow at High Reynolds Number," in U. Schumann and R. Friedrich (eds.), *Direct and Large Eddy Simulation of Turbulence*, (Notes on Numerical Fluid Mechanics, Vol. 15) Vieweg, Braunschweig (1986).

SCOTT, E. J., ECKERT, E. R. G., JONSSON, V. K., and YANG, JI-WU, "Measurements of Velocity and Concentration Profiles for Helium Injection into a Turbulent Boundary Layer Flowing over an Axial Circular Cylinder," University of Minnesota HTL-TR-55 (Feb. 1964).

SCOTT-WILSON, J. B., and CAPPS, D. S., "Wind Tunnel Observations of Boundary Layer Transition on Two Sweptback Wings at Mach 1.61," Rep. Aero. Res. Coun., London No. 17627 (1954).

SEIFF, A., "Examination of the Existing Data on the Heat Transfer of Turbulent Boundary Layers at Supersonic Speeds from the Point of View of Reynolds Analogy," NACA TN 3284 (1954).

SFEIER, A. A., "The Velocity and Temperature Fields of Rectangular Jets," *Int. J. Heat Mass Trans.*, Vol. 19, pp. 1289–1297 (1976).

SFORZA, P. M., STEIGER, M. H., and TRENTACOSTE, N., "Studies on Three-Dimensional Viscous Jets," *AIAA J.*, Vol. 4, pp. 800–806 (1966).

SHABAKA, I. M. M. A., and BRADSHAW, P., "Turbulent Flow Measurements in an Idealized Wing-Body Junction," *AIAA J.*, Vol. 19, pp. 131–132 (1981).

SHAH, R. K., and LONDON, A. L., "Laminar Flow Forced Convection in Ducts," in J. P. Hartnett and T. F. Irvine, Jr., (eds.), *Advances in Heat Transfer, Supplement 1*, Academic Press, New York (1978).

SHAPIRO, A. H., *The Dynamics and Thermodynamics of Compressible Fluid Flow*, The Ronald Press, New York (1953).

SHEETZ, N. W., JR., "Ballistics Range Boundary-Layer Transition Measurements on Cones at Hypersonic Speeds," in C. W. Wells (ed.), *Viscous Drag Reduction*, Plenum Press, New York (1969).

SHEN, S. F., "Calculated Amplified Oscillations in Plane Poiseuille and Blasius Flows," *J. Aero. Sci.*, Vol. 21, pp. 62–64 (1954).

SHEVELEV, YU. D., Numerical Calculation of the Three-Dimensional Boundary Layer in an Incompressible Fluid," *Fluid Dynamics*, Vol. 1, pp. 77–80 (1966).

SIEDER, E. N., and TATE, G. E., "Heat Transfer and Pressure Drop of Liquid in Tubes," *Ind. Eng. Chem.*, Vol. 28, pp. 1429–1435 (1936).

SIMPSON, R. L., "The Turbulent Boundary Layer on a Porous Wall," Ph.D. thesis, Stanford University (1968).

SIMPSON, R. L., "A Review of Two-Dimensional Turbulent Separated Flow Calculation Methods," in F. T. Smith (ed.), *Boundary Layer Separation*, Springer-Verlag, Heidelberg (1987).

SIMPSON, R. L., "Turbulent Boundary-Layer Separation," in J. L. Lumley, N. Van Dyke, and H. L. Reed (eds.), *Ann. Rev. Fluid Mech.*, Vol. 21, Annual Reviews, Inc., Palo Alto, CA (1989).

SIMPSON, R. L., CHEW, Y.-T., and SHIVAPRASAD, B. G., "The Structure of a Separating Turbulent Boundary Layer, Pt. 1. Mean Flow and Reynolds Stresses," *J. Fluid Mech.*, Vol. 113, pp. 23–51 (1981).

SITU, M., and SCHETZ, J. A., "New Mixing Length Model for Turbulent High-Speed Flows," *AIAA J.*, Vol. 29, pp. 872–873 (1991).

SIVASEGARAM, S., and WHITELAW, J. H., "The Prediction of Turbulent Supersonic Two-Dimensional Boundary Layer Flows," *Aeronaut. Q.*, pp. 274–294 (1971).

SKELLAND, A. H. P., "Momentum, Heat and Mass Transfer in Turbulent Non-Newtonian Boundary Layers," *AIChE J.*, Vol. 12, pp. 69–75 (1966).

SMITH, A. M. O., and CLUTTER, D. W., "Machine Calculation of Compressible Laminar Boundary Layers," *AIAA J.*, Vol. 3, pp. 639–647 (1965).

SMITH, A. M. O., JAFFE, N. A., and LIND, R. C., "Study of a General Method of Solution to the Incompressible Boundary Layer Equations," Douglas Aircraft Div., Rep. LB52949 (1965).

SMITH, P. D., "An Integral Prediction Method for Three-Dimensional Compressible Turbulent Boundary Layers," ARC R&M 3739 (1972).

SOWERBY, L., "Secondary Flow in a Boundary Layer," Rep. Aero. Res. Council, London No. 16832 (1954).

SQUIRE, H. B., "On the Stability for Three-Dimensional Disturbances of Viscous Fluid between Parallel Walls," *Proc. R. Soc. Lond.*, Vol. A142, pp. 621–628 (1933).

STEVENSON, T. N., "A Law of the Wall for Turbulent Boundary Layers with Suction or Injection," Cranfield Coll. Aero. Rep. 166 (1963).

STEWARTSON, K., "Multistructured Boundary Layers on Flat Plates and Related Bodies," C. S. Yih, (ed.) *Advances in Applied Mechanics,* Academic Press, New York, Vol. 14, pp. 145–239 (1974).

STEWARTSON, K., and WILLIAMS, P. G., "Self-Induced Separation," *Proc. R. Soc. London*, Ser. A, Vol. 312, pp. 181–206 (1969).

STEWARTSON, K., and WILLIAMS, P. G., "Self-Induced Separation, Pt. 2," *Mathematika*, Vol. 20, pp. 98–108 (1973).

STOKES, G. G., "On the Theories of Internal Friction of Fluids in Motion," *Trans. Camb. Philos. Soc.*, Vol. 8, pp. 287–305 (1845).

STOKES, G. G., "On the Effect of the Internal Friction of Fluids on the Motion of Pendulums," *Trans. Camb. Philos. Soc.*, Vol. 9 (1851).

STUART, J. T., Chapter 9 in L. Rosenhead (ed.), *Laminar Boundary Layers*, Clarendon Press, Oxford, 1963.

STÜPER, J., "Untersuchung von Reibungsschichten am fliegenden Flugzeug" [Flight Experiments of Boundary Layers on an Airplane], NACA TM 751 (1934).

SUTHERLAND, W., "The Viscosity of Gases and Molecular Force," *Phil. Mag.*, Ser. 5, pp. 507–531 (1893).

SVEHLA, R. A., and MCBRIDE, B. J., "FORTRAN IV Computer Program for Calculation of Thermodynamic and Transport Properties of Complex Chemical Systems," NASA TN D-7056 (1973).

TAKEMITSU, N., "An Analytic Study of the Standard K-ϵ Model," *J. Fluids Eng.*, Vol. 112, pp. 192–198 (1990).

TANNER, R. I., "Normal Stress Effects in Drag-Reducing Fluids," in C. W. Wells (ed.), *Viscous Drag Reduction*, Plenum Press, New York (1969).

TAYLOR, G. I., "The Transport of Vorticity and Heat through Fluids in Turbulent Motion," appendix by A. Fage and V. M. Faulkner, *Proc. R. Soc. Lond.*, Vol. A135, pp. 685–705 (1932); see also *Philos. Trans.*, Vol. A215, pp. 1–26 (1915).

THWAITES, B., "Approximate Calculation of the Laminar Boundary Layer," *Aeronaut. Q.*, Vol. 1, pp. 245–280 (1949).

TING, I., and LIBBY, P. A., "Remarks on the Eddy Viscosity in Compressible Mixing Flows," *J. Aerosp. Sci.*, Vol. 27, pp. 797–798 (1960).

TOLLMIEN, W., "Berechnung turbulenter Ausbreitungsvorgänge" [Calculation of Turbulent Jets], *ZAMM.*, Vol. 6, pp. 468–478 (1926).

TOLLMIEN, W., "Über die Enstehung der Turbulenz" [On the Production of Turbulence], *Nachr. Akad. Wiss. Göttingen*, Math. Phys. Klasse pp. 21–44 (1929). [Translated as "The Production of Turbulence," NACA TM 609 (1931).]

TOULOUKIAN, Y. S., and HO, C. Y., *Thermophysical Properties of Matter—The TPRC Data Series*, IFI/Plenum Data Co., New York, Vol. 3, 1970, p. 120; Vol. 6, 1970, p. 102; and Vol. 11, 1975, p. 94.

TRENTACOSTE, N., and SFORZA, P. M., "Further Experimental Results for Three-Dimensional Free Jets," *AIAA J.*, Vol. 5, pp. 855–891 (1967).

VANCOILLIE, G., "A Turbulence Model for the Numerical Simulation of Transitional Boundary Layers," *Proc. 2nd IUTAM Symp. Laminar-Turbulent Transition*, Novosibirsk, U.S.S.R. (1984).

VAN DRIEST, E. R., "Turbulent Boundary Layer in Compressible Fluids," *J. Aerosp. Sci.*, Vol. 18, pp. 145–160 (1951).

VAN DRIEST, E. R., "Investigation of Laminar Boundary Layer Compressible Fluids Using the Crocco Method," NACA TN 2597 (1952).

VAN DRIEST, E. R., "On Turbulent Flow near a Wall," *J. Aerosp. Sci.*, Vol. 23, pp. 1007–1012 (1956a).

VAN DRIEST, E. R., "The Problem of Aerodynamic Heating," *Aeronaut. Eng. Rev.*, Vol. 15, pp. 26–41 (1956b).

VAN DRIEST, E. R., and BLUMER, C. B., "Boundary Layer Transition, Free Stream Turbulence, and Pressure Gradient Effects," *AIAA J.*, Vol. 1, pp. 1303–1306 (1963).

VAN DRIEST, E. R., and BOISON, J. C., "Experiments on Boundary Layer Transition at Supersonic Speeds," *J. Aerosp. Sci.*, Vol. 24, pp. 885–899 (1957).

VAN DYKE, M., *Perturbation Methods in Fluid Mechanics*, Academic Press, New York, 1964.

VINCENTI, W. G., and KRUGER, C. H., JR., *An Introduction to Physical Gas Dynamics*, John Wiley and Sons, Inc., New York (1965).

VON KÁRMÁN, TH., "Mechanische Ähnlichkeit und Turbulenz" [Mechanical Similarity and Turbulence], *Nachr. Ges. Wiss. Göttingen, Math. Phys. Kl.*, p. 58 (1930), and *Proc. 3rd Int. Congr. Appl. Mech.*, Stockholm, Pt. I, 1930, p. 85; NACA TM 611 (1931).

VON KÁRMÁN, TH., "The Analogy between Fluid Friction and Heat Transfer," *Trans. ASME*, Vol. 61, pp. 705–710 (1939).

WAGNER, R. D., "Mean Flow and Turbulence Measurements in a Mach 5 Shear Layer," NASA TN D-7366 (1973).

WAHLS, R. A., BARNWELL, R. W., and DEJARNETTE, F. R., "Finite-Difference Outer-Layer, Analytic Inner-Layer Method for Turbulent Boundary Layers," *AIAA J.*, Vol. 27, pp. 15–21 (1989).

WALTRUP, P. J., and SCHETZ, J. A., "Supersonic Turbulent Boundary Layer Subjected to Adverse Pressure Gradients," *AIAA J.*, Vol. 11, pp. 50–57 (1973).

WALZ, A., "Ein neuer Ansatz für das Geschwindigkeitsprofil der laminaren Reibungsschicht" [A New Relation for the Velocity Profile in Laminar Shear Layers], *Ber. Lilienthal-Ges. Luftfahrtf.* No. 141, pp. 8–12 (1941).

WAZZAN, A. R., OKAMURA, T. T., and SMITH, A. M. O., "Spatial and Temporal Stability Charts for the Falkner-Skan Boundary-Layer Profiles," McDonnell-Douglas Rep. DAC-67086 (1968).

WEINSTEIN, A. S., OSTERLE, J. F., and FORSTALL, W., "Momentum Diffusion from a Slot Jet into a Moving Secondary," *J. Appl. Mech.*, pp. 437–443 (1956).

WELLS, C. S., JR., "Effects of Freestream Turbulence on Boundary-Layer Transition," *AIAA J.*, Vol. 5, pp. 172–174 (1967).

WERLÉ, H., private communication, 1982.

WILLIAMS, J. C. III, "Viscous Compressible and Incompressible Flow in Slender Channels," *AIAA J.*, Vol. 1, pp. 186–195 (1963).

WILSON, R. E., "Turbulent Boundary Layer Characteristics at Supersonic Speeds—Theory and Experiment," *J. Aerosp. Sci.*, Vol. 17, pp. 585–594 (1950).

WILSON, R. E., Secs. 13 and 14, *Handbook of Supersonic Aerodynamics*, NAVORD Rep. 1488, Vol. 5, U. S. Government Printing Office, Washington, D.C., 1966.

WOOLLEY, H. W., "Effect of Dissociation on Thermodynamic Properties of Pure Diatomic Gases," NACA TN 3270 (1955).

WU, J. C., "On the Finite Difference Solution of Laminar Boundary Layer Problems," *Proceedings of the 1961 Heat Transfer and Fluid Mechanics Institute*, Stanford University Press, Stanford, CA, 1961.

WYGNANSKI, I., and FIEDLER, H. E., "Some Measurements in the Self-Preserving Jet," *J. Fluid Mech.*, Vol. 38, pp. 577–612 (1969).

ZAHORSKI, S., *Mechanics of Viscoelastic Fluids*, Martinus Nijhoff Publ., The Hague (1982).

ZHUKAUSKAS, A., and SLANCHAUSKAS, A., *Heat Transfer in a Turbulent Liquid Flow*, Mintis, Vilnius, U.S.S.R., 1973.

INDEX

Adiabatic wall temperature, 128, 132, 134, 426
Adverse pressure gradient, 37, 40, 85, 164, 182–83, 220–24, 242, 292, 296, 302, 426, 438, 452–53, 457, 463, 484–86
Airfoil, 20–22, 85, 165, 191–92, 220–21
Algebraic turbulence model (*See* Reynolds stress, models, algebraic stress model)
Amplification rate, 187
Apparent viscosity, 23, 198
Axisymmetric body, (*See also* Body of revolution), 40, 63–64, 73
Axisymmetric flow, 40, 59–60, 63–64, 69, 71, 73, 462
  displacement thickness, 583
  equations, 40, 59–60, 63–64, 69, 71, 73
  jets (See Jets, round)
  stream function, 60
  wakes (See Wakes, round)

Backward difference (*See* Finite difference, backward form)
Baldwin-Lomax turbulence model (*See* Eddy viscosity, outer region models, Baldwin-Lomax model)
Bernoulli's equation, 10–11, 19, 33
Binary diffusion, 7
  coefficient, 7
    laminar, 7, 49–52, 70, 89–92, 144–49, 158–63
    modeling of, 70, 375–78, 434
    turbulent, 332, 375–78, 416, 433–34
Bingham plastic fluid, 23, 53–54, 65, 92–93, 324–26
Blasius:
  flat-plate solution, 81–84, 135, 182, 257, 470, 475
  skin friction law, 213–14
Body forces, 15, 33, 61–62
Body of revolution, 64
Body system of coordinates, 63–64, 69
Boundary and initial conditions:
  boundary layer edge, 3, 9–11, 31–32, 43, 56, 67–69, 74, 82, 87, 97–98, 102–3, 123, 132, 144–46, 162, 281, 304, 362, 364, 468, 471, 475
  downstream, 75
  fluid/solid interface, (*See also* Wall functions) 2–3, 9, 17, 31–32, 43, 46, 48–49, 67–69, 74, 77, 82, 87, 90–92, 97–98, 102–3, 123, 129–32, 144–48, 162, 208–10, 252, 257, 275–77, 285, 295, 301, 312–13, 318, 471, 475
  in numerical methods, 97–98, 102–3, 123, 477–79
  symmetry, 146, 313, 318
  in turbulent transport models (*See also* Wall functions), 262, 275–77, 281, 285, 295, 301, 304–5, 412
  3D flows, 464, 468, 471, 475, 477–79
  upstream, 36, 39–40, 47, 62, 67–68, 74–75, 82, 87, 97–98, 304–5, 412, 464, 477–79
Boundary layer:
  assumption (and approximation), 9–11, 16–19, 33, 66, 70, 282, 315, 322, 431
  concept, 9, 462–64
  concentration, 3, 50, 90–91, 147, 375–77
  equations,
    laminar, 19, 63–65, 69, 70–71
    reduction of Navier-Stokes equations to, 16–19
    turbulent, 236, 264, 282–84, 295, 299, 430–33, 454
    3D flows, 467–69
  thermal, 3, 45–47, 88–89, 140–41, 372–73
  thickness definitions, 3, 33–35, 53, 138, 223, 257, 434, 469
Bradshaw turbulence model (*See* Turbulent kinetic energy, models, Bradshaw model)
Bulk temperature, 315, 318
Bursting, 231–33
Busemann energy integral, 130

Catalytic wall, 162–63
Central (*and* centered) finite difference (*See* Finite difference, central form)
Channel flows, 322–23, 325–26, 335–36, 340–41, 350–52, 482–84
Chemically reacting flows (*See also* Real gas effects)
  laminar, 132, 138, 157–63
  turbulent, 454–56
Circuit analogy of Karplus, 106–7
Clauser boundary layer thickness, 257–58
Clauser eddy viscosity model (*See* Eddy viscosity, outer region models, Clauser model)

Closure problem for turbulent flows, 236, 298
Coles' wake function, 212–13, 241, 332–34
Coles' wake parameter, 212, 222–23
Concentration of species, 4, 7
  fluctuations of, 375–79
Conductivity, thermal (*See* Thermal conductivity)
Conservation of energy, 31, 44, 65–67
Conservation of mass, 16, 31–32, 48, 58–60
Conservation of mass of species, 49–50, 70
Conservation of momentum, 31–33, 48–50, 60–62
Continuity equation, 16, 58–60
Convective Mach number, 382–84
Convective speed, 382
Conversion factors:
  for thermal conductivity, 513
  for viscosity, 512
Couette flow, 77–78
Crank-Nicolson finite difference formulation, 101–2, 110, 479
Crocco integrals, 130–32, 144–46, 420, 423
Crocco transformation, 136
Crocco-Lees mixing theory, 168–69, 457–690
Curvature, velocity profile, 182

Daly-Harlow model (*See* Reynolds stress, models, Daly-Harlow model)
Damping, 79, 180, 249–51
Defect law, 207, 211–12, 220–22, 256–61, 265–67, 434–36
  in pipes, 332–34
  plot, 207, 211–12, 220–22, 256–61, 265–67, 434–36
  for temperature, 415–17
Developable surface, 469
Diffusion of mass (*See* Binary diffusion)
Diffusion, binary (*See* Binary diffusion)
Diffusion coefficient (*See* Binary diffusion coefficient)
Dilatant fluid, 5, 23
Direct numerical simulations (*of turbulent flow*), 303–4, 341, 411–12, 454
Displacement thickness, 34–35, 138, 166, 170, 259, 392–93, 434
  axisymmetric, 393
  compressible, 138, 434
  kinematic, 434
  3D flows, 469
Dissociation, 132, 157–63, 454–56
Donaldson model (*See* Reynolds stress, models, Donaldson model)
Drag, 1, 30, 194–96, 366, 385–86
Duct flow:
  non-circular, 311, 316–17, 335–36, 482
  heat transfer in, 316–17
  pressure drop in, 316–17
  square, 316–17, 341–47, 350–53

Eddies, 25, 229
Eddy conductivity (*See* Thermal conductivity, turbulent)
Eddy diffusivity (*See* Binary diffusion coefficient, turbulent)
Eddy viscosity:
  composite model for a boundary layer, 262–63
  definition, 238, 433
  inner region models, 247–55, 434
    Granville model, 255–56
    with injection or suction, 252–54
    Reichardt model, 250
    with roughness, 251
    Van Driest model, 249
    variable-density flow, 434
  models for jets and wakes, 386–98
    variable-density flow, 388–95
    3D flows, 508
  model for shear layers, 381
  outer region models, 256–61, 434–36
    Baldwin-Lomax model, 260–61, 436, 502
    Clauser model, 256–60, 434
    internal flows, 340–41, 348
    Kiss-Schetz model, 435–36
    Situ-Schetz model, 436, 505
    variable-density flow, 434–39
    3D flows, 497, 499, 502
  Prandtl model, 381, 387–89, 393
  pressure gradient effects, 255, 260, 267–68, 270–74
  relation to mixing length, 239
  3D boundary layer flows, 486, 497, 499, 502, 505
e^N method, 183–87, 327, 481
Effective velocity, 420–21
Ejections, 233
Elliptic partial differential equations, 74
Energy:
  internal (*See* Energy, thermal)
  mechanical, equation, 68
  thermal, 65
    conservation (*See* Conservation of energy)
    equation (*See* Energy equation)
  transfer by mass transfer, 71, 144, 433, 454–56
  turbulent kinetic (*See* Turbulent kinetic energy)
Energy cascade, 228
Energy equation, 65–69, 87, 127, 129–30, 136, 144, 158, 433, 467–68
  integral form (*See* Integral energy equation)
Enthalpy, definition of, 67, 71
Entrainment velocity, 213
Entrance region flows:
  laminar, 317–22,
  turbulent, 341–49
Equation of state, 74, 157, 454
Equilibrium pressure gradient (*See* Pressure gradient, equilibrium in turbulent flow)
Equilibrium flow, 158–59, 162, 454
Equilibrium constant, 158–59
Excess temperature, 87, 133
Explicit finite difference formulation, 97–100, 103–10, 477–79

Falkner-Skan wedge flows, 85–86, 89, 220–21, 260–61
Fanno line, 355
Favorable pressure gradient, 37, 40, 85, 164, 183, 220, 438, 452–53
Favre averaging (*See* Mass weighted averaging)
Fick's law, 7, 70

Film coefficient:
heat transfer, 47–48, 52, 128, 316, 416–18
based on enthalpy, 143
mass transfer (*and* diffusion), 50, 52, 416
Finite difference:
backward form, 96, 105
central form, 96, 104, 478
definition, 94–96
forward form, 96, 97
method:
Crank-Nicolson type, 101–2, 110, 479
explicit type, 97–100, 103–10, 477–79
implicit type, 100–3, 110–14, 117–18, 150–53, 184, 477–79
stability of, 99–100, 106–7, 479
in turbulent flows, 264–66
3D flows, 476–80
Finite element method, 118–23, 151
Flat plate flow:
laminar, 36–37, 45–48, 50–54, 81–84, 88, 90–93, 120, 135, 138–42, 150–54
hypersonic, 170–72
non-Newtonian fluid, 52–54, 90–93, 305
transition, 180–88, 190–93, 196–200
turbulent, 204–19, 224–33, 256–59, 265–67, 303–5, 414–24, 445–46
Fluctuation in turbulent flows:
concentration, 375–79
density, 382, 431
pressure, 282–83
temperature, 204, 371–75, 426–29
velocity (*See also* Turbulence intensity), 23–28, 224–38, 367–71, 380, 385, 426–29, 431–32, 492
Foreign gas injection, 89–91, 144–49, 375–79, 416–19, 439–42
Forward difference (*See* Finite difference, forward form)
Fourier's law, 6, 47, 69
for non-Newtonian fluids, 69
Freestream turbulence, 281, 290–91
Frictional heating, 44, 128
Friction factor (*See* Resistance coefficient)
Friction velocity, definition of, 206, 420
Frozen flow, 158, 162, 454
Fully developed flow, 312–17, 324–26
turbulent, 328–41
Fully rough flow, 215, 330

Galerkin method of weighted residuals, 120
Gradient transport formulations (*and* models), 238, 283, 297, 375, 492

Half-width (*and* radius), 366–74, 387–89, 506–10
Hanjalic-Launder model (*See* Reynolds stress, models, Hanjalic–Launder model)
Heat of formation (*and* dissociation), 71, 158
Heat equation, 97, 99
Heat flux:
laminar, 6, 66
modeling, 69
turbulent, 372, 394, 433
vector, 66
Heat (and energy) transfer:
laminar:
flat plate, 45–48, 87–89
high-speed flow, 128–29, 142–43, 148–49, 150–54, 160–63, 173–74
with injection, 89–91, 148–49
nonuniform wall temperature, 48–49, 89
non-circular ducts, 316–17
pipe flow, 315–16
entrance region, 317–22
pressure gradient effects, 88–89, 143
stagnation point, 142–43, 160–63, 173–74
transitional:
pipe flow, 329–31
turbulent:
flat plate, 416–19, 436–37, 455–56
with injection, 416–19
in jets, 372–75
pipe flow, 329–31, 341, 345
Heat transfer coefficient (*See* Film coefficient, heat transfer)
Heat transfer temperature, 415
Hot-wire anemometer, 25
Howarth problem, 107–8, 112, 123
Howarth-Dorodnitzin transformation, 134–35, 137
Hydraulic diameter, 316, 327, 329, 335
Hydrodynamic stability theory, 179–83, 327
Hypersonic interaction parameter, 171

Implicit finite difference formulation, 100–3, 110–14, 117–18, 150–53, 184, 477–79
Injection, 89–91, 144–49, 182–83, 216–19, 252–54, 274–76, 292, 297, 301–2, 375–79, 416–19, 439–42
Infinite swept wing, 474–76, 487–97, 499–500
Inner (*and* wall) region of a turbulent boundary layer, 207–11, 214–20, 224, 226–27, 247–56, 284–85, 294–97, 299, 332–38, 415–17, 434, 489–92, 499–502
Instability:
flow, 179–86, 481–82
numerical, 99–100, 106–7, 479
Integral energy equation, 43–48
Integral momentum equation:
jets and wakes, 364, 384
laminar, 31–43
turbulent, 240–42
3D flows, 469, 496
Integral species conservation equation, 49–52
Intermittency, 25–26, 224–26, 228–29, 495
factor, 265
Interpolation function, 118–21
Inviscid:
pressure distribution, 10–11, 33, 104, 310, 468
edge velocity distribution, 110–11, 33, 36, 39–40, 85, 116, 468
Isotropic fluid (and flow), 13, 486

Jets, 359–61
laminar:
planar, 363–65
round, 365
turbulent:
planar, 372–73, 384, 386–406, 408–10
round, 366–79, 382–85, 386–406, 408–10
3D jets, 506–10
Johnson-King model (*See* Turbulent kinetic energy, models, Johnson-King model)

Karplus stability analysis, 106–7
Kϵ models:
compressibilty corrected, 405–8, 445
constant-density boundary layer flows, 293–97
equation for ϵ (*and Z*), 293, 295, 299
with injection or suction, 297
internal flows, 340–41
jets and wakes, 403–8
Kϵ 2, 403–8
3D jets, 509–10
shear layers, 405–8
standard model, 295
transition prediction with, 188
variable-density boundary layer flows, 445, 449–51
3D boundary layer flows, 502–4
Kinematic displacement thickness, 434–35, 439–43
Kinematic viscosity, 6
Kutta condition, 19–22

Laminar sublayer, 304–6, 372–73, 378, 618–19, 628
Laminar shear stress:
modeling of, 4, 12–14, 62–63
non-Newtonian, 4–5, 22–23, 64–65
Large eddy simulations, 303–4
Law of mass action, 158
Law of the Wake, 212–13, 222–23, 241, 332–33, 423–24
3D flows, 489–92
Law of the Wall, 208–10, 247
with injection or suction, 216–19
internal flow, 332–33
with roughness, 214–15
for temperature (*See* Temperature law of the wall)
variable-density flow, 420–21, 423–26, 434
3D flows, 489–91, 502
Levy-Lees transformation (*and* coordinates), 136–37, 142, 149–50, 160, 436
Lewis number, 7, 144–46, 455–56
turbulent, 378
Logarithmic region (*See* Overlap region)

Mangler transformation, 73, 137
Mass, conservation (*See* Conservation of mass)
Mass flux:
laminar, 7
modeling of, 7, 70
turbulent, 375
Mass fraction, 7
Mass transfer:
laminar:
duct flow, 317
flat plate, 50–51, 89–92, 146–49
high-speed flow, 146–49, 160–63
pipe flow, 316,
entrance region, 322
stagnation point, 160–63
turbulent:
flat plate, 416–19
in jets, 375–79, 394–95, 401–5
pipe flow, 332–33
Mass transfer coefficient (*See* Film coefficient, mass transfer)
Mass-weighted averaging, 430–33, 445
Matched asymptotic expansions, 18
Mean flow variables in turbulent flow, definition, 25, 234, 430–33
Mean flow turbulence models (*See also* Eddy viscosity and Mixing length), 237–39
Mean free path, 6, 238
Mechanical energy equation, 68
Metric coefficients, 322–23, 467
Michel method for transition, 179
Mixing length:
composite model for a boundary layer, 262–63
definition, 238–39, 433
inner region models, 248–50
with injection or suction, 252–53
with roughness, 251
Van Driest model, 249–50, 646
variable-density flow, 420, 434, 436, 439, 445–48
models for jets and wakes, 385–89, 401–2
Prandtl model, 388
Taylor model, 388
model for shear layers, 380
outer region models, 256
internal flows, 339
variable-density flow, 434–36, 439, 445–48
relation to eddy viscosity, 239
relation to turbulent length scale, 284
pressure gradient effects, 239, 270–73
Situ-Schetz model, 436, 439, 444
von Karman model, 388, 423
3D flows, 499–502
Mole fraction, 71–73
Momentum, conservation (*See* Conservation of momentum)
Momentum equation(s), 16, 18–19, 62–65, 77, 87, 103–5, 130, 134, 136, 144, 182, 235, 433
3D flows, 467
Momentum integral equation (*See* Integral momentum equation)
Momentum integral method:
laminar, 30–43
Thwaites-Walz method, 38–40
axisymmetric, 40
compressible flow, 137–38
3D flows, 469, 475–76
turbulent, 239–44
Moses method, 241–44
3D flows, 469, 495–99
Momentum thickness, 33–35
3D flows, 469
Moody chart, 329–30
Moses method (*See* Momentum integral method, turbulent, Moses method)

Navier-Stokes equations, 9–16, 166, 350, 457, 464
Neutral stability, 180–82, 184–85, 327–28
Newtonian fluids, 4–5, 13, 63
Newton's law of cooling, 47
Non-Newtonian fluids, 4–5, 22–23, 52–54, 63–65, 92–93, 198, 200, 324–27
in turbulent flow, 305–6, 353–54
Normal pressure gradient, 10
Normal stresses, 12, 14
turbulent, 287, 335, 371
No-slip condition, 2
Numerical methods (*See also* Finite differences, method and Finite element method): 93–123, 149–56, 318–21, 325–26
turbulent flows, 264–67, 345–49, 436
3D flows, 476–80
Nusselt number:
definitions, 17

Nusselt number (*cont.*)
laminar flow:
flat plate, 17, 88
with injection, 91, 148–49
nonuniform wall temperature, 17
non-circular ducts, 316–17
pipe flow, 316, 320–21
turbulent flow, 329–32
Nusselt number for diffusion:
definition, 50
laminar flow:
flat plate, 50
with injection, 91
pipe flow, 316–17
relation to skin friction and Nusselt number, 52
turbulent flow, 332

Orr-Sommerfeld equation, 180
Outer region (and layer) of a turbulent boundary layer, 204–7, 210–13, 220–24, 256–62, 332, 423–26
Overlap region of a turbulent boundary layer, 210–11

Parabolic partial differential equations, 74, 97, 283
Parallel flows, 77–79, 312
Perfect gas, 74, 157
Perry-Schofield outer region law, 223–24
Pipe flow:
laminar:
entrance region, 317–22
fully developed, 312–17
heat transfer in, 315–17
mass transfer in, 316–17
non-Newtonian fluids, 324–27
transitional, 327–28
turbulent:
entrance region, 341–50
fully developed, 328–41
heat transfer in, 329, 332–33, 341, 345
mass transfer in, 332–33
non-Newtonian fluids, 353–54
Plane of symmetry flows, 479
Plastic fluid, 5, 23
Plate thermometer problem, 132–34
Pohlhausen:
method, 35–38
polynomial velocity profile, 37–38
pressure gradient parameter, 37, 192
Poiseuille flow (*See* Pipe flow, laminar, fully developed)
Polar plot for 3D boundary layers, 484, 488–89
Potential core, 359–60, 366, 375, 401, 506–7
Power law fluid, 23, 52–53, 64–65, 92–93, 324–27
Prandtl energy method (*See* Turbulent kinetic energy, models, Prandtl energy method)
Prandtl independence principle, 475
Prandtl number, 6–7, 128, 129, 132, 134, 316, 332, 416,
for turbulence quantities, 283, 293
turbulent, 345, 372, 394, 422
Pressure drop (and loss), 1, 312–13
laminar flow,
entrance region, 317–18, 320–21
fully developed, 312–14
non-circular ducts, 316–17
non-Newtonian fluids, 324
turbulent flow,
entrance region, 341, 344–48
fully developed, 329–30
non-Newtonian fluids, 353–54
Pressure gradient:
adverse (*See* Adverse pressure gradient)
equilibrium in turbulent flow, 222, 242, 260–61, 268, 270
favorable (*See* Favorable pressure gradient)
pipe flow (*See* pressure drop)
Pseudolaminar flow, 257–61, 435
Pseudoplastic fluid, 5, 23

Rayleigh line, 355–56
Real gas effects, 157–63, 454–56
Recovery factor:
laminar, 128, 132–33, 427
turbulent, 426–27
Recovery temperature (*See* Adiabatic wall temperature)
Reference temperature (*and* method), 128–29, 143
Reichardt eddy viscosity model (*See* Eddy viscosity, inner region models, Reichardt model)
Reichardt sublayer length scale, 250–51, 254–55
Relaminarization, 198
Resistance coefficient:
laminar, 313–14, 316–17
turbulent, 329–30
non-Newtonian fluids, 324–26
Reynolds analogy, 48, 416
Reynolds pipe flow experiments, 177–78
Reynolds number, 6
critical value of, 178, 327
Reynolds (and turbulent shear) stress:
convection of, 298
definition of, 224–26, 236–37
in compressible flow, 426, 430, 433
diffusion of, 298–99
dissipation of, 298
equation for, 298
models,
algebraic stress models, 301, 349–50
constant-density flows, 297–302
Daly-Harlow model, 301, 408–10
Donaldson model, 299–302, 408–10, 449–53
Hanjalic-Launder model, 298–302, 341–43
with injection or suction, 301–2
internal flows, 341–43, 349–50
jets and wakes, 408–10
shear layers, 408–10
variable-density flows
boundary layer flows, 449–53
pressure strain, 298
production of, 298–99
redistribution of, 298–99
3D flows, 492–94
Richardson number, 196
Roughness effects:
transition, 192–94, 197–200
turbulent flow, 214–15, 251, 329–30
Round-off errors, 99

Schmidt number, 7, 50, 316, 332
turbulent, 378
Second viscosity coefficient, 14
Secondary flows (*and* crossflow), 310–11, 316, 334–36, 341–43, 345–47, 349–52, 464–64, 471–72, 475–76
pressure-driven, 335, 463, 482–86
shear-driven, 335–36, 341–43, 482

Separation, boundary layer, 19–22, 30, 37–38, 42–43, 85–86, 220, 224, 242, 311
 by shocks, 164–68, 457–59
 3D flows, 465–66, 475, 484–86
Shape factor (*and* parameter), 38–39, 183, 205
Shear layers:
 laminar, 359–63
  high-speed flow, 363
  initial boundary layer effects, 362–63
 turbulent, 379–83, 405–11
  high-speed flow, 382–83, 405–8
  initial boundary layer effects, 381–83
Shear strain rate tensor, 12–13
Shear stress tensor, 12–14
Shock/boundary layer interaction:
 laminar, 164–68
 turbulent, 457–59
 3D flows, 505
Shooting, 83
Similar solutions:
 energy equation, 87–89
 jets, 363–65, 383–84
 momentum equation, (*See also* Blasius flat-plate solution and Falkner-Skan wedge flows), 79–86, 150–53
 non-Newtonian fluids, 92–93
 shear layers, 361–63, 379–82
 species conservation equation, 89–92, 146–49
 stagnation point, 85–86, 142–43, 160–63
 wakes, 385–86
 3D flows, 470–75
Skin friction coefficient:
 definition, 30
 laminar flow, 36, 37, 84, 86
  high-speed flow, 139, 148
  with injection, 148
  non-Newtonian fluids, 53–54, 92–93,
  pressure gradient effects, 86
  relation to Nusselt number and Nusselt number for diffusion, 52
  with viscous interaction, 172
  with vorticity interaction, 173
 measurement of, 208
 turbulent flow, 213–14, 240–41
  high-speed flow, 422–24, 449–52
  with injection, 219, 419
  pressure gradient effects, 223
  relation to Nusselt number, 416
  with roughness, 215
  3D flows, 489–90, 496–97
Slender channel equations, 322–23, 350
Species continuity equation, 70–71
Specific heat:
 constant pressure, 67
 constant volume, 69
Sphere, flow past, 24–25, 194–96
Spreading parameter, 380–83, 405–8
Spring/mass/damper problem, 17–18
Squire, theorem of, 180
Stability:
 laminar flow, 177–87
 numerical methods, 99–100, 107–8, 477–79
Stagnation point flow, 39, 40, 85, 86, 142–43, 160–63, 170–73
 3D bodies, 471–74
Stanton number:
 definition, 48
 laminar flow, 48
  high-speed flow, 154
  relation to skin friction coefficient, 52
 turbulent flow, 416
  with chemical reactions, 455–56
  flat plate, 416, 437
  high-speed flow, 154
  with injection, 419
  in pipes, 329, 341, 345
  relation to skin friction coefficient, 416
Stanton number for diffusion:
 definition, 52
 laminar flow, 52
 relation to skin friction coefficient, 52
 turbulent flow, 622
Stiffness matrix, 121, 124
Stokes' flows, 78–81, 249, 255
Stokes' hypothesis, 14
Stokes' law, 4
Stratification, density, 196–97
Streaks, 232–33
Stream function:
 axisymmetric flow, 60, 318–19
 compressible flow, 60, 134
 planar flow, 60, 81–82
 3D flows, 478
Streamline coordinates, 322
Sublayer, laminar (*See* Laminar sublayer)
Suction, 42–43, 90–91, 183
Superlayer, 213
Surface forces, 11, 61
Sutherland viscosity law, 8, 138, 162
Sweeps, 233

Temperature:
 bulk, 315–16, 318
 heat transfer (*See* Heat transfer temperature)
 reference, 128–29, 143
Temperature law of the wall, 415
Thermal boundary layer, 3
 thickness, 3
Thermal conductivity:
 conversion factors, 513
 laminar, 6–8, 513–15
 mixture, 72–73
 turbulent (*and* eddy), 345, 372
Thomas algorithm, 102–3, 112, 118, 123, 150
Three-dimensional flow, definition of, 462
Thwaites and Walz method (*See* Momentum integral method, Thwaites-Walz method)
Time averaging, 25, 27, 234, 430–31
Tollmein-Schlichting waves, 182, 481
Transformations:
 compressible flow, 134–37
 similarity, 79, 81–82, 85, 87, 90, 92, 146, 470–73
 to two-dimensional form, 73
Transition to turbulence, 176–78
 bluff bodies at low speeds, 194–96
 density stratification effects, 196–97
 freestream turbulence effects, 190–92, 197
 heating/cooling effects, 182–83, 197–99
 injection/suction effects, 182–83, 482
 internal flows, 177–78, 327–28
 nature of, 188–90
 non-Newtonian fluids, 198, 200
 prediction of, 179–88, 327–28, 481–82
 pressure gradient effects, 191–94, 194–96, 198
 roughness effects, 192–94, 197, 200
 supersonic flow, 197–200, 482
 3D flows, 480–82

Transpiration cooling:
laminar, 90–91, 146–49
turbulent, 416–19
Triple Deck formulation, 166–68
Truncation error, 99–101
Tubes, flow in (*See* Pipe flow)
Turbulence intensity:
axial, 27, 224–25, 229–30, 299–300, 335–38, 345–47, 367–69, 426–28, 449–51
spectra, 228, 335, 338, 370–71
mass-averaged, 431–32
normal (*and* transverse), 27, 224–25, 229–31, 299–300, 335–37, 345–47, 367, 449–51
tangential (*and* out-of-plane), 27, 224–25, 229–31, 299–300, 335–37, 345–47, 367, 449–51
Turbulent flow, nature of, 23–28, 213, 229–33
Turbulent kinetic energy:
balance, 229–30, 335–37, 371, 492–94
convection of, 228–30, 282, 335–37, 492–94
definition, 226
diffusion of, 228–30, 282–83, 335–37, 492–94
dissipation of, 228–30, 282, 335–37, 492–94
equation for, 282–84, 295, 299
models:
Bradshaw model, 282–83, 285–87, 399–402, 445–48, 502–3
with freestream turbulence, 291–92,
with injection or suction, 292
internal flows, 341, 351–52
jets and wakes, 399–402
Johnson-King model, 289–91, 445–49, 502
Prandtl energy method, 282–85, 292, 399–402, 445–46
variable-density flow, 445–49
3D flows, 502–3
profiles, 226–27, 335–37, 369, 492–93
production of, 228–30, 282–83, 335–37, 492–94
Turbulent length scale, 282–84, 293–94, 298–99, 403, 445
relation to mixing length, 284
Turbulent Prandtl number (*See* Prandtl number, turbulent)
Turbulent shear stress (*See* Reynolds stress)
Turbulent transport models (*See* Eddy viscosity, models; Mixing length, models; Turbulent kinetic energy, models; Kϵ models; Reynolds stress, models)

Unheated starting length problem, 45–47

Van Driest:
damping factor, 249–50
with injection and suction, 252–53
pressure gradient effects, 267
with roughness, 251
eddy viscosity model (*See* Eddy viscosity, inner region models, Van Driest model)
effective velocity (*See* Effective velocity)
mixing-length model (*See* Mixing length, inner region models, Van Driest model)
Van Driest I and II correlation laws, 420–26
Velocity defect, 34, 206
Viscosity:
apparent, 23
conversion factors, 512
definition, 4
kinematic, 6
laminar, 4–8, 513–14
measurement of, 4–5
mixture, 71–72
non-Newtonian fluid, 4–5, 23
second coefficient of, 14
turbulent (*See* Eddy viscosity)
Viscous/inviscid interaction:
laminar, 166–74
hypersonic, 169–74
turbulent, 457–60
vorticity, 171–74
Viscous throat, 169, 356, 459
von Karman mixing-length model (*See* Mixing length, von Karman model)
von Neumann stability analysis, 154–56
Vorticity parameter, 173

Wake:
Coles' law of (*See* Law of the wake)
function (*See* Coles' wake function)
parameter (*See* Coles' wake parameter)
Wakes, 359–61
laminar:
planar, 365–66
round (*and* axisymmetric), 366
near wake, 359–60
turbulent:
planar, 385–86, 388–91, 396–98
round (*and* circular and axisymmetric), 386, 398
Wall functions:
in Kϵ models, 295, 449, 502
in mean flow models, 277
in Reynolds stress models, 301
in TKE models, 285
3D flows, 502
Wall law (*See* Law of the Wall)
coordinates (or variables), 210
plot, 208–9, 214, 216–18, 250, 252–53, 265–66, 421, 491
Wall shear (*See* Skin friction coefficient)
Walz method (*See* Thwaites and Walz method)
Wave number of fluctuations, 226–28, 371
Wedge flows (*See* Falkner-Skan wedge flows)

Zone of dependence, 464, 476
Zone of influence, 464, 476